T0181079

Lecture Notes in Computer Science 13231

Stan Sclaroff · Cosimo Distante · Marco Leo ·
Giovanni M. Farinella · Federico Tombari (Eds.)

Image Analysis and Processing – ICIAP 2022

21st International Conference
Lecce, Italy, May 23–27, 2022
Proceedings, Part I

 Springer

Editors
Stan Sclaroff 🆔
Boston University
Boston, MA, USA

Cosimo Distante 🆔
National Research Council
Lecce, Italy

Marco Leo 🆔
National Research Council
Lecce, Italy

Giovanni M. Farinella 🆔
University of Catania
Catania, Italy

Federico Tombari 🆔
Technische Universität München
Garching, Germany

ISSN 0302-9743 ISSN 1611-3349 (electronic)
Lecture Notes in Computer Science
ISBN 978-3-031-06426-5 ISBN 978-3-031-06427-2 (eBook)
https://doi.org/10.1007/978-3-031-06427-2

This Springer imprint is published by the registered company Springer Nature Switzerland AG
The registered company address is: Gewerbestrasse 11, 6330 Cham, Switzerland

Preface

The International Conference on Image Analysis and Processing (ICIAP) is an established biennial scientific meeting promoted by the Italian Association for Computer Vision, Pattern Recognition and Machine Learning (CVPL - formerly GIRPR), which is the Italian IAPR Member Society, and covers topics related to theoretical and experimental areas of Computer Vision, Image Processing, Pattern Recognition, and Machine Learning with emphasis on theoretical aspects and applications. The 21st International Conference on Image Analysis and Processing (ICIAP 2022) was held in Lecce, Italy, during May 23–27, 2022 (postponed from 2021 due to the COVID-19 pandemic), in the magnificent venue of the Teatro Apollo, (http://www.iciap2021.org). It was organized by the Lecce Institute of Applied Sciences and Intelligent Systems, an Institute of CNR, the National Research Council of Italy. ICIAP 2022 was sponsored by NVIDIA, E4 Computer Engineering s.p.a., and ImageS s.p.a., and was endorsed by the Apulia Region and the Province of Lecce.

The conference covered both established and recent scientific trends, with particular emphasis on Video Analysis and Understanding, Pattern Recognition and Machine Learning, Deep Learning, Multiview Geometry and 3D Computer Vision, Image Analysis, Detection and Recognition, Multimedia, Biomedical and Assistive Technology, Digital Forensics and Biometrics, Image Processing for Cultural Heritage, and Robot Vision.

The ICIAP 2022 main conference received 297 paper submissions from all over the world, including Austria, Azerbaijan, Bangladesh, Belgium, Brazil, Canada, China, the Czech Republic, the UK, Finland, France, Germany, Greece, Hungary, India, Ireland, Japan, Latvia, Lebanon, Mongolia, New Zealand, the Netherlands, Italy, Pakistan, Peru, Poland, Portugal, Russia, Syria, Spain, South Korea, Sri Lanka, Luxembourg, South Africa, Sweden, Switzerland, Turkey, the United Arab Emirates, and the USA. Two rounds of submissions were introduced, independent from each other, in order to improve the quality of papers. Paper selection was carried out by 25 expert researchers who acted as area chairs, together with the International Program Committee and an expert team of reviewers. The rigorous peer-review selection process, carried out by three distinct reviewers for each submission, ultimately led to the selection of 162 high-quality manuscripts, with an overall acceptance rate of 54%.

The main conference program included 45 oral presentations, 145 posters, and five invited talks. The invited talks were presented by leading experts in computer vision and pattern recognition: Larry S. Davis, University of Maryland and Amazon (USA), Roberto Cipolla, University of Cambridge (UK), Dima Damen, University of Bristol (UK), and Laura Leal-Taixe, Technische Universität München (Germany).

ICIAP 2022 also included 11 tutorials and hosted 16 workshops, seven competitions, and a special session on topics of great relevance with respect to the state of the art. The tutorial and workshop organizers came from both industry and academia.

Several awards were presented during the ICIAP 2022 conference. The Best Student paper award was supported by the MDPI Journal of Imaging, and several other prizes were conferred under the Platinum level sponsorship provided by NVIDIA.

The success of ICIAP 2022 is credited to the contribution of many people. Special thanks should be given to the area chairs, who did a truly outstanding job. We wish to thank the reviewers for the immense amount of hard work and professionalism that went into making ICIAP 2022 a successful meeting. Our thanks also go to the Organizing Committee for their unstinting dedication, advice, and support.

We hope that ICIAP 2022 has helped to build a piece of the future, a future where technologies can allow people to live comfortably, healthily, and in peace.

May 2022

Cosimo Distante
Stan Sclaroff
Giovanni Maria Farinella
Marco Leo
Federico Tombari

Organization

General Chairs

Cosimo Distante   National Research Council, Italy
Stan Sclaroff   Boston University, USA

Technical Program Chairs

Giovanni Maria Farinella   University of Catania, Italy
Marco Leo   National Research Council, Italy
Federico Tombari   Google and TUM, Germany

Area Chairs

Lamberto Ballan   University of Padua, Italy
Francois Bremond   Inria, France
Simone Calderara   University of Modena and Reggio Emilia, Italy
Modesto Castrillon Santana   University of Las Palmas de Gran Canaria, Spain
Marco Cristani   University of Verona, Italy
Luigi Di Stefano   University of Bologna, Italy
Sergio Escalera   University of Barcelona, Spain
Luiz Marcos Garcia Goncalves   UFRN, Brazil
Javier Ortega Garcia   Universidad Autonoma de Madrid, Spain
Costantino Grana   University of Modena and Reggio Emilia, Italy
Tal Hassner   Facebook AML and Open University of Israel, Israel
Gian Luca Marcialis   University of Cagliari, Italy
Christian Micheloni   University of Udine, Italy
Fausto Milletarì   NVIDIA, USA
Vittorio Murino   Italian Institute of Technology, Italy
Vishal Patel   Johns Hopkins University, USA
Marcello Pelillo   Università Ca' Foscari Venice, Italy
Federico Pernici   University of Florence, Italy
Andrea Prati   University of Parma, Italy
Justus Piater   University of Innsbruck, Austria
Elisa Ricci   University of Trento, Italy
Alessia Saggese   University of Salerno, Italy
Roberto Scopigno   National Research Council, Italy

Filippo Stanco   University of Catania, Italy
Mario Vento   University of Salerno, Italy

Workshop Chairs

Emanuele Frontoni   Università Politecnica delle Marche, Italy
Pier Luigi Mazzeo   National Research Council, Italy

Publication Chair

Pierluigi Carcagni   National Research Council, Italy

Publicity Chairs

Marco Del Coco   National Research Council, Italy
Antonino Furnari   University of Catania, Italy

Finance and Registration Chairs

Maria Grazia Distante   National Research Council, Italy
Paolo Spagnolo   National Research Council, Italy

Web Chair

Arturo Argentieri   National Research Council, Italy

Tutorial Chairs

Alessio Del Bue   Italian Institute of Technology, Italy
Lorenzo Seidenari   University of Florence, Italy

Special Session Chairs

Marco La Cascia   University of Palermo, Italy
Nichi Martinel   University of Udine, Italy

Industrial Chairs

Ettore Stella   National Research Council, Italy
Giuseppe Celeste   National Research Council, Italy
Fabio Galasso   Sapienza University of Rome, Italy

North Africa Liaison Chair

Dorra Sellami   University of Sfax, Tunisia

Oceania Liaison Chair

Wei Qi Yan   Auckland University of Technology, New Zealand

North America Liaison Chair

Larry S. Davis   University of Maryland, USA

Asia Liaison Chair

Wei Shi Zheng   Sun Yat-sen University, China

Latin America Liaison Chair

Luiz Marcos Garcia Goncalves   UFRN, Brazil

Invited Speakers

Larry S. Davis   University of Maryland and Amazon, USA
Roberto Cipolla   University of Cambridge,UK
Dima Aldamen   University of Bristol, UK
Laura Leal-Taixe   Technische Universität München, Germany

Steering Committee

Virginio Cantoni   University of Pavia, Italy
Luigi Pietro Cordella   University of Napoli Federico II, Italy
Rita Cucchiara   University of Modena and Reggio Emilia, Italy
Alberto Del Bimbo   University of Firenze, Italy
Marco Ferretti   University of Pavia, Italy
Fabio Roli   University of Cagliari, Italy
Gabriella Sanniti di Baja   National Research Council, Italy

Endorsing Institutions

International Association for Pattern Recognition (IAPR)
Italian Association for Computer Vision, Pattern Recognition and Machine Learning
(CVPL)
Springer

Institutional Patronage

Institute of Applied Sciences and Intelligent Systems (ISASI)
National Research Council of Italy (CNR)
Provincia di Lecce
Regione Puglia

Contents – Part I

Multimedia

Deep Learning

Image Processing for Cultural Heritage

Robot Vision

Contents – Part II

Digital Forensics and Biometrics

Image Analysis, Detection and Recognition

Contents – Part III

Video Analysis and Understanding

Brave New Ideas

A Lightweight Model for Satellite Pose Estimation

Pierluigi Carcagnì(✉), Marco Leo, Paolo Spagnolo, Pier Luigi Mazzeo, and Cosimo Distante

CNR-ISASI, Ecotekne Campus via Monteroni snc, 73100 Lecce, Italy
pierluigi.carcagni@cnr.it

Abstract. In this work, a study on computer vision techniques for automating rendezvous manoeuvres in space has been carried out. A lightweight algorithm pipeline for achieving the 6 degrees of freedom (DOF) object pose estimation, i.e. relative position and attitude, of a spacecraft in a non-cooperative context using a monocular camera has been studied. In particular, the considered lite architecture has been never exploited for space operations and it allows to be compliant with operational constraints, in terms of payload and power, of small satellite platforms. Experiments were performed on a benchmark Satellite Pose Estimation Dataset of synthetic and real spacecraft imageries specifically introduced for the challenging task of the 6DOF object pose estimation in space. Extensive comparisons with existing approaches are provided both in terms of reliability/accuracy and in terms of model size that ineluctably affect resource requirements for deployment on space vehicles.

Keywords: Spacecraft pose estimation · 6DOF pose · Deep learning · Monocular vision · Space imagery

1 Introduction

Estimating the 6 degrees of freedom (6DOF) pose of space-borne objects (e.g., satellites, spacecraft, orbital debris) is a crucial step in many space operations such as docking, non-cooperative proximity tasks (e.g., debris removal), and inter-spacecraft communications (e.g., establishing quantum links). It has unique challenges that are not commonly encountered in the terrestrial setting. Due to the importance of the problem, the Advanced Concepts Team (ACT) at ESA recently held a benchmark competition called Kelvins Pose Estimation Challenge (KPEC)[1].

The 6DOF pose estimation of objects in images is a traditional computer vision task. Methods based on template matching [21] were initially used to this aim. Unfortunately, when an object undergoes occlusions or drastic illumination changes they become unreliable. To fix this issues local feature/keypoint matching approaches were introduced [11]. The extraction of the fiducial points for

---

[1] https://kelvins.esa.int/satellite-pose-estimation-challenge/home/.

S. Sclaroff et al. (Eds.): ICIAP 2022, LNCS 13231, pp. 3–14, 2022.
https://doi.org/10.1007/978-3-031-06427-2_1

the purpose of calculating the correspondences is carried out using handcrafted descriptors such as Harris corners or Canny edges, lines e.g. by Hough transform, or scale invariants such as SIFT, SURF and ORB [13]. The object poses are subsequently obtained by solving a Perspective-n-Point (PnP) problem [10]. The above mentioned methods based on handcrafted descriptors, tend to produce low-quality outputs in difficult conditions (weak or absent texture of the surfaces, strong lights, etc.) typical in an operational context in space.

To overcome these drawbacks, recent advances in pose estimation techniques, for terrestrial applications, have been based on deep learning (DL) algorithms instead. In general, these algorithms bypass the classic pipeline based on handcrafted features, and instead try, through the use of an appropriate deep convolutional neural network (DCNN), to learn, in an end-to-end way, the non-linear transformation between the two-dimensional space of the input image and the six-dimensional exposure space of the network output. Learning is carried out through appropriate supervised training.

Deep learning strategies have demonstrated robust behaviour under difficult operating conditions in terms of scene illumination and object surface texture. However, for efficient operations in terms of processed fps (frames per second) it is necessary to have appropriate hardware with power constraints as opposed to the classic approaches based on handcrafted descriptors.

Besides, they require a large amount of manual labels including the 2D keypoints, masks, 6D poses of objects, and other extra labels, which are usually very costly. Many recent 6DOF pose estimation methods exploited 3D object models to generate synthetic images for training because labels come for free. However, due to the domain shift of data distributions between real images and synthetic images, the network trained only on synthetic images fails to capture robust features in real images for 6DOF pose estimation [19]. Another effective pathway could be to combine the strength of deep neural networks and geometric optimisation for example by incorporating a perspective-n-point (PnP) solver in a deep neural architecture [3]. Most 6DOF pose estimation deep networks rely on an encoder-decoder architecture. To handle large scale variations for 6D object pose estimation they can rely on an additional object detection network or they exploit the inherent hierarchical architecture of the encoder network, which extracts features at different scales [6]. Among different deep architectures, the High-Resolution Network (HRNet) [15], initially introduced for human pose estimation [17], has also recently gathered very relevant results also for object detection and semantic segmentation. Differently from existing state-of-the-art frameworks, that first encode the input image as a low-resolution representation through a sub-network that is formed by connecting high-to-low resolution convolutions in series (e.g., ResNet, VGGNet), and then recover the high-resolution representation from the encoded low-resolution representation, HRNet maintains high-resolution representations through the whole process. The benefit is that the resulting representation is semantically richer and spatially more precise.

This scientific fervour and the increased accessibility of space platforms, and space data, recently pushed researchers to investigate the application of machine learning researches in space activities. One of these 'research frontiers' concerns the pose estimation of space objects for autonomous rendezvous or for capturing uncontrolled targets in debris removal operations. Spacecraft (vehicles designed for operation outside the earth's atmosphere) and satellites (objects that orbit a natural body) have two types of systems: payload, which comprises instruments that facilitate the primary purpose of the spacecrafts; and operations systems, which support the payload and allow it to reach, stay, and work in space. Modern space compute systems are moving towards shared/re-configurable, multicore systems which would be capable of running just lite DL models. Besides, another limiting factor for on-board compute is power [9].

As a consequence, not all the DL models developed for terrestrial operations can be exploited in space and then it is important to investigate how to find a good trade-off between model complexity and accuracy/reliability.

In this work, a study on computer vision techniques for automating rendezvous manoeuvres in space has been carried out. In particular, algorithms for estimating the 6DOF object pose estimation, i.e. relative position and attitude, of a spacecraft in a non-cooperative context using a monocular camera have been studied. The term "non-cooperative" implies that the target spacecraft does not have an active communication link or markers such as LEDs or reflectors useful for distance and attitude estimation. To this aim, in this work, a 6DOF system suited to be exploited on a Satellite Platform, in 50-kg micro-satellite class, has been developed. The paper introduces and assesses, for the specific challenging task of the 6DOf object pose estimation in space, a lite architecture inspired by the Lite-HRNet [18] recently introduced for human pose estimation. By our knowledge, this is the first attempt to exploit the lite-HRNet for space operations. Experiments were performed on the Spacecraft PosE Estimation Dataset (SPEED) [14], the first publicly available machine learning set of synthetic and real spacecraft imageries. Extensive comparisons with existing approaches are provided both in terms of reliability/accuracy and in terms of model size that ineluctably affect resource requirements for deployment on space vehicles [5].

The rest of the paper is organized as follows: Sect. 2 describes the problem, the proposed algorithmic pipeline introduced to address it and the dataset of space images used for experimental tests. Subsequently, Sect. 3 accurately describes experimental results and it provides a deep discussion about the advisability of the proposed trade-off between accuracy and computational requirement of resources with respect to existing works in the literature. Finally, Sect. 4 concludes the paper.

2 Methodology and Data

In this work, a PnP based pose estimation approach, exploiting keypoints extracted by means of deep learning techniques, has been employed [16]. There are extensive applications based on this approach in the literature, including action recognition, human-computer interaction, intelligent photo editing, pedestrian monitoring, etc. [7]. In particular, it is possible to divide the problem into two main categories: top-down methods and bottom-up methods. According to the top-down paradigm, the object of which to estimate the pose is first detected and then its pose is estimated by exploiting only the informative content of the region into the bounding box surrounding the identified object. The bottom-up paradigm directly regresses the positions of the keypoints belonging to the same object, or it detects all the keypoints in the scene and subsequently it groups the keypoints on the same object. Although the top-down paradigm is more expensive from a computational point of view, since in addition to the extraction of keypoints it requires a preliminary phase of detection of the object of interest, it is more accurate than the bottom-up paradigm [4].

2.1   Processing Pipeline

In this paper, a top-down paradigm has been exploited. The starting point was the pipeline introduced in [2] and depicted in Fig. 1. The algorithmic components of the pipeline have been modified in order to make the entire pipeline suitable for use with embedded systems, and then to be compliant with computational constraints for in space operations where power and resources are much more limited than for terrestrial tasks. The pipeline is model-based, i.e. it relies on the availability of a 3D model of the target object and the pose of the on-board camera has to be estimated with respect to the actual model configuration extracted from the acquired images. Hence, the pipeline consists of three main processing modules performing:

1. the detection of the object in the image;
2. the keypoints estimation in the detected bounding-box of the object;
3. the computation of the 2D-3D keypoints correspondences, i.e. between the available 3D points of the 3D model of the target and the estimated 2D ones of the detected object, and final pose estimation by means of a PnP (Perspective-n-Point) based algorithm.

   Taking into account the limited computational resources and energy consumption constraints on board of a spacecraft/satellite, the underlying idea of this work is to address the keypoints extraction task of the target vehicle by a 'lite' deep convolutional network in order to reduce the computational complexity and the size of the trained model, in terms of parameters and memory occupation.

   In this paper, a Lite-HRNet architecture [18] has been implemented and tested for the keypoints detection task, lowering this way the architectural complexity, compared to HRNet based approach exploited in [2], and therefore achieving the results of making the pipeline compliant with the computational resources available on board of the vehicle.

   The Lite-HRNet has been built-up following the same architectural strategies exploited for HRNet. Starting from a high-resolution convolutional subnetwork

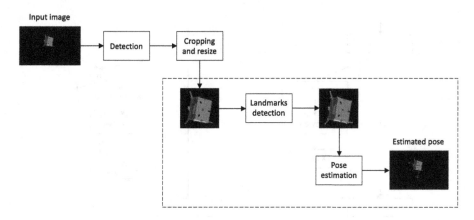

Fig. 1. Pose estimation pipeline.

as the first stage, high-to-low resolution subnetworks are gradually added and connected in parallel, keeping the high-resolution of the initial image through the whole process (Fig. 2).

Complexity is reduced by applying the efficient shuffle block, introduced with ShuffleNet [20], to the HRNet architecture and then using fewer layers and smaller widths. Moreover, the costly point-wise 1×1 convolution operation, heavily used in the original shuffle blocks, has been replaced by a lightweight unit, named conditional channel weighting. This allows, exploiting element-wise weighting operations, the architecture to obtain a linear complexity with respect to the number of channels instead of the quadratic complexity of the 1×1 convolution operations in the original implementation of the shuffle block. Finally, two lightweight functions have been introduced, a cross-resolution weighting function and a spatial weighting function, in order to compute the weight maps from all the channels across resolutions and for each resolution respectively compensating the role played by the point-wise 1×1 convolution operation (Fig. 3).

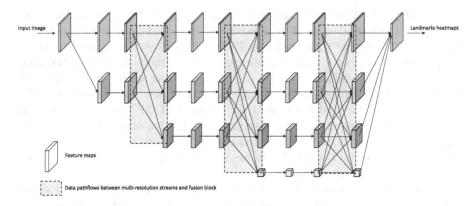

Fig. 2. Starting from an high-resolution convolutional subnetwork as first stage, high-to-low resolution subnetworks are gradually added and connected in parallel maintaining high-resolution through the whole process.

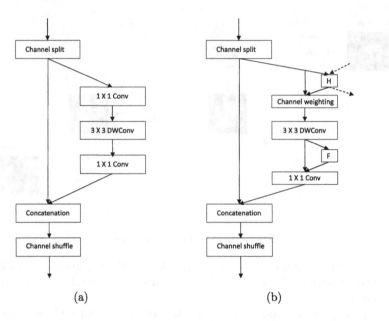

Fig. 3. (a) The shuffle block. (b) Conditional channel weighting block. H denotes the cross-resolution weighting function, F denotes the spatial weighting function. Weights distribution and representations from and to other resolutions are represented by the dotted lines.

In this work, in particular, the Lite-HRNet-18 implementation, where suffix 18 indicates the number of layers, has been exploited. A size of 768×768 pixels (the same of the solution in [2]) has been set for the input window and for the output heatmaps.

2.2   The Dataset

For training and validating the proposed pipeline, the SPEED (Spacecraft Pose Estimation Dataset) dataset [8] was exploited. It consists of 8-bit monochrome images in JPEG format with resolution 1920×1200 pixels. The dataset has three main folders of images: a folder containing 12000 synthetic images for training, a folder with 2998 similar synthetic images for testing and a folder with 300 real images of the Tango satellite mock-up, same format and resolution as the synthetic images. Ground truth data, in terms of 6DOF poses (position and orientation), is provided only for the images in the training set. Some of the training images are shown in Fig. 4. Images are particularly challenging due to the large variations for the satellite in terms of lighting condition, distance from the camera, orientation and background.

The camera model used for rendering the synthetic images is the same one as the actual camera used for capturing the 300 images of the mock-up. The related intrinsics camera parameters are: $resolution = 1920 \times 1200$ pixels, focal

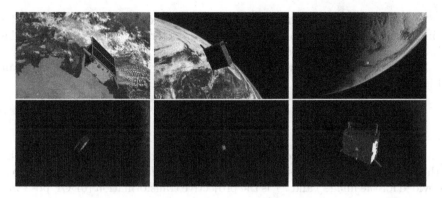

Fig. 4. Some Tango satellite images, at different distances from the camera, sampled from the SPEED dataset.

length $f = 17.6$ mm, Pixel pitch $p = 5.86$ μm/px, Horizontal $FoV = 35.452°$, Vertical $FoV = 22.595°$.

3 Experimental Results

Algorithmic pipeline implementation and testing were carried out using the Pytorch framework, and the Python language, in Ubuntu Linux 20.04 environment on a machine equipped with: Intel i7 processor; 64 GB of RAM; NVIDIA TITATN RTX GPU card with 24 GB of RAM.

In the first experimental phase the 3D model of the Tango satellite mockup has been estimated (3D model and landmarks annotations are not provided with SPEED dataset but only 6DOF ground truth for the training set). This estimation has been carried out by picking up 9 close-up images from the training folder. In each image, 11 keypoints were manually selected: they correspond to some of the strongest visual characteristics in the images which, moreover, are not occluded by other surfaces of the vehicle space.

Figure 5 shows the configurations of the 11 selected keypoints on 3 training images.

Fig. 5. Three configurations of the selected keypoints for the initial estimation of the 3D model of the Tango satellite mock-up.

Starting from the aforementioned 2D positions of the 11 keypoints in the 9 images and knowing the camera model parameters, the 3D structure of the Tango satellite has been reconstructed by multi-view triangulation.

In particular, multi-view triangulation was performed by minimizing an objective function

$$\sum_{i,j} \left\| p_{i,j} - \pi_{T_j^*}(\mathbf{x}_i) \right\|_2^2 \tag{1}$$

where $p_{i;j}$ denotes the 2D coordinates of the i-th landmark, obtained from the j-th image, and x_i the corresponding 3D landmark. T_j^* is the ground truth pose provided for the image j and π_T is the projective transformation (6DOF pose and intrinsic camera parameters are known) for the x_i 3D landmark onto the image plane. From Eq. 1 the 3D positions of the i selected landmarks were estimated in each image j.

Exploiting the 6DOF ground truth data in the dataset and knowing the estimated 3D model of the satellite, the ground truth 2D positions of landmarks for all training images were obtained by projecting x_i to the image plane by π_{T^*}. Finally, a bounding box was chosen in each image so that the 11 landmarks lie in it.

The 2D positions of the landmarks and the corresponding bounding boxes were then exploited to validate the proposed pipeline for 6DOF pose estimation.

The validation task was carried out by exploiting all available annotated images (i.e. the images in the training folder of the SPEED dataset) and a K-Fold Cross-Validation approach, with a number of folds equal to 6. In the cross-validation, 5 folds were used for training the Lite-HRNet. The input of the net were the patches obtained by cropping the original training image, around the available bounding boxes surrounding the landmarks and the corresponding 2D positions of the 11 selected landmarks. They were then resized to a common dimension of 768×768 pixels. In particular, the network was trained in order to regress 11 heatmaps, of the same size of the input patch, corresponding to the 11 selected landmarks. Ground truth heatmaps were generated as 2D normal distributions with mean equal to the ground truth 2D position of each landmark and standard deviation of 1 pixel.

The remaining fold was then used for validating the capability of the net to automatically estimate landmarks positions in unseen patches. This process was repeated 6 times changing the fold used for validation among the available ones.

The object poses in all the patches extracted for the 6 validation folds (12.000 images) were then obtained by solving a Perspective-n-Point (PnP) problem [10] exploiting the 2D-3D correspondences between the 11 predicted landmarks and the 3D structure model of the satellite mock-up.

Lite-HRNet was trained by scratch, for each of the 6 validation steps, employing the ADAM optimizer with starting learning rate = 0.001 (dropped by a 0.1 factor at the 120th and 170th epochs respectively), momentum = 0.9 and weight decay = 0.0001 parameters. A total number of 180 training epochs has been chosen.

A rotation and a translation error were finally computed. In particular, indicating with q^* and q the rotation quaternion ground truth and the estimated one, the rotation error E_r is defined as:

$$E_R = 2 \cdot cos^{-1}|q \cdot q^*|, \qquad (2)$$

and the translation error is defined as:

$$E_T = \frac{\|t - t^*\|_2}{\|t\|_2}, \qquad (3)$$

for an overall error:

$$E = E_R + E_T. \qquad (4)$$

The experienced mean errors on validation folds were $\overline{E}_R = 0.0302$, $\overline{E}_T = 0.0075$ and $\overline{E} = 0.377$ respectively. The $\overline{E}$ error is plotted in Fig. 6 versus the distance between the camera and the target. In particular, the figure reports the mean total error $\overline{E}$ on the y-axis, whereas the x-axis indicates distance ranges from the camera, expressed in meters, of the corresponding detected TANGO vehicle. The green bars correspond are related to the proposed pipeline whereas the orange ones correspond to the error achieved by the pipeline in [2], where the non-lite version of the HRNet was exploited. For a fair comparison, the pipeline in [2] was applied in the identical operating conditions as for the proposed algorithmic pipeline, i.e. without the final iterative refinement, in order to not add external bias in the evaluation of the benefit in using the lightweight version of the HRNet. As expected, the error decreased while the distance increased. It is worth noting that for short distances errors for the two pipelines are comparable. Of course, the gap between the performance of the lite version of HRNet increased at long distances since, on small targets, the landmark positioning failed.

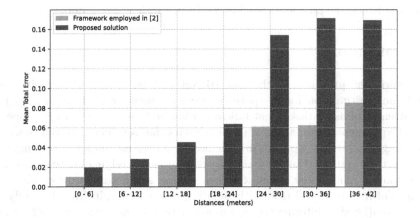

Fig. 6. Mean total error vs Distance of the target from the camera.

The numerical comparison between the proposed lite pipeline and the original one is reported in Table 1. The overall rotational and translation errors increased with a total error increasing from 0.0185 to 0.377. The rightmost column indicates the model size, in terms of number of network parameters (M), for both pipelines: it is worth noting that the original model had a size of 28.5 M whereas the proposed one has a size of 1.1 M (more than 25 times less).

Table 1. Errors and complexity comparison.

Approach	Mean rot. error	Mean trans. error	Mean total error	Network #Params
[2]	0.0141	0.0044	0.0185	28.5 M
Proposed	0.0302	0.0075	0.0377	1.1 M

Another relevant aspect to take into account is the average prediction time. The average time to predict the pose of the target vehicle by using the proposed pipeline is about 5 ms (i.e. 5.10 ms) whereas it is about 12 ms (i.e. 11.94 ms) by using the pipeline in [2].

Since, as largely stated in the introductory section, the computational load and the hardware resources are crucial points for space applications, the above numerical outcomes demonstrated that the introduce pipeline is particularly suited for the rendezvous process since it relies on a model with much fewer parameters allowing shorter computational time per image. To this aim, it is worth noting that a typical rendezvous process can be divided into several phases, including phasing of close-range rendezvous, final approaching, and docking. Relative navigation and control are mainly used in the close-range rendezvous phase, and orbit and attitude combined six-degree of freedom (6DOF) control is used in the final approaching phase [12]. Under this perspective, it is worth noting that, for close-range operations (up to about 20 m), the errors in pose estimations obtained with the Lite-HRNet are comparable with those gathered by state-of-the-art approaches relying on deep learning-based models.

4 Conclusions

This work has proven how the Lite-HRNet can be effectively exploited for 6DOF pose estimation for in space rendez-vous maneuvers. It represents the first attempt of using this recent full-resolution architecture for in space operations. Experimental validations on a benchmark dataset demonstrated that, for close range operations (up to 20 m), the errors in pose estimations are comparable with those gathered by state-of-the-art approaches relying on deep learning based models much more complex allowing, also, a half time of processing per image. Future works will deal with building a top-down pipeline that, relying on temporal/spatial filtering tricks, can alleviate also the computational load of the object detection step. Besides, a tool for generating synthetic datasets of photorealistic GAN-Generated scenes [1] will be introduced in order to train the pipeline for any target.

Acknowledgement. This work was supported in part by the Ministry of Education, University and Research under the grant PM3 AER01_01181 Modular Multi-Mission Platform.

References

1. Brock, A., Donahue, J., Simonyan, K.: Large scale gan training for high fidelity natural image synthesis. arXiv preprint arXiv:1809.11096 (2018)
2. Chen, B., Cao, J., Parra, A., Chin, T.J.: Satellite pose estimation with deep landmark regression and nonlinear pose refinement. In: Proceedings of the IEEE/CVF International Conference on Computer Vision Workshops (2019)
3. Chen, B., Parra, A., Cao, J., Li, N., Chin, T.J.: End-to-end learnable geometric vision by backpropagating PNP optimization. In: Proceedings of the IEEE/CVF Conference on Computer Vision and Pattern Recognition, pp. 8100–8109 (2020)
4. Cheng, Y., Wang, B., Yang, B., Tan, R.T.: Monocular 3d multi-person pose estimation by integrating top-down and bottom-up networks. In: Proceedings of the IEEE/CVF Conference on Computer Vision and Pattern Recognition, pp. 7649–7659 (2021)
5. Hu, X., Chu, L., Pei, J., Liu, W., Bian, J.: Model complexity of deep learning: A survey. arXiv preprint arXiv:2103.05127 (2021)
6. Hu, Y., Speierer, S., Jakob, W., Fua, P., Salzmann, M.: Wide-depth-range 6d object pose estimation in space. In: Proceedings of the IEEE/CVF Conference on Computer Vision and Pattern Recognition, pp. 15870–15879 (2021)
7. Jeon, M.H., Kim, A.: Prima6d: rotational primitive reconstruction for enhanced and robust 6d pose estimation. IEEE Robot. Autom. Lett. 5(3), 4955–4962 (2020)
8. Kisantal, M., Sharma, S., Park, T.H., Izzo, D., Märtens, M., D'Amico, S.: Satellite pose estimation challenge: Dataset, competition design, and results. IEEE Trans. Aerosp. Electron. Syst. 56(5), 4083–4098 (2020)
9. Kothari, V., Liberis, E., Lane, N.D.: The final frontier: deep learning in space. In: Proceedings of the 21st International Workshop on Mobile Computing Systems and Applications, pp. 45–49 (2020)
10. Lepetit, V., Moreno-Noguer, F., Fua, P.: Epnp: an accurate o (n) solution to the pnp problem. Int. J. Comput. Vision 81(2), 155 (2009)
11. Li, Z., Wang, G., Ji, X.: Cdpn: coordinates-based disentangled pose network for real-time rgb-based 6-dof object pose estimation. In: Proceedings of the IEEE/CVF International Conference on Computer Vision, pp. 7678–7687 (2019)
12. Luo, Y., Zhang, J., Tang, G.: Survey of orbital dynamics and control of space rendezvous. Chin. J. Aeronaut. 27(1), 1–11 (2014)
13. Ma, J., Jiang, X., Fan, A., Jiang, J., Yan, J.: Image matching from handcrafted to deep features: A survey. Int. J. Comput. Vision 129(1), 23–79 (2021)
14. Sharma, S., D'Amico, S.: Neural network-based pose estimation for noncooperative spacecraft rendezvous. IEEE Trans. Aerosp. Electron. Syst. 56(6), 4638–4658 (2020)
15. Sun, K., Xiao, B., Liu, D., Wang, J.: Deep high-resolution representation learning for human pose estimation. In: CVPR (2019)

16. Sundermeyer, M., Marton, Z.C., Durner, M., Brucker, M., Triebel, R.: Implicit 3d orientation learning for 6d object detection from rgb images. In: Proceedings of the European Conference on Computer Vision (ECCV), pp. 699–715 (2018)
17. Xiao, B., Wu, H., Wei, Y.: Simple baselines for human pose estimation and tracking. In: European Conference on Computer Vision (ECCV) (2018)
18. Yu, C., Xiao, B., Gao, C., Yuan, L., Zhang, L., Sang, N., Wang, J.: Lite-hrnet: a lightweight high-resolution network. In: Proceedings of the IEEE/CVF Conference on Computer Vision and Pattern Recognition, pp. 10440–10450 (2021)
19. Zhang, S., Zhao, W., Guan, Z., Peng, X., Peng, J.: Keypoint-graph-driven learning framework for object pose estimation. In: Proceedings of the IEEE/CVF Conference on Computer Vision and Pattern Recognition, pp. 1065–1073 (2021)
20. Zhang, X., Zhou, X., Lin, M., Sun, J.: Shufflenet: an extremely efficient convolutional neural network for mobile devices. In: Proceedings of the IEEE Conference on Computer Vision and Pattern Recognition, pp. 6848–6856 (2018)
21. Zhu, M., et al.: Single image 3d object detection and pose estimation for grasping. In: 2014 IEEE International Conference on Robotics and Automation (ICRA), pp. 3936–3943. IEEE (2014)

Imitation Learning for Autonomous Vehicle Driving: How Does the Representation Matter?

Antonio Greco[✉], Leonardo Rundo, Alessia Saggese, Mario Vento, and Antonio Vicinanza

Department of Information Engineering, Electrical Engineering and Applied Mathematics, University of Salerno, 84133 Fisciano, SA, Italy
{agreco,lrundo,asaggese,mvento,anvicinanza}@unisa.it

Abstract. Autonomous vehicle driving is gaining ground, by receiving increasing attention from the academic and industrial communities. Despite this considerable effort, there is a lack of a systematic and fair analysis of the input representations by means of a careful experimental evaluation on the same framework. To this aim, this work proposes the first comprehensive, comparative analysis of the most common inputs that can be processed by a conditional imitation learning (CIL) approach. With more details, we considered the combinations of raw and processed data—namely RGB images, depth (D) images and semantic segmentation (S)—to be assessed as inputs of the well-established Conditional Imitation Learning with ResNet and Speed prediction (CILRS) architecture. We performed a benchmark analysis, endorsed by statistical tests, on the CARLA simulator to compare the considered configurations. The achieved results showed that RGB outperformed the other monomodal inputs, in terms of success rate on the most popular benchmark NoCrash. However, RGB did not generalize well when tested on different weather conditions; overall, the best multimodal configuration was a combination of the RGB image and semantic segmentation inputs (i.e., RGBS) compared to the others, especially in regular and dense traffic scenarios. This confirms that an appropriate fusion of multimodal sensors is an effective approach in autonomous vehicle driving.

Keywords: Autonomous vehicle driving · Imitation learning · Conditional imitation learning · Benchmarking · CARLA

1 Introduction

Driving statistics confirm that distracted behavior is one of the main causes of the car accidents. Despite the effort of an experienced driver, keeping thorough attention on the road may be difficult, in particular during long travels [15]. This has motivated in the last decades a growing interest towards the design and development of autonomous vehicles. Anyway, nowadays this seems to be

S. Sclaroff et al. (Eds.): ICIAP 2022, LNCS 13231, pp. 15–26, 2022.
https://doi.org/10.1007/978-3-031-06427-2_2

feasible just in a limited set of scenarios, since the presence of dynamic obstacles in an unconstrained environment and the need to control the vehicle in real-time make this task particularly challenging. As pointed out in [20], the main techniques currently available for designing autonomous vehicles can be divided into modular and end-to-end (E2E) approaches. In a modular approach, the whole problem is partitioned into single tasks (e.g., data analysis, local planning, behavioral planning, motion planning). On the one hand, the advantage is the complete knowledge and control of each sub-task, thus allowing us to diagnose and solve possible driving errors. On the other hand, it requires a huge effort in the design of a system that takes into account any possible scenario. The E2E approaches drastically reduce the design effort by treating the whole driving problem as a single learning task despite a relevant amount of data would be necessary. In this field, we are experiencing a particular interest towards the imitation learning (IL) techniques, which allow us to train a policy learning from driving examples provided by an expert. For example, during the training phase, the model can be fed with raw data directly collected from sensors—such as an RGB image—or processed data—such as a semantic segmentation of the environment—and then it yields either the controlling value of the system—such as throttling, brake and steer values—or a set of waypoints that describe the local trajectory. The driving examples of the expert, also known as demonstrations, are a collection of pairs of inputs and outputs of the policy. The simplicity of data annotation and the possibility of training policies on recorded data are the most important advantages of IL [1,5]. Therefore, in this paper we focus on the approaches based on IL, which are attracting the interest of the scientific community in the latest years [14].

Relying upon the analysis of the literature, we can observe that most of the papers introduced new architectures or different inputs, but there is a lack of a systematic and fair analysis of the input representations by performing the experimental evaluation on the same framework. In addition, we note that the CIL technique is the most promising and investigated E2E approach in recent years. According to these observations, in this paper we aim to contribute to the state-of-the-art on multi-input CIL by proposing:

1. a comprehensive, comparative analysis of the most common inputs and their combinations that can be processed by a CIL approach [6]. To this end, we analyzed raw and processed data, namely RGB images, depth images and semantic segmentation;
2. a benchmark analysis, endorsed by appropriate statistical tests, on the CARLA simulator [9], to compare the considered configurations. We executed the analysis of 6 input configurations over 450 benchmark episodes for a total time of 135 h of testing time.

This work is organized as follows: Sect. 2 summarizes the literature background. Section 3 describes the dataset and benchmarks, as well as provides a complete description of the architectures and configurations used in our work. Section 4 presents and extensively discusses the results obtained. Finally, in Sect. 5 we draw some conclusions and future work.

2   Related Work

The most relevant literature approaches are described in this section.

One of the first examples was proposed by Bojarski *et al.* [2], who introduced the use of convolutional neural networks (CNNs) for imitation learning applied to autonomous vehicle driving. This method can only perform simple tasks, such as lane following, because it has not any command (from user or from automatic navigation systems) as input that can make a decision at an intersection, and so it cannot perform a navigation task. In [3], the authors showed the regions of an input image that contribute to the prediction of new actions. This approach attempts to provide an explanation to the output of the self-driving policy. This was the first approach to the explainability problem in this field. The recent work in [7] used attention mechanisms to achieve the same purpose. The authors highlighted that their method achieved results comparable with the state-of-art and, at the same time, the output is explainable.

A temporal sequence of data as input of a driving policy was introduced by Xu *et al.* [18] in which a recurrent neural network (RNN) was used to predict a moving path. The temporal information took into account the whole sequence of the performed actions but the RNN made the model computationally heavier. A step forward for the E2E architectures was proposed by Codevilla *et al.* [5] with the introduction of the conditional imitation learning (CIL) technique. This framework aims to solve the ambiguity at the intersections that the previous approaches suffer. It introduced lateral and longitudinal controls that perform the navigation task and defined the 'high-level command' input that enables the interaction with the navigation system. A modified version of this method was presented in [6]. The authors pointed out that the approach proposed in [5] exhibited limitations due to the bias of the adopted driving dataset. They proposed to add an input branch, as well as a deeper convolutional backend network, to use the measured speed as additional feature. This approach can reduce the inertia problem, i.e., the difficulty of restarting the vehicle after a stop for any reason. The policy does not directly model the causal signals, so it confuses the causes that lead to the stop command. Further works based on CIL methods exploit the effectiveness of the architecture using different learning methodologies. Chen *et al.* [4] presented, to the best of our knowledge, the state-of-art approach on simulator benchmarks, such as NoCrash [6] and CARLA [9]. The method was based on an agent trained with privileged information obtained from a simulator and then adopted as an expert for training a second agent without additional information. It outperformed the previous model but it is difficult to train in a real environment due to the presence of the privileged agent that should be trained using additional information that is not directly available in a real-world environment because it requires a huge effort for the labeling. The method proposed by Ohn *et al.* [13] was based on a multi-expert policy. It was trained using multiple driving policies with an IL strategy and the combination of the policies was then optimized *via* a task-driven refinement. From the one hand, it achieved the performance comparable to [4] but without requiring an additional image labeling. On the other hand, it needed a simulator to execute the task-driven refinement with an on-policy training.

The previous approaches assessed the performance of an end-to-end approach by using data acquired from a single sensor. Nevertheless, an autonomous vehicle is equipped with several sensors and all of them can contribute to the navigation task. The use of raw or processed data may be an additional design choice. An in-depth analysis on the representation of the input data is required to fully exploit the huge number of information extracted by the various sensors. In the recent literature, the exploitation of limits about the CIL technique by varying the input data has attracted a huge interest. In [1], a semantic segmentation (S) image was used as input of the network and the authors showed the effectiveness of this representation by varying the number of output classes and the resolution of the representation but they limited this analysis only to this type of input. The authors of [10] used a multimodal approach based on camera and Lidar. They proposed an algorithm to extract a Polar Grid View representation from the Lidar and then used a mid-fusion scheme.

In [17], an analysis of a CIL method with the use of an RGB camera with depth information was reported. The work focused on the different sensor-fusion methods between an RGB image and a depth image and demonstrated the effectiveness of an early-fusion scheme that used the four channel image composed of RGB image and depth image (RGBD) as input of the neural network. However, the best performing method showed the following constraint: the size of the RGB and depth images have to be the same to align them. The use of an RGBD image was typical also in [12], where the authors introduced the semantic segmentation of the scene as an additional task. They obtained an improvement of the performance compared to the early-fusion RGBD of [17] but their approach required additional effort for providing the semantic segmentation of input images.

From an overview of the previous methods, we notice that the integration of multiple inputs (such as the depth image or semantic segmentation) into the CIL method generally produces an increase of the driving ability, thus obtaining the best performance in the navigation without the dynamic obstacle task of the CARLA Benchmark. A further improvement of performance is obtained with the use of a deeper backend architecture and providing an additional task as the estimation of the semantic segmentation.

3   Materials and Methods

3.1   Dataset

The dataset CARLA100 [6] contains about 100 h of driving collected on Town01 of the CARLA simulator. From a first analysis of the dataset, we noticed that only 25 h provide RGB, semantic segmentation and depth images and, therefore, we selected only this subset of data for our experiments. We used also additional information, such as the speed measurement and the high-level command provided by the user. This last information expresses the intention of a user to take a specific direction at an intersection, thus it implicitly describes different behaviors.

The high-level commands are listed in what follows:

- `Follow Lane`: the user is sufficiently far from an intersection and it continues to follow the lane;
- `Left`: the user expressed the intention of turning to left at the next intersection;
- `Right`: the user expressed the intention of turning to right at the next intersection;
- `Go Straight`: the user expresses the intention of going straight at the next intersection.

We analyzed the distributions of high-level commands over the episodes of the dataset obtaining Fig. 1a, where we can observe that the data are balanced.

A weather type is defined for each episode. In the whole dataset, there are the following weather conditions:

- Clear Noon;
- After Rain Noon;
- Heavy Rain Noon;
- Clear Sunset.

We analyzed also the distribution of the weather conditions over the episodes of the dataset in Fig. 1b, which shows a quite even distribution among the weather conditions.

We selected 20 h from the whole dataset preserving the *a priori* distribution of the high-level commands that allows for discriminating overall different scenarios (turns, lane following). We divided the dataset into 15 h for the training set and the remaining 5 h for the validation set.

3.2 Benchmark

The used benchmark was NoCrash [6], that is composed of three distinct tasks: Empty, Regular Traffic, and Dense Traffic. The tasks aim at emulating different traffic levels from an empty to a densely-populated town.

For each task, six different weather conditions were defined: the same of the training set and two additional ones (namely rainy after rain, soft rain sunset). For each pair ⟨Task, Weather⟩, the benchmark contains 25 episodes. Each of them consists of a path that the vehicle should travel across a specific CARLA town. An episode is considered successful when the vehicle completes the path without collisions and in a given time. The final measure is the success rate (expressed as a percentage) of episodes for each task provided by the driving benchmark. For result comparability with the literature, we performed the evaluation in Town01 and Town02.

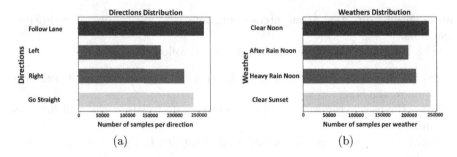

Fig. 1. Data distribution of the high-level commands (i.e., `Follow Lane`, `Left`, `Right`, `Go Straight`) and the weather conditions (i.e., Clear Noon, After Rain Noon, Heavy Rain Noon, Clear Sunset) over the samples of the dataset. **(a)** Data distribution of the high-level commands in the dataset; **(b)** Data distribution of the weather conditions in the dataset.

3.3   The Investigated CIL Method

The objective of imitation learning in the field of autonomous vehicle driving is the determination of a policy:

$$\pi : \mathcal{X} \to \mathcal{A}, \tag{1}$$

which links the input space $\mathcal{X}$, that can be composed of raw data collected from sensors, processed data or driving intentions, to a controlling space $\mathcal{A}$, defined in terms of driving commands or waypoints.

The policy is trained by minimizing a loss function $\mathcal{L}(\cdot, \cdot)$ between the model predictions and the demonstrations of an expert on the same input data. In this work, we adopt as output a three-dimensional controlling space composed of steering, throttling and braking commands, $\mathbf{a} = \langle s, a, b \rangle$. We also consider as additional task the prediction of the actual speed. The input space $\mathcal{X}$ depends on the specific configuration that we trained.

We considered the Conditional Imitation Learning with ResNet and Speed prediction (CILRS) architecture, introduced in [6], as our baseline, graphically represented in Fig. 2. It considers the predicted action $\mathbf{a}$ and the expert action $\mathbf{a}_{\text{expert}}$, and relies upon the loss function in Eq. (2), calculated for each sample:

$$
\begin{aligned}
\mathcal{L}(\mathbf{a}, \mathbf{a}_{\text{expert}}) &= \mathcal{L}\left(\langle v, s, a, b \rangle, \langle v_{\text{expert}}, s_{\text{expert}}, a_{\text{expert}}, b_{\text{expert}} \rangle\right) \\
&= w_1 \cdot ||v - v_{\text{expert}}|| + w_2 \cdot (\lambda_1 ||s - s_{\text{expert}}|| + \lambda_2 ||a - a_{\text{expert}}|| + \lambda_3 ||b - b_{\text{expert}}||),
\end{aligned} \tag{2}
$$

where w_1 and w_2 denote the weighting factors for the predicted speed and actual commands, respectively. Regarding the actual commands, $\lambda_1, \lambda_2, \lambda_3$ represent the weights for the terms s, a, b, respectively.

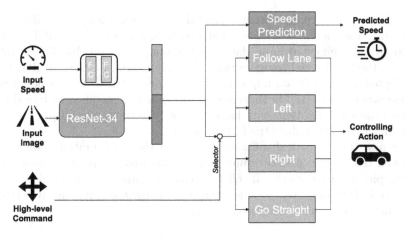

Fig. 2. General scheme of the CILRS architecture [6], which is used as our baseline for imitation learning. The investigated configurations can include as monomodal inputs: depth image, the estimation of a semantic segmentation, RGB image, or their multi-modal combinations. The network takes two additional inputs: (*i*) the current speed, and (*ii*) the high-level command that acts as a switch and can select the output values to use. The branches are: `Follow Lane`, `Left`, `Right` and `Go Straight`. This mechanism is adopted to predict the controlling action aimed at performing different behaviors at intersections.

Considering the per-sample loss function $\mathcal{L}(\cdot, \cdot)$ in Eq. (2) and assuming the dataset as a set composed of N $\langle \mathsf{Observation}, \mathsf{Action} \rangle$ pairs $\mathcal{D} = \{\langle \mathbf{o}_i, \mathbf{a}_i \rangle\}_{i=1}^{N}$, the objective of the imitation learning process is:

$$\underset{\theta}{\mathrm{minimize}} = \sum_{i=0}^{N} \mathcal{L}(\tilde{\pi}(\mathbf{o}_i; \theta), \pi_{\mathrm{expert}}(\mathbf{o}_i)), \tag{3}$$

where θ is the set of parameters of a function approximator $\tilde{\pi}(\mathbf{o}_i; \theta)$ of the expert's policy $\pi_{\mathrm{expert}}(\mathbf{o}_i)$ for the i-th sample.

3.4 Analysis of the Input Representation

In particular, we propose to investigate three monomodal and three multimodal images to represent the input in an autonomous vehicle:

1. Depth (D);
2. Semantic Segmentation (S);
3. RGB image (RGB);
4. RGB and Semantic Segmentation (RGBS);
5. RGB and Depth (RGBD);
6. RGB, Depth and Semantic Segmentation (RGBDS).

All the previous configurations were fed to the same architecture illustrated in Fig. 2. In case of the three monomodal configurations D, S and RGB—which provide a single image as input data—we did not modify our baseline architecture. For the multimodal configurations—namely RGBS, RGBD and RGBDS—we adopted an early-fusion approach that demonstrated to be the best fusion method according to [17]. In this case, the baseline was modified by introducing an extra convolutional layer with kernel size of 1×1 to keep as output the same input dimension and reduce the feature layers.

For assessing statistical differences between the configurations, we used a Wilcoxon test [16] on paired results (i.e., the distribution metric values for all the samples of the dataset). In all the tests, a significance level of 0.05 was considered. The computed p-values were then corrected using the Bonferroni-Holm method for multiple comparisons [11].

3.5   Training Procedure

The models were trained using the Adam optimizer with a batch size of 128 and an initial learning rate of 0.0002. We performed a min-max scaling, in the $[0, 1]$ range, of the input images. With the RGB image inputs, the following data augmentation transformations were applied: Random Hue Variation, Add Shadow, Add Fog, Darken, Brighten. The initial weights were pretrained on ImageNet [8]. In the loss function, the weighting factors were set as follows: $w_1 = 0.08, w_2 = 0.92$ and $\lambda_1 = 0.50, \lambda_2 = 0.45, \lambda_3 = 0.05$ according to [6].

All the models were implemented in Python using TensorFlow version 1.14.

4   Experimental Results

The methods were evaluated on the NoCrash [6] benchmark on Town01 and Town02. Table 1 shows the achieved results in terms of success rate.

The evaluation of the different configurations on the benchmark shows that among the monomodal inputs, D and S are not able to drive in a dense and new scenario. However, the Depth configuration did not generalize well and obtained the worst performance over all the configurations in presence of dynamic obstacles. This might be due to the lack of information on the surrounding environment, thus demonstrating difficulties to distinguish between the obstacles and the road. The semantic segmentation alone achieved low performance for all the benchmark tasks. The monomodal RGB configuration showed quite good driving abilities in an already known scenario but it did not generalize well in presence of different weather conditions. Although the adopted data augmentation should reduce this effect, over-specialization might still remain.

The multimodal inputs obtained overall the best results. Actually, they are able to effectively exploit the information conveyed by each sensor. From our analysis, the best combination of inputs is RGBS. It achieves the best performance on six tasks over the different scenarios against the four best values

Table 1. Results achieved by the investigated configurations. Each row contains the success rate (in percentage) of the episodes for each task in a specific scenario. The Training scenarios contain the results obtained on Town01 with the training weathers. The New Weather scenarios were obtained on Town01 with the testing weathers. The New Town scenarios were the results obtained on Town02 with the training weathers, while the New Town & Weather scenarios contain the results obtained on Town02 with the testing weathers. The highest performance for each setting is highlighted in bold.

Scenario	Task	D	S	RGB	RGBS	RGBD	RGBDS
Training	Empty	83	13	80	**97**	96	**97**
	Regular	59	21	70	**94**	83	86
	Dense	9	23	26	**58**	43	47
New Weather	Empty	80	10	92	96	92	**98**
	Regular	60	14	76	**94**	78	**94**
	Dense	10	26	44	**52**	40	38
New Town	Empty	11	8	27	**95**	69	90
	Regular	5	20	16	**83**	48	77
	Dense	0	8	6	29	16	**30**
New Town & Weather	Empty	10	10	8	74	48	**86**
	Regular	4	18	4	70	48	**80**
	Dense	0	6	2	**28**	10	26
Average	Empty	46.00	10.25	51.75	90.50	76.25	**92.75**
	Regular	32.00	18.25	41.50	**85.25**	64.25	84.25
	Dense	4.75	15.75	19.50	**41.75**	27.25	35.25

obtained by RGBDS. The evaluation of the average performance for each driving task confirms the previous trend. The RGBD configuration overcomes the monomodal inputs, but it is beyond the other multimodal combinations. Furthermore, we consider that the depth image can be obtained directly from a depth sensor, instead the semantic segmentation requires an additional computational effort for its estimation. Considering the trade-off between performance and prediction time, a well-established network is the Bilateral Segmentation Network (BiSeNet) V2 [19] takes on average 0.083 s on a NVIDIA Jetson Xavier AGX with a mean Intersection over Union (mIoU) of 72.6% for a prediction.

According to Table 1, all pairwise comparisons, based on the Wilcoxon tests, showed statistically significant differences ($p \ll 0.001$ after the Bonferroni-Holm correction).

We recorded the episodes performed by our best model on the CARLA Simulator, as shown in Fig. 3. From the recorded video, we identified two problems: the former is the jerky driving (i.e., non-smooth accelerations); the latter is related to the well-known inertia problem (i.e., when the vehicle stops or tends to remain stopped).

We tested the best configuration obtained on a real-size autonomous vehicle. We performed a fine-tuning on our real dataset acquired in our University campus and then we selected a set of well-known paths. From our experience, the autonomous vehicle is able to effectively perform the lane following task. Moreover, it stops in presence of pedestrians or other vehicles and it also reduces its speed at the crosswalks. However, we identified the same problems of the simulation environment: when its speed is 0, it is slow to restart, thus exhibiting the well-known inertia problem; on the other hand, we noted a jerky driving, making the experience of the human passenger not totally comfortable. A future direction of our research in this field may be the definition of a loss function that explicitly takes into account these negative effects on the driving task; currently, the inertia problem and the jerk are not considered in the adopted loss functions.

| (a) | (b) | (c) |

Fig. 3. Example frames extracted from the CARLA simulator during the execution of an episode of our best model. **(a)** The vehicle is stopped at the red traffic light. **(b)** The vehicle starts when the traffic light is green. **(c)** The vehicle stops beyond a car.

5 Conclusions

In this work, we presented the first comprehensive, comparative analysis of the most common monomodal inputs that can be processed by a conditional imitation learning approach, namely CILRS [6]. In particular, we considered raw and processed data, namely RGB, depth and semantic segmentation images, as well as their multimodal combinations.

The achieved results showed that RGB outperformed the other monomodal inputs, in terms of success rate on the most popular benchmark NoCrash. However, RGB alone did not generalize well when tested on different weather conditions. This confirms the limitations of a monomodal approach and that an appropriate fusion of multimodal sensors is an effective approach in autonomous vehicle driving. The achieved overall results showed that the best configuration was RGBS compared to the others. Interestingly, the development of CIL-based models using different input data might offer reliable systems in the case of sensor fault (e.g., depth image acquisition devices), by deploying the best performing CIL approaches, trained on different input data, on board.

Future work will be devoted to the extension of the current CIL approach proposed by Codevilla *et al.* [5,6], by injecting the driving rules into the learning procedure and fully exploiting the full potential of imitation learning. In real-world applications, we aim at assessing the semantic segmentation obtained by a CNN-based solution, such as BiSeNet V2 [19].

References

1. Behl, A., Chitta, K., Prakash, A., Ohn-Bar, E., Geiger, A.: Label efficient visual abstractions for autonomous driving. In: Proceedings of IEEE/RSJ International Conference on Intelligent Robots and Systems (IROS), pp. 2338–2345. IEEE (2020)
2. Bojarski, M., et al.: End to end learning for self-driving cars. arXiv preprint arXiv:1604.07316 (2016)
3. Bojarski, M., et al.: Explaining how a deep neural network trained with end-to-end learning steers a car. arXiv preprint arXiv:1704.07911 (2017)
4. Chen, D., Zhou, B., Koltun, V., Krähenbühl, P.: Learning by cheating. In: Proceedings of Conference on Robot Learning, pp. 66–75. PMLR (2020)
5. Codevilla, F., Müller, M., López, A., Koltun, V., Dosovitskiy, A.: End-to-end driving via conditional imitation learning. In: IEEE International Conference on Robotics and Automation (ICRA), pp. 4693–4700. IEEE (2018)
6. Codevilla, F., Santana, E., López, A.M., Gaidon, A.: Exploring the limitations of behavior cloning for autonomous driving. In: Proceedings of IEEE/CVF International Conference on Computer Vision (ICCV), pp. 9329–9338 (2019)
7. Cultrera, L., Seidenari, L., Becattini, F., Pala, P., Del Bimbo, A.: Explaining autonomous driving by learning end-to-end visual attention. In: Proceedings of IEEE/CVF Conference on Computer Vision and Pattern Recognition (CVPR) Workshops, pp. 340–341 (2020)
8. Deng, J., Dong, W., Socher, R., Li, L.J., Li, K., Fei-Fei, L.: ImageNet: a large-scale hierarchical image database. In: Proceedings of IEEE/CVF Conference on Computer Vision and Pattern Recognition (CVPR), pp. 248–255. IEEE (2009)
9. Dosovitskiy, A., Ros, G., Codevilla, F., Lopez, A., Koltun, V.: Carla: an open urban driving simulator. In: Conference on Robot Learning, pp. 1–16. PMLR (2017)
10. Eraqi, H.M., Moustafa, M.N., Honer, J.: Efficient occupancy grid mapping and camera-lidar fusion for conditional imitation learning driving. In: Proceedings of IEEE 23rd International Conference on Intelligent Transportation Systems (ITSC), pp. 1–7. IEEE (2020)
11. Holm, S.: A simple sequentially rejective multiple test procedure. Scand. J. Statist. **6**(2), 65–70 (1979). https://doi.org/10.2307/4615733
12. Huang, Z., Lv, C., Xing, Y., Wu, J.: Multi-modal sensor fusion-based deep neural network for end-to-end autonomous driving with scene understanding. IEEE Sensors J. MINO **21**(10), 11781–11790 (2020)
13. Ohn-Bar, E., Prakash, A., Behl, A., Chitta, K., Geiger, A.: Learning situational driving. In: Proceedings of IEEE/CVF Conference on Computer Vision and Pattern Recognition (CVPR), pp. 11296–11305 (2020)
14. Tampuu, A., Matiisen, T., Semikin, M., Fishman, D., Muhammad, N.: A survey of end-to-end driving: architectures and training methods. IEEE Trans. Neural Netw. Learn, Syst. (2020)
15. United States Department of Transportation: Risky Driving (2021). https://www.nhtsa.gov/. Accessed 18 Nov 2021

16. Wilcoxon, F.: Individual comparisons by ranking methods. Biometrics Bull. **1**(6), 196–202 (80–83). https://doi.org/10.2307/3001968
17. Xiao, Y., Codevilla, F., Gurram, A., Urfalioglu, O., López, A.M.: Multimodal end-to-end autonomous driving. IEEE Trans. Intell. Transp. Syst. (2020)
18. Xu, H., Gao, Y., Yu, F., Darrell, T.: End-to-end learning of driving models from large-scale video datasets. In: Proceedings of IEEE/CVF Conference on Computer Vision and Pattern Recognition (CVPR), pp. 2174–2182 (2017)
19. Yu, C., Gao, C., Wang, J., Yu, G., Shen, C., Sang, N.: BiSeNet V2: Bilateral network with guided aggregation for real-time semantic segmentation. Int. J. Comput. Vis. **129**(11), 3051–3068 (2021)
20. Yurtsever, E., Lambert, J., Carballo, A., Takeda, K.: A survey of autonomous driving: common practices and emerging technologies. IEEE Access **8**, 58443–58469 (2020)

LessonAble: Leveraging Deep Fakes in MOOC Content Creation

Ciro Sannino[1], Michela Gravina[1]([✉]) [iD], Stefano Marrone[1] [iD],
Giuseppe Fiameni[2] [iD], and Carlo Sansone[1] [iD]

[1] DIETI, University of Naples Federico II, Naples, Italy
ciro.sannino9@studenti.unina.it,
{michela.gravina,stefano.marrone,carlo.sansone}@unina.it
[2] NVIDIA AI Technology Center, Luxembourg, Italy
gfiameni@nvidia.com

Abstract. This paper introduces LessonAble, a pipelined methodology leveraging the concept of Deep Fakes for generating MOOC (Massive Online Open Course) visual contents directly from a lesson narrative. To achieve this, the proposed pipeline consists of three main modules: audio generation, video generation and lip-syncing. In this work, we use the NVIDIA Tacotron2 Text-to-Speech model to generate custom speech from text, adapt the famous First Order Motion Model to generate the video sequence from different driving sequences and target images, and modify the Wav2Lip model to deal with lip-syncing. Moreover, we introduce some novel strategies to support the use of markdown-like formatting to guide the pipeline in the generation of expression aware (i.e. curious, happy, etc.) contents. Despite the use and adaptation of third parties modules, developing such a pipeline presented interesting challenges, all analysed and reported in this work. The result is an extremely intuitive tool to support MOOC content generation.

Keywords: Deep fake · Tacotron 2 · Lip-sync · MOOC

1 Introduction

In recent years, Massive Open Online Courses (MOOCs) have spread exponentially, with their global market size expected to grow from USD 3.9 billion in 2018 to USD 20.8 billion by 2023[1], at a Compound Annual Growth Rate (CAGR) of 40.1% during the forecast period. This success is mostly due to the wide range of benefits they offer, such as the prospect to rethink a course content based on analytics and the opportunity to provide different course experiences through A/B testing to know which educational experience is more effective. MOOCs [9] are modern online courses for many participants at the same time ("massive"), without access restrictions ("open"), and in a course format (with video

---

[1] https://www.marketsandmarkets.com/Market-Reports/massive-open-online-course-market-237288995.html.

S. Sclaroff et al. (Eds.): ICIAP 2022, LNCS 13231, pp. 27–37, 2022.
https://doi.org/10.1007/978-3-031-06427-2_3

lectures and integrated tests). As a consequence, the students have the comfort of studying from home, self-paced learning, and much more.

An educational institution creates the learning content of a MOOC to teach and train students and experts. However, creating the MOOC content, i.e. a video lesson, often implies following a script (a text) already defined, with the author required to interpret it following every line of the defined text instead of recording a lesson on a wimp (as during a classical frontal lecture) to generate high-quality metadata (e.g. dubs) to support impaired students. Thus, despite this need tends to be extremely time-consuming, it is often a mandatory fair requirement. To take the best from this need, *in this work we introduce a pipelined methodology leveraging the concept of Deep Fakes, a recent AI-based approach to generate fake videos of a target subject, for generating MOOC (Massive Online Open Course) visual contents directly from that lesson script.* The idea is to realise a tool that supports the content generation by relieving the lecturer of the duty of both writing the lesson script and recording its lecture, automatically generating the latter from the former. To achieve this, the proposed pipeline consists of three main modules: the audio generation module, the video generation module and a lip-syncing step. Despite the proposed approach being general and realisable by using any approaches implementing the mentioned modules, in this work we:

- use the NVIDIA Tacotron2 [10] text-to-speech model to generate custom speech from text, by fine-tuning a pre-trained Deep Learning audio model on the target lecturer voice;
- adapt the famous First Order Motion Model [12] to generate the target video sequence from different driving sequences and target images, in order to make the resulting movie more natural and dynamic. We also introduced a simple trick to make the driving sequence switch smooth and duration independent;
- modify the famous Wav2Lip [7] model to deal with lip-syncing in our particular context.

Besides these three macroblocks, the proposed methodology also introduces a novel strategy to support the use of markdown-like formatting to guide the generation of expression aware (i.e. curious, happy, etc.) content. Despite the use and adaptation of third parties modules, developing such a pipeline presented interesting challenges, all analysed and reported in this work. The result is an extremely intuitive tool to support MOOC context generation leveraging for a good purpose Deep Fakes [5] a technology often associated with AI misuses, such as fake news, hoaxes and scams.

The rest of the paper is organised as follows: Sect. 2 introduces the Lesson-Able pipeline; Sect. 3 describes the application of the proposed solution and the experimental setup; finally Sect. 4 provides some conclusions.

2   Methodology

The proposed methodology is described following a top-down approach, introducing the LessonAble structure before describing each module and its implementation. The proposed work is a pipelined architecture designed to create, by sequential steps, the video lesson generating the voice, the video and lip-syncing the latter on the former. The conceived work is mainly focused on making the modules independent from each other, giving the possibility to replace or add new modules without the need for rewriting the whole architecture. Figure 1 shows how the modules of the pipeline interact in terms of input and output. The next sections detail each of the modules, focusing on the particular implementation choices we made in this paper. Nonetheless, it is worth noting that each module can be implemented by using different technologies (e.g., a different voice generator).

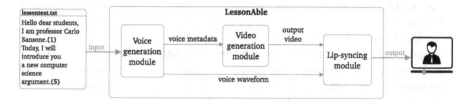

Fig. 1. The LessonAble pipeline consisting of three main modules: the *Voice generation*, the *Video generation* and the *Lip-syncing*. On the left, a lesson script is used as input to the voice generation module (Sect. 2.1). The voice module generates both a voice waveform and a voice metadata file containing the duration of each synthesized sentence and the markdown associated to it, to serve as input to the video generation module (Sect. 2.2). Finally, the lip-syncing module (Sect. 2.3) synchronises the lips in the generated video, based on the generated voice, producing the final lesson.

2.1 Voice Generation

The *Voice Generation* module aims to generate an audio file (e.g., in .wav format) reproducing the user's voice from a written text. To create a module able to generate audio in a target subject voice, it is needed to collect a dataset consisting of audio samples belonging to the target subject associated with their transcription. Besides the particular technology used to generate the voice, during the design of LessonAble we observed that the dataset collection stage can strongly affect the final quality. Indeed, although the formatting step often depends on the used text-to-speech technique, the dataset should always respect some characteristics, summarised as follows:

- **Gaussian like distribution on clips and text lengths**: the distribution of clip lengths should cover both short and long voice clips;
- **Mistake free**: any wrong or broken files should be removed. It is necessary to check annotations, compare transcript and audio length;
- **Noise free**: background noise might lead the voice module to struggle to learn;

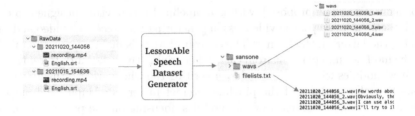

Fig. 2. Illustration of the Speech Dataset Generator working schema.

- **Naturalness of recordings**: the voice module can mime only the information within the dataset. Thus, to make the generated voice more natural it is important to have a dataset consisting of voice samples having tone and pitch differences (e.g., different punctuation);
- **Good phoneme coverage (for phoneme-based languages)**: the dataset must cover a good portion of phonemes, di-phonemes, and in some languages tri-phonemes. Depending on the case, if phoneme coverage is low, the model might have a hard time pronouncing novel difficult words.

In this work we leverage the NVIDIA Tacotron 2 [10] as Text-To-Speech (TTS) model to generate audios of the target subject, as in our preliminary analysis it outperformed all the other TTS systems, such as concatenate and parametric baseline ones. NVIDIA recommends using 16-bit audio with a sampling rate 22050 Hz. Indeed, this bit-depth provides a good signal to noise ratio and the sampling rate provides a good inference-speed-to-audio-quality ratio because it covers most of the human voice frequency range (80 Hz to 16 kHz). Audio with a higher bit-depth or sampling rate can be easily modified to match the described audio setup. Both training from scratch and fine-tuning can be exploited for Tacotron. To optimize training and memory consumption in large batches, it is recommended to use audio samples no longer than 10 s. Tacotron 2 recommends these prerequisites for its optimal usage; however, many users fail to achieve great results due to the dataset preparation and generation stage. Thus, to support the generation of a suitable dataset, we also realised the *LessonAble Speech Dataset Generator*[2] (Fig. 2). The tool pre-processes long audio/video files and the corresponding subtitles (in .srt format), generating voice samples and transcripts split into portions with a maximum length of 10 s, also meeting most of the aforementioned characteristics. To this aim, the tool iteratively splits the text into a set of sentences, each made up of a variable number of words and characters estimated so that they do not exceed the 10-second limit when pronounced.

---

[2] https://github.com/priamus-lab/LessonAble_SDG.

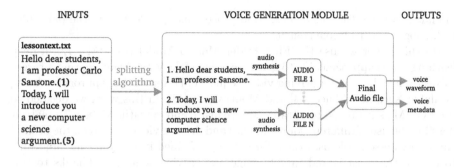

Fig. 3. The text is split into a set of sentences to provide audio lengths of at most 10 s. Each individual sentence is synthesised and then all the audio are merged to create a single output.

The consequence of having the training dataset consisting of several short samples is that also the whole lesson will be generated sentence by sentence, which will then be merged to obtain the final single large audio file (Fig. 3). It is also worth noting that the output of the *Voice Generation* module consists non only of the generated audio (i.e., the voice waveform), but also of an additional voice metadata file containing the duration of each single synthesised sentence and indications about the desired emphasis, to be used as input to the *Video Generation* module (Fig. 4).

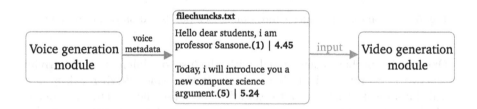

Fig. 4. Illustration of a voice metadata file. For each sentence, the file also reports the desired emphasis (in brackets) and the duration in seconds (after the pipe symbol), to be used by the *Video Generation* module.

2.2 Video Generation

Figure 4 depicts the video generation module using the voice metadata file previously generated by the voice generation module ("filechunks.txt" in the example) as input. This file contains *i*) the sentence (always ending with a dot), *ii*) the desired emphasis level (a number within brackets indicating which expression the user should have when pronouncing a certain sentence) and *iii*) the audio length in seconds (after a pipe symbol). The last two inputs are those actually needed by the video generation module, with the former used to guide the generation process and the latter fixing the generated video length. This structure is intended to make the video generation process independent from the particular technology used to implement it.

In this paper we use the First Order Motion Model (FOMM) [12], a deep learning based approach able to generate a deep fake video by animating a target subject image by using a driving video sequence. Despite other approaches being available, such as X2Face [14] and Monkey-Net [11], in this work we decided to use FOMM as it proved to obtain more realistic and versatile videos. To make it possible the use of different source images and driving videos based on the desired emphasis, LessonAble uses the information contained in the voice metadata file to choose the most suited target image and driving sequence. Thanks to this configuration, the algorithm is able to figure out which expression should be used in every specific part of the video lesson, as illustrated in Fig. 5. This allows us to make the generated lesson more natural, as the expression, movements, and so on change during the video based on the lesson text.

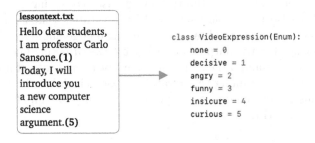

Fig. 5. An example of a video expression association based on the lesson text.

One of the problems associated with the use of different short driving sequences to generate a single longer video is that since driving videos have a fixed duration it is necessary to extend or decrease their length in order to match the audio pronunciation time. According to the frame per second (fps) of the driving sequence, LessonAble calculates the number of required frames by multiplying the audio duration (in seconds) by fps. In other words, denoting with spt the duration of a sentence's pronunciation, the number of required frames rf is computed as

$$rf = fps \cdot spt$$

If the pronunciation time of a sentence is greater than the duration of the driving sequence, the driving sequence is repeated in reverse order, to avoid glitches and spurious movements. The described process is iterated for every period or sentence with a different video expression. Finally, all the videos are merged together, obtaining the final output with the same duration of the audio file produced by the *Voice Generation* module. We refer to this process as *explicit expression-based video generation*.

2.3   Lip-syncing

The *Lip-syncing* module represents the final phase of the LessonAble pipeline. Different approaches have been proposed so far, including target-specific [3,13] as well as target-agnostic [4,8], introducing constrained and unconstrained talking face generation from speech. In this work we use an unconstrained method because it is independent of training data and thus can be applied to generic videos and audios. This approach is the best for our use case as any user input can be transformed to desired high-quality videos. Among all, we use the Wav2Lip library [7], a tool addressing the problem of lip-syncing a talking face video of an arbitrary identity by matching a target speech segment in a dynamic manner. Designed by the same authors of LipGan [8], it makes a step forward toward high-accuracy lip-syncing by using a more accurate discriminator and other smart tricks to make the synchronised video more natural and realistic.

3   Use Case

As a case of study, we generate a deep fake MOOC lesson for professor Carlo Sansone, from the University of Naples, Federico II. The content generation has been tested in two languages, English and Italian. For the *Voice Generation* module we performed fine-tuning for both languages, using the pre-trained model provided by NVIDIA for English and the ITAcotron2 [2] model for Italian. We performed the fine-tuning of both models with just fifteen minutes (for each language) of audio provided by Prof. Sansone, processing the raw data with the LessonAble Speech Data Generator. As described in Sect. 2.1, we just need a .mp4 or .mp3 file associated with a .srt file containing its transcription to generate a dataset ready to be used for fine-tuning. Every audio has a maximum duration of 10 s, saved as a waveform with a frequency 22050 Hz. A randomly selected 80% of the dataset was assigned to the training set while the remaining 20% to the test set. For the training, the number of epochs was set to 1000, a learning rate of 10^{-3}, weight decay of 10^{-6} and batch size of 64 and of 42 for English and Italian respectively. The rest of the parameters have not been changed from the configuration file proposed by Tacotron [10]. The main difference between Italian and English training is the 'cleaners' hyper-parameter, namely transformations (i.e., abbreviations, number expansions) that run over the input text at both training and validation time. Some cleaners are specific, such as the 'english_cleaners' used for the English language; for the Italian language we decided to use the 'phoneme_cleaners' [1], allowing simple phonemization of words and texts in many languages. Since the ITAcotron model is trained for multi speakers, it was first necessary to extract the speaker embedding before the training.

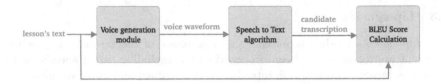

Fig. 6. Illustration of the audio evaluation process based on the BLUE Score.

To evaluate the output audio of the lesson, we used the Bilingual Evaluation Understudy (BLEU) score [6], a widely adopted metric for evaluating a generated sentence to a reference one. A perfect match results in a score of 1, whereas a perfect mismatch results in a score of 0. The score was developed for evaluating the predictions made by automatic machine translation systems. The BLEU score is quick and inexpensive to calculate and language independent, highly correlating with human evaluation. It works by counting matching n-grams in the candidate translation to n-grams in the reference text, where 1-gram or unigram would be each token, and a bigram comparison corresponds to each word pair. The comparison is made regardless of word order. In our case, the reference text is represented by the text of the lesson, while the candidate is calculated by entering the audio generated by LessonAble as input to a speech recogniser (Fig. 6). Based on a table shown in Fig. 7 and created by Google to evaluate machine-translated text, we can evaluate the synthesis of the audio generated by LessonAble: a good audio generation is the one with a BLEU greater than 40%.

BLEU Score	Interpretation
< 10	Almost useless
10 - 19	Hard to get the gist
20 - 29	The gist is clear, but has significant grammatical errors
30 - 40	Understandable to good translations
40 - 50	High quality translations
50 - 60	Very high quality, adequate, and fluent translations
> 60	Quality often better than human

The following color gradient can be used as a general scale interpretation of the BLEU score:

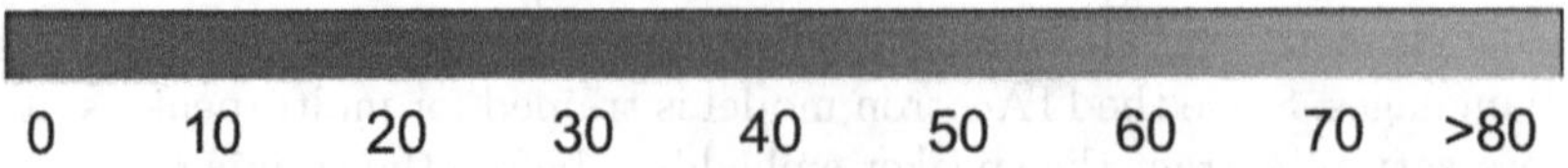

0   10   20   30   40   50   60   70   >80

Fig. 7. Percentage-based BLEU Score interpretation as suggested by Google.

The *Video Generation* and *Lip-syncing* modules use pre-trained models that require no further training. The provided source images and driving videos are used to create the video lesson. An example of output with a single source image and multiple driving videos is shown in Fig. 8, while the output of the generated lesson is available at this link. After every audio generation, we use the BLEU score. Based on 100 test cases of small lessons lasting at least 100 words, the average BLEU score of both the Italian and English models is above 0.52.

(a) Target image   (b) Driving videos, decisive (left) and curious (right).

Fig. 8. The video data inputs provided for the example.

4 Conclusion

In this work we have introduced LessonAble, a new approach to generate MOOC content based only on text. In particular, we proposed a pipelined approach consisting of three modules to *i*) generate the target lecturer custom fake voice, *ii*) generate the target lecturer Deep Fake video and *iii*) connecting both together by using lip-syncing. Despite in this work we use some specific approaches for each of these models, we want to highlight that the proposed methodology is general and can theoretically be used with any approach able to implement the specific module behaviour. Based on this, we argue that the pipeline's modules should be independent so that it can be easy to swap one model for a better one in the future. We invite readers to view the demonstration video at this *link*. We believe our efforts and our ideas in this approach can lead to new directions, such as the possibility of creating real fake MOOC content hubs. Furthermore, updating courses will be much easier as authors will only need to edit the text, while the video will be generated automatically.

Future research should focus on the application of expressions to the generated audio, providing the possibility to add slides during the lesson. Moreover, we plan to add more video expressions and to create voice models from scratch (for the Italian language). The result will be an extremely intuitive tool to support MOOC content generation leveraging for a good purpose Deep Fakes, a technology often associated with AI misuses, such as fake news, hoaxes and scams. Thus, a side effect of this work is also to further highlight that AI is only a (very powerful) tool, which misuse is to be associated with the user and not with the tool itself.

Code Base and Model Weights

The source code is available at https://github.com/priamus-lab/LessonAble, where we also provide the links to download the weights of the fine-tuned models for both Italian and English languages.

Acknowledgements. We acknowledge the CINECA award under the ISCRA initiatives, for the availability of high-performance computing resources and support within the projects IsC80_FEAD-D and IsC93_FEAD-DII. We also acknowledge the NVIDIA AI Technology Center, EMEA, for its support and access to computing resources, and the Federica Web Learning University center for providing professor Sansone's videos.

References

1. Bernard, M., Titeux, H.: Phonemizer: Text to phones transcription for multiple languages in python. J. Open Source Softw. **6**(68), 3958 (2021). https://doi.org/10.21105/joss.03958, https://doi.org/10.21105/joss.03958
2. Favaro, A., Sbattella, L., Tedesco, R., Scotti, V.: ITAcotron 2: transfering English speech synthesis architectures and speech features to Italian. In: Proceedings of The Fourth International Conference on Natural Language and Speech Processing (ICNLSP 2021), pp. 83–88. Association for Computational Linguistics, Trento, Italy, 12–13 Nov 2021. https://aclanthology.org/2021.icnlsp-1.10
3. Fried, O., et al.: Text-based editing of talking-head video. CoRR abs/1906.01524 (2019). http://arxiv.org/abs/1906.01524
4. Jamaludin, A., Chung, J.S., Zisserman, A.: You said that? synthesising talking faces from audio. Int. J. Comput. Vis. **127**, December 2019. https://doi.org/10.1007/s11263-019-01150-y
5. Nguyen, T.T., Nguyen, C.M., Nguyen, D.T., Nguyen, D.T., Nahavandi, S.: Deep learning for deepfakes creation and detection. CoRR abs/1909.11573 (2019). http://arxiv.org/abs/1909.11573
6. Post, M.: A call for clarity in reporting BLEU scores. CoRR abs/1804.08771 (2018). http://arxiv.org/abs/1804.08771
7. Prajwal, K.R., Mukhopadhyay, R., Namboodiri, V., Jawahar, C.V.: A lip sync expert is all you need for speech to lip generation in the wild. CoRR abs/2008.10010 (2020). https://arxiv.org/abs/2008.10010
8. Prajwal, K.R., Mukhopadhyay, R., Philip, J., Jha, A., Namboodiri, V., Jawahar, C.V.: Towards automatic face-to-face translation. CoRR abs/2003.00418 (2020), https://arxiv.org/abs/2003.00418
9. Reich, J.: Rebooting mooc research. Science **347**(6217), 34–35 (2015). https://doi.org/10.1126/science.1261627. https://www.science.org/doi/abs/10.1126/science.1261627
10. Shen, J., et al.: Natural TTS synthesis by conditioning wavenet on mel spectrogram predictions. CoRR abs/1712.05884 (2017). http://arxiv.org/abs/1712.05884
11. Siarohin, A., Lathuilière, S., Tulyakov, S., Ricci, E., Sebe, N.: Animating arbitrary objects via deep motion transfer. CoRR abs/1812.08861 (2018). http://arxiv.org/abs/1812.08861

12. Siarohin, A., Lathuilière, S., Tulyakov, S., Ricci, E., Sebe, N.: First order motion model for image animation. In: Wallach, H., Larochelle, H., Beygelzimer, A., d'Alché-Buc, F., Fox, E., Garnett, R. (eds.) Advances in Neural Information Processing Systems, vol. 32. Curran Associates, Inc. (2019). https://proceedings. neurips.cc/paper/2019/file/31c0b36aef265d9221af80872ceb62f9-Paper.pdf
13. Thies, J., Elgharib, M., Tewari, A., Theobalt, C., Nießner, M.: Neural voice puppetry: Audio-driven facial reenactment. CoRR abs/1912.05566 (2019). http:// arxiv.org/abs/1912.05566
14. Wiles, O., Koepke, A.S., Zisserman, A.: X2face: A network for controlling face generation by using images, audio, and pose codes. CoRR abs/1807.10550 (2018). http://arxiv.org/abs/1807.10550

An Intelligent Scanning Vehicle for Waste Collection Monitoring

Georg Waltner[1]([✉]), Malte Jaschik[2], Alfred Rinnhofer[2], Horst Possegger[1], and Horst Bischof[1]

[1] Institute of Computer Graphics and Vision, Graz University of Technology, 8010 Graz, Austria
waltner@icg.tugraz.at
[2] JOANNEUM RESEARCH Forschungsgesellschaft mbH, 8010 Graz, Austria
http://lrs.icg.tugraz.at

Abstract. While many industries have adopted digital solutions to improve ecological footprints and optimize services, new technologies have not yet found broad acceptance in waste management. In addition, past efforts to motivate households to improve waste separation have shown limited success. To reduce greenhouse gas emissions as part of a greater plan for fighting climate change, institutions like the European Union (EU) undertake strong efforts. In this context, developing intelligent digital technologies for waste management helps to increase the recycling rate and as a consequence reduces greenhouse gas emissions. Within this work, we propose an innovative computer vision system that is able to assess the residential waste in real-time and deliver individual feedback to the households and waste management companies with the aim of increasing recycling rates and thus reducing emissions. It consists of two core components: A compact scanning hardware designed specifically for rugged environments like the innards of a garbage truck and an intelligent software that applies a convolutional neural network (CNN) to automatically identify the composition of the waste which was dumped into the truck and subsequently delivers the results to a web portal for further analysis and communication. We show that our system can impact household separation behavior and result in higher recycling rates leading to noticeable reduction of CO_2 emissions in the long term.

Keywords: Convolutional neural networks · Deep learning · Computer vision · Cloud computing · Circular economy

1 Introduction

The world faces many ecological and economical challenges due to the steadily increasing global demand of resources, a process often referred to as the great

Powered by Austrian Climate- and Energy Fund - Project *DigiColl (877638)*.

acceleration [1]. Besides groundwater and air pollution the main issue is greenhouse gas emissions, among which carbon dioxide (CO_2) makes up the vast majority. Parts of this emissions are caused by household waste management, where waste must be transported, separated or incinerated. Each of these actions incurs considerable amounts of emissions. To reduce emissions and pollution in general, in 2015 the UN introduced the Agenda 2030 for Sustainable Development [2], where 17 global goals were defined, among them *paying special attention to air quality and municipal and other waste management* (Goal 11). One main pillar in waste management to increase recycling quotes is waste avoidance, with the aim of eliminating waste by keeping materials in use. This systematic approach is commonly referred to as circular economy. In that respect, several international efforts have been made [3–6]. More recently the European Green Deal [7] was announced, with ambitious goals like reduction of emissions by 40% until 2030 and to 0% until 2050.

One aspect in above programs is transforming local communities and the waste management industries for the digital era. The availability of cheap devices and sensors allows for broad deployment in urban areas and leads to so called smart cities. In this context, opportunities for Internet-of-Things (IoT) enabled waste management arise [8,9], mostly targeting the inefficiencies in transport and disposal in a general approach. A complementary and more specific approach is to target households to reduce the fraction of burned residential waste and keep valuable resources in the loop. This leads not only to less emissions but also lower costs for communities, giving monetary incentives to the households

Fig. 1. System overview. In the center of the waste monitoring system is the intelligent garbage truck that is equipped with a complex hardware set combined within a compact and sealed casing. The truck acquires images and sends them for processing to the cloud using a GSM module. After that, the AI model classifies the waste and calculates a waste distribution that is sent to the customer web portal, where the information is communicated to single households. (This figure has been created using and modifying resources from https://www.freepik.com/pch-vector.)

and communities as main target groups. Especially households play a critical role in increasing recycling quotes, as sorting the mixed residual waste later at the sorting plants is very inefficient and too expensive compared to burning the waste or landfilling. To achieve higher separation quotes, the awareness in households must be raised - a task that did not succeed in previous attempts, as they were either not targeted enough or required too much effort to reach a critical mass of households. One main factor that influences household waste separation is the individuals attitude [10], therefore constant qualified feedback is required to keep the individuals in the loop and raise the overall participation and engagement of citizens. With this feedback positive separation habits should be strengthened, especially in regards to the most problematic contaminants like batteries or plastics. There is common agreement that plastics and microplastics pose the biggest threat to our environment, polluting agricultural soil and marine ecosystems. After the import stop announced by China and Hong Kong in 2018 - who prior to that had imported over 70% of the worldwide share since 1992 [11], the problem of plastic recycling intensifies. In this concern, the EU has launched the European Strategy for Plastics in a Circular Economy [12] to achieve higher plastic waste recycling rates. The European targets for 2035 include a recycling rate of 75% for plastic waste and 65% for municipal waste, while limiting landfill to 10% at maximum [5]. As of now, while some countries like Austria, Belgium, Denmark, Germany, Luxembourg, Netherlands and Sweden are close to this target, the average of the EU-28 states remains at only 46.8% of recycled or composted municipal waste [13]. This reflects the different waste management strategies within the EU, where mostly in eastern countries incineration and landfilling dominate. The average EU recycling rate is comparable to the rate of the USA, as of now only few countries worldwide like Korea accomplish a higher rate of around 70% [14].

The proposed system focuses on the household level to target the recycling problem in an efficient and intelligent manner by automating the waste collection monitoring and providing real-time individual feedback. This is accomplished with an intelligent garbage truck that automatically acquires a set of images from the vehicle interior directly after waste containers are dumped inside. An RGB image is complemented with a multispectral image that delivers information up to near-infrared wavelengths (useful for organic waste) and depth information (useful for discrimination based on surface properties). The images are analyzed with artificial intelligence methods and the results are submitted to a web portal, where feedback is generated and passed on to the households. This feedback loop improves the waste separation habits, leads to increased recycling quotes and subsequently to reduced emissions. The proposed system is scalable and designed with cost efficiency in mind, to allow for broad and quick roll-out to communities for better waste management.

While there exist highly sophisticated computer vision supported systems in waste separation facilities, they not only come at a very high price, making them infeasible for deployment in large quantities, but also miss the link to single households. Only with the evaluation of every single garbage container directly at the source, i.e., after it is dumped into the garbage truck, the connection between household and separation habits can be made, which is essential for qualified feedback and change in behavior. Our intelligent vehicle enables exactly this: Providing the households individual feedback with the aim of behavior change at large scale.

2 Related Work

Coordinators in charge of waste management make continuous attempts to improve waste separation quotes. On one hand, they try to improve efficiency of the waste collection process itself, e.g. they focus on routing optimization [15–17] or waste quantity prediction [18–20]. Some approaches build on the use of IoT devices, e.g. for measuring fill levels [21,22] or weight and volume [23] of garbage bins. On the other hand, recent measurements taken in smart cities to reduce waste management emissions usually also involve the citizens and households. It was shown that putting containers for recyclables closer to households [24] has an impact on household level and can lead to better waste separation. An incentives system where waste pickup fees were reduced in exchange for better separation also prove effective [25]. Other than with structural improvements or incentives/penalties we involve households in a more direct fashion where they get instant feedback to raise awareness for their own recycling behavior. A scalable feedback approach cannot be based on manual monitoring, which is why our approach includes the use of an intelligent waste recognition module based on modern computer vision techniques. This computer vision method enables to extract information from waste images. Computer vision was previously dominated by handcrafted methods to extract the most distinct features from an image for content evaluation. Within the last decade Convolutional Neural Networks (CNN) replaced those handcrafted methods and showed astonishing performances over a variety of computer vision tasks. These networks no longer require manual design of features but instead learn to extract them from given data, where each data point (image) is associated with one or more labels. Since the introduction of AlexNet [26] research shifted from handcrafted feature design to network architecture design. Many CNN architectures have been proposed for a multitude of tasks like image classification, object detection or image segmentation. In this work we focus on the task of semantic image segmentation, where the CNN assigns a label to every single image pixel. Contrary to image classification, where the image is assigned a single label or object detection methods that locate objects in the image using bounding boxes (Fast R-CNN [27], Faster R-CNN [28], SSD [29], ...) or polygons (Mask R-CNN [30], ...), semantic segmentation allows for a much more detailed image analysis.

The few currently available computer vision approaches for waste recognition are often employed to decide in a binary fashion, if an image contains waste or not. Authors in [31] use a CNN to detect waste in urban images, binary classification of organic vs. recyclable waste is applied in [32]. Other works try to detect general waste classes like bags, bins or blobs from video streams [33].

Some works try to extract information about the waste composition directly from images. While not automated, in [34] authors estimate the physical composition of mixed waste using image based material analysis. Other approaches involve machine learning methods for classification of construction and demolition waste [35], classification of different plastic bottle types on a conveyor belt [36] or multi-class detection of single waste items [37]. Similar to our approach of classifying not only single objects but a mixture, authors in [38] use as sliding-window approach with a CNN classifier to generate waste label points in the image as input for a Gaussian clustering process. However they show results on simulated piles where objects do not overlap and are only partly identified, an oversimplified and unrealistic setting considering real world waste scenarios. As in our case, authors in [39] apply pixelwise segmentation, but only use two labels (waste and non-waste). Our approach in contrast classifies every single pixel of an real waste image into several classes and produces a much more detailed analysis of the composition. To the best of our knowledge, the only previous approach incorporating computer vision technology directly into waste vehicles is a road sweeper applying computer vision to control brushes, as a result saving electricity and reducing wearout [40].

3   Intelligent Waste Recognition System

At the heart of the proposed intelligent waste recognition system is the garbage truck, as shown in Fig. 1. The truck is equipped with a compact hardware housing that contains multiple cameras, illumination, a GPS device, a GSM router and an industrial PC for image acquisition. This acquisition system is mounted on the ceiling of the lifting unit on the rear of the truck and immediately starts the recording sequence after a garbage container was lifted and emptied to the hopper. The resulting dataset consists of images from different sensors and a GPS position. As soon as the acquisition finishes, the GSM module starts transferring the dataset to the cloud, where the different modalities are preprocessed into one image stack, which is passed on to an AI based image recognition system that automatically assesses the waste composition. The results are then delivered to a web portal that sends feedback via a SMS gateway to the households. In addition, authorized community waste management personnel can use the web portal to monitor separation status for whole streets, districts or communities.

4   Hardware Design

To reduce maintenance of the involved hardware parts, a robust custom designed stainless steel box with glass windows was built. This ensures that all parts can be mounted in a rigid fashion and image acquisition is reliable. The housing and all plugs must be protected against water damage during regular cleaning with a high-pressure cleaner. Furthermore, in case of unexpected problems the box can be quickly replaced and analyzed while operation can be continued. Inside the box is a multispectral snapshot camera that records 9 channels between 550 nm and 830 nm wavelength, a RGB camera and a dual monochrome camera stereo setup for depth. For a broad acceptance of the system in the market a convincing cost effectiveness is needed. The combination of a high resolution RGB camera, covering the visible range of wavelengths with a snapshot VISNIR multispectral camera is a good approach to get the maximum information gain [41]. As most multispectral cameras, especially if they include the short-wave infrared (SWIR) range, are far too expensive for mass deployment and a pushbroom scanning approach is not applicable in the truck setup, we opted for the proposed setup which allowed a fast realization and a field test with a higher number of installed systems. For a constant illumination covering all necessary wavelengths, several halogen floodlights are placed inside the box. To prevent cooling problems of the system and to extend the service life of the lamps, each of them is switched on only with a short pre-glow time for the exposures. To operate such systems in trucks with a power supply providing voltage peaks from nominal 24V up to 40V requires special solutions, since the intensity and wavelength distribution of the lamps is highly influenced by the supply voltage. In addition, switching on multiple lamps at the same time produces voltage drops, which causes the controlling PC to reboot. The rationale for using lamps, even though they are complicated to handle, was that at the time of the project start no sufficiently powerful multispectral LED illuminators covering the VISNIR range were available at affordable prices. The recording process is controlled by an industrial PC unit with a passive cooling system. A GSM module is connected to the PC for data transfers into the cloud. A systemic overview of the hardware composition is shown in Fig. 2, where also the power and trigger signals coming from the vehicle are depicted.

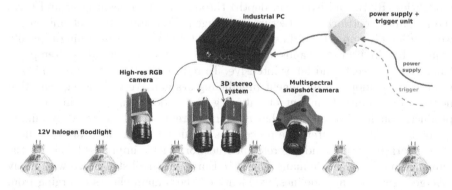

Fig. 2. Hardware overview. The hardware system consists of a stereo system, RGB and multispectral cameras, halogen floodlights and an industrial PC. The trigger signal and power supply are provided by the garbage truck.

5   Software Design

Upon arrival in the cloud, the data goes through a two-step procedure. First, the preprocessing instance takes all modalities, registers them so that pixels from different sensors are located at the same position and forwards them to the intelligent waste segmentation instance. In the second step, a convolutional neural network assigns a class to each pixel and outputs a class distribution over all pixels, where the background is ignored. This evaluation is then sent to the households along with a customized feedback. In the following, we describe the steps in more detail.

5.1   Data Preprocessing

The utilized sensors yield different data types which must be aggregated into one data stack per recording. First, the image modalities (3 RGB channels, 1 depth channel and 9 multispectral channels) are registered, warped and resized. For the RGB and multispectral images, an automated white balance correction to account for illumination changes is applied. The outcome of this preprocessing procedure is the input for the subsequent intelligent waste segmentation task.

5.2   Intelligent Waste Segmentation

To automatically extract the composition of waste from images a computer vision method is applied that generates waste composition statistics from waste images. Our model is a slightly modified DeepLabv3 [42] version that allows us to use the different modalities within the framework without adding too much computational overhead. We therefore apply early fusion on the different image modalities (RGB, multispectral, depth) through a compression layer and leave the remaining network architecture unchanged. To counter dataset imbalance, we additionally apply Focal Loss [43]. Successful CNN model training typically requires hundreds or thousands manually annotated images, which is often problematic as it is cost- and labor-intensive. To reduce the labeling effort and save resources, we adopt an intelligent prelabeling system. We train a segmentation network on a limited amount of data, e.g. a few dozen images. This network produces rough labels at first, that are further refined and corrected by human annotators. This is an iterative process, as the new human labels are then used for network retraining and the retrained network yields improved labels for the human annotators. Waste management in Europe is small-structured with many divergent separation guidelines, we identified the 6 main classes emerging from those guidelines for evaluation in our approach. For regional requirements, a finer differentiation might be needed. The results of the intelligent waste segmentation model are submitted to the web portal for analysis and communication.

5.3   Result Analysis and Communication

To close the loop to households, the customized feedback is generated from the web portal. Along with information about the recycling progress over time, the users receive detailed information about their waste compositions. This shows them which contaminants were wrongly placed in the waste bin and influences their future behavior.

6   Results

We verify both our intelligent CNN based waste segmentation model and the overall system impact. For the first part, to evaluate the performances of our CNN model, we curated a dataset of waste images and annotated them by hand as human groundtruth. The dataset contains the dominant 6 waste categories *organic*, *garbage_bag*, *paper*, *PET*, *plastic* and *residual*. We also add a *background* class for the garbage truck and an *ignore* class to mark image areas that cannot be identified and are left out during training. Our labeled dataset consists of 3107 images, where 2908 are used for training, 60 for model validation and 139 for evaluation of the final models. A big challenge is the small proportion of waste in relation to background as well as the imbalance in image area proportions for different classes (see Table 1 for details). We train several CNN models and evaluate them on the held-out test dataset. To evaluate the difference between

Table 1. Training and test dataset overview: 2908 and 139 images were used for training and testing the CNN models. Strong class imbalances makes waste segmentation a very hard task, as some classes only cover 2% or less image area on average.

Class	Img count	Img area (mean±std) [%]	Img area range [%]
background	2908/2908	71.67 ± 9.77	0.61 − 93.86
ignore	1882/2908	2.68 ± 5.65	0.00 − 67.06
organic	956/2908	1.79 ± 4.27	0.00 − 39.39
garbage_bag	2364/2908	6.92 ± 5.42	0.00 − 49.07
paper	2449/2908	2.04 ± 2.39	0.00 − 18.85
pet	1787/2908	0.62 ± 0.77	0.00 − 9.67
plastic	2844/2908	6.57 ± 5.33	0.00 − 33.91
residual	2908/2908	11.85 ± 5.29	5.02 − 58.13
a) train			
background	139/139	68.85 ± 10.21	39.03 − 100.00
ignore	72/139	1.78 ± 2.16	0.00 − 10.21
organic	96/139	19.84 ± 16.05	0.03 − 58.46
garbage_bag	93/139	6.39 ± 5.31	0.05 − 23.51
paper	82/139	3.10 ± 2.83	0.04 − 13.99
pet	59/139	0.87 ± 0.89	0.08 − 4.77
plastic	90/139	6.46 ± 5.08	0.05 − 32.46
residual	112/139	7.30 ± 6.47	0.00 − 35.78
b) test			

groundtruth (manual labels) and predictions (CNN output), we use two established metrics, accuracy (ACC) and Intersection-over-Union (IOU) scores and report mean/std over all images. The accuracy measures the amount of correctly labeled pixels, while the IOU measures how good the groundtruth and prediction areas overlap. Typically a score of 0.5 (50%) or higher is considered a good score. Evaluation results can be seen in Table 2. It is obvious, that the model perfectly learns to separate waste from the truck interior (*background* class). For the waste categories, results differ from 35% to 68% accuracy. This is due to the fact, that some classes are quite underrepresented in the training set (see Table 1) with *organic*, *paper* and *pet* covering 2% or less of the image area on average. A second issue is the subjective labeling due to the challenging nature of the data, where errors in the CNN predictions are often discussable. While the scores due to imperfect CNN segmentation masks are not optimal yet, it is often sufficient to find only parts of the contaminants for a qualitative feedback. If we weight the results with the image area proportions the different classes cover on average, the results are much better - the accuracy increases from 53.36% to 82.77%, the IOU from 43.84% to 77.22% (weighted average marked with * in Table 2). To get a better intuition for the numbers we show some qualitative results in Table 3. Due to the similar surface it is sometimes hard for the network to distinguish between *plastic* (yellow) and *garbage_bag* (light pink). Also, *pet* (turquoise) is a class with typically very small items where every inexact segmentation has large impact on the overall scores. On the other hand, our trained model is often able to spot contaminants that the human annotators missed during the labeling procedure. For example, in the lower right sample in Table 3 the CNN model correctly predicts organic waste (green) opposed to the annotated residual waste (grey) in the groundtruth.

Table 2. Results of our CNN model. Weighted results are marked with ᵂ, see text for details.

Class	ACC	IOU
background	98.35 ± 3.62	96.03 ± 5.15
organic	56.86 ± 35.91	53.24 ± 33.64
garbage_bag	68.42 ± 28.26	54.66 ± 26.24
paper	50.36 ± 28.77	40.13 ± 24.69
pet	35.61 ± 25.80	27.56 ± 20.60
plastic	49.97 ± 22.05	38.81 ± 18.80
residual	67.26 ± 26.01	40.32 ± 22.71
average	**53.36 ± 26.55**	**43.84 ± 25.41**
averageᵂ	**82.77 ± 14.06**	**77.22 ± 13.93**

Table 3. Qualitative results of our intelligent waste segmentation. Left: Input (RGB only for visualization), middle: CNN prediction, right: groundtruth labels. Best viewed digitally.

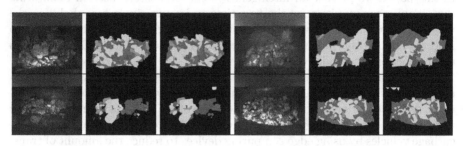

For the overall system evaluation, we monitored around 40 households over a period of 14 months. As Fig. 3 shows, the immediate household feedback improves separation rates by around 30% with ups and downs indicating seasonal fluctuations like increased organic waste around public holidays and reduced contaminants after improving communication strategies. An exact quantitative evaluation of resulting reductions in waste management costs and CO_2 emissions is part of an ongoing project. There are multiple dependencies on the collection area that must be taken into account (urban/rural, household income, education level, age structure, ...) to quantify the outcome of increased environmental awareness and separation behavior as result of our user-centric feedback component.

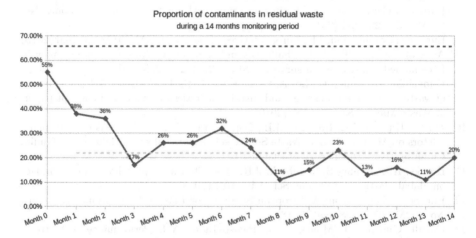

Fig. 3. Proportion of contaminants in residual waste over time. Households were monitored to evaluate the feedback impact. After communication start, the proportion went from 55% to 38% (blue line), showing the raised awareness. During the next year, it averaged at around 22% with seasonal fluctuations due to holidays etc. (yellow dashed line). The dashed grey line shows a nationwide average for comparison. (Color figure online)

7   Conclusion

In this paper, we present an intelligent system for waste collection monitoring that enables communities for the first time to involve single households in the waste management process. The real-time feedback allows the households to reflect on habits that might harm the common goal of reducing CO_2 emissions and a functioning circular economy.

At this stage the proposed system is designed after European waste management systems but adaptation for other parts of the world should be straightforward, as the collection of household waste is one of the main tasks in communities worldwide. In near future we plan to further enhance the hardware within the garbage vehicles by using edge computing devices to reduce the amount of transferred data and incorporate fast LED illumination for faster acquisition times.

References

1. Steffen, W., Broadgate, W., Deutsch, L., Gaffney, O., Ludwig, C.: The trajectory of the anthropocene: the great acceleration. J. Anthr. Rev. **2**(1), 81 (2015)
2. United Nations: Transforming our World: The 2030 Agenda for Sustainable Development. United Nations (2015)
3. Organisation for Economic Co-operation and Development: RE-CIRCLE: Resource Efficiency and Circular Economy (2018)
4. European Commission: Towards a Circular Economy: A Zero Waste Programme for Europe. No. COM(2014) 398 final/23 (2014)
5. European Commission: Closing the Loop - An EU Action Plan for the Circular Economy. No. COM(2015) 614 final (2015)
6. European Commission: A Monitoring Framework for the Circular Economy. No. COM(2018) 29 final (2018)
7. European Commission: The European Green Deal. No. COM(2019) 640 final (2019)
8. Anagnostopoulos, T., et al.: Challenges and opportunities of waste management in IoT-enabled smart cities: a survey. T-SUSC **2**(3), 275–298 (2017)
9. Esmaeilian, B., Wang, B., Lewis, K., Duarte, F., Ratti, C., Behdad, S.: The future of waste management in smart and sustainable cities: a review and concept paper. J. Waste Manag. **81**, 177–195 (2018)
10. Ma, J., Hipel, K.W., Hanson, M.L., Cai, X., Liu., Y.: An Analysis of Influencing Factors on Municipal Solid Waste Source-separated Collection Behavior in Guilin, China by Using the Theory of Planned Behavior. SCS 37, 336–343 (2018)
11. Brooks, A.L., Wang, S., Jambeck, J.R.: The Chinese import ban and its impact on global plastic waste trade. Sci. Adv. **4**(6) (2018)
12. European Commission: A European Strategy for Plastics in a Circular Economy. No. COM(2018) 028 final (2018)
13. Castillo-Giménez, J., Montañés, A., Picazo-Tadeo, A.J.: Performance in the treatment of municipal waste: are European union member states so different? Sci. Total Environ. **687**, 1305–1314 (2019)
14. Park, S., Lah, T.J.: Same material different recycling standards: comparing the municipal solid waste standards of the European Union, South Korea and the USA. Int. J. Environ. Waste Manag. **21**(1), 80–93 (2018)

15. A Heuristic Solution Method for Node Routing Based Solid Waste Collection Problems. J. Heuristics **19**(2), 129–156
16. Shah, P.J., Anagnostopoulos, T., Zaslavsky, A., Behdad, S.: A stochastic optimization framework for planning of waste collection and value recovery operations in smart and sustainable cities. J. Waste Manag. **78**, 104–114 (2018)
17. Aljoscha Gruler, Antoni Pérez-Navarro, L.C., Juan, A.A.: A simheuristic algorithm for time-dependent waste collection management with stochastic travel times. SORT **44**(2), 285–310 (2020)
18. Kannangara, M., Dua, R., Ahmadi, L., Bensebaa, F.: Modeling and prediction of regional municipal solid waste generation and diversion in Canada using machine learning approaches. J. Waste Manag. **74**, 3–15 (2018)
19. Hoque, M.M., Rahman, M.T.U.: Landfill area estimation based on solid waste collection prediction using ANN model and final waste disposal options. J. Clean. Prod. **256**, 120387 (2020)
20. Ghanbari, F., Kamalan, H., Sarraf, A.: An evolutionary machine learning approach for municipal solid waste generation estimation utilizing socioeconomic components. Arab. J. Geosci. **14**(2), 1–16 (2021)
21. Kumar, N., Swamy, C., Nagadarshini, K.N.: Efficient garbage disposal management in metropolitan cities using VANETs. JOCET **2**(3), 258–262 (2014)
22. Nirde, K., Mulay, P.S., Chaskar, U.M.: Iot based solid waste management system for smart city. In: Proceedings of ICICCS, pp. 666–669 (2017)
23. Sharmin, S., Al-Amin, S.T.: A cloud-based dynamic waste management system for smart cities. In: Proceedings of ACM DEV, pp. 1–4 (2016)
24. Rousta, K., Bolton, K., Dahlén., L.: A procedure to transform recycling behavior for source separation of household waste. Recycling **1**(1), 147–165 (2016)
25. Struk, M.: Distance and incentives matter: the separation of recyclable municipal waste. RCR **122**, 155–162 (2017)
26. Krizhevsky, A., Sutskever, I., Hinton, G.E.: Imagenet classification with deep convolutional neural networks. In: NeurIPS (2012)
27. Girshick, R.: Fast R-CNN. In: Proceedings of ICCV (2015)
28. Ren, S., He, K., Girshick, R., Sun, J.: Faster R-CNN: towards real-time object detection with region proposal networks. In: NeurIPS (2015)
29. Liu, W., et al.: SSD: single shot MultiBox detector. In: Leibe, B., Matas, J., Sebe, N., Welling, M. (eds.) ECCV 2016. LNCS, vol. 9905, pp. 21–37. Springer, Cham (2016). https://doi.org/10.1007/978-3-319-46448-0_2
30. He, K., Gkioxari, G., Dollár, P., Girshick, R.: Mask R-CNN. In: Proceedings of ICCV (2017)
31. Wang, Y., Zhang, X.: Autonomous garbage detection for intelligent urban management. In: MATEC Web Conference, vol. 232 (2018)
32. Toğaçar, M., Ergen, B., Cömert, Z.: Waste Classification using AutoEncoder Network with Integrated Feature Selection Method in Convolutional Neural Network Models. Measurement 153 (2019)
33. De Carolis, B., Ladogana, F., Macchiarulo, N.: YOLO TrashNet: garbage detection in video streams. In: Proceedings of EAIS, pp. 1–7 (2020)
34. Wagland, S.T., Veltre, F., Longhurst, P.J.: Development of an image-based analysis method to determine the physical composition of a mixed waste material. J. Waste Manag. **32**(2), 245–248 (2012)
35. Di Maria, F., et al.: Quality assessment for recycling aggregates from construction and demolition waste: an image-based approach for particle size estimation. J. Waste Manag. **48**, 344–352 (2016)

36. Wang, Z., Peng, B., Huang, Y., Sun, G.: Classification for plastic bottles recycling based on image recognition. J. Waste Manag. **88**, 170–181 (2019)
37. A Novel YOLOv3 Algorithm-Based Deep Learning Approach for Waste Segregation: Towards Smart Waste Management. Electronics 10(1) (2021],)
38. Wang, Y., Zhao, W.J., Xu, J., Hong, R.: Recyclable Waste Identification Using CNN Image Recognition and Gaussian Clustering. arXiv preprint arXiv:2011.01353 (2020)
39. Wang, T., Cai, Y., Liang, L., Ye, D.: A multi-level approach to waste object segmentation. Sensors **20**(14), 3816 (2020)
40. Donati, L., Fontanini, T., Tagliaferri, F., Prati, A.: An energy saving road sweeper using deep vision for garbage detection. Appl. Sci. **10**(22), 8146 (2020)
41. Rinnhofer, A.: Bestimmung von Wertstoffen im Restmüll, Leitfaden zur hyperspektralen Bildverarbeitung (2019)
42. Chen, L.C., Papandreou, G., Schroff, F., Adam, H.: Rethinking Atrous Convolution for Semantic Image Segmentation. arXiv preprint arXiv:1706.05587 (2017)
43. Lin, T.Y., Goyal, P., Girshick, R., He, K., Dollár, P.: Focal loss for dense object detection. In: Proceedings of CVPR, pp. 2980–2988 (2017)

Morphological Galaxies Classification According to Hubble-de Vaucouleurs Diagram Using CNNs

Pier Luigi Mazzeo$^{2(\boxtimes)}$ (ID), Antonio Rizzo^1, and Cosimo Distante^2 (ID)

^{1} Universitá del Salento, Via Monteroni sn, 73100 Lecce, Italy
^{2} ISASI - CNR c/o DHITECH, Via Monteroni sn, 73100 Lecce, Italy
pierluigi.mazzeo@cnr.it

Abstract. Galaxies morphology classification is a crucial task for studying their physical properties, formation and evolutionary histories. The large-scale surveys on universe has boosted the need to develop techniques for automated galaxies morphological classification. This paper proposes a system able to classify automatically galaxies according to the Hubble De Vaucouleurs diagram. We introduce a novel CNN architectures that for the first time was trained to automatically classify galaxies according to 26-classes Hubble-De Vaucouleurs scheme. We use Galaxy Zoo dataset, using the decision tree, to extract a labeled examples containing an even amount of images of each 26-classes. We also compared different CNN Backbones in order to assess obtained galaxies classification results. We obtain a balanced multi-class accuracy (BCA) of more than 80% in classifying all 26 Hubble-De Vaucouleurs galaxy categories.

Keywords: Convolutional neural networks · Galaxy morphology · Classification

1 Introduction

Nowadays 8 orbital optical telescopes and over 1000 professional observatories worldwide grabs many images a day for finding new stars, galaxies and planets. Galaxies show a different morphology over cosmic time [9], ranging from broad families of spirals to large elliptical. Morphology is linked to galaxy internal physical processes [27], such as dynamical and chemical evolution, star formation [18], and can reveal new knowledge from its evolutionary history such as mergers [20] and interactions with its environment [30]. A galaxy's morphology is commonly defined by visual inspection, and even if it is sufficient to recognize a huge range of morphological types in the nearby Universe [6], remains the intrinsic scalability issue of as future discoveries brings to far greater magnitudes of data.

The authors thank **Arturo Argentieri** for his technical support in the setup of the hardware used for network training and data processing.

Human based visual inspection of images or videos for classifying the apparent morphologies of galaxies is a laborious, time- consuming task. In the previous decade, thanked to the citizen science it has seen great success in classifying larger datasets [19,29,33], however it is important to highlight that this was achieved by increasing number of human classifiers, but the classification speed is remained unchanged. Modern telescopes continue to collect more and more images every day yield enormous amounts of data, i.e. only Euclid (orbital optical telescope) expected to pick up at least of 1 billion galaxies. The huge data quantity overwhelm the capacity of human volunteers [28], and developing and deploying of techniques for automated classification is necessary. Automated classification methodologies, particularly those using neural networks [12,13,36], increase the speed at which each galaxy images can be classified. These methodologies varying from Fourier analysis of luminosity profiles [24] and isophote decompositions [22], to statistical learning methods [31], random forest classifiers [5] and ensemble classifiers [2], which are all part of a larger group of machine learning methods [8]. Artificial Neural Network have been used in this field for a long time, but only in the last few years with growing field of deep learning seen increasing number of applications in astronomy [4]. CNNs have been successfully employed to detect quasars and gravitational lenses [25,26], to investigate bulge/disk dominance [14], detect stellar bars [1] and to classify radio morphologies [34]. CNNs have also been successfully used with simulations, including cosmological simulations [21] and mock surveys [23], as well as to realize tools for galaxy photometric profile analysis [32]. Recent studies have use CNNs for binary classification [14], or for classifying galaxies among general morphological shapes [36]. Fewer studies have looked at 3-way classification between distinct morphological types, though some works have explored classifying between ellipticals and barred/unbarred spirals [3], or between ellipticals, spirals and irregulars [10]. In [7] different CNN architectures are compared to distinguishing galaxies among four classes: elliptical, lenticular, spiral and irregular/miscellaneous. This work introduces for the first time, to the best of our knowledge, an automatic classifier based on CNN able to catalogue the galaxies according to the 26-classes Hubble-De Vaucouleurs diagram [11]. The Hubble-De Vaucouleurs scheme is one of the most complete turning fork diagram for galaxy classification. Furthermore in this paper different well-known CNN backbones are compared with a novel CNN architecture and all of them were trained on a subset of "galaxy-zoo" dataset, appropriately augmented. Galaxy Zoo [19] is a citizen science project in which about nearly one million galaxies are manually classified. We used a subset of the dataset used in [17] that contains more than 60k labelled images, and 80k unclassified ones for training and testing the CNNs. We obtained an balanced multi-class accuracy on the 26-classes up to 80%. This paper is organized as follows: in Sect. 2 is described the problem and the idea and how we have labelled the galaxy dataset. Section 3 summarizes the different network architecture we tested. In Sect. 4 are detailed and discussed the obtained results. Finally in Sect. 5 some conclusions are drawn.

2   Problem Statement

Figure 1 contains a Hubble-de Vaucouleurs classification scheme [11]. The de Vaucouleurs system retains Hubble's basic division of galaxies into ellipticals, lenticulars, spirals and irregulars. To complement Hubble's scheme, de Vaucouleurs introduced a more elaborate classification tuning system for spiral galaxies, based on three morphological characteristics: **Bars**: Galaxies are divided on the basis of the presence or absence of a nuclear bar. De Vaucouleurs introduced the notation SA to denote spiral galaxies without bars, complementing Hubble's use of SB for barred spirals. He also allowed for an intermediate class, denoted SAB, containing weakly barred spirals. **Lenticular**: galaxies are also classified as unbarred $SA0$ or barred $SB0$, with the notation $S0$ reserved for those galaxies for which it is impossible to tell if a bar is present or not (usually because they are edge-on to the line-of-sight). **Rings**: Galaxies are divided into those possessing ring-like structures (denoted '(r)') and those without rings (denoted '(s)'). So-called 'transition' galaxies are given the symbol (rs). **Spiral arms** As in Hubble's original scheme, spiral galaxies are assigned to a class based primarily on the tightness of their spiral arms. The de Vaucouleurs scheme extends the arms of Hubble's tuning fork to include several additional spiral classes: $Sd(SBd)$ - diffuse, broken arms made up of individual stellar clusters and nebulae; very faint central bulge $Sm(SBm)$ - irregular in appearance; no bulge component Im - highly irregular galaxy.

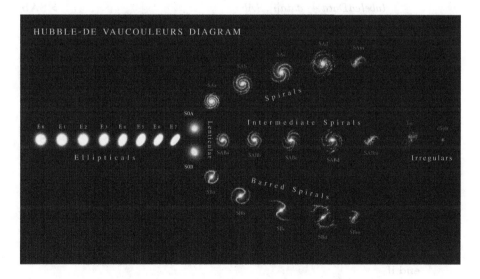

Fig. 1. Hubble-de Vaucouleurs tuning fork diagram

Algorithm 1. Galaxies decision tree pseudocode

$i \leftarrow 0$
while $i < len(data)$ **do**
   if $C_0 \leq 0.3$ **then**   ▷ Exclude all mergers
      if $V_y > 0.7$ **then**   ▷ Check for Elliptical **E**
         $label \leftarrow E\lceil (V_y * 10) - 10) * -1\rceil$
         $labeledData \leftarrow data[i], label$
      end if
      if $(C_9 > 0.7 || C_5 > 0.5)$ **then**
                                                   ▷ Check for Lenticulars **S0**
         if $(C_3 > 0.4 \& C_1 > 0.2)$ **then**
            $label \leftarrow S0B$ ▷ **S0B**
            $labeledData \leftarrow data[i], label$
         else
            if $((C_3 > 0.4 \& C_1 < 0.2)$ **then**
               $label \leftarrow S0A$ ▷ **S0A**
               $labeledData \leftarrow data[i], label$
            end if
         end if
      end if
      if $C_3 > 0.6 \& C_4 > 0.8$ **then**   ▷ Check for Spirals **SA**
         if $C_{10} > 0.5$ **then**
            $labeledData \leftarrow data[i], SAa$ ▷ **SAa**
         end if
         if $C_{10} > 0.5$ **then**
            $labeledData \leftarrow data[i], SAb$ ▷ **SAb**
         end if
         if $C_{11} > 0.4$ **then**
            $labeledData \leftarrow data[i], SAc$ ▷ **SAc**
         end if
         if $C_{11} > 0.5$ **then**
            $labeledData \leftarrow data[i], SAm$ ▷ **SAm**
         end if
         if $C_{10} > 0.4$ **then**
            $labeledData \leftarrow data[i], SAd$ ▷ **SAd**
         end if
         *   ▷ Similar code for Spirals **SAB** and **SB**
         *
         *
      if $C_4..C_8 > 0.7$ **then**   ▷ Check for Irregulars **Irr**
         $label \leftarrow Irr$
         $labeledData \leftarrow data[i], label$
      end if
   end if
   $i \leftarrow i + 1$

Table 1. Dataset composition after executing Algorithm 1

Elliptical	E0	E1	E2	E3	E4	E5	E6	E7
	41	565	1598	1312	1019	927	77	94
Lenticular	**S0A**	**S0B**						
	753	130						
Spirals	**SAa**	**SAb**	**SAc**	**SAd**	**SAm**			
	1216	710	823	116	188			
Barred Spirals	**SBa**	**SBb**	**SBc**	**SBd**	**SBm**			
	120	346	280	103	184			
Intermediate Spirals	**SABa**	**SABb**	**SABc**	**SABd**	**SABm**			
	217	401	432	103	205			
Irregulars	**Irr**							
	294							

2.1 Galaxy Zoo Dataset

The galaxy zoo project [19] proposed eleven questions to describe galaxies, in this work we use only a subset of them: i.e. we need only questions to classify galaxies, also the questions about how many arms a galaxy has have little meaning for classification. We transform each question in a condition $C_{0..11}$ and use them for galaxies categorization. In order to automatize this task we propose Algorithm 1 that with some nested-if allows to the galaxies images to be selected and assigned to right own categories.

Note that the threshold values can be optimized to gather more labeled images from the GalaxyZoo dataset, these values where set in an empirical way. Each class is determined through the execution of Algorithm 1 that contains conditions associated to threshold values that represents a probability to be or not part of a certain galaxy category (between 0 and 1). Algorithm 1 works as follows:

Elliptical Classes. Starting with Elliptical galaxies, we have 8 conditions going from a flat disk E0 to a squashed ellipse E7. To filter between all 7 levels the three conditions are used: one for E0 to E2, one for E3 to E5, E6 and one for E7 (i.e. for a particular galaxy in which the condition is equal to 0.9, the corresponding class is E1). If the value of the condition C_Y is greater than 0.7 then the equation to differentiate between classes is:

$$X_Y = \lceil ((C_Y * 10) - 10) * -1 \rceil \tag{1}$$

where X_Y is the number after 'E' of the corresponding conditions C_Y.

Lenticular Classes. For Lenticular galaxies S0B S0A we need condition C_9 and C_5 for checking how thick is the galaxy bulge. If Barred then question C_3 will be satisfied, vice versa for unbarred.

Spirals Classes. Spiral galaxies can be divided into 3 sub classes: barred (B), unbarred (A) and semi-barred (AB). For this class the condition C_4 will be strongly satisfied. We can define how barred a galaxy is by checking condition C_3, if is positive and above 0.5 then is barred, if both yes or no hangs over 0.5 then is semi-barred, finally if no is way above 0.5 then is unbarred.

Irregular Class. For irregular galaxy we just need to check condition from C_4 to C_8: if satisfied is pretty high we have an irregular galaxy.

After the categorization process executed through the Algorithm 1 galaxy dataset appears as detailed in Table 1.

2.2   Data Augmentation

After the nested-if process described in Algorithm 1 the dataset obtained is inevitably unbalanced, in order to make it balanced the best way is the data augmentation. The class with the most images is the pivot number, using randomly any combination of three transformations (random rotation, random noise, horizontal flip) each class with the number of images less than the pivot will be increased in size, in order to have all of them with the same image number. In Fig. 2 are showed some examples of image augmentation, a the end of this process the galaxy dataset has been increased up to 41598 images well balanced in each class.

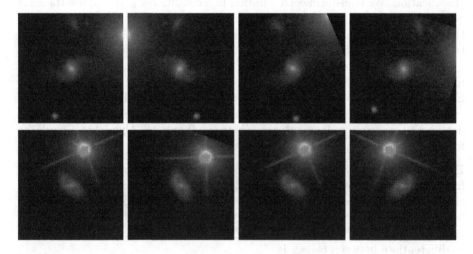

Fig. 2. Example of data augmentation: first column contains the original galaxy images (1st row **SBd** 2nd row **SBc**); other columns contain random rotation and noise and horizontal flip of the original galaxy images.

3   Method

In order to evaluate the best convolutional neural network able to solve the automatic galaxies classification we have compared different well known CNN backbones with the proposed architecture (see Fig. 3).

3.1   Proposed Architecture

The proposed architecture is principally composed by the concatenation of six convolution layers and 3 MaxPool as detailed in Table 2. A the and of the convolutional neural network is located a fully connected layer with 26-dimension output. Each row of Table 2 details the type of the Layers of the proposed CNN (see Fig. 3) with the input output and Kernel dimensions. As activation function we have used the 'LeakyRELU' to overcome the 'deadRELU' problem. This way we have created a pretty slim network with small complexity and low resource computational consuming impact.

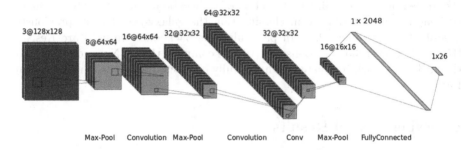

Fig. 3. Proposed CNN architecture.

Table 2. Layers configuration of the proposed CNN architecture

		Input (C×H×W)	Output (C×H× W)	Kernel (K×K)
L1	Conv	(3×128×128)	(8×128×128)	(21×21)
	Max Pool	(8×128×128)	(8×64×64)	(2×2)
L2	Conv	(8×64×64)	(16×64×64)	(15×15)
L3	Conv	(16×64×64)	(32×64×64)	(9×9)
	Max Pool	(32×64×64)	(32×32×32)	(2×2)
L4	Conv	(32×32×32)	(64×32×32)	(7×7)
L5	Conv	(64×32×32)	(32×32×32)	(5×5)
L6	Conv	(32×32×32)	(16×32×32)	(3×3)
	Max Pool	(16×32×32)	(16×16×16)	(2×2)
	FC	(1×1×4096)	(1×1×2048)	-
	FC	(1×1×2048)	(1×1×26)	-

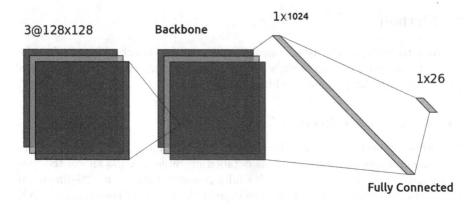

Fig. 4. Schema of CNN based on well-known backbones.

3.2   CNNs Based on Well-Known Backbones

We tested also three different well-known backbones in order to evaluate their accuracy performance in classification the galaxy-zoo dataset previously described (see Subsect. 2.1). In particular the backbones we used are ResNet [15], WideResNet [35] and Densenet [16]. To complete the CNN we concatenated the backbone block completed with two fully connected layers as schematized in Fig. 4.

4   Experimental Results

In order to test the different CNN described in Sect. 3 we split the galaxy dataset (obtained as described in Subsect. 2.1) in 3 subsets containing images, for training, validation and testing. For all the experimental sessions we have used the same subsets in order to compare, fairly, the obtained Balanced Multiclass Accuracy (BCA). Furthermore we have employed for all the tested CNN *Adam* as optimizer and *CrossEntropy* as loss function. The experimental sessions are conducted on an *NVIDIA GTX* 1080*Ti* GPU. The proposed CNN architecture described in Subsect. 3.1 has been trained from scratch on the galaxy dataset training set; instead the other three CNNs (see Subsect. 3.2) using different backbones that starting from weights 'pre-trained' on the IMAGENET dataset are fine-tuned on galaxy dataset training set. All the training sessions for all the tested CNNs are stopped after 100 epochs. Table 3 reported the obtained classification results in terms of F1-score for each of the 26 Galaxy class for the proposed CNN network architecture. The overall Balanced multiclass accuracy (BCA) reach the 78%. It can be noticed that some galaxy classes (i.e. E3, SABb) have the worst accuracy rate among all classes. This suggesting that a better dataset distribution could improve the overall accuracy, namely, using different threshold values during the nested-if process (see Algorithm 1) when galaxy dataset has being formatted. Table 4 summarized the results in terms

Table 3. F1-score performance for 26 Galaxies class for the proposed CNN Architecture

Elliptical	E0	E1		E2	E3	E4	E5	E6	E7
	0.97	0.68		0.64	0.33	0.57	0.64	0.95	0.99
Lenticular	**S0A**	**S0B**							
	0.83	0.54							
Spirals	**SAa**	**SAb**		**SAc**	**SAd**	**SAm**			
	0.91	0.90		0.76	0.64	0.90			
Barred Spirals	**SBa**	**SBb**		**SBc**	**SBd**	**SBm**			
	0.93	0.91		0.75	0.89	0.96			
Intermediate Spirals	**SABa**	**SABb**		**SABc**	**SABd**	**SABm**			
	0.91	0.51		0.59	0.66	0.94			
Irregulars	**Irr**								
Overall	**BCA**	**Weighted Avg**							
	0.78	0.78							

Table 4. CNN Architectures comparison for balanced multi-class accuracy (BCA) and network complexity (# parameters).

	BCA %	# Parameters
Resnet18 [15]	77	11.1M
Proposed	78	8.1M
WideResnet [35]	79	66.9M
Densenet121 [16]	81	6.9M

of Balanced Multiclass Accuracy (BCA) and of network complexity in terms of number of parameters, for each of the tested CNNs. As it can be noticed, all the networks reach good performance (up to 81%) in terms of classification accuracy, but the proposed architecture and the CNN based on Densenet121 backbones reach high performance with lower number of network parameters.

5   Conclusions

In this paper we faced with the problem of automatic morphological galaxies classification according to Hubble-de Vaucoulers diagram. We proposed, for the first time, an automatic classifier based on CNN able to classify galaxies in 26 different categories. We introduced an algorithm that based on a decision tree is able to assign a galaxies label. We proposed a pretty slim CNN architecture with small complexity and lower resource consuming impact. We tested different well known CNN backbones in order to assess obtained classification accuracy results. We successful demonstrated that all the tested networks reach classification accuracy up to 81% on the galaxy-zoo dataset, appropriately formatted. Future work will be addressed to test more convolutional neural network architectures on different and larger galaxies datasets.

References

1. Abraham, S., Aniyan, A.K., Kembhavi, A.K., Philip, N.S., Vaghmare, K.: Detection of bars in galaxies using a deep convolutional neural network. Monthly Notices Royal Astron. Soc. **477**(1), 894–903 (2018). https://doi.org/10.1093/mnras/sty627
2. Baqui, P.O., et al.: The minijpas survey: star-galaxy classification using machine learning. Astron. Astrophys. **645**, A87 (2021). https://doi.org/10.1051/0004-6361/202038986
3. Barchi, P., et al.: Machine and deep learning applied to galaxy morphology - a comparative study. Astron. Comput. **30**, 100334 (2020). https://doi.org/10.1016/j.ascom.2019.100334
4. Baron, D.: Machine learning in astronomy: a practical overview (2019)
5. Beck, M.R., et al.: Integrating human and machine intelligence in galaxy morphology classification tasks. Mon. Not. R. Astron. Soc. **476**(4), 5516–5534 (2018). https://doi.org/10.1093/mnras/sty503
6. Buta, R.J.: Galaxy morphology. In: Oswalt, T.D., Keel, W.C. (eds.) Planets, Stars and Stellar Systems Volume 6: Extragalactic Astronomy and Cosmology, chap. 1, pp. 1–90. Springer, Oxford (2013)
7. Cavanagh, M.K., Bekki, K., Groves, B.A.: Morphological classification of galaxies with deep learning: comparing 3-way and 4-way CNNs. Mon. Not. R. Astron. Soc. **506**(1), 659–676, 100334 (2021). https://doi.org/10.1093/mnras/stab1552
8. Cheng, T.Y.; et al.: Optimizing automatic morphological classification of galaxies with machine learning and deep learning using dark energy survey imaging. Mon. Not. R. Astron. Soc. **493**(3), 4209–4228, 100334 (2020). https://doi.org/10.1093/mnras/staa501
9. Conselice, C.J.: The evolution of galaxy structure over cosmic time. Ann. Rev. Astron. Astrophys. **52**(1), 291–337 (2014). https://doi.org/10.1146/annurev-astro-081913-040037
10. De La Calleja, J., Fuentes, O.: Machine learning and image analysis for morphological galaxy classification. Monthly Notices Roy. Astrono. Soc. **349**(1), 87–93 (2004). https://doi.org/10.1111/j.1365-2966.2004.07442.x
11. De Vaucouleurs, G.: Classification and Morphology of External Galaxies, pp. 275–310. Springer, Heidelberg (1959). https://doi.org/10.1007/978-3-642-45932-0-7
12. Dieleman, S., Willett, K.W., Dambre, J.: Rotation-invariant convolutional neural networks for galaxy morphology prediction. Mon. Not. R. Astron. Soc. **450**(2), 1441–1459 (2015). https://doi.org/10.1093/mnras/stv632
13. Domínguez Sánchez, H., Huertas-Company, M., Bernardi, M., Tuccillo, D., Fischer, J.L.: Improving galaxy morphologies for SDSS with deep learning. Mon. Not. R. Astron. Soc. **476**(3), 3661–3676, 100334 (2018). https://doi.org/10.1093/mnras/sty338
14. Ghosh, A., Urry, C.M., Wang, Z., Schawinski, K., Turp, D., Powell, M.C.: Galaxy morphology network: a convolutional neural network used to study morphology and quenching in 100,000 sdss and 20,000 candels galaxies. Astrophysical J. **895**(2), 112 (2020). https://doi.org/10.3847/1538-4357/ab8a47
15. He, K., Zhang, X., Ren, S., Sun, J.: Deep residual learning for image recognition. In: 2016 IEEE Conference on Computer Vision and Pattern Recognition (CVPR), pp. 770–778 (2016). https://doi.org/10.1109/CVPR.2016.90
16. Huang, G., Liu, Z., Weinberger, K.Q.: Densely connected convolutional networks, pp. 2261–2269 (2017)

17. Kaggle: Galaxy zoo - the galaxy challenge: Classify the morphologies of distant galaxies in our universe. https://www.kaggle.com/c/galaxy-zoo-the-galaxy-challenge/data
18. Kennicutt, R.C.: Star formation in galaxies along the hubble sequence. Ann. Rev. Astron. Astrophys. **36**(1), 189–231 (1998). https://doi.org/10.1146/annurev.astro.36.1.189
19. Lintott, C.J., et al.: Galaxy zoo: morphologies derived from visual inspection of galaxies from the sloan digital sky survey. Monthly Notices Roy. Astron. Soc. **389**(3), 1179–1189 (2008). https://doi.org/10.1111/j.1365-2966.2008.13689.x
20. Mihos, J.C., Hernquist, L.: Gasdynamics and starbursts in major mergers. Astrophys J **464**, 641 (1996). https://doi.org/10.1086/177353
21. Mustafa, M., Bard, D., Bhimji, W., Lukić, Z., Al-Rfou, R., Kratochvil, J.M.: CosmoGAN: creating high-fidelity weak lensing convergence maps using Generative Adversarial Networks. Computational Astrophysics and Cosmology **6**(1), 1–13 (2019). https://doi.org/10.1186/s40668-019-0029-9
22. Méndez-Abreu, J., Ruiz-Lara, T., Sánchez-Menguiano, L., de Lorenzo-Cáceres, A., Costantin, L., Catalán-Torrecilla, C., Florido, E., Aguerri, J.A.L., Bland-Hawthorn, J., Corsini, E.M., et al.: Two-dimensional multi-component photometric decomposition of califa galaxies. Astronomy & Astrophysics **598**, A32 (2017). https://doi.org/10.1051/0004-6361/201629525
23. Ntampaka, M., Eisenstein, D.J., Yuan, S., Garrison, L.H.: A hybrid deep learning approach to cosmological constraints from galaxy redshift surveys. Astrophys. J. **889**(2), 151, 100334 (2020). https://doi.org/10.3847/1538-4357/ab5f5e
24. Odewahn, S.C., Cohen, S.H., Windhorst, R.A., Philip, N.S.: Automated galaxy morphology: a fourier approach. Astrophys. J. **568**(2), 539–557 (2002). https://doi.org/10.1086/339036
25. Pasquet-Itam, J., Pasquet, J.: Deep learning approach for classifying, detecting and predicting photometric redshifts of quasars in the sloan digital sky survey stripe 82. Astron. Astrophys. **611**, A97, 100334 (2018). https://doi.org/10.1051/0004-6361/201731106
26. Schaefer, C., Geiger, M., Kuntzer, T., Kneib, J.P.: Deep convolutional neural networks as strong gravitational lens detectors. Astron. Astrophys. **611**, A2 (Mar 2018). https://doi.org/10.1051/0004-6361/201731201
27. Sellwood, J.A.: Secular evolution in disk galaxies. Rev. Mod. Phys. **86**, 1–46 (2014). https://doi.org/10.1103/RevModPhys.86.1
28. Silva, P.T., Cao, L.T., Hayes, W.B.: Sparcfire: enhancing spiral galaxy recognition using arm analysis and random forests. Galaxies (2018)
29. Simmons, B.D., et al.: Galaxy zoo: quantitative visual morphological classifications for 48 000 galaxies from candels. Monthly Notices Royal Astron. Soc. **464**(4), 4420–4447 (2016). https://doi.org/10.1093/mnras/stw2587
30. Sol Alonso, M., Lambas, D.G., Tissera, P., Coldwell, G.: Effects of galaxy interactions in different environments. Mon. Not. R. Astron. Soc. **367**(3), 1029–1038 (2006). https://doi.org/10.1111/j.1365-2966.2006.10020.x
31. Sreejith, S., et al.: Galaxy and mass assembly: automatic morphological classification of galaxies using statistical learning. Mon. Not. R. Astron. Soc. **474**(4), 5232–5258 (2017). https://doi.org/10.1093/mnras/stx2976
32. Tuccillo, D., Huertas-Company, M., Decenciére, E., Velasco-Forero, S., Domínguez Sánchez, H., Dimauro, P.: Deep learning for galaxy surface brightness profile fitting. Mon. Not. R. Astron. Soc. **475**(1), 894–909 (2017). https://doi.org/10.1093/mnras/stx3186

33. Willett, K.W., et al.: Galaxy zoo 2: detailed morphological classifications for 304,122 galaxies from the sloan digital sky survey. Monthly Notices Roy. Astron. Soc. **435**, 2835–2860 (2013)
34. Wu, C., et al.: Radio galaxy zoo: claran- a deep learning classifier for radio morphologies. Mon. Not. R. Astron. Soc. **482**(1), 1211–1230, 100334 (2018). https://doi.org/10.1093/mnras/sty2646
35. Zagoruyko, S., Komodakis, N.: Wide residual networks. In: Proceedings of the British Machine Vision Conference (BMVC), pp. 87.1-87.12. BMVA Press (2016)
36. Zhu, X.-P., Dai, J.-M., Bian, C.-J., Chen, Yu., Chen, S., Hu, C.: Galaxy morphology classification with deep convolutional neural networks. Astrophys. Space Sci. **364**(4), 1–15 (2019). https://doi.org/10.1007/s10509-019-3540-1

Biomedical and Assistive Technology

Pulmonary-Restricted COVID-19 Informative Visual Screening Using Chest X-ray Images from Portable Devices

Plácido L. Vidal[1,2]([✉])[iD], Joaquim de Moura[1,2][iD], Jorge Novo[1,2][iD], and Marcos Ortega[1,2][iD]

[1] Centro de investigación CITIC, Universidade da Coruña, Campus de Elviña, s/n, 15071 A Coruña, Spain
{placido.francisco.lizancos.vidal,joaquim.demoura, jnovo,mortega}@udc.es
[2] Grupo VARPA, Instituto de Investigación Biomédica de A Coruña (INIBIC), Universidade da Coruña, Xubias de Arriba, 84, 15006 A Coruña, Spain

Abstract. In the recent COVID-19 outbreak, chest X-rays were the main tool for diagnosing and monitoring the pathology. To prevent further spread of this disease, special circuits had to be implemented in the healthcare services. For this reason, these chest X-rays were captured with portable X-ray devices that compensate its lower quality and limitations with more deployment flexibility. However, most of the proposed computer-aided diagnosis methodologies were designed to work with traditional fixed X-ray machines and their performance is diminished when faced with these portable images. Additionally, given that the equipment needed to properly treat the disease (such as for life support and monitoring of vital signs) most of these systems learnt to identify these artifacts in the images instead of real clinically-significant variables. In this work, we present the first methodology forced to extract features exclusively from the pulmonary region of interest that is specially designed to work with these difficult portable images. Additionally, we generate a class activation map so the methodology also provides explainability to the results returned to the clinician. To ensure the robustness of our proposal, we tested the methodology with chest radiographs from patients

This research was funded by Instituto de Salud Carlos III, Government of Spain, DTS18/00136 research project; Ministerio de Ciencia e Innovación y Universidades, Government of Spain, RTI2018-095894-B-I00 research project, Ayudas para la formación de profesorado universitario (FPU), grant ref. FPU18/02271; Ministerio de Ciencia e Innovación, Government of Spain through the research project PID2019-108435RB-I00; Consellería de Cultura, Educación e Universidade, Xunta de Galicia, Grupos de Referencia Competitiva, grant ref. ED431C 2020/24 and through the postdoctoral grant contract ref. ED481B 2021/059; Axencia Galega de Innovación (GAIN), Xunta de Galicia, grant ref. IN845D 2020/38; CITIC, as Research Center accredited by Galician University System, is funded by "Consellería de Cultura, Educación e Universidade from Xunta de Galicia", supported in an 80% through ERDF Funds, ERDF Operational Programme Galicia 2014–2020, and the remaining 20% by "Secretaría Xeral de Universidades" (Grant ED431G 2019/01).

S. Sclaroff et al. (Eds.): ICIAP 2022, LNCS 13231, pp. 65–76, 2022.
https://doi.org/10.1007/978-3-031-06427-2_6

diagnosed with COVID-19, pathologies similar to COVID-19 (such as other types of viral pneumonias) and healthy patients in different combinations with three convolutional networks from the state of the art (for a total of 9 studied scenarios). The experimentation confirms that our proposal is able to separate COVID-19 cases, reaching a 94.7% ± 1.34% of accuracy.

Keywords: COVID-19 · Chest X-ray · CAD system · Class activation map · Deep learning · X-ray portable devices

1 Introduction

Coronavirus COVID-19 disease is an affection that primarily impacts in the pulmonary region [2], predominantly resulting in viral pneumonia. This results in the appearance of pathological structures visible through chest X-rays and other imaging techniques [19], present even in asymptomatic patients [4]. Amidst the global COVID-19 pandemic of 2020, these chest radiographs were one of the main means to diagnose the aforementioned pathology. For this reason, numerous automatic computer-aided diagnosis (CAD) methodologies were developed, aimed at reducing the subjectivity of the experts as well as the needed time and resources [3,7,14]. These proposals were designed to work with images from either fixed X-ray devices (which allow to obtain high-definition projection of the internal tissues in different angles and depths) or CT devices (which generate a defined single slice representation of the human body).

However, during medical emergencies and cross-contamination risk, the main mean to diagnose lung diseases is by using portable X-ray devices. These devices do not require a special platform nor room, so they can be used in dedicated circuits in hospitals and isolated areas (at the cost of lesser image quality and limited angles with worse capture conditions) [10]. However, this lesser image quality and limitations severely hinder the performance of methodologies not tested for them. For this reason, contributions specially designed to work with these portable devices were proposed, such as the work of de Moura *et al.* [13] with the objective of performing a screening, separating normal images from COVID-19 and from pathologies with COVID-19-like patterns. In this last work (and in some works based on fixed devices [1]), they also generate a gradient flow map so they can help the experts to assess what has the deep learning model analyzed in the images. Additionally, given the scarcity and complexity of the images extracted from these portable devices, works aimed to generate synthetic images have also been proposed, such as in the works of Morís *et al.* [11,12] generating synthetic images to palliate the lack of samples in this issue.

These methodologies (as shown by the gradient flow maps generated by the networks) were taking advantage of collateral information present in the radiographs to generate their prediction. That is, for example, if foreign artifacts commonly present in afflicted patients from certain diseases (such as respirators, pacemakers, heart rate sensors, etc.) are shown in the radiographs, these networks use this information to improve the metrics [16] and guess that an intubated patient must be afflicted by some kind of pulmonary disease or at least be at higher risk of being infected by the studied disease. While it is common for neural networks to take advantage of underlying statistical features commonly missed by human experts, this implies that the system will perform worse with different clinical protocols, stages of the pathology or even age groups as these artifacts used to their advantage may change.

For this reason, we propose the first fully automatic methodology to remove these factors by masking these extraneous artifacts and able to successfully work with images from these portable X-ray devices. This way, we obtain a robust methodology that only extracts information from the pulmonary region of interest, resulting in a system that is able to work in any stage of the progression of the patient and generate a prediction solely based on real clinical features. In addition, we take advantage of this mechanism that forces the attention of the network on the lung region to generate a representation that explains the results to the clinician. Thus, the expert will not only be able to verify that the network is obtaining the desired behavior, but may even discover factors overlooked by the expert that can help determine a better treatment and assist in their future monitoring. And, thanks to being able to work in these special scenarios with portable X-ray devices, can be seamlessly implemented in healthcare circuits specially designed to treat patients in a pandemic scenario. To the best of our knowledge, this proposal represents the only study designed to work with these portable devices able to extract information from exclusively the pulmonary regions, perform a classification of the results between COVID-19, healthy and similar pathologies to COVID-19, and generate an explainable representation of these classifications.

2   Materials

To train, test and validate our methodology we used a specific dataset composed by 2,071 radiographs from patients diagnosed with COVID-19, 716 from patients diagnosed with non-COVID-19 pathologies (but with very similar features to it), and 797 images from healthy patients. These images were acquired with an Agfa dr100E GE and Optima Rx200 portable X-ray devices. All these images were obtained by the radiology service of the Complexo Hospitalario Universidario de A Coruña (CHUAC) during real live clinical practice. The images with non-COVID-19 pathologies but with very similar patterns belong to types of viral pneuomonias or bacterial infections that leave a comparable trace and damage on the lungs of the afflicted. This is a specially interesting case, as the system need to discern these images from the very similar belonging to COVID-19 patients while also limited by the restrictions imposed by the portable X-ray devices. For the sake of balancing the dataset between the three available types of images, 716 images were randomly selected from each set to compose the final dataset.

3   Methodology

Our proposal is divided into two main steps presented in Fig. 1. In the first step, we use a specially designed methodology to extract the lung region of interest from portable chest radiographs (Sect. 3.1). Once we have delimited this region of interest, we mask the dataset and use it to train a convolutional neural network. Finally, using this trained network, we extract the gradient flow to generate an intuitive visualization of the relevant regions from the final classification (Sect. 3.2). This methodology will be followed with three different combinations of the labels that are available in the dataset to further evaluate the capabilities of the methodology and its behavior: Non-Healthy (COVID-19 + Pathologies similar to COVID-19) versus Healthy, COVID-19 versus non-COVID-19 (Pathologies similar to COVID-19 and Healthy), and every label as an independent class (COVID-19 versus Pathological versus Healthy).

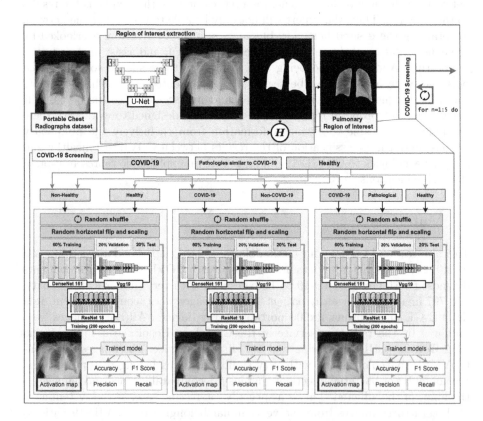

Fig. 1. Workflow for each repetition during the training of the proposed methodology.

3.1   Region of Interest Extraction

To extract the region of interest we will use a methodology that is robust to the artifacts and structures that are commonly present in portable chest radiographs. In our case, we will use a proposal based on the work of Vidal *et al.* [20]. This methodology performs a double-stage transfer learning: in the first step, the chosen network (in this case, an U-Net [15]) is adapted via transfer learning from magnetic resonance imaging for brain glioma segmentation [5] to general-purpose chest radiographs. Afterwards, the network undergoes a second transfer learning stage, where the network is further refined to be able to work also with radiographs extracted from portable devices (as well as with the pathologies considered in this work). With this specially designed segmentation methodology, we extract all the pulmonary regions of interest from the dataset. Finally, for each of these images, we apply the Hadamard product (H) to filter-out all artifacts not belonging to the lung area from the lower to the apical zone in the chest radiographs. This way, we remove any external artifacts the network could use to its advantage, artificially improving the metrics.

3.2   COVID-19 Screening

In this second step, we will train three different networks for each of the case studies considered in this work. Non-Healthy versus Healthy, COVID-19 versus non-COVID-19, and every label as an independent class (COVID-19 versus Pathological versus Healthy).

To train each of the networks, we will use the masked dataset of the previous step and three architectures from the state of the art: a ResNet18 [8], a Vgg19 [18] and a DenseNet161 [9]. Additionally, all these network architectures were pretrained with the ImageNet dataset [6] to further speed up the training and reduce the impact of the number of samples. Furthermore, to increase the effective number of available samples, a data augmentation strategy was used by randomly horizontally flipping and scaling the input lung radiographs. Finally, the dataset is randomly shuffled and divided into three sets: 60% of the images to be used for the training of the network, 20% for validation purposes of each epoch of training, and the remaining 20% to test the performance and robustness of the final models. Each model is set to be trained for a fixed number of 200 epochs, empirically determined to be enough to reach an stability point where no further improvements in the final metrics are achieved. From these 200 epochs, the model that achieved the best validation loss in a given epoch is the one chosen from the set. The models were trained using the cross-entropy as loss function, using as optimizer the Stochastic Gradient Descent (SGD) with a constant learning rate of 0.01 and a momentum of 0.9. Each network is trained independently and this process is repeated five times for each model and case study mentioned above in order to diminish the improbable selection bias and obtain more robust final metrics.

Finally, as our proposal is designed to help the clinicians with their diagnostic task, we extract the regions where the networks centered its attention. To do so, we use gradient-weighted class activation mapping in the network [17]. This strategy uses the gradient information flowing towards the last convolutional layer to determine the relevance of each neuron towards the generation of the classification, a generalization of Class Activation Mapping or CAM. This strategy is commonly used to assess the robustness of the predictions of a network, as it allows to remove models that take advantage on circumstantial features instead of actual relevant descriptors. However, once the models were verified to be valid, the information generated by the network is also useful to the clinicians, as we generate explainable results that can, even, help to discover new biomarkers hitherto unconsidered or determine if some elements present in the image are product of the quality of the X-ray images from portable devices/actual pathological structures.

4   Results and Discussion

Below we present the results that were obtained during the experimentation, divided into results during the training (Sect. 4.1) and the metrics of the final model and class activation maps (Sect. 4.2). Additionally, within each section, each of the three aforementioned cases of study will be discussed separately.

4.1   Training Results

The results for the training of the models studying the case Healthy versus Non-Healthy can be seen in Fig. 2. In this figure, we can see how this situation represents the easiest for the model to learn, as the accuracy during the training of all the three models quickly reached the maximum value. However, in this case, we can also see how the training loss quickly falls down while the validation loss stays on average on the same height. We see how the system is training for the initial variability, but quickly stabilises. This indicates that the issue at hand might be too easy for the methodology and a variable learning rate should be used. Because of this, we are perceiving slight traces of overfitting in the model in the final epochs (albeit we will only keep the one with the best validation loss nonetheless). This can also be seen in how the simpler architecture (Vgg19) is the one who obtains best validation accuracy and loss from the three tested models. This is normal given that healthy lungs do not present opacities, infiltrates and dense tissues present in the studied afflicted lungs.

On the other hand, in the results of the training for the study case of COVID-19 versus non-COVID-19 samples shown in Fig. 3, we can see how the results are significantly different. We are evaluating a more difficult situation than in the first scenario, and now the models present a more balanced progression between training and validation. In this case, we see how both the training loss and validation loss are proximal respective between the same model and no predominant signs of overfitting (as well as for significantly improved accuracy metrics during validation).

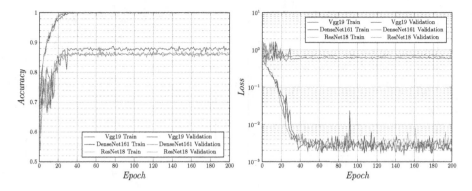

Fig. 2. Accuracy and loss during training and validation of the models for the experiment Healthy versus Non-Healthy.

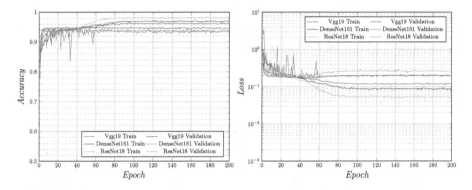

Fig. 3. Accuracy and loss during training and validation of the models for the experiment non-COVID-19 versus COVID-19.

Finally, in Fig. 4 we can see the training and validation results for the case study where we consider each class independently in the model targets. In this case, we observe how all the models ended achieving a compromise between the same models in the two previous experiments. As we will analyze in the following sections, the Healthy and the Pathological class probably present some overlapping in certain scenarios that is being exacerbated by the quality and limitations of the portable chest radiographs used in this work.

4.2 Test Results

The metrics for the first experiment (Healthy versus Non-Healthy) are shown in Table 1. In this case, we can confirm what we saw in Fig. 2, where the Vgg19 model is the one obtaining the best overall results. We can see how, in this case, the simpler model is obtaining the best results of the experiment (albeit satisfactory nonetheless in all scenarios). Additionally, we can see how the ResNet18

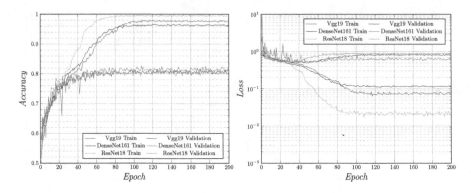

Fig. 4. Accuracy and loss during training and validation of the models for the experiment healthy versus pathological versus COVID-19.

has particularly high standard deviation in several metrics confirming what we stated in the previous sections: the network has some traces of overfitting.

Table 1. Test results for the experiment evaluating healthy (H) versus pathological (P) and COVID-19 (C) classes.

Architecture	Accuracy	Class	Precision	Recall	F1-Score
Densenet161	87.2% ± 2.53%	**H**	85.0 % ± 2.76%	88.8% ± 4.58%	87.0% ± 3.33%
		P+C	89.2% ± 2.89%	85.6% ± 1.62%	87.4% ± 1.85%
ResNet18	85.7% ± 1.63%	**H**	85.8% ± 3.71%	89.4% ± 4.80%	86.2% ± 1.94%
		P+C	86.6% ± 4.18%	84.2% ± 4.40%	85.2% ± 1.47%
Vgg19	89.6% ± 1.29%	**H**	87.6% ± 1.74%	91.8% ± 1.72%	89.8% ± 1.47%
		P+C	91.8% ± 1.83%	87.4% ± 1.36%	89.6% ± 1.29%

On the other hand, in the second experiment (COVID-19 versus non-COVID-19) shown in Table 2 we see how the networks were able to obtain outstanding results. Moreover, in this case, we can highlight the performance of the DenseNet161 in terms of recall for the non-COVID-19 class and precision for the COVID-19 class. Additionally, we can observe how, like in the previous experiment, the ResNet18 architecture stands out for its higher than average standard deviation. Nonetheless, the DenseNet161 in this case would be the preferable model, as it obtained the highest metrics even in the F1 Score that is resilient to the unbalances of the dataset (unlike the accuracy, for example). Additionally, its densely connected blocks that confer the network an extra layer of self-supervision are preventing the network from falling into overfitting like the ResNet18.

Finally, in Table 3 we can see the results for the last experiment, where we train the networks with all the classes independently. In this case, we can see

Table 2. Test results for the experiment evaluating healthy (H) and pathological (P) versus COVID-19 (C) classes.

Architecture	Accuracy	Class	Precision	Recall	F1-Score
Densenet161	94.7% ± 1.34%	H+P	93.0% ± 1.10%	99.4% ± 0.8%	96.0% ± 1.10%
		C	98.4% ± 2.24%	86.4% ± 2.42%	91.8% ± 1.94%
ResNet18	94.4% ± 1.06%	H+P	93.2% ± 1.47%	98.6% ± 1.74%	95.8% ± 0.75%
		C	97.2% ± 3.49%	85.8% ± 3.76%	91.6% ± 2.94%
Vgg19	94.3% ± 1.43%	H+P	93.2% ± 1.17%	98.6% ± 1.36%	95.8% ± 1.17%
		C	96.8% ± 2.79%	86.2% ± 1.33%	91.0% ± 1.67%

how, in all three networks the COVID-19 class stands out from the rest, being easily distinguished from the rest. On the other hand, the precision of the healthy class and the recall of the pathological class suffer a bit when separated. In this case, we see how the pathological class and the healthy are intermixing results in some cases. This may be due to the fact that some of the patients could have been tested with an early onset stage of the diseases similar to COVID-19 that, in conjunction with the difficulties imposed by the portable X-ray devices, present very similar patterns to healthy images. This way, when these two classes are separated (such as in this experiment and in the first one) they return mixed results. For this reason, the desirable models in these cases are the ones generated in the second experiment. In this case, the pathological and the healthy classes are combined and the confusing patterns, instead of penalizing the model, help it to achieve outstanding better results.

Table 3. Test results for the experiment evaluating healthy (H) versus pathological (P) versus COVID-19 (C) classes.

Architecture	Accuracy	Class	Precision	Recall	F1-Score
Densenet161	83.4% ± 2.06%	H	77.2% ± 4.07%	81.4% ± 3.38%	79.2% ± 2.99%
		P	80.4% ± 3.88%	78.8% ± 4.12%	79.4% ± 2.73%
		C	93.8% ± 2.64%	90.2% ± 1.72%	91.8% ± 0.40%
ResNet18	81.8% ± 1.56%	H	72.4% ± 5.39%	82.6% ± 3.55%	77.0% ± 2.83%
		P	81.2% ± 1.72%	76.8% ± 3.87%	78.8% ± 1.60%
		C	95.0% ± 2.68%	85.8% ± 3.19%	90.0% ± 1.10%
Vgg19	81.3% ± 2.57%	H	72.8% ± 6.85%	80.4% ± 4.17%	75.8% ± 3.19%
		P	82.0% ± 3.03%	76.8% ± 8.33%	79.0% ± 3.74%
		C	92.4% ± 3.77%	87.8% ± 3.82%	89.8% ± 2.57%

For this reason, the DenseNet161 model from the second experiment is the one that we will use to generate the class activation maps. In Fig. 5, we can see four chest radiographs from portable devices as an example of the class activation maps generated with our proposal. As the reader can see, the proposed methodology has achieved satisfactory results, being able to successfully discern COVID-19 patients from healthy/pathological samples. Additionally, as shown in that same figure, both for the non-COVID-19 cases and COVID-19 cases, thanks to the proposed masking strategy, the system is only focusing on lung regions. While on non-COVID-19 images the lung regions remain clean, COVID-19 detections are centered on the affected lung regions (such as the bronchus/bronchioles in the leftmost COVID-19 image). Thus, all the performed predictions are purely based on the analysis of lung structures and not other artifacts (such as the wires present in the leftmost healthy radiography).

Non-COVID-19   **COVID-19**

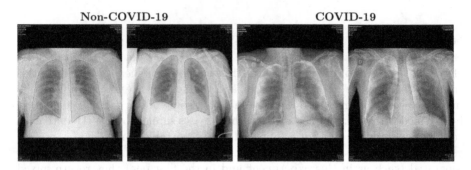

Fig. 5. Examples of generated maps with the DenseNet161 trained to differentiate non-COVID-19 patients from COVID-19 ones.

5 Conclusions

In this work we have presented a methodology capable of working with portable X-ray images, critical for healthcare systems in emergency scenarios thanks to their flexibility (at the cost of inferior capture quality). Our system has been able to successfully differentiate COVID-19 from COVID-19-like pathologies and from images of healthy patients, even with the difficulties imposed by the capture quality of these portable devices. In addition, we have generated the class activation maps of the chosen model, which has demonstrated that our methodology of masking the region of interest in these portable device images ensures that the model is focused exclusively on the pulmonary region. This means that our system is not taking advantage of information foreign to the domain (such as the presence of pacemakers, respirators, cardiac pulse meters, etc.) and able to work independently of the established protocol. Moreover, these class activation maps of the network are useful to assist the clinical expert in his diagnostic task by providing an explainable representation of what is detected by the system. Thus, the expert can not only determine the validity of the results, but may also discover factors that went unnoticed during the overloaded clinical practice of crowded emergency services. As a result, the system will provide robust measurements regardless of clinical procedure, patient status or stage of the pathology development.

As future work, we plan to further improve the experiments to reduce the impact of overfitting in some of the models with a dynamic loss paradigm (to improve the resolution of the gradient descent) and early stopping strategies, as well as the implementation of a data augmentation strategy by synthesis of new samples by using, for example, the cycle-consistent generative adversarial approaches presented in the state of the art. This would aid to complement the variability of the dataset. Finally, as mentioned, some of the samples presented are probably from early onset stages of different pathologies with similar symptoms to COVID-19. A study with clinicians to reverse-engineer the models and discover these early markers in images could be also of interest.

References

1. Explainable deep learning for pulmonary disease and coronavirus covid-19 detection from x-rays. Comp. Methods Programs Biomed. **196**, 105608 (2020). https://doi.org/10.1016/j.cmpb.2020.105608
2. Aguiar, D., Lobrinus, J.A., Schibler, M., Fracasso, T., Lardi, C.: Inside the lungs of COVID-19 disease. Int. J. Legal Med. **134**(4), 1271–1274 (2020). https://doi.org/10.1007/s00414-020-02318-9
3. Alom, M.Z., Rahman, M.M.S., Nasrin, M.S., Taha, T.M., Asari, V.K.: Covid mtnet: Covid-19 detection with multi-task deep learning approaches (2020)
4. Bandirali, M., Sconfienza, L.M., Serra, R., Brembilla, R., Albano, D., Pregliasco, F.E., Messina, C.: Chest radiograph findings in asymptomatic and minimally symptomatic quarantined patients. Radiology **295**(3), E7–E7 (2020). https://doi.org/10.1148/radiol.2020201102
5. Buda, M., Saha, A., Mazurowski, M.A.: Association of genomic subtypes of lower-grade gliomas with shape features automatically extracted by a deep learning algorithm. Comput. Biol. Med. **109**, 218–225 (2019). https://doi.org/10.1016/j.compbiomed.2019.05.002
6. Deng, J., Dong, W., Socher, R., Li, L.J., Li, K., Fei-Fei, L.: Imagenet: a large-scale hierarchical image database. In: 2009 IEEE Conference on Computer Vision and Pattern Recognition, pp. 248–255. IEEE (2009)
7. Fan, D.P., Zhou, T., Ji, G.P., Zhou, Y., Chen, G., Fu, H., Shen, J., Shao, L.: Inf-net: automatic COVID-19 lung infection segmentation from CT images. IEEE Trans. Med. Imaging **39**(8), 2626–2637 (2020). https://doi.org/10.1109/tmi.2020.2996645
8. He, K., Zhang, X., Ren, S., Sun, J.: Deep residual learning for image recognition. CoRR abs/1512.03385 (2015). http://arxiv.org/abs/1512.03385
9. Huang, G., Liu, Z., Weinberger, K.Q.: Densely connected convolutional networks. CoRR abs/1608.06993 (2016). http://arxiv.org/abs/1608.06993
10. Jacobi, A., Chung, M., Bernheim, A., Eber, C.: Portable chest x-ray in coronavirus disease-19 (COVID-19): a pictorial review. Clin. Imaging **64**, 35–42 (2020). https://doi.org/10.1016/j.clinimag.2020.04.001
11. Morís, D.I., de Moura, J., Novo, J., Ortega, M.: Cycle generative adversarial network approaches to produce novel portable chest x-rays images for covid-19 diagnosis. In: ICASSP 2021–2021 IEEE International Conference on Acoustics, Speech and Signal Processing (ICASSP). IEEE, June 2021. https://doi.org/10.1109/icassp39728.2021.9414031

12. Morís, D.I., de Moura Ramos, J.J., Buján, J.N., Hortas, M.O.: Data augmentation approaches using cycle-consistent adversarial networks for improving COVID-19 screening in portable chest x-ray images. Expert Syst. Appl. **185**, 115681 (2021). https://doi.org/10.1016/j.eswa.2021.115681
13. de Moura, J., et al.: Deep convolutional approaches for the analysis of covid-19 using chest x-ray images from portable devices. IEEE Access **8**, 195594–195607 (2020). https://doi.org/10.1109/ACCESS.2020.3033762
14. de Moura, J., Novo, J., Ortega, M.: Fully automatic deep convolutional approaches for the analysis of covid-19 using chest x-ray images. medRxiv, May 2020. https://doi.org/10.1101/2020.05.01.20087254
15. Ronneberger, O., Fischer, P., Brox, T.: U-net: Convolutional networks for biomedical image segmentation. CoRR abs/1505.04597 (2015). http://arxiv.org/abs/1505.04597
16. Sadre, R., Sundaram, B., Majumdar, S., Ushizima, D.: Validating deep learning inference during chest x-ray classification for COVID-19 screening. Sci. Rep. **11**(1), August 2021. https://doi.org/10.1038/s41598-021-95561-y
17. Selvaraju, R.R., Cogswell, M., Das, A., Vedantam, R., Parikh, D., Batra, D.: Grad-cam: Visual explanations from deep networks via gradient-based localization. In: 2017 IEEE International Conference on Computer Vision (ICCV), pp. 618–626 (2017). https://doi.org/10.1109/ICCV.2017.74
18. Simonyan, K., Zisserman, A.: Very deep convolutional networks for large-scale image recognition. CoRR abs/1409.1556 (2015). http://arxiv.org/abs/1409.1556
19. Velavan, T.P., Meyer, C.G.: The COVID-19 epidemic. Tropical Med. Int. Health **25**(3), 278–280 (2020). https://doi.org/10.1111/tmi.13383
20. Vidal, P.L., de Moura, J., Novo, J., Ortega, M.: Multi-stage transfer learning for lung segmentation using portable x-ray devices for patients with COVID-19. Expert Syst. Appl. **173**, 114677 (2021). https://doi.org/10.1016/j.eswa.2021.114677

Comparison of Different Supervised and Self-supervised Learning Techniques in Skin Disease Classification

Loris Cino[1], Pier Luigi Mazzeo[2(✉)] (iD), and Cosimo Distante[2] (iD)

[1] Universitá del Salento, Via Monteroni sn, 73100 Lecce, Italy
[2] ISASI - CNR c/o DHITECH, Via Monteroni sn, 73100 Lecce, Italy
`pierluigi.mazzeo@cnr.it`

Abstract. For years now, The International Skin Imaging Collaboration has been providing datasets of dermoscopic images. Several studies show that dermoscopy provides improved diagnostic accuracy, in comparison to standard photography. Excellent results have been obtained that even exceed human performance. In this paper we broke the state of the art for the dataset provided for the ISIC 2019 challenge. In this work were compared the performance of various convolutional networks, various data augmentations, and various cost functions and optimizers. Results obtained using transfer learning from IMAGENET were compared with the performance obtained using BYOL (bootstrap your own latent), a self-supervised technique. Moreover, it has been demonstrated that self-supervised learning techniques can be used in this field improving the performance of the network compared to training from scratch. We were obtained a balanced multiclass accuracy (BCA) of 87% in the test and validation dataset and with a top-2 accuracy of 97%.

Keywords: Skin disease classification · CNN · Self-supervised learning

1 Introduction

The early diagnosis of melanoma can guarantee, almost with certainty, the survival of the patient. In last years dermoscopy has been of fundamental importance, a technique of image acquisition that by eliminating the light reflected from the skin, allows a more in-depth visual analysis. Unfortunately, a timely diagnosis is not always possible due to the large number of patients who have to visit the dermatologists. For this reason making available to doctors a set of tools to automate, even if partially, the diagnosis process would help. Over the years many solutions have been proposed to solve this problem, starting from some more traditional image processing solutions to techniques based on neural networks [1,2,10,23], as well as mixed techniques [12]. A common solution in the works that use neural network with better performance is to used to

The authors thank **Arturo Argentieri** for his technical support in the setup of the hardware used for network training and data processing.

start the training with pretrained weight, usually on IMAGENET [8], because in ISIC datasets there are not enough images to train a deep neural network from scratch.

An alternative to transfer learning could be self-supervised learning [15]. The network is trained using pseudo labels that are automatically generated for a predefined pretext task without involving any human annotation. By doing that the cost of making a dataset significantly drops because the human annotation is the cost-expensive phase, also considering that, in this task, annotation is made by doctor and requires special tools. The network is trained in two stages: pretext task and downstream task. The first stage involves pseudo-labeled data. Pretext tasks can be splitted into several categories based on which kind of information use. Recent works use a set of random augmentation to create two different versions of the same image and then train the network to produce the same representation for the two versions of the image. Downstream task is used for train the network for real task.

Main contribution of this work is focused on an investigation of different supervised and self-supervised techniques applied to the automatic skin disease classification domain. Different convolutional neural networks (CNN) performance are evaluated varying loss functions, optimizer and augmented training data. We, also, compared the performance of the network pre-trained on IMAGENET dataset with those obtained with the BYOL (Bootstrap Your Own Latent) self-supervised technique. We demonstrated that the applied self-supervised methodology improves the network performance comparing them with the same network trained from scratch. This paper is organized as follows: in Sect. 2 a brief look at related work is given; Sect. 3 explains adopted methodology for automatic skin disease classification; Sect. 4 describes the experiments that have been conducted; finally in Sect. 5 some conclusion remarks are drawn.

2   Related Work

There are two different subsection of related works on self-supervised and on ISIC datasets [6,12].

2.1   Self-supervised Learning

A well known work in self-supervised learning is PIRL [17], pretext invariant representation learning. This paper figure it out that during pretext task is better train the network to produce similar representation of the same image but transformed in different ways to make the network learn more meaningful features. In PIRL has been demonstrate that using only one augmentation produce a hidden space that is strongly correlated to the pretext task used. But using more augmentations the last layer produces a better representation compared to others layers. Their work focused on using the jigsaw puzzle augmentation to generate more versions of the same image. They introduce contrastive learning in the self-supervised field and the use of a memory bank to prevent the network to collapse to a trivial solution. The problem of contrastive learning is that require a lot of memory to get good performance and makes the architecture hard to implement because mini batch-SGD makes difficult to have enough negative example without having a batch size that is practical infeasible. So, the feature space of many images was cached in a memory bank. SimCLR [4] further extend this technique by introducing a set of sequential random augmentations. As PIRL, the loss forces the network to produce a close representation on different versions of the same image and far representation of different images. But in this case the authors introduce a learnable non-linear transformation between representation and loss. Their studies show that this transformation substantially improves the quality of learned representation of the layer before it. This non-linear transformation is implemented by a fully connected neural network called MLP, multi-layer projection head. They also show that some data augmentations perform better that others as pretext task, but better results are reached when a set of augmentations is used. When using only one augmentation, although the network is completely able to recognise pair of images, no good representation is learnt. Moreover, in that paper has been proved that contrastive learning benefits of large batch size and more training epochs than standard supervised. SimCLR do not use a memory bank but only confront image in the same batch. In SimCLRv2 [5] was introduced a further teacher-student network to increase the performance of the network. A popular alternative is BYOL [11], bootstrap your own latent. It keeps the sequence of random augmentations but in this case the network is trained to guess the prediction of a target network for a given image but augmented in different ways. The set of random augmentations are the same of SimCLR. The target network has the same architecture of the main one, but the weights are calculated by the moving average of the online network's weight. This prevents collapse to the trivial solution although it is not so clear why this happens, probably because undesirable equilibria are unstable. The performance got are very similar to both version of SimCLR but it is easier to implement and require much less memory. They use a MLP head as SimCLR but do not use contrastive learning.

2.2 On ISIC 2019 Dataset

Several studies have been made on this dataset, many of them show promising results. A CNN's performance has been also compared to that of physician. The results were surprising, that neural network outperforms most of the dermatology except the ones with more than 10 years of experience [23]. One work on this dataset is skin Deep [3], this paper achieved an average weighted recall score of 0.72 with a VGG-19 [24] net with a some fully connected layer on top. The VGG net was pretrained on IMAGENET and the weight frozen, only the additional fully connected layers were fine-tuned. As loss and optimizer are used respectively cross entropy loss and Adam whit a learning rate of $7e-4$ and a weight decay of $5e-6$, several values of learning rate and weight decay were tested. The network was trained using a batch size of 128 and for 20 epochs. The authors tried to reduce reduced over-fitting by using Dropout in the fully

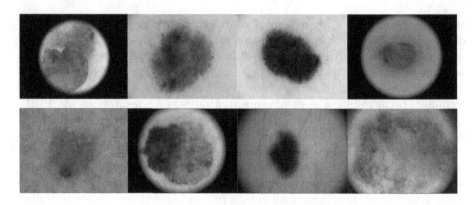

Fig. 1. Sample ISIC 2019 images.

connected layers with a p of 0.2. In systematic Investigation on Deep Architectures for Automatic Skin Lesions Classification [1] 23 CNN were tested on this task showing that the family of EfficientNet [21] performs better than other common CNN. Ensembles of neural network were also tested. In this work each image has been randomly cropped and resized to a resolution of 224 × 224 pixels. Then have been applied several augmentations such as: horizontal and vertical image flipping; contrast, saturation flipping; contrast, saturation, brightness, and colour hue modifications; image rotation up to 90 °C. For training the CNNs, the SGD optimizer was used with learning rate of 0.001, momentum of 0.9, and weight decay of 0.001. In addition, an early stopping strategy of 22 epochs on the validation set accuracy and a max number of 200 training epochs have been chosen. The weighted cross entropy is used to reduce the impact of the imbalanced dataset and it has been shown that works better than other technique such as data replication. The best performance is reached with RegNetY16 [22]. Deep-learning approach in the study of skin lesions [10] reached a 78,5% BCA with a ResNet50 [13]. In this paper has been done a deep study over CNNs, batch size, optimizers, learning rate, number of epochs and number of layers to freeze to get better performance. The images were resized at size 512 × 512, and the train-validation-test splitting was used with respectively 80%-10%-10%. During the process the metadata is also used to help the network to perform better. The metadata include age, sex, and other information about the patient. This data has been pre-processed to get an array of 19 values. Two training approach were tried: train the network with all images and training the network with balanced subset of the dataset using data replication to balance the dataset. The first method got better performance. Many pretrained on IMAGENET networks were tested: VGG169 [24], NASNetMobile10, ResNet50 [13]. The last one was the one with best results, this network has been used throughout subsequent experiments. To obtain the best possible model, each hyperparameter should be adjusted independently, using a grid search, but the optimization time would been too long so a greedy approach was used (Fig. 1).

3 Method

The best hyperparameters configuration has been found after many time consuming experimental sessions. Some experiments with big image size took days with two Nvidia Titan RTX with 24 GB of RAM, and all the possible combination of hyperparameters taken months. We used 'weighted cross-entropy' as loss function, 'ReduceLROnPlateau' as learning rate scheduler and as optimizer '$MADGRAD$' [19]. For all these hyperparameters were made several experiments for searching the configuration that gives a better performance on BCA, balanced multiclass accuracy. The dataset has been shuffled and then divided into train-validation- test with a 80%-10%-10% splitting. The results showed the performance can significantly increase outperforming the referred state-of-the-art. The configuration with higher BCA is, then used for the self-supervised learning experiments.

3.1 Loss and Optimizer

Then, starting from the configuration with better performance proposed in [1] various experiment has been done on losses and optimizers. Best results were obtained with weighted cross entropy, a standard cross entropy but with the addition of a term that use a weight, that is a vector with an element for each class, to penalize less the wrong classification of the most common class and penalize more the rarest class. This is done to make the network learn to recognise all the classes with the same importance. Standard cross entropy, Focal Loss [7] and a Weighted Focal Loss were used but they got a lower BCA score. Another technique used was Label Smoothing [16], this is a technique that use soft targets that are a weighted average of the hard targets and the uniform distribution over labels, that theoretically would nicely address the problem of overfitting and the problem of overconfidence of the network that, in a problem difficult like skin issues classification, could be relevant. But, despite the theoretical background, using label did not improve performance. For optimizer has been tested stochastic gradient descend, RADAM that stands for rectified ADAM and MADGRAD [19], a relatively new optimization method in the family of AdaGrad adaptive gradient methods. As expected, each of these optimizers required a tuning phase for tuning hyperparameter as learning rate, weight decay and momentum. The hyperparameters were very different from optimizers from optimizers this made the experiment very long and complex. The best configuration was the optimizer MADGRAD with learning rate of 0.00025, momentum 0.9 and weight decay 0. The MADGRAD optimizer was the hardest to use, even a small variation of learning rate can make the network unstable or kill the gradient.

3.2 Choosing CNN and Input Image Size

Many architectures were tested from different families of convolutional neural network ResNext [13], Resnet [20] and EfficientNet [21]. The last one has a recommended image size for each neural network, for example EfficientNet-B4 should works with image of size 380×380 pixels. In reverse, Resnet family has not a recommended image size. Figure 2 summarizes the performance obtained with different CNN and input image size.

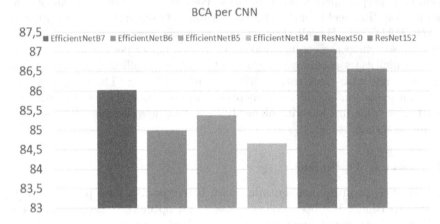

(a) Comparison of test dataset BCA of different convolutional neural network.

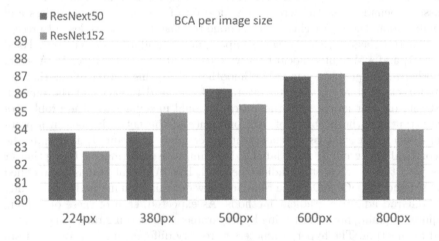

(b) Comparison of test dataset BCA of different image size.

Fig. 2. Performance accuracy evaluation comparing different CNN varying the input image size.

As can be seen in Fig. 2a the best performing neural network have a bigger image size so was made an experiment to understand how image size impact on the performance. Many image sizes were tried: 224px, 380px, 500px, 600px and 800px. As Fig. 2b shows the performance increases as image size increases.

The difference in performance is due to the fact that making the image smaller probably create some artefact that destroy the pattern of the various skin issues. With ResNet152 the best performance is get with 600px images size.

3.3 Data Augmentations

In this work has been used mainly geometrical transformation: random rotation, random horizontal and vertical flip, and affine transformation in general as suggest in the and Data Augmentation for Skin Lesion Analysis [25] but after several experiments the results showed that is better not to use random centre cropping and to use augmentation that are slightly different compared the ones suggested in that paper. The parameters of the affine transformation were scale $= (0.7, 1.7)$, that is the scaling factor interval; and shear $= (-30, 30)$. Brightness and contrast have been modified sampling from a distribution of 0.2 of variance instead, hue and saturation the variance was 0.05. This augmentation has been found after many trial and error experiments, testing the most common augmentation used with convolutional neural network. For example, random grayscale augmentation, that is quite common, does not worked in this case because deteriorate the network performance. This is probably due to the fact that the colour is important for disease classification, indeed is one of the major differences between Nevus and Melanoma. Then, to further improve the network performance, were tried two others augmentation: gaussian noise and gaussian blur none of these augmentations improve overall performance. In Fig. 3 are presented the obtained performance in terms of BCA for different sets of augmented data.

4   Experimental Results

In Table 1 are summarised the performance in term of accuracy of the proposed approach. The best result was obtained with a ResNext50 [13] using images of size 800 × 800px. As can be seen in Table 1a, the network is able to classify quite well every class especially nevus (NV), this is reasonable because it is the most common class in the dataset, and vasculitis (VASC), even though in the dataset there are few images of this class. This probably means that this kind of disease has a particular pattern that the network recognize easily. The classes that are most difficult to classify for the network are dermatofibroma (DF), that is the less frequent class in the dataset, and actinic keratoses (AK). This could be due to a high interclass similarities of AK with seborrheic keratoses (BKL) and basal cell carcinoma (BCC), these classes are three or four times more common in the dataset, therefore the network is lead to predict one of these two classes. The mean precision of the network is 0.85, the balance multiclass accuracy is 87.8% while the mean F1-score is 0.86. These results are better than the state of the art of this classification task as shown in Table 1b, previous results reached were, using k-fold cross validation, with a RegNetY16GF [22]. The best performance got are F1-score 81.9 ± 0.6, precision 83.4 ± 0.4, recall 80.6 ± 0.8. Moreover, the

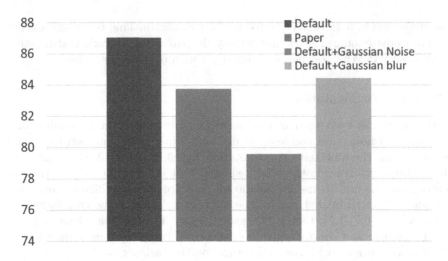

Fig. 3. Comparison of test dataset BCA of different sets of data augmentations.

network has a top-2 accuracy of 97% and a top 3 accuracy of 99%, this means that the network recognizes the issues even when it misses the prediction.

Table 1. Summarize and Comparison of accuracy performance.

(a) Precision, Recall and F1-Score for each class (Resnext50 - 800 × 800 px).

Skin Issue	Precision	Recall	F1-Score
MEL	89	80	84
NV	93	96	94
BCC	90	93	92
AK	79	79	79
BKL	85	83	84
DF	73	90	81
VASC	96	1	98
SCC	90	90	90

(b) Medium Precision, Recall and F1-Score of some papers.

Paper	Precision	Recall	F1-Score
SkinDeep [3]	73.26	72.88	72.91
Filipescu et al. [10]	–	78.11	–
Carcagni et al. [1]	81.9	83.4	80.6
Ours	**85**	**87**	**86**

4.1 Self-supervised Experiments

Selfsupervised learning is a new field of research. Most of the works are done on common and well-known dataset like IMAGENET [8] and CIFAR10 [18]. Appling this technique with a dataset like the ISIC 2019 dataset [6,12] is a challenging task for many reasons. To use a self-supervised method in melanoma detection has been chosen BYOL [11] architecture because it is relatively easy to implement and require much less memory and less computationally capability compared to other self-supervised techniques. To evaluate the quality of the features learnt during the self-supervised pretext task the network has been

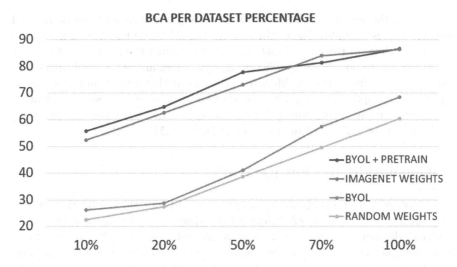

Fig. 4. Comparison of BCA of BYOL task started from pretrained weight, BYOL pretext task started from random weights, finetuning from IMAGENET, random initializzated network.

finetuned using different portions of the dataset 10%, 20%, 50%, 70% and the whole dataset. In Fig. 4 are summarized obtained results comparing the same architecture but pre-trained with IMAGENET and with same network initialized with random weights. First, to make sure that our reproduction of BYOL framework works properly, a test on CIFAR10 has been done. Two backbones have been used for this experiment EfficientNet-B4 [21] and ResNext50 [13], both get similar result to the BYOL's paper results. The experiments on this ISIC dataset were made in two different directions: demonstrate that self-supervised approach can be used in skin issues classification and try to apply this method to break the state-of the art in this field. So two groups of experiments were done, one starting pretext task with IMAGENET weight, to try to break the state-of-art, and the another one starting from random weights to prove the usefulness of self-supervised technique in this task. Using the setup that showed best results with supervised experiments, two networks were trained. One using pretrained weight on IMAGENET and one with weights trained using BYOL pretext. In this case the pretext task was made starting from pretained weights also. Use finetuning before a pretext task, was never done before. The results shows that four time out of five the BYOL pretext task improve the network performance, but the gain is up to 4%. The gain is so little that can be due to chance. More experiments are needed to prove that BYOL can be used in this scenario. The same experiment was repeated but starting the pretext task with random weights initialization of the network. Image size of 600px is used for both pretext and downstream task. In this experiment starting with the network's weights obtained with BYOL pretext task gives always better performance compared with random initialized network, up to 8%. But the balanced accuracy of the network is far from the accuracy got starting from pretrained weights. This is probably due to the fact that in IMAGENET there are millions of images, while, in the case of melanoma, even adding the images that are not annotated we do not reach even 30000 images, there are not enough images to train a network from scratch. Moreover, this result shows that supervised technique can be applied successfully in particular task as skin issues classification that is innovation in self-supervised technique because self-supervised learning has been applied only on standard dataset such as IMAGENET.

5   Conclusion and Future Work

This work, collected a preliminary results especially for the application of self-supervised technique that outperformed the state of the art for the ISIC 2019 challenge dataset. Future work will be addressed to study different loss functions, such as Center Loss [18] that has good performance due to its capability of making the features separable but also discriminative. In this task, where the images are very similar each other, discriminative features could improve the performance of the network. Label Smoothing [16] did not increase the classification accuracy, this because it was not weighted and the performance drops despite the theoretical background was promising. A weighted version of Label smoothing could be used to improve performance. Few families of convolutional neural network have been tried, several like VGG [24] net and DenseNet [14], that are widely used in this classification task, were not tested at all. In addition, Transformers [9] should be considered, the have shown great ability in image classification task. Bigger improvement can be obtained using patient's metadata. Information such age, sex and location of the disease has been proved to be useful in dermatology. Another option could be use ensamble of neural networks to have a more accurate prediction. But for a reliable network more images and image of skin disease of other skin colour must be collected.

References

1. Carcagni, P., Leo, M., Celeste, G., Distante, C., Cuna, A.: A systematic investigation on deep architectures for automatic skin lesions classification. In: 2020 25th International Conference on Pattern Recognition (ICPR), pp. 8639–8646. IEEE Computer Society, Los Alamitos, January 2021. https://doi.org/10.1109/ICPR48806.2021.9412789

2. Carcagnì, P., Leo, M., Cuna, A., Mazzeo, P.L., Spagnolo, P., Celeste, G., Distante, C.: Classification of skin lesions by combining multilevel learnings in a DenseNet architecture. In: Ricci, E., Rota Bulò, S., Snoek, C., Lanz, O., Messelodi, S., Sebe, N. (eds.) ICIAP 2019. LNCS, vol. 11751, pp. 335–344. Springer, Cham (2019). https://doi.org/10.1007/978-3-030-30642-7_30

3. Chen, L., Siraj, F., Su, J., Wong, C., Wong, M.: Skindeep: diagnosing dermatological images via computer vision. University of Michigan, Technical report (2021)

4. Chen, T., Kornblith, S., Norouzi, M., Hinton, G.: A simple framework for contrastive learning of visual representations. In: III, H.D., Singh, A. (eds.) Proceedings of the 37th International Conference on Machine Learning. Proceedings of Machine Learning Research, vol. 119, pp. 1597–1607. PMLR (13–18 Jul 2020). https://proceedings.mlr.press/v119/chen20j.html

5. Chen, T., Kornblith, S., Swersky, K., Norouzi, M., Hinton, G.E.: Big self-supervised models are strong semi-supervised learners. In: Larochelle, H., Ranzato, M., Hadsell, R., Balcan, M.F., Lin, H. (eds.) Advances in Neural Information Processing Systems. vol. 33, pp. 22243–22255. Curran Associates, Inc. (2020). https://proceedings.neurips.cc/paper/2020/file/fcbc95ccdd551da181207c0c1400c655-Paper.pdf

6. Combalia, M., et al.: Bcn20000: Dermoscopic lesions in the wild (2019)

7. Defazio, A., Jelassi, S.: Adaptivity without compromise: a momentumized, adaptive, dual averaged gradient method for stochastic optimization (2021)

8. Deng, J., Dong, W., Socher, R., Li, L.J., Li, K., Fei-Fei, L.: Imagenet: a large-scale hierarchical image database. In: 2009 IEEE Conference on Computer Vision and Pattern Recognition, pp. 248–255 (2009)

9. Dosovitskiy, A., et al.: An image is worth 16x16 words: Transformers for image recognition at scale. In: 9th International Conference on Learning Representations, ICLR 2021, Virtual Event, Austria, May 3–7, 2021. OpenReview.net (2021)

10. Filipescu, S.G., Butacu, A.I., Tiplica, G.S., Nastac, D.I.: Deep-learning approach in the study of skin lesions. Skin Res. Technol. **27**, 931–939 (2021)

11. Grill, J.B., et al.: Bootstrap your own latent - a new approach to self-supervised learning. In: Larochelle, H., Ranzato, M., Hadsell, R., Balcan, M.F., Lin, H. (eds.) Advances in Neural Information Processing Systems. vol. 33, pp. 21271–21284. Curran Associates, Inc. (2020), https://proceedings.neurips.cc/paper/2020/file/f3ada80d5c4ee70142b17b8192b2958e-Paper.pdf

12. Gutman, D.A., Codella, N.C.F., Celebi, M.E., Helba, B., Marchetti, M.A., Mishra, N.K., Halpern, A.: Skin lesion analysis toward melanoma detection: A challenge at the international symposium on biomedical imaging (ISBI) 2016, hosted by the international skin imaging collaboration (ISIC). CoRR abs/1605.01397 (2016). http://arxiv.org/abs/1605.01397

13. He, K., Zhang, X., Ren, S., Sun, J.: Deep residual learning for image recognition. In: 2016 IEEE Conference on Computer Vision and Pattern Recognition (CVPR), pp. 770–778 (2016)

14. Huang, G., Liu, Z., Weinberger, K.Q.: Densely connected convolutional networks. In: 2017 IEEE Conference on Computer Vision and Pattern Recognition (CVPR), pp. 2261–2269 (2017)

15. Jing, L., Tian, Y.: Self-supervised visual feature learning with deep neural networks: a survey. IEEE Trans. Pattern Anal. Mach. Intell. **43**, 4037–4058 (2021)

16. Lin, T.Y., Goyal, P., Girshick, R.B., He, K., Dollár, P.: Focal loss for dense object detection. 2017 IEEE International Conference on Computer Vision (ICCV), pp. 2999–3007 (2017)

17. Misra, I., van der Maaten, L.: Self-supervised learning of pretext-invariant representations. In: 2020 IEEE/CVF Conference on Computer Vision and Pattern Recognition (CVPR), pp. 6706–6716 (2020)

18. Müller, R., Kornblith, S., Hinton, G.E.: When does label smoothing help? In: NeurIPS (2019)

19. Perez, F., Vasconcelos, C.N., Avila, S., Valle, E.: Data augmentation for skin lesion analysis. In: OR 2.0/CARE/CLIP/ISIC@MICCAI (2018)

20. Radosavovic, I., Kosaraju, R.P., Girshick, R.B., He, K., Dollár, P.: Designing network design spaces. In: 2020 IEEE/CVF Conference on Computer Vision and Pattern Recognition (CVPR), pp. 10425–10433 (2020)

21. Simonyan, K., Zisserman, A.: Very deep convolutional networks for large-scale image recognition. CoRR abs/1409.1556 (2015)

22. Tan, M., Le, Q.: EfficientNet: Rethinking model scaling for convolutional neural networks. In: Chaudhuri, K., Salakhutdinov, R. (eds.) Proceedings of the 36th International Conference on Machine Learning. Proceedings of Machine Learning Research, vol. 97, 09–15 June 2019, pp. 6105–6114. PMLR. https://proceedings.mlr.press/v97/tan19a.html

23. Tschandl, P., et al.: Expert-level diagnosis of nonpigmented skin cancer by combined convolutional neural networks. JAMA Dermatol. **155**, 58–65 (2019)

24. Wen, Y., Zhang, K., Li, Z., Qiao, Y.: A discriminative feature learning approach for deep face recognition. In: ECCV (2016)

25. Xie, S., Girshick, R.B., Dollár, P., Tu, Z., He, K.: Aggregated residual transformations for deep neural networks. In: 2017 IEEE Conference on Computer Vision and Pattern Recognition (CVPR), pp. 5987–5995 (2017)

Unsupervised Deformable Image Registration in a Landmark Scarcity Scenario: Choroid OCTA

Emilio López-Varela[1,2](✉) [ID], Jorge Novo[1,2] [ID],
José Ignacio Fernández-Vigo[3,4] [ID], Francisco Javier Moreno-Morillo[3] [ID],
and Marcos Ortega[1,2] [ID]

[1] VARPA Group, Biomedical Research Institute of A Coruña (INIBIC),
University of A Coruña, A Coruña, Spain
{e.lopezv,jnovo,mortega}@udc.es
[2] CITIC-Research Center of Information and Communication Technologies,
University of A Coruña, A Coruña, Spain
[3] Department of Ophthalmology, Hospital Clínico San Carlos,
Instituto de Investigación Sanitaria (IdISSC), Madrid, Spain
[4] Centro Internacional de Oftalmología Avanzada, Madrid, Spain

Abstract. Recent advances in OCTA allow the imaging of blood flow deeper than the retinal layers at the level of the choriocapillaris (CC), where a pattern of small dark areas represents the absence of flow, called flow voids. The distribution of flow voids can be used as a biomarker to diagnose and monitor the progression of relevant pathologies or the efficacy of applied treatments. A pixel-to-pixel comparison can help to carry out this monitoring effectively, although in order to carry out this comparison, the used images must be perfectly aligned. CC images are characterized by their granularity, presenting numerous and complex local deformations, so a deformable registration is necessary to carry out a reliable comparison. However, CC OCTA images also present a characteristic absence of visually significant anatomical structures. This landmark scarcity hardens drastically the identification of points of interest to achieve an accurate registration. Based on this context, we designed a methodology to accurately perform this deformable registration in this challenging scenario. Hence, we propose a convolutional neural network

This research was funded by Instituto de Salud Carlos III, Government of Spain, DTS18/00136 research project; Ministerio de Ciencia e Innovación y Universidades, Government of Spain, RTI2018-095894-B-I00 research project; Ministerio de Ciencia e Innovación, Government of Spain through the research project with reference PID2019-108435RB-I00; Consellería de Cultura, Educación e Universidade, Xunta de Galicia, Grupos de Referencia Competitiva, grant ref. ED431C 2020/24; Axencia Galega de Innovación (GAIN), Xunta de Galicia, grant ref. IN845D 2020/38; CITIC, Centro de Investigación de Galicia ref. ED431G 2019/01, receives financial support from Consellería de Educación, Universidade e Formación Profesional, Xunta de Galicia, through the ERDF (80%) and Secretaría Xeral de Universidades (20%). Emilio López Varela acknowledges its support under FPI Grant Program through PID2019-108435RB-I00 project.

model trained by unsupervised learning to register images in a real clinical scenario, being obtained at different time instants from patients with central serous chorioretinopathy (CSC) treated with photodynamic therapy. Our methodology produces superior alignment to those achieved with other proven methods, helping to improve the monitoring of the efficacy of photodynamic therapy applied to patients with CSC. Our robust and adaptable methodology can also be exploited in other similar scenarios of complex registrations with anatomical landmark scarcity.

Keywords: Ophthalmology · OCTA imaging · Choriocapillaris · Deformable image registration · Flow voids · Convolutional neural networks

1   Introduction

Optical coherence tomographic angiography (OCTA) is a noninvasive imaging modality characterized by its capability to show a detailed visualization of the retinal vascularity. OCTA uses the variation in signal intensity of the OCT image over time as a contrast mechanism to obtain images [10,15]. Recent advances in OCTA allow imaging of blood flow deeper than the retinal layers, at the level of the choriocapillaris (CC), providing new dynamic information about choroidal physiology. OCTA is currently the only noninvasive modality available for imaging the CC in a clinical setting. Also, CC flow is not adequately visualized in traditional angiography so OCTA is the only usable alternative. When OCTA is used to image the choroidal innermost thickness, a granular image is obtained [24] showing a pattern of bright areas, representing flow, and a pattern of small dark regions showing areas of no flow called flow voids. Several examples of this type of image can be seen in Fig. 1. Several studies [6,8,19] reported a close correlation between abnormal flow void distribution and multiple retinal and choroidal diseases such as age-related macular degeneration, diabetic retinopathy, glaucoma, etc. Therefore, this different distribution of flow voids can be used as a biomarker to diagnose and monitor the progression of certain pathologies. In addition, it can also be used to monitor the efficacy of an applied treatment, as is the case, for example, in the treatment of central serous chorioretinopathy (CSC) [12].

In order to carry out an accurate and objective monitoring over time of the evolution of a pathology or the effectiveness of a treatment, it is necessary to perform an adequate comparison of the different progressive obtained images. A pixel-to-pixel comparison can adequately reflect these changes in the flow voids distribution over time as long as the images are properly aligned. CC images are characterized by a very rough and grainy appearance, presenting numerous complex local deformations that make pixel-to-pixel comparisons unfeasible and inappropriate. In order to carry out this type of comparison, these local deformations must be corrected as much efficiently and accurately as possible without distorting the original image by means of a deformable registration.

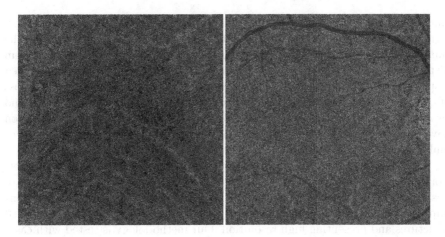

Fig. 1. Two examples of OCTA cc images.

However, there are several factors that hinder this registration process. Firstly, the roughness of the images discussed above, which gives rise to numerous complex deformations. Secondly, the characteristic absence of visually significant anatomical structures. This landmark scarcity hardens drastically the identification of points of interest to achieve an accurate registration. Lastly, the regular use of high-resolution imaging (1024 × 1024 pixels) in the clinical setting. The use of high-resolution images complicates the alignment process in terms of time and accuracy. To the best of our knowledge, there are no works that try to solve this deformable image registration problem in this novel image modality.

From the perspective of deformable registration in other medical imaging modalities, several traditional optimization methods such as B splines [22] or dense vector fields [25] deal with matching pairs of images. Unfortunately, these methods often require considerable time and computer resources to register a given pair of images. In the state of the art, several contributions have proposed the use of neural networks to perform medical image registration issues [9,17]. There are some perspectives that use unsupervised training that does not rely on any ground truth directly, although it can be used in a complementary way [4,18]. In general, these works propose the use of a convolutional neural network such as U-net [21] and a spatial transformation function [14] that warps images to one another. These architectures proved to be useful in solving registration tasks in modalities such as magnetic resonance imaging (MRI) where there are clearly defined anatomical structures that facilitate the registration process in a natural(due to the presence of characteristic points on the image) or guided way (as complementary ground truth). In contrast, CC OCTA images are characterized by their granularity, and unlike other imaging modalities as MRI, there do not present clear and defined anatomical structures, making it complicated to accurately align images. In addition, these architectures are designed to work with relatively small image resolutions (typically 256 × 256). The use of high-resolution images poses several problems, the main one being the loss of model efficiency due to factors such as insufficient receptive field. To adequately perform a registration task, the receptive field of the convolutional kernel in the smallest layer of the network must be at least as large as the expected maximum displacement. An increase in image resolution causes the maximum expected displacement to be larger than the cases of smaller images, resulting in a loss of quality in the registration. To increase the receptive field of the network and solve this problem, several works in related tasks such as semantic segmentation have used techniques such as dilated convolutions [29], wider networks [28] or multiscale images [30].

In this work, a novel approach is proposed using a convolutional neural network trained by unsupervised learning to register CC OCTA images. The proposal aims at achieving an efficient registration by overcoming the limitations imposed by the characteristics of using rough images, without clearly defined structures and presenting high resolution. Our methodology is tested with cases of a real clinical study about monitoring the response to a treatment.

2   Materials and Methods

2.1   Dataset

The dataset consists of a total of 821 CC OCTA images (1024 × 1024 pixels) from 52 patients with chronic CSC obtained using the Zeiss Plex Elite capture device. All the images were acquired by two well-trained clinical experts, being preliminarily used in a clinical study to evaluate the efficacy of studying changes in CC and choroid (CH) flow signal voids as a biomarker for monitoring the response of photodynamic therapy applied to CSC. Within the dataset, there are two different types of images, CC slab and CH slab images. For each patient, images were obtained at different times during treatment. The time instants correspond to pre-treatment, 2 to 4 days after treatment, 1 month after treatment, 3 months after treatment and 6 months after treatment. An example of the first 4 time instants for the CC image modality and the CH modality is shown in Fig. 2.

2.2   Network Architecture

We created an architecture (PyTorch 1.6 [20]) composed of convolutional layers and a spatial transform. The input of our network consists of the reference image and the image to be registered. The network is divided into 3 branches that use the input images at different scales (low, medium and high resolution images). In each of these branches, convolutions are applied with kernels of size 3 × 3 and a stride of 2. Each convolution is followed by a batch normalization layer that helps the convergence of the model acting as a regulating factor [13] and a LeakyReLU layer. The low resolution branch allows to make convolutions in a fast and efficient way, being able to accumulate several consecutive convolutions which increases the receptive field. In order to further increase the receptive

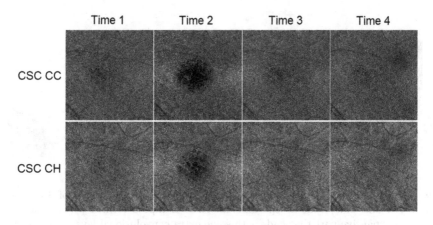

Fig. 2. Images for the first 4 temporal instants of a patient.

field, the last convolution of this branch is applied using a dilation. This branch allows us to obtain most of the image features in a coarse but efficient way. This allows the independent medium and high resolution branches to concentrate specifically on fine details that refine the final result without having to provide a complete representation of all the features that make up the image so that far fewer convolutions are needed. This reduces the number of model parameters that must be optimized, which reduces the risk of overfitting and aids in the convergence [5]. The 3 branches are merged using transpose convolutions, also using skip connections. Several extra convolutions are used to refine the result at the final resolution. The output of this plus the image to be registered serve as input to the spatial transformer which produces the final registered image. Figure 3 shows an overview of the neural architecture.

2.3 Training Details

The dataset was randomly divided into training (533 images), validation (128) and test (160) subsets. All the sets were independent from each other and all the images of a patient belonged to a single set. Local normalized cross correlation [3], a popular metric that is robust to intensity variations, was used as loss function. Network parameters were initialized using the He et al. [11] method. A batch size of 5 was used as it offered the best results in previous tests. As optimizer, we used the stochastic gradient descent with an initial learning rate of 0.001. It used a dynamic learning rate that was reduced by a factor of 0.7 if the loss of validation did not fall after 40 epochs. Early stopping was performed based on the validation loss. To make the model more robust and avoid overfitting, an exhaustive data augmentation process has also been applied [7]. Some transformations were grouped so that only one from the group could be applied at a time. Table 1 shows all the transformations applied to the images.

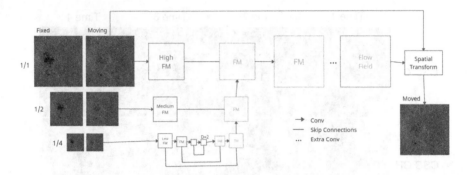

Fig. 3. Overview of the neural architecture. The network is composed of 3 branches (green, red and blue) that accept different sizes of inputs formed by convolutional layers, an expansive part that ends with some extra convolutional layers (yellow) and finally a spatial transform that outputs the registration image. FM = Feature maps. (Color figure online)

Table 1. Transformations applied to fixed and moving images in each batch. From each group only one transformation can be applied at a time.

Groups	Transformations
Group 1	Coarse Dropout
Group 2	Elastic Transform, Piecewise Affine
Group 3	Shift Scale Rotate
Group 4	Horizontal Flip
Group 5	Vertical Flip
Group 6	Random Rotate 90°
Group 7	Blur, Gaussian Blur, Motion Blur, Median Blur
Group 8	CLAHE, Brightness Contrast
Group 9	Gauss Noise, Image Compression, Multiplicative Noise

2.4   Baseline Methods

In this work, we selected representative baseline methods to compare and remark the suitable performance of the proposal. In particular, we use as first baseline Symmetric Normalization (Syn) [1], one of the best performing registration algorithms [16]. We use the version of this algorithm in the publicly available software package Advanced Normalization Tools (ANTs) [2], using mutual information (Syn) and cross-correlation (SynCC) as optimization metrics. We also tested other registration algorithms belonging to this package, such as the time-varying diffeomorphism using mean square metrics (TVMSQ) or some simple affine transformations.

2.5   Evaluation Metric

As metrics to quantitatively evaluate the alignment of this proposal, we selected different complementary statistics that are frequently used in these issues. For example, MSE is a traditional and simple method for measuring point distances between two images. The two images are compared pixel by pixel and the average of the square of the difference between the error of the two images is calculated. Other types of metrics such as SSIM [26] are also commonly used to evaluate image quality, being a type of metric that correlates well with human visual perception, as it evaluates structural differences between images by comparing local statistics rather than measuring point distances. In particular, we considered: MSE, NRMSE, SSIM, MSSIM [27], and VIF [23].

3   Results and Discussion

3.1   Quantitative Evaluation

Table 2 shows the results obtained for the different similarity metrics by the registration algorithms that were tested. In the table header, the value before registration is shown below the metric name and the time is given in seconds. All the experiments were performed using an intel(R) Core(TM) i5-6300HQ processor and NVIDIA GeForce GTX 950M graphics. As can be seen, our proposal achieves the best values for all the tested similarity metrics, demonstrating the ability to achieve better image alignment than the other tested algorithms. Also, the registration efficiency of our method is superior to the rest of the tested algorithms involving lower execution times than the rest of the tested methods using a CPU or even better exploiting GPU capabilities.

Table 2. Performance of different registration methods and the proposal in terms of time and image similarity. The header of each metric: the pre-registration value.

Methods	SSIM 0.0931	MSSIM 0.2103	VIF 0.0394	MSE 1482.1	NRMSE 0.3180	CPU seconds	GPU seconds
Affine	0.1564	0.3901	0.0527	1384.7	0.3068	361.2	
TVMSQ	0.1130	0.2230	0.0390	1437.3	0.3132	15787.6	
Syn	0.2920	0.5772	0.0828	1109.3	0.2744	1372.8	
SynCC	0.4452	0.6659	0.1180	965.3	0.2550	47772.6	
Our model	**0.7228**	**0.7113**	**0.1694**	**473.0**	**0.1796**	**233.7**	**38.1**

3.2   Qualitative Evaluation

While the used metrics show that our proposal is efficient in achieving image alignment, we must check if our method is able to align the images obtained at different time instants in such a way that no strange deformations are produced, the noise produced by a bad alignment is mitigated whereas the truly significant differences between the images are maintained. We must also check if our method allows a pixel-to-pixel comparison and that this ultimately produces a benefit for the clinician. Figure 4 shows illustrative results of applying the different registration algorithms to the same pair of images. Column one shows the fixed image, column two the registered image and column three the pixel by pixel differences between the fixed and registered images with a color code where the red represents a darker value in the fixed image, blue a darker value in the moving image and white the same value. Two values whose intensity value is less than 20 units apart are counted as equal as being not representative. As can be seen, our method substantially improves the visualization of pixel-to-pixel differences compared to the other tested methods. In addition, it manages to register the images without producing strange and undesired deformations.

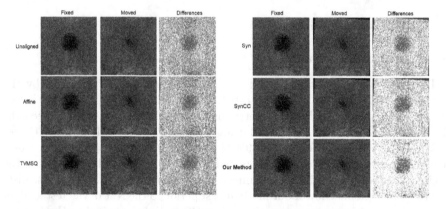

Fig. 4. Comparison of the registration of the different tested algorithms. Column 1, the fixed image; column 2, the image after the registration (except the first row, where the default unregistered images are shown); and column 3, pixel by pixel differences between the fixed and the moving image. The difference image shows in red the darkest areas in the first image, in blue the darkest areas in the second image and in white the areas that are considered equal in both images. (Color figure online)

As can be seen in Fig. 5, our proposal is able to eliminate a large part of the noise resulting from local deformations without altering the truly significant changes in the image such as changes in the distribution of the flow voids. Regarding the clinical case that was used to test our methodology, this produces an improvement in the monitoring of the efficacy of photodynamic therapy treatment in patients with CSC, and can also be used as a complementary preliminary step to other computational and clinical analyses. Although our experiment was based on the monitoring of photodynamic therapy in patients with CSC, our methodology is expandable to other pathologies associated with this type of image. Therefore, our methodology has a direct clinical relevance. Also, our proposal presents the potential of being applied to other complex medical image modalities, specially to other depths of OCTA imaging, of a great complexity and interest.

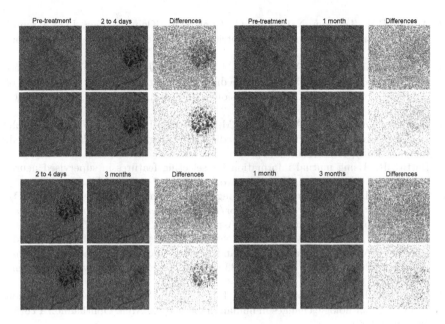

Fig. 5. Images and pixel-to-pixel differences obtained with our proposal for different time instants of the same patient. First row of each set, the fixed image and the unregistered moving image; second row, the fixed image and the registered image.

4 Conclusions

In this work, we have presented a robust registration methodology that employs a convolutional neural network trained by unsupervised learning to register CC OCTA images. The quantitative and qualitative obtained results have demonstrated that our methodology is capable of achieving efficient and effective image registration in a current clinical problem such as CSC treatment monitoring, overcoming the limitations imposed by the characteristics of using high-resolution, rough images without clearly defined anatomical structures. Furthermore, our methodology can be applied to other pathologies associated with this type of imaging and being adapted to other medical imaging modalities of great complexity and lack of representative visual anatomical structures. As future work, a more detailed ablation study of the proposed architecture could be performed to see which of its components most affect the performance of the architecture.

References

1. Avants, B.B., Epstein, C.L., Grossman, M., Gee, J.C.: Symmetric diffeomorphic image registration with cross-correlation: evaluating automated labeling of elderly and neurodegenerative brain. Med. Image Anal. **12**(1), 26–41 (2008)

2. Avants, B.B., Tustison, N.J., Song, G., Cook, P.A., Klein, A., Gee, J.C.: A reproducible evaluation of ants similarity metric performance in brain image registration. Neuroimage **54**(3), 2033–2044 (2011)

3. Balakrishnan, G., Zhao, A., Sabuncu, M.R., Guttag, J., Dalca, A.V.: An unsupervised learning model for deformable medical image registration. In: Proceedings of the IEEE Conference on Computer Vision and Pattern Recognition, pp. 9252–9260 (2018)

4. Balakrishnan, G., Zhao, A., Sabuncu, M.R., Guttag, J., Dalca, A.V.: Voxelmorph: a learning framework for deformable medical image registration. IEEE Trans. Med. Imaging **38**(8), 1788–1800 (2019)

5. Battiti, R.: Using mutual information for selecting features in supervised neural net learning. IEEE Trans. Neural Networks **5**(4), 537–550 (1994)

6. Bhutto, I., Lutty, G.: Understanding age-related macular degeneration (AMD): relationships between the photoreceptor/retinal pigment epithelium/Bruch's membrane/choriocapillaris complex. Mol. Aspects Med. **33**(4), 295–317 (2012)

7. Buslaev, A., Iglovikov, V.I., Khvedchenya, E., Parinov, A., Druzhinin, M., Kalinin, A.A.: Albumentations: fast and flexible image augmentations. Information **11**(2) (2020). https://doi.org/10.3390/info11020125, https://www.mdpi.com/2078-2489/11/2/125

8. Cao, J., McLeod, D.S., Merges, C.A., Lutty, G.A.: Choriocapillaris degeneration and related pathologic changes in human diabetic eyes. Arch. Ophthalmol. **116**(5), 589–597 (1998)

9. Cao, X., et al.: Deformable image registration based on similarity-steered CNN regression. In: Descoteaux, M., Maier-Hein, L., Franz, A., Jannin, P., Collins, D.L., Duchesne, S. (eds.) MICCAI 2017. LNCS, vol. 10433, pp. 300–308. Springer, Cham (2017). https://doi.org/10.1007/978-3-319-66182-7_35

10. De Carlo, T.E., Romano, A., Waheed, N.K., Duker, J.S.: A review of optical coherence tomography angiography (octa). Int. J. Retina and Vitreous **1**(1), 5 (2015)

11. He, K., Zhang, X., Ren, S., Sun, J.: Delving deep into rectifiers: surpassing human-level performance on imagenet classification. In: Proceedings of the IEEE International Conference on Computer Vision, pp. 1026–1034 (2015)

12. Ho, M., et al.: Analysis of choriocapillaris perfusion and choroidal layer changes in patients with chronic central serous chorioretinopathy randomised to micropulse laser or photodynamic therapy. Br. J. Ophthalmol. **105**, 555–560 (2020)

13. Horwath, J.P., Zakharov, D.N., Megret, R., Stach, E.A.: Understanding important features of deep learning models for segmentation of high-resolution transmission electron microscopy images. NPJ Comput. Mater. **6**(1), 1–9 (2020)

14. Jaderberg, M., Simonyan, K., Zisserman, A., Kavukcuoglu, K.: Spatial transformer networks. arXiv preprint arXiv:1506.02025 (2015)

15. Jia, Y., et al.: Split-spectrum amplitude-decorrelation angiography with optical coherence tomography. Opt. Express **20**(4), 4710–4725 (2012). https://doi.org/10.1364/OE.20.004710, http://www.opticsexpress.org/abstract.cfm?URI=oe-20-4-4710

16. Klein, A., Andersson, J., Ardekani, B.A., Ashburner, J., Avants, B., Chiang, M.C., Christensen, G.E., Collins, D.L., Gee, J., Hellier, P., et al.: Evaluation of 14 nonlinear deformation algorithms applied to human brain MRI registration. Neuroimage **46**(3), 786–802 (2009)

17. Krebs, J., et al.: Robust non-rigid registration through agent-based action learning. In: Descoteaux, M., Maier-Hein, L., Franz, A., Jannin, P., Collins, D.L., Duchesne, S. (eds.) MICCAI 2017. LNCS, vol. 10433, pp. 344–352. Springer, Cham (2017). https://doi.org/10.1007/978-3-319-66182-7_40

18. Li, H., Fan, Y.: Non-rigid image registration using fully convolutional networks with deep self-supervision. arXiv preprint arXiv:1709.00799 (2017)

19. Lutty, G., Grunwald, J., Majji, A.B., Uyama, M., Yoneya, S.: Changes in choriocapillaris and retinal pigment epithelium in age-related macular degeneration. Mol. Vis. **5**(35), 35 (1999)

20. Paszke, A., et al.: Pytorch: an imperative style, high-performance deep learning library. In: Wallach, H., Larochelle, H., Beygelzimer, A., d'Alché-Buc, F., Fox, E., Garnett, R. (eds.) Advances in Neural Information Processing Systems, vol. 32, pp. 8024–8035. Curran Associates, Inc. (2019). http://papers.neurips.cc/paper/9015-pytorch-an-imperative-style-high-performance-deep-learning-library.pdf

21. Ronneberger, O., Fischer, P., Brox, T.: U-Net: convolutional networks for biomedical image segmentation. In: Navab, N., Hornegger, J., Wells, W.M., Frangi, A.F. (eds.) MICCAI 2015. LNCS, vol. 9351, pp. 234–241. Springer, Cham (2015). https://doi.org/10.1007/978-3-319-24574-4_28

22. Rueckert, D., Sonoda, L.I., Hayes, C., Hill, D.L., Leach, M.O., Hawkes, D.J.: Non-rigid registration using free-form deformations: application to breast MR images. IEEE Trans. Med. Imaging **18**(8), 712–721 (1999)

23. Sheikh, H.R., Bovik, A.C.: Image information and visual quality. IEEE Trans. Image Process. **15**(2), 430–444 (2006)

24. Spaide, R.F., Fujimoto, J.G., Waheed, N.K.: Image artifacts in optical coherence angiography. Retina (Philadelphia, Pa.) **35**(11), 2163 (2015)

25. Thirion, J.P.: Image matching as a diffusion process: an analogy with Maxwell's demons. Med. Image Anal. **2**(3), 243–260 (1998)

26. Wang, Z., Bovik, A.C., Sheikh, H.R., Simoncelli, E.P.: Image quality assessment: from error visibility to structural similarity. IEEE Trans. Image Process. **13**(4), 600–612 (2004)

27. Wang, Z., Simoncelli, E.P., Bovik, A.C.: Multiscale structural similarity for image quality assessment. In: The Thrity-Seventh Asilomar Conference on Signals, Systems & Computers, vol. 2, pp. 1398–1402. IEEE (2003)

28. Wu, Z., Shen, C., Van Den Hengel, A.: Wider or deeper: revisiting the resnet model for visual recognition. Pattern Recogn. **90**, 119–133 (2019)

29. Yu, F., Koltun, V.: Multi-scale context aggregation by dilated convolutions. arXiv preprint arXiv:1511.07122 (2015)

30. Zhao, H., Qi, X., Shen, X., Shi, J., Jia, J.: ICNet for real-time semantic segmentation on high-resolution images. In: Ferrari, V., Hebert, M., Sminchisescu, C., Weiss, Y. (eds.) ECCV 2018. LNCS, vol. 11207, pp. 418–434. Springer, Cham (2018). https://doi.org/10.1007/978-3-030-01219-9_25

Leveraging CycleGAN in Lung CT Sinogram-free Kernel Conversion

Michela Gravina[1]([✉]) [iD], Stefano Marrone[1] [iD], Ludovico Docimo[2] [iD],
Mario Santini[3] [iD], Alfonso Fiorelli[3] [iD], Domenico Parmeggiani[2] [iD],
and Carlo Sansone[1] [iD]

[1] DIETI, University of Naples Federico II, Naples, Italy
{Michela.Gravina,Stefano.Marrone,Carlo.Sansone}@unina.it
[2] DAMSS, Università della Campania "L.Vanvitelli", Caserta, Italy
{Ludovico.Docimo,Domenico.Parmeggiani}@unicampania.it
[3] Department of Translational Medical Sciences, Università della Campania
"L.Vanvitelli", Caserta, Italy
{Mario.Santini,Alfonso.Fiorelli}@unicampania.it

Abstract. Cancer screening guidelines recommend annual screening
with low-dose Computed Tomography (CT) for high-risk groups to
reduce lung cancer mortality. Unfortunately, lung CT effectiveness can be
strongly impacted by the considered reconstruction kernel. This selection
is (almost) final, implying that it is no longer possible to change the used
reconstruction kernel once applied, unless a sinogram for the conversion
is available. The aim of this paper was to introduce a new sinogram-free
kernel conversion in the contest of lung CT imaging. In particular, we
wanted to define a procedure able to deal with different acquisition protocols, able to be used in an unpaired images scenario. To this aim, we
leveraged a CycleGAN, considering the CT kernel conversion task as a
style transfer problem. Results show that the CT kernel conversion can
be effectively addressed as a style transfer problem.

Keywords: Kernel conversion · Deep Learning · CycleGAN

1 Introduction

According to the World Health Organization (WHO), lung cancer is one of the
most frequent diseases, causing death estimated for nearly 1.59 million people
per year [9]. It is strictly associated with the consumption of tobacco prod-
ucts, showing an increase of 2% per year in its worldwide incidence. Cancer
screening guidelines [13] recommends annual screening with low-dose Computed
Tomography (CT) for high-risk groups to reduce cancer mortality. Indeed, still
today early diagnosis is considered the key to reducing mortality and increasing the chances for a successful treatment. Over the years, CT imaging has
demonstrated great potential in cancer detection, providing information about
the presence and the stage of a tumour. From a technical perspective, it is a

radiological technique that uses computer-processed combinations of multiple X-ray measurements, taken from different angles, to provide volumetric acquisitions organised in a series of sectional 2D images (slices) of the body allowing to distinguish the various organs and tissues based on their density. This procedure results in high-resolution images, that are commonly used in a wide set of medical practices, including to guide biopsy procedures, local treatments (cryotherapy, radiofrequency ablation, etc.) or to assess whether a cancer is responding to treatment or not.

One of the main advantages of CT scans is in its high versatility which makes CT suited for analysing both hard tissues (i.e. bones [3]) and soft ones (i.e. organs [10]). The key for this flexibility is in the mathematical process used to reconstruct the tomographic images from the acquired multi-angle X-ray projections. Despite several reconstruction techniques exist, they all aim at backprojecting the X-ray data to the image domain by using a convolutional filter [7], most commonly referred as *reconstruction kernel* (or simply kernel). The aim of such a kernel is to adjust the frequency components of the X-day data before the reconstruction, in order to reduce blurring. As a consequence, the used kernel plays a crucial role in the reconstruction process, resulting to be among the most important aspects impacting the quality of the final image and, in turn, the needed radiation dose. Despite different kernels being available for reconstructing images from specific anatomical districts, they can all be roughly gathered into smooth kernels, usually generating low-noise but low-resolution images, and sharp kernels, usually generating higher resolution images (most commonly affected by noise).

The selection of the most suited kernel should thus be based on the expected use for the acquired image, considering the trade-off between the image quality required for the examination and the delivered radiation dose. This becomes problematic in contexts where the anatomical district under examination encompasses tissues that require different kernels for best outcomes. A suited example for this is the chest, with researchers analysing the effects associated with the use of different kernels in the case of CT images of lungs. Unfortunately, this selection is (almost) final, implying that it is no longer possible to change the used reconstruction kernel once applied, unless a sinogram is available.

As the radiomics methodology becomes more established and with the spread of Deep Learning (DL) in several biomedical imaging tasks, researchers are analysing if, and to what extent, the used reconstruction kernel can affect automatic CT image processing and whether DL, and in particular deep Convolutional Neural Networks (CNN), can be used to cope with sinogram-free kernel conversion. In [2] the authors show how it is possible to learn mutual relationships between different kernel types of smooth and sharp kernels. For the study, the paper uses a super-resolution CNN enhanced by squeeze-and-excitation residual blocks and task-specific losses. Despite the model does not consider the sinogram, it is important to highlight that it has been trained on images pairs obtained by reconstructing patients' CT raw data (i.e. the sinogram) using the desired kernels. In both [5] and [1], the authors show that radiomics features can be widely affected by the used reconstruction kernel. The latter work also proves that it is possible to perform post-reconstruction image conversion by using a deep Convolutional Neural Network (CNN), trained on coupled residual images, to make the extracted features more stable across kernels.

Despite the cited papers report promising results, the main limitation is in the need for training pairs reconstructed by using different kernels, not usually available in clinical contexts or in retrospective studies. Moreover, the considered models may result to be very task and dataset oriented. Thus, to cope with these problems, in this work we propose to perform sinogram-free kernel conversion in lung CT by using a Generative Adversarial Network (GAN), trained on a publicly available dataset. Among the different available GANs architectures, we make use of CycleGAN [14], a particular model designed to perform unpaired image-to-image translation. To the best of our knowledge, this is the first work using CycleGAN in lung CT kernel conversion. The rest of the paper is organised as follows: Sect. 2 introduces the proposed methodology; Sect. 3 describes the involved dataset and the experimental setup; Sect. 4 shows the obtained results; finally Sect. 5 provides some conclusions.

2   Methodology

In this work we propose a lung CT kernel conversion methodology consisting of two main steps: the *Pre-processing*, used to prepare data belonging to different patients and the *Kernel Conversion*, using a CycleGAN to perform the actual adaptation of images with different reconstruction kernels.

2.1   Pre-processing

Besides the used reconstruction kernel, CT images can also vary for other aspects, such as the slice thickness, the acquisition size, etc., resulting in slices with diverse pixel spacing. Thus, in order to make our approach general enough to be used with different acquisition protocols, as a first step we implement a pre-processing stage during which each CT series is re-sampled with a bi-linear interpolation to have a spacing of 1 mm in every direction (x, y, z). Moreover, as a chest CT acquisition includes different organs, there is also the need for a lung segmentation module in order to allow the conversion module to focus only on lung parenchyma. As we want this step to be as protocol independent as possible, the lungs are segmented by using the properties of the Hounsfield Scale: each voxel in the CT series is clipped to a maximum value of 4096 and to a minimum value of -1024, and then normalized in $[0, 1]$ before using the central slice to derive the threshold to use for distinguishing the pixel belonging to the lung tissue from the others. K-means algorithm, with $k = 2$, is used to determine the aforementioned threshold. The result of this step is a binary lung mask, including only lung parenchyma. This mask is further refined by considering only the largest regions obtained and applying dilation and erosion operations on them. The obtained mask is used to extract the lung voxels. For each CT volume, the slices representing the central centimetre are extracted, resulting in a total of 10 slices per acquisition. Finally, CT images are normalized in the range $[-1, 1]$ to prepare data for the implemented CycleGan. It is worth noting that the described preprocessing is applied to the images of all the considered kernels.

2.2   Kernel Conversion

CycleGAN is a technique for training image-to-image translation models from a source domain (X) to a target domain (Y) *in the absence of paired examples* via a GAN architecture. The aim is to learn a map between the two domains X and Y given training samples belonging to them. Distributions of data belonging to X and Y are referred as $x \sim p_{data}(x)$ and $y \sim p_{data}(y)$ respectively. As proposed in [14], the CycleGAN includes two mappings $G : X \rightarrow Y$ and $F : Y \rightarrow X$, that are implemented with two generator networks sharing the same architecture. In addition two discriminators D_X and D_Y are introduced, where the former aims at distinguishing between images x and translated images $F(y)$ while the latter is used for differentiating between y and $G(x)$. Each discriminator provides as output the probability that the input images belong to the real distribution of the target domain. The generator, inspired by the network proposed in [6], is a fully convolutional network (FCN), able to process images having different sizes. Its architecture (Fig. 1) consists of three main components: the Encoder, the Transformer and the Decoder. The first performs dimensionality reduction through three convolutional layers. The size of the feature maps remains constant in the Transformer component, which consists of nine Residual Blocks, whose aim is to extract features to perform the domain translation. In a Residual Block, the standard batch normalization is replaced with an Instance Normalization, as proposed in [12]. The main idea is that the result of kernel conversion from one image to another should not depend on the contrast of the source image. Therefore the generator network should remove this kind of information. The Instance Normalization is then introduced to efficiently implement contrast normalization. In opposition to batch normalization, it applies normalization to each instance instead of the whole batch and it is also applied in the inference phase. Finally, the Decoder uses three transpose convolution operations to generate an output with the same size as the input image. The Discriminator is a PatchGaN [4]. Differently from a standard CNN, which produces a single scalar value as output for binary classification, a PatchGaN provides a two-dimensional matrix of size $N \times N$. Each pixel of the output refers to a patch of $M \times M$ size of the input image and represents its probability to belong to the real distribution of the target domain. In other words, each pixel is assigned a receptive field of $M \times M$. The classification probability is then computed as the average of the matrix produced as output. Similar to the generator, the Discriminator is an FCN resulting in a network with fewer parameters than standard CNN that can be used for images of varying and relatively large sizes. The Discriminator architecture (Fig. 2) consists of four convolutional layers followed by Instance

Normalization and Leaky ReLU as activation function, and a final convolution layer with a sigmoid function.

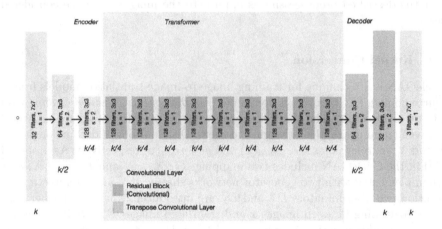

Fig. 1. Generator architecture consisting of three main components: the Encoder, the Transformer and the Decoder. The size of the output feature map in each layer is represented by k.

During the training of the implemented CycleGan, the objective function consists of two main terms: the *adversarial losses* for matching the distribution of generated images to the data distribution in the target domain; and *cycle consistency* losses to prevent the learned mappings G and F from contradicting each other. The adversarial losses (adv) are applied to both mapping functions, using a least-square loss, as suggested in [8], to solve the vanishing gradients problem that may occur considering the standard loss used for GAN. More in detail, for the mapping function $G : X \rightarrow Y$, with D_Y discriminator, the objective function is defined as follows:

$$L_{adv}(G, D_Y, X, Y) = \begin{cases} \mathbb{E}_{y \sim pdata(y)}[(D_Y(y) - 1)^2] + \mathbb{E}_{x \sim pdata(x)}[(D_Y(G(x))^2] \\ \mathbb{E}_{x \sim pdata(x)}[(D_Y(G(x)) - 1)^2] \end{cases}$$
(1)

where the first line is referred to D_Y that aims to distinguish between the real samples y and the generated images $G(x)$, while the second equation is used for G that tries to generate images $(G(x))$ similar to those belonging to Y. A similar adversarial loss $L_{adv}(F, D_X, Y, X)$ is introduced for the mapping $F : Y \rightarrow X$ and its discriminator D_X. Moreover, the learned mapping functions should be cycle-consistent: it should be verified that $x \rightarrow G(x) \rightarrow F(G(x)) \approx x$ and $y \rightarrow F(y) \rightarrow G(F(y)) \approx y$. Based on this observation, the *cycle consistency loss* (*cyc*) is defined as follows:

$$L_{cyc}(G, F) = \mathbb{E}_{x \sim pdata(x)}[\|F(G(x)) - x\|_1] + \mathbb{E}_{y \sim pdata(y)}[\|G(F(y)) - y\|_1] \quad (2)$$

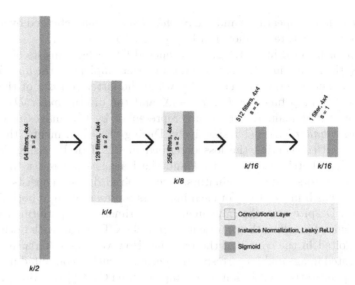

Fig. 2. Discriminator architecture consisting of convolutional layers followed by Instance Normalization and Leaky ReLU as activation function, and a final convolution layer with a sigmoid function to provide classification probabilities. The size of the output feature map in each layer is represented by k

To improve the stability of adversarial training, the Image Pool technique is used, as suggested in [11], to solve the problem that the discriminator only focuses on the last generated images when its weights are updated. This lack of memory not only causes divergence in training but make the generator reintroduce artefacts that the discriminator has forgotten about. Since any generated image during the entire training procedure is *fake* for the discriminator, the latter should be able to correctly classify all the images provided by the generator. Based on this observation, the discriminator is updated using a history of previously generated images rather than only the ones in the current iteration, that is in the minibatch of the training process.

3   Experimental Setup

In this work we use the publicly available Lung-PET-CT-Dx[1] dataset, consisting of 898 CT-images belonging to patients with suspicion of lung cancer (adenocarcinoma, small cell carcinoma, large cell carcinoma, squamous cell carcinoma). The CT slice interval varies from 0.625 mm to 5 mm. Scanning mode includes plain, contrast and 3D reconstruction. CT images are acquired with different scans using several reconstruction kernels. We focus on two reconstruction kernels with different characteristics: the *Lung kernel*, which gives images a

---

[1] https://wiki.cancerimagingarchive.net/pages/viewpage.action?pageId=70224216.

particularly sharp appearance and the *Standard kernel* characterized by a noticeably 'smoother' texture, as shown in Fig. 5a and 5b respectively.

As aforementioned in Sect. 2.2, the proposed CycleGan consists of two coupled GAN: the former includes the generator G, that implement the map function $G : X \rightarrow Y$ and the discriminator D_Y, while the latter consists of the generator F, for the map function $F : Y \rightarrow X$ and the discriminator D_X. In our experiments the domain X and Y are represented by CT-images using *Lung kernel* and *Standard Kernel* respectively. During the experiments, the CycleGan is trained considering the loss described in Eqs. 1 and 2 using Adam as optimizer. The batch size is set to 4 and the learning rate is set to $3 \cdot 10^{-4}$ for both generators and discriminators. The involved dataset consists of *paired CT-images*, which means that for each image its representation in both domains is available. Despite this, the implemented experiments are performed with an *unpaired dataset* in which the representation of the CT-images with both kernels is not exploited in the course of the training. However, the difference between each generated image and its corresponding ground truth is computed in order to provide a quantitative evaluation of the implemented CycleGAN. More in detail, we consider different similarity indexes that aim to measure the quality of the generated images. Denoting with I and I_g the ground-truth and the generated image respectively, with $\tilde{I}$ and $\tilde{I}_g$ the mean of the images, with (p,q) the coordinates of a pixel in the image and with N the number of pixels, the implemented similarity indexes are detailed in Table 1 and are defined as follows:

- **Normalized Cross-correlation (NCC)**, measuring the correlation between the pixel intensity of the generated image and the corresponding ground-truth. A value near zero implies no correlation between the images;
- **Sum of Squared (SSD)**, calculated as the mean of the squares of the differences between the pixel intensities of two images. A value near 0 implies a good similarity;
- **Sum of absolute differences (SAD)**, computed as the average of the absolute differences between the pixel intensities of the two images. Similar to SSD, a value near 0 means a good similarity;
- **Cosine similarity (CS)**, the cosine of the angle between images. Similar to NCC, a value near zero means that there is no similarity between images.

To estimate the generalization error we use a hold-out split, considering a randomly selected 25% of the dataset for testing purposes.

4   Results

Table 2 reports the performance of the proposed approach. In both map functions, the proposed metrics show very promising results. The SSD and SAD values are very close to 0, while the NNC and CS report a high correlation score. Figure 4 shows the training curves representing the loss functions monitored during the training. More in detail, Fig. 4a reports the L_{adv} for the generators G and F, that is the second expression in the Eq. 1 representing $L_{adv}(G, D_Y, X, Y)$

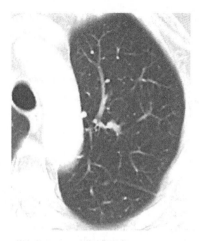
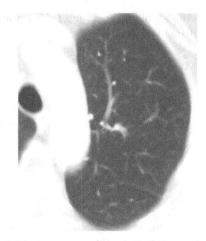

(a) CT image using *Lung kernel*   (b) CT image using *Standard kernel*

Fig. 3. Examples of the considered reconstruction kernels: the *Lung kernel* (5a) gives images a particularly sharp appearance and the *Standard kernel* (5b) is characterized by a noticeably 'smoother' texture.

Table 1. Definition of the implemented similarity indexes: I and I_g are the ground truth and the generated image respectively; $\tilde{I}$ and $\tilde{I}_g$ are the mean of the images; p, q are the coordinates of a pixel in the image; N is the number of pixels.

Similarity index	Math formula
$NCC(I, \hat{I})$	$\frac{\sum(I(p,q)-\tilde{I})(I_g(p,q)-\tilde{I}_g)}{\sqrt{\sum(I(p,q)-\tilde{I})^2 \sum(I_g(p,q)-\tilde{I}_g)^2}}$
$SSD(I, \hat{I})$	$\frac{1}{N}\sum_N (I(p,q) - I_g(p,q))^2$
$SAD(I, \hat{I})$	$\frac{1}{N}\sum_N |I(p,q) - I_g(p,q)|$
$CS(I, \hat{I})$	$\frac{\sum I(p,q)I_g(p,q)}{\sqrt{\sum I(p,q)^2 \sum I_g(p,q)^2}}$

Table 2. Results of the proposed methodology for the image-to-image translation models from a source domain (X) to the target domain (Y) and vice versa. The domain X and Y are represented by CT-images using *Lung kernel* and *Standard Kernel* respectively.

Mapping function	SSD	SAD	NCC	CS
$G : X \rightarrow Y$	0.00499	0.02322	0.81721	0.99719
$F : Y \rightarrow X$	0.00593	0.02730	0.78917	0.99668

and $L_{adv}(F, D_X, Y, X)$ respectively while the Fig. 4b shows the L_{adv} for the discriminators D_X and D_Y corresponding to the first part of the Eq. 1. Finally, the Fig. 4c reports the loss defined in Eq. 2 representing $L_{cyc}(G, F)$ computed for $G : X \rightarrow Y$ and $F : Y \rightarrow X$ separately. The training curves suggest that losses become stable after about 1200 iterations, confirming good convergence of the implemented model.

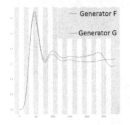 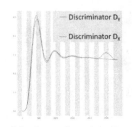 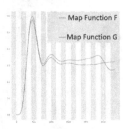

(a) L_{adv} for the generators G and F corresponding to the the second expression in the Equation 1.

(b) L_{adv} for the discriminators D_X and D_Y corresponding to the first part of the Equation 1.

(c) $L_{cyc}(G, F)$ computed for $G : X \rightarrow Y$ and $F : Y \rightarrow X$ separately representing the Equation 2.

Fig. 4. Training Curves for the loss functions monitored during the training.

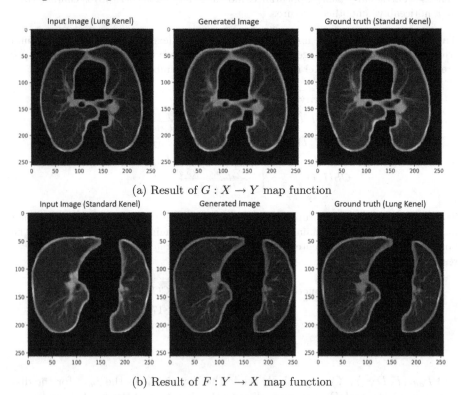

(a) Result of $G : X \rightarrow Y$ map function

(b) Result of $F : Y \rightarrow X$ map function

Fig. 5. Qualitative assessment of the proposed methodology. Input CT image (left), generated image (center), ground truth image (right)

Finally, Fig. 5 can be used to make a qualitative assessment of the proposed methodology. More in detail, the Fig. 5a shows the result of the map function $G : X \rightarrow Y$, that translate the CT images from *Lung Kernel* to *Standard Kernel*, while the Fig. 5b reports an example of the inverse map, $F : Y \rightarrow X$. When moving from *Lung Kernel* to *Standard Kernel*, the generator reduces the contrast in the image, making it sharper. In contrast, during the inverse translation, the network learns how to perform the image enhancement highlighting the details.

5   Conclusion

The aim of this paper was to introduce a new sinogram-free kernel conversion in the contest of lung CT imaging. In particular, we wanted to define a procedure able to deal with different acquisition protocols and suitable to be used in an unpaired images scenario. To this aim, we leveraged a CycleGAN, considering the CT kernel conversion task as a style transfer problem. To measure the effectiveness of the proposed approach, we considered a publicly available dataset, using our approach to transfer CT images from a domain X, consisting of acquisitions that use a *Lung* reconstruction kernel, to a target domain Y, representing images reconstructed by using a *Standard* kernel. We trained the used Cycle-GAN proposing an adversarial loss (L_{adv}) (Eq. 1) based on the least-squares loss function and using the Instance Normalization and Image Pool techniques to efficiently implement contrast normalization and to improve the stability of adversarial training respectively. The results (Table 2) are very promising, confirming the GAN as an effective techniques for image-to-image translation with the aim of reducing the heterogeneity arising from the use of different protocols in CT acquisitions. To the best of our knowledge, this is the first work using CycleGAN in lung CT kernel conversion.

Acknowledgments. The authors gratefully acknowledge the support of NVIDIA Corporation with the donation of the Titan Xp GPU used for this research, as well as the availability of the Calculation Centre SCoPE of the University of Naples Federico II and its staff. This work is part of the "Synergy-net: Research and Digital Solutions against Cancer" project (funded in the framework of the POR Campania FESR 2014-2020 - CUP B61C17000090007).

References

1. Denzler, S., et al.: Impact of CT convolution kernel on robustness of radiomic features for different lung diseases and tissue types. Br. J. Radiol. **94**(1120), 20200947 (2021)
2. Eun, D.i., Woo, I., Park, B., Kim, N., Seo, J.B., et al.: Ct kernel conversions using convolutional neural net for super-resolution with simplified squeeze-and-excitation blocks and progressive learning among smooth and sharp kernels. Comput. Methods Programs Biomed. **196**, 105615 (2020)
3. Ibragimov, B., Xing, L.: Segmentation of organs-at-risks in head and neck CT images using convolutional neural networks. Med. Phys. **44**(2), 547–557 (2017)

4. Isola, P., Zhu, J.Y., Zhou, T., Efros, A.A.: Image-to-image translation with conditional adversarial networks. In: Proceedings of the IEEE Conference on Computer Vision and Pattern Recognition, pp. 1125–1134 (2017)
5. Jin, H., Heo, C., Kim, J.H.: Deep learning-enabled accurate normalization of reconstruction kernel effects on emphysema quantification in low-dose ct. Phys. Med. Biol. **64**(13), 135010 (2019)
6. Johnson, J., Alahi, A., Fei-Fei, L.: Perceptual losses for real-time style transfer and super-resolution. In: Leibe, B., Matas, J., Sebe, N., Welling, M. (eds.) ECCV 2016. Perceptual losses for real-time style transfer and super-resolution, vol. 9906, pp. 694–711. Springer, Cham (2016). https://doi.org/10.1007/978-3-319-46475-6_43
7. Kak, A.C., Slaney, M.: Principles of computerized tomographic imaging. In: SIAM (2001)
8. Mao, X., Li, Q., Xie, H., Lau, R.Y., Wang, Z., Paul Smolley, S.: Least squares generative adversarial networks. In: Proceedings of the IEEE International Conference on Computer Vision, pp. 2794–2802 (2017)
9. Serj, M.F., Lavi, B., Hoff, G., Valls, D.P.: A deep convolutional neural network for lung cancer diagnostic. arXiv preprint arXiv:1804.08170 (2018)
10. Shimizu, A., Ohno, R., Ikegami, T., Kobatake, H., Nawano, S., Smutek, D.: Segmentation of multiple organs in non-contrast 3d abdominal CT images. Int. J. Comput. Assist. Radiol. Surg. **2**(3), 135–142 (2007)
11. Shrivastava, A., Pfister, T., Tuzel, O., Susskind, J., Wang, W., Webb, R.: Learning from simulated and unsupervised images through adversarial training. In: Proceedings of the IEEE Conference on Computer Vision and Pattern Recognition. pp. 2107–2116 (2017)
12. Ulyanov, D., Vedaldi, A., Lempitsky, V.: Instance normalization: the missing ingredient for fast stylization. arXiv preprint arXiv:1607.08022 (2016)
13. Wender, R., et al.: American cancer society lung cancer screening guidelines. CA: Cancer J. Clin. **63**(2), 106–117 (2013)
14. Zhu, J.Y., Park, T., Isola, P., Efros, A.A.: Unpaired image-to-image translation using cycle-consistent adversarial networks. In: Proceedings of the IEEE International Conference on Computer Vision, pp. 2223–2232 (2017)

Investigating One-Class Classifiers to Diagnose Alzheimer's Disease from Handwriting

Antonio Parziale[1,2]([envelope]) [ID], Antonio Della Cioppa[1,2,3] [ID], and Angelo Marcelli[1,2] [ID]

[1] DIEM, University of Salerno, Via Giovanni Paolo II 132, 84084 Fisciano, SA, Italy
{anparziale,adellacioppa,amarcelli}@unisa.it
[2] University of Salerno Unit, CINI, National Laboratory of Artificial Intelligence and
Intelligent Systems, Fisciano, SA, Italy
[3] ICAR-CNR, Via P. Castellino 111, 80131 Naples, Italy

Abstract. The analysis of handwriting and drawing has been adopted since the early studies to help diagnose neurodegenerative diseases, such as Alzheimer's and Parkinson's. Departing from the current state-of-the-art methods that approach the problem of discriminating between healthy subjects and patients by using two- or multi-class classifiers, we propose to adopt one-class classifier models, as they require only data by healthy subjects to build the classifier, thus avoiding to collect patient data, as requested by competing techniques. In this framework, we evaluated the performance of three models of one-class classifiers, namely the Negative Selection Algorithm, the Isolation Forest and the One-Class Support Vector Machine, on the DARWIN dataset, which includes 174 subjects performing 25 handwriting/drawing tasks. The comparison with the state-of-the-art shows that the methods achieve state-of-the-art performance, and therefore may represent a viable alternative to the dominant approach.

Keywords: Alzheimer's disease diagnosis · Handwriting analysis · One-class classifiers · Negative selection algorithm · Isolation forest · One-class support vector machine

1 Introduction

Neurodegenerative diseases affect millions of people worldwide, and among them, Alzheimer's disease (AD) is the most frequent one. In 2015 it affected more than 40 million people worldwide, and it is estimated this number will increase to 131.5 million by 2050 [33]. There is no cure for the disease, but the decline can be somehow managed during its progression by proper treatments, with the aim of ensuring to the patients an autonomous lifestyle as long as possible. The best results, however, can be reached with an early diagnosis, so that the treatment can be initiated at the very beginning of the disease. This creates a critical need for conducting periodic screening of the population at risk of developing the disease, as suggested by national and international health organizations.

In this context, there has been an increasing interest in developing methodologies and tools to help the early diagnosis of Alzheimer's disease [25,37]. Most of the approaches exploit biomarkers, such as clinical, imaging, biochemical, and genetic, but their predicting values have been questioned [22,23], they are either expensive, invasive, time-consuming and need tools and materials mostly available in hospitals and specialized labs.

The analysis of handwriting dynamics might provide a cheap and non-invasive method for evaluating the disease progression [18], since handwriting requires a precise and properly coordinated control of the body [32], which, in turn, envisages the cognitive and motor functions that are impaired by the insurgence of the diseases. Cheap and widely available graphic tablets have then been introduced to administer various handwriting and/or drawing tests and to record kinematic and dynamic information of the performed movements, and several studies on the neural processes involved in motor learning and execution, as well as the deterioration of motor processes involved in writing and drawing in patients, have highlighted the most distinctive features of patient's movements [2,19,28,35,36,38–40]. Based on these findings, machine learning methods combining a variety of tasks, features, and classifiers to discriminate between healthy subjects and AD patients have been proposed in the literature [3,7,9,10,26,27,41]. All of them, however, aim at modelling both populations and then use the learned models for telling apart samples produced by healthy subjects and AD patients. Departing from this dominant point of view, we approach the problem as a case of anomaly detection. After all, the disease affects different brain areas to various extents, that in turn may produce a large variety of effects on motor control functions, and eventually result in anomalous variations of the kinematic and dynamics of the movements with respect to the one observed in healthy subjects. In addition, our experience in collecting a large dataset has shown that acquiring patient samples is a very demanding task, for various reasons. First of all, patients are reluctant to disclose their pathology to others because of both societal and self-stigma [13]. Patients very often need assistance to move around or perform other tasks and thus need to be reached at their living places. Administering the test, moreover, may need the assistance of the caregiver or even specialized medical personnel. Those mentioned above are possible reasons why there are a few datasets publicly available, and most of them include a few dozen subjects, are often unbalanced and the two populations may differ significantly in their characteristics [11,30,31].

In this framework, we have turned our attention toward the one-class classifier models, because they have been designed and used to deal with anomaly detection, and because they do not require patient samples to train the classifier. We present here the results of an empirical study aimed at evaluating the performance of three one-class algorithms on the problem at hand, namely the negative selection algorithm (NSA), the isolation forest (iForest) and the one-class support vector machine (OCSVM). The first one has been considered because it was specifically conceived for anomaly detection, while the others because they are based on the same paradigms of random forest and support vector machine, that are two of the most used and top-performing models adopted in other studies for the automatic diagnosis of neurodegenerative diseases.

The paper is organized as follows. In Sect. 2, we review previous works on using NSA, iForest and OCSVM to support the automatic diagnosis of diseases, while in Sect. 3, we summarize the main features of each algorithm for implementing the learning process. Section 4 illustrates the experiments we performed to evaluate the effectiveness of the proposed system and reports the obtained results. Finally, discussion, conclusions and future research directions are discussed in Sect. 5.

2 Related Works

There is a rich body of literature on the analysis of handwriting with pattern recognition methods and machine learning tools to discriminate people affected by Alzheimer's or Parkinson's disease from healthy controls [10, 41].

The analysis of the state-of-art reveals that a fundamental aspect to take care of to diagnose a neurodegenerative disorder through handwriting is the selection of handwriting tasks to administer to subjects. The work reported in [6] addresses specifically the protocol for data acquisition that was adopted for collecting the dataset used in the experimental part of this paper. The proposed protocol includes different writing tasks, namely copy of words, letters, and sentences, and the authors aimed at evaluating the kinematic properties of both on-paper and in-air movements performed by the subject. Two different feature sets have been extracted on the handwriting data collected according to that protocol [4,9]. Both feature sets were used to train different classifiers and ensembles of classifiers. The classification accuracy obtained with the feature set proposed in [4] varied task by task and a mean accuracy of 84.7% was reached with an XGB classifier and a RFE feature selection method. The best results obtained with the feature set described in [9], the one adopted in this paper, was 91.43% and it was reached by a multiclassifier that combined the outcomes of 9 Random Forest working on the feature sets extracted by as many different tasks.

Decision interpretability is another important aspect of the diagnosis of neurodegenerative disorder through machine learning tools since a clear and entirely explicable process can be easily understood and adopted by physicians. In [27], the problem of automatic diagnosis of Parkinson's disease through the analysis of handwriting and drawing samples is addressed and different machine learning techniques are compared by evaluating their classification accuracy and their degree of interpretability. The authors of the study have considered Decision Tree (DT), Random Forest (RF) and Support Vector Machine (SVM) model to build the classifier, and the results they reported show that DT achieves good results by providing a high level of interpretability. In [26], the authors use Cartesian Genetic Programming (CGP) as a classification method and show that it performs better in terms of both accuracy and interpretability. Eventually, in

[3] the use of Grammatical Evolution was investigated to generate decision rules that are even more interpretable by the physicians.

As with regards to the use of the algorithms we have considered in this study in medical applications, to the best of our knowledge, they have never been applied to the diagnosis of neurodegenerative disease. We found three works adopting the negative selection algorithm (NSA) and its variants in medical applications [16]. A variant of the original NSA, the real value NSA (RNSA) has been applied to the diagnoses of epilepsy, breast cancer, and liver pathologies. In [1], RNSA was used to generate the detectors and a particle swarm optimization to push them to explore the space of epileptic signals and maintain diversity and generality among them. The method compared favourably to other state-of-the-art methods proposed in the literature for EEG signals discrimination. A variant of the RNSA called V-detector [20], which supports variable size of detector radius, is adopted in [21] to randomly generate the sets of detectors with the aim of maximizing the coverage area of each detector. Experimental results on Breast Cancer Wisconsin and BUPA Liver Disorder datasets show that the V-Detector can achieve the highest detection rates and lowest false alarm rates in comparison with Artificial Neural Network and Sequential Minimal Optimization.

The iForest algorithm [24] has been tested on four medical datasets for diagnostic purposes, namely Arrhythmia, Annthyroid, Mammography, and Breast Cancer Wisconsin, all showing a great unbalance between the two classes, with the anomaly class (i.e. the class of interest) accounting from a tiny 2% of the samples in case of the Mammography dataset, up 35% for the Breast Cancer Wisconsin. The performance have been evaluated in terms of accuracy by using the AUC metric and compared with state-of-the-art algorithms currently used for anomaly detection (including unsupervised Random Forest). The results reported show that iForest was the top-performing method on all the medical datasets (and, overall, on ten over twelve datasets), with the AUC ranging between 0.80 for the Arrhythmia dataset and 0.99 for the Breast Cancer Wisconsin.

OCSVM has been used for tumor segmentation and detection in medical images, and for nosocomial infection detection. Zhang et al. [42] have addressed medical image segmentation by reformulating the issue as a one-class learning problem, and developed a user-friendly tumor segmentation method implemented by an OCSVM, because of its ability to learn the nonlinear distribution of the tumor data without using any prior knowledge. Experimental results obtained from patients' medical images showed that the proposed segmentation method outperforms a two-class SVM segmentation method. Leveraging on the same feature, i.e. the ability to learn nonlinear distribution of tumor data without prior knowledge. OCSVM have been applied to tumor detection as well. In [15] a method called immune feature weighted SVM (IFWSVM) was proposed to detect tumors in MR images. The immune algorithm was introduced in searching for the optimal feature weights and the values of the OCSVM parameters simultaneously. Theoretical analysis and experimental results showed that IFWSVM has better performance than conventional OCSVM. In [8] OCSVM was adopted to identify patients with one or more infections by using clinical and other data because of its intrinsic ability to deal with class (infected/not infected) imbalance. The infected subjects are then identified as samples that deviate significantly from the normal profile. The experimental results reported by the authors showed that whereas standard 2-class SVMs reached a sensitivity of 50.6%, the one-class approach increased sensitivity to as much as 92.6%, suggesting OCSVMs can provide an effective and efficient way of overcoming data imbalance in classification problems.

3   One-class Classifiers

3.1   RNSA

The Negative Selection Algorithm was firstly introduced by Stephanie Forrest and her colleagues in 1994 [12] by drawing an analogy with the way the biological immune system learns to discriminate body cells from foreign cells and pathogens. NSA starts with a set of self antigens S, and proceeds by generating a set of detectors, D, that only recognize the complement of S. A detector d is said to recognize a self antigen if its affinity is below a given threshold σ, called detector size. Once a detector that does not recognize any of the self antigens of S is found, it is stored in D, and the process is iterated until a given number of detectors is discovered. These detectors are eventually applied to new data in order to classify them as being self or non-self: samples falling within a distance from a detector smaller than the detector's size are considered as non-self. Thus, when used for pattern recognition, it implements a prototype-based approach, where the prototypes of one class are learned by using only the samples of the others. In this study, we adopt the NSA variant named real value negative selection algorithm (RNSA) [14] that allows representing detectors and antigens by real-valued vectors, so that the affinity could be evaluated by using Euclidean distance. Moreover, we adopt the canonical RNSA implementation, i.e. the detectors are generated randomly in the representation space and they all have the same, fixed size. Self antigens represent samples from healthy subjects, while detectors represent the prototypes of AD patients to be learned. Both are represented by a real-valued feature vector, normalized into the feature space $[0, 1]^n$, where n denotes the number of features, and the affinity between samples and detectors is represented by the Euclidean distance. Thus, there are 2 parameters, the number N of detectors and the detector size σ, that need to be set depending on the problem at hand.

3.2   Isolation Forest

Isolation Forest (iForest) is an anomaly detection algorithm that, differently from other methods, focuses on anomaly isolation, i.e. how far a data point is from the rest of the data, rather than normal instance profiling [24]. iForest doesn't utilize distance or density measures to detect anomalies but it isolates observations by randomly selecting a feature and then randomly setting a split value between the maximum and minimum values of the selected feature. iForest, similar to a Random Forest, builds an ensemble of binary trees named Isolation Trees (iTrees). Anomalies are those instances that have short average path lengths on the iTrees in the forest. iForest shows similar performance when it is trained only with normal data or with normal data and anomalies. Two are the main hyper-parameters of this method: the number of trees to build and the number of samples to draw from the training set to train each iTree.

3.3   One-class Support Vector Machine

The classical SVM algorithm builds hyperplanes in a multidimensional space to separate samples belonging to different classes. One-class SVM is an extension of the classical algorithm to the one-class classification problem [34]. The algorithm maps the feature vectors in a new feature space through a kernel function in a way that the origin of the space represents the only member of the second class. Then, it separates the new feature vectors from the origin with a hyperplane at the maximum distance from the origin. Two are the hyper-parameters of this method: the kernel and the scalar parameter nu, which is the margin of the One-Class SVM and it corresponds to the probability of finding a new, but regular, observation outside the frontier.

4   Experimental Results

4.1   Dataset

We used DARWIN [5], a public dataset that includes handwriting samples executed by 174 individuals, 89 of which are AD patients and the remaining ones are healthy subjects. All subjects were recruited so as to match the patients and the control group for age, education level, type of occupation, and gender.

Each subject was involved in 25 different motor tasks, which were grouped into four categories, in increasing order of difficulty: graphic tasks, copy and reverse copy tasks, memory tasks, dictation.

The samples were recorded by using a Bamboo Wacom tablet, which allows reproducing the pen-and-paper setting and simultaneously digitizing the handwriting result.

Each handwriting sample is represented by 18 features aimed at describing the time of execution, mean speed, acceleration and jerk both in-air and on-paper, mean pressure and the tremor visible in the trace. Thus, each subject was represented by a feature vector of 450 real values obtained by the concatenation of the 25 feature vectors extracted from the data recorded during the execution of each task.

4.2   Features Selection

Features play a key role in creating predictive models. A large number of features may lead to overfitting and, at the same time, will cause the curse of dimensionality, i.e., the dimensions of search space for the problem will increase.

Among the feature selection methods that have been proposed in the literature, univariate feature-ranking methods have the lowest computational complexity [17], and are therefore the elective choice to deal with applications that involve thousands of features, as in our case.

Pair-wise ranking methods evaluate the degree of dependency between each feature and the target concept, one feature at a time. Several different criteria have been used to conduct pair-wise dependency evaluation, based on estimation of correlation, uncertainty, discriminative power of and between the features, or hypothesis test. In this work, we have used a variant of the Minimum Redundancy feature selection algorithm [29], to find out highly correlated features and remove them from the dataset. The method finds out all the pairs of features whose absolute Pearson correlation index is larger than a given threshold τ. Then, the Mutual Information Score (MIS) between each selected feature and the class is computed. Eventually, the features are selected one by one in descending order of MIS and the features they are correlated to are eliminated. In this way, the features that have the least correlation with each other and the highest MIS scores with respect to the class are collected.

To set the value for τ, we observed that there are groups of features that are strongly correlated, as for instance the kinematic features describing the velocity and acceleration profiles, or the kinematic and pressure features. Considering that reducing the number of features is beneficial in terms of computational cost for the NSA algorithm, and it can contribute to reducing the complexity of the test in case some of the tasks do not contribute any feature to the final pool, we have used a set of values for τ ranging from 0.2 to 0.5 with a step of 0.1 with the aim to determine the best set of correlated features to pass the first step of the algorithm described above. With such a choice, the obtained feature set includes 28, 38, 88 and 121 features for $\tau = \{0.2, 0.3, 0.4, 0.5\}$, respectively.

4.3 Performance Evaluation

In the experiments described below, we have used 80% of the healthy subjects for the training set and the remaining 20% along with 100% of the patients' samples for the test set. Moreover, for each classier, we have performed 15 independent runs by selecting at random the samples of the healthy subjects to be included in the training set. Then, the final performance figures have been obtained by averaging over all the runs.

To be used, RNSA requires that the number N of detectors and the detector size σ are set. We performed a preliminary tuning phase to search for a number of detectors that was suitable for the 5 feature spaces that we investigated. At the end of this phase, we set $N = 4,000$.

As regards the detector size σ, since it depends on the feature space size, we have computed the distances among the healthy subjects' samples, recording the maximum value. Considering that in our feature space the maximum distance between two points is $\sqrt{n}$, that a detector assumes the shape of an n-sphere, and that larger detectors have a higher probability of covering self samples and therefore being discarded, thus favouring the exploration of the feature space, we restricted the search for the values of σ in the range $[\frac{\sqrt{n}}{2} - 1.0, \frac{\sqrt{n}}{2}]$ with step 0.1, so as to have the smallest detectors of the same size of the hypercube including all the samples of the healthy subjects, and the largest ones being the largest possible ones.

Both for iForest and OCSVM a 5-fold grid search was performed to select the best set of hyper-parameters for each configuration of the feature set and for each iteration. For OCSVM, the grid search aimed at minimizing the hinge loss, which is a metric that considers only prediction errors and it is used in maximal margin classifiers such as support vector machines. For iForest, the grid search aimed at maximizing the classification accuracy. Table 1 reports the hyper-parameters that were varied and the corresponding range of variation.

Table 2 reports the average performance of all the methods in terms of weighted accuracy, sensitivity, specificity and their standard deviation. Given the imbalance of the test set in representing the two classes, we suggest comparing the performance of the classifiers in terms of weighted classification accuracy, i.e. the mean value between the classification accuracy computed on the samples handwritten by healthy people and the accuracy computed on the samples handwritten by the patients. In this case the weighted accuracy corresponds also to the mean value between sensitivity and specificity.

When the original feature set is not reduced ($\tau = 1.0$), the OCSVM shows a greater sensitivity value but the RSNA performs better in terms of weighted accuracy thanks to the specificity equal to 100%. In all the other scenarios, when the feature selection is applied, the RNSA outperforms the other two methods both in terms of weighted accuracy and sensitivity. We can notice that the RSNA performance worsen significantly when the dimension of the feature set increased from 121 to 450, while its performance are almost comparable when the number of features varies between 28 and 121. This deterioration of the performance suggests that the number of detectors must be varied when the dimension of the feature space goes beyond a certain value. Eventually, we can notice that when the number of features varies between 28 and 121 the iForest performs slightly better than the OCSVM.

The comparison between the results in Table 2 for the case $\tau = 1.0$ and those reported in [9], where 6 different classifiers were tested on the same dataset, shows that OCSVM and iForest achieved performance similar or better than RF (80.00%) and much better than SVM (68.57%), and that RNSA outperforms the Gaussian Naive Bayesian (85.71%) that was the top-performing classifier among those considered in that work.

Table 1. Range of the hyper-parameters of iForest and OCSVM tuned with a grid search approach

Classifier	Parameter	Range of variation
iForest	Number of trees	[50, 700] step: 50
	Max number of samples	[20, 40, 60, 80, 100]%
	Bootstrap	True
OCSVM	kernel	[rbf, linear]
	Gamma (rbf only)	Powers of ten in $[10^{-8}, 10]$
	nu	[0.05, 1] step: 0.05

Table 2. Performance of OCSVM, iForest and RNSA as the number of features varies. For the RNSA, the values of σ that exhibits the best performance have been reported while the number of detectors is equal to 4,000.

τ	# of features	Method	Weighted accuracy (%)	Sensitivity (%)	Specificity (%)
1.0	450	OCSVM	82.12 ± 10.50	96.40 ± 2.17	67.84 ± 22.65
		iForest	79.77 ± 6.67	89.74 ± 5.53	69.80 ± 15.06
		RNSA ($\sigma = 9.7$)	89.59 ± 2.30	79.18 ± 4.61	100.00 ± 0.00
0.5	121	OCSVM	76.06 ± 14.36	94.08 ± 5.75	58.04 ± 33.85
		iForest	84.83 ± 5.75	88.09 ± 5.83	81.56 ± 11.09
		RNSA ($\sigma = 5.1$)	98.45 ± 0.89	97.68 ± 1.67	99.22 ± 2.07
0.4	88	OCSVM	76.15 ± 10.37	88.39 ± 7.16	63.92 ± 27.22
		iForest	78.74 ± 6.08	79.85 ± 7.64	77.65 ± 15.91
		RNSA ($\sigma = 4.4$)	97.88 ± 1.57	96.55 ± 2.53	99.22 ± 3.04
0.3	58	OCSVM	74.76 ± 8.57	77.38 ± 13.32	72.16 ± 28.43
		iForest	83.35 ± 3.79	89.06 ± 4.10	77.65 ± 9.22
		RNSA ($\sigma = 3.6$)	98.10 ± 2.28	98.13 ± 2.35	98.04 ± 4.80
0.2	28	OCSVM	73.78 ± 5.74	70.71 ± 10.65	76.86 ± 17.29
		iForest	76.30 ± 6.25	78.88 ± 4.14	73.72 ± 12.75
		RNSA ($\sigma = 2.6$)	97.12 ± 1.89	94.23 ± 3.78	100.00 ± 0.00

5   Discussion and Conclusions

Our research aims at diagnosing Alzheimer's disease through a cheap and easy to administer test based on handwriting analysis. The motivation behind this study was to overcome the problem of collecting patients' samples to train the system. Patients' samples are very hard to acquire and therefore represents one of the major obstacles to adopting machine learning methods for the automatic diagnosis of Alzheimer's disease. Their scarcity, moreover, leads to unbalanced datasets, that negatively affect the learning process of the classifier.

We have proposed to deal with these issues by means of one-class classifiers, which need only samples produced by healthy subjects in order to recognise samples drawn by patients. In this paper, we compared the performance of Negative

Selection Algorithm, One-Class Support Vector Machine and Isolation Forest, three different methods mainly adopted to detect anomalies in a dataset.

The results showed that all the methods we have considered in this study exhibit similar or better performance than state-of-the-art methods on the largest publicly available dataset for AD diagnosis. They also show that, among them, the Negative Selection Algorithm exhibited better performance in terms of sensitivity and specificity compared to the other two one-class classification methods. Moreover, NSA has obtained better results than those reported in another study [9] on the same dataset, when other machine learning tools were trained on both the classes. The most remarkable result, however, is that in the best configuration ($\tau = 0.3$, $\sigma = 3.6$, $N = 4,000$) the RNSA outperformed the best results reported in the literature by a much more complex architecture that combines the outputs of 9 RF classifiers by the majority vote decision rule.

The result of our experiments confirms that the motor variability induced by Alzheimer's disease can be characterized by properly modelling the handwriting of a population of healthy subjects. In particular, it has been of paramount importance to collect data from healthy subjects that matched the patient population in terms of demographic features as age, level of education, sex, nutrition habits, anamnestic data, and so forth. All together, they provide robust evidence that handwriting recorded with a tablet following a well-defined test is a valid digital biomarker for Alzheimer's disease that can be easily implemented and adopted for periodical screening of the population at risk of developing the disease.

In our future investigations, we will aim at defining if it is possible to choose a smaller number of tasks and features that allow to achieve the best performance, so as to simplify the test administration. Eventually, we plan to vary the NSA algorithm by exploiting different approaches to define the number of detectors and their radius.

References

1. Ba-Karait, N.O., Shamsuddin, S.M., Sudirman, R.: Eeg signals classification using a hybrid method based on negative selection and particle swarm optimization. In: Proceedings of the 8th International Conference on Machine Learning and Data Mining in Pattern Recognition, pp. 427–438 (2012)
2. Broderick, M.P., Van Gemmert, A.W., Shill, H.A., Stelmach, G.E.: Hypometria and bradykinesia during drawing movements in individuals with Parkinson's disease. Exp. Brain Res. **197**(3), 223–233 (2009)
3. Cavaliere, F., Della Cioppa, A., Marcelli, A., Parziale, A., Senatore, R.: Parkinson's disease diagnosis: towards grammar-based explainable artificial intelligence. In: 2020 IEEE Symposium on Computers and Communications (ISCC), pp. 1–6 (2020)
4. Cilia, N.D., D'Alessandro, T., De Stefano, C., Fontanella, F., Molinara, M.: From online handwriting to synthetic images for Alzheimer's disease detection using a deep transfer learning approach. IEEE J. Biomed. Health Inform. **25**(12), 4243–4254 (2021)

5. Cilia, N.D., De Gregorio, G., De Stefano, C., Fontanella, F., Marcelli, A., Parziale, A.: Diagnosing Alzheimer's disease from on-line handwriting: a novel dataset and performance benchmarking. Eng. Appl. Artif. Intell. **111**, 104822 (2022). https://doi.org/10.1016/j.engappai.2022.104822
6. Cilia, N.D., De Stefano, C., Fontanella, F., Di Freca, A.S.: An experimental protocol to support cognitive impairment diagnosis by using handwriting analysis. Procedia Comput. Sci. **141**, 466–471 (2018)
7. Cilia, N.D., De Stefano, C., Fontanella, F., Molinara, M., Scotto Di Freca, A.: Using handwriting features to characterize cognitive impairment. In: Ricci, E., Rota Bulò, S., Snoek, C., Lanz, O., Messelodi, S., Sebe, N. (eds.) ICIAP 2019. LNCS, vol. 11752, pp. 683–693. Springer, Cham (2019). https://doi.org/10.1007/978-3-030-30645-8_62
8. Cohen, G., Hilario, M., Sax, H., Hugonnet, S., Pellegrini, C., Geissbuhler, A.: An application of one-class support vector machines to nosocomial infection detection. In: MEDINFO 2004, pp. 716–720. IOS Press (2004)
9. De Gregorio, G., Desiato, D., Marcelli, A., Polese, G.: A multi classifier approach for supporting Alzheimer's diagnosis based on handwriting analysis. In: Del Bimbo, A., et al. (eds.) ICPR 2021. LNCS, vol. 12661, pp. 559–574. Springer, Cham (2021). https://doi.org/10.1007/978-3-030-68763-2_43
10. De Stefano, C., Fontanella, F., Impedovo, D., Pirlo, G., di Freca, A.S.: Handwriting analysis to support neurodegenerative diseases diagnosis: a review. Pattern Recogn. Lett. **121**, 37–45 (2019)
11. Drotár, P., Mekyska, J., Rektorová, I., Masarová, L., Smékal, Z., Faundez-Zanuy, M.: Evaluation of handwriting kinematics and pressure for differential diagnosis of Parkinson's disease. Artif. Intell. Med. **67**, 39–46 (2016)
12. Forrest, S., Perelson, A.S., Allen, L., Cherukuri, R.: Self-nonself discrimination in a computer. In: Proceedings of 1994 IEEE Computer Society Symposium on Research in Security and Privacy, pp. 202–212 (1994)
13. Gautier, S., Rosa-Neto, P., Morais, J.a., Webster, C.: World Alzheimer Report 2021: Journey through the diagnosis of dementia. ADI, London, UK (2021)
14. Gonzalez, F., Dasgupta, D., Kozma, R.: Combining negative selection and classification techniques for anomaly detection. In: Proceedings of the 2002 Congress on Evolutionary Computation. CEC 2002, vol. 1, pp. 705–710 (2002)
15. Guo, L., Zhao, L., Wu, Y., Li, Y., Xu, G., Yan, Q.: Tumor detection in MR images using one-class immune feature weighted SVMs. IEEE Trans. Magn. **47**(10), 3849–3852 (2011)
16. Gupta, K.D., Dasgupta, D.: Negative selection algorithm research and applications in the last decade: a review (2021)
17. Huang, S.H.: Supervised feature selection: a tutorial. Artif. Intell. Res. **4**(2), 22–37 (2015)
18. Impedovo, D., Pirlo, G., Vessio, G.: Dynamic handwriting analysis for supporting earlier Parkinson's disease diagnosis. Information **9**(10), 247 (2018)
19. Jankovic, J.: Parkinson's disease: clinical features and diagnosis. J. Neurol. Neurosur. Psychiatry **79**(4), 368–376 (2008)
20. Ji, Z., Dasgupta, D.: V-detector: an efficient negative selection algorithm with "probably adequate" detector coverage. Inf. Sci. **179**(10), 1390–1406 (2009)
21. Lasisi, A., Ghazali, R., Herawan, T.: Chapter 11 - application of real-valued negative selection algorithm to improve medical diagnosis. In: Al-Jumeily, D., Hussain, A., Mallucci, C., Oliver, C. (eds.) Applied Computing in Medicine and Health, pp. 231–243. Morgan Kaufmann, Boston (2016)

22. Le, W., Dong, J., Li, S., Korczyn, A.D.: Can biomarkers help the early diagnosis of Parkinson's disease? Neurosci. Bull. **33**(5), 535–542 (2017)
23. Li, T., Le, W.: Biomarkers for Parkinson's disease: how good are they? Neurosci. Bull. **36**(2), 183–194 (2020)
24. Liu, F.T., Ting, K.M., Zhou, Z.H.: Isolation-based anomaly detection. ACM Trans. Knowl. Discovery Data (TKDD) **6**(1), 1–39 (2012)
25. Myszczynska, M.A., et al.: Applications of machine learning to diagnosis and treatment of neurodegenerative diseases. Nat. Rev. Neurol. **16**, 440–456 (2020)
26. Parziale, A., Senatore, R., Della Cioppa, A., Marcelli, A.: Cartesian genetic programming for diagnosis of Parkinson disease through handwriting analysis: performance vs interpretability issues. Artif. Intell. Med. **111**, 101984 (2021)
27. Parziale, A., Della Cioppa, A., Senatore, R., Marcelli, A.: A decision tree for automatic diagnosis of Parkinson's disease from offline drawing samples: experiments and findings. In: Ricci, E., et al. (eds.) Image Analysis and Processing - ICIAP 2019, pp. 196–206 (2019)
28. Parziale, A., Senatore, R., Marcelli, A.: Exploring speed-accuracy tradeoff in reaching movements: a neurocomputational model. Neural Comput. Appl. **32**, 13377–13403 (2020)
29. Peng, H., Long, F., Ding, C.: Feature selection based on mutual information criteria of max-dependency, max-relevance, and min-redundancy. IEEE Trans. Pattern Anal. Mach. Intell. **27**(8), 1226–1238 (2005)
30. Pereira, C.R., Weber, S.A.T., Hook, C., Rosa, G.H., Papa, J.P.: Deep learning-aided Parkinson's disease diagnosis from handwritten dynamics. In: 2016 29th Conference on Graphics, Patterns and Images, pp. 340–346, October 2016
31. Pereira, C.R., et al.: A new computer vision-based approach to aid the diagnosis of Parkinson's disease. Comput. Methods Programs Biomed. **136**, 79–88 (2016)
32. Precup, R.E., Teban, T.A., Albu, A., Borlea, A.B., Zamfirache, I.A., Petriu, E.M.: Evolving fuzzy models for prosthetic hand myoelectric-based control. IEEE Trans. Instrum. Meas. **69**(7), 4625–4636 (2020)
33. Prince, M., Wimo, A., Guercet, M., Ali, G.C., Wu, Y.T., Prina, M.: World Alzheimer Report 2015: The Global Impact of Dementia. ADI, London, UK (2015)
34. Schölkopf, B., Williamson, R.C., Smola, A.J., Shawe-Taylor, J., Platt, J.C., et al.: Support vector method for novelty detection. In: NIPS, vol. 12, pp. 582–588. Citeseer (1999)
35. Senatore, R., Marcelli, A.: A neural scheme for procedural motor learning of handwriting. In: International Conference on Frontiers on Handwriting Recognition, pp. 659–664. Springer (2012)
36. Senatore, R., Marcelli, A.: A paradigm for emulating the early learning stage of handwriting: performance comparison between healthy controls and Parkinson's disease patients in drawing loop shapes. Hum. Mov. Sci. **65**, 89–101 (2019)
37. Tanveer, M., et al.: Machine learning techniques for the diagnosis of Alzheimer's disease: a review. ACM Trans. Multimedia Comput. Commun. Appl. **16**(1s), 1–35 (2020)
38. Teulings, H.L., Contreras-Vidal, J.L., Stelmach, G.E., Adler, C.H.: Parkinsonism reduces coordination of fingers, wrist, and arm in fine motor control. Exp. Neurol. **146**(1), 159–170 (1997)
39. Teulings, H.L., Stelmach, G.E.: Control of stroke size, peak acceleration, and stroke duration in parkinsonian handwriting. Human Mov. Sci. **10**(2–3), 315–334 (1991)
40. Van Gemmert, A., Adler, C.H., Stelmach, G.: Parkinson's disease patients undershoot target size in handwriting and similar tasks. J. Neurol. Neurosur. Psychiatry **74**(11), 1502–1508 (2003)

41. Vessio, G.: Dynamic handwriting analysis for neurodegenerative disease assessment: a literary review. Appl. Sci. **9**(21), 4666 (2019)
42. Zhang, J., Ma, K.K., Er, M.H., Chong, V.: Tumor segmentation from magnetic resonance imaging by learning via one-class support vector machine. In: International Workshop on Advanced Image Technology, pp. 207–211 (2004)

Learning Unrolling-Based Neural Network for Magnetic Resonance Imaging Reconstruction

Qiunv Yan, Li Liu$^{(\boxtimes)}$, and Lanyin Mei

Birentech Research, Shanghai 200000, China
{qnyan,lliu}@birentech.com

Abstract. Accelerated magnetic resonance imaging (MRI) based on neural networks is an effective solution for fast MRI reconstruction, producing competitive performance in restoring the image domain from its undersampled measurements. However, most existing works rely on convolutional neural networks (CNNs), which are limited by the inherent locality in capturing the long-distance dependency. In this work, we propose a UNet-like Transformer network (UTrans) that is capable of mapping the measurements back to image domain, resulting in an efficient MRI reconstruction. To better capture the non-local features, window-based self-attention operators are adopted to replace the convolutional layers in both encoder and decoder branches of UTrans. Inspired by unrolled optimization approaches, we apply a recurrent block to integrate the forward measurement operator and UTrans to unroll the iterative reconstruction. In the unrolling framework, UTrans served as a regularizer for image reconstruction with limited data. Finally, we replace feed forward network (FFN) module of the window-based self-attention operators with layer-fixed FFN (LF-FFN) whose parameters in the first hidden layer are obtained by random initialization and are fixed, with those in the second layer being updated in the usual fashion. Experiments on fastMRI indicate that the proposed method can attain improved reconstruction results with high performance on limited measurements with fewer network parameters.

Keywords: Underdetermined system · Image inversion · Transformer · Unrolled-network · MRI reconstruction

1 Introduction

Accelerated magnetic resonance imaging (MRI) is a typical instance of solving underdetermined image inversion problem to recover a vectorized image from k-space undersampled measurements of the form $f = Am + \varepsilon$, where $A : \mathbb{C}^N \to \mathbb{C}^M$ is the Fourier transform with undersampling, $f \in \mathbb{C}^M$ is the under-sampled data, $m \in \mathbb{C}^N$ is the image to be reconstructed, and ε represents Gaussian noise [1].

Existing methods based on unrolled networks for fast MRI restoration have proven their feasibilities in solving underdetermined inverse problems [2, 3, 22–24]. In an unrolled network, the amount of blocks are analogous to the amount of iterations. Additionally, learning the regularizer in this framework by fixing these blocks can improve

S. Sclaroff et al. (Eds.): ICIAP 2022, LNCS 13231, pp. 124–136, 2022.
https://doi.org/10.1007/978-3-031-06427-2_11

the convergence speed of the network. However, the ability of using different blocks to learn a regularizer and accelerate network training varies. Thus, how to learn an effective regularizer is an inherent problem in unrolling.

The approaches based on non-unrolling [4–6] directly use the general network architecture to learn the mapping from measurements to images. In the past, UNet [5] and its variants [6, 26] have successfully been applied to inverse problems for MR imaging on fastMRI challenge. In fact, most of these reconstruction approaches rely on CNNs, leading to limited performance due to the inherent local feature dependency.

Recent works have shown tremendous success of Transformers in vision tasks [8–12, 26]. Swin Transformer builds hierarchical feature maps using shifted windows, which is compatible with a broad range of vision tasks [9]. Then, the adapted architecture named Swin Transformer V2 [10] sets new records on four representative vision benchmarks (ImageNet-V2, COCO, ADE20K and Kinetics-400). More recently, transformers integrated into U-shaped architecture have been employed for biomedical image segmentation [11, 12, 26]. Since there are relatively less works on image reconstruction based on the Transformers, the potential advantages of transformers for image reconstruction remain to be verified. To this end, we aim to design a powerful transformer model that is capable of learning a regularizer in the unrolling optimization framework, to improve performance in underdetermined image reconstruction tasks.

This paper proposes an unrolling-based network for image reconstruction from under-sampled measurements as illustrated. Motivated by the Swin Transformer [9], our method proposes UTrans, consisting of a symmetric encoder-decoder architecture and skip connections as depicted in Fig. 1. To learn fine-grained and long-distance feature representation, the window-based self-attention operators are adopted to replace the convolutional layers in both encoder and decoder branches in UTrans. Another key contribution of this paper is applying a recurrent structure to integrate the forward measurement operator and UTrans as shown in Fig. 2. Moreover, we replace original FFN in the window-based self-attention operators with LF-FFN. Extensive experiments on fastMRI indicate that the proposed approach can obtain more competitive reconstruction results with high performance with fewer network parameters and limited measurements.

Our key contributions can be summarized as:

(1) A UNet-like Transformer architecture (UTrans), consisting of symmetric encoder-decoder architecture and skip connections, is presented to learn non-local feature representation.
(2) We introduce an unrolling block for image inversion from under-sampled measurements, which leverages UTrans and the forward measurement operator for leaning a regularizer within this structure.
(3) It is found in the experiment that using LF-FFN in Swin Transformer block is more effective than MLP in the reconstruction results. This method can reduce the amount of model parameters.

2 Related Work

2.1 Deep Learning for MR Imaging

It is worth noting that traditional MRI technique has a slow imaging modality, and thus several techniques have been proposed to accelerate the reconstruction of high-quality images. Compared with conventional algorithms [13–16], deep learning approaches [17, 18] are more suitable for medical images and complex images, where small abnormal details are much more important than overall features. Ronneberger et al. [5] explained that encoder-decoder CNNs can learn the mapping of higher-level representations from inputs. Subsequently, UNet [5] has successfully been applied to image inversion. For example, Zbontar et al. [20] apply UNet to capture spatial information for inverse problems related to MR imaging, or Jin et al. [19] who utilize a similar architecture of encoder-decoder CNN with skip connections as refinement to their solution for medical imaging reconstruction tasks. Heckel et al. used deep decoder [21] to reconstruct multi-coil datasets of the fastMRI challenge [6] and obtain decent performance. However, these methods are all based on CNN architectures, which struggle to capture global knowledge due to the inherent locality of the convolution operation.

While transformer architectures can capture the non-local features by modeling long-distance dependency. Motivated by these reasons, transformers have been applied to computer vision tasks including biomedical images. It is worth mentioning that a UNet-like pure Transformer, called Swin-UNet [11], is designed for medical image segmentation, using rearrange operation to expand the resolution in decoder. Although, transformers integrated into UNet have been employed for biomedical image segmentation [11, 12, 26], there are fewer image inversion tasks based on pure transformer architecture. Our proposed UTrans merges the improved Swin Transformer blocks to UNet-like architecture for non-local feature representation learning for image inversion. Furthermore, MTrans proposed by Feng et al. fuse the informative features from MR imaging scans of different modalities [7]. However, this approach adds auxiliary modality as prior knowledge for the target modality to improve the quality of the reconstructed images. In contrast, our approach is a single-modal approach based on the pure transformer architecture. Without fusing the extra information as prior knowledge, UTrans can learn a regularizer for inverse problems of which we are unaware during training.

2.2 Unrolling-Based Deep Learning Approaches

Most of the unrolling-based deep learning approaches are built on the unrolled optimization framework, where all free parameters and functions are learned via training [2, 3, 22–24]. Yang et al. proposed a model-based unrolling method [2], in place of proximal operators in unrolled optimization algorithms to learn the regularization parameters of CS-MRI reconstruction. Another model-based unrolling architecture (MoDL) solved the data consistency problem (DC) combining prior regularization [3]. The iterative threshold reduction network (ISTA-Net) unrolls ISTA as a regularizer to a deep neural network, in order to learn the image transformation and parameters involved in the original ISTA algorithm [22]. For the above unrolling optimization approaches, the number of blocks is akin to the number of training iterations [23]. By fixing the number of blocks and learning a regularizer in the network, this reconstruction method can correspondingly reduce the number of iterations for training [1]. The ability of using different blocks to learn a regularizer in the framework and accelerate the network training varies. Thus, we design a powerful block which is capable of learning a regularizer to accelerate MRI reconstruction.

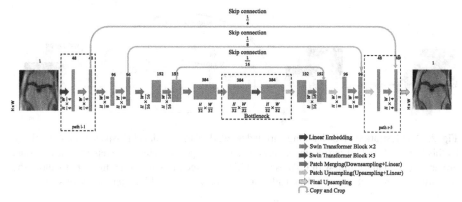

Fig. 1. The architecture of UTrans includes encoder, decoder, bottleneck and symmetric skip connections. Window-based self-attention operators are adopted to replace the convolutional layers in each component.

3 Approach

3.1 Overall Architecture

As illustrated in Fig. 2, our proposed unrolling-based architecture incorporates the forward measurement operator and UTrans (seen in Fig. 1) into multiple blocks for MRI reconstruction. Consider the MAP estimation Eq. (1) where the regularizer $r(\cdot)$ (the negative logarithm prior of m) is convex, in this case, the common algorithm for optimizing (1) is proximal gradient descent [25], whose iterations are:

$$m^{(n+1)} = P\left(m^{(n)} - \eta A^*\left(Am^{(n)} - f\right)\right) \tag{1}$$

where $\mathcal{P}(z) = \underset{m}{\operatorname{argmin}} p\left\{\frac{1}{2}\|m - z\|^2 + r(m)\right\}$ represents the proximal operator corresponding to the regularizer $r(\cdot)$, and the step-size parameter η is a free parameter learnt by the unrolled network. We take as this reconstruction network the nth iterate of proximal gradient descent $m^{(n)}$ starting from the given initialization $m^{(0)}$ as shown in Fig. 2. Subsequently, we transform this into a learning reconstruction network by replacing all instances of the proximal operator $\mathcal{P}(\cdot)$ with UTrans ($\mathcal{P}_\theta(\cdot)$) mapping from measurements to images. The basic unit of UTrans is Swin Transformer block (window-based self-attention operators), which mainly includes window based self-attention and shifted window based self-attention [9] to capture the non-local features. As shown in Fig. 1, the architecture of UTrans is consist of symmetric encoder-decoder, bottleneck and skip connections. In the encoder, patch merging layers are form a hierarchical feature representation. Symmetrically, the patch upsampling layer in the decoder is designed for upsampling and halving the number of feature channels. Finally, a linear projection layer is applied to the output of the decoder to keep the consistent resolution ($W \times H$) and number of channel (1) of the feature maps to the zero-filling images.

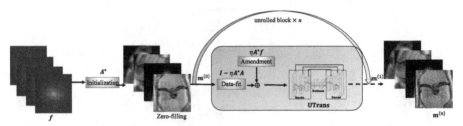

Fig. 2. An illustration of the learning unrolled network. Sampled data in k-space is first initialized by this network to obtain zero-filling images. Then our proposed architecture replaces the proximal gradient descent algorithm with UTrans and utilizes the recurrent block architecture to unroll the forward measurement operator A and adjoint operator A^* into multiple iteration layers.

3.2 Improved Swin Transformer Blocks of UTrans

In window based self-attention (W-MSA) module and shifted window based self-attention (SW-MSA) module [9], we propose the layer-fixed FFN (LF-FFN) to reduce the amount of parameters. For LF-FFN, the parameters in the first hidden layer are fixed, with those in the second layer being updated as usual. The improved module is built by replacing the standard FFN in multi-head self-attention (MSA).

Encoder of UTrans
First, the measurement of MRI is spilt into non-overlapping patches by patch partition module which patch size is 4. Regard as a "token", the feature of each patch is set to the concatenation of the grayscale values of the raw pixels, so the feature dimension of each patch turns into 16. And a linear embedding layer maps the dimension of this raw feature to an arbitrary dimensions C. Then, $\frac{H}{4} \times \frac{W}{4}$ tokens are fed into the improved Swin Transformer blocks to perform representation learning. These improved blocks maintain the feature dimension whilst leaving the output resolution unchanged, together with the linear embedding layer known as "Path 1-1". At each downsampling step, patch merging layer connects the features of each group of 2×2 adjacent patches accordingly, meanwhile, a linear layer is used to adjust the dimension of the 4C cascade feature to 2C. The $\frac{H}{8} \times \frac{W}{8}$ tokens with 2C dimension are served as the input of the improved Swin Transformer blocks to perform feature transformation. This first block of patch merging layer and feature transformation is referred to "Path 1-2". Such procedure is repeated

once, as "Path l-3", with output resolution of $\frac{H}{16} \times \frac{W}{16}$. Finally, this patch merging layer is implemented before the bottleneck, resulting the output resolution of $\frac{H}{32} \times \frac{W}{32}$.

Bottleneck of UTrans

To decrease the depth of the deep Transformer and minimize the amount of model parameters, as illustrated in Fig. 1, two asymmetric improved Swin Transformer blocks are used to build the bottleneck to learn deep feature representation. In the bottleneck, the feature dimension and resolution remain unchanged.

Decoder of UTrans

The decoder has a similar structure to the encoder and forms a bilaterally symmetrical structure within it. For the decoder, we use upsampling layers to expand the extracted features. The patch upsampling layer includes a bilinear interpolation layer and a linear layer. They are designed to reshape the feature maps of adjacent dimensions into higher-resolution feature maps (2 times upsampling), and correspondingly reduce the feature dimension to half of the current dimension, respectively. The output feature ($\frac{H}{32} \times \frac{W}{32} \times 8C$) of the bottleneck first passes through the upsampling layer in the decoder, the feature resolution becomes $\frac{H}{16} \times \frac{W}{16}$. Simultaneously, a linear layer is applied to this feature to reduce the current dimension ($8C \rightarrow 4C$). Then, the $\frac{H}{16} \times \frac{W}{16}$ tokens are sent to the Swin Transformer block to perform representation learning. The Swin Transformer block also remains the feature dimension and resolution unchanged, and together with the patch upsampling layer are called "path r-1". Such process is repeated twice, named as "path r-2", "path r-3". Finally, a special patch upsampling layer is applied to the output of path r-3. The difference is that the linear layer will be applied to the feature ($H \times W \times C$), which aims to map the feature dimension $H \times W \times C$ to the MRI grayscale image dimension. To reduce the loss of spatial information caused by compression and downsampling, we connect the shallow features and deep features by skip connections to compensate. Similar with UNet [5], skip connections are used to fuse multi-scale features from downsampling features of encoder and upsampling features of decoder.

4 Experiments

In the following, we first introduce the dataset and baseline models used in our experiments, and explain the implementation details. Then, we analyze and summarize the results of the comparative experiment. Finally, we conducted ablation experiments to investigate the effectiveness of our proposed strategy.

4.1 Datasets and Baseline Models

We use a raw MRI dataset to evaluate our proposed unrolling-based network. FastMRI stems originally from a collaborative research project between Facebook AI Research (FAIR) and NYU Langone Health, furthermore, currently acts as the largest open-access raw MR image dataset [20]. We choose knee singlecoil train, knee singlecoil val and knee singlecoil test as the experimental datasets. In the experiments, peak-signal-to-noise ratio (PSNR), structural similarity index (SSIM) and normalized mean square error (NMSE) are used to evaluate the performance of our proposed method. We compare our network with following algorithms: UNet [5], Swin-UNet [9], unrolling approach based on UNet and unrolling approach based on Swin-UNet, provided by fastMRI [20] for MRI reconstruction. For accelerating MRI, the undersampled k-space data is produced by $\hat{y} = M \odot y$, where M is the binary mask operator and $\odot$ represents element-wise multiplication. In this reconstruction task, we use a random mask with pattern 4-fold acceleration [20] to select a subset of the k-space points for fastMRI. All compared methods are retrained with their default setting.

4.2 Implementation Details

Our proposed model is implemented in Python 3.8 and Pytorch 1.7.0 with eight NVIDIA A100 GPUs and 40 GB of memory per card. The input Magnetic resonance image size and patch size are 256×256 and 4, respectively. In the improved Swin Transformer blocks, window size is set to 8 and the arbitrary dimensions C is 48. We employ an Adam optimizer with the initial learning rate of 8e−4 and a weight decay of 1e−3, and train our model over 100 epochs. We directly use the training weight of Swin Transformer [10] on ImageNet to initialize the encoder of our proposed model.

4.3 Result and Evaluation

We evaluate our reconstruction results by calculating the PSNR, SSIM and NMSE between the reconstructed images and fully sampled ground truth images. As illustrated in Table 1, we show the reconstruction results over different baselines on fastMRI. The first two rows display the reconstruction results by the non-unrolled CNN and Transformer methods, while the last three provide results of the unrolled networks with three blocks including ours. The non-unrolling methods, whether based on CNN or Transformer, are not as effective as unrolling-based approaches as shown in Table 1. The improvements made by our approach are significantly higher than the two other unrolled

Table 1. The reconstruction results (with standard deviation) of SSIM, PSNR and NMSE on different methods under random 4× undersampling pattern [20].

fastMRI	Random		
Method	PSNR	SSIM	NMSE
UNet	26.2±1.00	0.62 ± 0.08	0.042 ± 0.015
Swin-UNet	27.1±0.80	0.68 ± 0.06	0.030 ± 0.008
Unrolling-based UNet	29.5±0.65	0.72 ± 0.06	0.029 ± 0.015
Unrolling-based Swin-UNet	30.5±0.43	0.73 ± 0.03	0.028 ± 0.033
Unrolling-based UTrans	31.2±0.35	0.74 ± 0.07	0.029 ± 0.015

methods, achieving nearly 31.2 dB in PSNR and nearly 0.74 in SSIM on the fastMRI dataset.

We provide this reconstructed images on the fastMRI dataset in Fig. 3. Column 2 in Fig. 3 shows that the zero-filling reconstruction produce obvious aliasing artifacts and confusing anatomical details. Compared with zero-filling, the above methods can improve reconstruction to a different extent. However, unrolling-based approaches can produce the lowest reconstruction error and better preserves important details. This advanced performance is attributed to the superiority that unrolling optimization network can effectively capture prior knowledge form MR image information during training. In particular, our approach produces high-quality reconstructed images with clearer details and less structural loss than unrolling-based UNet. Thus, we can conclude that UTrans can predict the missing high-frequency content from the low-pass filtered image, since the Transformer can extract non-local features and perform feature representation more effectively. On the other hand, our approach surpasses the unrolling-based Swin-UNet, we obverse that the patch upsampling layers play a pivotal role in improving the performance of image reconstruction. In Sect. 4.4, ablation analysis has confirmed our conjecture.

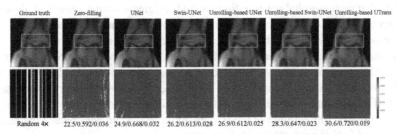

Fig. 3. Comparison of the reconstruction results of various methods on fastMRI with three unrolled blocks. The reconstructed image (up) and error map (down) provide corresponding quantitative measurements in PSNR/SSIM/NMSE. The detailed information in reconstructed image can be compared in the yellow boxes. The more obvious the error, the worse the reconstruction effect. (Color figure online)

4.4   Ablation Analysis

Here, we present several ablation studies designed to understand the individual components, such as the number of unrolling iterations and the efficiency of upsampling. We ablate such important elements in the proposed approach using fastMRI dataset.

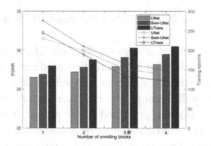

Fig. 4. Comparison of PSNR and training convergence epochs with other methods based on different unrolling blocks. Best configuration with an asterisk.

Number of Unrolling Blocks

Ablation study of the number of blocks on MRI reconstruction and training convergence epochs are summarized in Fig. 4. To qualitatively analyze the impact of the number of blocks on reconstruction results, we also display visual results on fastMRI with error maps in Fig. 5. As we can see, to a different extent, appropriately increasing the number of proximal gradient descent iterations can improve the reconstruction performance. We speculate that such unrolled optimization methods originate from a numerical algorithm with convergence or asymptotic convergence guarantee, and therefore may provide more information about the relationship between network structure and performance. In addition, it was found that as the number of blocks in unrolling architecture increases, the convergence speed of the network will also increase. However, it is obviously unreasonable to blindly expand the number of blocks, because this will greatly increase the amount of model parameters and computational complexity. This unrolling approach incorporating the configuration of the framework in Fig. 4 can both reduce the number of blocks and further promote the performance.

Efficiency of Patch Upsampling

Corresponding to the patch merging layer in the encoder, we present a patch upsampling layer in the decoder to achieve upsampling and feature dimension decrease. In order to explore the efficiency of the proposed patch upsampling layer, we perform the ablation experiments of our unrolling approach based on UTrans with pixel shuffle, transposed convolution, rearrange (Swin-UNet) and patch upsampling layer on the fastMRI dataset with three unrolled blocks. Experimental results shown in Fig. 6 indicate that our unrolling-based architecture based on UTrans combined on the patch upsampling layer can attain the best effectiveness of MRI reconstruction. We speculate that the decoder

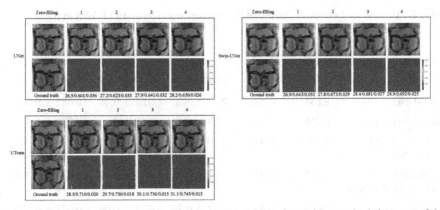

Fig. 5. Comparison of different number of unrolling blocks for various methods in terms of the reconstruction results on fastMRI. The reconstructed image (up) and error map (down) provide corresponding quantitative measurements in PSNR/SSIM/NMSE. The more obvious the error, the worse the reconstruction effect.

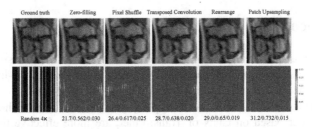

Fig. 6. Comparison of different approaches of upsampling in terms of the reconstruction results on fastMRI. The reconstructed image (up) and error map (down) provide corresponding quantitative measurements in PSNR/SSIM/NMSE. The more obvious the error, the worse the reconstruction effect.

of UTrans is more effective in restoring the spatial information of the feature maps than Swin-UNet.

Efficiency of LF-FFN

Next, we investigate the influence of LF-FFN, which replaces original FFN module in transformer. We record the reconstruction results of Swin-UNet, unrolling-based Swin-UNet and our method on fastMRI in Table 2. As can be seen, all above methods provided with LF-FFN can attain slightly better performance than original FFN module, while the amount of parameters are reduced. Our architecture with LF-FFN brings nearly 0.5–1 upgrade in PSNR with slightly smaller model size and FLOPs.

Table 2. Ablation study on the influence of reconstruction results using LF-FFN in transformers with three unrolling blocks.

fastMRI	#param	FLOPs	Random		
Method			PSNR	SSIM	NMSE
Swin-UNet	8.8M	1.2G	27.1 ± 1.00	0.62 ± 0.80	0.042 ± 0.015
Swin-UNet+LF-FFN	6.2M	0.8G	27.8 ± 0.80	0.68 ± 0.56	0.030 ± 0.008
Unrolling-based Swin-UNet	26.4M	3.6G	29.0 ± 0.35	0.70 ± 0.06	0.029 ± 0.009
Unrolling-based Swin-UNet+LF-FFN	18.6M	2.4G	29.5 ± 0.65	0.71 ± 0.12	0.028 ± 0.005
Unrolling-based UTrans	24.9M	3.5G	31.1 ± 0.46	0.71 ± 0.09	0.027 ± 0.003
Unrolling-based UTrans+LF-FFN	17.1M	2.6G	31.4 ± 0.67	0.72 ± 0.03	0.029 ± 0.004

5 Conclusion

This paper presents a learning unrolling-based network for MRI reconstruction from under-sampled measurements. In particular, we focus on exploring non-local features in imaging for accelerated MRI reconstruction. We conducted extensive experiments on fastMRI under random $4 \times$ undersampling pattern. The results shows that the unrolling, whether based on CNN or Transformer, are more effective than the non-unrolled approaches. Ablation studies have proven the importance and effectiveness of the core components in this network, such as unrolled optimization method, upsampling approach and LF-FFN. The proposed unrolled optimization network can obtain higher quality performance with an increased convergence speed. Our approach achieves high-quality reconstruction images with clearer details and less structural loss than unrolling-based Swin-UNet, due to the decoder of UTrans being more effective in restoring the spatial information of the feature maps. Furthermore, it can be seen in the reconstruction results that the LF-FFN in the Swin Transformer module can not only reduce the scale of the network but also improve the reconstruction performance. Regarding inference speed, our architecture with LF-FFN has more advantages, though inference time is generally not an evaluation metric to assess the quality of MRI reconstruction. In the future, we will explore the impact of non-random initialization (e.g. Gaussian distribution) of the first hidden layer in LF-FFN on the learning unrolling-based network. Furthermore, extensive experiments on the fastMRI under different settings of undersampling patterns will be conducted. We hope this work provides a promising direction for future research on MRI reconstruction using Transformer-based networks.

References

1. Ongie, G., Jalal, A., Metzler, C.A., Baraniuk, R.G., Dimakis, A.G., Willett, R.: Deep learning techniques for inverse problems in imaging. IEEE J. Sel. Areas Inf. Theory **1**(1), 39–56 (2020). https://doi.org/10.1109/JSAIT.2020.2991563

2. Yang, Y., Sun, J., Li, H., Xu, Z.: ADMM-CSNet: a deep learning approach for image compressive sensing. IEEE Trans. Pattern Anal. Mach. Intell. **42**(3), 521–538 (2020). https://doi.org/10.1109/TPAMI.2018.2883941

3. Aggarwal, H., Mani, M., Jacob, M.: MoDL: model-based deep learning architecture for inverse problems. IEEE Trans. Med. Imaging **38**, 394–405 (2019)

4. Akcakaya, M., Moeller, S., Weingartner, S., Ugurbil, K.: Scan-specific robust artificial-neural-networks for k-space interpolation (RAKI) reconstruction: database-free deep learning for fast imaging. Magn. Reson. Med. **81**, 439–453 (2019). https://doi.org/10.1002/mrm.27420

5. Ronneberger, O., Fischer, P., Brox, T.: U-net: convolutional networks for biomedical image segmentation. In: Navab, N., Hornegger, J., Wells, W.M., Frangi, A.F. (eds.) MICCAI 2015. LNCS, vol. 9351, pp. 234–241. Springer, Cham (2015). https://doi.org/10.1007/978-3-319-24574-4_28

6. https://fastmri.org/leaderboards/challenge/2020

7. Feng, C., Yan, Y., Chen, G., Fu, H., Xu, Y., Shao, L.: Accelerated Multi-Modal MR Imaging with Transformers. arXiv:abs/2106.14248 (2021)

8. Dosovitskiy, A., et al.: An Image is Worth 16x16 Words: transformers for Image Recognition at Scale. arXiv:abs/2010.11929 (2021)

9. Liu, Z., et al.: Swin Transformer: Hierarchical Vision Transformer using Shifted Windows. arXiv:abs/2103.14030 (2021)

10. Liu, Z., et al.: Swin Transformer V2: Scaling Up Capacity and Resolution. arXiv:abs/2111.09883 (2021)

11. Cao, H., et al.: Swin-Unet: Unet-like Pure Transformer for Medical Image Segmentation. arXiv:abs/2105.05537 (2021)

12. Wang, W., Chen, C., Ding, M., Yu, H., Zha, S., Li, J.: TransBTS: multimodal brain tumor segmentation using transformer. In: de Bruijne, M., et al. (eds.) MICCAI 2021. LNCS, vol. 12901, pp. 109–119. Springer, Cham (2021). https://doi.org/10.1007/978-3-030-87193-2_11

13. Haldar, J.P., Hernando, D., Liang, Z.: Compressed-sensing MRI with random encoding. IEEE Trans. Med. Imaging **30**, 893–903 (2011)

14. Daubechies, I., Defrise, M., Mol, C.D.: An iterative thresholding algorithm for linear inverse problems with a sparsity constraint. Commun. Pure Appl. Math. **57**, 1413–1457 (2003)

15. Seewann, A., et al.: MRI characteristics of atypical idiopathic inflammatory demyelinating lesions of the brain. J. Neurol. **255**, 1–10 (2007)

16. Xu, S., Zeng, S., Romberg, J.: Fast compressive sensing recovery using generative models with structured latent variables. In: ICASSP 2019 - 2019 IEEE International Conference on Acoustics, Speech and Signal Processing (ICASSP), pp. 2967–2971 (2019)

17. Zhu, B., Liu, J.Z., Cauley, S.F., Rosen, B.R., Rosen, M.S.: Image reconstruction by domain-transform manifold learning. Nature **555**(7697), 487–492 (2018)

18. Mardani, M., et al.: Deep generative adversarial neural networks for compressive sensing MRI. IEEE Trans. Med. Imaging **38**, 167–179 (2019)

19. Jin, K.H., McCann, M.T., Froustey, E., Unser, M.A.: Deep convolutional neural network for inverse problems in imaging. IEEE Trans. Image Process. **26**, 4509–4522 (2017)

20. Zbontar, J., et al.: fastMRI: an open dataset and benchmarks for accelerated MRI. arXiv:abs/1811.08839(2018)

21. Heckel, R., Hand, P.: Deep decoder: concise image representations from untrained non-convolutional networks. arXiv:abs/1810.03982(2019)

22. Hammernik, K., et al.: Learning a variational network for reconstruction of accelerated MRI data. Magnetic Resonance in Medicine, 79. Zhang, J. and Bernard Ghanem. "ISTA-Net: Interpretable Optimization-Inspired Deep Network for Image Compressive Sensing. In: 2018 IEEE/CVF Conference on Computer Vision and Pattern Recognition (2018), pp. 1828–1837 (2018)

23. Cheng, J., Wang, H., Ying, L., Liang, D.: Model learning: primal dual networks for fast MR imaging. arXiv:abs/1908.02426(2019)

24. Schlemper, J., Caballero, J., Hajnal, J., Price, A., Rueckert, D.: A deep cascade of convolutional neural networks for dynamic MR image reconstruction. IEEE Trans. Med. Imaging **37**, 491–503 (2018)

25. Combettes, P.L., Pesquet, J.: Proximal splitting methods in signal processing. In: Fixed-Point Algorithms for Inverse Problems in Science and Engineering (2011)

26. Chen, J., et al.: TransUNet: Transformers Make Strong Encoders for Medical Image Segmentation. arXiv:abs/2102.04306(2021)

Machine Learning to Predict Cognitive Decline of Patients with Alzheimer's Disease Using EEG Markers: A Preliminary Study

Francesco Fontanella[1(✉)], Sonia Pinelli[1], Claudio Babiloni[2], Roberta Lizio[2], Claudio Del Percio[2], Susanna Lopez[2], Giuseppe Noce[2], Franco Giubilei[2], Fabrizio Stocchi[3], Giovanni B. Frisoni[4], Flavio Nobili[5], Raffaele Ferri[6], Tiziana D'Alessandro[1], Nicole Dalia Cilia[7], and Claudio De Stefano[1]

[1] University of Cassino and Southern Lazio, Cassino, FR, Italy
`fontanella@unicas.it`
[2] University of Rome "La Sapienza", Rome, Italy
[3] IRCCS San Raffaele Pisana, Rome, Italy
[4] Geneva University, Geneva, Switzerland
[5] IRCCS Ospedale Policlinico San Martino, Genoa, Italy
[6] I.R.C.C.S. Oasi Maria SS., Troina, EN, Italy
[7] University of Enna "Kore", Enna, Italy

Abstract. Alzheimer's disease causes most of dementia cases. Although currently there is no cure for this disease, predicting the cognitive decline of people at the first stage of the disease allows clinicians to alleviate its burden. Clinicians evaluate individuals' cognitive decline by using neuropsychological tests consisting of different sections, each devoted to testing a specific set of cognitive skills. In this paper, we present the results of a preliminary study aimed at assessing the ability of machine learning based tools to predict the cognitive decline of Alzheimer's patients using features extracted from EEG records at resting state. We tested seven classification schemes in predicting nine scores, provided by different sections of four neuropsychological tests. The experimental results demonstrated that at least three of these scores allows EEG-based features to be effective in predicting the cognitive decline of Alzheimer's patients by using machine learning tools.

1 Introduction

Neurodegenerative diseases belonging to severe cognitive deficits and disabilities (dementia) represent a major threat to human health. These diseases are age-dependent and are increasingly prevalent (million patients worldwide according to World Health Organization) in part because the elderly population has increased in recent years. Alzheimer's disease (AD) accounts for 60–80% of all cases of dementia [10]. Neurodegeneration of brain neurons is induced by an

S. Sclaroff et al. (Eds.): ICIAP 2022, LNCS 13231, pp. 137–147, 2022.
https://doi.org/10.1007/978-3-031-06427-2_12

abnormal accumulation of amyloid proteins extracellularly and phospho tau proteins intracellularly. New diagnostic criteria for research recommend the in-vivo measures of those amyloid and tau proteins in the brain by cerebrospinal fluid (CSF) and positron emission tomography, although they are invasive or expensive.

The typical primary clinical manifestation is a deficit in episodic memory, often associated with an impairment in visuospatial and frontal functions. AD patients are monitored using neuropsychological tests which help clinicians to assess their cognitive status and monitor the progression of the disease. These tests evaluate an individual's cognitive abilities through questionnaires consisting of different sections, each devoted to testing a specific set of cognitive skills. These tests provide a final score, obtained summing up the scores from the different sections they are made of.

Many clinical trials are in progress to discover an effective disease-modifying cure for AD. Waiting for that discovery, noninvasive and cost-effective prescreening methodologies are important to select old persons at risk of AD for the mentioned procedures. In this line, EEG is non-invasive, easy and faster to use and able to differentiate the severity of dementia at a lower cost than other imaging devices. EEG data recorded in condition of resting state allow clinicians to detect brain rhythm abnormalities, generally correlated with the severity of cognitive impairment [14]. Therefore, EEG signal analysis provides useful indications on the patterns of brain activity and allows the prediction of the stage of dementia [9,12]. In particular, the slowing of the rhythms in the EEG signals of AD patients can be explained by a gain of the activity in the theta and delta frequency ranges, and a reduction of the activity in the alpha and beta frequency ranges. The reduction of complexity in the EEG temporal patterns can be explained by a modification of the neural network architecture observed in AD patients due to loss of neurons and functional interaction alteration which make the activity of the brain more predictable, more regular, and simpler than in healthy people. Therefore, this difference can be exploited to distinguish the EEG signals of healthy people from those of people affected by AD or other pathologies (e.g., epilepsy).

Although EEG and other approaches (e.g.neuroimaging) have been demonstrated to be an effective tool to support clinicians in the neurophysiological evaluation of patients at risk of AD, its use for predicting the rate of cognitive decline in AD patients at 1-year follow-up is still a challenging task. This prediction would be of paramount importance in the framework of Precision Medicine to concentrate the resources for patients' managements on the cases with the worse predicted clinical outcome. In the last few years, machine learning is demonstrating to be a viable way to solve a wide spectrum of real-world problems [6–8]. Regarding the prediction of cognitive decline in AD patients, to date, most studies using machine learning methods are based on magnetic resonance image (MRI) and positron emission tomography (PET) data [11,13].

In this paper, we present the results of a preliminary study aimed at assessing the ability of machine learning to predict the progression of cognitive decline at

1-year follow up in patients with MCI due to Alzheimer's disease (ADMCI) using features extracted from EEG records. To assess this progression, we considered the scores of the cognitive tests mentioned above and tested seven widely-used and well-known machine learning algorithms on a dataset containing features extracted from the EEG biomarkers in ADMCI participants.

The remaining of the paper is organized as follows: Sect. 2 introduces the features we extracted from the EEG records; Sect. 3 describe the tests used to characterize the cognitive status of the participants to our study; Sect. 4 reports the experimental results; finally, Sect. 5 concludes the paper.

2 EEG Features

The Central Nervous System (CNS) generally consists of nerve cells and glia cells, which are located between neurons. Each nerve cell consists of axons, dendrites, and cell bodies. Nerve cells respond to stimuli and transmit information over long distances. In the human brain each nerve is connected to approximately 10,000 other nerves, mostly through dendritic connections. The activities in the CNS are mainly related to the synaptic currents transferred between the junctions (called synapses) of axons and dendrites, or dendrites and dendrites of cells [17]. When brain cells (neurons) are activated, local current flows are produced. EEG measures mostly the currents that flow during synaptic excitations of the dendrites of many pyramidal neurons in the cerebral cortex. Only large populations of active neurons can generate electrical activity recordable on the head surface. Between electrode and neuronal layers current penetrates through skin, skull and several other layers. Weak electrical signals detected by the scalp electrodes are massively amplified, and then displayed on paper or stored to computer memory.

Most of the useful information about the functional state of a human brain lies in five major brain waves distinguished by their different frequency bands. These frequency bands are:

- **delta band** (0–4 Hz): The brainwaves show the greatest amplitude and slowest frequency. They typically center around a range of 1.5 to 4 cycles per second. They never go down to zero because that would mean that you were brain dead. But, deep dreamless sleep would take you down to the lowest frequency. Typically, 2 to 3 cycles a second. When a person goes to bed read for a few minutes before attempting sleep, is likely to be in low beta.
- **theta band** (3.5–7.5 Hz): Theta brainwaves, are typically of even greater amplitude and slower frequency. This frequency range is normally between 5 and 8 cycles a second. A person who has taken timeoff from a task and begins to daydream is often in a theta brainwave state. A person who is driving on a freeway, and discovers that they can't recall the last five miles, is often in a theta state-induced by the process of freeway driving. The repetitious nature of that form of driving compared to a country road would differentiate a theta state and a beta state in order to perform the driving task safely.

- **alpha band** (7.5–13 Hz): The next brainwave category in order of frequency is alpha. Where beta represented arousal, alpha represents non-arousal. Alpha brain waves are slower, and higher in amplitude. Their frequency ranges from 9 to 14 cycles persecond. A person who has completed a task and sits down to rest is often in an alpha state. A person who takes time out to reflect or meditate is usually in an alpha state. A person who takes a break from a conference and walks in the garden is often in an alpha state;
- **beta band** (13–26 Hz): When the brain is aroused and actively engaged in mental activities, it generates beta waves. These beta waves are of relatively low amplitude, and are the fastest of the four different brainwaves. The frequency of beta waves ranges from 15 to 40 cycles a second. Beta waves are characteristics of a strongly engaged mind. A person in active conversation would be in beta. A debater would be in high beta. A person making a speech, or a teacher, or a talk show host would all be in beta when they are engaged in their work;
- **gamma band** (26–70 Hz): Gamma waves are fast oscillations and are usually found during conscious perception. Due to small amplitude and high contamination by muscle artifacts, gamma waves are underestimated and not widely studied as compared to other slow brain waves. High gamma activity at temporal locations is associated with memory processes. Research studies reported that gamma activity is involved in attention, working memory, and long-term memory processes. Gamma activity is also involved in psychiatric disorders such as schizophrenia, hallucination, Alzheimer's disease, and epilepsy.

The EEG signal is measured as the potential difference between an electrode active and a reference electrode with the aid of a third electrode, known as an earth electrode. A sufficient number of electrodes may be 1 to 256 (or more in near future) placed on EEG cap for easy application. To provide ionic current and to reduce contact impedance between the electrode surface and the scalp, EEG gel or paste must be used together for proper skin preparation. In biopotential measurements, the most important point is preserving the biosignal originality. The contact impedance should be between $1\,k\Omega$ to $10\,k\Omega$ to record an accurate signal. Less than $1\,k\Omega$ contact impedance indicates a possible shortcut between electrodes; on the other hand, impedance greater than $10\,k\Omega$ can cause distorting artifacts.

The EEG activity was recorded while the participants were relaxed with eyes closed and seated on a comfortable reclined chair in a silent room with dim lights. Instructions encouraged the participants to experience quiet wake fulness with muscle relaxation, no voluntary movements, no talking, and no development of systematic goal-oriented mentalization during the EEG recording. Rather, a quiet wondering mode of mentalization was kindly required. The participants, including the ADMCI patients, did not experience any significant difficulties following those instructions. In all participants, the (eyes-closed) EEG recordings lasted about 3–5 min. Considering all clinical recording units, the EEG data were recorded with a sampling frequency of 128–512 Hz and related antialiasing

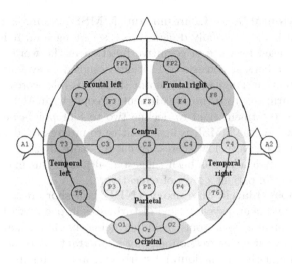

Fig. 1. Electrode montage.

bandpass between 0.01 Hz and 60–100 Hz. The electrode montage included 19 scalp monopolar sensors placed following the 10–20 System (i.e., O1, O2, P3, Pz, P4, T3, T5, T4, T6, C3, Cz, C4, F7, F3, Fz, F4, F8, Fp1, and Fp2;, see Fig. 1). A frontal ground electrode was used, while cephalic or linked earlobe electrodes were used as electric references according to local methodological facilities and standards. Electrode impedances were kept below 5 kΩ. Vertical and horizontal electro-oculographic (EOG) potentials (0.3–70 Hz bandpass) were recorded to control eye movements and blinking.

We used as features the following markers: De-P, De-O, Th-P, Th-O, A2-P, A2-O, A2-T, A3-P, A3-O. We used as features also gender, age, and education.

3   Cognitive Tests and Scales

In the following subsections we describe the tests and scales we used to assess the cognitive status of the participants.

– The **Alzheimer's Disease Assessment Scale (ADAS)** is a two part scale designed to assess cognitive and non-cognitive symptoms of AD. It is one of the most frequently used scales in clinical trials.
  The part which measures cognitive faculties is known as the ADAS-Cog. There are seven performance items and four clinician-rated items: coving memory, orientation, language and praxis. Possible scores range from 70 (severe impairment) to 0 (no impairment). The part that measures non-cognition contains ten clinician-rated items covering agitation, depressed mood, psychosis (delusions and hallucinations), attention/concentration and changes in weight.

- The **Mini-Mental State Examination (MMSE)** is a measure of cognitive function which was originally designed as a screening tool. It is perhaps the most widely used measure of cognitive function in the world and it is also used by researchers. It consists of 11 items designed to evaluate orientation, memory, attention, language and motor skills. Possible scores range from 30 (severe impairment) to 0 (no impairment). People with Alzheimer's disease not receiving treatment tend to decline two to four points per year on the MMSE. However, the MMSE is less sensitive to progressive decline in people who already have very low scores. A cut-off score of 23 for the presence of cognitive impairment has been suggested although this may vary slightly depending on the patient's level of education.
- The **Neuropsychiatric Inventory (NPI)** is a short test used to assess behavioural and neuropsychological disturbances in people with dementia. It measures a range of symptoms including delusions, hallucinations, dysphoria, anxiety, agitation/aggression, euphoria, aberrant motor activity, apathy, irritability, night-time behavioural disturbances and eating disturbances. The severity and frequency of symptoms are measured separately.
- The **Auditory Verbal Learning Test (AVLT)** [15] is one of the most widely used word learning tests in clinical research and practice. Five presentations of a 15-word list (list A) are given, each followed by attempted recall. This is followed by a second 15-word list (list B), followed by recall of list A, and delayed recall and recognition are also tested. AVLT allows clinicians to measure the rate of learning, as opposed to recall of a single stimulus, or series of stimuli.

4   Experimental Results

The experiments presented in this study were performed using the "ADMCI" dataset, which comes from a national archive, formed by clinical, neuropsychological, anthropometric, genetic, and EEG data. The dataset also contains age, gender, and education for each participant. The study cohort consisted of 64 participants in AD or MCI aged between 56 and 80 years. All experiments were performed with the informed and overt consent of each participant or caregiver, in line with the Code of Ethics of the World Medical Association (Declaration of Helsinki) and the standards established by the local Institutional Review Board. Generally, enrollment criteria included the availability of a caregiver, written informed consent of participant and caregiver, no need for 24-h care, and the absence of other physical or neurological causes of dementia-like symptoms. All patients underwent a routine laboratory assessment, measurement of serum vitamin B12 and folic acid levels, as well as serologic (HIV, Lues) and thyroid testing. The status of AD/MCI was based on the "positivity" to one or more of the following biomarkers: $A\beta_{1-42}$/phospho-tau ratio ($A\beta_{42}$/p-tau) in the CSF, FDG-PET, and structural MRI of the hippocampus, parietal, temporal, and posterior cingulate regions. The "positivity" was judged by the physicians in charge for releasing the clinical diagnosis to the patients, according to the local diagnostic routine of the participating clinical units.

To evaluate the cognitive status of the participants we used the following scores:

- MMSE (MM in the following);
- ADAS (AS);
- NPI: Delayed word recall (DL), Word recall 1 (R1), Word recall 2 (R2), and Word recall 3 (R3);
- AVLT: Immediate (TI), Delayed recall (TD), and Recall (TR);

For each score, a participant was labeled as "stable" or "decliner", if the value of that score after twelve months was the same or worse.

To explore the possibility of using EEG markers to predict individuals'cognitive status we tested seven well-known and widely used classifiers: Decision Tree (DT) [3], K-Nearest Neighbor (KNN) [1], Logistic Regression (LR) [4], Multilayer Perceptron (MLP) [16], Random Forest (RF) [2], Simple Logistic (SLO), and Support Vector Machine (SVM) [5]. To assess the performance of the above-mentioned classifiers we used the five-fold cross-validation strategy. Furthermore, to reduce the bias introduced by the random shuffling of data, we performed twenty runs. The results reported in the following were computed by averaging those obtained by the twenty runs performed.

Table 1 shows the classification results achieved in terms of overall accuracy (ACC), sensitivity (True Positive Rate, TPR), and specificity (True Negative Rate, TNR). Since for some participants not all scores were available, for each score the table also shows the actual number of Decliner/Stable (Dec/Sta) participants available for that score. From the table we can observe that the performance achieved varied widely. Looking at the average performance, we can see that the best trade-off between TPR and TNR was obtained with the scores whose datasets were balanced (e.g. TD and TI), whereas with the imbalanced datasets (e.g. R2, R3, and TR) we achieved good performance only in recognizing stable patients (TNR). Excluding the performance achieved on the TR score by using SLO and SVM (poor trade-off between TPR and TNR) the best performance was achieved by SVM on the MM score. In this case we achieved the best result in recognizing cognitive declining patients (TPR, 84.1%). From the table we can also observe that we achieved good results in recognizing stable patients (TNR) only with poor performance in terms of TPR.

In order to compare the performance achieved using the nine scores considered, we have plotted the best and average classification results (See Fig. 2). In terms of accuracy, from Fig. 2 we can see that apart of TR (with bad TPR-TNR trade-off, see above) the best accuracy ranged between 60% and 70%, whereas the average showed a wider variation and did not reflect the same trends of the best results. As for TPR, we can observe that MM, AS, and TI allowed us to achieve performance greater than 70%. It is also worth noting that this trend was replicated by the average performance, confirming the effectiveness of EEG-based features in predicting cognitive decline. Finally, looking at the TNR results we can see that their trend is complementary to that of TPR.

To compare the effectiveness of the six classifiers used we plotted the number of times each of them achieved the best performance. This plot is shown in

Table 1. Classification results. For each score the best result is shown in bold.

Score	R1			R2			R3		
Dec/Sta	25/25			20/40			20/40		
	ACC	TPR	TNR	ACC	TPR	TNR	ACC	TPR	TNR
DT	60.2	48.4	68.6	63.8	26.5	82.5	62.8	30.5	79.0
KNN	**63.8**	46.4	76.3	62.7	**42.0**	73.0	61.0	30.5	76.3
LOG	57.7	47.2	65.1	61.7	40.0	72.5	58.8	37.0	69.8
MLP	60.7	**52.0**	66.9	60.5	41.5	70.0	58.5	**37.5**	69.0
RF	53.8	32.8	68.9	65.7	22.5	87.3	62.8	22.5	83.0
SLO	57.8	25.6	80.9	66.1	16.7	**89.5**	66.1	27.8	**84.2**
SVM	59.3	20.8	**86.9**	**69.2**	33.3	83.8	**66.7**	29.4	83.8
Average	59.0	39.0	73.3	64.2	31.8	79.8	62.4	30.7	77.9
Score	MM			AS			DL		
Dec/Sta	34/30			21/36			34/29		
	ACC	TPR	TNR	ACC	TPR	TNR	ACC	TPR	TNR
DT	67.6	69.7	65.2	59.7	67.1	51.3	64.9	37.6	80.8
KNN	66.2	75.6	55.2	54.1	62.9	44.0	61.6	28.1	81.1
LOG	62.1	68.2	54.8	50.3	54.4	45.7	63.3	49.0	71.7
MLP	61.3	65.0	56.9	55.0	58.2	51.3	62.1	**49.5**	69.4
RF	64.1	73.8	52.8	**65.0**	**72.4**	**56.7**	65.3	33.3	83.9
SLO	64.1	67.6	60.0	57.0	62.1	51.3	**67.7**	36.7	85.8
SVM	**70.5**	**84.1**	54.5	49.2	66.8	29.3	65.4	22.2	**88.2**
Average	65.1	72.0	57.0	55.8	63.4	47.1	64.3	36.6	80.1
Score	TD			TI			TR		
Dec/Sta	30/30			28/32			8/40		
	ACC	TPR	TNR	ACC	TPR	TNR	ACC	TPR	TNR
DT	58.8	55.3	62.3	58.0	49.3	65.6	79.2	37.5	87.5
KNN	49.8	46.0	53.7	**66.8**	**65.0**	68.4	77.1	37.5	85.0
LOG	45.5	49.3	41.7	57.5	56.4	58.4	71.7	**38.8**	78.3
MLP	55.0	58.7	51.3	55.7	58.9	52.8	73.1	20.0	83.8
RF	50.2	53.3	47.0	59.5	58.9	60.0	79.2	25.0	90.0
SLO	49.7	25.0	**74.3**	50.5	25.4	**72.5**	**83.3**	25.0	**95.0**
SVM	**61.8**	**73.0**	50.7	46.3	22.5	67.2	**83.3**	25.0	**95.0**
Average	53.0	51.5	54.4	56.3	48.1	63.6	78.1	29.8	87.8

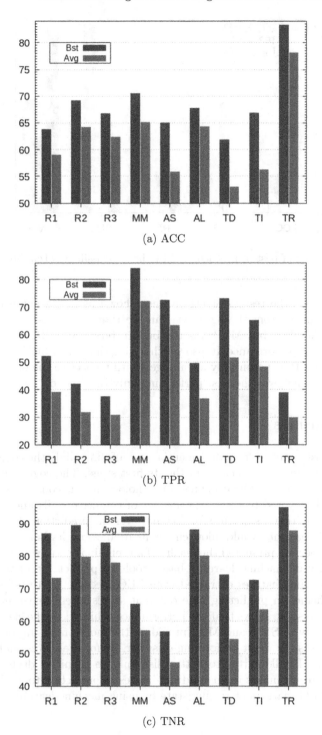

Fig. 2. Overall classification results.

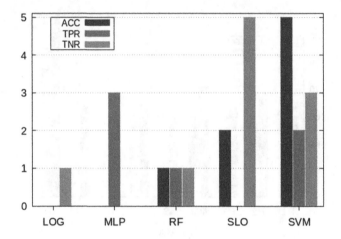

Fig. 3. Number of best results achieved by the six classifiers taken into account.

Fig. 3. Note that the results of DT are not shown as it did not achieved any best performance. From the figure we can see that SVM was the overall best performing classifier, but MLP was the best in terms of TPR, even if did not achieve any best performance. It also worth noting that RF, which is an ensemble based classifier, thus potentially advantaged with respect to the other classifiers taken into account, was the best performing only three times.

5   Conclusions

AD is the most common form of dementia, and affects mainly the cognitive skills of the people affected, especially in the its first stage. The cognitive decline of AD patients is monitored by using neuropsychological tests consisting of different sections, each devoted to test a specific set of cognitive skills and providing a score. Currently there is no cure for AD, however the prediction of the cognitive decline of AD patients would allow clinicians to alleviate its burden.

In this paper, we presented the results of a preliminary study aimed at assessing the ability of machine learning based tools to predict cognitive decline of AD patients using features extracted from EEG records. We tested seven classification schemes in predicting nine different scores after twelve months. The experimental results, shows that the performance achieved by the classifiers used varied widely, with SVM and MLP providing the most promising performance, whereas DT was the worst performing classifier. As for the scores analyzed, MM MM, AS, and TI allowed us to better discriminate cognitive declining people (TPR). This result makes them good candidates as prediction tools through the analysis of EEG-markers at resting state by using machine learning.

References

1. Bishop, C.M.: Pattern Recognition and Machine Learning (Information Science and Statistics). Springer, Heidelberg (2006)
2. Breiman, L.: Random forests. Mach. Learn. **45**(1), 5–32 (2001)
3. Breiman, L., Friedman, J., Stone, C.J., Olshen, R.A.: Classification and Regression Trees. CRC Press, Boca Raton (1984)
4. le Cessie, S., van Houwelingen, J.C.: Ridge estimators in logistic regression. J. Roy. Stat. Soc.: Ser. C (Appl. Stat.) **41**(1), 191–201 (1992)
5. Chang, C.C., Lin, C.J.: LIBSVM: a library for support vector machines. ACM Trans. Intell. Syst. Technol. **2**, 27:1-27:27 (2011)
6. Cilia, N., De Stefano, C., Fontanella, F., Marrocco, C., Molinara, M., Scotto Di Freca, A.: An end-to-end deep learning system for medieval writer identification. Pattern Recogn. Lett. **129**, 137–143 (2020). https://doi.org/10.1016/j.patrec.2019.11.025
7. Cilia, N.D., De Gregorio, G., De Stefano, C., Fontanella, F., Marcelli, A., Parziale, A.: Diagnosing Alzheimer's disease from on-line handwriting: a novel dataset and performance benchmarking. Eng. Appl. Artif. Intell. **111**, 104822 (2022). https://doi.org/10.1016/j.engappai.2022.104822
8. De Stefano, C., Ferrigno, L., Fontanella, F., Gerevini, L., Molinara, M.: Evolutionary computation to implement an IoT-based system for water pollution detection. SN Comput. Sci. **3**(2), 1–15 (2022)
9. Fiscon, G., et al.: Combining EEG signal processing with supervised methods for Alzheimer's patients classification. BMC Med. Inform. Decis. Mak. **18**(1), 1–10 (2018)
10. Gauthier, S., Rosa-Neto, P., Morais, J., Webster, C.: World Alzheimer report 2021 journey through the diagnosis of dementia (2021). https://www.alzint.org/u/World-Alzheimer-Report-2021.pdf
11. Grueso, S., Viejo-Sobera, R.: Machine learning methods for predicting progression from mild cognitive impairment to Alzheimer's disease dementia: a systematic review. Alzheimers Res. Ther. **13**(1) (2021)
12. Hampel, H., et al.: Perspective on future role of biological markers in clinical therapy trials of Alzheimer's disease: a long-range point of view beyond 2020. Biochem. Pharmacol. **88**(4), 426–449 (2014)
13. James, C., Ranson, J., Everson, R., Llewellyn, D.: Performance of machine learning algorithms for predicting progression to dementia in memory clinic patients. JAMA Netw Open **4**(12) (2021)
14. Jeong, J.: EEG dynamics in patients with Alzheimer's disease. Clin. Neurophysiol. **115**(7), 1490–1505 (2004)
15. Rey, A.: L'examen clinique en psychologie. Presses universitaires de France (1964)
16. Rumelhart, D.E., Hinton, G.E., Williams, R.J.: Learning representations by back-propagating errors. Nature **323**(6088), 533–536 (1986)
17. Sanei, S., Chambers, J.: EEG Signal Processing. Wiley, Hoboken (2007)

Improving AMD Diagnosis
by the Simultaneous Identification
of Associated Retinal Lesions

José Morano[1,2]([✉]) [iD], Álvaro S. Hervella[1,2][iD], José Rouco[1,2][iD], Jorge Novo[1,2][iD],
José Ignacio Fernández-Vigo[3,4][iD], and Marcos Ortega[1,2][iD]

[1] Centro de Investigación CITIC, Universidade da Coruña, A Coruña, Spain
{j.morano,a.suarezh,jrouco,jnovo,mortega}@udc.es
[2] VARPA Research Group, Instituto de Investigación Biomédica de A Coruña
(INIBIC), Universidade da Coruña, A Coruña, Spain
[3] Department of Ophthalmology, Hospital Clínico San Carlos,
Instituto de Investigación Sanitaria (IdISSC), Madrid, Spain
[4] Department of Ophthalmology, Centro Internacional de Oftalmología Avanzada,
Madrid, Spain

Abstract. Age-related Macular Degeneration (AMD) is the predominant cause of blindness in developed countries, specially in elderly people. Moreover, its prevalence is increasing due to the global population ageing. In this scenario, early detection is crucial to avert later vision impairment. Nonetheless, implementing large-scale screening programmes is usually not viable, since the population at-risk is large and the analysis must be performed by expert clinicians. Also, the diagnosis of AMD is considered to be particularly difficult, as it is characterized by many different lesions that, in many cases, resemble those of other macular diseases. To overcome these issues, several works have proposed automatic methods for the detection of AMD in retinography images, the most widely used modality for the screening of the disease. Nowadays, most of these works use Convolutional Neural Networks (CNNs) for the binary classification of images into AMD and non-AMD classes. In this work, we propose a novel approach based on CNNs that simultaneously performs AMD diagnosis and the classification of its potential lesions. This latter secondary task has not yet been addressed in this domain, and provides complementary useful information that improves the diagnosis performance and helps understanding the decision. A CNN model is trained using retinography images with image-level labels for both AMD and lesion presence, which are relatively easy to obtain. The experiments conducted in several public datasets show that the proposed approach improves the detection of AMD, while achieving satisfactory results in the identification of most lesions.

Keywords: Medical imaging · Deep learning · Ophthalmology ·
Age-related Macular Degeneration

S. Sclaroff et al. (Eds.): ICIAP 2022, LNCS 13231, pp. 148–159, 2022.
https://doi.org/10.1007/978-3-031-06427-2_13

1 Introduction

Age-related Macular Degeneration (AMD) is a degenerative disorder affecting the macula. Nowadays, it is the most frequent cause of blindness in developed countries, specially in people over 60 years old [21]. Worldwide, an estimated 8.7% of blindness cases are caused by AMD [10], and it is expected to increase due to the global population ageing.

Conventionally, the clinical classification of AMD comprises 5 classes related to its developmental stage [10]: (1) no apparent ageing changes, (2) normal ageing changes, (3) early AMD, (4) intermediate AMD and (5) late AMD. The late stage is generally divided into *dry* and *wet* AMD types, of which the first is the most common (about 90% of the people diagnosed with AMD present the dry type) [10]. All these stages are characterized by the presence of certain lesions within a distance of two optic disc diameters from the fovea of either eye. [1,10]. Thus, early AMD is characterized by drusen, intermediate AMD, by pigmentary abnormalities (associated to drusen), wet AMD, by choroidal neovascularization and pigment epithelial detachment, and dry AMD, by geographic atrophy. Other less common signs of wet AMD are hemorrhages and exudates in or around the macula. Also, it has been reported that untreated choroidal neovascularization occasionally cause a disciform scar under the macula.

Ophthalmologists can identify the presence of AMD by examining retinography (also called color fundus photography) images and optical coherence tomography (OCT) images [20]. OCT is more appropriate for accurate grading on diagnosed patients. Retinography, instead, is more convenient for large-scale screening and early detection programmes for the at-risk population, due to its affordability and widespread availability. However, implementing this type of programmes for diseases like AMD is usually not viable, since the population at risk is large and the analysis of color fundus images is challenging, as AMD is characterized by many different lesions that, in many cases, resemble those of other macular disorders [16]. This forces such analysis to be performed either by clinical experts or by people specifically trained for that task. Moreover, the visual interpretation of the images can be subjective, and there may be relevant differences between the diagnoses of different analysts. All this motivates the research on automatic diagnostic methods [15,20].

Despite some methods have approached the grading of AMD using retinography images [2,18], the objective of most methods is the diagnosis of referable AMD [15,20], as it is the most relevant issue in screening. Referable AMD comprises intermediate and late AMD, but not early AMD. The diagnosis of referable AMD is the focus of the work herein described. The common approach in the state of the art for AMD diagnosis follows a machine learning-based binary AMD/non-AMD classification. While early works were based on classical methods with ad hoc features [14,15], most recent works apply Convolutional Neural Networks (CNNs) [3,7,12,18,19], as in this work. These neural network approaches have explored the use of ad hoc CNN architectures [18], ensembles of these networks [7,19], or standard CNN classification architectures [3,12]. Moreover, while ImageNet pretraining is common when using standard CNNs [3,20], other kinds of self-supervised pretraining were also successfully applied [12]. All of these are common approaches in the diagnosis of ocular diseases [7,19,20].

In this work, we approach the AMD detection using an ImageNet pretrained CNN [17]. In contrast to previous approaches, we propose the use of image-level lesion labels along with diagnosis labels to train a CNN that simultaneously identifies AMD and its associated retinal lesions. This simultaneous task was not yet addressed in this domain. Some previous works have approached AMD diagnosis through the identification of certain lesions, such as drusen [22] or geographic atrophy [13]. In these works, the lesions are first segmented in the image, and then, the AMD diagnosis is derived from that segmentation maps. Differently, we aim at detecting the image-level presence of a wide variety of lesions associated with AMD, along with the overall AMD diagnosis, in a multi-task learning setting.

The proposed setting has several advantages. First, it allows to incorporate useful information that conveniently complements the diagnosis, increasing the feedback received by the models during the training. This multi-task feedback of highly related tasks can help the models to generalize better and to improve the diagnostic performance [4]. Second, the lesion presence information provided by the models is of clinical interest, as it can be indicative of the stage of AMD or the presence of other diseases distinct of AMD. Moreover, due to the direct link between the lesions and the diagnosis, the lesion information can also help to better understand the decisions made by the automatic system, improving its explainability. Since the model outputs both the diagnosis and the lesion detection, it is easier for clinicians to understand the final diagnosis. Last, the benefits of this approach are achieved without much extra effort from clinicians for the creation of the training datasets, since the image-level lesion presence identification (lesion labels) is implicit in the diagnosis assessment. Furthermore, these type of image-level annotations are commonly available in medical records.

To evaluate the proposed approach, we trained a CNN using retinographies and image-level labels from a public dataset, and evaluated its performance in other two public datasets without using any lesion information nor further refinement of the CNN. The provided ablation study and comparison experiments clearly demonstrate that the proposed approach achieves state-of-the-art performance, and surpasses the performance of the traditional AMD diagnosis approach, while achieving satisfactory results in the identification of most lesions. Also, a more detailed analysis of the networks outcomes proves that the proposed approach contributes to the explainability of the models.

2   Materials and Methods

This work is focused on the simultaneous detection of AMD and the identification of its associated retinal lesions from retinography images. To perform this joint task, we train the single CNN depicted in Fig. 1 end-to-end, from raw RGB retinographies. This joint identification involves predicting the presence/absence of AMD along with the presence/absence of N lesions for each

Fig. 1. Proposed approach for the simultaneous detection of AMD and the identification of its associated retinal lesions.

image (AMD + lesions, A+L). An image can be associated with any number of lesions between 0 and N, and the presence of lesions does not necessarily imply the presence of AMD, as they may belong to a different pathology. Thus, the network outputs correspond to $N + 1$ independent image-level detectors.

2.1 Prediction Loss

To train the networks using the proposed approach, we use a loss function that combines the diagnosis error and the lesion detection error by means of a weighted sum. Formally, the loss is defined as

$$\mathcal{L}_{total} = \mathcal{L}_{diagnosis} + \alpha \, \mathcal{L}_{lesions} \ , \tag{1}$$

where α is the weight controlling the relative importance of the diagnosis and the lesion losses, which are defined as

$$\mathcal{L}_{diagnosis} = \mathcal{L}\left(\mathbf{f(r)}_1, \mathbf{d}\right) \ , \tag{2}$$

$$\mathcal{L}_{lesions} = \frac{1}{N} \sum_{i=1}^{N} \mathcal{L}\left(\mathbf{f(r)}_{i+1}, \mathbf{l}_i\right) \ , \tag{3}$$

where $\mathbf{f(r)}$ denotes the predicted network output for retinography $\mathbf{r}$, $\mathbf{d}$ the target AMD diagnosis, $\mathbf{l}_i$ the target for the identification of lesion i, N the number of lesions, and $\mathcal{L}$ a base binary classification loss.

In the proposed approach the base loss $\mathcal{L}$ is Binary Cross-Entropy, and the number of lesions is $N = 5$. The considered lesions are *drusen, exudate, hemorrhage, scar* and *others* (unknown). Also, to emphasize the importance of AMD diagnosis, we assign twice as much weight to it in the loss by setting $\alpha = 0.5$.

2.2 Network Architecture

The proposed CNN architecture is based on the VGG-13 backbone [17], as depicted in Fig. 1. The convolutional trunk, denoted as VGG in Fig. 1, corresponds to the convolutional blocks in VGG-13 before the last pooling layer. This allows to reuse the pre-trained weights in ImageNet classification as initialization. Moreover, this convolutional block can be directly applied to images of arbitrary size. The rest of the original VGG-13 network is replaced by a custom pooling and classification head. First, we add an N channel 1×1 convolutional layer with ReLU activation to reduce the 512 channels of the previous layers into N channels—one channel per considered lesion. This is intended to provide local activation maps of approximately 1/16 of the original image size on each dimension. Then, these N maps are reduced to a fixed size of 31×31 using an adaptive max pooling. Finally, the output layer consists on a fully-connected layer of $N + 1$ sigmoid units corresponding to each lesion identification and the AMD diagnosis.

2.3 Data

For the experiments conducted in this work, we employed the publicly available iChallenge-AMD [6], ARIA [5] and STARE [9] datasets. iChallenge-AMD is used both to train and validate the models in the lesion identification and AMD detection tasks. ARIA and STARE are only used to validate the models in AMD detection.

iChallenge-AMD. The iChallenge-AMD dataset [6] is composed of 400 images, of which 89 are from patients with AMD. The size of some images is 2124×2056 pixels, and 1444×1444 pixels for others. All the images are manually labeled as AMD or non-AMD. The reference standard for AMD presence is based on the retinographies themselves and other complementary information, such as optical coherence tomography (OCT) and visual field. This information, however, was not released with the dataset and it is not available. Along with the AMD labels, 118 of the images include pixel-level lesion annotations. Each lesion is labeled with one of the following classes: drusen (61 images), exudate (38), hemorrhage (19), scar (13) and others (17). The presence of a lesion does not necessarily indicate the presence of AMD, and vice versa. Although the dataset description does not indicate it, we have found that at least 125 of its images belong to the same eye as another image in the dataset. This issue is taken into account when partitioning the data into training and test to avoid having images of the same patient in both sets.

ARIA. The public ARIA dataset [5] is composed of 143 color fundus images from patients with AMD (23), diabetic retinopathy (59) and without any disease (61). Image sizes are 768×576 pixels. To validate our models, we use the 23 images labeled as AMD and the 61 images from healthy patients.

STARE. The publicly available STARE dataset [9] contains 397 retinographies, of size 700×605, from both healthy and pathological patients, with an associated comment indicating its diagnosis. To validate our approach, we use the 46 images labeled as AMD and the 36 images labeled as "Normal".

In Fig. 2, representative examples of retinography images from the iChallenge-AMD, ARIA and STARE datasets are provided.

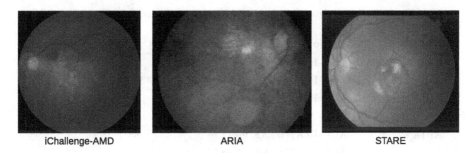

iChallenge-AMD   ARIA   STARE

Fig. 2. Representative examples of retinography images from the iChallenge-AMD, ARIA and STARE datasets.

2.4   Training Details

For both training and inference, we rescale the iChallenge-AMD dataset images to a fixed width of 720 pixels. The ARIA and STARE images are preserved at their original resolution, since their width is similar. To mitigate data scarcity and avoid overfitting, we use online data augmentation. On each epoch, random transformations are applied to the input images. These transformations consist of combinations of random intensity and color variations, slight affine transformations (scaling, shearing and rotation) and horizontal and vertical flipping.

To train the different models, we use the Adam optimization algorithm [11]. The parameters of the algorithm were empirically set as follows. Initial learning rate: $\alpha = 1 \times 10^{-6}$; decay rates: $\beta_1 = 0.9$, $\beta_2 = 0.999$. The learning rate remains constant during the training, which has a fixed duration of 200 epochs.

The convolutional layers of the networks are initialized to the parameter values of a VGG13 model pre-trained in ImageNet, while the added layers are initialized using the He et al. [8] method with uniform distribution.

3   Results and Discussion

To evaluate the impact of including lesion identification feedback, we compare the performance of the models trained with the proposed approach (A+L) with the performance of the models trained with the traditional approach, which only involves predicting the presence of AMD (AMD-only, A-O). For this end, we trained the same models using both approaches in the iChallenge-AMD dataset. In addition, to take into account the stochasticity of the networks training—magnified by the scarcity of annotated data—we performed 5 repetitions of 2-fold cross-validation with randomly created folds for each considered alternative. All these models were then also evaluated in the STARE and ARIA datasets.

The quantitative evaluation of the models was performed using Precision-Recall (PR) and Receiver Operating Characteristic (ROC) analyses. Since a total of 10 repetitions is made for each model, we computed the mean curves by merging each operating point, and computed Area Under the Curve (AUC)

Table 1. AMD detection results. All values indicate percentages.

Dataset	Metric	A+L	A-O	Li et al. [12]	Mookiah et al. [14]
iChallenge-AMD	AUC-ROC	94.01	93.47	77.19	–
	AUC-PR	86.22	84.84	–	–
STARE	AUC-ROC	86.41	83.85	–	100
	AUC-PR	88.25	86.11	–	–
ARIA	AUC-ROC	92.52	90.11	–	85
	AUC-PR	84.92	81.88	–	–

Table 2. Lesions detection results for the A+L approach on iChallenge-AMD.

	Drusen	Exudate	Hemorrhage	Scar	Others
AUC-ROC (%)	71.43	85.87	85.67	89.29	41.61

values for each curve. Table 1 reports the AUC values of the mean ROC and PR curves for the A+L and the A-O approaches in the AMD detection task. In the same table, we include the state-of-the-art works that report their results on the iChallenge-AMD, STARE and ARIA datasets. On the other hand, Fig. 3 depicts the mean ROC and PR curves of the same models in the iChallenge-AMD, ARIA and STARE datasets, with their corresponding AUC. Also, two examples of correctly classified images of the iChallenge-AMD, along with their corresponding lesion activation maps using the A+L and A-O approaches, are shown in Fig. 4. Lastly, Table 2 shows, for the A+L alternative, the AUC values of the mean ROC curves for lesion identification.

As it can be observed in Table 1 and Fig. 3, the best results are achieved by the A+L models. The difference between these models increases for STARE and ARIA datasets, used only for validation. This demonstrates that incorporating lesion identification feedback helps the models to generalize better. In addition, both the A+L and the A-O approaches clearly outperform the works in the state of the art using the iChallenge-AMD and ARIA datasets. In STARE, however, this is not the case, and the work by Mookiah et al. [14] is the one that yields the best results. In this regard, however, it is worth noting that the results shown in [14] correspond to a model directly trained in STARE, while ours correspond to a model trained on a different dataset (iChallenge-AMD), without further refinement in STARE.

As observed in Fig. 4, the A+L CNN local activation maps clearly differ between pathological and healthy images. In the latter case, the overall highest values along the whole image belong to a single map. Differently, for pathological images, these maximum values are distributed among several maps. Furthermore, these values are commonly placed in areas where the lesions occur. This is the same for all the models of the A+L approach. Thus, in a sense, the activations of the last map (bottom right corner) represent the healthy areas of the image,

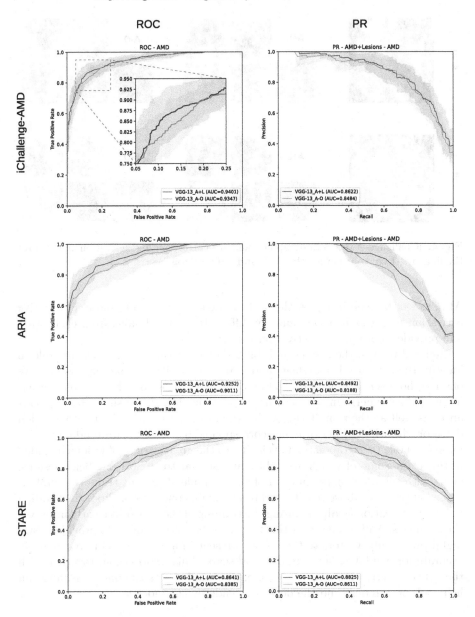

Fig. 3. Mean ROC and PR curves for AMD detection for the networks trained using A+L and A-O in the iChallenge-AMD, ARIA and STARE datasets.

while the activations of the other maps indicate the presence of a lesion. This information is of clinical interest, as it indicates the areas in which the model estimates that a lesion is placed. Also, it helps to better understand the decisions made by the model, increasing its explainability and facilitating its evaluation.

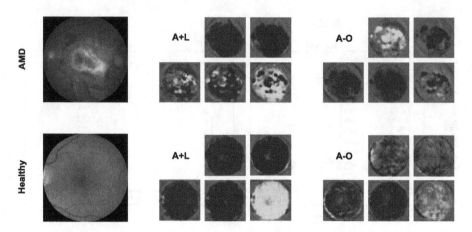

Fig. 4. iChallenge-AMD retinographies and its corresponding activation maps after the final 1×1 convolution for the A+L and A-O approaches.

Also, as can be seen in Fig. 4, the A+L approach gives rise to more informative activation maps, as it approximately differentiates the lesions from each other and provides more precise activation regions.

Related to this, the results in Table 2 show that the A+L models are able to identify the exudate, hemorrhage and scar lesions with satisfactory results. We observe, however, a remarkable lower performance for the 'drusen' and 'other' lesions. The low performance on drusen is explained by their usual subtle appearance, as well as their similarity with exudates. In the case of the other lesion detector, the achieved performance corresponds to a system with random behavior and unable to extrapolate. This is due to the great variety of lesions included in this class, some of which are subtle and similar to the other labeled lesions, along with the poor representation of such complexity in the training samples.

In general, the above results lead to two main conclusions. First, that adding lesion identification feedback during the training of the networks clearly improves the diagnosis. And second, that this task also allows the models to provide meaningful probability vectors of the lesions present in an image, being particularly accurate for exudates, hemorrhages and scars. This information, together with the network activation maps, conveniently complements the diagnosis, and can help to better understand the decisions made by the models.

4   Conclusions

In this work, we have proposed a novel approach for improving the AMD diagnosis by the simultaneous identification of its associated retinal lesions. To this end, we trained a CNN using the retinography images and the image-level annotations (both for lesions and diagnosis) from the publicly available iChallenge-AMD dataset. The experiments performed in this and other two public datasets

(STARE and ARIA) demonstrate that the proposed approach surpasses the traditional approach in AMD detection, while achieving satisfactory results regarding the identification of most lesions. The information resulting from this latter task (the vector of lesion probabilities and the network activation maps) conveniently complements the diagnosis, and can be useful to better understand the decisions made by the model. Moreover, collecting the data needed for this task does not imply much extra effort from clinicians, since the identification of lesions is integrated into the diagnostic process. Apart from this, there are currently many datasets that, while not providing this type of annotations, do provide pixel-level lesion annotations for a subset of its data, from which the image-level lesion labels can be easily obtained. In light of the results herein presented, we think that the proposed methodology could be successfully applied in most of these scenarios.

Notwithstanding, our approach also presents some potential issues for further improvement. The most important one is related to lesion identification. Based on the provided results, the capability of our model to identify the selected lesions could be substantially improved, particularly for the most difficult ones (drusen and others). In this regard, one of the aspects that could help to improve the performance is the addition of more data, since our approach is trained with very few examples. On the other hand, the activation maps that are provided by our approach, although useful, are not linked to the lesions, which complicates its analysis. For this to be the case, new architectural designs need to be explored. Both issues represent interesting fields for further research.

Acknowledgments. This work was funded by Instituto de Salud Carlos III, Government of Spain, and the European Regional Development Fund (ERDF) of the European Union (EU) through the DTS18/00136 research project; Ministerio de Ciencia e Innovación, Government of Spain, through RTI2018-095894-B-I00 and PID2019-108435RB-I00 research projects; Axencia Galega de Innovación (GAIN), Xunta de Galicia, ref. IN845D 2020/38; Xunta de Galicia and the European Social Fund (ESF) of the EU through the predoctoral grant contracts ref. ED481A-2017/328 and ref. ED481A 2021/140; Consellería de Cultura, Educación e Universidade, Xunta de Galicia, through Grupos de Referencia Competitiva, grant ref. ED431C 2020/24; CITIC, Centro de Investigación de Galicia ref. ED431G 2019/01, is funded by Consellería de Educación, Universidade e Formación Profesional, Xunta de Galicia, through the ERDF (80%) and Secretaría Xeral de Universidades (20%).

References

1. Bird, A.C., et al.: An international classification and grading system for age-related maculopathy and age-related macular degeneration. Surv. Ophthalmol. **39**(5), 367–374 (1995). https://doi.org/10.1016/S0039-6257(05)80092-X
2. Burlina, P.M., Joshi, N., Pacheco, K.D., Freund, D.E., Kong, J., Bressler, N.M.: Use of deep learning for detailed severity characterization and estimation of 5-year risk among patients with age-related macular degeneration. JAMA Ophthalmol. **136**(12), 1359–1366 (2018). https://doi.org/10.1001/jamaophthalmol.2018.4118

3. Burlina, P.M., Joshi, N., Pekala, M., Pacheco, K.D., Freund, D.E., Bressler, N.M.: Automated grading of age-related macular degeneration from color fundus images using deep convolutional neural networks. JAMA Ophthalmol. **135**(11), 1170–1176 (2017). https://doi.org/10.1001/jamaophthalmol.2017.3782

4. Caruana, R.: Multitask learning. Mach. Learn. **28**(1), 41–75 (1997). https://doi.org/10.1023/A:1007379606734

5. Farnell, D.J.J., et al.: Enhancement of blood vessels in digital fundus photographs via the application of multiscale line operators. J. Franklin Inst. (2008). https://doi.org/10.1016/j.jfranklin.2008.04.009

6. Fu, H., et al.: ADAM: automatic detection challenge on age-related macular degeneration (2020). https://doi.org/10.21227/dt4f-rt59

7. González-Gonzalo, C., et al.: Evaluation of a deep learning system for the joint automated detection of diabetic retinopathy and age-related macular degeneration. Acta Ophthalmol. **98**(4), 368–377 (2020). https://doi.org/10.1111/aos.14306

8. He, K., Zhang, X., Ren, S., Sun, J.: Delving deep into rectifiers: surpassing human-level performance on imagenet classification. In: Proceedings of the 2015 IEEE International Conference on Computer Vision (ICCV). ICCV, Washington, DC, USA, pp. 1026–1034 (2015). https://doi.org/10.1109/ICCV.2015.123

9. Hoover, A.D., Kouznetsova, V., Goldbaum, M.: Locating blood vessels in retinal images by piecewise threshold probing of a matched filter response. IEEE Trans. Med. Imaging **19**(3), 203–210 (2000). https://doi.org/10.1109/42.845178

10. Kanski, J.J., Bowling, B.: Clinical Ophthalmology: A Systematic Approach, 7th edn. Elsevier Health Sciences, New York (2011)

11. Kingma, D.P., Ba, J.L.: Adam: a method for stochastic optimization. In: Bengio, Y., LeCun, Y. (eds.) Proceedings of the 3rd International Conference on Learning Representations (ICLR) (2015)

12. Li, X., Jia, M., Islam, M.T., Yu, L., Xing, L.: Self-supervised feature learning via exploiting multi-modal data for retinal disease diagnosis. IEEE Trans. Med. Imaging **39**(12), 4023–4033 (2020). https://doi.org/10.1109/TMI.2020.3008871

13. Liefers, B., et al.: A deep learning model for segmentation of geographic atrophy to study its long-term natural history. Ophthalmology **127**(8), 1086–1096 (2020). https://doi.org/10.1016/j.ophtha.2020.02.009

14. Mookiah, M.R.K., et al.: Automated detection of age-related macular degeneration using empirical mode decomposition. Knowl.-Based Syst. **89**, 654–668 (2015). https://doi.org/10.1016/j.knosys.2015.09.012

15. Pead, E., et al.: Automated detection of age-related macular degeneration in color fundus photography: a systematic review. Surv. Ophthalmol. **64**(4), 498–511 (2019). https://doi.org/10.1016/j.survophthal.2019.02.003

16. Saksens, N.T., et al.: Macular dystrophies mimicking age-related macular degeneration. Prog. Retin. Eye Res. **39**, 23–57 (2014). https://doi.org/10.1016/j.preteyeres.2013.11.001

17. Simonyan, K., Zisserman, A.: Very deep convolutional networks for large-scale image recognition. In: Bengio, Y., LeCun, Y. (eds.) Proceedings of the 3rd International Conference on Learning Representations (ICLR) (2015)

18. Tan, J.H., et al.: Age-related macular degeneration detection using deep convolutional neural network. Futur. Gener. Comput. Syst. **87**, 127–135 (2018). https://doi.org/10.1016/j.future.2018.05.001

19. Ting, D.S.W., et al.: Development and validation of a deep learning system for diabetic retinopathy and related eye diseases using retinal images from multiethnic populations with diabetes. JAMA **318**(22), 2211–2223 (2017). https://doi.org/10.1001/jama.2017.18152

20. Ting, D.S., et al.: Deep learning in ophthalmology: the technical and clinical considerations. Prog. Retin. Eye Res. **72**, 100759 (2019). https://doi.org/10.1016/j.preteyeres.2019.04.003

21. Wong, W.L., et al.: Global prevalence of age-related macular degeneration and disease burden projection for 2020 and 2040: a systematic review and meta-analysis. Lancet Glob. Health **2**(2), e106–e116 (2014). https://doi.org/10.1016/S2214-109X(13)70145-1

22. Zheng, Y., et al.: An automated drusen detection system for classifying age-related macular degeneration with color fundus photographs. In: 2013 IEEE 10th International Symposium on Biomedical Imaging, pp. 1448–1451 (2013). https://doi.org/10.1109/ISBI.2013.6556807

Eye Diseases Classification Using Deep Learning

Patrycja Haraburda[✉] and Lukasz Dabała

Warsaw University of Technology, Warsaw, Poland
patrycjaharaburda97@gmail.com, Lukasz.Dabala@pw.edu.pl

Abstract. Eye disease recognition is a challenging task, which usually requires years of medical experience. In this work, we conducted research that can be a start for the most versatile solution. We tried to solve the problem of the classification of different eye diseases using neural networks. The first step of this work consists of gathering all publicly available eye disease datasets and preprocessing them to make the experiments as generalized as possible. This led to the creation of a dataset composed of over 30,000 images. The aim was to teach the model the actual symptoms of the diseases instead of adjusting the results to a given part of the dataset. Several deep convolutional neural networks were used as feature extractors and they were combined with the Synergic Deep Learning model. We conducted experiments on the data and were able to achieve promising results.

Keywords: Classification · Neural networks · Medical images · Diabetic retinopathy · Glaucoma · Cataract

1 Introduction

1.1 Motivation

According to the World report on vision provided by the World Health Organization [4] in 2019, there were at least 2.2 billion people in the world affected by visual impairment or vision loss. The report also claims that the fates of around 1 billion of those people could have been prevented if they were provided a proper diagnosis and treatment. A crucial note is that not all people with a condition will have notable vision impairment. Data shows that only 5.3% of the population with age-related macular degeneration experience a severe state of the disease, causing vision impairment or vision loss. Similarly, it happens for only 10.9% of patients with diagnosed glaucoma.

At the moment of writing this paper, there are no available automated solutions for a general diagnosis of eye conditions [22] on the market. In the literature it is possible to find methods for recognizing individual eye diseases, but not cases with combined eye conditions. One of the reasons is the lack of standardization in medical data records, which makes it a difficult problem for computer-aided

S. Sclaroff et al. (Eds.): ICIAP 2022, LNCS 13231, pp. 160–172, 2022.
https://doi.org/10.1007/978-3-031-06427-2_14

systems. However, as shown before, there is a huge need for such recognition systems. Many people are developing different kinds of eye impairment. Moreover, a lot of them are not provided with access to appropriate medical care. Hence, a potential solution to this problem would be of great benefit.

1.2   Overview of the Study

This study undertook the challenge of creating the versatile model. It faced problems that appeared during the research and implementation. One of them is the fact that retinal photographs have a high variance of both symptoms and characteristics of camera model properties. For this reason, in order to create a universal model for classification, a large and diverse dataset is required. In this study, we managed to collect over 30,000 images coming from 18 different datasets. What is more, as every disease is manifesting in different eye structures, for each there should be performed separate segmentation algorithm. In our work, we proposed the optic disc segmentation method as a part of the preprocessing step for glaucoma. In addition, medical images have characteristics that make a classification of them a challenging task for regular neural networks. Because of that, we have used SDL architecture that is particularly designed for this kind of data. Next, we performed various trials, both for individual and combined datasets. This allowed us to make an assumption regarding the division of the images into training and testing data.

2   Related Work

2.1   Diabetic Retinopathy

Medical literature is abundant in papers with efficient Diabetic Retinopathy (DR) classification with accuracy reaching nearly 100%. Authors in [26] presented a Synergic Deep Learning (SDL) model for fundus images classification. It classified images as either healthy or one of 3 stages of Diabetic Retinopathy. Many works tried to utilize Convolutional Neural Networks (CNN) [9] for the task. They were mainly used for feature extraction [14].

K. Shankar, Y. Zhang et al. [27] developed the HPTI-v4 model to classify different stages of DR. The solution consisted of segmentation and classification processes. The former was developed with histogram features extraction and the latter with Inception v4. Model hyperparameters were tuned utilizing Bayesian optimization techniques.

Imran Qureshi et al. [25] considered the Active Deep Learning (ADL) CNN model. Additionally, the solution provided an interpretation layer yielding detailed information about the symptoms of the disease.

2.2 Glaucoma

The next group of publications that are relevant for this study relates to Glaucoma classification with the usage of CNN. One of them [15] used standard CNN models like VGG19, ResNet50, DENet, and GoogleLeNet and verified the impact of the transfer learning technique with successful outcomes.

Some of the studies explored the possibilities of optic disc usage for a better classification. These tasks were divided into two steps: finding the proper region and classifying the images. Authors in [21] tried to use a classic approach: Hough transform and ResNet101 model. Classic approaches have to be really well-tuned. Therefore, the authors in [6], proposed an automated solution using Regions and Convolutional Neural Network. A study in [11], presents how a CNN can be used to find pixels that are the most probable to belong to the optic disc. Furthermore, the authors compared 9 different architectures to classify images, among others: VGG16, VGG19, InceptionV3, and ResNet50.

2.3 Cataract

In comparison to other diseases, there are few works related to cataract classification. In one of them [29] authors extract textural features of the images. The collected data is trained using an SVM classifier into four stages: normal, slight, medium, and severe.

The authors of [12] use ResNet18 and gray level co-occurrence matrix for feature extraction. In this work, images were labeled into six levels of the illness with an SVM classifier.

3 Datasets

Medical images are very difficult to obtain because they require patient and health facility consents and they often need expensive hardware to store and process. Although there are many restrictions on making medical images public, there exist some datasets that facilitate exploration of the area of diseases recognition. In this work, we describe available data sources depicting eye illnesses.

Table 1 presents publicly available fundus image datasets. Each row contains the name of the datasets as well as the number of images for each medical condition and the number of health ("normal") pictures. Moreover, images with unsatisfactory quality have been manually removed. That includes underexposed and overexposed images, but also images where the optic disc was not visible. Despite being irrelevant for e.g. Diabetic Retinopathy (DR) recognition, the optic disc is crucial for glaucoma. The goal of the database was to make it universal and allow the classification of all selected illnesses.

4 Synergic Deep Learning

SDL model was firstly proposed in [30]. It was designed to solve problems related to medical images. That is problems with high inter-class similarity and intra-class variance. It was first used to classify medical images modalities and reached

Table 1. List of publicly available eye disease datasets. Numbers in cells describe how many images of a given disease are available in the given dataset.

Dataset name	Diabetic retinopathy	Glaucoma	Normal	Cataract
DiaretDB0 [20]	108	–	22	–
DiaretDB1 [19]	84	–	–	–
DRHagis [17]	10	10	–	–
EyePACS [8]	9317	–	25810	–
HRF [7]	15	15	15	–
IDRID [24]	348	–	–	–
JSIEC [3]	106	–	38	–
MESSIDOR [10]	654	460	–	–
ODIR [2]	849	114	1674	216
ACRIMA [11]	396	–	–	–
BinRushed [5]	195	–	–	–
KAGGLE [1]	101	–	300	100
Magrabia [5]	–	94	–	–
ORIGA [31]	–	168	482	–
RIM-ONE [13]	–	227	–	–
REFUGE [23]	–	40	–	–
DRIVE [28]	–	–	33	–
STARE [18]	–	–	58	–
SUM	*12183*	*1128*	*28432*	*316*

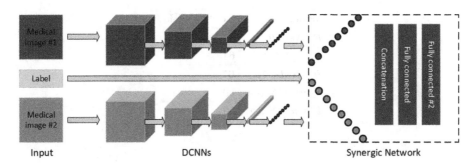

Fig. 1. SDL architecture

state-of-the-art accuracy. Therefore, it is a good candidate for the problem of disease classification on fundus images.

Model architecture is presented in Fig. 1. It is composed of two deep convolutional neural networks (DCNN) models, which can be of any architecture, and Synergic Network (SN), consisting of concatenation, learning, and output layers. The input for the model is a pair of batches. Each of the batches is processed by a separate DCNN. Outputs are passed to the SN together with a label retrieved from the following equation:

$$y_s(Z_A, Z_B) = \begin{cases} 1 & \text{if} \quad y_A = y_B \\ 0 & \text{if} \quad y_A \neq y_B \end{cases} \tag{1}$$

where, y_s - synergic input label, Z_A - DCNN-A model output, Z_B - DCNN-B model output, y_A - DCNN-A input label, y_B - DCNN-B input label

A pair of images from the same class will result in label 1, while images from different classes, return label 0. Outputs of the DCNN are concatenated inside SN and processed by learning layers. An additional learning signal is passed through DCNNs based on the output.

The architecture facilitates learning different patterns belonging to the same class. They can be both symptoms of a disease as well as parameters of a camera. On the other hand, specific lesions, like hemorrhages can be present on images depicting different illnesses. What is more, they can be captured with the same device, which will cause similar luminance and zoom magnitude. All those problems are addressed with the SDL model and particularly with the synergic signal.

Many pre-trained models have been tested as DCNNs. This includes ResNet-50, ResNet-101, ResNet-152, EfficientNet-b0 and EfiicientNet-b4. The experiments showed that ResNet-152 achieves the best results in terms of accuracy.

Table 2 presents scores achieved by models for the final, multilabel dataset classification (as described in Sect. 6).

Table 2. Comparison of scores achieved by different models

Model	Accuracy (%)
ResNet-50	61
ResNet-101	64
ResNet-152	**67**
EfficientNet-b0	59
EfficientNet-b4	50

5   Segmentation

Segmentation of the area surrounding the optic disc was performed for the glaucomatous images. Previous research in [11] and [16] showed that optic area segmentation significantly improves classification accuracy. It is due to the fact, that the disease is mostly present in that region. An image should be cropped to an area twice as big as the radius of the optic disc. In the literature, there are available many methods for segmentation. However, for the convenience of further processing, we have proposed our method.

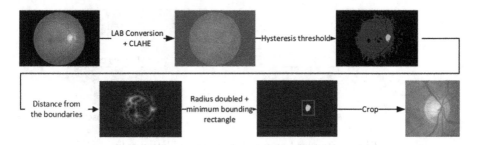

Fig. 2. Optic Disc Segmentation algorithm

The segmentation algorithm is presented in Fig. 2. The main steps consist of finding optic disc (OD) location based on pixel intensities and their neighborhood. When the OD is located, the original image is cropped to the minimum bounding rectangle. After all images were processed, samples were manually evaluated and only images with perfect segmentation were incorporated into the training dataset. There were two main requirements for the data to be considered well segmented. Firstly, it had to have 100% IOU between ground truth and segmented areas of the OD. Secondly, the cropped area had to be approximately two times bigger than the radius of the OD. Conditions were strict in order to provide the best quality of the data.

6   Experiments Results

Multiple experiments with various combinations of the input data were carried out in this study. The next subsections, show the impact of the division of the original datasets on the outcomes. It is particularly important to understand where the differences come from. This knowledge is used for the construction of the generalized disease classification model. The same hyperparameters were used in all the experiments. These were stochastic gradient descent optimizer and cross entropy loss criterion. The training was carried during 5 epochs.

All images were resized to have 224×224 pixels and were preprocessed with the following techniques: Contrast Limited Adaptive Histogram Equalization (CLAHE), which was performed for contrast improvement and normalization, which ensured the same range of colors in all images and had a positive impact on the learning process.

6.1   Initial Approach

The preliminary approach to the problem assumed division of the dataset accordingly to the scheme presented in Fig. 3. The images from the initial datasets were included in both training and validation datasets. The target variable consisted of 4 classes: Age Related Macular Degeneration (AMD), Diabetic Retinopathy (DR), glaucoma, and normal. Results are presented in Table 4. The division of the datasets is shown in Table 3.

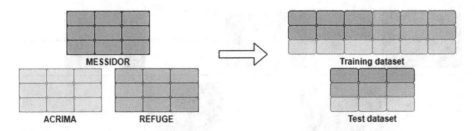

Fig. 3. Standard approach in train-validation split

Table 3. Division of datasets for classes: AMD, DR, glaucoma, normal

Class	Training dataset	Validation dataset
amd	ARIA, Baidu	Baidu
dr	IDRID, MESSIDOR	IDRID
glaucoma	KAGGLE, Magrabia, REFUGE	Magrabia, REFUGE
normal	KAGGLE, JSIEC	KAGGLE

Table 4. Results for classes: AMD, DR, glaucoma, normal

Class	Accuracy (%)	Sensitivity (%)	Specificity (%)
amd	94	81.8	97.8
dr	99	100	99.65
glaucoma	79	76	92.95
normal	82	100	94
AVERAGE	89		

Results attained for the individual eye conditions are outstanding, especially for classes DR (99% accuracy) and AMD (94% accuracy). Nevertheless, they cannot be considered reliable. The distribution of the images does not simulate a real-life scenario. In this approach, the model is tested on pictures taken with the same device (with the same parameters) as it was trained on. This will hardly ever occur in real cases. What is more, such a strategy does not guarantee, that the model properly recognizes an illness. In fact, it may be performing classification based on camera properties. These issues are addressed in further trials.

6.2   Individual Diseases Classification

In this study, experiments were conducted aiming at the classification of individual diseases against healthy ones. ODIR dataset is composed of a large number of images taken with different cameras. Therefore, it was chosen for validation purposes in all experiments. To achieve better generalization of the solution, data augmentation operations were applied, namely horizontal flip and rotation by 10°.

Diabetic Retinopathy. The training for DR was done using datasets listed in Table 5. The total number of training images, after data augmentation reached 2,000 images. The final model achieved 66% accuracy, 67% sensitivity, and 65.7% specificity.

Table 5. Division of datasets for classes: DR, normal

Class	Training dataset	Validation dataset
dr	IDRID, JSIEC, MESSIDOR, eyePACS	ODIR
normal	JSIEC, eyePACS, ORIGA, KAGGLE, STARE	ODIR

Glacuoma. The training for glaucoma was performed according to data presented in Table 6. The total number of training images after data augmentation was equal to nearly 3,000. Images were preprocessed with the algorithm presented in Sect. 5, so they were depicting only the region around the optic disc. The final method reached 77% accuracy, 75% sensitivity, and 78% specificity, which is a huge improvement compared to the outcomes without segmentation, (52% accuracy).

Table 6. Division of datasets for classes: glaucoma, normal

Class	Training dataset	Validation dataset
glaucoma	REFUGE, KAGGLE, ORIGA, BinRushed, MESSIDOR	ODIR
normal	JSIEC, ORIGA, KAGGLE, STARE	ODIR

Cataract. Classification of the cataract versus normal was carried on data displayed in Table 7. The number of training images after augmentation was near 1,500. The model results were: 87% accuracy, 85% sensitivity, and 88% specificity.

Table 7. Division of datasets for classes: cataract, normal

Class	Training dataset	Validation dataset
cataract	KAGGLE	ODIR
normal	JSIEC, ORIGA, KAGGLE, STARE	ODIR

The results justify the statement, that the model can appropriately classify such illnesses as DR, glaucoma, and cataract. During the experiments, AMD was also analyzed. However, the literature states that fundus photography is not the appropriate examination method for this type of illness. Therefore AMD fundus images were excluded from further processing and combined model were built based only on the three above-mentioned eye impairments.

6.3   Multiple Diseases Classification

In the final solution, datasets were split according to the scheme in Fig. 4. Detailed information is presented in Table 8. In order to achieve better results, transfer learning was applied. Firstly, the model was trained on a 4-label dataset during 4 epochs. Next, it was fine-tuned on cropped, glaucomatous images. Then, it was further fine-tuned with the same parameters, during 4 epochs on full images containing glaucoma, diabetic retinopathy, cataract symptoms, or none of them. The procedure improved accuracy for glaucomatous images from 49% to 65%. The final outcome is shown in Table 9.

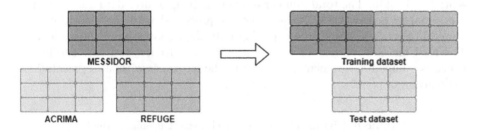

Fig. 4. Proper approach for train-validation split

Table 8. Division of datasets for classes: DR, glaucoma, cataract and normal

Class	Training dataset	Validation dataset
dr	IDRID, JSIEC, MESSIDOR, eyePACS	ODIR
glaucoma	KAGGLE, REFUGE, ORIGA, BinRushed, MESSIDOR	ODIR
normal	KAGGLE, JSIEC, ORIGA, STARE, eyePACS	ODIR
cataract	KAGGLE	ODIR

Table 9. Results for classes: DR, glaucoma, cataract, normal

Class	Accuracy (%)	Sensitivity (%)	Specificity (%)
dr	73	70	81
glaucoma	65	69	93
cataract	66	73	87
normal	73	67	89
AVERAGE	70		

The final results are lower than those presented in Sect. 6.1. The difference in accuracy is over 20% points. Therefore, the new results confirm that dataset division highly affects accuracy.

This shows how important is the understanding of the characteristics of the problem. This issue also highlights the pitfalls of neural networks, which behave as black boxes. More attention should be put into the analysis of a model itself and its activation regions. Furthermore, the complexity of medical images should be also considered. Many factors and attributes impact the final image and they should all be thoroughly analyzed and reviewed.

6.4   Comparison to Standard DCNN Model

Multiple trials were performed to compare the final solution with the standard ResNet-152 model. The training was conducted with the same hyperparameters and on the same dataset. In terms of overall accuracy, the results were similar. The standard model achieved 67% accuracy (see Table 10). The advantage of the SDL model is visible in the accuracy of individual classes, which are much more scattered for standard ResNet. The difference between classes with the lowest and highest accuracy is over 50% points. In comparison, in the SDL model accuracies of individual classes oscillate around 70%, which makes the final output more credible. This proves the utility of the chosen model for the problem of eye illness classification.

Table 10. Regular ResNet-152, results for classes: DR, glaucoma, cataract, normal

Class	Accuracy (%)	Sensitivity (%)	Specificity (%)
dr	55	60	85
glaucoma	37	93	89
cataract	89	68	95
normal	72	64	88
AVERAGE	67		

7   Discussion

To create a universal model for eye condition classification, each of the diseases needs to be considered separately. Each illness is demonstrated in different eye structures and because of that requires a different preprocessing method. Moreover, the same modality of images may not be a good choice for all the cases. A model taking versatile input data is worth consideration. Furthermore, patients are commonly diagnosed by physicians who have access to not only their eye images but also medical records and family history. Despite the fact, that such data is not present in publicly available datasets, it could be of great use.

Before implementing the solution in practice, the results should be discussed and analyzed with the doctors. Therefore, in the potential first phase, the solution can serve as physician support and can be verified and tuned to achieve better results. After the first phase, the certification process can be started in order to implement this solution and facilitate the work of doctors.

8   Conclusions

There are many available works in the literature that present individual eye disease classification. Despite that, no papers aim at the creation of the versatile model, which will be able to distinguish between many illnesses. The model was implemented and described in our study. In the first step, we managed to collect over 30,000 images from 18 separate datasets, which were all publicly available. What is more, we have used SDL architecture which was previously proven to properly address medical images classification issues. A crucial observation was made and verified in terms of model learning. Testing should be performed on images taken with a camera that was not previously seen by the model. Such testing is useful for demonstrating whether a universal pattern of the illness was extracted. It also exhibits the potential risk that a model learned a pattern of a dataset itself. Finally, we managed to achieve promising results, with 70% accuracy for 4 classes (healthy, diabetic retinopathy, glaucoma, and cataract). Still, the solution can be evolved further. Output classes can also be divided based on the stage of the illness. Next, particular symptoms can be marked on the image. There is still plenty of room for improvements.

Acknowledgements. The work was supported by the Foundation for Supporting the Development of Radiocommunication and Multimedia Techniques, which provided a scholarship to help with the research process.

References

1. Cataract image dataset (2019). https://www.kaggle.com/jr2ngb/cataractdataset/
2. International competition on ocular disease intelligent recognition, July 2019. https://odir2019.grand-challenge.org/
3. Joint shantou international eye centre (jsiec) (2019). https://www.kaggle.com/linchundan/fundusimage1000
4. World report on vision (2019). https://www.who.int/publications/i/item/9789241516570
5. Almazroa, A., Alodhayb, S., et al.: Retinal fundus images for glaucoma analysis: the riga dataset. In: Medical Imaging 2018: Imaging Informatics for Healthcare, Research, and Applications, vol. 10579, p. 105790B. SPIE (2018)
6. Bajwa, M.N., Malik, M.I., et al.: Two-stage framework for optic disc localization and glaucoma classification in retinal fundus images using deep learning. BMC Med. Inf. Dec. Making **19**(1), 1–16 (2019)
7. Budai, A., Bock, R., et al.: Robust vessel segmentation in fundus images. Int. J. Biomed. Imaging **2013** (2013)

8. Cuadros, J., Bresnick, G.: Eyepacs: an adaptable telemedicine system for diabetic retinopathy screening. J. Diabetes Sci. Technol. **3**(3), 509–516 (2009)

9. Das, S., Kharbanda, K., et al.: Deep learning architecture based on segmented fundus image features for classification of diabetic retinopathy. Biomed. Sig. Process. Control **68**, 102600 (2021)

10. Decencière, E., Zhang, X., et al.: Feedback on a publicly distributed database: the messidor database. Image Anal. Stereol. **33**(3), 231–234 (2014)

11. Diaz-Pinto, A., Morales, S., et al.: CNNs for automatic glaucoma assessment using fundus images: an extensive validation. Biomed. Eng. Online **18**(1), 1–19 (2019)

12. Dong, Y., Wang, Q., Zhang, Q., Yang, J.: Classification of cataract fundus image based on retinal vascular information. In: Xing, C., Zhang, Y., Liang, Y. (eds.) ICSH 2016. LNCS, vol. 10219, pp. 166–173. Springer, Cham (2017). https://doi.org/10.1007/978-3-319-59858-1_16

13. Fumero, F., Alayón, S., et al.: Rim-one: an open retinal image database for optic nerve evaluation. In: 2011 24th International Symposium on Computer-Based Medical Systems (CBMS), pp. 1–6. IEEE (2011)

14. Gayathri, S., Gopi, V.P., Palanisamy, P.: A lightweight CNN for diabetic retinopathy classification from fundus images. Biomed. Signal Process. Control **62**, 102115 (2020)

15. Gómez-Valverde, J.J., Antón, A., et al.: Automatic glaucoma classification using color fundus images based on convolutional neural networks and transfer learning. Biomed. Opt. Express **10**(2), 892–913 (2019)

16. Haleem, M.S., Han, L., et al.: Regional image features model for automatic classification between normal and glaucoma in fundus and scanning laser ophthalmoscopy (SLO) images. J. Med. Syst. **40**(6), 132 (2016)

17. Holm, S., Russell, G., et al.: Dr hagis-a fundus image database for the automatic extraction of retinal surface vessels from diabetic patients. J. Med. Imaging **4**(1), 014503 (2017)

18. Hoover, A., Goldbaum, M.: Stare public online database. http://www.ces.clemson.edu/~ahoover/stare/

19. Kälviäinen, R., Uusitalo, H.: Diaretdb1 diabetic retinopathy database and evaluation protocol. In: Medical Image Understanding and Analysis, vol. 2007, p. 61 (2007)

20. Kauppi, T., Kalesnykiene, V., et al.: Diaretdb0: Evaluation database and methodology for diabetic retinopathy algorithms. In: Machine Vision and Pattern Recognition Research Group, Lappeenranta University of Technology, Finland, vol. 73, pp. 1–17 (2006)

21. Li, F., et al.: Deep learning-based automated detection of glaucomatous optic neuropathy on color fundus photographs. Graefes Arch. Clin. Exp. Ophthalmol. **258**(4), 851–867 (2020). https://doi.org/10.1007/s00417-020-04609-8

22. Lu, W., Tong, Y., et al.: Applications of artificial intelligence in ophthalmology: general overview. J. Ophthalmol. (2018)

23. Orlando, J.I., Fu, H., et al.: Refuge challenge: a unified framework for evaluating automated methods for glaucoma assessment from fundus photographs. Med. Image Anal. **59**, 101570 (2020)

24. Porwal, P., Pachade, S., et al.: Indian diabetic retinopathy image dataset (idrid): a database for diabetic retinopathy screening research. Data **3**(3), 25 (2018)

25. Qureshi, I., Ma, J., Abbas, Q.: Diabetic retinopathy detection and stage classification in eye fundus images using active deep learning. Multimedia Tools Appl. **80**(8), 11691–11721 (2021). https://doi.org/10.1007/s11042-020-10238-4

26. Shankar, K., Sait, A.R.W., et al.: Automated detection and classification of fundus diabetic retinopathy images using synergic deep learning model. Pattern Recogn. Lett. **133**, 210–216 (2020)
27. Shankar, K., Zhang, Y., et al.: Hyperparameter tuning deep learning for diabetic retinopathy fundus image classification. IEEE Access **8**, 118164–118173 (2020)
28. Staal, J., Abràmoff, M.D., et al.: Ridge-based vessel segmentation in color images of the retina. IEEE Trans. Med. Imaging **23**(4), 501–509 (2004)
29. Zhang, H., Niu, K., et al.: Automatic cataract grading methods based on deep learning. Comput. Methods Programs Biomed. **182**, 104978 (2019)
30. Zhang, J., Xia, Y., et al.: Classification of medical images and illustrations in the biomedical literature using synergic deep learning. CoRR abs/1706.09092 (2017)
31. Zhang, Z., Yin, F.S., et al.: Origa-light: An online retinal fundus image database for glaucoma analysis and research. In: 2010 Annual International Conference of the IEEE Engineering in Medicine and Biology. pp. 3065–3068. IEEE (2010)

A Two-Step Radiologist-Like Approach for Covid-19 Computer-Aided Diagnosis from Chest X-Ray Images

Carlo Alberto Barbano[1]([✉])(iD), Enzo Tartaglione[1,2](iD), Claudio Berzovini[3],
Marco Calandri[1], and Marco Grangetto[1](iD)

[1] University of Turin, Turin, Italy
{carlo.barbano,enzo.tartaglione,marco.calandri,marco.grangetto}@unito.it
[2] LTCI, Télécom Paris, Institut Polytechnique de Paris, Paris, France
[3] Azienda Ospedaliera Citta della Salute e della Scienza, Turin, Italy
cberzovini@cittadellasalute.to.it

Abstract. Thanks to the rapid increase in computational capability during the latest years, traditional and more explainable methods have been gradually replaced by more complex deep-learning-based approaches, which have in fact reached new state-of-the-art results for a variety of tasks. However, for certain kinds of applications performance alone is not enough. A prime example is represented by the medical field, in which building trust between the physicians and the AI models is fundamental. Providing an explainable or trustful model, however, is not a trivial task, considering the black-box nature of deep-learning based methods. While some existing methods, such as gradient or saliency maps, try to provide insights about the functioning of deep neural networks, they often provide limited information with regards to clinical needs.

We propose a two-step diagnostic approach for the detection of Covid-19 infection from Chest X-Ray images. Our approach is designed to mimic the diagnosis process of human radiologists: it detects objective radiological findings in the lungs, which are then employed for making a final Covid-19 diagnosis. We believe that this kind of *structural* explainability can be preferable in this context. The proposed approach achieves promising performance in Covid-19 detection, compatible with expert human radiologists. Moreover, despite this work being focused Covid-19, we believe that this approach could be employed for many different CXR-based diagnosis.

Keywords: Deep learning · Chest x-ray · Radiological findings · Covid-19

This work has received funding from the European Union's Horizon 2020 research and innovation programme under grant agreement No. 825111, DeepHealth Project.

Supplementary Information The online version contains supplementary material available at https://doi.org/10.1007/978-3-031-06427-2_15.

S. Sclaroff et al. (Eds.): ICIAP 2022, LNCS 13231, pp. 173–184, 2022.
https://doi.org/10.1007/978-3-031-06427-2_15

1   Introduction

Early Covid-19 diagnosis is a key element for proper treatment of the patients and prevention of the spread of the disease. Given the high tropism of Covid-19 for respiratory airways and lung epythelium, identification of lung involvement in infected patients can be relevant for treatment and monitoring of the disease. Virus testing is currently considered the only specific method of diagnosis. The US Center for Disease Control (CDC) recommends collecting and testing specimens from the upper respiratory tract (nasopharyngeal and oropharyngeal swabs) or from the lower respiratory tract when available (bronchoalveolar lavage, BAL) for viral testing with reverse transcription polymerase chain reaction (RT-PCR) assay [1]. Testing on BAL samples provides higher accuracy, however this test is uncomfortable for the patient, possibly dangerous for the operator due to aerosol emission during the procedure and cannot be performed routinely. Nasopharingeal swabs are easily executable and affordable and current standard in diagnostic setting; their accuracy in literature is influenced by the severity of the disease and the time from symptoms onset and is reported up to 73.3% [25]. Current position papers from radiological societies (Fleischner Society, SIRM, RSNA) [1,10,16] do not recommend routine use of imaging for Covid-19 diagnosis; however, it has been widely demonstrated that, even at early stages of the disease, chest x-rays (CXR) can show pathological findings.

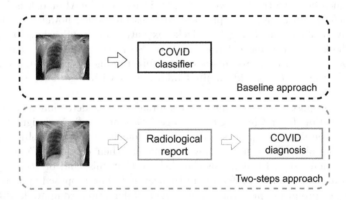

Fig. 1. Comparison between standard approaches for Covid-19 diagnosis and our proposed two-step method.

In the last year, many works attempted to tackle this problem, proposing deep learning-based strategies [3,12,18,20,21]. All of the proposed approaches include some elements in common:

- the images collected during the pandemic need to be augmented with non-Covid cases from publicly available datasets;
- some standard pre-processing is applied to the images, like lung segmentation using U-Net [15] or similar models [20] or converting the pixels of the CXR scan in Hounsfield units;

– the deep learning model is trained to predict the final diagnosis using state-of-the-art approaches for deep neural networks.

Despite some very optimistic results [3,12,13,19,21], the proposed approaches exhibit significant limitations that deserve further analysis. For example, a recent work [20] showed that augmenting Covid-19 datasets with negative cases from publicly-available datasets can inject dangerous biases, where the trained model learn to discriminate different data sources rather than actual radiological features related to the disease. These unwanted effects are difficult to spot when using a *black box* model like deep neural networks, without having control on the decision process.

In this work, we propose an explainable approach, mimicking the radiologists' decision process. We break-down the Covid-19 diagnosis problem into two separate sub-problems. First, we train a model to detect anomalies in the lungs. These anomalies are widely known and, following [6], they comprise 14 objective radiological observations which can be found in lungs. Then, the employ this information to train a decision tree model, such that the Covid-19 diagnosis is fully explainable and grounded to objective radiological findings (Fig. 1). Mimicking the radiologist's decision is more robust to biases and aims at building trust for the physicians and patients towards the AI tool, which can be useful for a more precise COVID diagnosis. Thanks to the collaboration with the radiology units of Città della Salute e della Scienza di Torino (CDSS) and San Luigi Hospital (SLG) in Turin, we collected the COvid Radiographic images DAta-set for AI (CORDA), comprising both positive and negative COVID cases as well as a ground truth on the human radiological reporting, and it currently comprises almost 1000 CXRs.

The rest of the paper is structured as follows. Section 2 introduces and describes the datasets used in this work; Sect. 3 discusses the possible radiological reports, objective findings in the radiographies; Sect. 4 discusses the approach thanks to, from the radiological reports generated, the COVID diagnosis is being elaborated; Sect. 5 discusses the results achieved and finally Sect. 6 draws the conclusions.

2   Datasets

In this section we introduce the datasets that we employed for building our proposed method. As shown in recent works such as [20], augmenting COVID datasets with negative cases from publicly-available datasets (such as ChestXRay or RSNA) can drive the models towards spurious correlations, such as discriminating only between healthy and unhealthy lung, or also be influenced by information related to the image acquisition sites. This is why we focus on the recognition of objective pathologies and radiological findings (CheXpert), which we then use for training a model to elaborate the final Covid-19 diagnosis (CORDA). We now provide a short description of the main datasets, additional information can be found in the supplementary material.

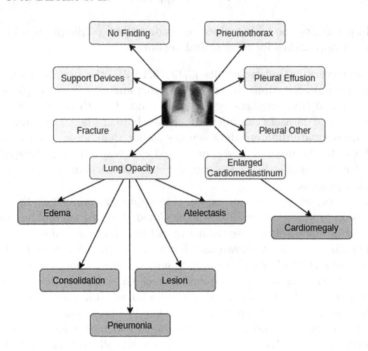

Fig. 2. Radiological findings provided by CheXpert, conforming to the medical known categorization [6].

CheXpert. This is a large dataset comprising about 224k CXRs. This dataset consists of 14 different observations on the radiographic image: differently from many other datasets which are focused on disease classification based on clinical diagnosis, the main focus here is "chest radiograph interpretation", where anomalies are detected [9]. The learnable radiological findings are summarized in Fig. 2.

ChestXRay. This dataset contains 5857 X-ray images collected at the Guangzhou Women and Children's Medical Center, Guangzhou, China. In this dataset, three different labels are provided: normal patients (1583), patients affected by bacterial pneumonia (2780) and affected by viral pneumonia (1493). This dataset is granted under CC by 4.0 and is part of a work on Optical Coherence Tomography [11].[1]

RSNA. Developed for the RSNA Pneumonia Detection Challenge, this dataset contains pneumonia cases found in the NIH Chest X-ray dataset [22].

---

[1] https://data.mendeley.com/datasets/rscbjbr9sj/2/files/f12eaf6d-6023-432f-acc9-80c9d7393433.

It comprises 20672 normal CXR scans and 6012 pneumonia cases, for a total of 26684 images.[2]

CORDA. This dataset was created for this study by retrospectively selecting chest x-rays performed at a dedicated Radiology Unit in CDSS and at SLG in all patients with fever or respiratory symptoms (cough, shortness of breath, dyspnea) that underwent nasopharyngeal swab to rule out COVID-19 infection. Patients' average age is 61 years (range 17–97 years old). It contains a total of 898 CXRs and can be split by different collecting institution into two similarly sized subgroups: CORDA-CDSS, which contains a total of 447 CXRs from 386 patients, with 150 images coming from COVID-negative patients and 297 from positive ones, and CORDA-SLG, which contains the remaining 451 CXRs, with 129 COVID-positive and 322 COVID-negative images. Including data from different hospitals at test time is crucial to doublecheck the generalization capability of our model.

The data collection is still in progress, with other 5 hospitals in Italy willing to contribute at time of writing. We plan to make CORDA available for research purposes according to EU regulations as soon as possible.

3   Radiological Report

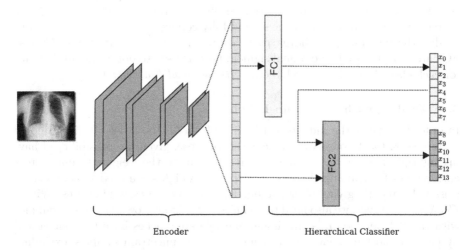

Fig. 3. Radiological report framework. After the convolutional encoder extracts features from the CXR, the hierarchical classifier provides outcome for the different lung pathologies.

In this section we describe our proposed method for detecting and classifying radiological findings from CXRs. For this task, we leverage the large scale dataset

---

[2] https://www.kaggle.com/c/rsna-pneumonia-detection-challenge.

CheXpert, which contains annotation for different kinds of common radiological findings that can be observed in CXR images (like opacity, pleural effusion, cardiomegaly, etc.). Given the high heterogeneity and the large size of CheXpert, its use is perfect for our purposes: in fact, once the model is trained on this dataset, there is no need to fine-tune it for the Covid-19 diagnosis, since it will already extract objective radiological findings.

CheXpert provides 14 different types of observations for each image in the dataset. For each class, the labels have been generated from radiology reports associated with the studies with NLP techniques, conforming to the Fleischner Society's recommended glossary [6], and marked as: negative (N), positive (P), uncertain (U) or blank (N/A). Following the relationship among labels illustrated in Fig. 2, as proposed by [9], we can identify 8 top-level pathologies and 6 child ones.

3.1 Architecture

As the backbone of our model, we use the widely known ResNet [7] and DenseNet [4] convolutional architecture. The encoder is followed by a fully connected classifier. The classifier architecture that we design reflects the hierarchy of the different lung pathologies presented in Fig. 2. As shown in Fig. 3, the classifier is constructed by stacking two fully-connected layers, and makes use of connectivity patterns similar to "dense connections" as proposed by [8]. The first fully-connected layer (*FC1*) is used to classify the 8 top-level classes from the extracted features. Output logits from FC1 are then concatenated with the extracted image features, and the second fully-connected layer (*FC2*) is used to predict the remaining 6 children pathologies. Finally, the logits from FC1 and FC2 are concatenated, and a final sigmoid layer is used to obtain the probability for each class. We call this architecture *Hierarchical Classifier* (HC).

3.2 Dealing with Uncertain Labels

In order to extract the radiological findings from CXRs, a deep learning model is trained on the 14 observations. For this purpose, given the possibility of having multiple findings in the same CXR, we employ the weighted binary cross entropy loss for training the model. Typically, weights are used to compensate class unbalancing, giving higher importance to less represented classes. Within *CheXpert*, however, we also need to tackle another issue: how to treat the samples with the U label. Towards this issue, multiple approaches have been suggested by [9]. The most popular is to just ignore all the uncertain samples, excluding them from the training process and considering them as N/A.

We propose to include the U samples in the learning process, mapping them to the maximum uncertainty (probability 0.5 to be P or N). Then, we assign a weight to the P and N outcomes:

$$w_n = \begin{cases} 1 + S_n^+ / S_n^- & \text{if } y_n = 0 \\ 1 + S_n^- / S_n^+ & \text{if } y_n = 1 \\ 1 & \text{if } y_n = 0.5 \end{cases} \tag{1}$$

where S_n^- and S_n^+ respectively represent the cardinality of negative and positive samples for the n-th class. These weights are then plugged in the weighted loss entropy loss to be minimized:

$$\mathcal{L}_n = -w_n \cdot [y_n \log(x_n) + (1 - y_n) \log(1 - x_n)] \tag{2}$$

Hence, uncertain samples will have a lower influence during the training process, while being pushed either towards 0 or 1 by the higher weight certain samples in the same class. All of the remaining blank labels are ignored when computing the BCE loss, considering them as missing labels.

Table 1 shows a performance comparison between the standard approach as proposed by [9] and our proposal (U-label use), for 5 salient radiological findings, using the same setting as in [9]. We observe an overall improvement in the performance, which is expected by the inclusion of the U-labeled examples. For all our experiments, we will use models trained using the U labeled samples.

Table 1. Performance (AUC) for a DenseNet-121 trained on CheXpert.

Method	Atelectasis	Cardiomegaly	Consolidation	Edema	Pleural Eff.
Baseline [9]	0.79	**0.81**	0.90	0.91	0.92
U-label	**0.83**	0.79	**0.93**	**0.93**	**0.93**

4 COVID Diagnosis

The second step of the proposed approach is building the model which can actually provide a clinical diagnosis for COVID. We freeze the model obtained from Sect. 3 and use its output as input features to train a new binary classifier on the CORDA dataset. We test two different types of classifiers: a decision tree (Tree) and a two-layers fully-connected classifier (FC). The decision tree is trained on the probabilities output of the radiological reports, using the standard CART Algorithm implementation provided by the Python scikit-learn [14] package. The fully-connected classifier, comprising one hidden layer of size 512 and the output layer, is instead trained on the encoder latent space. The reason is that training it on the output probabilities would result in a loss of explainability compared to the decision tree, hence it makes more sense to maximize the achievable performance by training on the full latent space as discussed in Sect. 5.

5 Experiments

In this section we compare the COVID diagnosis generalization capability through a direct deep learning-based approach (baseline) and our proposed two-step diagnosis, where first we detect the radiological findings, and then we

Table 2. Results for COVID diagnosis. Abbreviations: ResNet-18 (RN-18), DenseNet-121 (DN-121), Pretrain dataset (Pretrain), chest X-Ray (CXR), CheXpert (ChXp), Train dataset (Train), CORDA-CDSS (CDSS), CORDA-SLG (SLG). We denote fully-explainable methods with [†].

Method	Baseline [20]			Two-step			
Backbone	RN-18	RN-18	RN-18	RN-18	DN-121	DN-121	DN-121
Classifier	FC	FC	FC	FC	FC	Tree[†]	FC
Pretrain	-	RSNA	CXR	ChXp			ChXp
Train	CDSS			CDSS			SLG
Sensitivity	**0.56**	0.54	0.54	0.69	0.72	**0.77**	0.79
Specificity	0.58	**0.80**	0.58	0.73	**0.78**	0.60	0.82
BA	0.57	**0.67**	0.56	0.71	**0.75**	0.68	0.81
AUC	0.59	**0.72**	0.67	0.76	**0.81**	0.70	0.84

discriminate patients affected by COVID using a decision tree-based diagnosis (Tree) or a deep learning-based classifier from the radiological findings (FC). The performance is tested on a subset of *patients* not included in the training/validation set. The assessed metrics are: balanced accuracy (BA), sensitivity, specificity and area under the ROC curve (AUC). For all of the methods we adopt a 70%-30% train-test split. For the deep learning-based strategy, SGD is used with a learning rate 0.01 and a weight decay of 10^{-5}. All of the experiments were run on NVIDIA Tesla T4 GPUs using PyTorch 1.4.[3]

Table 2 compares the standard deep learning-based approach [20] to our two-step diagnosis. Baseline results are obtained pre-training the model on some of the most used publicly-available datasets. We observe that the best achievable performance is very low, consisting in a BA of 0.67. A key takeaway is that trying to directly diagnose diseases such as COVID-19 from CXRs might be currently infeasible, probably given the small dataset sizes and strong selective bias in the datasets. We can clearly see how the two-step method outperforms the direct diagnosis: using the same network architecture (ResNet-18 as backbone and a fully-connected classifier on top of it), we obtain a significant increase in all of the assessed metrics. Even better results are achieved by using a DenseNet-121 as backbone and the fully-connected classifier.

Focusing on the fully-explainable decision tree method, we found that a maximum depth of 4 gave the best results in terms of model complexity and generalization ability.

Figure 4 graphically shows the learned decision tree (whose performance is shown in Table 2): this provides a very clear interpretation for the decision process. From the clinical and radiological perspective, these data are consistent with the COVID-19 CXR semiotics that radiologists are used to deal with. The edema feature, although unspecific, is strictly related to the interstitial

---
[3] The source code is available at https://github.com/EIDOSlab/covid-two-step.

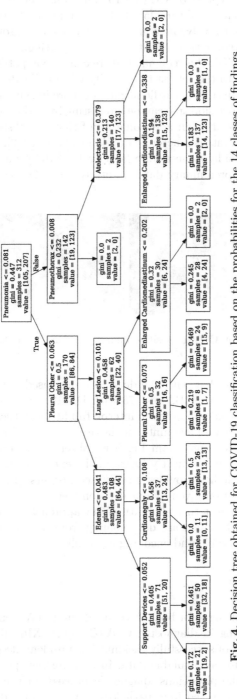

Fig. 4. Decision tree obtained for COVID-19 classification based on the probabilities for the 14 classes of findings.

involvement that is typical of COVID-19 infections and it has been largely reported in the recent literature [5]. Indeed, in recent COVID-19 radiological cal papers, interstitial involvement has been reported as ground glass opacity appearance [24]. However this definition is more pertinent to the CT imaging setting rather than CXR; the "edema" feature can be compatible, from the radiological perspective, to the interstitial opacity of COVID-19 patients. Furthermore, the not irrelevant role of cardiomegaly (or more in general enlarged cardiomediastinum) in the decision tree can be interesting from the clinical perspective. In fact, this can be read as an additional proof that established cardiovascular disease can be a relevant risk factor to develop COVID-19 [2]. Moreover, it may be consistent with the hypotheses of a larger role of the primary cardiovascular damage observed on preliminary data of autopsies of COVID-19 patients [23].

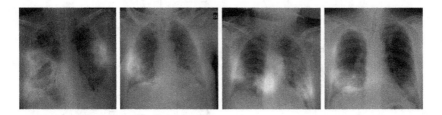

Fig. 5. GradCAM on COVID-positive images obtained from Densenet121+FC.

Although it is true that with the deep learning-based approach we observe a boost in the performance (BA of 0.75 with DN-121+FC vs 0.68 with Tree), this is the result of a trade-off between interpretability and discriminative power. Using Grad-CAM [17] we have hints on the area the model focused on to take the final diagnostic decision. From Fig. 5 we observe that on COVID-positive images, the model seems to mostly focus on the expected lung areas. However this kind of interpretability provides very limited insights when compared to the clinical-based explanation given by our decision tree.

Finally, to further test the reliability of our approach, we used our strategy also on CORDA-SLG (which are data coming from a different hospital structure), reaching comparable and encouraging results.

6   Conclusions

One of the latest challenges for both the clinical and the AI community has been applying deep learning in diagnosing COVID from CXRs. Recent works suggested the possibility of successfully tackling this problem, despite the currently small quantity of publicly available data. In this work we propose a multi-step approach, close to the physicians' diagnostic process, in which the final diagnosis is based upon detected lung pathologies. We performed our experiments on CORDA, a COVID-19 CXR dataset comprising approximately 1000 images.

All of our experiments have been carried out bearing in mind that, especially for clinical applications, explainability plays a major role for building trust in machine learning algorithms, although better interpretability can come at the cost of a lower prediction accuracy.

References

1. ACR recommendations for the use of chest radiography and computed tomography (CT) for suspected COVID-19 infection. https://www.acr.org/
2. ESC Guidance for the Diagnosis and Management of CV Disease During the COVID-19 Pandemic. (2020). https://www.escardio.org/Education/COVID-19-and-Cardiology/ESC-COVID-19-Guidance
3. Apostolopoulos, I.D., Bessiana, T.: Covid-19: automatic detection from x-ray images utilizing transfer learning with convolutional neural networks. arXiv preprint arXiv:2003.11617 (2020)
4. De Fauw, J., et al.: Clinically applicable deep learning for diagnosis and referral in retinal disease. Nat. Med. **24**(9), 1342–1350 (2018)
5. Guan, W., et al.: Clinical characteristics of coronavirus disease 2019 in China. N. Engl. J. Med. **382**(18), 1708–1720 (2020)
6. Hansell, D.M., Bankier, A.A., MacMahon, H., McLoud, T.C., Muller, N.L., Remy, J.: Fleischner society: glossary of terms for thoracic imaging. Radiology **246**(3), 697–722 (2008)
7. He, K., Zhang, X., Ren, S., Sun, J.: Deep residual learning for image recognition. In: Proceedings of the IEEE Conference on Computer Vision and Pattern Recognition, pp. 770–778 (2016)
8. Huang, G., Liu, Z., Van Der Maaten, L., Weinberger, K.Q.: Densely connected convolutional networks. In: Proceedings of the IEEE Conference on Computer Vision and Pattern Recognition, pp. 4700–4708 (2017)
9. Irvin, J., et al.: Chexpert: a large chest radiograph dataset with uncertainty labels and expert comparison. In: Proceedings of the AAAI Conference on Artificial Intelligence, vol. 33, pp. 590–597 (2019)
10. Italian Radiology Society: Utilizzo della Diagnostica per Immagini nei pazienti Covid 19. https://www.sirm.org/
11. Kermany, D., Zhang, K., Goldbaum, M.: Labeled optical coherence tomography (OCT) and chest x-ray images for classification. Mendeley Data **2** (2018)
12. Narin, A., Kaya, C., Pamuk, Z.: Automatic detection of coronavirus disease (covid-19) using x-ray images and deep convolutional neural networks. arXiv preprint arXiv:2003.10849 (2020)
13. Oh, Y., Park, S., Ye, J.C.: Deep learning Covid-19 features on CXR using limited training data sets. IEEE Trans. Med. Imaging **39**(8), 2688–2700 (2020)
14. Pedregosa, F., et al.: Scikit-learn: machine learning in Python. J. Mach. Learn. Res. **12**, 2825–2830 (2011)
15. Ronneberger, O., Fischer, P., Brox, T.: U-net: convolutional networks for biomedical image segmentation. In: Navab, N., Hornegger, J., Wells, W.M., Frangi, A.F. (eds.) MICCAI 2015. LNCS, vol. 9351, pp. 234–241. Springer, Cham (2015). https://doi.org/10.1007/978-3-319-24574-4_28
16. Rubin, G.D., Ryerson, C.J., Haramati, L.B., Sverzellati, N., et al.: The role of chest imaging in patient management during the Covid-19 pandemic: a multinational consensus statement from the Fleischner society. RSNA Radiol. (2020). https://doi.org/10.1148/radiol.2020201365

17. Selvaraju, R.R., Cogswell, M., Das, A., Vedantam, R., Parikh, D., Batra, D.: Grad-cam: visual explanations from deep networks via gradient-based localization. In: Proceedings of the IEEE International Conference on Computer Vision, pp. 618–626 (2017)
18. Sethy, P.K., Behera, S.K.: Detection of coronavirus disease (Covid-19) based on deep features (2020)
19. Sitaula, C., Hossain, M.B.: Attention-based VGG-16 model for Covid-19 chest x-ray image classification. Appl. Intell. **51**(5), 2850–2863 (2021)
20. Tartaglione, E., Barbano, C.A., Berzovini, C., Calandri, M., Grangetto, M.: Unveiling Covid-19 from chest x-ray with deep learning: a hurdles race with small data. Int. J. Environ. Res. Publ. Health **17**(18), 6933 (2020). https://doi.org/10.3390/ijerph17186933
21. Wang, L., Lin, Z.Q., Wong, A.: Covid-net: a tailored deep convolutional neural network design for detection of Covid-19 cases from chest x-ray images. Sci. Rep. **10**(1), 1–12 (2020)
22. Wang, X., Peng, Y., Lu, L., Lu, Z., Bagheri, M., Summers, R.: Chestx-ray8: hospital-scale chest x-ray database and benchmarks on weakly-supervised classification and localization of common thorax diseases. In: 2017 IEEE Conference on Computer Vision and Pattern Recognition (CVPR), pp. 3462–3471 (2017)
23. Wichmann, D., et al.: Autopsy findings and venous thromboembolism in patients with Covid-19: a prospective cohort study. Ann. Internal Med. (2020)
24. Wong, H.Y.F., et al.: Frequency and distribution of chest radiographic findings in Covid-19 positive patients. Radiology, 201160 (2020)
25. Yang, Y., et al.: Laboratory diagnosis and monitoring the viral shedding of 2019-NCOV infections. medRxiv (2020)

UniToChest: A Lung Image Dataset for Segmentation of Cancerous Nodules on CT Scans

Hafiza Ayesha Hoor Chaudhry[1], Riccardo Renzulli[1], Daniele Perlo[2],
Francesca Santinelli[3], Stefano Tibaldi[3], Carmen Cristiano[3],
Marco Grosso[3], Giorgio Limerutti[3], Attilio Fiandrotti[1(✉)],
Marco Grangetto[1], and Paolo Fonio[1]

[1] University of Turin, Turin, Italy
attilio.fiandrotti@unito.it
[2] Fondazione Ricerca Molinette Onlus, Turin, Italy
[3] Città della Salute e della Scienza di Torino, Turin, Italy

Abstract. Lung cancer has emerged as a major causes of death and
early detection of lung nodules is the key towards early cancer diagnosis and treatment effectiveness assessment. Deep neural networks achieve
outstanding results in tasks such as lung nodules detection, segmentation
and classification, however their performance depends on the quality of
the training images and on the training procedure. This paper introduces
UniToChest, a dataset consisting Computed Tomography (CT) scans of
623 patients. Then, we propose a lung nodules segmentation scheme relying on a convolutional neural architecture that we also re-purpose for a
nodule detection task. The experimental results show accurate segmentation of lung nodules across a wide diameter range and better detection
accuracy over a traditional detection approach. The datasets and the
code used in this paper are publicly made available as a baseline reference.

Keywords: Medical image segmentation · Deep learning · U-Net ·
Dataset · Chest CT scan · Lung nodules · DeepHealth

1 Introduction

Lung cancer has become the leading cause of death for men and women in 2021,
surpassing breast and prostate cancer [19]. With such a low survival rate of 14–
15% at late stages of lung cancer, detecting and monitoring malign cancerous
nodules is the key towards better recovery rates [7]. Conventionally, first a thoracic *Computed Tomography* (CT) scan of the lungs generates high resolution
images of the chest structures [16]. The same procedure is used to monitor the
growth of lung nodules over time, as an indicator of the success of the treatment
or as a warning in case of a sudden volume change [13].

© The Author(s), under exclusive license to Springer Nature Switzerland AG 2022
S. Sclaroff et al. (Eds.): ICIAP 2022, LNCS 13231, pp. 185–196, 2022.
https://doi.org/10.1007/978-3-031-06427-2_16

Manual lung nodules analysis is time-consuming, so *Computer-Aided Diagnosis* (CAD) systems are commonly employed for the detection and segmentation tasks. Over the past decade, several systems based on traditional or deep learning based image processing techniques have been proposed for the detection and segmentation of lung nodules [8,10,21]. Differences in size and shape of the nodules, age and gender of the patients, imaging device model and brand along with the similarity between nodules and their surrounding, make this a challenging task. Most of the methods rely on supervised approaches, so an important factor towards precise segmentation and detection is the training dataset quality. Images and relative annotations in fact often lack in terms of quality or quantity or both, due to the cost of acquiring and annotating the images by a radiologist. While some methods attempt to cope with small sample sizes [22] or noisy labels [6], a good training set remains of paramount importance. Finally, releasing annotated medical images to the public requires abiding by the privacy protection laws, which includes making sure that neither the images nor the annotations leak any sensitive information.

This paper present a twofold contribution towards accurate lung nodule detection and segmentation.

First, we present UniToChest [15], a dataset collected and annotated by Radiology Unit in Città della Salute e della Scienze Hospital within the framework of the EU-H2020 *DeepHealth* project[1]. The dataset includes 306440 lung cancer screening thoracic computed tomography (CT) scans of 623 patients. Each patient file contains diagnostic lung cancer CT scan images and associated segmentation masks for the annotated lesions. This dataset is the largest of its kind with most diversity in lesions (lung nodule) size.

Second, we propose a complete nodule detection and segmentation pipeline designed around a convolutional neural network. Namely, we first segment nodules using an autoencoder with skip connection that we train in a fully supervised way. Next, we recast the nodule detection task as a segmentation problem, showing better performance than a baseline nodule detector.

2   Background and Related Works

In this section we first provide the medical background relevant to the understanding of this work, next we review existing techniques in pulmonary nodules detection highlighting the main limitations that prompted this research.

Computed Tomography (CT) scan is a medical imaging procedure that uses a computer linked to an x-ray machine to grab series of pictures of areas of the inner body. The pictures are taken from different angles and are used to create 3-dimensional views of tissues and organs. Sometimes to increase the chances of seeing diseases, a drug called contrast medium is used which is injected into the venous circulation to make the blood vessels opaque and reveal neoplastic lesions. Pulmonary nodules are small, focal, radiographic opacities that may be solitary

---

[1] https://deephealth-project.eu/.

or multiple. A classic solitary pulmonary nodule (SPN) [4] is a single, spherical, well-circumscribed, radiographic opacity measuring less than or equal to 30 mm in diameter and is surrounded completely by aerated lung. The SPN is a coined term that in the past described solitary nodules detected incidentally by chest radiography (CXR). Today, most nodules are detected by computed tomography (CT). The detailed CT images frequently identify more than one nodule, or enlarged lymph nodes. Indeterminate nodules are those that do not possess features clearly associated with a benign etiology, such as a benign pattern of calcification or stability on imaging for >2 years. On CT scans, a nodule appears as a rounded or irregular opacity, well or poorly defined, measuring up to 3 cm in diameter. Advances in chest imaging and the increased use of CT as a diagnostic modality have lead to incidental identification of many small pulmonary nodules. The vast majority of nodules detected on CT are sub-centimeter based on early lung screening trials (61%–89%). The overwhelming majority of these are benign. The prevalence of pulmonary nodules changes significantly across studies. This variation stems from the inconsistency among studies in method, enrolled population, and reporting results. Most lung nodules are detected incidentally on CXR or CT scans obtained for other purposes. The actual risk for malignancy in sub-centimeter nodules is lower than the predicted risk based on clinical and radiographic criteria for pulmonary nodules [11]. The risk factor varies with the nodule diameter, staying under 35% for nodules below to 1 cm in diameter and exceeding 97% for nodules above 3 cm. For this reason, accurate nodule segmentation is of paramount importance to estimate its malignancy probability. Methods for the detection and segmentation of lung nodules can be categorized into traditional and learning-based.

Traditional methods rely on handcrafted feature extraction [2], often coupled with shallow classifiers or regressors. The main problem with such techniques is the manually designing feature extractors time consuming activity and features may be tailored to some specific dataset.

Learning-based methods usually rely on a type of artificial neural networks known as *Convolutional Neural Networks (CNN)* [9]. The underlying idea is to let the convolutional layers learn feature extractors that maximize some loss metric on an annotated dataset rather than handcrafting the feature extractor. Such architectures include millions of learnable parameters and represent the state of the art in a number of medical applications today [20]. In particular, the U-Net [17] architecture is designed around an autoencoder topology with skip connections and represents a standard for semantic segmentation tasks. Due to the amount of learnable parameters they include, their performance strongly depends on the amount of data available for training, prompting the collection of large annotated datasets. The *Lung Image Database Consortium image collection (LIDC-IDRI)* dataset [1] is the largest publicly available dataset for the detection and segmentation of lung nodules. *LUNA16* (Lung Nodule Analysis 2016) [18] is a segmentation challenge that uses a subset of *LIDC-IDRI*. The *LIDC-IDRI* dataset contains 7371 nodules annotated by atleast 3 out of 4 radiologists performing the study. For nodules greater than 3 mm the complete volumetric nodule boundary is given as Region of Interests (ROIs) [12]. Whereas for nodules less than 3 mm only the centroid point(x,y,z) is provided as ROI instead of whole nodule boundary, which makes detecting the smaller nodules harder.

3 The UniToChest Dataset

The *UniToChest* dataset has been collected within the EU-H2020 *DeepHealth* [3, 14] project and consists of about than 300k lung CT scans of pulmonary lungs from 623 different patients. The scans are in DICOM format and each scan comes with a manually annotated segmentation mask in black and white PNG format, both being 512×512 in size. The slice thickness of CT scans ranges from 1.25 to 6.5 mm and the pixel spacing from 0.41 to 0.97 mm. A comparison with similar datasets in Table 3 shows that our dataset has more nodules and from a wider diameter range especially at the top end. In fact, for most patients are available images collected over multiple exams over the years including late stages as shows in Fig. 1a. The dataset contains data collected from a gender-balanced population (377 males and 246 females) and spanning across a wide rage of ages (from 7 to 9, most of the population being between 60 and 80), as in Fig. 1b For many comparable datasets, the images come from a single acquisition device that may hide some specific bias; conversely, our dataset includes images acquired using 10 different devices as in Fig. 1c. For each and every image, the radiologist inspected the image for nodules and, where found, each nodule was manually segmented across multiple slices. Finally, compliance with the UE regulation on privacy is guaranteed since any sensitive information (name, birth date, identity) was carefully removed from images and annotations.

Table 1. Dataset population for the three splits we provide

Splits	Number of patients	Male	Female	Average age
Training	498	303	195	66
Validation	62	39	23	68
Test	63	35	28	65
Total	623	377	246	66

The total number of nodules in the malignant CT scans of our dataset surpasses any publicly available dataset. The distribution of overall nodule diameter in our dataset is represented in Fig. 1a, and a detail description of nodule diameter with respect to splits made in our experiment can be seen in Table 2.

For the purpose of training a neural network, we split the dataset into training, validation and test set randomly as 80-10-10 of patients. We maintain data consistency across multiple splits by assigning a single split to each patient.

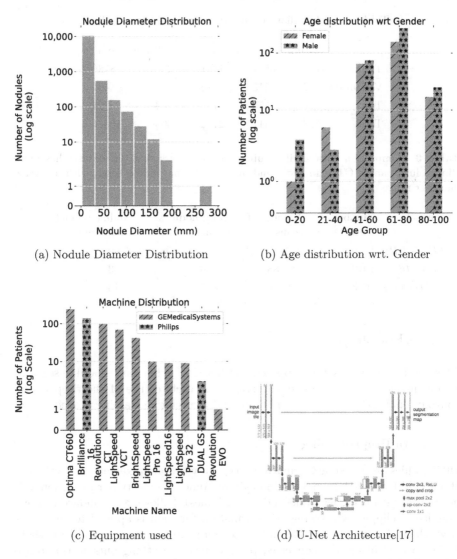

(a) Nodule Diameter Distribution   (b) Age distribution wrt. Gender

(c) Equipment used   (d) U-Net Architecture[17]

Fig. 1. Dataset distributions show no sign of bias with respect to nodule diameter range, patient gender and machines used. The U-Net architecture is also displayed for reference

The data population with respect to the splits is summarized in Table 1 All the three sets (train, validate, and test) have a 60 to 40 ratio between the number of male and female patients.

Table 2. Nodule diameter distribution across three splits

	<3 mm	<10 mm	<30 mm	>30 mm	Total
Train	149	6527	1861	249	8786
Validation	7	315	116	23	461
Test	21	575	195	33	824
Total	177	7417	2172	305	10071

Table 3. Comparison with similar public dataset shows that our dataset has more clinical lung cancer CT scan slices and annotated lung nodule count with a diverse diameter range

Dataset	Number of patients	Number of scans	Total nodules count	Nodule diameter range (mm)
LIDC-IDRI	1010	244527	7371	2–69
LUNA16	1010	888	1836	3–33
UniToChest	623	306440	10071	1–136

4 Methodology

This section describes the proposed method for pulmonary nodules segmentation, including the preprocessing stage, the architecture of the deep neural convolutional architecture we rely upon and the relative training procedures.

4.1 Data Preprocessing

DICOM files produced by CT machines typically contain pixel intensity values in Hounsfield Units (HU), i.e. they indicate radiometric density per pixel (low values indicating air, higher values indicating bones). Following a standard medical practice, a clipped windowing transformation function is applied to such density values. The window width and center indicate the range of the Hounsfield Units covered inside the converted pixel values, everything outside this range will be equivalent to either zero or one. According to standard practice, we have used a window width of 1600 and a window center of −500 to account for the radiometric density of body structures actually useful for nodule detection.

4.2 Network Architecture

Our approach relies on the U-Net implementation [17] in Fig. 1d. The encoder consists of 5 convolutional layers with max-pooling for featuremap downsampling. As in other convolutional architectures, as the size of the featuremaps shrinks the number of featuremaps increases by a two factor. The decoder includes 5 convolutional layers followed by an upconvolutions, where the size of the featuremaps increases while their number decreases at each layer. A number of encoder and decoder layers are matched with skip connections, where the feature maps generated by the respective encoder layer is concatenated with the output of decoder layer, enabling the precise learning and localization of image object by allowing different tradeoffs between semantic level and spatial accuracy of the featuremaps.

4.3 Training Procedure

The training method is fully supervised and consists in randomly initializing the network weights (*from scratch*) and then training the network for nodule segmentation minimizing the loss between the network output and the segmentation mask relative to the input image. As for similar segmentation tasks, we minimize the *DICE* loss since it has a derivative allowing for error gradient backpropagation and minimizing the dice loss amount to maximizing the IoU between predicted and ground truth mask. As a preliminary stage, we found beneficial pretraining the network over the LIDC dataset prior to the training on UniToChest. The rationale behind this pretraining is to have the network learning additional features from the LIDC dataset that may be possibly useful when trained for segmentation on UniToChest. Next, the network is trained over UniToChest train set until the *Intersection over Union* (IoU) score as measured over UniToChest validation set did not improved for 50 epochs. For this training, only scans with one or more nodules have been considered, since other scans we experimentally verified do not bring any useful information for segmentation. The CT slices are provided in input to the network in batches of 5, as that enabled a reasonable tradeoff between memory footprint and performance. We found beneficial resorting to on the fly data augmentation during the training to avoid overfitting to the training data. The augmentation technique we used consist in cropping random patches from the slices and performing random flips and rotations (the very same transformations are also applied to the corresponding segmentation mask). The optimizer used in our experiment is Adam with an initial learning rate of 0.001 and weight decay of 0.0001. The whole architecture has been implemented in PyTorch and is available on github.[2]

5 Results and Discussion

In this section, we first experiment over the UniToChest dataset with the neural network based method described in the previous section for nodule segmentation. Next, we repurpose and the same method for nodules detection, with particular attention to the tradeoff between sensitivity and specificity. All results are relative to UniToChest test set, i.e. images that the network has never seen at training time.

---

[2] https://github.com/deephealthproject/UC4_pipeline.

5.1   Nodules Segmentation

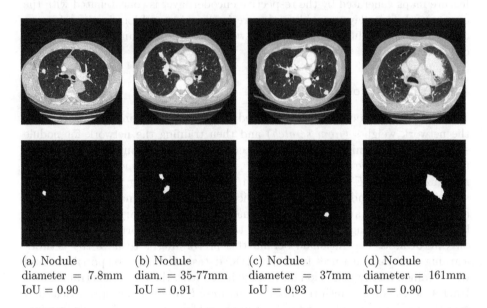

(a) Nodule
diameter = 7.8mm
IoU = 0.90

(b) Nodule
diam. = 35-77mm
IoU = 0.91

(c) Nodule
diameter = 37mm
IoU = 0.93

(d) Nodule
diameter = 161mm
IoU = 0.90

Fig. 2. Segmentation results over different nodule diameter range; regions in yellow represent the overlap between predicted and annotated segmentation masks. (Color figure online)

As a first experiment, we evaluate the performance of the U-Net trained as in the previous section at nodule segmentation. For this task, we consider only positive slices from the test set, i.e. slices with at least one nodule. Figure 3 shows the *Intersection over Union* (IoU) and DICE scores for both the network pretrained as in the previous section and a reference network that was trained from scratch. The pretraining improves both segmentation accuracy (about +10% IoU) and convergence speed. As further experiment, we tested the network pretrained on LIDC only over UniToChest dataset, the top IoU settling at about 43% as a proof of the benefit yield by pretraining.

Table 4 correlates the average IoU and DICE scores with the nodule diameter size for both the cases without and with pretraining. The number of nodules in the sub 10 mm bin is approximately equal to the number of nodules in the above 10 mm bin. The average IoU is as large as 61% and even on the nodules having a diameter of less than 3 mm, we achieve an average IoU of 59%. We hypothesize that the above 10 mm bin benefits the most from the pretraining because LIDC contains nodules ranging mainly in the 10 mm to 50 mm range.

Finally, Fig. 2 shows some samples of the segmentation mask predicted by the network (bottom row) for some sample test images (top row). Red pixels represent false negatives, green pixel false positives and yellow pixels correctly segmented pixels: most of the pixels are correctly segmented, a few errors only remaining at the borders of the nodule.

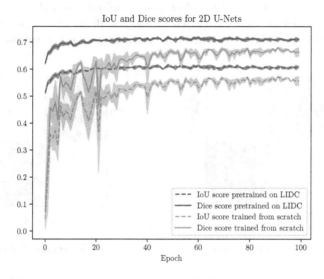

Fig. 3. Segmentation scores for the first 100 epochs: pretraining improves asymptotic IoU, plus the network converges faster.

Table 4. Segmentation accuracy: pretraining improves accuracy, especially for nodule sizes most represented in the dataset used for pretraining

Nodule diameter	From scratch (not pretrained)		Pretrained on LIDC	
	IoU	DICE	IoU	DICE
<10 mm	0.575	0.69	0.59	0.69
>10 mm	0.58	0.69	0.615	0.72

5.2   Detection

Next, we evaluate the performance of the same U-Net network trained for segmentation on a nodule detection task, this time considering both positive and negative test scans. We provide each scans in input to the network and we count the number of white pixels in the predicted segmentation mask. If such number is greater than zero, the slice is labeled as positive, negative otherwise. Figure 4a (left) shows that the network achieves a sensitivity of 0.95 and specificity of 0.80. We investigate whether the specificity value could be increased further, finding a balance between sensitivity and specificity, since in many medical trials the aim is also to reduce the number of false positives. For this reason, we finetune the network adding to the train set 10% of the negative training samples drawn at random. The confusion matrix on the right shows that the specificity improves from 0.8 to 0.95, reducing the number of false positives.

As a baseline reference, we also trained a binary classifier based on a ResNet18 [5] pretrained over ImageNet to discriminate each slice as positive or negative. We achieved a sensitivity of 0.74 and specificity of 0.82, so the number of false positives is higher than our segmentation-based detection method.

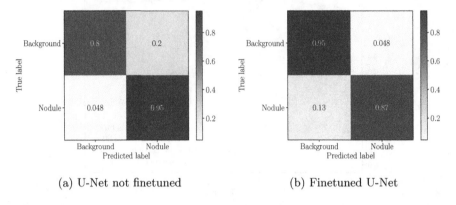

(a) U-Net not finetuned   (b) Finetuned U-Net

Fig. 4. Normalized confusion matrices at nodule detection for a U-Net network trained at segmentation. The network finetuned on negative slices has better tradeoff between sensitivity and specificity

6   Conclusion and Future Works

This paper presented UniToChest, a CT scan lung nodules dataset, that is among the largest of its kind and boasts a diversity of patient ages, acquisition machines and nodules diameter. We proposed a U-Net based architecture that yield promising results at both detection and segmentation of lung nodules. Future research directions of this work include exploiting the thee-dimensional information of nodules across neighboring slices.

Acknowledgements. This work has received funding from the European Union's Horizon 2020 research and innovation programme under grant agreement No 825111, DeepHealth Project.

References

1. Armato, S.G., et al.: The lung image database consortium (LIDC) and image database resource initiative (IDRI): a completed reference database of lung nodules on CT scans. Med. Phys. **38**(2), 915–931 (2011). https://doi.org/10.1118/1.3528204, https://wiki.cancerimagingarchive.net/display/Public/LIDC-IDRI

2. Chaudhary, A.H., Ikhlaq, J., Iftikhar, M.A., Alvi, M.: Blood cell counting and segmentation using image processing techniques. In: Khan, F., Jan, M.A., Alam, M. (eds.) Applications of Intelligent Technologies in Healthcare. EICC, pp. 87–98. Springer, Cham (2019). https://doi.org/10.1007/978-3-319-96139-2_9

3. DeepHealth: Deep-learning and HPC to boost biomedical applications for health (2019). https://deephealth-project.eu/

4. Gaga, M., et al.: Lung nodules: a comprehensive review on current approach and management. Ann. Thoracic Med. **14**(4), 226 (2019). https://doi.org/10.4103/atm.atm_110_19

5. He, K., Zhang, X., Ren, S., Sun, J.: Deep residual learning for image recognition. In: Proceedings of the IEEE Conference on Computer Vision and Pattern Recognition, pp. 770–778 (2016)

6. Karimi, D., Dou, H., Warfield, S.K., Gholipour, A.: Deep learning with noisy labels: exploring techniques and remedies in medical image analysis. Med. Image Anal. **65**, 101759 (2020). https://doi.org/10.1016/j.media.2020.101759

7. Knight, S.B., Crosbie, P.A., Balata, H., Chudziak, J., Hussell, T., Dive, C.: Progress and prospects of early detection in lung cancer. Open Biol. **7**(9), 170070 (2017). https://doi.org/10.1098/rsob.170070

8. LeCun, Y., Bengio, Y., Hinton, G.: Deep learning. Nature **521**(7553), 436–444 (2015)

9. LeCun, Y., Bottou, L., Bengio, Y., Haffner, P.: Gradient-based learning applied to document recognition. Proc. IEEE **86**(11), 2278–2324 (1998)

10. Liu, H., et al.: A cascaded dual-pathway residual network for lung nodule segmentation in CT images. Physica Med. **63**, 112–121 (2019). https://doi.org/10.1016/j.ejmp.2019.06.003

11. MacMahon, H., et al.: Guidelines for management of incidental pulmonary nodules detected on CT images: from the Fleischner society 2017. Radiology **284**(1), 228–243 (2017). https://doi.org/10.1148/radiol.2017161659

12. McNitt-Gray, M.F., et al.: The lung image database consortium (LIDC) data collection process for nodule detection and annotation. Acad. Radiol. **14**(12), 1464–1474 (2007)

13. Mozley, P.D., et al.: Measurement of tumor volumes improves RECIST-based response assessments in advanced lung cancer. Transl. Oncol. **5**(1), 19–25 (2012). https://doi.org/10.1593/tlo.11232

14. Oniga, D., et al.: Applications of AI and HPC in health domain. In: HPC, Big Data, AI Convergence Toward Exascale: Challenge and Vision (Chap. 11). CRC Press, Taylor & Francis Group, Boca Raton (2021). ISBN 9781032009841

15. Perlo, D., et al.: UniToChest (2022). https://doi.org/10.5281/zenodo.5797912

16. Puderbach, M., Kauczor, H.U.: Can lung MR replace lung CT? Pediatr. Radiol. **38**(S3), 439–451 (2008). https://doi.org/10.1007/s00247-008-0844-7

17. Ronneberger, O., Fischer, P., Brox, T.: U-net: convolutional networks for biomedical image segmentation. In: Navab, N., Hornegger, J., Wells, W.M., Frangi, A.F. (eds.) MICCAI 2015. LNCS, vol. 9351, pp. 234–241. Springer, Cham (2015). https://doi.org/10.1007/978-3-319-24574-4_28

18. Setio, A.A.A., et al.: Validation, comparison, and combination of algorithms for automatic detection of pulmonary nodules in computed tomography images: the luna16 challenge. Med. Image Anal. **42**, 1–13 (2017). https://doi.org/10.1016/j.media.2017.06.015, https://luna16.grand-challenge.org/

19. Siegel, R.L., Miller, K.D., Fuchs, H.E., Jemal, A.: Cancer statistics. Cancer J. Clin. **71**(1), 7–33 (2021). https://doi.org/10.3322/caac.21654

20. Ulku, I., Akagunduz, E.: A survey on deep learning-based architectures for semantic segmentation on 2D images. arXiv preprint arXiv:1912.10230 (2019)
21. Wu, J., Qian, T.: A survey of pulmonary nodule detection, segmentation and classification in computed tomography with deep learning techniques. J. Med. Artif. Intell. **2**(8), 1–12 (2019). https://doi.org/10.21037/jmai.2019.04.01
22. Zhang, P., Zhong, Y., Deng, Y., Tang, X., Li, X.: A survey on deep learning of small sample in biomedical image analysis. arXiv preprint arXiv:1908.00473 (2019)

Optimized Fusion of CNNs to Diagnose Pulmonary Diseases on Chest X-Rays

Valerio Guarrasi[1,2(✉)] and Paolo Soda[1]

[1] Unit of Computer Systems and Bioinformatics, Department of Engineering,
University Campus Bio-Medico of Rome, Rome, Italy
{valerio.guarrasi,p.soda}@unicampus.it
[2] Department of Computer, Control, and Management Engineering,
Sapienza University of Rome, Rome, Italy

Abstract. Since the beginning of the COVID-19 pandemic, more than 350 million cases and 5 million deaths have occurred. Since day one, multiple methods have been provided to diagnose patients who have been infected. Alongside the gold standard of laboratory analyses, deep learning algorithms on chest X-rays (CXR) have been developed to support the COVID-19 diagnosis. The literature reports that convolutional neural networks (CNNs) have obtained excellent results on image datasets when the tests are performed in cross-validation, but such models fail to generalize to unseen data. To overcome this limitation, we exploit the strength of multiple CNNs by building an ensemble of classifiers via an optimized late fusion approach. To demonstrate the system's robustness, we present different experiments on open source CXR datasets to simulate a real-world scenario, where scans of patients affected by various lung pathologies and coming from external datasets are tested. Promising performances are obtained both in cross-validation and in external validation, obtaining an average accuracy of 93.02% and 91.02%, respectively.

Keywords: COVID-19 · X-ray · Convolutional neural networks · Fusion of classifiers

1 Introduction

Since the start of the pandemic, the [1] recorded more than 350 million cases and 5 million deaths caused by the acute respiratory syndrome COVID-19. To control and reduce the spread of the pandemic, different testing modalities, like the reverse transcriptase-polymerase chain reaction (RT-PCR), have been introduced to validate the presence of the virus in patients.

Further to laboratory tests, there have also been efforts to use medical images as a means to diagnose COVID-19 pneumonia [2], mainly using computed tomography (CT) and chest X-ray (CXR) scans. The choice of the imaging modality carries pros and cons. Thanks to its high specificity and its facility of recognizing the different stages of the pathology, CT is the key modality for diagnosing lung pathologies. However, on CT scans it is hard to differentiate between COVID-19 positive patients and those affected by other lung pathologies [2]. Moreover, with CT scanning there is a high risk of contamination for both patients and clinicians, since the cleaning procedure of the scanners is not trivial. Conversely, although CXR has less sensitivity than CT, it is more used for its cost-effectiveness, compactness and limited cross-infection. With the CXR modality there is also the possibility of using portable scanners, useful in emergency care units or at the patients' house, facilitating the control of the virus also in underdeveloped countries.

Over the last decade, deep-learning (DL) has demonstrated to be one of the best solutions to overcome challenges coming from multiple fields of study since it can extrapolate information from data useful for the task at hand [7,22]. Therefore, during the COVID-19 pandemic, researchers have developed DL models able to diagnose COVID-19 on CXR. The state-of-the-art has focused mainly on two classification tasks. The first detects COVID-19 pneumonia in a binary classification task distinguishing between images of patients suffering from COVID-19 and those not affected by this disease, including healthy subjects and those affected by other pneumonia. This task is shortly referred to as COVID-19 vs. non-COVID-19 in the following. The second aims to discriminate images of patients affected by COVID-19 pneumonia, other types of pneumonia and healthy subjects shortly named COVID-19 vs. Pneumonia vs. Healthy hereinafter. Providing a survey of the work on these tasks is out of the scope of this contribution, but the interested readers can refer to [31] for further details. However, it is worth noting that the analysis of the literature reveals a major limitation: such models do not reflect a real-world where patients affected by different lung diseases, further to pneumonia, arrive at hospitals and are scanned for diagnosis. For instance, a model trained on the non-COVID-19 vs. COVID-19 task, where the non-COVID-19 class includes only healthy patients, is not useful in this scenario since the model is not specific to the COVID-19 diseases. Similarly, in the Healthy vs. Pneumonia vs. COVID-19 task, the algorithm learns how to classify between patients affected by COVID-19, healthy patients and pneumonia, but it is not able to detect other lung diseases. These motivations go hand-in-hand with clinical motivations, which state that it is important to detect if a patient is healthy or is affected by pulmonary disease, discriminating between a generic pulmonary disease, COVID-19 pneumonia and other types of pneumonia [5] since each therapy is different [21].

In the literature, the few papers which extend the 2-class and the 3-class classification tasks to other pulmonary diseases are few in number. In [21] the authors used a pre-trained CNN which uses texture descriptors of CXR images [8,30] to recognize different types of pneumonia: COVID-19, SARS, MERS, Pneumocystis, Streptococcus, Varicella and healthy cases. In [9] the authors proposed a CNN model similar to the InceptionNetV3 that screens COVID-19 positive cases from other types of Pnuemonia, Tubercolosis patients and healthy cases, using a CXR dataset [8,18,25,30]. Finally, in [3] the authors presented a trans-

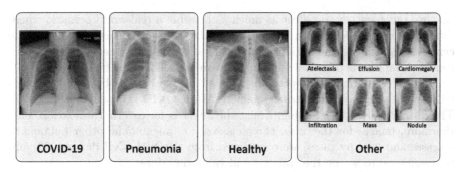

Fig. 1. Example of CXR scans for each class.

fer learning approach working with a pre-trained CNN on a CXR dataset [8,30] to discriminate between four classes: COVID-19 pneumonia, other pneumonia, other diseases and healthy patients.

A general limitation of most approaches processing CXR scans for classification goals is that they do not externally validate the models on never-seen data because only a simple hold-out or a cross-validation (CV) scheme are usually used to compute the performance. This favorably biases the performance concerning a real-world scenario where CXR scans come from different scanners and hospitals, which makes non-trivial the generalization of the model. Such issue is also confirmed by our results reported in the next sections: we find that state-of-the-art CNNs have high performance when tested in CV, but they drop when tested on an external data source. To overcome this limitation, and also to deal with the need of extending the 2- and 3-class task to more classes, in this paper we present a method to algorithmically build an ensemble of pre-trained CNNs that performs a 4-class classification task on CXR scans, where we have patients affected by COVID-19, other pneumonia, other lung diseases and healthy subjects, shortly referred as COVID-19 vs. Pneumonia vs. Other vs. Healthy. Figure 1 shows examples of images belonging to those four classes. The extension to a 4-class scenario may seem straightforward, but given that many lung diseases are collected in the new class, it extends the capabilities of the system, which now can work with a vast number of lung conditions. To make the DL model usable in clinical practice, we present an approach that performs well not only in CV but is also robust to external validation.

The manuscript is organized as follows: Sect. 2 presents the datasets used for training and testing, Sect. 3 shows the proposed method and explains the experiments followed, and Sect. 4 shows and discusses the results obtained, and finally Sect. 5 provides the concluding remarks.

2 Materials

The scientific community has focused on gathering various COVID-19 open-access datasets. Among them, here we focus on those containing CXR images that we put together to reflect as much as possible a real-world scenario where different lung diseases are studied and where the scans are collected from multiple centers, augmenting the variance in the data. As also discussed in [21], inter-center variability is a crucial step [9] to make the algorithms more robust.

We collected images of patients affected by COVID-19 by exploiting two COVID-19 multi-centric datasets, namely AIforCOVID [27] and COVIDX [29]. The former is used for training, whilst the latter is for external validation. Furthermore, images for the other three classes (i.e. pneumonia, other pulmonary diseases and healthy cases) were retrieved from the NIH CXR dataset [30] and they are used to set up the training and validation datasets.

The AIforCOVID dataset [27] is composed of anteroposterior and posteroanterior views of 1100 COVID-19 positive patients, with a mild or severe outcome collected from six different Italian hospitals.

The well-known COVIDX dataset [29] is composed of both COVID-19 positive and negative CXR scans: here we retrieved 16690 scans of the positive class since the non-COVID-19 cases came from the NIH CXR dataset [30].

The NIH CXR dataset [30] contains 112120 CXR images in anteroposterior view collected from NIH clinical center's internal PACS systems: it includes 60361 scans of healthy cases, 1431 scans of patients affected by pneumonia and 50328 scans of cases affected by other lung pathologies, which are atelectasis, cardiomegaly, effusion, infiltration, mass, nodule, and pneumothorax. To have a balanced dataset for training, we randomly selected 1100 images for each of the three classes. The remaining 108820 scans were used for external validation.

To sum up, the dataset used for experiments performed in CV is composed of 1100 scans for each of the four classes, whereas the one used in external validation accounts for 16690 scans for the COVID-19 class, 331 scans for the pneumonia class, 49228 scans for the other lung pathologies class, and 59261 scans for the healthy class.

3   Methods

Our DL approach works with CXR images to perform a 4-class classification task, which discriminates between COVID-19 cases, pneumonia cases, healthy patients and patients affected by other lung diseases, and it algorithmically builds an optimized late fusion ensemble of multiple pre-trained CNNs. The idea stems from observing that today many pre-trained CNNs are available, permitting researchers and practitioners to explore many different deep architectures by exploiting transfer learning even when the available dataset would not permit training from scratch. Furthermore, once several CNNs have been trained, a question arises: is it better to pick the CNN with the best performance on a validation set or to explore the possibility to build an ensemble of CNNs? Indeed, it is well known that in many cases, ensembles of classifiers combined in late fusion have provided better performance than single learners [14]. This happens since fusing multiple models provides complementary and more powerful data representation, and the success of such a mixture relies on having diverse classifiers [14] offering different and complementary points of view to the ensemble.

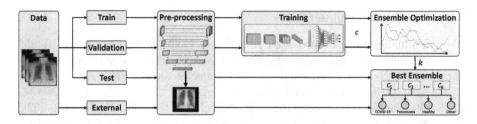

Fig. 2. Schematic representation of our proposal.

Moreover, in the case we opt for the ensemble, there is another question: which are the CNNs to be included in the ensemble? To this end, denoting with n and k the number of available CNNs and the number of CNNs to included in the ensemble, a researcher should explore $\gamma = \sum_{k=2}^{n} \binom{n}{k}$ combinations to find which is the best one. But putting together the CNNs with the largest performance not always retrieves the best ensemble. This happens because the CNNs should provide wrong classifications on the same samples: this phenomenon can be measured by the diversity score, which measures how much the classifications returned by a mixture of classifiers vary on a set of data. In this respect, here we present a multi-objective solution to this search that returns the ensemble and therefore the set of CNNs, maximizing accuracy and diversity scores on a validation set. Figure 2 shows the whole pipeline that is further described in the next subsections.

3.1 Pre-processing

As a first step, there is the need to align the CXR images because they are collected from different centers: the goal is to obtain a cropped image of the bounding box containing the patient's lung, excluding unnecessary regions of the scan. This is performed using a U-Net [23] trained on the Montgomery County CXR collection [18] and the Japanese Society of Radiological Technology repository [25] with a total of 7717 CXR scans of non-COVID-19 patients, which extrapolates the mask of the lung pixels. Given the mask, the cropped image is the minimum squared bounding box containing both lungs.

The U-Net was trained for 100 epochs using a binary cross-entropy loss function and an Adam optimizer after resizing the images to a $3 \times 256 \times 256$ normalized tensor. To prevent overfitting, we applied a random augmentation, which consists of a random rotation ($\pm 25°$), random horizontal and vertical shift (± 25 pixels), and random zoom (0–0.2%). To remove any artifact, we selected the top two biggest segmented regions representing the lungs. We also assessed the U-Net performance by running a 5-fold CV on the two aforementioned datasets that returned an average Dice score equal to 96.32%.

3.2 Training of Single CNNs

We individually trained and tested 20 different CNNs with a stratified 10-fold CV where the train-validation-test split is 70-20-10%. They are well-known state-of-the-art CNNs [20] pre-trained on the ImageNet dataset [10]: AlexNet [19], VGG11, VGG13, VGG16, VGG19 [26], GoogLeNet [28], ResNet18, ResNet34, ResNet50, ResNet101, ResNet152 [15], WideResNet50 [33], ResNeXt50 [32], SqeezeNet1(0), SqeezeNet1(1) [17], DenseNet121, DenseNet161, DenseNet169, DenseNet201 [16], and MobileNetV2 [24].

After the alignment phase, the images are resized to a $3 \times 224 \times 224$ tensor and normalized. To prevent overfitting, during training a random augmentation is performed: random horizontal and vertical random shift (± 7 pixels), flip along the vertical axis, random rotation ($\pm 45°$) and elastic transformation ($\alpha = 20-40$, $\sigma = 7$). All the CNNs are trained using the cross-entropy as loss function, with a maximum of 300 epochs and an early stopping of 25 epochs on the validation set. We used a batch size of 32 and used stochastic gradient descent as optimizer with an initial learning rate of 0.001 and a momentum of 0.9, a learning rate scheduler with a step size of 7, and $\gamma = 0.1$.

3.3 Ensemble Optimization

As already mentioned, the composition of the ensemble is determined by maximizing both the accuracy and the diversity scores provided by the ensemble itself on a validation set. While the accuracy ACC is uniquely defined, the diversity can be measured using different scores heuristically set, which are divided into pairwise and non-pairwise measures, although the former usually perform better than the latter [6]. In this work, the used the pairwise double-fault score $DF_{i,j}$, which is the proportion of samples miss-classified by the classifiers i and j. For a team of c classifiers, the averaged double-fault $\overline{DF}$ over all pairs of classifiers is given by $\overline{DF} = \frac{2}{c(c-1)} \sum_{i=1}^{c-1} \sum_{j=i+1}^{c} DF_{i,j}$. Both ACC and $\overline{DF}$ range in $[0, 1]$, and the higher the values, the more accurate and diverse the models. In practice, given c classifiers collected in the set $\mathbf{C} = \{C_i\}_{i=1}^{c}$ our method looks for the combination of $k \leq c$ models maximizing both the accuracy and the double-fault score $(\overline{DF})$ on a validation set among all the θ possible combinations which are collected in the set $\mathbf{\Theta} = \{\Theta_j\}_{j=1}^{\theta}$, where Θ_j denotes one of the possible mixture of classifiers. The method returns $\hat{\Theta}$ containing the set of k classifiers from $\mathbf{C}$ that constitute the best ensemble so that $\hat{\Theta} = \{C_i \in \Theta_j | j = \arg\min_{\Theta_j \in \mathbf{\Theta}} F\}$, where F is the objective function defined as $F = (1 - ACC(\Theta_j))^2 + (1 - \overline{DF}(\Theta_j))^2$. Let us also notice that, as proofed in [12,13], solving this two-objective minimization problem corresponds to finding the Pareto optimum of the optimization problem that has a unique solution.

Furthermore, the method can work with any aggregation rule combining the outputs provided by the single classifier in the ensemble. In this respect, here we use the majority voting rule, which assigns the most common label among the classifications since it has demonstrated to be the most performing in many

Table 1. Results of single CNNs in CV.

Rank	CNN	ACC	Recall			
			COVID-19	Pneumonia	Other	Healthy
1	VGG13	89.87 ± 0.14	96.34 ± 0.07	87.24 ± 0.16	86.53 ± 0.20	89.35 ± 0.20
2	MobileNetV2	89.80 ± 0.11	97.56 ± 0.04	87.83 ± 0.18	84.38 ± 0.26	89.42 ± 0.16
3	VGG11	89.55 ± 0.01	98.78 ± 0.23	87.25 ± 0.03	84.52 ± 0.08	87.63 ± 0.12
4	DenseNet121	88.27 ± 0.05	100.0 ± 0.00	84.44 ± 0.14	81.44 ± 0.20	87.21 ± 0.30
5	DenseNet201	88.20 ± 0.16	98.78 ± 0.25	85.55 ± 0.18	82.74 ± 0.23	85.74 ± 0.22
6	ResNet34	88.11 ± 0.13	97.56 ± 0.26	85.00 ± 0.02	83.62 ± 0.12	86.26 ± 0.01
7	ResNet101	87.43 ± 0.12	98.78 ± 0.11	83.62 ± 0.25	82.31 ± 0.24	85.00 ± 0.08
8	DenseNet161	87.02 ± 0.04	98.78 ± 0.03	83.89 ± 0.08	80.89 ± 0.05	84.52 ± 0.19
9	GooLeNet	86.76 ± 0.13	98.78 ± 0.22	80.15 ± 0.08	82.19 ± 0.04	85.91 ± 0.13
10	SqueezeNet1(1)	86.61 ± 0.19	96.34 ± 0.27	82.46 ± 0.29	82.31 ± 0.28	85.31 ± 0.22
11	VGG16	86.32 ± 0.11	98.78 ± 0.11	76.30 ± 0.02	83.74 ± 0.03	86.47 ± 0.11
12	WideResNet50	86.25 ± 0.03	97.56 ± 0.04	81.41 ± 0.06	78.41 ± 0.01	87.63 ± 0.15
13	SqueezeNet1(0)	86.14 ± 0.14	97.56 ± 0.20	78.79 ± 0.19	82.92 ± 0.25	85.29 ± 0.17
14	AlexNet	85.75 ± 0.24	97.56 ± 0.14	82.22 ± 0.14	79.23 ± 0.09	84.00 ± 0.28
15	ResNeXt50	85.53 ± 0.28	98.78 ± 0.04	81.61 ± 0.21	79.31 ± 0.15	82.40 ± 0.14
16	VGG19	85.48 ± 0.13	97.56 ± 0.07	76.72 ± 0.26	82.31 ± 0.30	85.31 ± 0.10
17	DenseNet169	85.13 ± 0.04	98.78 ± 0.06	80.28 ± 0.09	79.42 ± 0.11	82.04 ± 0.18
18	ResNet18	84.80 ± 0.08	100.0 ± 0.00	75.74 ± 0.17	80.31 ± 0.14	83.13 ± 0.23
19	ResNet152	84.77 ± 0.15	97.56 ± 0.22	80.01 ± 0.28	77.15 ± 0.19	84.36 ± 0.19
20	ResNet50	83.83 ± 0.11	96.34 ± 0.12	80.31 ± 0.16	77.13 ± 0.04	81.55 ± 0.25

applications [4]. To prevent any tie, we considered only odd values of k in $[3, 20]$, resulting in 9 combinations.

Notice also that, to prevent any bias, the optimization is performed on a validation set without intersections with the test and external validation sets.

4 Results and Discussions

Tables 1, 2, 3 and 4 show all the results achieved in terms of accuracy and recall for each of the four classes. On the one hand, Tables 1 and 2 present the performance attained by each of the 20 CNNs when the experiments were performed in 10 fold CV and on the external dataset, respectively. On the other hand, in the case of CV, each row of Table 3 reports the scores achieved by the ensemble returning the minimum F among all the possible mixture of classifiers in Θ; note that such ensembles were built considering only odd values of k to avoid any ties in the final decision. Furthermore, for any k, we fixed the ensemble in Table 3 and applied it to the external dataset: the corresponding results are shown in Table 4. All such four tables also show in the first column the rank of each row.

Table 2. Results of single CNNs in external validation.

Rank	CNN	ACC	Recall			
			COVID-19	Pneumonia	Other	Healthy
2	VGG13	82.83 ± 0.95	90.26 ± 0.87	80.46 ± 0.93	81.49 ± 0.58	82.24 ± 0.65
1	MobileNetV2	83.68 ± 0.91	91.84 ± 0.57	83.16 ± 0.73	77.65 ± 0.60	84.35 ± 0.45
3	VGG11	82.70 ± 0.52	93.98 ± 0.40	82.07 ± 0.68	78.44 ± 0.89	80.26 ± 0.85
6	DenseNet121	81.64 ± 0.97	94.18 ± 0.79	78.85 ± 0.80	74.74 ± 0.96	80.06 ± 0.66
7	DenseNet201	81.49 ± 0.59	93.97 ± 0.92	78.18 ± 0.60	78.01 ± 0.59	80.17 ± 0.61
5	ResNet34	82.37 ± 0.77	90.75 ± 0.67	80.17 ± 0.80	76.66 ± 0.82	80.83 ± 0.73
4	ResNet101	82.69 ± 0.49	93.01 ± 0.42	77.11 ± 0.69	75.76 ± 0.44	78.32 ± 0.77
8	DenseNet161	80.97 ± 0.45	93.70 ± 0.95	77.10 ± 0.52	74.57 ± 0.56	79.92 ± 0.79
11	GooLeNet	80.86 ± 0.52	93.78 ± 0.80	74.22 ± 0.57	75.02 ± 0.53	79.69 ± 0.52
13	SqueezeNet1(1)	80.13 ± 0.98	90.27 ± 0.65	76.87 ± 0.92	76.84 ± 0.80	78.93 ± 0.51
9	VGG16	80.94 ± 0.94	91.82 ± 0.92	71.33 ± 0.74	76.41 ± 0.45	79.07 ± 0.49
12	WideResNet50	80.42 ± 0.61	91.97 ± 0.76	75.04 ± 0.71	73.35 ± 0.64	82.94 ± 0.63
10	SqueezeNet1(0)	80.94 ± 0.69	91.12 ± 0.61	74.28 ± 0.55	76.64 ± 0.97	78.80 ± 0.65
16	AlexNet	79.47 ± 0.74	92.68 ± 0.79	76.44 ± 0.86	74.28 ± 0.87	78.21 ± 0.60
17	ResNeXt50	79.15 ± 0.71	93.01 ± 0.77	74.71 ± 0.51	73.97 ± 0.42	77.49 ± 0.87
20	VGG19	78.22 ± 0.79	91.89 ± 0.41	70.65 ± 0.52	76.95 ± 0.47	77.91 ± 0.45
15	DenseNet169	79.50 ± 0.85	91.86 ± 0.75	75.17 ± 0.87	72.03 ± 0.59	75.31 ± 0.81
18	ResNet18	78.70 ± 0.91	94.68 ± 0.75	68.98 ± 0.80	72.93 ± 0.79	78.37 ± 0.43
14	ResNet152	80.01 ± 0.76	92.02 ± 0.48	74.31 ± 0.88	70.34 ± 0.77	78.49 ± 0.68
19	ResNet50	78.70 ± 0.53	89.26 ± 0.90	73.78 ± 0.64	71.06 ± 0.70	74.12 ± 0.90

In Table 1 we notice that the models provide satisfactory performance in CV, but Table 2 reveals that they do not generalize well to images belonging from a cohort different from the one used for training, as the drop of 5–8% in accuracy suggests. Similar behavior occurs for the recalls of each class. This observation is strengthened by the fact that the results in external validation are attained on a set of thousands of images, much larger than the set used for training. This finding also confirms previous work showing that DL suffers from this limitation in several bio-medicine applications [11]. This behavior also occurs for the models used by the authors in [3], which worked with the AlexNet, VGG16, and ResNet50.

Let us now focus on the results attained by the optimized ensembles (Tables 3 and 4). We notice in all the cases in Table 3 that the accuracy is larger than 90% and the ensembles, whatever the number of single CNNs used, always outperform the results provided by the single deep networks. This suggests that the ensemble of classifiers successfully exploits the diversity introduced by the different CNNs. Furthermore, the ensemble is robust to external validation since its performance drops to a lesser extent, i.e. around 2–3%, for any k. Among all the ensembles, the combination with the highest accuracy in both the experiments has $k = 3$, showing that the best choice of k can be obtained prior to

Table 3. Results of ensembles in CV.

Rank	k	ACC	Recall			
			COVID-19	Pneumonia	Other	Healthy
1	3	93.02 ± 0.04	98.78 ± 0.21	90.40 ± 0.15	88.51 ± 0.25	94.71 ± 0.13
4	5	91.60 ± 0.22	100.0 ± 0.00	89.38 ± 0.26	87.68 ± 0.07	89.33 ± 0.23
2	7	92.09 ± 0.27	98.78 ± 0.21	89.38 ± 0.26	87.68 ± 0.07	92.59 ± 0.07
3	9	92.08 ± 0.15	98.78 ± 0.21	90.40 ± 0.15	88.51 ± 0.25	90.42 ± 0.29
8	11	90.53 ± 0.25	98.78 ± 0.21	89.38 ± 0.26	87.68 ± 0.07	86.52 ± 0.15
9	13	90.33 ± 0.22	98.78 ± 0.21	89.38 ± 0.26	87.68 ± 0.07	85.46 ± 0.21
7	15	90.77 ± 0.23	98.78 ± 0.21	90.40 ± 0.15	88.51 ± 0.25	85.46 ± 0.21
6	17	91.28 ± 0.17	97.56 ± 0.30	91.42 ± 0.06	89.75 ± 0.28	86.52 ± 0.15
5	19	91.31 ± 0.17	97.56 ± 0.30	91.42 ± 0.06	89.75 ± 0.28	86.52 ± 0.15

Table 4. Results of ensembles in external validation.

Rank	k	ACC	Recall			
			COVID-19	Pneumonia	Other	Healthy
1	3	91.02 ± 0.86	96.24 ± 0.82	87.83 ± 0.42	85.62 ± 0.51	92.60 ± 0.84
5	5	88.84 ± 0.58	97.81 ± 0.91	86.99 ± 0.52	85.46 ± 0.45	86.58 ± 0.86
3	7	89.94 ± 0.96	95.99 ± 0.57	86.42 ± 0.40	85.26 ± 0.89	90.49 ± 0.58
2	9	90.03 ± 0.73	96.44 ± 0.83	87.89 ± 0.95	85.84 ± 0.50	88.17 ± 0.57
9	11	87.53 ± 0.74	96.44 ± 0.58	86.58 ± 0.98	85.50 ± 0.85	83.68 ± 0.87
8	13	87.73 ± 0.85	96.40 ± 0.66	87.29 ± 0.55	85.29 ± 0.51	83.12 ± 0.52
6	15	88.51 ± 0.45	95.79 ± 0.63	88.28 ± 0.83	86.07 ± 0.98	82.90 ± 0.84
7	17	88.44 ± 0.93	94.66 ± 0.94	89.25 ± 0.66	86.93 ± 0.42	84.48 ± 0.52
4	19	89.21 ± 0.68	94.90 ± 0.47	88.79 ± 0.73	87.07 ± 0.41	84.31 ± 0.57

the external validation. This combination is composed of the VGG11, ResNet34, and DenseNet161 CNNs, which do not correspond to the top 3 single models.

Furthermore, we assess if the performance of the CNNs and of the ensembles are different. To this goal, rather than focusing on the best models, we run the pairwise t-test between the distributions of each performance score. In other words, both in CV and external validation, we compare the performance scores between the single CNNs and the ensembles, finding that they are statistically different (p-value ≤ 0.05). In particular, this result holds not only for the global accuracy but also for the recall of each class. The statistical significance of the performance differences is also confirmed when comparing the best CNN and the best ensemble.

Finally, we deepen how the optimization works on the validation set. To this goal, the left panel in Fig. 3 shows the values of accuracy and diversity computed for any of the θ ensembles we tested. Straightforwardly, the best ensemble

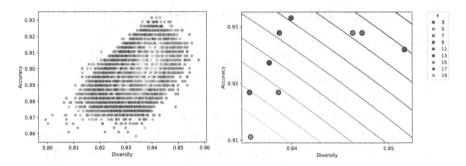

Fig. 3. Left panel: diversity vs accuracy plot of all the possible Θ_j for all ks. Right panel: diversity vs accuracy plot of the optimal $\hat{\Theta}$ for each k. The objective function F is plotted in red, the darker the color the better the value. (Color figure online)

is the one lying in the top right corner of this plot. Furthermore, we notice that the lowest diversity values correspond to the lowest accuracy, confirming the empirical observation that any mixture of classifiers should include learners making errors on different samples. Observing the colors, we notice that there is a concentric scheme as k rises, showing that ensembles of lower values of k have a higher range of accuracy and diversity. This suggests that randomly picking three or even more CNNs and including them in an ensemble does not guarantee to get larger performance than using one of the CNNs. To further prove the importance of having the diversity in the objective function F we also performed an ablation experiment where its contribution is neglected: this means that the composition of the best ensemble is determined only by maximizing the accuracy. The results, not shown here for space reason, reveal that for any k the performance of an ensemble built using our F is better than those attained maximizing only the accuracy. Let us now focus on the right panel, which zooms the plot close to the top right corner, showing level curves that correspond to points where F is constant. As already reported, the two-objective optimization problem is solved by an ensemble of 3 classifiers. We also notice that sub-optimal performances are attained not by other ensembles with other three classifiers but, rather, by ensembles with more classifiers. Nevertheless, the positions of the colored circles confirm that maximizing one of the two scores is not enough to get the best performance. Indeed we note that the diversity drops as the accuracy increases. Furthermore, as the number of classifiers in the mixture increases, the multi-objective function F drops, and in some cases also the diversity and the accuracy drop. This empirically suggests that ensuring diversity while keeping large accuracy becomes more difficult as the number of classifiers in the ensemble increases.

5   Conclusions

In this manuscript we presented an approach to build an optimized ensemble combining several CNNs via a late fusion approach. The goal is to obtain a classifier robust to CXR scans of multiple pulmonary diseases coming from multiple data sources, as it happens in the real world. In an effort to deploy the solution in practice, the results on the one side show that our approach is able to generalize to unseen data, overcoming the limits of single classifiers. On the other side, the rankings shown in Tables 3 and 4 reveal that the best ensemble in CV is also the best in the external validation, an observation that does not hold in the case of single CNNs, confirming the robustness of the method. Future works are directed towards the external validation on other public, as well as to extend the number of classes.

Acknowledgements. This work is partially funded by: POR CAMPANIA FESR 2014–2020, AP1-OS1.3 project "Protocolli TC del torace a bassissima dose e tecniche di intelligenza artificiale per la diagnosi precoce e quantificazione della malattia da COVID-19" CUP D54I20001410002; EU project "University-Industrial Educational Centre in Advanced Biomedical and Medical Informatics (CeBMI) No. 612462-EPP-1-2019-1-SK-EPPKA2-KA"; "AI against COVID-19 Competition", organized by IEEE SIGHT Montreal, Vision and Image Processing Research Group of the University of Waterloo, and DarwinAI Corp., and sponsored by Microsoft.

References

1. Worldmeters coronavirus. https://www.worldometers.info/coronavirus/. Accessed 01 Feb 2022
2. Aljondi, R., et al.: Diagnostic value of imaging modalities for COVID-19: scoping review. J. Med. Internet Res. **22**(8), e19673 (2020)
3. Basu, S., et al.: Deep learning for screening COVID-19 using chest x-ray images. In: 2020 IEEE Symposium Series on Computational Intelligence (SSCI), pp. 2521–2527. IEEE (2020)
4. Brown, G., et al.: Diversity creation methods: a survey and categorisation. Inf. Fusion **6**(1), 5–20 (2005)
5. Brunese, L., et al.: Explainable deep learning for pulmonary disease and coronavirus COVID-19 detection from x-rays. Comput. Methods Programs Biomed. **196**, 105608 (2020)
6. Cavalcanti, G.D., et al.: Combining diversity measures for ensemble pruning. Pattern Recogn. Lett. **74**, 38–45 (2016)
7. Cipollari, S., et al.: Convolutional neural networks for automated classification of prostate multiparametric magnetic resonance imaging based on image quality. J. Magn. Reson. Imaging **55**(2), 480–490 (2022)
8. Cohen, J.P., et al.: COVID-19 image data collection: prospective predictions are the future. arXiv preprint arXiv:2006.11988 (2020)
9. Das, D., Santosh, K.C., Pal, U.: Truncated inception net: COVID-19 outbreak screening using chest x-rays. Phys. Eng. Sci. Med. **43**(3), 915–925 (2020). https://doi.org/10.1007/s13246-020-00888-x
10. Deng, J., et al.: ImageNet: a large-scale hierarchical image database. In: 2009 IEEE Conference on Computer Vision and Pattern Recognition, pp. 248–255. IEEE (2009)
11. Futoma, J., et al.: The myth of generalisability in clinical research and machine learning in health care. Lancet Digit. Health **2**(9), e489–e492 (2020)

12. Guarrasi, V., et al.: A multi-expert system to detect COVID-19 cases in x-ray images. In: 2021 IEEE 34th International Symposium on Computer-Based Medical Systems (CBMS), pp. 395–400. IEEE (2021)
13. Guarrasi, V., et al.: Pareto optimization of deep networks for COVID-19 diagnosis from chest x-rays. Pattern Recogn. **121**, 108242 (2022)
14. Hansen, L.K., et al.: Neural network ensembles. IEEE Trans. Pattern Anal. Mach. Intell. **12**(10), 993–1001 (1990)
15. He, K., et al.: Deep residual learning for image recognition. In: Proceedings of the IEEE Conference on Computer Vision and Pattern Recognition, pp. 770–778 (2016)
16. Huang, G., et al.: Densely connected convolutional networks. In: Proceedings of the IEEE Conference on Computer Vision and Pattern Recognition, pp. 4700–4708 (2017)
17. Iandola, F.N., et al.: SqueezeNet: AlexNet-level accuracy with 50x fewer parameters and <0.5MB model size. arXiv preprint arXiv:1602.07360 (2016)
18. Jaeger, S., et al.: Two public chest x-ray datasets for computer-aided screening of pulmonary diseases. Quant. Imaging Med. Surg. **4**(6), 475 (2014)
19. Krizhevsky, A.: One weird trick for parallelizing convolutional neural networks. arXiv preprint arXiv:1404.5997 (2014)
20. Litjens, G., et al.: A survey on deep learning in medical image analysis. Med. Image Anal. **42**, 60–88 (2017)
21. Pereira, R.M., et al.: COVID-19 identification in chest x-ray images on flat and hierarchical classification scenarios. Comput. Methods Programs Biomed. **194**, 105532 (2020)
22. Pouyanfar, S., et al.: A survey on deep learning: algorithms, techniques, and applications. ACM Comput. Surv. (CSUR) **51**(5), 1–36 (2018)
23. Ronneberger, O., Fischer, P., Brox, T.: U-net: convolutional networks for biomedical image segmentation. In: Navab, N., Hornegger, J., Wells, W.M., Frangi, A.F. (eds.) MICCAI 2015. LNCS, vol. 9351, pp. 234–241. Springer, Cham (2015). https://doi.org/10.1007/978-3-319-24574-4_28
24. Sandler, M., et al.: MobileNetV2: inverted residuals and linear bottlenecks. In: Proceedings of the IEEE Conference on Computer Vision and Pattern Recognition, pp. 4510–4520 (2018)
25. Shiraishi, J., et al.: Development of a digital image database for chest radiographs with and without a lung nodule: receiver operating characteristic analysis of radiologists' detection of pulmonary nodules. Am. J. Roentgenol. **174**(1), 71–74 (2000)
26. Simonyan, K., et al.: Very deep convolutional networks for large-scale image recognition. arXiv preprint arXiv:1409.1556 (2014)
27. Soda, P., et al.: AIforCOVID: predicting the clinical outcomes in patients with COVID-19 applying AI to chest-x-rays. An Italian multicentre study. Med. Image Anal. **74**, 102216 (2021)
28. Szegedy, C., et al.: Going deeper with convolutions. In: Proceedings of the IEEE Conference on Computer Vision and Pattern Recognition, pp. 1–9 (2015)
29. Wang, L., et al.: Covid-net: a tailored deep convolutional neural network design for detection of COVID-19 cases from chest x-ray images. Sci. Rep. **10**(1), 1–12 (2020)
30. Wang, X., et al.: ChestX-ray8: hospital-scale chest x-ray database and benchmarks on weakly-supervised classification and localization of common thorax diseases. In: Proceedings of the IEEE Conference on Computer Vision and Pattern Recognition, pp. 2097–2106 (2017)

31. Wynants, L., et al.: Prediction models for diagnosis and prognosis of COVID-19 infection: systematic review and critical appraisal. Brit. Med. J. **369** (2020)
32. Xie, S., et al.: Aggregated residual transformations for deep neural networks. In: Proceedings of the IEEE Conference on Computer Vision and Pattern Recognition, pp. 1492–1500 (2017)
33. Zagoruyko, S., et al.: Wide residual networks. arXiv preprint arXiv:1605.07146 (2016)

High/Low Quality Style Transfer for Mutual Conversion of OCT Images Using Contrastive Unpaired Translation Generative Adversarial Networks

Mateo Gende[1,2]([✉])[ID], Joaquim de Moura[1,2][ID], Jorge Novo[1,2][ID], and Marcos Ortega[1,2][ID]

[1] Centro de investigación CITIC, Universidade da Coruña, A Coruña, Spain
{m.gende,joaquim.demoura,jnovo,mortega}@udc.es
[2] Grupo VARPA, Instituto de Investigación Biomédica de A Coruña (INIBIC), Universidade da Coruña, A Coruña, Spain

Abstract. Recent advances in artificial intelligence and deep learning models are contributing to the development of advanced computer-aided diagnosis (CAD) systems. In the context of medical imaging, Optical Coherence Tomography (OCT) is a valuable technique that is able to provide cross-sectional visualisations of the ocular tissue. However, OCT is constrained by a limitation between the quality of the visualisations that it can produce and the overall amount of tissue that can be analysed at once. This limitation leads to a scarcity of high quality data, a problem that is very prevalent when developing machine learning-based CAD systems intended for medical imaging. To mitigate this problem, we present a novel methodology for the unpaired conversion of OCT images acquired with a low quality extensive scanning preset into the visual style of those taken with a high quality intensive scan and vice versa. This is achieved by employing contrastive unpaired translation generative adversarial networks to convert between the visual styles of the different acquisition presets. The results we obtained in the validation experiments show that these synthetic generated images can mirror the visual features of the original ones while preserving the natural tissue texture, effectively increasing the total number of available samples that can be used to train robust machine learning-based CAD systems.

This research was funded by Instituto de Salud Carlos III, Government of Spain, DTS18/00136 research project; Ministerio de Ciencia e Innovación y Universidades, Government of Spain, RTI2018-095894-B-I00 research project; Ministerio de Ciencia e Innovación, Government of Spain through the research project with reference PID2019-108435RB-I00; Consellería de Cultura, Educación e Universidade, Xunta de Galicia, Grupos de Referencia Competitiva, grant ref. ED431C 2020/24, predoctoral grant ref. ED481A 2021/161 and postdoctoral grant ref. ED481B 2021/059; Axencia Galega de Innovación (GAIN), Xunta de Galicia, grant ref. IN845D 2020/38; CITIC, Centro de Investigación de Galicia ref. ED431G 2019/01, receives financial support from Consellería de Educación, Universidade e Formación Profesional, Xunta de Galicia, through the ERDF (80%) and Secretaría Xeral de Universidades (20%).

S. Sclaroff et al. (Eds.): ICIAP 2022, LNCS 13231, pp. 210–220, 2022.
https://doi.org/10.1007/978-3-031-06427-2_18

Keywords: Optical coherence tomography · Generative adversarial networks · Style transfer · Synthetic images

1   Introduction

According to the World Health Organisation, more than 2.2 billion people suffer from vision impairment. Of these, at least 1 billion have a condition that could have been prevented or is yet to be addressed. Conversely, a considerable proportion of people with eye conditions who receive timely diagnosis and treatment do not develop blindness [25]. This motivates the need for accurate and accessible diagnostic tools that can detect these diseases in their early stages.

Recent advances in medical imaging techniques, plus the development of new and improved machine learning algorithms, are contributing to build Computer-Aided Diagnosis (CAD) systems that can help to recognise and prevent these conditions. These algorithms allow deep learning models to be trained directly from annotated data. This greatly simplifies the development of CAD systems at the cost of requiring vast amounts of training images that, unfortunately, are not always readily available. The problem of data scarcity affects many domains of application of deep learning [20], but it is especially prevalent in medical imaging [13], due to the sensitive nature of the data and its acquisition costs.

In the context of medical imaging, Optical Coherence Tomography (OCT) is a non-invasive imaging technique that can obtain volumetric digitalisations of the tissue of the eyes. Due to its ability to visualise cross-sections of relevant pathological structures, OCT has been used extensively to diagnose ocular diseases [18,19] such as diabetic macular edema [24] or glaucoma [9,23]. While OCT-based diagnosis is usually performed by the clinician visually inspecting several images, this task is considerably labour-intensive and time-consuming, as well as subjective in nature.

To mitigate this problem and to aid the experts in their undertaking, many OCT-based CAD systems make use of machine learning algorithms and deep learning models [3,15,22], all while achieving results that are equal to or better than board-certified specialists [6,11,13,21]. The functions of such systems can range from the detection and visualisation of pathological structures [5], to the segmentation and measurement of anatomical parts of the eye that are relevant for diagnosis [4] or the automatic diagnosis of patients in screening tests [3]. However, as mentioned above, the development process of these systems requires a considerable amount of data.

To obtain the images, an OCT scanner sweeps the retina of the patient via interferometry with a low coherence beam of light, obtaining samples for each point and usually creating a visualisation based on the average of said samples. The number of samples that is taken has a direct effect on the image quality, with longer scans over a smaller area of the retina providing higher quality images than those that are taken over a larger surface, while the latter allow for a wider portion of the eye to be analysed at once. To simplify the operation, manufacturers typically provide different scanning presets so that the clinicians

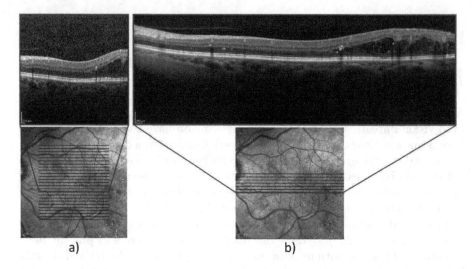

Fig. 1. Example of two slices from different scanning settings taken from the same location. a) *"Macular Cube"* preset. b) *"Seven Lines"* preset.

can choose which kind of scan they want to perform based on the needs and the condition of the patient, as well as the time they have available. This usually tends to result in either small, high-quality volumes that cover a long and narrow part of the retina or noisier, lower-resolution images that can cover most of the retina surface. Figure 1 shows an example of the two most representative scan presets used in clinical practice, displaying a *"Macular Cube"* preset where 25 OCT slices are sampled at a low rate, resulting in low resolution, noisy images of the whole retina and a *"Seven Lines"* preset where only 7 slices are sampled at a higher rate and quality over a narrow strip of tissue. These two kinds of presets are the most prevalent in medical services, reference works and publicly available datasets.

This compromise between OCT slice quality and sampled area places a limit on the amount of high quality data that can be obtained in a single session, in this way contributing to the problem of data scarcity that was mentioned above. This problem has been approached by several authors by applying super-resolution or reducing the speckle noise present in the images, for reference [1,8,12,26], with varying degrees of success at enhancing the quality of extensive OCT scans. Despite these results, however, none of these works have addressed the visual differences that exist between images acquired with different scanning presets. Moreover, a machine learning system that has been trained with images acquired with a particular preset may not be able to perform as well when applied to data from another configuration due to the visual differences between images of different presets.

In this work, we present a novel fully automatic methodology to address the issue of data scarcity in OCT imaging. By making use of a Contrastive

Unpaired Translation Generative Adversarial Network (CUT-GAN) architecture, the visual style of images acquired with a superior quality preset can be transferred to data obtained with a more extensive, lower quality preset and vice versa. This way, the total number of available samples can be effectively multiplied, contributing to mitigate this problem of data scarcity that is so prevalent when developing OCT-based CAD systems. Furthermore, this methodology has the potential to create multi-preset datasets which can then be used to train CAD systems in a robust manner, without the need to obtain samples acquired with every preset the system may encounter in service. To the best of our knowledge, this proposal represents the only study designed specifically for the mutual conversion of OCT samples acquired with different scanning presets most commonly used in the health services.

2 Methodology

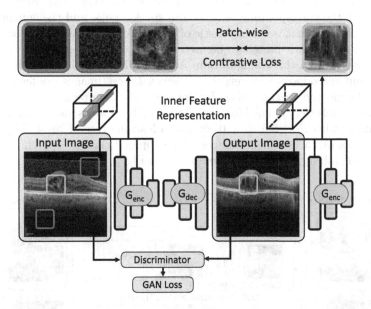

Fig. 2. Patch-wise contrastive learning for OCT images. Contrastive loss is obtained by comparing the inner representations of patches extracted from the input and output images.

To convert the images between the two scanning presets, a CUT-GAN architecture was used [17]. This architecture is able to perform unpaired image-to-image translation between two classes by training in a patch-wise manner while trying to preserve the mutual information in both input and output images. By training this architecture on our two classes corresponding to different acquisition presets, we can obtain models that are able to perform a "style transfer", conferring the visual features of any preset to images taken with another.

This can be achieved by employing a patch-wise contrastive loss. As an image is propagated through the network, patches of its inner representation, taken as the output of the encoder part of the generative network, are compared with the inner representation of the corresponding synthetic generated image. Similar inner representations will minimise this contrastive loss. This way, we can leverage the potential of the encoder part of the network to capture features that are common to both presets, such as the choroid or the location of the inner limiting membrane, while taking advantage of the ability of the decoder part to synthesise preset-specific features such as speckle noise or detailed tissue texture. This process is illustrated in Fig. 2.

For this purpose, we trained a model to make the conversion of images acquired with the *"Macular Cube"* preset into those of a visual quality matching the *"Seven Lines"* configuration. This way, the more numerous, lower quality *"Macular Cube"* images can be converted to the visual style of the *"Seven Lines"* preset, with the corresponding reduction in speckle noise and the enhancement in tissue visibility that is characteristic of this more intensive scan, allowing the generation of synthetic images to be used for over-sampling as if they were originally acquired with this configuration. Additionally, we trained another model to carry out the opposite conversion of *"Seven Lines"* images, conferring them the visual style and speckle noise of *"Macular Cube"* samples.

These models were trained for a maximum of 400 epochs, using Adam [10] as an optimiser with $\beta_1 = 0.5, \beta_2 = 0.999$ and a learning rate of $2e-4$. These were used to generate the synthetic images of the opposed class by applying them to the original training samples, as represented in Fig. 3.

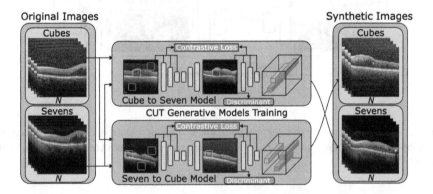

Fig. 3. Diagram showing the synthetic image generation methodology. By training two CUT generative models, we can then transform original *"Macular Cube"* and *"Seven Lines"* images into synthetic images of the opposed class.

3   Results and Discussion

To train and validate the CUT-based synthetic image generator models, we used 240 OCT slices acquired from different patients in accordance with the Declaration of Helsinki, approved by the local Ethics Committee of Investigation from A Coruña/Ferrol (2014/437) the 24th of November, 2014. The platform used to acquire the images was a Heidelberg SPECTRALIS® OCT scanner configured with the two aforementioned presets shown in Fig. 1. In particular, 120 images belonged to the *"Macular Cube"* class with lower overall quality and resolution, while the remaining 120 were acquired with the *"Seven Lines"* preset. These images were resized to a resolution of 256 × 256 pixels and then used to train both of the CUT-GAN models. The loss curves for the 400 epochs of training are shown in Fig. 4.

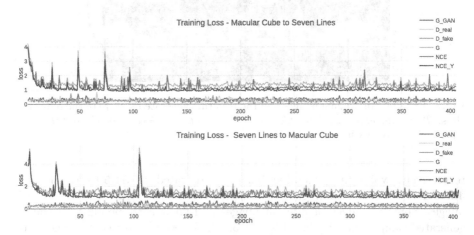

Fig. 4. Training losses for both generative models. *G_GAN*: generator GAN loss, *D_real*: discriminator loss for real images, *D_fake*: discriminator loss for fake images, *G*: generator loss, *NCE*: Noise Contrastive Estimation loss for images of the original class, *NCE_Y*: NCE loss for images of the target class.

To evaluate and assess the quality of the synthetic data, an experiment was performed using the Blind/Referenceless Image Spatial QUality Evaluator BRISQUE [14], an automatic image quality evaluator commonly used in medical imaging literature for similar purposes [2,27,28]. BRISQUE does not require a reference image to perform a comparison unlike other metrics such as Structural Similarity Index Measure or Peak Signal to Noise Ratio. Instead, when analysing a single image, it returns a score that is indicative of the perceptual quality of said image. The higher the perceived quality, the lower the BRISQUE score.

In the experiment, the quality score of every original image was assessed and compared to that of the synthetic ones. The original *"Cube"* and *"Seven"* images achieved an average BRISQUE score of **80.36 ± 25.05** and **42.42 ± 9.29**

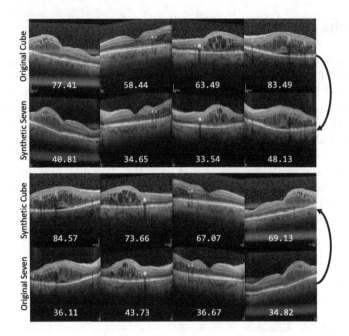

Fig. 5. Examples of original images and the resulting generated picture of the opposite class. *Top*: *"Macular Cube"* to *"Seven Lines"*. *Bottom*: *"Seven Lines"* to *"Macular Cube"*. Numbers in white indicate the BRISQUE score.

respectively, while the generated images were rated at **84.59 ± 15.47** and **42.91 ± 7.94**, correspondingly, to synthetic generated *"Cube"* and *"Seven"* images. Some examples of original images and their corresponding synthetic generated samples are displayed in Fig. 5. These results show that the generative networks are able to mimic the visual features of the target classes whilst preserving the original tissue structure. Since the available dataset is unpaired, no reference image-based quality assessment can be performed.

The generative adversarial network training process involves the use of a discriminator that distinguishes between original and synthetic images. The aim of this discriminator is to guide the training process by providing a loss component which penalises generated images that do not mimic the visual features of the original images correctly. However, this discriminator is intentionally simple, in order to allow the generative model to progressively adjust during training. Furthermore, the training process forces the discriminator to progressively learn the visual features that distinguish original and synthetic images. This leaves this discriminator biased towards the training process, which aims to a relative stability in terms of losses, as can be seen in Fig. 4. Because of its simplicity and bias, this discriminator is a poor candidate for assessing the separability of the synthetic generated images. In an effort to more objectively assess the validity of the synthetic samples, an independent separability test was performed. The aim of this separability test is to verify whether the synthetic generated images display the visual features that characterise each of the presets. In this test, a classifier is first trained to classify the original images according to the visual features that they display, learning the characteristics that distinguish each preset. When tested on the synthetic generated images, these should be classified according to the visual features of the preset they are converted to, instead of those corresponding to the scanning preset that was used to acquire them. By using a Densely Connected Convolutional Network [7] architecture, which has demonstrated its advantages for classification tasks in medical imaging [16], we can achieve an objective validation of the separability of the synthetic data.

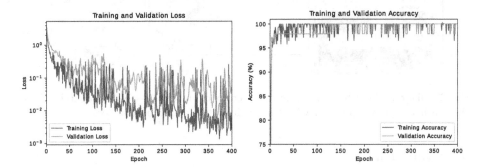

Fig. 6. Training and validation losses in logarithmic scale and accuracies for the DenseNet-121 classifier. The model with the lowest validation loss is marked with a red star. (Color figure online)

In this separability test, a DenseNet-121 classifier was trained to discern between *"Macular Cube"* and *"Seven Lines"* images from the original dataset. This dataset was partitioned with 60% of the samples being used for training, 20% for validation and the remaining 20% to test the model. The model was trained for a maximum of 400 epochs on the training samples, and the checkpoint which performed best on the validation set was selected for testing. During this training, we used Adam [10] as an optimiser, with a learning rate of 2e−4, $\beta_1 = 0.5$, and $\beta_2 = 0.999$. The training and validation loss and accuracy curves are displayed in Fig. 6. The loss curves are represented in logarithmic scale in order to provide a better visualisation.

Afterwards, the synthetic images that were generated by the CUT-GAN models were assigned to their transferred class and tested with this classifier, comparing the results with those obtained with the test that used original images. This process, and the results that were obtained, are represented in Fig. 7. These satisfactory results show that the generative models are able to synthesise images that can successfully display the visual characteristics of the target class.

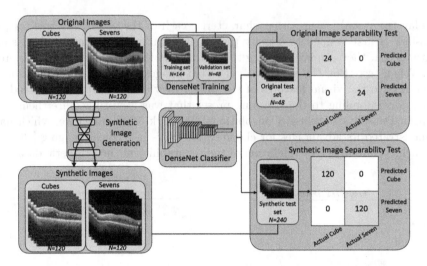

Fig. 7. Diagram describing the tests that were performed to assess the separability of the synthetic images, together with the resulting test confusion matrices.

4   Conclusions

Almost half of the population that suffers from a form of visual impairment is under-diagnosed or affected by a preventable condition. OCT imaging can be used to create CAD systems that help to diagnose and assess these pathologies. However, the necessary compromise between image quality and the amount of tissue that can be analysed in a single session leads to a scarcity of high quality data.

In this work, we present a fully automatic methodology for the mutual "style transfer" of OCT images taken from the two most widely used scanning presets, constituting the first example of such conversion in the literature. This method is able to transform noisy, low quality *"Macular Cube"* OCT scans into cleaner *"Seven Lines"*-quality scans while preserving tissue texture, as well as to perform the inverse transformation. The obtained test results show that these synthetic images achieve the visual features of their assigned class, with a very similar BRISQUE score when compared with original images. Additionally, the results that were produced by the separability test indicate that these synthetic generated images are clearly separable into their respective target classes. The ability to mutually transfer the style of multiple presets has the potential to produce multi-preset datasets with which to train more robust CAD systems. Moreover, the potential of this methodology is not limited to OCT imaging, as it is applicable to many other imaging techniques such as angiographies or retinographies.

As future work, our plans encompass the validation of this synthetic image generation methodology with human experts, verifying if ophthalmologists are able to distinguish the original images from the synthetic generated ones. Additionally, we plan to evaluate this generative methodology on screening tests by supplying single-preset datasets with synthetic generated images converted to other presets, assessing the benefits these may provide on the training of robust CAD systems designed for the screening of ocular pathologies.

References

1. Apostolopoulos, S., et al.: Automatically enhanced OCT scans of the retina: a proof of concept study. Sci. Rep. **10**(1) (2020). https://doi.org/10.1038/s41598-020-64724-8
2. Chaabouni, A., Gaudeau, Y., Lambert, J., Moureaux, J.M., Gallet, P.: Subjective and objective quality assessment for h264 compressed medical video sequences. In: 2014 4th International Conference on Image Processing Theory, Tools and Applications (IPTA), pp. 1–5 (2014). https://doi.org/10.1109/IPTA.2014.7001922
3. Cheung, C.Y., Tang, F., Ting, D.S.W., Tan, G.S.W., Wong, T.Y.: Artificial intelligence in diabetic eye disease screening. Asia-Pac. J. Ophthalmol. (2019). https://doi.org/10.22608/apo.201976
4. Fu, H., et al.: Joint optic disc and cup segmentation based on multi-label deep network and polar transformation. IEEE Trans. Med. Imaging **37**(7), 1597–1605 (2018). https://doi.org/10.1109/tmi.2018.2791488
5. Gende, M., De Moura, J., Novo, J., Charlón, P., Ortega, M.: Automatic segmentation and intuitive visualisation of the epiretinal membrane in 3D OCT images using deep convolutional approaches. IEEE Access **9**, 75993–76004 (2021). https://doi.org/10.1109/ACCESS.2021.3082638
6. Gulshan, V., et al.: Development and validation of a deep learning algorithm for detection of diabetic retinopathy in retinal fundus photographs. JAMA **316**(22), 2402 (2016). https://doi.org/10.1001/jama.2016.17216
7. Huang, G., Liu, Z., Maaten, L.V.D., Weinberger, K.Q.: Densely connected convolutional networks. In: 2017 IEEE Conference on Computer Vision and Pattern Recognition (CVPR), pp. 2261–2269. IEEE, July 2017. https://doi.org/10.1109/cvpr.2017.243
8. Huang, Y., et al.: Simultaneous denoising and super-resolution of optical coherence tomography images based on generative adversarial network. Opt. Express **27**(9), 12289 (2019). https://doi.org/10.1364/oe.27.012289
9. Kamalipour, A., Moghimi, S.: Macular optical coherence tomography imaging in glaucoma. J. Ophthalmic Vis. Res. (2021). https://doi.org/10.18502/jovr.v16i3.9442
10. Kingma, D.P., Ba, J.: Adam: a method for stochastic optimization. In: Bengio, Y., LeCun, Y. (eds.) 3rd International Conference on Learning Representations, ICLR 2015, San Diego, CA, USA, 7–9 May 2015, Conference Track Proceedings (2015). http://arxiv.org/abs/1412.6980
11. Lee, J.H., Kim, Y.T., Lee, J.B., Jeong, S.N.: A performance comparison between automated deep learning and dental professionals in classification of dental implant systems from dental imaging: a multi-center study. Diagnostics **10**(11), 910 (2020). https://doi.org/10.3390/diagnostics10110910
12. Li, M., Idoughi, R., Choudhury, B., Heidrich, W.: Statistical model for oct image denoising. Biomed. Opt. Express **8**(9), 3903–3917 (2017)
13. Litjens, G., et al.: A survey on deep learning in medical image analysis. Med. Image Anal. **42**, 60–88 (2017). https://doi.org/10.1016/j.media.2017.07.005

14. Mittal, A., Moorthy, A.K., Bovik, A.C.: No-reference image quality assessment in the spatial domain. IEEE Trans. Image Process. **21**(12), 4695–4708 (2012). https://doi.org/10.1109/TIP.2012.2214050

15. de Moura, J., Novo, J., Ortega, M.: Deep feature analysis in a transfer learning-based approach for the automatic identification of diabetic macular edema. In: 2019 International Joint Conference on Neural Networks (IJCNN). IEEE, July 2019. https://doi.org/10.1109/ijcnn.2019.8852196

16. Nugroho, K.A.: A comparison of handcrafted and deep neural network feature extraction for classifying optical coherence tomography (OCT) images. In: 2018 2nd International Conference on Informatics and Computational Sciences (ICICoS), pp. 1–6 (2018). https://doi.org/10.1109/ICICOS.2018.8621687

17. Park, T., Efros, A.A., Zhang, R., Zhu, J.-Y.: Contrastive learning for unpaired image-to-image translation. In: Vedaldi, A., Bischof, H., Brox, T., Frahm, J.-M. (eds.) ECCV 2020. LNCS, vol. 12354, pp. 319–345. Springer, Cham (2020). https://doi.org/10.1007/978-3-030-58545-7_19

18. Puliafito, C.A., et al.: Imaging of macular diseases with optical coherence tomography. Ophthalmology **102**(2), 217–229 (1995)

19. Schmitt, J.: Optical coherence tomography (OCT): a review. IEEE J. Sel. Top. Quant. Electron. **5**(4), 1205–1215 (1999). https://doi.org/10.1109/2944.796348

20. Shorten, C., Khoshgoftaar, T.M.: A survey on image data augmentation for deep learning. J. Big Data **6**(1) (2019). https://doi.org/10.1186/s40537-019-0197-0

21. Ting, D.S.W., et al.: Development and validation of a deep learning system for diabetic retinopathy and related eye diseases using retinal images from multiethnic populations with diabetes. JAMA **318**(22), 2211 (2017). https://doi.org/10.1001/jama.2017.18152

22. Ting, D.S.W., et al.: Artificial intelligence and deep learning in ophthalmology. Br. J. Ophthalmol. **103**(2), 167–175 (2018). https://doi.org/10.1136/bjophthalmol-2018-313173

23. Triolo, G., Rabiolo, A.: Optical coherence tomography and optical coherence tomography angiography in glaucoma: diagnosis, progression, and correlation with functional tests. Ther. Adv. Ophthalmol. **12**, 251584141989982 (2020). https://doi.org/10.1177/2515841419899822

24. Vujosevic, S., et al.: Diabetic macular edema with neuroretinal detachment: OCT and OCT-angiography biomarkers of treatment response to anti-VEGF and steroids. Acta Diabetol. **57**(3), 287–296 (2019). https://doi.org/10.1007/s00592-019-01424-4

25. World Health Organization: World Report on Vision. World Health Organization (2019). https://www.who.int/publications/i/item/9789241516570

26. Xu, M., Tang, C., Hao, F., Chen, M., Lei, Z.: Texture preservation and speckle reduction in poor optical coherence tomography using the convolutional neural network. Med. Image Anal. **64**, 101727 (2020). https://doi.org/10.1016/j.media.2020.101727

27. Yu, S., Dai, G., Wang, Z., Li, L., Wei, X., Xie, Y.: A consistency evaluation of signal-to-noise ratio in the quality assessment of human brain magnetic resonance images. BMC Med. Imaging **18**(1) (2018). https://doi.org/10.1186/s12880-018-0256-6

28. Zhang, Z., et al.: Can signal-to-noise ratio perform as a baseline indicator for medical image quality assessment. IEEE Access **6**, 11534–11543 (2018). https://doi.org/10.1109/access.2018.2796632

Real-Time Respiration Monitoring of Neonates from Thermography Images Using Deep Learning

Simon Lyra$^{(\boxtimes)}$ ⓘ, Ines Groß-Weege, Steffen Leonhardt ⓘ, and Markus Lüken ⓘ

Chair for Medical Information Technology,
Helmholtz Institute for Biomedical Engineering, RWTH Aachen University,
Aachen, Germany
lyra@hia.rwth-aachen.de

Abstract. In this work, we present an approach for non-contact automatic extraction of respiration in infants using infrared thermography video sequences, which were recorded in a neonatal intensive care unit. The respiratory signal was extracted in real-time on low-cost embedded GPUs by analyzing breathing-related temperature fluctuations in the nasal region. The automatic detection of the patient's nose was performed using the Deep Learning-based YOLOv4-Tiny object detector. Additionally, the head was detected for movement tracking. A leave-one-out cross validation showed a mean intersection over union of 79% and a mean average precision of 93% for the detection algorithm. Since no clinical reference was provided, the extracted respiratory activity was validated for video sequences without motion artifacts using Farnebäck's Optical Flow algorithm. A mean MAE of 8.5 breaths per minute and a mean F_1-score of 80% for respiration detection were achieved. The model inference on NVIDIA Jetson modules showed a performance of 32 fps on the Xavier NX and 62 fps on the Xavier AGX. These outcomes showed promising results for the real-time extraction of respiratory activity from thermography recordings of neonates using Deep Learning-based techniques on embedded GPUs.

Keywords: Deep Learning · Camera-based respiration monitoring · NICU · Infrared thermography

1 Introduction

The earlier an infant is born, the higher the risk of death or serious disability [27]. Every year, one out of ten neonates is born too early [13]. Due to the immature organ system and different courses of complications and diseases, approximately 1 million neonates die annually [15]. The medical treatment of premature infants in neonatal intensive care units (NICUs) is provided using an incubator, which enables the isolation of the patient from harmful ambient influences, e.g., pathogenic agents or noise. Further, it regulates ambient conditions such as temperature, humidity, and oxygen inside, which is crucial for proper development of the patients [6].

During neonatal care, continuous monitoring of vital signs such as heart rate or respiratory rate (RR) is necessary for the assessment of the patient's health status. Today, all measurement techniques use adhesive sensors, which require direct contact with the patient's skin. Due to the immaturity of the skin, the attached sensors can involve a great risk, because the regular replacement can lead to medical adhesive-related skin injuries (MARSI) [19]. Therefore, the non-contact acquisition of vital signs is investigated by research groups worldwide to provide an alternative for wired patient monitoring in neonatal care.

Next to e.g., radar-based techniques [8,9], camera-based methods experienced an increase in popularity due to wide and low-cost accessibility of imaging devices [12]. While RGB cameras can be used for remote extraction of heart rate or respiratory activity [7], infrared thermography (IRT) devices enable the acquisition of thermoregulatory process by measuring the skin temperature of a patient [10]. Further, the respiration-related temperature changes in the nasal region can be used for breathing detection, which is the focus of this work. In Fig. 1, a heatmap of the data used in this work is presented, where the breathing-related temperature differences can be obtained in the nostrils. During the inspiration, an inflow of (colder) ambient air can be seen, while the expiration shows an outflow of heated air.

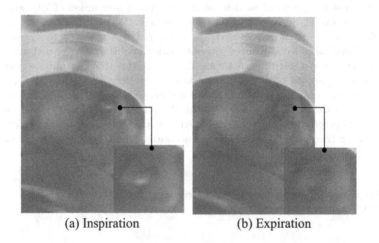

     (a) Inspiration   (b) Expiration

Fig. 1. Heatmap of IRT frames during a) inspiration and b) expiration.

In 2011, Abbas et al. introduced an IRT-based technique to measure temperature deviations around the nostrils for neonates using continuous wavelet transform [1]. Here, a temperature deviation of approx. $0.27\,^{\circ}\text{C}$ was obtained in the nasal region during respiratory activity. Since the nasal region of interest (ROI) was manually defined in the first frame and not changed in position, Pereira et al. improved the approach in 2017 by implementing a tracking algorithm [23]. One year later, the same research group further developed their method and published an algorithm, which automatically detected ROIs containing the respiratory signal [24]. Since a costly high-definition IRT device was used for measurement, Lorato et al. presented studies to show feasibility of measuring respiration in the same ROI using a thermopile array [16] and several low-cost FLIR Lepton IRT devices [17,18]. Although the described methods showed high accuracies for the RR, the real-time capability of the CPU-based algorithms was not the focus, so a clinical implementation could be difficult.

In recent years, Deep Learning (DL)-based algorithms revolutionized real-time image processing in various medical fields [22]. The GPU-based methods not only enable advanced accuracies for challenges such as ROI tracking or segmentation [11], but also show increased computation performance compared to classical approaches, which allows real-time processing even on embedded devices [20]. However, the literature in the field of DL-based real-time extraction of respiration in the neonatal context is limited. In this work, we will present a DL-based algorithm for respiration monitoring of neonates using the YOLOv4-Tiny object detector to track the nasal region for signal extraction in thermography video data recorded in a NICU. A patient-wise leave-one-out cross validation was conducted for analyzing the model accuracies of a pre-trained model (using adult data) and a model trained from scratch. Further, the inference performance was examined on NVIDIA Jetson devices to show real-time capability on embedded low-cost GPUs, which shows promising results for clinical applications.

2   Materials and Methods

2.1   Experimental Setup and Dataset

A video dataset including 19 patients was recorded at the NICU of Saveetha Medical College and Hospital, Chennai, India, while the trial was approved by the institutional ethics committee of Saveetha University (SMC/IEC/2018/03/067). The gestational ages were between 29 and 40 weeks, with actual ages from 37 h to 56 days post birth. Their weights were between 1500 g and 3010 g. Written informed consent was obtained from the parents of all patients. A camera setup was used for recording, which was equipped with the infrared thermography (IRT) camera VarioCAM HD head 820 S (InfraTec, Germany). The thermography sequences were acquired with a temporal resolution of 10 Hz and a format of $1,024 \times 768$ IR pixels. The neonates were recorded with a thermal sensitivity of 20 mK for 10 min using a mobile workstation.

As illustrated in Fig. 2, the neonates were placed next to the camera setup in an open incubator. The setup was attached to a stable stand. Since the camera setup was equipped with additional RGB and monochrome cameras, the measurement area was illuminated using organic LED panels. The benefits of using this light technology for remote vital sign monitoring were described in [21].

For this study, a subset of data from 10 neonatal patients was used for training and validation of the detection model by randomly sampling 100 IRT images

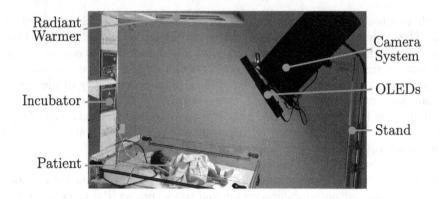

Fig. 2. Measurement setup placed above the neonate.

from every subject. Although the accuracies of detection models using only small IRT datasets were shown to be high in previous work [20], the influence of a pre-training step was analyzed using a subset of the thermal infrared face database from Kopaczka et al. [14]. This dataset provides IRT images from the upper torso and facial regions of adult subjects, while the same IRT camera with equal resolution was used for recording as in our study. Since (to the best of our knowledge) there are no available IRT datasets of neonates or neonatal faces, the adult dataset is used to generally learn facial features in human faces. As shown in Table 1, a subset of 2000 images containing 90 different subjects was used for the pre-training step. This pretrained model was then used in a transfer-learning step to further fine-tune the detector on the neonatal features.

Table 1. Dataset sampling for detector training and validation.

Dataset	Subjects	Images
Neonatal (ours)	10	1000
Kopaczka et al. [14]	90	2000

2.2 Data Preprocessing

In a first step, the neonatal images were converted from floating point 32-bit temperature values to 8-bit integers for training. Since an initial normalization of the images showed a low contrast (see Fig. 3 (a)), a histogram equalization was conducted to increase the global contrast. Although the applied histogram equalization showed a higher contrast in the IRT frames, Fig. 3 (b) reveals the problem of saturated image areas with relative high temperatures, which could complicate the feature extraction. Therefore, the contrast limited adaptive histogram equalization (CLAHE) implementation of OpenCV in Python was used.

In contrast to the standard implementation, the CLAHE algorithm reduces contrast boosting of noise and saturation of bright image areas by applying the histogram equalization to every cell of an M×N grid of the input frame. A contrast limit ensures that pixels of saturated areas are evenly assigned to other adjacent cells [26]. The result of the applied CLAHE method using the original image as input can be observed in Fig. 3 (c). As expected, the pixels in the facial region were not saturated and the overall contrast was increased.

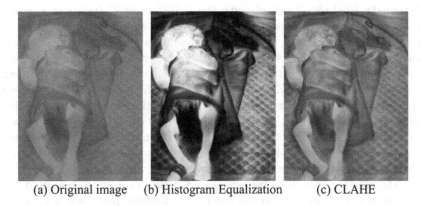

(a) Original image   (b) Histogram Equalization   (c) CLAHE

Fig. 3. Comparison of (a) original image with (b) applied standard histogram equalization and (c) applied contrast limited adaptive histogram equalization.

The dataset for pre-training the model was already provided as 8-bit integer frames. In contrast to the neonatal recording setup, where the heating element of the open incubator increased the background (mattress) temperature for some subjects and therefore decreased the thermographic contrast, Kopaczka et al. recorded the data in front of a cold wall [14]. This ensured a higher contrast of the images. In order of equalizing both the adult and neonatal datasets, random noise and blur was added to the adult recordings before applying the CLAHE.

All frames were normalized and labelled with ground truth bounding boxes for the training step of the detector. This was performed using the tool Yolo_mark [3]. In each frame, the head and nose were marked with a bounding box (if available). While the predicted nose should be used for temperature extraction in the nasal region, the position of the head should indicate the movement of the subjects. During the study, no gold-standard reference data was recorded for respiration, so the video data was manually labelled by evaluating the thorax movement in a 250×250 px ROI using Farnebäck's Optical Flow algorithm [5]. This algorithm detects the pixel intensity changes between two consecutive frames, which is caused by the movement of objects. As a result, it computes the magnitude and direction of displacement and returns an optical flow vector field. In the process of creating reference data, physical activity prevented a proper annotation process due to an indistinguishable overlap to the breathing-related motion. Therefore, both the head and nose motions, which were later tracked using the bounding box position of the prediction model, were automatically used to classify the activity level of the subject. To allow a validation of the proposed method, only signal sections without movements (deviation of bounding box centers less than 5 px between two frames) were analyzed.

2.3   Detector Training and Validation

For model training and inference, the YOLOv4-Tiny object detection algorithm in Python by Bochkovskiy et al. with the CSPDarknet53 backbone was used with an image size of 416 × 416 px [4]. This one-stage prediction algorithm processes the detection as a regression problem: a pre-trained feature extractor is tuned using a custom dataset, so the specific characteristics of the input image can be applied to a classification and regression stage for finding trained objects. In contrast to the larger model version YOLOv4, which showed high accuracies but poor real-time capability on embedded GPUs, the tiny model reaches high inference performance with a general loss in accuracy [4]. However, previous work showed that YOLOv4-Tiny can be beneficial for simple detection problems in IRT images [20]. Further, real-time performance was demonstrated on embedded GPUs. Thus, this model was used in this work.

For the training step, a high-performance desktop computer was used, running Ubuntu 18.04 and featuring an Intel Xeon E5-2620 processor, an NVIDIA Quadro RTX5000, and 64 GB RAM. For the training process, the GPU was deployed in combination with CUDA 11.0, cuDNN 8.0.5, and OpenCV 4.4.0. To compare the results of a pre-trained model with the outcome of a detector trained from scratch (see Sect. 2.1), two training cycles were performed. For the pre-trained cycle, the initial weights were trained for 4000 epochs using the labelled images from [14]. In a next step, the neonatal dataset was prepared for a patient-wise 10-fold leave-one-out cross-validation (LOOCV) to quantify the model accuracy. Therefore, in every fold, the data of one patient was defined as test set while the rest was used for training. Although the computational effort of a LOOCV is higher than for a standard cross validation, it is recommended for smaller datasets to address overfitting. However, in the case of varying data for particular folds, a higher variance should be expected in the evaluation [2].

The LOOCV was conducted for both the pre-trained and initial model weights. While the latter weights were trained for 4000 epochs, the pre-trained model weights were only fine-tuned in 1500 epochs using a reduced learning rate. In all training processes, early stopping was conducted to prevent overfitting by monitoring the validation loss. After the training iterations, the mean average precision (mAP), intersection over union (IoU) and F_1 score were computed. While the mAP describes the accuracy of a detector using all classes, the IoU provides information about the overlap of ground truth and predicted bounding boxes. Finally, the F_1-score represents the harmonic mean of sensitivity and precision. The results for the training processes are presented in Sect. 3.1.

2.4 Respiration Extraction

The respiration extraction was performed in real-time on different GPUs using Python. To ensure a proper evaluation of the approach and exclude overfitting, it was conducted patient-wise using the model weights of the fold in which the specific patient was in the test set. In a first step, the (raw) input images were loaded and converted into 8-bit in order to enable the inference using the detection algorithm. Before both nose and head were predicted from the model, the images were normalized and the CLAHE method was applied (see Fig. 4). While the predicted bounding box for the nose was used to crop the nasal region and extract the mean value from the original temperature frame, the head box was tracked for movement quantification. If more than one object was found for the labels, only this with the highest confidence score was chosen.

After the video sequence was processed, a 5th-order Butterworth band pass (BP) filter with cutoff frequencies of 0.3 Hz and 1.4 Hz was used to filter noise and also allow an evaluation of tachypnea episodes [25]. As illustrated in Fig. 4, a peak detection was conducted on the filtered signal to determine all local maxima. Afterwards, the peaks were counted using a sliding window with a window length of 10 s to extract the RR. In order to compare the extracted RR with the custom reference data, the signals received from the optical flow algorithm were processed identically.

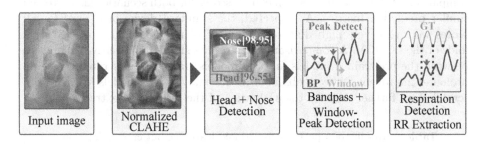

Fig. 4. Algorithm overview for the extraction of respiration.

For further evaluating the detection performance of respiration cycles, common detection metrics such as precision, recall, and further the F_1 score were computed by investigating the peaks in the extracted signal. Thus, a predicted peak between two minima in the reference signal was defined as a true positive, while a missing peak was considered as a false negative. Further, several peaks between two reference minima were specified as false positive. This allowed a profound evaluation of the prediction performance. After the signal analysis, the inference of the extraction algorithm was conducted on different embedded GPUs to show real-time capability on low-cost modules.

2.5   Real-Time Feasibility on Embedded GPUs

The low-cost system-on-modules Jetson AGX Xavier (approx. 700\$, November 2021) and the less performance version Jetson Xavier NX (approx. 400\$, November 2021) (NVIDIA, Santa Clara, USA) were used for inference to show the feasibility of applying the trained detector in combination with embedded GPU systems. Both development boards provide a 64-bit CPU, an NVIDIA Volta GPU, including tensor cores and 16 GB (AGX Xavier) and 8 GB (Xavier NX) of RAM. Low-cost and high-performing portable devices can be implemented to provide efficient embedded systems for real-time camera-based monitoring. Both modules were selected to determine the performance of inference and usability. The results of a detailed performance analysis will be presented in Sect. 3.3.

3   Results and Discussion

3.1   Detector Accuracy

In Table 2, the results of the patient-wise 10-fold LOOCV for both the pre-trained weights and those trained from scratch are presented. While the metrics for folds 2 and 5 revealed lower prediction accuracies, an analysis of the remaining subjects showed promising results, which resulted in mean mAPs of 0.93 for the model trained from scratch and 0.92 for the pre-trained case, respectively. The comparison of the mean IoUs and mean F_1 scores for both scenarios also revealed similar prediction metrics. As expected, the standard deviations (SDs) for all scores were relatively high due to the impact of poor outcomes of individual folds of the LOOCV. A comparison of the results showed no decisive differences between the pre-trained models and the models trained from scratch. Since no profound improvements were observed in the analysis, the models learned from scratch were further used for the evaluation of the respiration extraction.

Table 2. Results of a patient-wise 10-fold leave-one-out cross-validation.

Pre-train	Metric @ 0.5	Fold										Mean	SD
		1	2	3	4	5	6	7	8	9	10		
No	mAP	0.99	0.66	0.98	0.98	0.78	0.97	0.99	0.99	0.98	0.99	**0.93**	0.11
	IoU	0.85	0.67	0.79	0.81	0.82	0.80	0.80	0.82	0.76	0.83	0.79	0.04
	F_1	0.99	0.73	0.75	0.99	0.80	0.96	0.99	0.99	0.94	0.99	**0.91**	0.10
Yes	mAP	0.99	0.61	0.95	0.99	0.75	0.98	0.99	0.99	0.98	0.99	0.92	0.12
	IoU	0.87	0.71	0.81	0.82	0.77	0.80	0.80	0.83	0.81	0.78	**0.80**	0.04
	F_1	0.99	0.45	0.71	0.98	0.82	0.98	0.99	0.99	0.93	0.99	0.88	0.17

The lack of improvement from pre-trained weights can be explained by the complexity factor of the detection process. Since only two labels were predicted in the (initially high-definition) IRT images, the sole use of the neonatal dataset was sufficient to train well performing models. Another reason for the comparable results could be the extracted features for both head and nose. Although fundamental differences were assumed for the features between adult and neonatal relative geometry, the obvious IRT features of the head (round structure in front of a colder background) and nose (mostly cold spot in the face) could cause a related weight adjustment and, therefore, lead to equal conclusions. At the same time, both the pre-trained and the models trained from scratch showed similar results for folds 2 and 5. The poor outcome for these folds can be explained by a strongly decreased contrast due to an extensive use of the radiant warmer during the measurement, which heated the background.

In total, the presented results showed that the YOLOv4-Tiny object detector could be trained from scratch on a comparable small neonatal IRT dataset to predict the head and the nose of the subjects.

3.2 Respiration Extraction

In Table 3, the results for the respiration extraction and the detection of breathing cycles are presented. Although the YOLOv4-Tiny detector accuracy was conducted on all 10 folds, the motion analysis revealed continuous activity of the subjects from folds 9 and 10 during the entire video sequences and further motion disturbed sections in the remaining folds, so no reference signal could be extracted. Thus, only sequences without artifacts were used for evaluation. However, the exclusion of motion artifacts for camera-based respiration analysis was also conducted in literature [17].

Table 3. Results for respiration detection and MAE for RR extraction.

Metric	Fold								Mean	SD
	1	2	3	4	5	6	7	8		
Precision	0.92	0.94	0.94	0.99	0.95	0.87	0.94	0.90	0.93	0.04
Recall	0.73	0.73	0.68	0.60	0.73	0.63	0.70	0.79	0.71	0.05
F_1	0.82	0.82	0.79	0.75	0.83	0.72	0.80	0.84	0.80	0.04
MAE [BPM]	9.3	6.3	14.0	4.0	8.6	7.4	12.2	6.6	8.5	3.0

While the (averaged) mean average error (MAE) revealed a value of 8.5 breaths per minute (BPM), the metrics for precision and the F_1-score showed promising results for breathing detection. Next to the precision, which considers true and false positives, the recall gives conclusions about the predicted true positives and false negatives. While a high precision could be observed, the mean recall indicates that some reference peaks were not found by the extraction algorithm, which could be explained due to undetected temperature changes in the nose region. Since both nostrils were not visible in all frames, shallow breathing could be missed by the algorithm, because no intense heat flow was generated.

Although the presented results were promising for the DL-based respiration extraction, the method was validated using custom reference data measured from breathing-related thorax movements using Farnebäck's Optical Flow algorithm. Therefore, the validation results need to be discussed carefully and analyzed in future work with proper gold-standard reference.

3.3   Inference Performance

In Table 4, the results of a performance analysis using a Desktop GPU and two low-cost embedded GPU modules are presented. While the inference speed describes the performance of the detection step, the total time provides information about the efficiency of the entire algorithm, including data loading, preprocessing, and extraction of the respiration signal.

Table 4. Mean performance on different GPU platforms.

Platform	Inference [fps]	Total [fps]
Jetson Xavier NX	52	32
Jetson AGX Xavier	103	62
Quadro RTX5000	233	137

As it can be observed, all GPUs revealed real-time capability for the detection step and also the algorithm including data processing. While the desktop GPU showed a performance of 137 fps, the AGX Xavier was able to process the method with 62 fps. Although the Xavier NX module showed a loss in performance, it still allowed real-time respiration monitoring.

4   Conclusion and Outlook

In this work, we presented an approach for IRT-based respiration extraction for neonates using the YOLOv4-Tiny object detector to show real-time performance on embedded GPUs. While the respiratory activity was measured using the mean temperature from the detected nose ROI, the predicted head was used for movement detection. The extracted signal was compared to a reference measured from respiration-related thorax movement. Next to the neonatal dataset, adult thermography images were used for a pre-training step to compare the detection accuracy to a model learned from scratch. During a patient-wise 10-fold LOOCV, no major differences between both model types were observed, so the pre-trained model was neglected. Since the validation resulted in high mean accuracies, the pre-training step could be impractical due to the simplicity of a two-class detection problem. However, the accuracy of the models will be further improved in future work by additional neonatal data.

The evaluation of the respiration extraction for motion-less video sequences showed promising results, although the validation needs to be discussed carefully since no clinical reference data was provided. However, the outcomes proved the functionality of the method, so upcoming datasets with gold-standard reference for respiration can be used to further validate the monitoring technique.

A final analysis of the inference speed on embedded NVIDIA Jetson modules showed real-time capability for both embedded GPUs. While 62 fps could be achieved on the AGX Xavier, the less performant version Xavier NX demonstrated a performance of 32 fps for the algorithm including data processing. The promising outcomes showed the real-time capabilities of the YOLOv4-Tiny object detector deployed on embedded GPUs for the camera-based extraction of respiratory activity in neonates using an IRT dataset. We are confident that IRT cameras in combination with DL-based methods on embedded GPUs contribute to an enhanced use of wire-less vital sign measurement. The clinical application of this technique could contribute to reduce adhesive sensors, especially in neonatal care, and, therefore, minimize the risk of medical adhesive-related skin injuries.

Acknowledgments. We acknowledge the effort from the authors of the YOLOv4 object detection algorithm and Kopaczka et al. for providing the adult IR dataset. The authors gratefully acknowledge financial support provided by German Research Foundation [Deutsche Forschungsgemeinschaft, LE 817/32-1].

References

1. Abbas, A.K., Heimann, K., Jergus, K., Orlikowsky, T., Leonhardt, S.: Neonatal non-contact respiratory monitoring based on real-time infrared thermography. Biomed. Eng. Online **10**(1), 1–17 (2011)
2. Berrar, D.: Cross-validation. In: Ranganathan, S., Gribskov, M., Nakai, K., Schönbach, C. (eds.) Encyclopedia of Bioinformatics and Computational Biology, pp. 542–545. Academic Press, Oxford (2019)
3. Bochkovskiy, A.: Yolo_mark - github repository (2019). https://github.com/AlexeyAB/Yolo_mark. Accessed 06 Dec 2021
4. Bochkovskiy, A., Wang, C.Y., Liao, H.Y.M.: Yolov4: optimal speed and accuracy of object detection. arXiv: 2004.10934 (2020)
5. Farnebäck, G.: Two-frame motion estimation based on polynomial expansion. In: Bigun, J., Gustavsson, T. (eds.) SCIA 2003. LNCS, vol. 2749, pp. 363–370. Springer, Heidelberg (2003). https://doi.org/10.1007/3-540-45103-X_50
6. Ferris, T., Shepley, M.: The design of neonatal incubators: a systems-oriented, human-centered approach. J. Perinatol. **33**, S24–S31 (2013)
7. van Gastel, M., Stuijk, S., de Haan, G.: Robust respiration detection from remote photoplethysmography. Biomed. Opt. Express **7**(12), 4941–4957 (2016)
8. Gleichauf, J., et al.: Automated non-contact respiratory rate monitoring of neonates based on synchronous evaluation of a 3D time-of-flight camera and a microwave interferometric radar sensor. Sensors **21**(9), 2959 (2021)
9. Greneker, E.: Radar sensing of heartbeat and respiration at a distance with applications of the technology. In: Radar 97 (Conf. Publ. No. 449), pp. 150–154 (1997)

10. Heimann, K., Jergus, K., Abbas, A.K., Heussen, N., Leonhardt, S., Orlikowsky, T.: Infrared thermography for detailed registration of thermoregulation in premature infants. J. Perinat. Med. 1–8 (2013)

11. Hoog Antink, C., et al.: Fast body part segmentation and tracking of neonatal video data using deep learning. Med. Biol. Eng. Comput. **58**(12), 3049–3061 (2020). https://doi.org/10.1007/s11517-020-02251-4

12. Hoog Antink, C., Lyra, S., Paul, M., Yu, X., Leonhardt, S.: A broader look: camera-based vital sign estimation across the spectrum. Yearb. Med. Inform. **28**(01), 102–114 (2019)

13. Kinney, M., Howson, C., McDougall, L., Lawn, J.: Executive summary for born too soon: the global action report on preterm birth. March of Dimes, PMNCH (World Health Organization) (2012)

14. Kopaczka, M., Kolk, R., Schock, J., Burkhard, F., Merhof, D.: A thermal infrared face database with facial landmarks and emotion labels. IEEE Trans. Instrum. Meas. 1–13 (2019)

15. Liu, L., et al.: Global, regional, and national causes of under-5 mortality in 2000–15: an updated systematic analysis with implications for the sustainable development goals. LANCET **388** (2016)

16. Lorato, I., Bakkes, T., Stuijk, S., Meftah, M., De Haan, G.: Unobtrusive respiratory flow monitoring using a thermopile array: a feasibility study. Appl. Sci. **9**(12), 2449 (2019)

17. Lorato, I., et al.: Multi-camera infrared thermography for infant respiration monitoring. Biomed. Opt. Express **11**(9), 4848–4861 (2020)

18. Lorato, I., et al.: Towards continuous camera-based respiration monitoring in infants. Sensors **21**(7), 2268 (2021)

19. Lund, C.H., et al.: Neonatal skin care: clinical outcomes of the AWHONN/NANN evidence-based clinical practice guideline. J. Obstet. Gynecol. Neonatal Nurs. **30**(1), 41–51 (2001)

20. Lyra, S., et al.: A deep learning-based camera approach for vital sign monitoring using thermography images for ICU patients. Sensors **21**(4) (2021)

21. Lyra, S., Paul, M.: Organic led panels for pulse rate measurement using photoplethysmography imaging. In: Proceedings of the International Student Scientific Conference Poster, pp. BI04 1–4 (2019)

22. Maier, A., Syben, C., Lasser, T., Riess, C.: A gentle introduction to deep learning in medical image processing. Z. Med. Phys. **29**(2), 86–101 (2019)

23. Pereira, C.B., et al.: Estimation of respiratory rate from thermal videos of preterm infants. In: 2017 39th Annual International Conference of the IEEE Engineering in Medicine and Biology Society (EMBC), pp. 3818–3821. IEEE (2017)

24. Pereira, C.B., et al.: Noncontact monitoring of respiratory rate in newborn infants using thermal imaging. IEEE Trans. Biomed. Eng. **66**(4), 1105–1114 (2018)

25. Reuter, S., Moser, C., Baack, M.: Respiratory distress in the newborn. Pediatr. Rev. **35**(10), 417 (2014)

26. Reza, A.: Realization of the contrast limited adaptive histogram equalization (clahe) for real-time image enhancement. VLSI Sig. Process. **38**, 35–44 (2004)

27. Shelov, S.P., Altmann, T.R. (eds.): Caring for Your Baby and Young Child, 5th edn. American Academy of Pediatrics, Itasca (2005)

Improving Colon Carcinoma Grading
by Advanced CNN Models

Marco Leo[1]([✉])(iD), Pierluigi Carcagnì[1], Luca Signore[2], Giulio Benincasa[4,5](iD),
Mikko O. Laukkanen[3](iD), and Cosimo Distante[1,2](iD)

[1] CNR-ISASI, Ecotekne Campus via Monteroni snc, 73100 Lecce, Italy
`marco.leo@cnr.it`
[2] Universitá del Salento, Via Monteroni sn, 73100 Lecce, Italy
[3] CNR-IEOS, Via Pansini 5, 80131 Naples, Italy
[4] Pineta Grande Hospital, 81030 Caserta, Italy
[5] Italo Foundation, Via Desenzano 14, 20146 Milan, Italy

Abstract. Cancer ranks as a leading cause of death and an important
barrier to increasing life expectancy in every country of the world. For
this reason, there is a great requirement for developing computer-aided
approaches for accurate cancer diagnosis and grading that can overcome the problem of intra- and inter-observer inconsistency and thereby
improve the accuracy and consistency in the cancer detection and treatment planning processes. In particular, the studies exploiting deep learning for automatic grading of colon carcinoma are still in infancy since
the works in the literature did not exploit the most advanced models
and methodologies of machine learning and systematic exploration of
the most promising available convolutional networks is missing. To fill
this gap, in this work, the most performing convolutional architectures in
classification tasks have been exploited to improve colon carcinoma grading in histological images. The experimental proofs on the largest publicly
available dataset demonstrated marked improvement with respect to the
leading state of the art approaches.

Keywords: Colon carcinoma · Artificial intelligence · Deep learning ·
Optimally designed networks · Histological diagnosis

1 Introduction

Cancer ranks as a leading cause of death and an important barrier to increasing
life expectancy in every country of the world[1]. Through early diagnosis, there is
a great chance of recovery and thereby reduce the mortality rate. Traditionally,
tissue samples were reviewed by pathologists under a microscope, which is time-
consuming and prone to error [18]. In the last decade, the availability of whole

---

[1] World Health Organization (WHO). Global Health Estimates 2020: Deaths by
Cause, Age, Sex, by Country and by Region, 2000–2019. WHO; 2020. Accessed
March 8, 2021.

The original version of this chapter was revised: The name of the author Pierluigi
Carcagnì has been corrected. The correction to this chapter is available at
https://doi.org/10.1007/978-3-031-06427-2_65

slide digital scanners and a large amount of storage space brought to born of a new scientific field namely Digital Pathology (DP) that is the process by which histology slides are digitized to produce high-resolution images. Unfortunately, interpreting histological images is not easy anyway: the sensitivity is affected by image quality and the observer's level of expertise and, besides, from the multiple tissue types and characteristics. For this reason, there is a great requirement for developing a computer-aided accurate cancer diagnosis system that can overcome the problem of intra- and inter-observer inconsistency and thereby improve the accuracy and consistency in the cancer detection, grading, and, subsequently, in treatment planning processes [5,7,22]. Artificial intelligence (AI), especially recently developed computer vision methodologies based on deep learning, carry tremendous potential in assisting clinical diagnosis of classification, semantic and instance segmentation, detection, and grading of cancer [10] In particular, recent computer vision methodologies based on deep learning can help in several tasks such referred as classification, semantic and instance segmentation, detection, and grading. There are several studies already published addressing the automatic analysis of diseases of different tissues such as brain, cervical, breast, lung, prostate, stomach, skin, and colon [8] but, unfortunately, there are still several open issues to be addressed before reaching an accuracy level to be really trusted [1].

Colorectal carcinoma (CRC) is a well-characterized heterogeneous disease induced by different tumorigenic modifications in colon cells. The computer-based analysis of colon digital histological images involves different tasks which have been well investigated in the last decade [10,21]: normalization of histological staining (with the aim of matching stain colors with a given template in order to reduce appearance e of stained histology samples coming from different laboratories) [6], segmentation of cells (e.g., nuclei and glands segmentation), classification of tissues (e.g. tumor, stroma, adipose tissue) [28], detection of cancer progression (lymphocytes and mitosis counting) [25] and tumor stage prediction and survival period prediction using personal (age, gender, adjuvant therapy, medical illness, smoking history, coffee consumption, alcohol consumption, physical activity, etc.) and tumor-related histo-pathology parameters (gross appearance, circumferential involvement, tumor size, histologic type, etc.) [13]. Unfortunately less efforts have been spent for automatic grading of CRC that is the key elements of colorectal cancer diagnostics [20]. The grading of the cancer is diagnosed based on the differentiation of the cells, i.e. determining the abnormalities in the cellular phenotype. Colon cancer is usually divided into three grades: well-differentiated (low grade), moderately differentiated (intermediate grade), and poorly differentiated (high grade) [11]. Cancers representing higher grade or poor differentiation tend to grow and spread faster.

Grading describes how the cancer cells look compared to normal, healthy cells. Knowing the grade gives healthcare team an idea of how quickly the cancer may be growing and how likely it is to spread. The accurate diagnosis to predict the prognosis is crucial in selecting the therapy and estimating the response to treatment [19].

There are a few works specifically designed for colon cancer grading since there are some challenges i.e. extracting knowledge from limited datasets of labelled and biased data (the labelling task requires high clinical skills and it is time consuming) and getting rid of intra-class variance and inter-class similarity derived from the continuum existing from the different grading levels.

Relevant attempts to face the above challenges rely on two different strategies:

1. classifying the informative content of a single patch (in general of size 224 × 224 pixels) and then counting the predictions to label to whole image (patch-based approaches).
2. learning contextual information. This can be achieved by enlarging the patch size and introducing additional modules in the network architecture. This leads to architectures that are domain specific (context-aware approaches).

Early patch based works relied on handcrafted feature [2, 27] eventually using a multi-level analysis (from coarse to fine) by using an SVM based combination of weak classifiers such as linear and RBF. A contribution that deserves mention used shape specific parameters to perform a two-class classification of images into normal or cancerous tissue and a three-class classification into normal, low grade cancer, and high grade cancer [3]. In particular two different feature sets (commonly referred to as BAM-1 and BAM-2) were employed and a support vector machine (SVM) classifier was trained using shape features. Recently deep learning-based approaches demonstrated their superiority in colon cancer grading. CNNs are generally used for representation learning from small image patches (e.g. 224 × 224) extracted from digital histology images due to computational and memory constraints. Some of them made use of an intermediate tissue classification. They get advantage of the classification step but, on the other hand, grading depends on cancer type [29]. In general, these kind of approaches require to intelligently combine patch-level classification results and model the fact that not all patches will be discriminative. This challenge can be addressed by a max-voting or by training a decision fusion model to aggregate patch-level predictions given by patch-level CNNs [15]. The first end to end deep learning based approach for just grading was proposed in [4]: the authors combine convolutional and recurrent architectures to train a deep network to predict colorectal cancer outcome based on images of tumour tissue samples.

Some authors considered the possibility to embed contextual information in order to improve accuracy. The first attempt of introducing contextual information was in [24] that introduced a computational model that utilises feature sharing across scales and learns dependencies between scales using long-short term memory (LSTM) unit. In [30] a novel cell-graph convolutional neural network (CGC-Net) that converts each large histology image into a graph, where each node is represented by a nucleus within the original image and cellular interactions are denoted as edges between these nodes according to node similarity. A novel framework for context-aware learning of histology images has been recently proposed in [23]. The framework first learns the Local Representation by a CNN (LR-CNN) and then aggregate the contextual information through a representation aggregation CNN (RA-CNN).

Generally speaking, dividing a histological section high power field (HPF) into single patches, analyzing the single patches to predict the whole image would reduce staining artifacts causing variance inside one single section and between different histological sections. Combining several patches to one large patch would increase the quantity of contextual information, eventually increasing the accuracy of the tissue-wide diagnosis. On the other side, using patch-based approaches it makes possible to directly exploit network models designed for the classical task of image classification. In order to learn contextual information, some specific architectures have to be introduced instead and designing choices have to be manually done as for example the size of the area to be considered as a whole and the configuration of additional modules.

In this work, the recent advances in machine learning combining Densely Connected Convolutional Network (DenseNet) [17], a family of models, called EfficientNets [26] and Squeeze and Excitation networks [16] are exploited, at the best of our knowledge for the first time, to improve the diagnosis of colon adenocarcinoma. Lastly, the current study introduces optimally designed network model selected among those in a selected space [19] for classification. The developed algorithm demonstrated marked improvement in CRC diagnosis on the largest publicly available dataset annotated for carcinoma grading.

The rest of the paper is organized as follow: the Sect. 2 describes the considered convolutional architectures and the used publicly available dataset whereas the Sect. 3 describes experimental results and accuracy gathered in CRC grading. Finally, Sect. 4 concludes the paper.

2 Methods and Data

In this section the CNN architectures introduced for CRC grading are listed and briefly described. In particular, both manually and optimally automatically designed architectures were exploited. They are all described in the Subsect. 2.1 whereas the Subsect. 2.2 will introduce the datasets of annotated histological images used for testing.

2.1 Advanced Deep Network Architectures

The development of the deep learning diagnosis tool was done using a workstation equipped with Intel(R) Xeon(R) E5-1650 @ 3.20 GHz CPU, one GeForce GTX 1080 Ti with 11 GB of RAM GPU, Ubuntu 16.04 Linux operating system. Several advanced CNNs architectures have been trained in an end-to-end way and subsequently analyzed in order to establish how each particular architectural choice affects the classification results. The architectures considered in this work are those that have introduced the most significant and distinctive features in the recent years and that better performed in the ImageNet Large Scale Visual Recognition Challenge (ILSVRC) [9]. In particular only the architectures by which the vanishing gradient issue have been taken into consideration since the considered problem requires very deep architecture to extract distinctive features. At the same time, the high generalization capability and low memory footprint during inference in related problems have been also used in the selection process [12]. We can list, according to their date of introduction respectively, in: ResNet [14], DenseNet [17], SENet [16], EfficientNet [26]. The *ResNet* architecture addresses vanishing gradient and training degradation (accuracy saturation) problems by introducing a *deep residual learning* approach where a residual mapping, conceived by means of skip-connections, of each few stacked layers of the entire network is learnt. Exploiting the approach of skip-connections introduced with ResNet, in the DenseNet architecture each layer is connected to every other layer in a feed-forward fashion. In particular, for each layer, the feature-maps of all preceding layers are used as inputs, and its own feature-maps are used as inputs into all subsequent layers. In contrast to ResNets, the features are not added together before being passed as input to the generic layer, but rather concatenated allowing a strengthen feature propagation and alleviating the vanishing-gradient problem for very deep implementations. Unlike the networks described above, with the SENet architecture investigated strategies in order to better capture spatial correlations between features. In particular they introduced a new architectural unit, named *Squeeze-and- Excitation (SE)* block able to explicitly model the interdependencies of the convolutional channels. This makes available the global information to the network necessary to learn to emphasise informative features and suppress less useful ones. A different investigation has been followed regarding the Efficientnet architecture. It uses a compound coefficient to uniformly scale network width, depth, and resolution. Starting from a mobile-size baseline architecture created ad hoc (Efficient-B0) and following the proposed scaling strategy, a family of eight architectures has been built up with increasing grade of complexity. All the networks were modified to be adapted to a three classes inference problem. Finally, some networks designed by exploring a design space where populations of networks can be parameterized [19] were exploited as well for CRC grading. The idea is that, instead of designing a single best model under specific settings, tools from classical statistics can be employed in order to study the behaviour of populations of models and discover general design principles. That can provide information on the design of the network and, unlike a single model optimized for a specific scenario, it is more likely to generalize to new settings. According to the nomenclature in [19], the networks obtained in the design space are called RegNet (Regular Networks). Different models having different computational loads and epochs were generated. Models were generated with (called RegNetY) and without (called RegNetX) a Squeeze-and-Excitation (SE) block. To identify any specific generated model the corresponding FLOPs regime was used in its name. Hence, for instance, in the following, RegNetY-400MF means that a model with SE block and a computational regime of 400 mega-FLOPs was built.

For CNNs training, a fine-tuning strategy has been performed with starting models all pre-trained on the same Imagenet dataset. Moreover, data augmentation has been applied to the original data in terms of operations of: horizontal and vertical image flipping; rotation with a value of $\pm 45°$ and $\pm 90°$; shearing between $-20°$ and $20°$. Finally, the SGD optimizer was employed, with learning rate = 0.001, momentum = 0.9 and weight decay = 0.001 parameters, and setting an early stopping strategy of 22 epochs on validation set and a max number of 100 training epochs.

2.2 The Datasets: CRC and Extended CRC

One of the most used dataset for adenocarcinoma grading task is CRC-Dataset [3]. It is comprised of visual fields extracted from 38 Hematoxylin and Eosin stained whole-slide images (often abbreviated as H&E stained WSIs) of colorectal cancer cases and consists of 139 visual fields with an average size of 4548×7520 pixels obtained at 20× magnification. These visual fields are classified into three different classes (normal, low grade, and high grade) based on the organization of glands in the visual fields by the expert pathologist. Recently the CRC dataset has been extended with more visual fields extracted from another 68 H&E stained WSIs using the same criteria. This additional set of images has been named Extended version of CRC-Dataset[2]. The extended colorectal cancer (Extended CRC) dataset consists of 300 visual fields with an average size of 5000×7300 pixels. Table 1 reports a detailed distribution of the labels (i.e. Normal, Low Grade and High Grade) of the visual fields for both datasets. Labels correspond to the annotations provided by expert pathologists (i.e. the ground truth). The extended CRC dataset has been selected as benchmark to test the different approaches investigated in Sect. 3.

Table 1. Distribution of visual fields of different classes in CRC and extended CRC datasets.

Dataset	Normal	Low Grade	High Grade	Total
CRC	71	33	35	139
Extended CRC	120	120	60	300

3 Experimental Results on the Extended CRC Dataset

In this section the experimental results, gathered by using the different machine learning strategies described in Sect. 2 in order to analyze histological visual fields in the extended CRC dataset, are described. A 3-fold cross validation for a fair comparison of the proposed method with the method presented in the literature was followed. The fold splitting provided together with the extended CRC was used (this make as fairest as possible comparison with other approaches). All visual fields extracted from one case only lies in one fold and one fold is used for training, one for validation (hyper-parameter tuning) and one for the

---

[2] https://warwick.ac.uk/fac/sci/dcs/research/tia/data/extended_crc_grading/.

testing. Following the common way of operating of the approaches in the literature in this specific research field, two classification problems were addressed: the binary problem of distinguish between normal and tumor tissues, and the 3-class problem of grading the issues as normal, low grade, and high grade. For each visual field, non overlapping patches of size 224×224 pixels were extracted and given as input to the subsequent training (batch size was set to 16) and classification steps. After this process, the gathered patch distribution per fold and class is reported in Table 2. Background class refers to white regions, whose patches have 95% of pixels with average radiometric value (along the three color channels) higher than 235 (their percentage is about 11% of the total number of extracted patches). These patches were not considered for further analysis since their informative content is not relevant.

Table 2. Patch distribution per fold and class.

	No Tumor	Low Grade	High Grade	Background
Fold 1	20911	28298	13084	8799
Fold 2	22430	28042	12412	8768
Fold 3	22879	28388	13495	6302

In all the experiments on CRC extended dataset, the same training, validation, and test splits as in [24] were used for a fair comparison with the existing methods. For the same reason, both binary and ternary classification performance are reported where binary classification means that examples with intermediate and high grades have been put together and considered as a unique class against the class including only examples of lower-grade cancer. This could help to provide significant outcomes for the sub-task of differentiating between normal and tumor tissues. Two evaluation metrics were used: average accuracy and weighted accuracy. The average accuracy refers to the percentage of visual fields classified correctly whereas weighted accuracy is the sum of accuracy of each class weighted by the number of samples in that class.

In particular, for each fold j in range $[1, K]$ ($k = 3$ in the following experiments), *average accuracy* is computed as the average of:

$$acc_j = \frac{\sum_{i=1}^{c} TP_i}{\sum_{i=1}^{c} Ni} \tag{1}$$

Similarly, *weighted accuracy* is computed as the average of:

$$w\_acc_j = \frac{\sum_{i=1}^{C} \frac{TP_i}{N_i}}{C}. \tag{2}$$

where C indicated the number of classes (2 or 3), N_i is the of elements in the class i and TP_i is the number of true positive for the class i.

As already stated, in this experimental phase a set of advanced neural architectures have been exploited for the first time for colon carcinoma grading.

Table 3 reports average and weighted classification for the binary and 3-classes problems on the extended CRC dataset. It is worth noting that ResNet50 was a model already tested for this task. It has been used as a pivot to check if the data handling process was correctly implemented. As reported in the Sect. 3.1 gathered scores were coherent with those in [23] and this corroborate the validity of the test of the never tested models. In particular, EfficientNet-b2 model reported the highest accuracy scores for both the binary and three-classes problems. Anyway results gathered by using DenseNet121 and by other EfficientNet configurations were very encouraging as well.

Table 3. Results on the extended CRC dataset by using advanced deep learning architectures.

Model	Average (%) (Binary)	Weighted (%) (Binary)	Average (%) (3-classes)	Weighted (%) (3-classes)
DenseNet121	94.98 ± 2.14	95.69 ± 1.99	87.24 ± 2.94	83.33 ± 2.04
EfficientNet-B0	93.63 ± 0.94	93.80 ± 1.10	85.89 ± 3.64	83.55 ± 3.54
EfficientNet-B1	95.64 ± 1.23	94.79 ± 1.15	85.89 ± 3.64	83.56 ± 3.39
EfficientNet-B2	**96.99 ± 2.94**	**96.65 ± 3.11**	**87.58 ± 3.36**	**85.54 ± 2.21**
EfficientNet-B3	96.65 ± 2.05	96.22 ± 2.22	86.57 ± 2.68	83.31 ± 1.82
EfficientNet-B4	95.31 ± 1.24	94.36 ± 1.27	84.89 ± 2.91	82.44 ± 1.84
EfficientNet-B5	95.98 ± 1.62	95.66 ± 1.72	87.57 ± 3.37	84.98 ± 3.80
EfficientNet-B7	95.98 ± 1.62	95.36 ± 1.68	86.90 ± 3.01	84.41 ± 2.78
ResNet-50	94.96 ± 0.79	95.45 ± 1.20	86.57 ± 2.43	80.60 ± 1.73
ResNet-152	95.64 ± 0.94	95.82 ± 1.01	84.22 ± 4.58	79.99 ± 4.13
Se-ResNet50	93.30 ± 2.47	93.14 ± 2.54	84.89 ± 3.02	81.63 ± 2.08
RegNetY-200MF	92.97 ± 3.73	93.87 ± 2.92	83.90 ± 0.76	80.54 ± 1.03
RegNetY-400MF	93.97 ± 2.94	93.99 ± 3.11	84.23 ± 2.62	81.92 ± 1.74
RegNetY-800MF	93.65 ± 4.77	94.15 ± 4.17	84.24 ± 1.63	81.10 ± 1.41
RegNetY-4.0GF	95.64 ± 0.94	95.37 ± 1.52	84.55 ± 2.57	81.36 ± 1.43
RegNetY-6.4GF	94.31 ± 2.48	94.26 ± 2.15	86.57 ± 2.12	83.58 ± 2.21
RegNetY-8.0GF	91.95 ± 2.15	92.19 ± 2.40	82.55 ± 1.70	80.81 ± 2.06
RegNetY-12GF	93.97 ± 2.93	94.28 ± 2.93	84.22 ± 2.41	82.21 ± 3.09
RegNetY-16GF	94.97 ± 1.62	94.24 ± 2.08	85.22 ± 3.93	83.29 ± 3.45
RegNetY-32GF	94.64 ± 2.49	94.55 ± 2.79	84.56 ± 2.68	81.65 ± 2.39

Results obtained using optimally designed models were satisfying as well, but not at the top of accuracy. The importance of using optimally designed models could be anyway pointed out by observing training time. To this end, the Table 4 reports the average training time for all the considered convolutional architectures.

It emerged that some configurations of RegNetY, e.g. the 6.4GF, can get classification accuracy very close to the best performing architectures even saving more than 30% of time for training.

Table 4. Average Training Time (ATT), in minutes, for the deep learning architectures exploited for experiments on extended CRC dataset.

Model	ATT	Model	ATT
DenseNet121	746.67	SE-ResNet50	449.67
EfficientNet-B0	224.33	RegNetY-200MF	101.33
EfficientNet-B1	452.67	RegNetY-400MF	207.33
EfficientNet-B2	477.67	RegNetY-800MF	209.67
EfficientNet-B3	480.67	RegNetY-4.0GF	272.67
EfficientNet-B4	518.00	RegNetY-6.4GF	337.00
EfficientNet-B5	677.67	RegNetY-8.0GF	459.33
EfficientNet-B7	1188.00	RegNetY-12GF	563.69
ResNet50	276.00	RegNetY-16GF	699.33
ResNet152	493.67	RegNetY-32GF	1219.67

3.1 Comparisons to Leading Approaches in the Literature

In this subsection, the best pipelines introduced in this work, according to the results in Table 3, have been compared with the leading approaches in the literature for the task of colon carcinoma grading on the extended CRC dataset[3]. It is possible to observe that the proposed solutions outperformed most of the previous ones both for 2-classes and 3-classes classification tasks. In particular, for the 3-classes task, i.e. the most challenging and interesting for grading purposes, the introduced strategy based on EfficientNet-B2 model achieved the best classification scores. The performance for the binary classification task were only slightly lower than the ones obtained by introducing contextual information as proposed in [23]. It is worth noting that in [23] the context-aware network is a complex architecture comprising different blocks. In particular, the contextual information is learning, by exploiting an attentive mechanism, starting from local features extracted by another CNN as well. Experiments carried out in this paper proved that an easier patch-based approach, if implemented by properly exploiting modern CNN architectures can be effective as well, and for the 3-class grading can work even better.

---

[3] Data for previous works were taken from original papers.

Table 5. Comparisons of best performing pipelines introduced in this paper with leading approaches in the literature.

Model	Average (%) (Binary)	Weighted (%) (Binary)	Average (%) (3-classes)	Weighted (%) (3-classes)
Proposed solutions				
EfficientNet-B2	96.99 ± 2.94	96.65 ± 3.11	**87.58 ± 3.36**	**85.54 ± 2.21**
RegNetY-4.0GF	95.64 ± 0.94	95.37 ± 1.52	84.55 ± 2.57	81.36 ± 1.43
RegNetY-6.4GF	94.31 ± 2.48	94.26 ± 2.15	86.57 ± 2.12	83.58 ± 2.21
Previous works				
ResNet50 [23]	95.67 ± 2.05	95.69 ± 1.53	86.33 ± 0.94	80.56 ± 1.04
LR + LA-CNN [23]	**97.67 ± 0.94**	**97.64 ± 50.79**	86.67 ± 1.70	84.17 ± 84.17
CNN-LSTM [24]	95.33 ± 2.87	94.17 ± 3.58	82.33 ± 2.62	83.89 ± 2.08
CNN-SVM[15]	96.00 ± 0.82	96.39 ± 1.37	82.00 ± 1.63	76.67 ± 2.97
CNN-LR[15]	96.33 ± 1.70	96.39 ± 1.37	86.67 ± 1.25	82.50 ± 0.68

Besides, it is worth noting that patches having a fixed size of 224 × 224 and a batch size of 16 were used in all the experiments reported in this manuscript. Compared architectures used different patch and batch sizes (for instance patches of 1792 × 1792 pixels and a batch of 64 images were used in [23]). Hence, it is possible to conclude that introduced models have been able to extract knowledge from a minor amount of data e to generalize it, making this way the effectiveness of the training process more independent from setup parameters.

4 Conclusion

In this work, the most performing convolutional architectures in object classification tasks have been exploited to improve colon carcinoma grading in histological images. Results on the largest publicly available dataset demonstrated a substantial improvement with respect to leading state-of-the-art approaches. In our opinion, this is a highly important result since the substantial improvement of the accuracy in grading the extracted patches make it possible to develop robustly and objective (reduced bias from different operator expertise and evaluation circumstances) computational tools to support healthcare professionals in the challenging task of carcinoma diagnosis and therapy planning. Future works will deal with the use of ensembles of networks in order to further improve classification accuracy. Besides, transformer architectures, that work as an attention mechanism to highlight the most discriminative areas will be evaluated for this classification task.

Acknowledgment. The authors would like to thank Mr. Arturo Argentieri from CNR-ISASI Italy for his technical contribution on the multi-GPU computing facilities and in setting up and maintain the publicly available dataset.

References

1. Ahmad, Z., Rahim, S., Zubair, M., Abdul-Ghafar, J.: Artificial intelligence (AI) in medicine, current applications and future role with special emphasis on its potential and promise in pathology: present and future impact, obstacles including costs and acceptance among pathologists, practical and philosophical considerations. a comprehensive review. Diagn. Pathol. **16**(1), 1–16 (2021)
2. Altunbay, D., Cigir, C., Sokmensuer, C., Gunduz-Demir, C.: Color graphs for automated cancer diagnosis and grading. IEEE Trans. Biomed. Eng. **57**(3), 665–674 (2009)
3. Awan, R., et al.: Glandular morphometrics for objective grading of colorectal adenocarcinoma histology images. Sci. Rep. **7**(1), 1–12 (2017)
4. Bychkov, D., et al.: Deep learning based tissue analysis predicts outcome in colorectal cancer. Sci. Rep. **8**(1), 1–11 (2018)
5. Carcagnì, P., et al.: Classification of skin lesions by combining multilevel learnings in a DenseNet architecture. In: Ricci, E., et al. (eds.) ICIAP 2019. LNCS, vol. 11751, pp. 335–344. Springer, Cham (2019). https://doi.org/10.1007/978-3-030-30642-7_30
6. Ciompi, F., et al.: The importance of stain normalization in colorectal tissue classification with convolutional networks. In: 2017 IEEE 14th International Symposium on Biomedical Imaging (ISBI 2017), pp. 160–163. IEEE (2017)
7. Das, A., Nair, M.S., Peter, S.D.: Computer-aided histopathological image analysis techniques for automated nuclear atypia scoring of breast cancer: a review. J. Digit. Imaging **33**(5), 1091–1121 (2020)
8. Debelee, T.G., Kebede, S.R., Schwenker, F., Shewarega, Z.M.: Deep learning in selected cancers' image analysis - a survey. J. Imaging **6**(11), 121 (2020)
9. Deng, J., Dong, W., Socher, R., Li, L.J., Li, K., Fei-Fei, L.: ImageNet: a large-scale hierarchical image database. In: 2009 IEEE Conference on Computer Vision and Pattern Recognition, pp. 248–255. IEEE (2009)
10. Deng, S., et al.: Deep learning in digital pathology image analysis: a survey. Front. Med. 1–18 (2020)
11. Fleming, M., Ravula, S., Tatishchev, S.F., Wang, H.L.: Colorectal carcinoma: pathologic aspects. J. Gastrointest. Oncol. **3**(3), 153 (2012)
12. Gertych, A., et al.: Convolutional neural networks can accurately distinguish four histologic growth patterns of lung adenocarcinoma in digital slides. Sci. Rep. **9**(1), 1–12 (2019)
13. Gupta, P., et al.: Prediction of colon cancer stages and survival period with machine learning approach. Cancers **11**(12), 2007 (2019)
14. He, K., Zhang, X., Ren, S., Sun, J.: Deep residual learning for image recognition. In: Proceedings of the IEEE Conference on Computer Vision and Pattern Recognition, pp. 770–778 (2016)
15. Hou, L., Samaras, D., Kurc, T.M., Gao, Y., Davis, J.E., Saltz, J.H.: Patch-based convolutional neural network for whole slide tissue image classification. In: Proceedings of the IEEE Conference on Computer Vision and Pattern Recognition, pp. 2424–2433 (2016)
16. Hu, J., Shen, L., Sun, G.: Squeeze-and-excitation networks. In: Proceedings of the IEEE Conference on Computer Vision and Pattern Recognition, pp. 7132–7141 (2018)
17. Huang, G., Liu, Z., Van Der Maaten, L., Weinberger, K.Q.: Densely connected convolutional networks. In: Proceedings of the IEEE Conference on Computer Vision and Pattern Recognition, pp. 4700–4708 (2017)

18. Kong, B., Li, Z., Zhang, S.: Toward large-scale histopathological image analysis via deep learning. In: Biomedical Information Technology, pp. 397–414. Elsevier, Amsterdam (2020)

19. Radosavovic, I., Kosaraju, R.P., Girshick, R., He, K., Dollár, P.: Designing network design spaces. In: Proceedings of the IEEE/CVF Conference on Computer Vision and Pattern Recognition, pp. 10428–10436 (2020)

20. Rathore, S., Hussain, M., Ali, A., Khan, A.: A recent survey on colon cancer detection techniques. IEEE/ACM Trans. Comput. Biol. Bioinf. **10**(3), 545–563 (2013)

21. Salvi, M., Acharya, U.R., Molinari, F., Meiburger, K.M.: The impact of pre-and post-image processing techniques on deep learning frameworks: a comprehensive review for digital pathology image analysis. Comput. Biol. Med. 104129 (2020)

22. Saxena, S., Gyanchandani, M.: Machine learning methods for computer-aided breast cancer diagnosis using histopathology: a narrative review. J. Med. Imaging Radiat. Sci. **51**(1), 182–193 (2020)

23. Shaban, M., et al.: Context-aware convolutional neural network for grading of colorectal cancer histology images. IEEE Trans. Med. Imaging **39**(7), 2395–2405 (2020). https://doi.org/10.1109/TMI.2020.2971006

24. Sirinukunwattana, K., Alham, N.K., Verrill, C., Rittscher, J.: Improving whole slide segmentation through visual context - a systematic study. In: Frangi, A.F., Schnabel, J.A., Davatzikos, C., Alberola-López, C., Fichtinger, G. (eds.) MICCAI 2018. LNCS, vol. 11071, pp. 192–200. Springer, Cham (2018). https://doi.org/10.1007/978-3-030-00934-2_22

25. Swiderska-Chadaj, Z., et al.: Learning to detect lymphocytes in immunohistochemistry with deep learning. Med. Image Anal. **58**, 101547 (2019)

26. Tan, M., Le, Q.: EfficientNet: rethinking model scaling for convolutional neural networks. In: International Conference on Machine Learning, pp. 6105–6114. PMLR (2019)

27. Tosun, A.B., Kandemir, M., Sokmensuer, C., Gunduz-Demir, C.: Object-oriented texture analysis for the unsupervised segmentation of biopsy images for cancer detection. Pattern Recogn. **42**(6), 1104–1112 (2009)

28. Tsai, M.J., Tao, Y.H.: Deep learning techniques for the classification of colorectal cancer tissue. Electronics **10**(14), 1662 (2021)

29. Vuong, T.L.T., Lee, D., Kwak, J.T., Kim, K.: Multi-task deep learning for colon cancer grading. In: 2020 International Conference on Electronics, Information, and Communication (ICEIC), pp. 1–2. IEEE (2020)

30. Zhou, Y., et al.: CGC-net: cell graph convolutional network for grading of colorectal cancer histology images. In: 2019 IEEE/CVF International Conference on Computer Vision Workshop (ICCVW), pp. 388–398 (2019). https://doi.org/10.1109/ICCVW.2019.00050

Multimedia

Frame Adaptive Rate Control Scheme for Video Compressive Sensing

Fuma Kimishima[(✉)] [iD], Jian Yang [iD], Thuy T. T. Tran [iD], and Jinjia Zhou [iD]

Faculty of Science and Engineering, Hosei University, Tokyo, Japan
{fuma.kimishima.3c,jian.yang.4f}@stu.hosei.ac.jp,
jinjia.zhou.35@hosei.ac.jp

Abstract. Measurement coding compresses the output of compressive image sensors to improve the image/video transmission efficiency. In these coding systems, rate control plays a vital role. The major purpose of rate control is to determine the quantization parameters (or quantization stepsizes) to control the bitrate under available bandwidth (bits limitation) while maximizing the image/video quality. However, most of the existing rate control algorithms apply iterations to find the best quantization parameters, so it suffers from a long processing time and can't efficiently support video processing. This paper presents a frame adaptive rate control scheme for measurement coding. Firstly, the initialized quantization parameter (QP) of the first frame is determined by the triangle quantization method. Moreover, frame adaptive QP adjustment is proposed to refine the QP for each frame. As a result, this work improves the video quality up to 1.56 dB PSNR and reduces the processing time up to 53% compared to the state-of-the-art.

Keywords: Compressive sensing · Rate control · Quantization · Measurement coding

1 Introduction

Recently, with the rapid development of image processing techniques, the number of diverse cameras has increased in our daily life. In order to transfer all obtained signal information, enormous electric power is required, which adversely affects the environment. Therefore, it is significant to compress the data before transmission.

Compressive Sensing: Compressive sensing (CS) is one of the attractive data sampling methods, which can obtain the data in low dimensions. Thus CS has the ability to represent the original data X length of N as measurement data Y length of M, which is the linear combination of the pixel if M is always less than N. This means that CS can compress and sample at the same time. It is expressed as follows.

$$Y = \Phi X \tag{1}$$

S. Sclaroff et al. (Eds.): ICIAP 2022, LNCS 13231, pp. 247–256, 2022.
https://doi.org/10.1007/978-3-031-06427-2_21

where Φ is a M × N measurement matrix. Although CS is efficient, this method still has the problem. That is to sample only once from the entire frame. The large sampling range leads to high computational complexity. The method to solve this problem is the block-based CS (BCS) [5], which divides the frame into several smaller square blocks and samples from each block. Because the sampling area is smaller, sampling is easier and allows data to be read faster.

Measurement Coding: Measurement coding which further compresses the output of CS is useful to obtain a higher compression ratio. Many works [8–10,13] employ prediction based coding algorithms. The difference (residual) between the current measurement and predicted measurement are quantized to reduced data size. Scalar quantization (SQ) is a common quantization method [1], which divides the signal to be quantized by the quantization step. The quantization step depends on the quantization parameters QP. The larger QP is, the more color can be expressed and the original image can be reproduced faithfully. Quantized discrete cosine transform (QDCT) increases the coefficient of the low-frequency part and brings the coefficient of the high-frequency part closer to zero because the human eye doesn't care about the lost high-frequency component, so the result of compression is satisfactory [4]. SQ with differential pulse code modulation (DPCM) subtract the previous block that was quantized from the current one and quantize this little information [7]. The result of this method is better than only SQ with regard to the quality of image.

Rate Control in Measurement Coding: In prediction-based measurement coding systems, rate control plays an important role. The major purpose of rate control is to determine the quantization parameters (or quantization stepsizes) to control the bitrate under available bandwidth (BL: bits limitation) while maximizing the image/video quality. Wan et al. applied a 2-step rate control algorithm. In the first step, the QP candidates are roughly decided, and then in the second step, optimal QP that makes bpp less than BL [10] is searched by iteration. These iteration-based rate control methods repeat the calculation a lot to find the optimum QP so the amount of calculation is large and only be able to handle images.

Main Contributions of This Work: This paper has the following main contributions: 1) This is the first-rate control scheme for video compressive sensing system; 2) We effectively use the quantization and bit-rate information from the neighboring frame to predict the quantization parameter of the current frame, so the iteration is avoided and the processing time is greatly reduced. 3) By further accurately adjusting the quantization parameter of each block, the reconstructed video quality is greatly improved. Experimental result show that this work improves up to 1.56 dB PSNR and the processing time reduction up to 53% compared to the state-of-the-art.

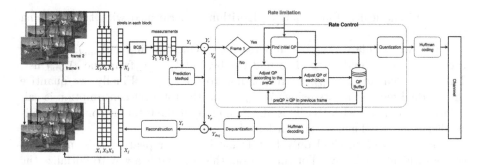

Fig. 1. The entire system for video compressive sensing

2 Proposed Frame Adaptive Rate Control Scheme

2.1 Overall System

Our proposal applies a rate control algorithm using BCS and intra prediction as shown in Fig. 1. On the sampling side, we divide the entire frame into $B \times B$ blocks to obtain BCS measurements. After BCS, the measurement of the current block is predicted by intra prediction and be subtracted from the best prediction candidate. These residual measurements are quantized. The rate control part gives quantization parameters to enable communication. To reconstruct the video, the basic pursuit with Inverse fast Walsh-Hadamard transform (IFWHT) is used.

2.2 Triangle Quantization for QP Initialization

Redundancy can be reduced by intra prediction, but there is potential to further compress. Residual measurements are quantized by SQ to change a scalar value that can be stored with a fixed memory size. The calculation of SQ is expressed as follows:

$$\Delta = \frac{2^n - 1}{2^{QP}} \tag{2}$$

$$y_q = \frac{y_r}{\Delta} \tag{3}$$

$$y_{deq} = y_q \times \Delta + 0.5 \tag{4}$$

where n is the number of bits, Δ is the quantization stepsizes and y_r is the residual measurement. The numerator of Δ means the capacity that can store the residual measurements. If the residual measurement is larger than $2^n - 1$, it is cut to that value. The larger QP is, the smaller Δ is, and quantized measurement denoted by y_q is closer to y_r so the amount of data to transfer is higher. Thus the purpose of rate control is to find optimal QP whose amount of data don't exceed the BL. We propose the rate control algorithm based on triangle threshold method [12], which enables encoder to select the best QP from

QP_{list} automatically. Rate control algorithm is performed with the following Algorithm1.

First, we calculate Δ_{cand} until it reaches 0. Once we find a Δ_{cand}, we apply the triangle threshold method to find the best $QP_{triangle}$ from the Δ_{cand} list. Then we calculate $\Delta_{triangle}$ corresponding to $QP_{triangle}$. Finally, we quantize the residual measurements using $\Delta_{triangle}$ and the total bit consumption is calculated to compare with BL. This algorithm stop if the total bit consumption is smaller than BL. Then, y_q is transferred.

We apply this method only first frame because repeating this method is computationally intensive. For this reason, this method is a way to initialize the QP to realize rate control for the video.

Algorithm 1. QP initialization algorithm

Input: BL
Output: y_q
  Initialization : QP = 0
  while 1 **do**
    $\Delta_{cand}[QP] = \frac{2^n - 1}{2^{QP}}$
    if $\Delta_{cand}[QP] == 0$ **then**
      break()
    end if
    sort(Δ_{cand})ASC
    $QP_{triangle} = Triangle(\Delta_{cand}[QP], size(\Delta_{cand}))$
    $\Delta_{triangle} = \frac{2^n - 1}{2^{QP_{triangle}}}$
    $y_q = \frac{y_r}{\Delta_{triangle}}$
    $H[QP_{triangle}] = Entropy(y_q)$
    if $H[QP_{triangle}] <= BL$ **then**
      end
    end if
  end while

2.3 Frame Adaptive QP Adjustment and Block-Based QP Refinement

There is an important property for achieving the rate control using the QP of the first frame. It means adjacent frames are similar. In other words, in order to find the optimum parameter QP_c for the current frame, we search for rate controllable parameter from a value close to the previous frame one QP_p. It can be expressed by the following equation:

$$QP_c = QP_p + 1 - (i \times 0.5) \qquad (5)$$

where i is initialized with 0. We add 1 to QP_p and reduce it by 0.5 until total bit consumption is less than BL.

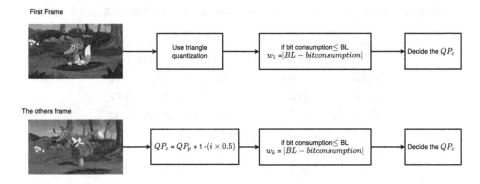

Fig. 2. How to determine the appropriate QP and BL for each frame

There is a difference between the finally obtained bit consumption and BL. Video data can be transferred efficiently if that difference is small. The sum of this difference and BL is the BL of the next frame. It can be represented as follows:

$$w_k = |BL - bitconsumption| \tag{6}$$

$$BL_{k+1} = BL_k + w_k \tag{7}$$

where k is the number of frames and w_k is the difference. As shown in Fig. 2, by changing BL for each frame as well as QP, rate control that can efficiently transfer with a small amount of calculation is realized.

After QP is decided for the current frame, block-based QP refinement is further processed. At the time of compression, quantization is performed block by block, and a round function that round to a closer integer value is used. The range when $y_q \times \Delta$ is subtracted from residual measurement is from $-\frac{QP}{2}$ to $\frac{QP}{2}$. If this difference is large, a lot of information is lost so we give this block measurement a little bigger QP. But if QP is changed too much, the quality of reconstructed image is deteriorate [6], so we change it a little. Expressed in a mathematical formula, it looks like this:

$$QP_c = QP_c + 0.1 - (i \times 0.01) \tag{8}$$

where i is initialized with 0. If rate control isn't realized when decreasing by 0.01, QP_c is applied to all blocks.

3   Results and Comparison

Two datasets WILDTRACK [3] and MCL-JCV [11] are used for testing. All experiments are implemented with MATLAB_2021a. The proposed coding framework is based on BCS and intra-prediction. We use the inverse discrete fast Walsh-Hadamard transform (IFHWT) and L1 minimization to recover the sparse measurement. l1eq_ algorithm in the L1-magic toolbox [2] is used to do the L1 minimization.

Table 1. Comparison with existing works with QP = 15, SR = 0.5 and n = 16

Images	[7]		[8]		This work	
	PSNR	bpp	PSNR	bpp	PSNR	bpp
Baby	27.30	1.71	21.52	1.77	32.16	1.71
Bird	30.71	1.77	21.86	1.82	33.66	1.77
Butterfly	24.77	2.32	15.38	2.38	26.26	2.32
Head	31.48	1.62	25.87	1.63	35.91	1.62
Woman	27.85	1.81	19.71	1.86	32.58	1.81
Bridge	26.73	2.22	21.02	2.15	28.71	2.22
Lenna	27.49	1.67	22.17	1.81	31.13	1.67
Man	27.01	2.12	20.26	2.18	28.36	2.11
Pepper	27.69	1.84	20.80	1.90	31.50	1.84
Baboon	26.32	2.13	21.99	2.06	27.77	2.12
Barbara	26.33	1.94	20.47	2.09	27.87	1.93
Coastguard	28.60	1.93	24.34	1.50	33.01	1.94
Face	31.48	1.62	25.91	1.63	35.89	1.61
Comic	24.81	2.41	18.12	2.37	26.87	2.41
Flowers	26.12	2.08	19.56	2.19	27.31	2.08
Foreman	26.59	1.78	20.43	1.87	31.84	1.77
Monarch	25.56	1.75	20.09	1.81	26.63	1.74
Ppt3	23.07	1.67	17.72	1.74	24.24	1.67
Zebra	23.37	2.26	18.39	2.12	24.42	2.26
Average	27.01	1.93	20.82	1.94	29.80	1.93

3.1 Quantization Performance Comparison

We firstly compare the QP initialization algorithm with the existing measurement coding algorithms EUSIPCO [7] and DCC'19 [8]. EUSIPCO [7] is a block-based quantization method that predicts the current block from the previous block. DCC'19 [8] have a subsequent measurement coding system and use scalar quantization. Table 1 illustrates the PSNR and bpp comparison results with quantization parameter QP = 15, sampling rate SR = 0.5, B = 16, the number of bits n = 16, and resize all the grayscale test images to 256×256. On average, our algorithm achieves an average PSNR of 29.80 dB and an average bpp of 1.93 bpp. We report the improvement of up to 7.66% bpp reduction in image Barbara and an improvement of 12.87 dB PSNR in image Woman. Compared to the previous works, this work obtains the highest PSNR while keeping a similar compression ability.

3.2   Rate Control Performance Comparison

The purpose of our experiment is to control the rate while changing the QP for each frame. For video rate control, it doesn't matter if a frame bpp is larger than BL, and the average from the beginning to that frame should be less than BL. Figure 4 illustrates the average bpp up to each frame for videoSRC04_30 dataset fixed sampling rate SR = 0.5, the block size B = 16, and BL set to 0.5 bpp. We use the same measurement matrix and intra prediction, therefore only the rate control algorithm is compared. As shown in Fig. 4, rate control can't work if QP is fixed in all frames. Compared to the state-of-the-art TMM'21 [10], it turns out that our result is closer to BL and never exceeded that value.

Table 2. Comparing the video quality and processing time with existing work under B = 16

SR = 0.5	BL = 0.5				BL = 1.0			
Sequences	TMM'21 [10]		This work		TMM'21 [10]		This work	
	PSNR	Time (min)	PSNR	Time (min)	PSNR	Time (min)	PSNR	Time (min)
WILD_TRACK	27.61	17.0	27.5	14.4	31.32	17.8	31.54	14.5
videoSRC20_28	33.28	34.2	33.55	31.2	36.55	68.3	36.62	42.2
videoSRC02_16	41.06	45.7	41.17	34.4	43.02	66.8	43.02	39.6
videoSRC16_06	41.99	33.1	42.08	25.6	42.60	34.3	42.56	26.0
videoSRC30_20	38.97	89.9	40.32	48.6	40.18	120.0	41.59	56.4
videoSRC04_30	24.72	29.2	24.85	24.7	27.23	38.1	27.26	30.7
Average	34.61	41.5	34.91	29.8	36.82	57.6	37.10	34.9
SR = 0.75	BL = 0.5				BL = 1.0			
WILD_TRACK	25.90	22.3	26.3	15.8	31.13	21.4	31.18	17.3
videoSRC20_28	31.92	44.5	32.08	41.7	37.09	69.3	37.20	48.3
videoSRC02_16	39.60	58.0	39.81	53.0	43.16	74.1	43.18	51.9
videoSRC16_06	41.90	50.5	42.01	30.7	43.18	44.2	43.15	36.1
videoSRC30_20	38.44	68.3	39.44	55.2	40.47	117.6	42.03	70.4
videoSRC04_30	23.58	37.8	24.20	34.0	27.49	47.4	27.46	39.3
Average	33.56	46.9	33.97	38.4	37.09	62.3	37.37	43.9

We compare our algorithm with TMM'21 [10] only in the rate control part. Table 2 shows the comparison result of average PSNR for the entire frame and the processing time for all test videos with BL = 0.5, 1.0, and SR = 0.5 and 0.75. As shown in Table 2, when SR = 0.5, we report up to 53% minute reduction for videoSRC30_20 when BL = 1.0 and an improvement of 1.41 dB PSNR in videoSRC30_20 when BL = 1.0. When SR = 0.75, our algorithm achieves a reduction of 40% processing time in terms of minute and an improvement of 1.56 dB for videoSRC30_20 when BL = 1.0. Figure 3 shows the visual quality comparison for the first frame to fourth frames compared with only the rate control part of TMM'21 [10]. This work has better visual quality than TMM'21 [10] under the same BL constrain.

254   F. Kimishima et al.

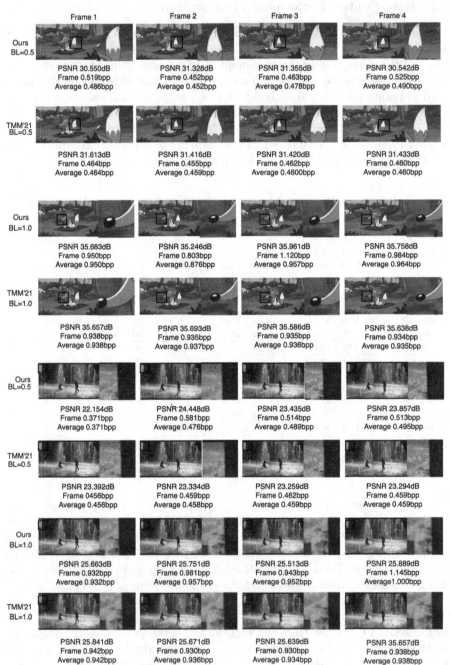

Fig. 3. The visual comparison for the first four frames of videoSRC20_28 and videoSRC04_30 video from the top to the bottom row with SR = 0.5 and B = 16

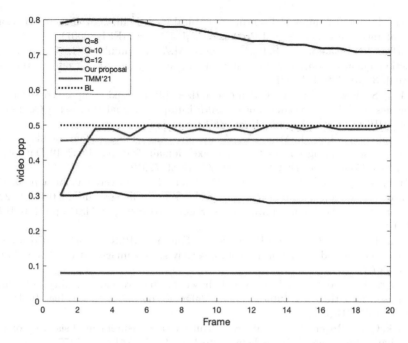

Fig. 4. Comparing the bitrate of each frame with fixed QP algorithms, and iteration based algorithms (TMM'21 [10]. Test data: videoSRC04_30 with SR = 0.5, B = 16 and BL = 0.5

4   Conclusion

In this paper, we propose the frame adaptive rate control scheme for compressively sensed videos. The system applies the triangle threshold based quantization (QP) for the first frame. The quantization parameters of the following frames are determined by referring to bitrate and QP of the previous frames. Experimental results show our proposed algorithm can realize the rate control, with the reduction of the processing time up to 53%, 1.56 dB increase in PSNR compared to the state-of-the-art.

References

1. Boufounos, P.T.: Universal rate-efficient scalar quantization. IEEE Trans. Inf. Theory **58**(3), 1861–1872 (2011)
2. Candes, E., Romberg, J.: l1-magic: recovery of sparse signals via convex programming, vol. 4, p. 14 (2005). www.acm.caltech.edu/l1magic/downloads/l1magic.pdf
3. Chavdarova, T., et al.: Wildtrack: a multi-camera HD dataset for dense unscripted pedestrian detection. In: Proceedings of the IEEE Conference on Computer Vision and Pattern Recognition, pp. 5030–5039 (2018)
4. Docef, A., Kossentini, F., Nguuyen-Phi, K., Ismaeil, I.R.: The quantized DCT and its application to DCT-based video coding. IEEE Trans. Image Process. **11**(3), 177–187 (2002)

5. Gan, L.: Block compressed sensing of natural images. In: 2007 15th International Conference on Digital Signal Processing, pp. 403–406. IEEE (2007)
6. Laska, J.N., Boufounos, P.T., Davenport, M.A., Baraniuk, R.G.: Democracy in action: quantization, saturation, and compressive sensing. Appl. Comput. Harmon. Anal. **31**(3), 429–443 (2011)
7. Mun, S., Fowler, J.E.: DPCM for quantized block-based compressed sensing of images. In: 2012 Proceedings of the 20th European Signal Processing Conference (EUSIPCO), pp. 1424–1428. IEEE (2012)
8. Peetakul, J., Zhou, J., Wada, K.: A measurement coding system for block-based compressive sensing images by using pixel-domain features. In: 2019 Data Compression Conference (DCC), pp. 599–599. IEEE (2019)
9. Tran, T.T., Peetakul, J., Pham, C.D., Zhou, J.: Bi-directional intra prediction based measurement coding for compressive sensing images. In: 2020 IEEE 22nd International Workshop on Multimedia Signal Processing (MMSP), pp. 1–6. IEEE (2020)
10. Wan, R., Zhou, J., Huang, B., Zeng, H., Fan, Y.: APMC: adjacent pixels based measurement coding system for compressively sensed images. IEEE Trans. Multimedia (2021)
11. Wang, H., et al.: MCL-JCV: a JND-based h. 264/avc video quality assessment dataset. In: 2016 IEEE International Conference on Image Processing (ICIP), pp. 1509–1513. IEEE (2016)
12. Zack, G.W., Rogers, W.E., Latt, S.A.: Automatic measurement of sister chromatid exchange frequency. J. Histochem. Cytochem. **25**(7), 741–753 (1977)
13. Zhou, J., Zhou, D., Guo, L., Takeshi, Y., Goto, S.: Measurement-domain intra prediction framework for compressively sensed images. In: 2017 IEEE International Symposium on Circuits and Systems (ISCAS), pp. 1–4. IEEE (2017)

Shot-Based Hybrid Fusion for Movie Genre Classification

Tianyu Bi[✉], Dimitri Jarnikov, and Johan Lukkien

Department of Mathematics and Computer Science, Eindhoven University
of Technology, P.O. Box 513, 5600 MB Eindhoven, The Netherlands
{t.bi.1,d.s.jarnikov,j.j.lukkien}@tue.nl

Abstract. Multi-modal fusion methods for movie genre classification
have shown to be superior over their single modality counterparts. However, it is still challenging to design a fusion strategy for real-world scenarios where missing data and weak labeling are common. Considering
the heterogeneity in different modalities, most existing works design late
fusion strategies that process and train models per modality, and combine the results at the decision level. A major drawback in such strategies
is the potential loss of across-modality dependencies, which is important
for understanding audiovisual contents. In this paper, we introduce a
Shot-based Hybrid Fusion Network (SHFN) for movie genre classification. It consists of single-modal feature fusion networks for video and
audio, a multi-modal feature fusion network working on a shot-basis,
and finally a late fusion part for video-level decisions. An ablation study
indicates the major contribution from video and the performance gain
from the additional modality, audio. The experimental results on the
LMTD-9 dataset demonstrate the effectiveness of our proposed method
in movie genre classification. Our best model outperforms the state-of-
the-art method by 5.7% on AUPRC(micro).

Keywords: Multi-modal fusion · Multi-label classification · Movie
genre classification

1 Introduction

Movie genre describes the style of a movie. With the development of online movie
services, movie genre classification plays an important role in many applications
such as movie indexing, searching, and recommendation systems [1]. Typically,
a movie is labelled by one or multiple genres, which makes it a multi-label classification problem. However, movie genres are immaterial concepts that cannot
be easily identified in a frame like an object-related class or pinpointed to a
sequence of frames like an action recognition class [11]. This inexact labeling
[16] of movie genres and the multi-modal nature of movies make it a challenging
task to identify genres in a movie.

Recently, deep learning methods became state-of-the-art in movie genre classification [3,4,9,12]. Different modalities such as video, audio, image, text, and

© The Author(s), under exclusive license to Springer Nature Switzerland AG 2022
S. Sclaroff et al. (Eds.): ICIAP 2022, LNCS 13231, pp. 257–269, 2022.
https://doi.org/10.1007/978-3-031-06427-2_22

their combinations have been used in identifying movie genres. Wei and Hung [3] use deep neural networks to identify movie genres from poster images. Simoes et al. [9] propose a CNN-Motion model that classifies movie genres from video frames. Wehrmann and Barros [12] propose a CTT-MMC model that learns movie genres from video and audio modalities. Ali and Pinar [4] use plot summaries with a bidirectional LSTM in movie genre prediction.

Although the existing methods perform well, there are still limitations that need to be addressed. First, these methods employ a pretrained image classifier to extract visual features per video frame and use them as video representations. Although these features can be later fused to learn temporal information, the spatial correlation information among frames is lost. A more effective visual representation learning method is required to capture such information for movies. Second, for most of the multi-modal methods, visual and audio representations are learned separately from entire video and audio tracks before fusion and prediction, which is late fusion strategy [8]. This fusion strategy suffers from losing information of mutual correlation between two modalities [15]. To learn correlations among multi-modals, a more effective fusion strategy such as model-level fusion is required [5].

Apart from that, real-world scenarios for movie genre classification have not been well studied in the previous works. For example, there may not be sufficient time and resources to process the entire content to get movie genres. In that case, the algorithm should be able to give reliable early predictions using sufficient movie content. Besides, there could be missing data from one modality at some part of the content and the multi-modal streams may not be always synchronized. The absence of a modality and the assumption of strict synchrony often lead to corruption and performance degradation [13,14]. We need a fusion method that can combine asynchronous modalities and effectively learn multi-modal representations to achieve comparable performance with synchronous fusion methods.

To address these issues, we design a new multi-model fusion architecture using deep learning neural networks that can learn high-quality video and audio representations, effectively combine multiple modalities, and flexibly work in different scenarios.

In this paper, we propose a Shot-based Hybrid Fusion Network (SHFN) for movie genre classification. It consists of three components: video-net, audio-net, and multi-modal-net. The video-net generates video representations from video frame-groups and learns long-term information from a sequence of frame-groups (a video shot) by visual feature fusion. Each frame-group contains a certain number of video frames. Similarly, the audio-net generates audio representations from audio frames and employs audio feature fusion on a group of audio frames (an audio shot) to learn temporal information. The video shot and audio shot can be set up properly to have the same size. In this way, video and audio features can be combined on a shot-basis. The combination can be a synchronous or an asynchronous way. Next, the multi-modal-net combines the shot-based visual and audio features to make shot-level predictions and conducts late fusion to get the video-level results. Since our method is shot-based, it is flexible and scalable to different input lengths of the movie. We also introduce an asynchronous multi-modal data augmentation mechanism to enable model training in asynchronous or missing data scenarios. In the experiments, we use LMTD-9 [11] as the dataset to evaluate our method for movie genre classification. The experimental results show that our SHFN outperforms state-of-the-art methods, and is applicable in several real-world scenarios.

There are two main contributions. First, we design a deep learning hybrid fusion network that learns and combines multi-modal representations flexibly in a shot-based way. This method enables early predictions by running on the first few shots rather than running through the entire content. Second, we introduce an effective data augmentation mechanism that can be applied to missing data and video-audio asynchronous scenarios and further improve the performance. Extensive experiments are conducted to investigate configurations for representation learning and fusion.

The paper is organized as follows. Section 2 describes our multi-modal fusion approach in detail. Section 3 presents the experiments and results. Section 4 ends this paper with a conclusion and future works.

2   Shot-Based Hybrid Fusion Networks

In this section, we introduce the components of our SHFN in detail. First, we illustrate the architecture of the video-net. Then, we explain the design of the audio-net. Finally, we discuss multi-modal fusion and multi-label classification.

2.1   Video-Net

Inspired by the VRFN in [1], we build the video-net in SHFN for learning and fusing video representations. Figure 1 shows the architecture of the video-net. It consists of two components: Video Feature Learning Network (VFLN) and Video Feature Fusion Network (VFFN). VFLN is used to learn visual representations from short video segments and VFFN is used to fuse features from several video segments to learn long-term information from the video.

We use a pretrained Inflated 3D-ConvNet (I3D) for VFLN because it has good performance on many video tasks [2]. We leverage transfer learning to create high-quality video representations. Kinetics [2] is selected as the large-scale video dataset to pretrain the I3D network. The video representations learned from VFLN can capture characteristic features from a short sequence of frames and that is good enough for general video tasks such as action recognition. However, movie genre is an immaterial high-level concept that cannot be easily pinpointed to one frame or even a sequence of frames like a general object-related class [11]. Capturing movie genres from a movie may require the understanding of lower-level concepts such as scenes, objects, human actions, moods, and their relationships from a longer period of the content. To achieve this, we create the

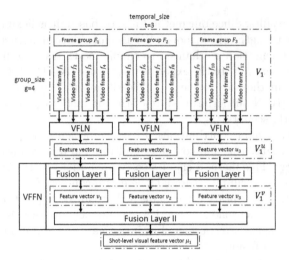

Fig. 1. Video-Net: visual representations are learned from a video shot.

VFFN using C3D-LSTM to conduct visual feature fusion [8] on the representations of short video segments. As a variation of CNN-RNN [10], C3D-LSTM is capable of learning temporal information from sequence data.

The video-net works in a shot-based way. It takes one video shot as one input unit. In our work, a video shot is a video segment extracted from the video content with a certain shot size. A video shot can be represented by a fixed-length sequence of successive video frames. To preprocess and fetch a video shot, we use a shot-level data fetcher. It first preprocesses the raw video frames to the required format and then groups the frames sequentially into video shots.

As shown in Fig. 1, to fetch the input of VFLN, the data fetcher collects video frames into frame-groups (F_i) and combines frame-groups to form video shots. Here, the number of video frames in a frame-group is set to g, which is called the group_size. Since VFLN takes one frame-group each time as the input unit, group_size is a hyperparameter that we can tune for VFLN. Another hyperparameter is temporal_size t, which defines how many frame-groups are contained in a video shot. Since VFFN takes one video shot as the input unit, we can tune temporal_size for optimizing VFFN.

Formally, a video shot V_i is fetched as the input of the video-net. It is represented by a list of frame-groups $[F_{i \cdot t}, F_{i \cdot t+1}, ..., F_{(i+1)t-1}]$, in which each frame-group is represented by a list of video frames (e.g., $F_i = [f_{i \cdot g}, f_{i \cdot g+1}, ..., f_{(i+1)g-1}]$). Thus, in total the video shot V_i at the input stage consists of $g \times t$ video frames. Using VFLN, each input F_i is transformed into an intermediate video representation u_i. Now the video shot is represented by $V_i^u = [u_{i \cdot t}, u_{i \cdot t+1}, ..., u_{(i+1)t-1}]$. Next, VFFN takes the video shot representation V_i^u as the input data and conducts visual feature fusion in two steps. In the first step, it fuses the 3d feature vector u_i into a 1d feature vector v_i using C3D and Flatten (Fusion Layer I) to learn temporal features within a frame-group. The

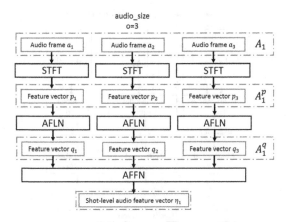

Fig. 2. Audio-Net: audio representations are learned from an audio shot.

video shot at this stage is represented by $V_i^v = [v_{i \cdot t}, v_{i \cdot t+1}, ..., v_{(i+1)t-1}]$. In the second step, LSTM (Fusion Layer II) is used to learn temporal features from the video shot representation V_i^v. The outcome is the shot-level feature vector μ_i that will be the visual input data representing the video shot V_i for multi-modal feature fusion.

2.2 Audio-Net

In SHFN, the audio-net is responsible for processing audio data, generating intermediate representations, and fusing audio features. Figure 2 shows the architecture of the audio-net, which consists of a Short-Time Fourier Transform (STFT) feature extractor, an Audio Feature Learning Network (AFLN), and an Audio Feature Fusion Network (AFFN). STFT is used to extract mel spectrogram features from audio data. AFLN is a VGGish network that generates intermediate audio representations from each processed temporal slice of an input audio shot. The VGGish network is a VGG-like audio classification model initialized with weights that are pretrained on a large-scale audio event detection dataset, AudioSet [6]. This dataset consists of 632 audio event classes with more than 2 million sound clips [6]. This guarantees the capability of the audio-net in generating high-quality audio representations. Finally, the audio representations from AFLN are fed to AFFN for feature fusion and learning long-term information from the audio. We compare three different methods (LSTM, Reshape and Merge) for AFFN in the experiments.

Similar to the video-net, the audio-net is also working in a shot-based way. An audio shot is an audio segment extracted from the audio content with a certain duration. The audio shot size is set according to the video shot size so that they have the same duration. We use the same shot-level data fetcher to preprocess raw audio data and make audio shots for the audio-net. As shown in Fig. 2, the data fetcher collects o audio frames as one audio shot, where o is

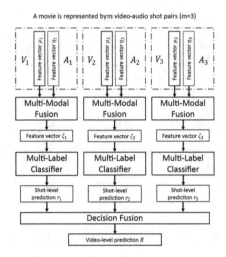

Fig. 3. The workflow of multi-modal feature fusion, classification, and decision fusion.

called the audio_size, which is the input size of audio-net. This hyperparameter can be tuned to optimize audio-net. The hyperparameters g, t, and o need to be tuned properly to have the same shot size on video and audio.

Formally, the audio-net takes the audio shot A_i as input, which is represented by a list of successive audio frames $[a_{i \cdot o}, a_{i \cdot o+1}, ..., a_{(i+1)o-1}]$. Each audio frame a_i is then processed by STFT to extract a mel spectrogram feature vector p_i. Now the audio shot is represented by $A_i^p = [p_{i \cdot o}, p_{i \cdot o+1}, ..., p_{(i+1)o-1}]$. Next, the AFLN takes each feature vector p_i as input data and generates an intermediate audio representation q_i. The outcome audio shot is represented by $A_i^q = [q_{i \cdot o}, q_{i \cdot o+1}, ..., q_{(i+1)o-1}]$. Finally, AFFN applies feature fusion methods on these representations and learns a shot-level audio feature vector η_i. It is the audio representation for the given audio shot A_i and is used together with the shot-level visual representation μ_i in multi-modal fusion and classification.

2.3 Multi-modal Feature Fusion and Decision Fusion

We apply multi-modal fusion on video and audio data for movie genre classification. As mentioned before, the movie content is divided into shots in pre-processing. Visual and audio features are computed from video shots and audio shots. Multi-modal-net conducts multi-modal fusion on each video-audio shot pair. The two modalities are correlated in a shot relationship, and a shot-level multi-modal representation is learned in this step.

The next step is classification. Movie genre classification is a multi-label classification problem. We use Binary Relevance (BR), a widely-used method for multi-label classification in deep learning. BR is a problem transformation method, which uses C binary classifiers for C classes [7]. In our network, the

BR module is a Dense layer with C Sigmoid neurons. Each Sigmoid neuron corresponds to a movie genre class.

The final step is decision fusion. In the previous steps, a movie is represented by a collection of video and audio shot pairs, and the classification is conducted on each shot pair. We need decision fusion to get the final prediction result from all movie shots. In our decision fusion strategy, the element-wise average is calculated for the multi-label classes from the shot-level predictions.

Formally, a movie can be represented by m video-audio shot pairs. From each shot pair (V_i, A_i), the video-net and audio-net can learn shot-level representations μ_i and η_i respectively from V_i and A_i. Then, the visual-audio representation ζ_i is learned from μ_i and η_i by multi-modal feature fusion. Now the movie is represented by a collection of shot-level multi-modal representations $[\zeta_1, \zeta_2, ..., \zeta_m]$. Next, the multi-label classifier predicts on each shot representation (ζ_i) and obtains the shot-level result (r_j). The final prediction result is calculated by decision fusion as follows: $R = decision\_fusion([r_1, r_2, ..., r_m])$. Figure 3 shows the workflow of multi-modal fusion, multi-label classification, and decision fusion.

3 Experiments and Results

3.1 Dataset and Data Preprocessing

We choose LMTD-9 [11] from publicly available datasets because it has a sufficient amount of data and reliable baselines for comparison. It contains over 4000 movie trailers: 2861 for training, 374 for validation, and 772 for testing. There are nine movie genres: action, adventure, comedy, crime, drama, horror, romance, SciFi, and thriller. A trailer can have one, two, or three genres as its labels.

We preprocess video and audio data for experiments. To enable transfer learning on VFLN, we use the same video preprocessing configurations as in the pretrained I3D model. In video preprocessing, we extract frames at 25 frames per second. For each frame, we resize the shorter side to 256 and crop a 224×224 piece from the center of the image. The pixel values are normalized to $[-1.0, 1.0]$. We design a data fetcher to make batches of data from the dataset for training and validation. Depending on the group_size, temporal_size, and batch_size, it loads data from b_{size} batches of trailers and fetches 1 video shot each time from each trailer. A video shot has t video segments (frame-groups) and each segment contains g frames. Therefore, the size of the mini-batch is $b_{size} \times g \times t$. Each test video is separated into video shots in test mode, and the data fetcher loads $1 \times g \times t$ frames as one video shot for shot-level inference.

We follow the audio preprocessing steps of AudioSet [6] and use the same configurations to enable transfer learning from the pretrained AudioSet model to our AFLN. Each audio frame is a 960 ms segment. STFT is applied on each segment with a 10 ms step and a 25 ms window. We obtain a log-mel spectrogram with 96×64 bins from each frame, where 64 is the mel-spaced frequency bins. Next, the data fetcher selects b_{size} trailers and fetches 1 audio shot for each trailer. An audio shot contains o (audio_size) audio frames. Thus, the size of an

Table 1. Settings for SHFN variations.

Multi-modal fusion	Concatenate, Maximum, Average
Audio fusion	LSTM, Reshape, Maximum, Average
Training	Synchronous, Asynchronous
LSTM_Size	32, 64, 128, 256, 512, 1024
(Temporal_Size, Audio_Size)	(3,8) (6,16) (9,24)

audio mini-batch is $b_{size} \times o$. Each test audio is separated into audio shots as the input data for the audio-net in test mode.

To train the models, we use both synchronous and asynchronous data augmentation mechanisms. The synchronous one needs the alignment of the input video shot and audio shot. We set up the hyperparameters (group_size, temporal_size, and audio_size) properly to get the same duration on video and audio. Each video shot contains three frame-groups (temporal_size = 3) with 64 frames in each frame-group (group_size = 64), while each audio shot consists of 8 audio frames (audio_size = 8). In this way, the video and audio shots are both 7.68 s and they can be easily aligned and extracted from the same position of the movie trailer. In the asynchronous approach, we use the same settings except that the video and audio shots are extracted from arbitrary locations of the same content.

3.2 Models Variations for Evaluation

In the experiment, we design several variations of SHFN models using different architecture settings. Table 1 illustrates the options for each of the architecture settings including multi-modal fusion, audio fusion, training data, LSTM_size, temporal_size, and audio_size. The multi-modal fusion has three options: *Concatenate*, *Maximum*, and *Average*. The visual and audio features are either concatenated or merged by maximum/average pooling in this step. The options of audio fusion include *LSTM*, *Reshape*, *Maximum*, and *Average*. The *Reshape* layer concatenates the feature vectors into a longer one, while *Maximum* and *Average* merge multiple feature vectors into a single one. The *LSTM* fuses the feature vectors temporally and outputs a new feature vector. For all the combinations of these options, different LSTM_sizes, temporal_sizes, and audio_sizes are tested in the experiments.

3.3 Experiment Settings and Evaluation Metrics

We implement our networks using Keras and TensorFlow with 1 Nvidia TITAN Xp GPU. To avoid overfitting, we use a Dropout layer and set the drop rate to 0.5. We apply the EarlyStopping callback to stop training when the monitored metric (val_loss) stops improving. For all networks, we set the batch_size to 64 in training stage. We use Adam with a learning rate of 0.001. The multi-label

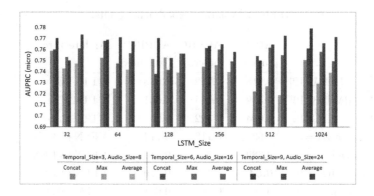

Fig. 4. Compare SHFN models with different hyperparameter settings.

binary cross-entropy loss function $L(y, \hat{y}) = \sum_{i=1}^{C}(-\frac{1}{N}\sum_{j=1}^{N}(y_{ij} * \log(\hat{y_{ij}}) + (1 - y_{ij}) * \log(1 - \hat{y_{ij}})))$ is used to optimize our networks.

To compare our method with the previous work, we use the same evaluation metrics. Area Under Precision-Recall Curve (AUPRC) was applied in [11] for evaluating multi-label classification. AUPRC calculates a real number from the Precision-Recall (PR) curve, in which the precision and recall values are calculated from different thresholds. We use three variations of AUPRC: AUPRC(macro), AUPRC(micro), and AUPRC(weigthed). AUPRC(macro) is calculated on each class with an unweighted average. AUPRC(micro) is computed globally on all examples. AUPRC(weighted) is performed on each class with a weighted average. For all these metrics, the larger the results, the better the performance.

3.4   Results and Discussion

Figure 4 illustrates the AUPRC(micro) results for SHFN variations. Each SHFN model is set up with different combinations of the hyperparameters: multi-modal fusion methods, temporal_size, audio_size, and LSTM_size. Here, we use the default C3D-LSTM for video fusion and use LSTM as the audio fusion method. The LSTM_size is tuned and applied for the LSTM layer in both video-net and audio-net. From the results, we can see the trend that the performance increases when using a larger temporal_size and audio_size. It is likely that the model captures more important information for classification from a larger group of video and audio segments.

To investigate the effectiveness of each module in our network, we conduct ablation experiments that compare the performance of SHFN models with Video-Only models and Audio-Only models. The Video-Only models and Audio-Only models are the video-nets and audio-nets from the SHFN and are trained alone for prediction. Table 2 shows the evaluation results. Here, t and o represent temporal_size and audio_size. We see a similar trend in Video-Only and Audio-Only methods in that the performance increases when we use a larger input size. We

Table 2. Ablation experiment results.

Models	AUPRC (macro)	AUPRC (micro)	AUPRC (weighted)	Architecture
CTT-MMC-A [12]	0.618	0.712	0.683	CTT+FC
CTT-MMC-B [12]	0.599	0.704	0.661	CTT+2FC
CTT-MMC-C [12]	0.624	0.722	0.697	CTT+Maxout
Video-Only($t = 3$)	0.634	0.718	0.714	VFLN+VFFN
Video-Only($t = 6$)	0.629	0.719	0.710	VFLN+VFFN
Video-Only($t = 9$)	**0.645**	0.735	0.720	VFLN+VFFN
Audio-Only($o = 8$)	0.568	0.688	0.667	LSTM(1024)
Audio-Only($o = 16$)	0.571	0.698	0.673	LSTM(256)
Audio-Only($o = 24$)	0.582	**0.717**	0.681	LSTM(512)
SHFN($t = 3, o = 8$)	0.653	0.758	0.737	Concat+LSTM(32)
SHFN($t = 6, o = 16$)	0.659	0.766	0.739	Concat+LSTM(64)
SHFN($t = 9, o = 24$)	**0.661**	**0.778**	**0.742**	Concat+LSTM(1024)

Table 3. Compare Sync and Async training mechanism.

Models	Input Size	AUPRC (macro)	AUPRC (micro)	AUPRC (weighted)	Training
Concat+LSTM(32)	(3,8)	0.653	0.758	0.737	Sync
Concat+LSTM(32)	(3,8)	0.663	0.763	0.743	Async
AVG+MAX	(6,16)	0.657	0.767	0.736	Sync
AVG+MAX	(6,16)	0.666	0.763	0.742	Async
MAX+AVG	(9,24)	0.673	0.773	0.748	Sync
MAX+AVG	(9,24)	0.672	0.784	0.751	Async

notice that our Video-Only models outperform the Audio-Only models, which is reasonable since the visual data contains more important information for classification. The Video-Only($t = 9$) improves AUPRC(macro) by 3.4% compared with CTT-MMC-C [12], the state-of-the-art video-only method. The Audio-Only models perform well, and the best one (Audio-Only($o = 24$)) even outperforms the previous video-only methods (CTT-MMC-A [12] and CTT-MMC-B [12]) on AUPRC(micro). These results demonstrate the effectiveness of representation learning and single-modal fusion in the video-net and audio-net. When it comes to multi-modal fusion, there is a significant improvement in model performance. Compared with Video-Only($t = 9$), the SHFN($t = 9, o = 24$) improves AUPRC(micro) by 5.9%, which shows the complementary contribution of the audio modality.

Table 4. Using partial data in inference, we compare the prediction results in AUPRC(micro) for using 1/4, 2/4, 3/4, and 4/4 (entire) of the movie trailer data.

Model	Position	1/4	2/4	3/4	4/4
MAX+AVG(9,24)	Start	0.687	0.737	0.765	0.784
MAX+AVG(9,24)	End	0.739	0.776	0.783	0.784
MAX+AVG(9,24)	Middle	0.741	0.765	0.776	0.784
CTT-MMC-TN [12]	–	–	–	–	0.742

Table 3 compares the performance of synchronous (Sync) and asynchronous (Async) training mechanisms. In the experiments, we select three typical SHFN models. They are the model architectures that achieve good performance in each (temporal_size, audio_size) setting: (3,8), (6,16), and (9,24). We can see that with the use of our shot-based hybrid fusion networks, the Async training mechanism can achieve comparable and even a bit better performance compared with the Sync way. A potential reason is that the model could capture supplementary information in the multi-modal features with an unfixed combination of visual and audio data. It demonstrates that our method is applicable in a data asynchronous scenario.

Finally, we would like to investigate how well our models can perform using much fewer resources in prediction. We make 12 different test sets, and each contains a different portion (1/4, 2/4, 3/4, 4/4) of movie trailer data that are extracted from one of three positions (Start, End, Middle) of the movie trailer. We use MAX+AVG(9,24) in this test, which uses *Maximum* for multi-modal fusion and *Average* for audio feature fusion. The results are summarized in Table 4. With more data in inference, the performance increases reasonably. When we use a portion of movie trailer data from the start of the content, it achieves much lower results compared with the predictions using the end and middle parts of the content. The reason could be that the first few movie trailer segments contain more noises, such as genre-free content. Compared with state-of-the-art methods, our methods (End and Middle) achieve much better performance, even when using only half of the data for inference. This demonstrates that our method is more flexible in resource usage and is applicable in a missing data scenario.

4   Conclusion

In this paper, we proposed SHFN for movie genre classification. This network consists of a video-net, an audio-net, and a multi-modal-net. The video-net and audio-net generate characteristic features in a shot basis and conduct temporal feature fusion from single-modal data. The multi-modal-net combines the shot-level visual and audio features, conducts multi-label classification, and predicts movie genres. The experimental results show that our models outperform state-of-the-art models using fewer data in inference. It demonstrates the effectiveness and efficiency of our method. In future work, we suggest investigating different representation learning and feature fusion methods and including more modalities (e.g., image and text) in predictions.

References

1. Bi, T., Jarnikov, D., Lukkien, J.: Video representation fusion network for multi-label movie genre classification. In: 2020 25th International Conference on Pattern Recognition (ICPR), pp. 9386–9391. IEEE (2021)
2. Carreira, J., Zisserman, A.: Quo Vadis, action recognition? A new model and the kinetics dataset. In: Proceedings of the IEEE Conference on Computer Vision and Pattern Recognition, pp. 6299–6308 (2017)
3. Chu, W.T., Guo, H.J.: Movie genre classification based on poster images with deep neural networks. In: Proceedings of the Workshop on Multimodal Understanding of Social, Affective and Subjective Attributes, pp. 39–45 (2017)
4. Ertugrul, A.M., Karagoz, P.: Movie genre classification from plot summaries using bidirectional LSTM. In: 2018 IEEE 12th International Conference on Semantic Computing (ICSC), pp. 248–251. IEEE (2018)
5. Gadzicki, K., Khamsehashari, R., Zetzsche, C.: Early vs late fusion in multimodal convolutional neural networks. In: 2020 IEEE 23rd International Conference on Information Fusion (FUSION), pp. 1–6. IEEE (2020)
6. Gemmeke, J.F., et al.: Audio set: an ontology and human-labeled dataset for audio events. In: 2017 IEEE International Conference on Acoustics, Speech and Signal Processing (ICASSP), pp. 776–780. IEEE (2017)
7. Gibaja, E., Ventura, S.: A tutorial on multilabel learning. ACM Comput. Surveys (CSUR) **47**(3), 1–38 (2015)
8. Mangai, U.G., Samanta, S., Das, S., Chowdhury, P.R.: A survey of decision fusion and feature fusion strategies for pattern classification. IETE Techn. Rev. **27**(4), 293–307 (2010)
9. Simões, G.S., Wehrmann, J., Barros, R.C., Ruiz, D.D.: Movie genre classification with convolutional neural networks. In: 2016 International Joint Conference on Neural Networks (IJCNN), pp. 259–266. IEEE (2016)
10. Wang, J., Yang, Y., Mao, J., Huang, Z., Huang, C., Xu, W.: CNN-RNN: a unified framework for multi-label image classification. In: Proceedings of the IEEE Conference on Computer Vision and Pattern Recognition, pp. 2285–2294 (2016)
11. Wehrmann, J., Barros, R.C.: Convolutions through time for multi-label movie genre classification. In: Proceedings of the Symposium on Applied Computing, pp. 114–119 (2017)
12. Wehrmann, J., Barros, R.C.: Movie genre classification: a multi-label approach based on convolutions through time. Appl. Soft Comput. **61**, 973–982 (2017)
13. Wimmer, M., Schuller, B., Arsic, D., Radig, B., Rigoll, G.: Low-level fusion of audio and video feature for multi-modal emotion recognition. In: Proceedings of the 3rd International Conference on Computer Vision Theory and Applications VISAPP, Funchal, Madeira, Portugal, pp. 145–151 (2008)
14. Wu, C.H., Lin, J.C., Wei, W.L.: Survey on audiovisual emotion recognition: databases, features, and data fusion strategies. APSIPA Trans. Sign. Inf. Process. **3** (2014)

15. Zeng, Z., Pantic, M., Roisman, G.I., Huang, T.S.: A survey of affect recognition methods: audio, visual, and spontaneous expressions. IEEE Trans. Pattern Anal. Mach. Intell. **31**(1), 39–58 (2008)
16. Zhou, Z.H.: A brief introduction to weakly supervised learning. Natl. Sci. Rev. **5**(1), 44–53 (2018)

Landmark-Guided Conditional GANs for Face Aging

Xin Huang[1] and Minglun Gong[2]([⊠])

[1] Memorial University of Newfoundland, St. John's, NL A1C 5S7, Canada
xhuang@mun.ca
[2] University of Guelph, Guelph, ON N1G 2W1, Canada
minglun@uoguelph.ca

Abstract. Face aging, which alters a person's facial photo to the appearance at a different age, is a popular topic in multimedia applications. Recently, conditional Generative Adversarial Networks (cGANs) have achieved visually impressive progress in this area. However, generating a convincing aging appearance while preserving the person's identity is still a challenging task. In this paper, we propose a novel Landmark-Guided cGAN (LGcGAN), which not only synthesizes texture changes related to aging, but also alters facial structures accordingly. We adapt a built-in attention mechanism to emphasize the most discriminative regions relevant to aging and minimize changes that affect personal identity and background. Conditioned with age vectors, the primal cGAN in our symmetric network converts input faces to target ages, and the dual cGAN inverts the previous task, which feeds synthesized target faces back to the original input age scope for enhancing age consistency. Both qualitative and quantitative experiments show that our method can generate appealing results in terms of image quality, personal identity, and age accuracy.

Keywords: Face aging · Conditional GANs · Dual-learning · Landmark attention

1 Introduction

Aging is a gradual and continuous process over time, which contains rich and complex alterations in facial features. Automatically rendering a given face under different ages is a problem of interests in Computer Vision. Face aging has wide applications in areas such as identifying a missing child at different ages [4], special effects in entertainment [3] and person identification [18].

In the last decade, there are two traditional types of methods for face aging: the prototype-based methods [8,21] and the physical model-based methods [15,16]. In the prototype-based methods, faces are grouped according to different ages and an average face is constructed as the prototype for each age group. Aging patterns between groups are learned and texture differences are transferred for synthesizing an aged face. However, synthesized aging faces always lose personal identity information, resulting in unrealistic visual results. Physical model-based methods apply a more complex parameter model to describe variations of winkles, muscles, hair colors, skin, etc. [16]. These methods heavily depend on face aging sequence images of the same individual, which are time-consuming to collect. Besides, these traditional methods learn a single mapping between two input age groups and hence cannot transfer an input image to an arbitrary age group.

With the success of deep convolution networks in image generation [6, 7], Generative Adversarial Networks (GANs) based models are developed as powerful tools for age progression [13, 20]. Features of generated face images are controlled by the age-condition in the conditional GANs (cGANs) [12], which can dramatically reduce artifacts and produce more appealing aging effects of the rendered images. In the last two decades, many recent face aging studies attempt to supervise generation results by addressing attribute manipulation [10], identity consistency [20] and age editing [22]. To capture the age-irrelevant features which impact individuals differently in their oldness, Liu et al. [9] propose a multimodal based on face disentanglement.

The problem of facial aging is far from being solved though, due to the difficulties of formulating the complex aging mechanism and the diversity of aging patterns. Most existing works have the following limitations: i) they need long-range sequential labeled faces of the same person for training, which are very rare in existing datasets; ii) they focus on handling texture changes (i.e., wrinkles) and ignore structural variations related to aging, making them ineffective for handling large age spans; and iii) they preserve personal identity through minimizing the differences between inputs and synthesized results, which often leads to blurry artifacts and insufficient variations. To tackle the aforementioned limitations, we propose a novel Landmark-Guided conditional Generative Adversarial Network (LGcGAN) for face aging. Inspired by CycleGAN [24], LGcGAN performs both a primal task (face aging) and a dual task (face reconstruction) simultaneously.

Specifically, the primal architecture of our proposed LGcGAN consists of three modules: a cGAN module with a built-in attention, a landmark prediction module, and a conditional discriminator. The inverse dual face aging framework for face rejuvenation has the same components with previous architecture. The generator G (both primal and dual) receives an input image and a target age code, and learns an attention-content mask in a similar way as in [14]. The attention mask learns the modified facial regions which underline aging diverse effects, whereas the content color mask learns pixels which focus on the person identity and constant features. The final output of the generator is a combination of the attention masks and content masks. Since the attention mechanism only modifies the regions relevant to face aging, it preserves the background and the identity of the person well. Furthermore, to supervise long-term age progression, the generator is not only conditioned by age labels but also conditioned by facial landmark code to enhance the different facial structures, features of head's sense organs and poses. To encourage the synthesized faces fall into the target age group, we send the generated aged faces to a pre-trained age classifier and add an age classification loss to the objective.

The dual task of input face reconstruction is treated as applying age transform on the result image using the age of input image as condition. If a face generated by the primal task preserves the person's identity well and is consistent with the target age group, then the reconstruction step should transform the aging results back to the input image. Hence, a reconstruction loss can be defined here to measure the error. Additionally, the discriminator D consists of an unconditional discriminator and a conditional discriminator which is combined with an age vector, aiming to make the generated face be more photo-realistic and guarantee the synthesized face lies in the target age group.

The main contributions of this study are: i) We develop a novel symmetric network for face aging, which incorporates both age progression (primal) and face reconstruction (inverse) operations. The primal and inverse reconstruction processes enhance both the quality of synthesized face images and the closeness of the appearance with the target age. ii) The generator is conditioned with an external landmark attention which can help to generate faces with more precise facial structures and poses because the facial skeleton changes when people go through large age span. iii) We highlight a built-in attention-content mechanism to better reflect the dynamic regions relevant to face aging with the result that aging accuracy and identity consistency can be simultaneously achieved. Besides, the ghost artifacts between adjacent age groups can be significantly reduced. iv) Extensive qualitative and quantitative experiment results on the UTKFace dataset demonstrate the ability of the proposed method in rendering effective aging effects in terms of image quality, personal identity, and age accuracy.

2   Related Work

In the early days of face aging research, Todd et al. [17] model face growth by revised cardioidal strain transformation in a computable geometric progress. Suo et al. [16] mechanically highlight the revolution of physical features such as muscles, wrinkles and hair. However, those algorithms are computational expensive and rely on large amount of paired image sequences of the same person.

Recently, Wang et al. [19] propose a recurrent face aging framework which can model the intermediate transition states, thus the face growth between adjacent age groups is more smooth. It requires sequential face images of the same person at different ages, which limits its applications. Zhang et al. [23] apply conditional adversarial auto-encoder (CAAE), which transforms the input image to manifold to simulate muscle sagging and gets rid of the requirement of paired training samples while preserving the personality. However, reconstructing the face image using only the age condition lacks sufficient constraints. This transformation process ignores more basic information of various images, causing the generated images unnatural. Wang et al. [20] employ cGANs and a pre-trained identity-preserved module to better maintain identity consistency, while the representation ability of discriminator is insufficient. Although they show some positive results, the rendered faces are blurry. These works all ignore the fact that the facial skeleton structure undergoes dramatical changes with aging from young to old, which results in the failure of reconstructing realistic faces in some specific cases (e.g., rendering a 90 years old face from a baby face or rendering a 5 years old baby face from a senior person.). To handle the above issues, our method uses the target age condition to generate the aging results, which are then used to reconstruct faces in the original face age group. Facial landmarks are employed to guide both synthesis processes.

3   The Proposed Methods

It is important to find a mapping between age features and face features for inferring faces with desired ages. This problem can be considered as a cGAN problem and solved by minimizing the distance between ground truth face distributions in target age groups and generated ones.

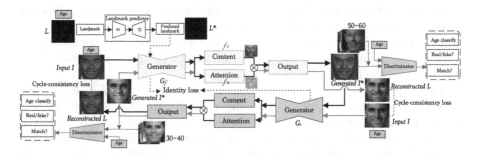

Fig. 1. The pipeline of the proposed LGcGAN for face aging. The black flowchart shows the primal cGAN G_f for aging face generation and the red flowchart shows the dual cGAN G_r for face reconstruction. G_f and G_r share parameters W_G. Based on the target age condition, an input face image is first transformed to the target ages before being reconstructed based on the initial input age condition. (Color figure online)

A high-level illustration scheme of our proposed architecture is shown in Fig. 1. We divide multiple sets of training faces with different ages into N non-overlapping groups: $\mathbb{I}_1, \mathbb{I}_2, \ldots, \mathbb{I}_N$ ($N = 10$). The images in different age groups can be different person. These 10 age groups corresponding to age 0–5,6–10,11–20, 21–30, 31–40, 41–50, 51–60, 61–70, 71–80 and 80+, respectively. Our goal is to learn an age progression model G_f for the primal task and a face regeneration model G_r for the dual task. G_f and G_r have the same structure and share the same parameters. Given a facial image I with age group $A_s \in \Re^{h \times w \times 10}$ that I belongs to a target age condition $A_t \in \Re^{h \times w \times 10}$, where h and w indicate the height and width of an intermediate feature map, 10 is the age group number. Similar to a 10-dim one-hot age vector, we reshape the vector as a 10-channel tensor and each channel represents a specific age group. Model G_f

aims to predict a synthesized face $I^* = G_f(I, A_t)$ of the input person that falls in target age group A_t. The generated face I^* should not only belong to target age group A_t but also preserve the original person's identity. The regenerated face $I_r = G_r(G_f(I, A_t), A_s)$ should be as much the same as possible with the input face I and it should lie in the input age group A_s. The detailed components are described in Subsect. 3.1 and the objective functions are presented in Subsect. 3.2.

3.1   Network Architecture

As shown in Fig. 1, our primal face aging network consists of three dominating modules: i) a generator G_f that is trained to generate a synthesized face I^* with target age A_t; ii) an associated discriminator D aims to make I^* looks realistic and guarantee I^* lie in target age group A_t; iii) target aging face landmark predictor works as an attention mechanism for the face generation. The dual procedure, which regenerates face image I_r to match the original face I, has the same 3 modules as the primal network.

Landmark Attentions: Most of the existing works have limited ability for handling facial structure changes and hence can handle age progression between adults only (e.g., between 20 and 80 years old). To overcome this limitation, we pre-train an external landmark attention model as a landmark predictor, which is imposed on the internal multi-scale features.

To be specific, we first retrieve a source landmark feature $L \in \mathbb{R}^{2 \times n}$ from the input face image I using 2D face alignment [1], where n is fixed at 68 in this work. The pre-trained landmark predictor $\mathcal{G}_L$ is then used to convert L to L^* that falls in the target age group L^{GT}. The landmark predictor has an encoder (E)-decoder (D) structure with four fully connected (FC) layers at both parts. The final landmark prediction is achieved by fusing the age feature vectors with the latent features of the source landmark. The training loss for $\mathcal{G}_L$ is $\mathcal{L}_{lmk} = \|E(L^*) - E(L)\|_2^2 + \|L^* - L^{GT}\|_2^2$.

The attention mechanism helps to precisely emphasize the variational structure areas, thus helping produce more sophisticated features in a fine-grained fashion. Specifically, to calculate the latent attention layers, we integrate original and predicted landmarks features to two hidden feature layers h_i and h_j in the image generation process. The overall attention features can be formulated as:

$$f_{attn1} = \sigma \left(f_c \left(H_l^1 \left(L^* \right) \right) \ominus \left(f_c \left(H_l^1 \left(L \right) \right) \right) \right); f_{attn2} = \beta \left(H_l^2 \left(H_l^1 \left(L^* \right) \oplus \left(H_l^1 \left(L \right) \right) \right) \right) \tag{1}$$

where H_l^1 is convolutional encoder to produce latent landmark vectors of landmarks. The difference between two latent landmark features forming a first attention maps f_{attn1} with size of $N_L \times 16 \times 16$. H_l^2 is the second convolutional encoder which can encode the concatenated landmark feature values to attention feature maps f_{attn2} with size of $N_L \times 32 \times 32$.

Generator: Basic GAN-based methods for face aging firstly apply numerous down-sampling convolutional layers to learn the high-level feature distributions, and then forward the feature maps into multiple up-sampling convolutional layers to render the final image. In our work, given an input RGB face image $I \in \mathbb{R}^{h \times w \times 3}$ under an arbitrary age group $\mathbf{A}_s \in \mathbb{R}^{1 \times 10}$ and a target age group $\mathbf{A}_t \in \mathbb{R}^{1 \times 10}$, we need to pad the one-hot label $\mathbf{A}_t$ into $\mathbb{R}^{h \times w \times 10}$. Then, we form the input of generator as a concatenation vector $(I, \mathbf{A}_t) \in \mathbb{R}^{h \times w \times (3+10)}$.

One key ingredient of our approach is to make G focus on those regions of the image that are relevant to face aging and keep the remaining information unchanged to preserve identity consistency. For this purpose, we embedded a built-in attention mechanism to the generator, which can disentangle the discriminative devise objects and the consistent part by producing an attention feature mask f_A and a content color mask f_C instead of regressing a full image. The final generated image can be obtained as:

$$I^* = G(I, \mathbf{A}_t, L^*, L) = (1 - f_A) \otimes f_C + f_A \otimes I \tag{2}$$

where $\otimes$ denotes the element-wise product, where $f_C = G(I, \mathbf{A}_t) \in \mathbb{R}^{h \times w \times 3}$, and $f_A = G(I, \mathbf{A}_t) \in \{0, \dots, 1\}^{h \times w \times 1}$. The mask f_C indicates how much the original image will contribute at each pixel location, f_A determines how much the aging condition will contribution to the changes of the final image.

Discriminator: Discriminator aims to distinguish between the generated image (fake) and its ground truth (real). This combined discriminator D_G consists of a conditional and an unconditional discriminator, aiming to promote the realistic of generated faces and guide the synthesized face lies in the target age group. The conditional discriminator concatenated with reshaped age condition A_t to the fifth convolution layer, which corresponds the age condition in generator network G_f.

3.2   Objective Functions

The loss function includes four terms: 1) an adversarial loss proposed by Gulrajani et al. [5], which pushes the distribution of the generated images to the distribution of the training images; 2) an age classification loss for encouraging accurate age identification for the generated facial images using an age classifier; 3) an identity feature loss, which helps to preserve the identity information for the generated fake face samples; and 4) a cycle-consistency loss [24], which further constrains the input and output faces correspond to the same person.

Adversarial Loss: The discriminator D_G is trained alternately with the generator. The objective of the discriminators consists of a visual realism adversarial loss and an age-face paired consistency adversarial loss. Mathematically, it is defined as:

$$\begin{aligned} \mathcal{L}_{D_G} = &-\tfrac{1}{2}\mathbb{E}_{I^{GT} \sim p_{I_{GT}}}\left[\log\left(D_G\left(I^{GT}\right)\right)\right] - \tfrac{1}{2}\mathbb{E}_{I^* \sim p_{I^*}}\left[\log\left(1 - D_G\left(I^*\right)\right)\right] \\ &- \tfrac{1}{2}\mathbb{E}_{I^{GT} \sim p_{I_{GT}}}\left[\log\left(D_G\left(I^{GT}, s_t\right)\right)\right] - \tfrac{1}{2}\mathbb{E}_{I^* \sim p_{I^*}}\left[\log\left(1 - D_G\left(I^*, s_t\right)\right)\right] \end{aligned} \tag{3}$$

$$\mathcal{L}_G = \mathbb{E}_{I^* \sim p_{\text{data}}(I^*)}\left[\log D_G\left([A_t, I^*]\right)\right] + \mathbb{E}_{I \sim p_{\text{data}}(I)}\left[\log\left(1 - D_G\left([A_t, G_f(I)]\right)\right)\right] \tag{4}$$

where I^{GT} is from the real image distribution p_I^{GT}. s_t is reshaped age condition A_t. G tries to minimize the adversarial loss objective $\mathcal{L}_{D_G}$ while D_G tries to maximize it. The target of G is to produce a facial image I^* that looks similar to the images from GT, while D_G aims to distinguish between rendered face images I^* and real images I. A similar adversarial loss of Eq. 3 for inverse mapping is defined as:

$$\mathcal{L}_{\mathcal{R}D_G} = -\tfrac{1}{2}\mathbb{E}_{I\sim p_I}\left[\log\left(D_G\left(I\right)\right)\right] - \tfrac{1}{2}\mathbb{E}_{I_r\sim p_{I_r}}\left[\log\left(1 - D_G\left(I_r\right)\right)\right] \\ - \tfrac{1}{2}\mathbb{E}_{I\sim p_I}\left[\log\left(D_G\left(I, s_r\right)\right)\right] - \tfrac{1}{2}\mathbb{E}_{I_r\sim p_{I_r}}\left[\log\left(1 - D_G\left(I_r, s_r\right)\right)\right] \tag{5}$$

$$\mathcal{L}_{\mathcal{R}G} = \mathbb{E}_{I_r\sim p_{\text{data}}\,(I_r)}\left[\log D_G\left(\left[A_t, I_r\right]\right)\right] + \mathbb{E}_{I^*\sim p_{\text{data}}(I^*)}\left[\log\left(1 - D_G\left(\left[A_t, G_r(I^*)\right]\right)\right)\right] \tag{6}$$

Age Classification Loss: Besides reducing the image adversarial loss, the generator must also reduce the age error by the age classifier D_A. The age classification loss is defined with two components: an age estimation loss with fake images used to optimize G, and an age estimation loss of real images used to learn the age classifier D_A. This loss $\mathcal{L}_{cls}\left(G, D_A, \boldsymbol{I}, \boldsymbol{A}_t, \boldsymbol{A}_s\right)$ is computed as:

$$\mathcal{L}_{cls} = \mathbb{E}_{\boldsymbol{I}\sim\mathbb{P}_I}\left[\ell\left(D_A\left(G\left(\boldsymbol{I}, \boldsymbol{A}_t\right)\right), \boldsymbol{A}_t\right) + \ell\left(D_A\left(\boldsymbol{I}\right), \boldsymbol{A}_s\right)\right] \tag{7}$$

where $\boldsymbol{A}_s$ is the source age condition of the input image I, $\ell(\cdot)$ corresponds to a softmax loss. Similarly, in the dual procedure, the loss function is defined as:

$$\mathcal{L}_{\mathcal{R}cls} = \mathbb{E}_{\boldsymbol{I}^*\sim\mathbb{P}_{I^*}}\left[\ell\left(D_A\left(G\left(\boldsymbol{I}^*, \boldsymbol{A}_t\right)\right), \boldsymbol{A}_t\right) + \ell\left(D_A\left(\boldsymbol{I}^*\right), \boldsymbol{A}_s\right)\right] \tag{8}$$

Identity Feature Loss: Adversarial loss and age classification loss only drive the generator to generate samples that follow the target data distribution. Hence, the generated images may not looks like the same person. This problem can be solved by a perceptual loss to shorten the distance between input image and output image in the same lower feature space where the feature layers are good at keeping the content. To better preserve the personal identity when perform aging, we use an identity feature loss:

$$\mathcal{L}_{\text{id}} = \sum_{I\in p_I(I)} \left\|\varPhi(I) - \varPhi\left(G\left(I \mid A_t\right)\right)\right\|^2 \tag{9}$$

Here $\varPhi(\cdot)$ represents the features extracted by the conv5 layer in our generative network.

Cycle-Consistency Loss: As we regenerate the face in a primal-dual loop, the rejuvenated face image I_r should be as similar to the input image I as possible. This is enforced by using the following lost function:

$$\mathcal{L}_{\text{cycle}}\left(G_f, G_r\right) = \mathbb{E}_{I\sim p_{\text{data}}\,(I)}\left[\left\|G_r\left(G_f(I)\right) - I\right\|_1\right] \\ + \mathbb{E}_{I^*\sim p_{\text{data}}\,(I^*)}\left[\left\|G_f\left(G_r(I^*)\right) - I^*\right\|_1\right] \tag{10}$$

where the regenerated image $I_r = G_r\left(G_f(I)\right)$ is compared against the input image I.

Full Objective: Overall, to generate desired facial image I^* with the target age and the same person identity, we linearly combine the aforementioned losses:

$$\mathcal{L}_G = \lambda_1 \mathcal{L}_G + \lambda_2 \mathcal{L}_{\mathcal{R}G} + \lambda_3 \mathcal{L}_{\text{cls}} + \lambda_4 \mathcal{L}_{\mathcal{R}\text{cls}} + \lambda_5 \mathcal{L}_{\text{id}} + \lambda_6 \mathcal{L}_{\text{cycle}} \quad (G_f, G_r) \quad (11)$$

$$\mathcal{L}_D = \mathcal{L}_{D_G} + \mathcal{L}_{\mathcal{R}D_G} \quad (12)$$

where $\mathcal{L}_{\mathcal{R}\cdot}$ is the loss produced by the face regeneration procedure. λ is a set of hyper-parameters that control the relative importance of each term.

4   Experiments

We evaluate our proposed model both qualitatively and quantitatively on a large public dataset UTKFace [2], which contains over 20,000 face images from 0 to 116 years old. All the images are annotated with ages. There are large variations in pose, illumination, expression, and even style in this dataset. The data is unpaired and non-sequential because there are no continuous images for the same person. We split UTKFace into two parts: 90% for training and the rest for test.

4.1   Implementation Details

Basics: Our model is implemented on PyTorch3 and tested on a single Nvidia GeForce GTX 1080 Ti GPU and the memory is 50GiB. We apply batch normalization, set a fixed learning rate of 0.0002, and use Adam algorithm as the optimizer.

Architecture Details: As for the generator G, it transforms an input RGB image with size $128 \times 128 \times 3$ and an age condition to a $16N_i \times 16 \times 16$ tensor by 4 groups of convolution- Batchnorm-Relu layers and 6 residual blocks, where $N_i = 16$ is the depth of latent image feature maps. The stride of each convolution layer is 2 and the kernel size is 3×3. The residual block includes 2 convolution

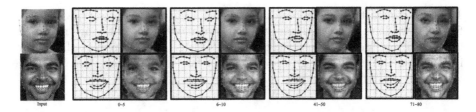

Fig. 2. Conditional landmark visualization. The input faces are shown in far left and have landmarks retrieved by 2D face alignment [1] displayed on the top. The remaining columns show the landmark prediction results (left) and the corresponding rendered aging faces (right) for different age groups, respectively. One can notice that the eyes are bigger and the faces are chubbier at young ages, suggesting the effectiveness of the proposed landmark attention module.

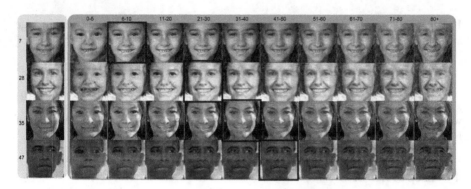

Fig. 3. Results of our proposed LGcGAN for age progression. The first column shows the original faces marked with their true ages on the left. The remaining 10 columns show the results synthesized for different age target groups (indicated at the top of each column). Generated faces are realistic in aspects of age and facial feature. The ones generated for the respective input age groups (highlighted in yellow boxes) have appealing similarity with the input faces. (Color figure online)

layers, followed by a zero-padding layer, a stride-1 convolution layer and a batch-normalization layer. We reshape the age condition as a 10-dim one-hot vector to a 10-channel tensor because we divided ages into 10 groups as aforementioned. The reshaped tensor further concatenates the latent image feature. Each channel of the conditioned tensor indicates a specific age group. Through the pre-trained landmark prediction model, we can directly get a predicted landmark with the target age group. Then we feed the predicted landmark vector into a landmark encoder which has a linear layer with Relu activation and a Conv layer, producing landmark features with size $N_i \times 16 \times 16$. Both age condition and landmark condition tensor are embedded to the image feature tensor after the residual blocks.

4.2   Qualitative Comparison

To better demonstrate the superiority in aging accuracy and preserving identity features, we have compared our methods with two state-of-the-art methods: CAAE [23] and Identity-Preserved cGAN (IPCGAN) [20]. We remove the gender information and use 10 age groups instead for fair comparisons.

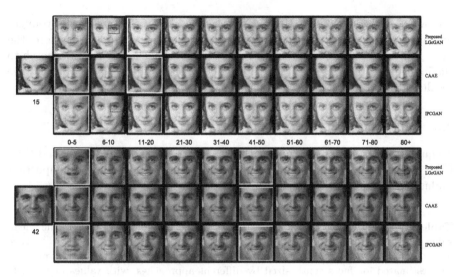

Fig. 4. Comparisons between results synthesized by CAAE [23], IPCGAN [20], and our LGcGAN. Two input faces images of different genders and age groups are used (shown on the left. The white boxes highlight the age groups that the initial faces belong to. Yellow boxes show that our method tend to generate rounder eyes and shorter faces at early ages. Overall, faces reconstructed by LGcGAN yield more distinct features and realistic aging effects, see for example areas highlighted green boxes. In comparison, eyes generated by CAAE and IPCGAN highlighted in the red boxes are blurry and have more artifacts. (Color figure online)

Facial landmarks can reflect the changes in the facial skeleton that occur with aging and hence have impact on facial appearance. The landmark changes should be dramatic from children to adults and the changes between adults and senior faces are not obvious because the skeleton itself undergoes minimal changes with aging after the growth and development of a person. Figure 2 illustrates the the effect of facial landmark prediction. For the two input faces, their facial landmarks are retrieved by 2D face alignment [1]. Changes to these landmark locations are predicted for different age groups, which guide the synthesized faces to follow. Hence, the external landmark attention contributes the accuracy of face progression.

Synthesized face manifold of our proposed LGcGAN is shown in Fig. 3. The generated face images not only resemble faces in the target age groups but also well preserve the personal identity over a long age span. It is clear that our method is efficient to simulate translations between age groups and synthesize elderly or young face images with high visual fidelity.

Performance comparison with prior works is shown in Fig. 4. It shows that the proposed LGcGAN generates more visually convincing faces for different ages, providing better age accuracy and identity consistency. Age accuracy is reflected by more precise facial structure and reasonable appearance changes. In comparison, the CAAE fails to generate the elder looking and the results are overly smooth, whereas IPCGAN lacks the ability to transfer the younger looking, as eye shapes and facial profiles do not change much. Furthermore, the faces generated using our model for the same agegroups as inputs best match to the input faces. The result of CAAE lost the wrinkleson the male's face and the results of IPCGAN introduced unnatural gray color to the female's face.

4.3   Quantitative Comparison

To quantitative evaluate our approach, we compare it with CAAE [23] and IPC-GAN [20] on both aging accuracy and identity preservation, which are important evaluation metrics in face aging.

Table 1. Estimated ages between real and synthesized faces. "Real" is the mean value of ages estimated for real photos from different age groups of the UTKFace datasets,which serves as ground truth. The following rows show the mean values of ages estimated for faces synthesized by different approaches, with values in brackets showing their absolute differences from the ground truth. Best results (smaller discrepancy)are shown in boldface. * represents the models that we re-trained on 10 age groups.

Age group	21–30	31–40	41–50	51–60	61–70	71–80	80+
Generic	25.03	35.01	45.12	54.63	65.40	73.66	87.29
CAAE* [23]	24.31(**0.72**)	32.43(2.58)	42.21(2.91)	51.49(3.14)	60.17(5.23)	70.57(3.09)	82.68(4.61)
IPCGAN* [20]	22.74(2.29)	31.74(3.27)	39.93(5.19)	50.04(4.59)	58.32(7.08)	68.42(5.24)	80.33(6.96)
Ours	26.18(**1.15**)	36.91(**1.10**)	44.68(**1.44**)	51.79(**2.84**)	62.52(**2.88**)	71.05(**1.61**)	88.24(**0.95**)

Aging Accuracy: The aging accuracy problem can be transformed to age distribution estimation. To evaluate the age accuracy of synthesized faces, we use young faces as input to generate elder faces. On UTKFace, faces with age under or equal to 20 are considered as testing samples, and the target aging faces in other seven age groups are generated. For fair comparison, the Face++ API [11] is then used to estimate ages for generated results. Since the algorithms are trained using the UTKFace datasets, we first use Face++ API to estimate all real faces in different age groups of the datasets. The mean age of each group serves as the respective ground truth age value. The mean value of all generated fake faces in the same age group is then compared with the ground truth. The less discrepancy between the two, the better synthesis accuracy in terms of aging effects. As shown in Table 1, our method out performs the two baseline approaches in all but the 21–30 age group, demonstrating the effectiveness of our network.

Table 2. Face identify verification results. The top shows the verification confidences among the input images and results synthesized by LGcGAN. The bottom compares the verification rates among three methods, with best verification rates shown in bold

Age group	21–30	31–40	41–50	51+
Verification confidence				
10–20	95.76	94.78	94.65	93.28
21–30	–	95.74	94.54	93.77
31–40	–	–	95.12	94.32
41–50	–	–	–	94.64
Verication rate				
CAAE [23]	87.05	81.07	73.36	60.25
IPCGAN [20]	**100**	**100**	**100**	**100**
ours	**100**	**100**	**100**	**100**

Identity Preservation: To determine whether the identities of input faces have been properly preserved during face aging process, face verification check is also performed. Scores from Face++ [11] are used to evaluate not only the similarity between the input faces and the synthesized age-progressed faces, but also the synthesized age-progressed faces in different age groups. We again select images with age under 20 as the input and generate images in other four groups (21–30,31–40,41–50,51+). Results of face verification experiments are shown in Table 2. The top part of Table 2 shows verification confidence between ground truth testing faces and their aging elder ones generated by LGcGAN, and the bottom part is verification rates of three methods which indicate the capability in preserving identity.

The above two tests show that IPCGAN is good at preserving identity information but lacks ability of generating aging faces with accurate target age appearance. CAAE performs better than IPCGAN in age accuracy but fails to preserve personal identity. Our LGcGAN consistently outperforms CAAE and IPCGAN in both evaluation metrics.

5   Conclusion

In this paper, we propose a novel Landmark-Guided Conditional GAN (LGc-GAN) for face aging. Based on an external, built-in attention mechanism and a primal-dual framework for both face generation and reconstruction, our model can learn the transition pattern at different ages and perform well in preserving the personal identity. Both qualitative and quantitative experiments validate the effectiveness of our approach and its advantages over the existing ones. Future work will include using the hand-crafted landmarks from the original database, which can bring higher accuracy results; Additionally, we believe that landmark based attention modules give better solutions in other face synthesis problems such as facial expression dynamics and facial image-image translations.

References

1. Bulat, A., Tzimiropoulos, G.: How far are we from solving the 2D & 3D face alignment problem? (and a dataset of 230,000 3D facial landmarks). In: Proceedings of the IEEE International Conference on Computer Vision, pp. 1021–1030 (2017)
2. Chen, B.-C., Chen, C.-S., Hsu, W.H.: Cross-age reference coding for age-invariant face recognition and retrieval. In: Fleet, D., Pajdla, T., Schiele, B., Tuytelaars, T. (eds.) ECCV 2014. LNCS, vol. 8694, pp. 768–783. Springer, Cham (2014). https://doi.org/10.1007/978-3-319-10599-4_49
3. Fu, Y., Guo, G., Huang, T.S.: Age synthesis and estimation via faces: A survey. IEEE Trans. Pattern Anal. Mach. Intell. **32**(11), 1955–1976 (2010)
4. Grother, P.J., Ngan, M.L., Hanaoka, K.K.: Ongoing face recognition vendor test (FRVT) part 2: Identification (2018)
5. Gulrajani, I., Ahmed, F., Arjovsky, M., Dumoulin, V., Courville, A.C.: Improved training of Wasserstein GANs. In: Advances in Neural Information Processing Systems, pp. 5767–5777 (2017)
6. Huang, X., Wang, M., Gong, M.: Hierarchically-fused generative adversarial network for text to realistic image synthesis. In: 2019 16th Conference on Computer and Robot Vision (CRV), pp. 73–80. IEEE (2019)
7. Isola, P., Zhu, J.Y., Zhou, T., Efros, A.A.: Image-to-image translation with conditional adversarial networks. In: Proceedings of the IEEE Conference on Computer Vision and Pattern Recognition, pp. 1125–1134 (2017)
8. Kemelmacher-Shlizerman, I., Suwajanakorn, S., Seitz, S.M.: Illumination-aware age progression. In: Proceedings of the IEEE Conference on Computer Vision and Pattern Recognition, pp. 3334–3341 (2014)
9. Liu, L., Wang, S., Wan, L., Yu, H.: Multimodal face aging framework via learning disentangled representation. J. Vis. Commun. Image Rep. **83**, 103452 (2022)
10. Liu, Y., Li, Q., Sun, Z., Tan, T.: A 3 GAN: an attribute-aware attentive generative adversarial network for face aging. IEEE Trans. Inf. Forensics Secur. **16**, 2776–2790 (2021)
11. Megvii, I.: Face++ research toolkit (2013)
12. Mirza, M., Osindero, S.: Conditional generative adversarial nets. arXiv preprint arXiv:1411.1784 (2014)
13. Pham, Q., Yang, J., Shin, J.: Semi-supervised FaceGAN for face-age progression and regression with synthesized paired images. Electronics **9**(4), 603 (2020)
14. Pumarola, A., Agudo, A., Martinez, A.M., Sanfeliu, A., Moreno-Noguer, F.: GANimation: anatomically-aware facial animation from a single image. In: Proceedings of the European Conference on Computer Vision (ECCV), pp. 818–833 (2018)
15. Ramanathan, N., Chellappa, R.: Modeling age progression in young faces. In: 2006 IEEE Computer Society Conference on Computer Vision and Pattern Recognition (CVPR 2006), vol. 1, pp. 387–394. IEEE (2006)
16. Suo, J., Zhu, S.C., Shan, S., Chen, X.: A compositional and dynamic model for face aging. IEEE Trans. Pattern Anal. Mach. Intell. **32**(3), 385–401 (2009)
17. Todd, J.T., Mark, L.S., Shaw, R.E., Pittenger, J.B.: The perception of human growth. Sci. Am. **242**(2), 132–145 (1980)
18. Wang, H., Gong, D., Li, Z., Liu, W.: Decorrelated adversarial learning for age-invariant face recognition. In: Proceedings of the IEEE/CVF Conference on Computer Vision and Pattern Recognition, pp. 3527–3536 (2019)
19. Wang, W., et al.: Recurrent face aging. In: Proceedings of the IEEE Conference on Computer Vision and Pattern Recognition, pp. 2378–2386 (2016)

20. Wang, Z., Tang, X., Luo, W., Gao, S.: Face aging with identity-preserved conditional generative adversarial networks. In: Proceedings of the IEEE Conference on Computer Vision and Pattern Recognition, pp. 7939–7947 (2018)
21. Yang, H., Huang, D., Wang, Y., Wang, H., Tang, Y.: Face aging effect simulation using hidden factor analysis joint sparse representation. IEEE Trans. Image Process. **25**(6), 2493–2507 (2016)
22. Yao, X., Puy, G., Newson, A., Gousseau, Y., Hellier, P.: High resolution face age editing. In: 2020 25th International Conference on Pattern Recognition (ICPR), pp. 8624–8631. IEEE (2021)
23. Zhang, Z., Song, Y., Qi, H.: Age progression/regression by conditional adversarial autoencoder. In: Proceedings of the IEEE Conference on Computer Vision and Pattern Recognition, pp. 5810–5818 (2017)
24. Zhu, J.Y., Park, T., Isola, P., Efros, A.A.: Unpaired image-to-image translation using cycle-consistent adversarial networks. In: Proceedings of the IEEE International Conference on Computer Vision, pp. 2223–2232 (2017)

Introducing AV1 Codec-Level Video Steganography

Lorenzo Catania[1], Dario Allegra[1]([✉]), Oliver Giudice[1,2],
Filippo Stanco[1], and Sebastiano Battiato[1]

[1] Image Processing Laboratory, Department of Mathematics and Computer Science,
University of Catania, Catania, Italy
lorenzo.catania@phd.unict.it, {allegra,giudice,fstanco}@unict.it
[2] Applied Research Team, IT Department, Bank of Italy, Rome, Italy
battiato@unict.it

Abstract. Steganography is the ancient art of concealing messages into data. High research interest has grown over the last years, however, techniques in literature are only focused on standard and in some way legacy multimedia formats (e.g., H.264). Moreover, most video steganography techniques are based on concealing data into contents of each frame employing many strategies. In this paper, a codec-level video steganography technique is presented on the novel AV1: a royalty-free video compression format proposed by the *Alliance for Open Media (AOM)*. The proposed method is based on the alteration of intra-prediction angles and, differently from other solutions, it works along the compression process, by allowing the encoder to reduce possible distortions caused by the messages to be hidden. The effectiveness of the technique was demonstrated by hiding up to 1024 characters into an highly compressed video of 40 s maintaining an average *Peak Signal-to-Noise Ratio* value of 37.53 dB.

Keywords: Steganography · Information hiding · Video encoding · Image coding

1 Introduction

Video data are daily consumed by people through analogical translations captured by perceptive organs and interpreted by our brain, a process which reduces the sensibility in respect to the range of all possible emitted signals. Typically, any kind of data should be compressed to optimise the storage usage, and the design of encoding algorithms is of utter importance, since the temporal redundancy between frames is notable and may be properly reduced. As a consequence, the compressed version of a video may lose part of its original information. In order to obtain a greater quality it's necessary to sacrifice some compression efficiency and, vice-versa, to generate a smaller sized output video, the encoder must carry out a quality drop.

S. Sclaroff et al. (Eds.): ICIAP 2022, LNCS 13231, pp. 284–294, 2022.
https://doi.org/10.1007/978-3-031-06427-2_24

The objective of digital steganography is to hide pieces of information, called *secret data*, inside a wider lake of information called *cover data* [13]. In most scenarios, the secret message is relatively small w.r.t. the cover data size, as the injection of secret data implies an alteration of the original media which leads to distortions that, if consistent, may be visible or audible, therefore revealing the presence of an hidden message [2,14]. Multimedia data is suitable for being used as cover data in steganography applications, as small variations are difficult to be perceived by observers. In particular videos are even better and more and more often chosen in this contexts given their larger dimension and bigger tolerance to distortions.

Techniques in literature [9] are mainly focused on standard and somewhat legacy multimedia formats (e.g., H.264). Moreover, most video steganography techniques are based on concealing data into video contents by employing many strategies mostly taking into account spatial and temporal information of single and/or range of frames. In this paper, a codec-level video steganography technique is presented on the novel AV1: a royalty-free video compression format proposed by the *Alliance for Open Media (AOM)*. At the best of the authors' knowledge, there are no similar solutions in literature able to perform video steganography on AV1. Moreover, the proposed technique is based on the alteration of the intra-prediction angles and, being codec-level, it works during the compression process and not on the image raw data. This allowed to reduce possible distortions caused by the messages to be hidden.

The choice of AV1 as target format is due to both the open-minded philosophy that characterizes the project and its recent birth and emergence. This standard, in fact, is the result of the combined work of the industry top players; it is open source and royalty-free and its first release is very recent, going back to 2018. Some criticism are often made about this format, in particular regarding the reference implementations not being production-ready and poor hardware support in respect to other standards. However, AV1 is a promising initiative for the industry and a partially unexplored ground for research purposes. Another reason AV1 is a fruitful field for steganography is its aim to become a widespread format on the web, through which the majority of communications takes place, including the concealed ones.

The proposed solution has been evaluated on various videos and demonstrated to achieve the best performance on short videos with static camera shots; however, it demonstrated to obtain appreciable quality results also in other settings.

The main contributions of this paper can be summarized as follows:

- a steganography system that works on the emerging AV1 video format, presented in 2018 by the consortium named *Alliance for Open Media*. As of the time of writing and limited to the knowledge of authors, this is the first publicly discussed technique of steganography over AV1 videos;
- an approach for the insertion of hidden messages that, differently from other works, is codec-level that means it is applied during the compression process, exploiting the encoding steps in order to reduce artifacts;

– an open source implementation which is publicly available at the following GitHub repository: https://github.com/av1stego.

The remainder of this paper is organized as follows: Sect. 2 presents related works; Sect. 3 introduces the proposed method while Sect. 4 details experimental results. Finally, Sect. 5 concludes the paper with a summary and discussion about future works.

2   Related Works

The amount of possible approaches to staganography over digital data is large, like extensively exposed by S. Dhawan and R. Gupta [6] in 2021. However, each method is unique on its own because it's conceived to be applied to an established kind of data and, eventually, to ship another precise typology of information as well. A comprehensive review of video steganography, which is probably the most common nowadays, can be found in [12].

Some techniques work on uncompressed video frames, while others are executed in harmony with a video compression algorithm. The former may be typically easier to generalise, but the latter are widely used because it is a requirement for a steganography system to cooperate with an encoder; actually, in practical applications it is very rare to work with uncompressed video data, especially if they have to be transmitted.

The most simple method of hiding information into a video is altering the spatial domain, usually represented by the pixels of each frame. A common technique consists in Least Significant Bit coding (LSB), which consists in modifying the least significant bits to match sections of the secret message, such that a decoder is able to read those bits and reconstruct the hidden data [5, 16–18]. Techniques based on this method are simple and effective, but they are not robust against statistical attacks and other similar attempts to detect or erase covert data [10]. Note that not all values need to be modified, as sometimes the original last bit matches the secret bit, therefore the message may be transmitted correctly without altering cover data, although this is unlikely.

Another important category of steganographic solutions is given by the so called transform-based techniques [9]. In these methods, video data are processed to pass to a different domain than the spatial one, then information is hidden using the transformed values as cover data. These kind of solutions are more complicated than the ones based on LSB over spatial values and other similarly trivial approaches, and trade a higher resistance to detection, erasure attempts and statistical attacks thanks to the higher complexity in design and implementation. An example of those kind of steganographic solution was given by Patel et al. in 2013 [7], which exploits an Integer Wavelet transform on the video data and applies symmetric key encryption as well on the secret message before embedding it in the video using LSB coding.

The growing number of encoding standards had a collateral impact on the field of digital steganography as well. Different raw and uncompressed video data share a lot of similarities, although the way they are stored may be different.

```
Secret message: 0110
Sequence of angle values: 2,1,3,5,6,4,0,5,2,3

offset = 0, padding = 0
Transmitted signal: 2,1,3,4,6,4,0,5,2,3

offset = 2, padding = 0
Transmitted signal: 2,1,2,5,6,5,0,5,2,3

offset = 0, padding = 1
Transmitted signal: 2,1,3,5,6,5,0,4,2,3

offset = 2, padding = 1
Transmitted signal: 2,1,2,5,6,5,0,5,2,2
```

Fig. 1. Examples of various injections of the same secret message into a sequence of angles values and four different *offset-padding* pairs: (0,0),(2,0),(0,1),(2,1). Note that the *6* value is always skipped, independently from the parameters used

Also, even though different compression processes can share some aspects, they output very different data and this makes hard building an efficient steganography system which works well along different encoding processes. For these reasons, a consistent amount of literature steganography techniques are format-based [1,4,11,19], i.e. they exploit one or more specific part of a given video encoding format, implying they are very efficient on videos compressed using that standard, but their usage is limited to the contexts where the reference format is feasible. Also, they are rarely trivial to be generalised, precisely because their functioning is bound to details that are proper of the format. Another minor aspect of format-based video steganography methods, mostly related to the research, is that almost all of such methods work on common standards like H264/AVC, like in [3] which proposes a video steganography method applied to the intra-prediction mode phase of H264. Although some concepts may be reused in other environments, it is very difficult to make proper generalisations about. Analogously, Bhautmage et al. [15] proposes a steganography algorithm that works on videos encoded in AVI containers. Even though this method is based on a bits manipulation technique, it is introduced specifically for that format, denoting which, sometimes, methods are presented as format-specific even when it is not strictly necessary.

In the next section, a codec-level video steganography technique is proposed on the novel AV1 working during the compression process and not on the image raw data.

3   Steganography on AV1 Codec-level

In order to correctly perform the information hiding operation, the part of the format taken into account is the AV1's intra-prediction process part. The idea is to inject information directly into the directional intra-prediction data. Each entry is composed of a base angle value named *intra mode* or *nominal angle* and an additional offset to be applied named *pAngle*. The angles are expressed in degrees. Therefore, the cover data is the sequence of *pAngle* values, as those are fine-grained values for which small variations do not considerably alter the video. The stream of secret bits (the message) can be embedded into this sequence by LSB coding. It has to be noted that there are 7 possible angle offsets: $\{-9, -6, -3, 0, 3, 6, 9\}$. Those are encoded by mapping them into 8-bit unsigned integers, thus the set of symbols that could be finally transmitted is: $\{0, 1, 2, 3, 4, 5, 6\}$. Angles encoded as 6 are excluded from the process as they'd be embedded with value 1 and would cause an error. In particular, the last bit is forced to be *0* or *1* by respectively adding 1 to odd values and subtracting 1 from even values. A graphical representation of the insertion process is reported in Fig. 1.

To minimize the distortion, the injection operations were distributed over the angle values, skipping some of them based on two parameters the user is allowed to select:

- *offset*: the count of values from the start of the sequence to the first angle with an injected bit.
- *padding*: the number of values skipped after each injection operation on an angle.

The chosen pair *offset-padding* is required to extract the original message (Fig. 1).

4   Experiments and Evaluation

The proposed method has been implemented by forking an alternative encoder, *rav1e*, adding support for hidden data insertion, and the reference decoder *aomdec*, which is part of the standard codec's library, adding an extra pass for hidden data extraction. The encoder *rav1e* has been chosen over the reference encoder, which is *aomenc*, because of its faster running times.

After processing each video to insert some hidden data, the efficiency of the implementation has been assessed by encoding and decoding the same video with vanilla versions of *rav1e* and *aomdec*, figuring how much the information injection process has impacted on the output size and distortion in respect to the original. In addition to measures collection, a validation phase on each test is performed as well.

The secret messages transmitted are strings and the stop symbol is encoded as the null character. All of the audio tracks were removed from the videos and each of them was converted to a raw format, YUV, before being fed to the evaluation pipeline.

Each test is formed by a combination of a video, a message and the padding and offset parameters passed to the encoder. The experiments were subdivided into sets of tests named test cases, made up of tests that share the same context, or have characteristics in common. The evaluation is made by analysing, in each test case, the variation of the resulting file's measures over the shifting of a single parameter. The sizes are reported in bytes, while *padding* and *offset* are expressed in number of angle values skipped, reminding that in this particular

Table 1. Results of inserting small and average random strings into a small video

Padding	Offset	Enriched size	Message size	Enrich cost	Integrity
0 bits	0 bits	371468 bytes	25 bytes	145 bytes	OK
4 bits	0 bits	405223 bytes	25 bytes	33900 bytes	OK
0 bits	0 bits	373599 bytes	45 bytes	2276 bytes	OK
4 bits	0 bits	438522 bytes	45 bytes	67199 bytes	OK
0 bits	0 bits	435972 bytes	175 bytes	64649 bytes	OK
4 bits	0 bits	447178 bytes	175 bytes	75855 bytes	ERROR

implementation each angle value hosts one secret bit. Any configuration which is not explicitly specified has the default value assigned to it by each of the tools.

These measures were used to assess the proposed method efficiency:

– *Naive video size*: The size of the video encoded using an unmodified version of *rav1e*.
– *Enriched video size*: The size of the video encoded using a fork implementation of *rav1e* and containing an hidden message
– *Message size*: The size of the hidden message.
– *Enrich cost*: The difference between the *enriched video size* and the *naive video size*.
– *Integrity*: The result of the integrity check. *OK* means the message has been stored entirely into the video, while *ERROR* reports that something went wrong. This control is performed by extracting the injected data with a modified version of *aomdec* and comparing this data with the secret message that was supposed to be hidden.

4.1   Solution Capabilities

Most of the experiments were focused on evaluating how much information is possible to store in a given video using the proposed method. Capability is not only one of the most important quality measures for a steganography application, but it also represents the maximum size that a message must have to be hidden without losses into a given video, which is a requirement of any successful steganographic transmission.

Small and Average Random Strings in Short Videos. In this experiment random strings of different sizes are hidden in short videos. The first trial was made by injecting these strings into *Me at the zoo*, the first video ever uploaded on YouTube, depicting Jawed Karim, YouTube's co-founder, at the zoo.

In one case, the message is too big for being stored in the video and the integrity check fails, which means that it is not possible to recover the secret message. However, the principal issue of these tests is that the output video

Table 2. Results of inserting small and average random strings into a VIRAT ground small video

Padding	Offset	Enriched size	Message size	Enrich cost	Integrity
0 bits	0 bits	1196786 bytes	25 bytes	−1006 bytes	OK
4 bits	0 bits	1197984 bytes	25 bytes	192 bytes	OK
0 bits	0 bits	1197378 bytes	45 bytes	−414 bytes	OK
4 bits	0 bits	1196518 bytes	45 bytes	−1274 bytes	OK
0 bits	0 bits	1197119 bytes	175 bytes	−673 bytes	OK
4 bits	0 bits	1202238 bytes	175 bytes	4446 bytes	OK
0 bits	0 bits	1199674 bytes	350 bytes	1882 bytes	OK
4 bits	0 bits	1207271 bytes	350 bytes	9479 bytes	OK

Table 3. Results of an inserting an average *lorem ipsum* in a short surveillance camera video

Padding	Offset	Enriched size	Message size	Enrich cost	Integrity
0 bits	0 bits	1196697 bytes	128 bytes	−1095 bytes	OK
16 bits	0 bits	1206503 bytes	128 bytes	8711 bytes	OK
0 bits	64 bits	1197280 bytes	128 bytes	−512 bytes	OK
16 bits	64 bits	1209575 bytes	128 bytes	11783 bytes	OK
0 bits	0 bits	1197585 bytes	256 bytes	−207 bytes	OK
16 bits	0 bits	1211517 bytes	256 bytes	13725 bytes	ERROR
0 bits	64 bits	1198055 bytes	256 bytes	263 bytes	OK
16 bits	64 bits	1211260 bytes	256 bytes	13468 bytes	ERROR

dimension is overwhelmingly larger in respect to the original one, as a consequence detecting that some extra data is present becomes trivial (see Table 1).

Another test was made using the same messages, but conceived into a short *VIRAT* [8] ground footage of 41 s.

The performance measured in this video is superior (see Table 2). Messages are always transmitted correctly and the enrichment cost is never exaggerated, in some cases the enriched video is even smaller than the original one. Although the two videos have a similar length, the increased capacity of the latter may be due to the considerably higher resolution.

An Average *Lorem Ipsum* in a Small Video. These tests were made by pushing a larger *lorem ipsum* to stress out the capacity of the same VIRAT [8] CCTV video pointing to the ground that was used in the last trial, chosen as cover data for this experiment as well (see Table 3).

We note that, intuitively, the *padding* parameter causes much more repercussions over the results of *offset* while requiring more space to store the full message.

4.2   Visual Quality Preservation

The pin around what the field of steganography spins is the conceived transmission of a message, as a consequence the presence of such a message must be out of sight. The largest sign that a video has been altered is the occurrence of visual glitches. It is practically impossible to not add any visible alteration in a video during the process, but one of the objectives of any stego technique should be to make these distortions as rare and less suspicious as possible. If a footage is evidently rigged, the secrecy of the message is compromised (see Fig. 2). In each test reported in this section the message has been transmitted entirely and correctly, if not further specified.

Injecting a *lorem ipsum* message in a VIRAT [8] ground video has produced some evident artifacts that last for a bunch of frames and then disappear once the block is decoded using a new angle value.

Fig. 2. Visible artifact during the insertion of a 256 characters long *lorem ipsum* message into a VIRAT ground footage. In the top-right image, a padding of 4 values is used, which makes the smaller artifact slightly less evident than the one on the top-left image that has no padding. The image on the bottom is the original frame.

Table 4. Peak Signal-to-Noise Ratio results of the insertion of various *lorem ipsum* into a VIRAT ground footage. When the maximum value is infinite (*inf*) it means that at least one of the frames has not been altered at all.

Size	Pad	Off	Y	U	V	Avg	Min	Max
256 bytes	0 bits	0 bits	39.22 dB	54.14 dB	55.37 dB	40.92 dB	33.58 dB	inf
256 bytes	4 bits	0 bits	39.12 dB	53.11 dB	54.72 dB	40.81 dB	35.27 dB	inf
1024 bytes	0 bits	0 bits	33.58 dB	50.27 dB	51.76 dB	35.30 dB	31.31 dB	inf
1024 bytes	4 bits	0 bits	35.83 dB	50.46 dB	52.38 dB	37.53 dB	34.63 dB	39.90

The key measure to understand if the video may have visual defects is the minimum PSNR. In this case, there is a drop in respect to the average value, revealing that at least one frame has suffered relevant alterations 4.

The chart in Fig. 3 reports the evolution of PSNR values in respect to the message size when a *lorem ipsum* is hidden in a VIRAT [8] video. No padding or offset were added to the insertion process. Cases where it is not possible to inject the whole message into the video are omitted.

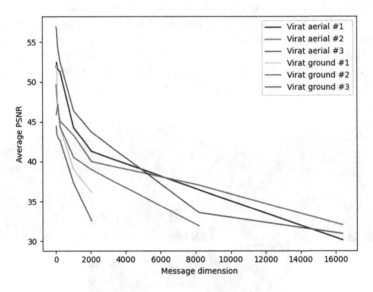

Fig. 3. Measurement of PSNR over message size variation about the insertion of a lorem ipsum message on various VIRAT [8] footages.

4.3   Discussion

The experimental trials performed have shown some limitations that are somehow intrinsic in the process described. The capability is not huge as the technique alters the intra-prediction angle values, that are present in a limited number in any video with an average compression ratio due to the high temporal redundancy reduction. Another deficit of the method is given by the presence of artefacts, sometimes eye-naked visible, mainly due to low or none padding. This sudden variation makes the distortion very evident and may therefore reveal the presence of the secret data.

Another aspect that may be problematic in some practical uses is the lack of an encryption process which is integral part of the technique. This means that if an attacker is able to retrieve the position of the hidden bits, and the secret message has not been encrypted previously, this information is exposed. Therefore, if it's possible to transmit the decryption secret over another secure way, an additional application along the steganography process of cryptography may result convenient.

Table 5 reports a resume of the implementation performance. In cases where the PSNR of a channel was not altered at all, the value was set to the maximum non-infinite value obtaine during the experiments. Note that the cost per byte is lower than 1 because the encoder applies entropy coding on hidden data as well, which confirms that in most of the cases this approach is convenient.

Table 5. Average performance in term of PSNR for channels Y, U and V

Cost per byte	0.941
Y PSNR	41.918
U PSNR	61.005
V PSNR	62.352

5   Conclusions and Future Works

The video steganography technique presented in this paper has proven to perform well on a wide range of settings, contributing to the development of the state-of-the-art free format AV1 as a pioneering proposal of digital steganography working on this standard. The reference implementation, used to perform the experiments reported, provides an effective configuration system that makes possible to tune the process for each specific video structure, by adding padding and offset to the insertion of the hidden bits.

Due to the recently born interest about on AV1, which is being adopted by big players of the industry like Netflix, there may be a lot of further improvements to the product of this research. For example, the inclusion of a smart policy to decide whether to inject a bit or not in a given angle value, with decisions based on the video structure, may lead to much more containment of video alteration artefacts.

Despite the current limitations it brings, the presented method is fully viable in any adapt practical application, like demonstrated by the results of the trials performed, and represents a promising point of start for the research on steganography that leans on this emerging video encoding standard.

References

1. Yao, Y., Zhang, W., Yu, N., Zhao, X.: Defining embedding distortion for motion vector-based video steganography. Multimedia Tools Appl. **74**(24), 11163–11186 (2014). https://doi.org/10.1007/s11042-014-2223-8
2. Battiato, S., Giudice, O., Paratore, A.: Multimedia forensics: discovering the history of multimedia contents. In: Proceedings of the 17th International Conference on Computer Systems and Technologies 2016, pp. 5–16 (2016)

3. Cao, M., Tian, L., Li, C.: A secure video steganography based on the intra-prediction mode (IPM) for h264. Sensors **20**(18), 5242 (2020)
4. Cao, Y., Zhang, H., Zhao, X., Yu, H.: Video steganography based on optimized motion estimation perturbation. In: Proceedings of the 3rd ACM Workshop on Information Hiding and Multimedia Security, pp. 25–31 (2015)
5. Dasgupta, K., Mandal, J.K., Dutta, P.: Hash based least significant bit technique for video steganography (HLSB). Int. J. Secur. Priv. Trust Manage. (IJSPTM) **1**(2), 1–11 (2012)
6. Dhawan, S., Gupta, R.: Analysis of various data security techniques of steganography: a survey. Inf. Secur. J. Glob. Perspect. **30**(2), 63–87 (2021)
7. Patel, et al.: Lazy wavelet transform based steganography in video. In: 2013 International Conference on Communication Systems and Network Technologies, pp. 497–500 (2013)
8. Oh, S., et al.: A large-scale benchmark dataset for event recognition in surveillance video (2011)
9. Manjula, G.R., Sushma, R.B.: Video steganography: a survey of techniques and methodologies. Available at SSRN 3851241 (2021)
10. Gomez, T.S., Guillermo, A., Francia, I.I.I.: Steganography obliterator: an attack on the least significant bits. In: InfoSecCD Conference 2006, September 2006
11. Gujjunoori, S., Amberker, B.B.: DCT based reversible data embedding for mpeg-4 video using HVS characteristics. J. Inf. Secur. Appl. **18**(4), 157–166 (2013)
12. Liu, Y., Liu, S., Wang, Y., Zhao, H., Liu, S.: Video steganography: a review. Neurocomputing **335**, 238–250 (2019)
13. Mstafa, R.J., Elleithy, K.M.: Compressed and raw video steganography techniques: a comprehensive survey and analysis. Multimedia Tools Appl. **76**(20), 21749–21786 (2016). https://doi.org/10.1007/s11042-016-4055-1
14. Piva, A.: An overview on image forensics. ISRN Signal Process. **2013**, 22 (2013)
15. Dahatonde, A., Bhautmage, P., Jeyakumar, A.: Advanced video steganography algorithm. Int. J. Eng. Res. Appl. (IJERA) **3**, 1641–1644 (2013)
16. Sampat, V., Dave, K., Madia, J., Toprani, P.: A novel video steganography technique using dynamic cover generation. In: National Conference on Advancement of Technologies-Information Systems & Computer Networks (ISCON-2012), Proceedings published in Int J of Comput Appl (IJCA). Citeseer (2012)
17. Swathi, A., Jilani, S.A.K.: Video steganography by LSB substitution using different polynomial equations. Int. J. Comput. Eng. Res. **2**(5), 1620–1623 (2012)
18. Yadav, P., Mishra, N., Sharma, S.: A secure video steganography with encryption based on LSB technique. In: 2013 IEEE International Conference on Computational Intelligence and Computing Research, pp. 1–5. IEEE (2013)
19. Zhang, M., Guo, Y.: Video steganography algorithm with motion search cost minimized. In: 2014 9Th IEEE Conference on Industrial Electronics and Applications, pp. 940–943. IEEE (2014)

Deep Learning

Efficient Transfer Learning for Visual Tasks via Continuous Optimization of Prompts

Jonathan Conder$^{(\boxtimes)}$ (iD), Josephine Jefferson(iD), Nathan Pages(iD),
Khurram Jawed(iD), Alireza Nejati(iD), and Mark Sagar(iD)

Soul Machines, Auckland, New Zealand
jonathan.conder@soulmachines.com

Abstract. Traditional methods for adapting pre-trained vision models to downstream tasks involve fine-tuning some or all of the model's parameters. There are a number of trade-offs with this approach. When too many parameters are fine-tuned, the model may lose the benefits associated with pre-training, such as the ability to generalize to out-of-distribution data. But, if instead too few parameters are fine-tuned, the model may be unable to adapt effectively for the tasks downstream. In this paper, we propose *Visual Prompt Tuning* (VPT) as an alternative to fine-tuning for Transformer-based vision models. Our method is closely related to, and inspired by, prefix-tuning of language models [22]. We find that, by adding additional parameters to a pre-trained model, VPT offers similar performance to fine-tuning the final layer. In addition, for low-data settings and for specialized tasks, such as traffic sign recognition, satellite photo recognition and handwriting classification, the performance of Transformer-based vision models is improved with the use of VPT.

Keywords: Computer vision · Few-shot · Fine-tuning · Prompt engineering · Prefix-tuning · CLIP · Transformers · Vision transformers

1 Introduction

Transfer learning [30] is a method for training neural network models on new tasks, starting from parameters that have been trained for other tasks. This allows the network to leverage knowledge common to both the original and the new tasks [42], and is of particular use when applying general models trained on large amounts of data in more specific contexts.

Several approaches to transfer learning are commonly used. For problems where training data is plentiful, the entire network can be trained on the new

Supplementary Information The online version contains supplementary material available at https://doi.org/10.1007/978-3-031-06427-2_25.

task. When data is scarce, however, this approach may increase generalization error [17] as the network "forgets" some of the knowledge it learned originally. For such tasks, the network can be used as the core of a larger model with additional components (such as a classifier network that converts the output features of the core network into probability vectors), and these components can be trained while keeping the core network frozen.

A recent finding in the domain of Natural Language Processing (NLP) has been that large-scale pre-trained models can be adapted to new tasks *without* additional training. This is done by prompting the model, during inference time, with some appropriate text [3].

For example, a language model pre-trained on a large text corpus can be made to summarize a body of text by prepending the sentence "Provide a summary of the following text", or simply appending "TL;DR:". In this way, the problem of adapting a network to a new task becomes a problem of manually *engineering* a good prompt for that task.

Applying this concept to computer vision, recent methods such as CLIP [29] have used joint contrastive training to encode mappings from text and images into a common feature space. This allows prompt engineering to be used for computer vision tasks, in addition to NLP. For instance, a "type of pie" classifier can be constructed by ranking images based on the distance of the feature vector to the feature vectors of "a photo of a blueberry pie" or "a photo of a meat pie."

While this approach is quite flexible, there are two downsides to prompt engineering:

1. Finding the best prompts can be a matter of trial and error.
2. It is limited to text domains.

Recently, AutoPrompt [34] and prefix-tuning (PT) [22] were proposed as solutions to the first issue. AutoPrompt finds a sequence of textual tokens through discrete optimization. In contrast, PT optimizes a number of vector "tokens" in the *input feature* space, which is continuous (unlike the text token space). Thus, the prefix can be optimized using standard gradient descent methods. This approach has been shown to perform at least as well as fine-tuning and adapter tuning [13,31], even with fewer trained parameters.

Inspired by the success of these techniques, we aim to address the second issue. In the visual domain, the most widely used models, traditionally, are convolutional neural networks (CNNs), but several recent works have explored applying Transformers [38] to computer vision tasks. One route to using transformers in visual tasks is to feed the feature vectors produced by a CNN into a Transformer [4]. As an alternative, Vision Transformer (ViT) models [8] avoid using CNNs altogether, by passing (linear projections of) a grid of image patches directly to a Transformer. The ViT approach has demonstrated better performance than conventional CNNs if the training dataset is sufficiently large [8], which aligns with the fact that Transformer models lack the inductive biases of CNNs. Unlike the text setting, it is not clear to us whether manually engineered images can be used as prompts in the visual domain. However, it is straightforward to adapt PT for use with ViT models, since they are very similar to text-based Transformers.

There are good reasons to suspect that prompts could improve the performance of Transformers on visual tasks. One comes from the experience of optical illusions involving color, where the color of one part of an image can alter the perception of color in another part. Another is related to the Transformer architecture. Since Transformers multiply their inputs with each other, it has been hypothesized they are good at learning *contextual representations* [14]. In other words, the representation of an input token is modulated by the other tokens. Recent work in prompt engineering suggests that prompts serve to *locate* a particular task in the space of all tasks the model has learned [32]. A Transformer trained on diverse visual data will learn a variety of tasks, such as recognizing both photographs and sketches of particular objects. Prompting it can then "prime" the network to solve a task more relevant to a specific domain.

In this work, we explore the performance of PT when applied to a large ViT model. Our method adds prompt vectors as additional inputs to ViT models, alongside image patches which have been linearly projected and combined with a positional embedding (see Fig. 1). We call our method *visual prompt tuning* (VPT), as the term "prefix" can be misleading when applied to non-sequential input data, like image patches. The Transformer architecture allows us to optimize these prompts using gradient descent, without modifying or removing any of the ViT parameters.

Our hypothesis is that VPT may improve transfer learning for visual tasks. In what follows, we evaluate the performance of VPT on a number of image classification benchmarks. For comparison, we measured the performance of the ViT model without additional training (zero-shot), a linear classifier trained on the output features of the ViT model, and a linear classifier combined with VPT. We found that using VPT always improved performance, and in many cases VPT alone outperformed the linear classifier. The tasks which gained the most from VPT were those for which data was scarce, and those which differed significantly from the pre-training task.

2 Methods

2.1 Pre-training

We chose the ViT component of CLIP [29] as the pre-trained model to use for our experiments.[1] The full CLIP model consists of an image encoder (in our case, a ViT model) and a text encoder; which both output real-valued vectors (with the same shape). To classify an image using CLIP, one can encode it and compare the resulting vector with several encoded text labels, using cosine similarity. Similarly, one can classify a string of text in terms of a set of image "labels". CLIP has been trained on 400 million pairs of images with related text, using a contrastive loss function. This is the average of the cross entropy loss of the classifiers described above, for each minibatch.

---

[1] Specifically ViT-B/32, which is available at https://github.com/openai/CLIP.

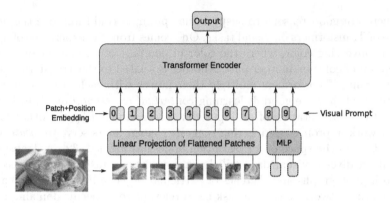

Fig. 1. Vision Transformer (ViT) with Visual Prompt Tuning. During VPT, only the parameters shown with a dotted border are trained.

The main advantage CLIP has over other ViT models is that it can classify images given any number of text labels, without additional fine-tuning. This allows us to evaluate VPT on its own, without altering the final layer of the model. It has also been trained on a large collection of diverse data, which is likely to improve transfer learning performance for real-world tasks.

2.2   Fine-Tuning

We evaluated the following approaches to fine-tuning the base CLIP model. The two non-VPT approaches have already been benchmarked [29], but we sought to independently reproduce them here.

Zero-shot (ZS). This approach does not train any existing or additional parameters. Given an image, we encoded it using CLIP and compared the result with a fixed set of encoded text labels (depending on the benchmark in question). We based our labels on those used in the original CLIP benchmarks [29, Fig. 21], but there was guesswork involved in naming some classes (especially for EuroSAT and GTSRB).

The classification scores could have been improved by using several labels for each class and averaging the corresponding feature vectors [29, §3.1.4]. Prefix-tuning [22] the labels is another possibility. However, for the purposes of evaluating VPT, we opted for the single label approach.

Visual Prompt Tuning (VPT). This approach is illustrated by Fig. 1. The first layer of the CLIP image encoder is a strided convolution, which effectively breaks the input image into a grid of patches, flattens the resulting tensors into vectors and projects each of these into a lower-dimensional space using a learned linear transformation. After that, the encoder adds a learned positional embedding to each vector. Ordinarily, these vectors, together with the learned "class" embedding, are the only inputs to the Transformer proper.

For VPT, we provided additional inputs (the "prompt") to the Transformer, bypassing the convolution and the positional embedding. This did not require architectural changes to the Transformer itself, because it is agnostic to the number of inputs.[2] The prompt can be trained directly using gradient descent. One can also use another network, such as a multi-layer perceptron (MLP), to generate the prompt from trainable input vectors. The latter approach has been shown to improve results for prefix-tuning [22]. We typically trained an MLP as well, but sometimes also added a positional embedding to its outputs. Note that the MLP and positional embedding are only needed for training; at inference time, the generated prompts are fixed, so the same pre-computed prompts can be used for all input images.

To use this modified model as a classifier, we compared the Transformer output with the encoded text labels from the zero-shot approach. We also considered prefix-tuning the text encoder (at the same time as VPT), but in preliminary experiments the small (if any) improvement in performance was not worth the large increase in training time.

Training a Linear Classifier. For this approach, we replaced the final layer of the CLIP image encoder (a linear projection) so that its output dimension matched the number of classes for a given benchmark.

We used Adam [16] to train this layer, as opposed to the L-BFGS algorithm (which was used for original benchmarks), as the PyTorch [27] implementations of L-BFGS we tried were unable to compete with the other approaches, and the scikit-learn [28] implementation was unsuitable for end-to-end training when combining this approach with VPT. We found that the PyTorch implementation of Adam (on a single GPU) performed similarly to, and converged faster than, the scikit-learn implementation of L-BFGS (on a multicore CPU). Incidentally, VPT is slower than both of these approaches, because gradients propagate through the entire ViT model, and one can no longer cache the CLIP-encoded images.

Combining VPT with a Linear Classifier (VPT+L). This approach is the same as VPT, but instead of using encoded text labels, we replaced the final layer of the CLIP image encoder and trained that together with the prompt.

2.3 Datasets

The datasets we experimented with are broadly similar to those used for the original CLIP benchmarks [29, Table 9]. However, we did not use (or recreate) Country211, Hateful Memes, Kinetics700, KITTI or PASCAL VOC 2007. For

---

[2] It does change the number of outputs, but CLIP only uses the one corresponding to the "class" embedding.

Table 1. Hyperparameters used for visual prompt tuning. Each column represents a distinct hyperparameter selection.

Trainable vectors	8	8	16	16
Dimension of trainable vectors	32	768	768	4
Positional embedding	No	No	No	Yes
Parameters	24,832	595,968	602,112	15,424
Learning rate	0.001	0.01	0.01	0.01

ImageNet, we randomly sampled some WordNet synsets and used a script to download around 600 images per class. We rendered SST2 ourselves. For UCF101 we used only the middle frame of each video.

We used provided train/validation/test splits when available. For Birdsnap we only used the images which were available online at the time, and sampled 4 images per class as a test set. We used a random selection of 30 images per class as the test set for Caltech-101. To generate CLEVR Counts, we sampled 2000 training/validation images and 500 test images; the remaining images were discarded. Similarly, for EuroSAT we sampled only 1000 training/validation images and 500 test images per class. We used a random selection of 70 images per class as the test set for ReSISC45. For STL10, SUN397 and UCF101, which provide multiple splits, we chose the first one. If a dataset did not provide a validation set, we simply used 10% of the training data. See the supplementary material for detailed references and split sizes.

We used mean per class accuracy as our evaluation metric for Caltech-101, FGVC-Aircraft, and both Oxford datasets; overall accuracy was used for the remaining datasets. In the discussion below (e.g. on graph axes), we slightly abuse language and use "accuracy" (and "error rate") as generic labels for these per-dataset metrics (and their complements).

2.4 Training Procedure

We trained each model on 2 Quadro RTX 8000 cards using automatic mixed precision, an initial learning rate ranging from 0.01 to 0.001, and a batch size of 512. It took a total of 3 weeks (averaging 51 min per run for few-shot classification, and 88 min per run for ordinary classification).

For Caltech 101, CIFAR-100, and Oxford Flowers, we experimented with a wide variety of VPT hyperparameters. We found that training the prompt vectors directly lead to poor performance. On the other hand, using an MLP to generate the prompt was no better than a single fully-connected (FC) layer. The best performing choices, recorded in Table 1, were then used for visual prompt tuning across all the datasets. For instance, in the leftmost case, each prompt vector was generated by passing one of eight vectors through a linear map $\mathbb{R}^{32} \to \mathbb{R}^{768}$. In the rightmost case, we instead used sixteen vectors in $\mathbb{R}^4$, and added the result to one of sixteen "positional embedding" vectors (in $\mathbb{R}^{768}$).

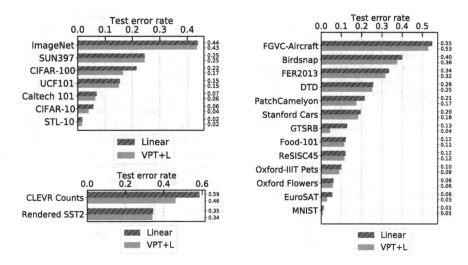

Fig. 2. Comparison of the test error rate of the Linear and VPT+L (VPT with Linear classifier, see Sect. 2.2) methods, on general-purpose classification datasets (top left), specialized classification datasets (right) and non-classification datasets (bottom left).

We used cross entropy as the loss function. Once the validation loss plateaued, we reduced the learning rate by a factor of 10. Training was stopped if the validation metric (usually accuracy) had not improved for 15 epochs. We considered including the validation set in the training data for a final session, reusing the best known hyperparameters, but found the performance difference (on the test set) to be negligible in experiments.

For few-shot classification, we only validated once every 10 epochs (as the validation sets were much larger than the new training sets), and we only used the best known hyperparameters for each dataset.

3   Results

Our attempts to replicate the original zero-shot and linear classifier benchmarks for CLIP yielded slightly different results, for a number of possible reasons. For example, some of our datasets (or the train/validation/test splits) did not match the originals exactly. For the zero-shot approach, we may have labelled some classes differently. Also, our linear classifiers were trained differently (to facilitate combining them with VPT).

3.1   Classification

We divided the datasets qualitatively into three categories: general-purpose classification (ImageNet [33], CIFAR-10 [19], CIFAR-100 [19], SUN397 [40,41],

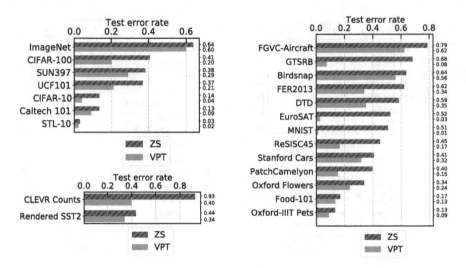

Fig. 3. Comparison of the test error rate of the ZS and VPT methods, on general-purpose classification datasets (top left), specialized classification datasets (right) and non-classification datasets (bottom left).

UCF101 [36], STL-10 [7], and Caltech 101 [23]), specialized classification (FGVC-Aircraft [24], GTSRB [37], Birdsnap [1], FER2013 [10], DTD [6], EuroSAT [11,12], MNIST [21], ReSISC45 [5], Stanford Cars [18], PatchCamelyon [9,39], Oxford Flowers [25], Oxford Pets [26], Food 101 [2]), and specialized tasks that weren't classification tasks (CLEVR Counts [15,29] and Rendered SST2 [29,35]).

Figure 2 presents the test error rate using the best per-dataset hyperparameter selections for the Linear and VPT+L methods (see Sect. 2.2, and supplementary information for hyperparameter values). As can be seen, in the general-purpose classification sets, VPT only offers a clear advantage for CIFAR-100 and CIFAR-10. However, for the specialized classification tasks, VPT improves accuracy for many of the datasets, especially EuroSAT and GTSRB, whereas e.g. DTD and Oxford Flowers do not seem to benefit from VPT.

Thus, we see a general pattern of VPT improving the performance more for tasks that are domain specific. In particular, tasks where the training images differ substantially from natural images and other images likely to appear in the CLIP training set. As regards CIFAR-100 and CIFAR-10 benefiting from VPT, the images in those two datasets have a much lower resolution than those typically seen on the internet. VPT also offers a performance advantage for CLEVR Counts, however the baseline performance is already poor (~60% error rate), thus accuracy with VPT is still relatively low.

Figure 3 shows the test error rate for the best per-dataset hyperparameter selections for the ZS and VPT methods (see Sect. 2.2, and supplementary information for hyperparameter values). Here, the advantages of VPT are more pronounced, since the ZS method does not use the training data. The trend of VPT offering larger improvements for specialized datasets is also more clear,

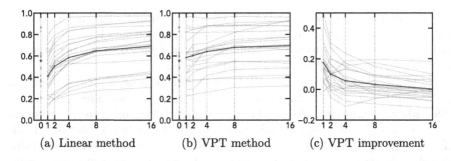

(a) Linear method   (b) VPT method   (c) VPT improvement

Fig. 4. Test accuracy (vertical axis) vs. number of labelled examples per class (horizontal axis) when using the Linear or VPT methods. The blue lines are the average of the accuracies over all the datasets (light gray lines). The zero-shot CLIP baseline is indicated by star. (Color figure online)

especially for the EuroSAT and MNIST datasets, where VPT takes the error rate from nearly 50% to almost state-of-the-art.

3.2 Few-Shot Classification

Figure 4a presents the test accuracy of the linear classifier method (see Sect. 2.2) when trained on only 1, 2, 4, 8 or 16 images per class. The test accuracy values reported at 0 are for the ZS method.

We observe that one-shot training of a linear classifier does not outperform the zero-shot method, except for a few datasets. For Oxford Pets and RenderedSST2, even 16-shot training underperforms. These results are coherent with the original benchmarks [29, Fig. 6], which found that (on average) 4 images per class were required for a few-shot linear classifier to match zero-shot performance.

Figure 4b shows the test accuracy of the VPT method (see Sect. 2.2) in the context of few-shot learning. Here, one-shot learning outperforms the zero-shot baseline in most cases. This demonstrates that VPT is a more robust approach to few-shot transfer learning than the linear classifier method.

Figure 4c compares the few-shot performance of the VPT and linear classifier methods directly. For all but one task, VPT outperforms the linear classifier method in the one-shot setting, and by about 20% on average. When more data is available the gap becomes smaller (as one might expect from Fig. 2 and Fig. 3). Overall VPT outperforms the Linear method when data is scarce.

4 Discussion

4.1 Applications

One of our goals in pursuing this line of research is to develop systems that learn in a human-like way. Transfer learning is a promising avenue, as people leverage existing knowledge when learning to recognize new objects [20].

Our benchmarks show that the VPT approach to transfer learning is effective for tasks which are highly specialized, and those for which data is scarce. In other cases, the time it takes to tune a prompt might not be worth the performance improvement over simply training a linear classifier.

Other possible applications of VPT are similar to those of prefix-tuning [22, §8]. For instance, visual prompts could be used by a cloud-based provider to efficiently run classifiers for several organizations at the same time, or even different users within the same organization. One could even employ multiple levels of visual prompt tuning: for instance, part of the prompt could improve traffic sign classification, and another part could be tuned for traffic signs in a specific country.

4.2 Hyperparameters for VPT

When tuning hyperparameters (see Table 1), we found that using a fully-connected network to generate prompts struck the best balance between performance and parameter count. For about half of the datasets, a layer with hundreds of inputs gave the best results. As few as four inputs worked well for many of the other datasets, after we added a "positional embedding."

Adding this layer changes the number of trainable parameters without affecting the network architecture at inference time. However, it does alter the way the prompt is initialized, which could contribute to the performance gap compared with tuning the prompt directly.

Another possibility is that the parameters of the fully-connected layer are updated more efficiently. Since they contribute to more than one prompt vector, these parameters may be weighted more heavily in attention calculations. This could allow general features of the downstream task to be learned quickly, while allowing the other parameters to update more slowly. It is likely that all the vectors in a good prompt have something in common. For example, when prefix-tuning a language model, it helps to initialize the prefix with real words, even if the words are unrelated to the task at hand [22]. To close the performance gap it might suffice to vary the learning rate more frequently, or use different rates for different parts of the prompt.

4.3 Conclusion

In this paper we have presented visual prompt tuning, a transfer learning method that preserves the weights of the model but fine-tunes on tasks by adding an auxiliary prompt input. For tasks that are specific and relatively out-of-domain, this method shows improved results over the linear classifier method. Performance is not significantly decreased for more general tasks, or in-domain tasks. Compared to the zero shot method, results are significantly improved. In addition, in data-scarce (few shot) domains, generalization is improved over the linear method. VPT can also be used in combination with other methods, to further improve performance on certain tasks. In conclusion, we believe VPT presents a valuable method for adapting transformer-based vision models to new domains and tasks.

References

1. Berg, T., et al.: Birdsnap: large-scale fine-grained visual categorization of birds. In: IEEE Conference on Computer Vision and Pattern Recognition, pp. 2019–2026 (2014). https://doi.org/10.1109/CVPR.2014.259
2. Bossard, L., et al.: Food-101 - mining discriminative components with random forests. In: European Conference on Computer Vision, pp. 446–461 (2014). https://doi.org/10.1007/978-3-319-10599-4_29
3. Brown, T., et al.: Language models are few-shot learners. In: Advances in Neural Information Processing Systems, vol. 33, pp. 1877–1901 (2020)
4. Carion, N., et al.: End-to-end object detection with transformers. In: European Conference on Computer Vision, pp. 213–229. Springer (2020). https://doi.org/10.1007/978-3-030-58452-8_13
5. Cheng, G., et al.: Remote sensing image scene classification: benchmark and state of the art. In: Proceedings of the IEEE, vol. 105, pp. 1865–1883 (2017). https://doi.org/10.1109/JPROC.2017.2675998
6. Cimpoi, M., et al.: Describing textures in the wild. In: IEEE Conference on Computer Vision and Pattern Recognition, pp. 3606–3613 (2014). https://doi.org/10.1109/CVPR.2014.461
7. Coates, A., et al.: An analysis of single-layer networks in unsupervised feature learning. In: Proceedings of the Fourteenth International Conference on Artificial Intelligence and Statistics, vol. 15, pp. 215–223 (2011)
8. Dosovitskiy, A., et al.: An image is worth 16 × 16 words: transformers for image recognition at scale. In: International Conference on Learning Representations (2021)
9. Ehteshami Bejnordi, B., et al.: Diagnostic assessment of deep learning algorithms for detection of lymph node metastases in women with breast cancer. JAMA **318**, 2199–2210 (2017). https://doi.org/10.1001/jama.2017.14585
10. Goodfellow, I.J., et al.: Challenges in representation learning: a report on three machine learning contests. Neural Netw. **64**, 59–63 (2015). https://doi.org/10.1016/j.neunet.2014.09.005
11. Helber, P., et al.: Introducing EuroSAT: a novel dataset and deep learning benchmark for land use and land cover classification. In: IEEE International Geoscience and Remote Sensing Symposium, pp. 204–207 (2018). https://doi.org/10.1109/IGARSS.2018.8519248
12. Helber, P., et al.: EuroSAT: a novel dataset and deep learning benchmark for land use and land cover classification. IEEE J. Selected Top. Appl. Earth Observ. Remote Sens. **12**, 2217–2226 (2019). https://doi.org/10.1109/JSTARS.2019.2918242
13. Houlsby, N., et al.: Parameter-efficient transfer learning for NLP. In: Proceedings of the 36th International Conference on Machine Learning, vol. 97 (2019)
14. Jayakumar, S.M., et al.: Multiplicative interactions and where to find them. In: International Conference on Learning Representations (2019)
15. Johnson, J., et al.: CLEVR: a diagnostic dataset for compositional language and elementary visual reasoning. In: IEEE Conference on Computer Vision and Pattern Recognition, pp. 1988–1997 (2017). https://doi.org/10.1109/CVPR.2017.215
16. Kingma, D.P., Ba, J.: Adam: a method for stochastic optimization. In: International Conference on Learning Representations (2015)
17. Kouw, W.M., Loog, M.: An introduction to domain adaptation and transfer learning. Delft University of Technology, Technical report (2018)

18. Krause, J., et al.: 3D object representations for fine-grained categorization. In: IEEE International Conference on Computer Vision Workshops, pp. 554–561 (2013). https://doi.org/10.1109/ICCVW.2013.77
19. Krizhevsky, A.: Learning multiple layers of features from tiny images. University of Toronto, Technical report (2009)
20. Lake, B., et al.: One shot learning of simple visual concepts. In: Proceedings of the Annual Meeting of the Cognitive Science Society 33 (2011)
21. Lecun, Y., et al.: Gradient-based learning applied to document recognition. In: Proceedings of the IEEE, pp. 2278–2324 (1998). https://doi.org/10.1109/5.726791
22. Li, X.L., Liang, P.: Prefix-tuning: optimizing continuous prompts for generation. In: Proceedings of the 59th Annual Meeting of the Association for Computational Linguistics and the 11th International Joint Conference on Natural Language Processing (Volume 1: Long Papers), pp. 4582–4597 (2021). https://doi.org/10.18653/v1/2021.acl-long.353
23. Fei-Fei, L., et al.: Learning generative visual models from few training examples: an incremental Bayesian approach tested on 101 object categories. In: Conference on Computer Vision and Pattern Recognition Workshop, pp. 178–178 (2004). https://doi.org/10.1109/CVPR.2004.383
24. Maji, S., et al.: Fine-grained visual classification of aircraft. arXiv preprint (2013)
25. Nilsback, M., Zisserman, A.: Automated flower classification over a large number of classes. In: 2008 Sixth Indian Conference on Computer Vision, Graphics & Image Processing, pp. 722–729 (2008). https://doi.org/10.1109/ICVGIP.2008.47
26. Parkhi, O.M., et al.: Cats and dogs. In: IEEE Conference on Computer Vision and Pattern Recognition, pp. 3498–3505 (2012). https://doi.org/10.1109/CVPR.2012.6248092
27. Paszke, A., et al.: PyTorch: an imperative style, high-performance deep learning library. In: Advances in Neural Information Processing Systems, vol. 32, pp. 8024–8035 (2019)
28. Pedregosa, F., et al.: Scikit-learn: machine learning in Python. J. Mach. Learn. Res. **12**, 2825–2830 (2011)
29. Radford, A., et al.: Learning transferable visual models from natural language supervision. In: ICML (2021)
30. Rawat, W., Wang, Z.: Deep convolutional neural networks for image classification: a comprehensive review. Neural Comput. **29**, 2352–2449 (2017). https://doi.org/10.1162/neco_a_00990
31. Rebuffi, S.A., et al.: Learning multiple visual domains with residual adapters. In: Advances in Neural Information Processing Systems, vol. 30 (2017)
32. Reynolds, L., McDonell, K.: Prompt programming for large language models: beyond the few-shot paradigm. In: Extended Abstracts of the 2021 CHI Conference on Human Factors in Computing Systems (2021). https://doi.org/10.1145/3411763.3451760
33. Russakovsky, O., et al.: ImageNet large scale visual recognition challenge. Int. J. Comput. Vis. **115**(3), 211–252 (2015). https://doi.org/10.1007/s11263-015-0816-y
34. Shin, T., et al.: AutoPrompt: eliciting knowledge from language models with automatically generated prompts. In: Proceedings of the 2020 Conference on Empirical Methods in Natural Language Processing, pp. 4222–4235 (2020). https://doi.org/10.18653/v1/2020.emnlp-main.346
35. Socher, R., et al.: Recursive deep models for semantic compositionality over a sentiment treebank. In: Proceedings of the 2013 Conference on Empirical Methods in Natural Language Processing, pp. 1631–1642 (2013)

36. Soomro, K., et al.: UCF101: a dataset of 101 human actions classes from videos in the wild. University of Central Florida, Technical report (2012)
37. Stallkamp, J., et al.: Man vs. computer: benchmarking machine learning algorithms for traffic sign recognition. Neural Netw. **32**, 323–32 (2012). https://doi.org/10. 1016/j.neunet.2012.02.016
38. Vaswani, A., et al.: Attention is all you need. In: Advances in Neural Information Processing Systems, vol. 30 (2017)
39. Veeling, B.S., et al.: Rotation equivariant CNNs for digital pathology. In: Medical Image Computing and Computer Assisted Intervention, pp. 210–218 (2018). https://doi.org/10.1007/978-3-030-00934-2_24
40. Xiao, J., et al.: SUN database: large-scale scene recognition from abbey to zoo. In: IEEE Conference on Computer Vision and Pattern Recognition (2010). https:// doi.org/10.1109/CVPR.2010.5539970
41. Xiao, J., Ehinger, K.A., Hays, J., Torralba, A., Oliva, A.: SUN database: exploring a large collection of scene categories. Int. J. Comput. Vis. **119**(1), 3–22 (2014). https://doi.org/10.1007/s11263-014-0748-y
42. Zamir, A.R., et al.: Taskonomy: disentangling task transfer learning. In: IEEE Conference on Computer Vision and Pattern Recognition, pp. 3712–3722 (2018). https://doi.org/10.1109/CVPR.2018.00391

Continual Learning with Neuron Activation Importance

Sohee Kim⬤ and Seungkyu Lee(✉)⬤

Department of Computer Science and Engineering, Kyunghee University, 1732,
Deogyeong-daero, Giheung-gu, Yongin-si, Gyeonggi-do, Republic of Korea
{soheekim,seungkyu}@khu.ac.kr

Abstract. Continual learning is a concept of online learning with multiple sequential tasks. One of the critical barriers of continual learning is that a network should learn a new task keeping the knowledge of old tasks without access to any data of the old tasks. We propose a neuron activation importance-based regularization method for stable continual learning regardless of the order of tasks. We conduct comprehensive experiments on existing benchmark data sets to evaluate not just the stability and plasticity of our method with improved classification accuracy also the robustness of the performance along the changes of task order.

Keywords: Continual learning · Neuron importance

1 Introduction

Continual learning is a sequential learning scheme on multiple different tasks. New tasks do not necessarily consist of only existing classes of previous tasks nor statistically similar instances of existing classes. In challenging situations, new tasks may consist of mutually disjoint classes or existing classes with unseen types of instances in previous tasks. One of the main challenges is learning such new tasks without catastrophic forgetting existing knowledge of previous tasks. Researchers have proposed diverse continual learning approaches to achieve both stability (remembering past tasks) and plasticity (adapting to new tasks) of their deep neural networks from sequential tasks of irregular composition of classes and varying characteristics of training instances. Since the training of a neural network is influenced more by recently and frequently observed data, the neural network forgets what it has learned in prior tasks without continuing access to them in the following tasks. Continual learning model has to adapt to a new task without access to some or entire classes of past tasks while it maintains acquired knowledge from the past tasks [19]. In addition, the continual learning model has to be evaluated with arbitrary order of tasks since the order of tasks is not able to be fixed nor predicted in real applications. It is required to function consistently regardless of the order of tasks.

Supplementary Information The online version contains supplementary material available at
https://doi.org/10.1007/978-3-031-06427-2_26.

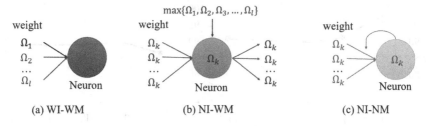

Fig. 1. Three different (Importance-Measurement) ways (a) Weight Importance ($\Omega_1 \sim \Omega_l$) by respective Weight Measurement (b) Neuron Importance (Ω_k) by Weight Measurements. The maximum value of weight importance out of ($\Omega_1 \sim \Omega_l$) is assigned to neuron importance (Ω_k). And then all weights connected to the neuron get the same importance of the neuron. (c) Neuron Importance (Ω_k) by Neuron Measurement, where l and Ω indicate weight index and importance of either weight or neuron respectively. The proposed method belongs to NI-NM.

There are three major categories in prior continual learning approaches; 1) architecture modification of neural networks [17,18,21], 2) rehearsal using sampled data from previous tasks [3,16], and 3) regularization freezing significant weights of a model calculating the importance of weights or neurons [1,2,4,7,8,10,14,15,22,23]. Most recent methods have tackled the problem with fundamental regularization approaches that utilize the weights of given networks to the fullest. The basic idea of regularization approaches is to constrain essential weights of prior tasks not to change. In general, they alleviate catastrophic interference with a new task by imposing a penalty on the difference of weights between the prior tasks and the new task. The extent of the penalty is controlled by the significance of weights or neurons in solving a certain task using respective measurements.

As Fig. 1 illustrates, weight importance can be decided by three different (Importance-Measurement) ways. WI-WM (Weight Importance by Weight Measurement) [2,10,15,22,23] calculates weight importance based on the measurement of the corresponding weight as described in Fig. 1a. Elastic weight consolidation (EWC) [10] estimates parameter importance using the diagonal of the Fisher information matrix equivalent to the second derivative of the loss. Synaptic intelligence (SI) [22] measures the importance of weights in an online manner by calculating each parameter's sensitivity to the loss change while it trains a network. Unlike SI [22], Memory aware synapses (MAS) [2] assesses the contribution of each weight to the change of a learned function. Variational Continual Learning (VCL) [15], a Bayesian neural network-based method, decides weight importance through variational inference. Bayesian Gradient Descent (BGD) [23] finds posterior parameters assuming that the posterior and the prior distributions are Gaussian. To mitigate the interference across multiple tasks in continual learning, weight importance-based approaches let each weight have its weight importance. However, in the case of convolutional neural networks, since a convolutional filter makes one feature map that can be regarded as one neuron, those weights should have the same importance. NI-WM (Neuron Importance by Weight Measurement) calculates neuron importance based on the measurement of all weights. Weight importance is redefined as the importance of its connected neuron [1]. Uncertainty-regularized Continual Learning (UCL) [1] measures weight importance by its

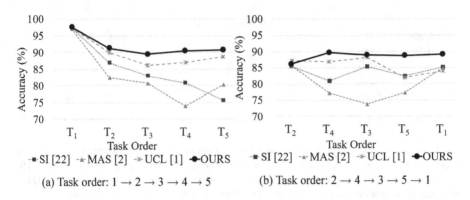

(a) Task order: $1 \rightarrow 2 \rightarrow 3 \rightarrow 4 \rightarrow 5$ (b) Task order: $2 \rightarrow 4 \rightarrow 3 \rightarrow 5 \rightarrow 1$

Fig. 2. Classification accuracy of continual learning on Split Cifar10. SI [22], MAS [2] and UCL [1] show critical changes in their performance as the order of tasks changes.

uncertainty indicating the variance of weight distribution. It claims that the distribution of essential weights for past tasks has low variance, and such stable weights during training a task are regarded as important weights not to forget. As illustrated in Fig. 1b, it suggests neuron-based importance in neural networks. NI-NM (Neuron Importance by Neuron Measurement) calculates neuron importance based on the measurement of the corresponding neuron [8].

Weight importance is defined as the importance of its connected neuron. [8] exploits proximal gradient descents using a neuron importance. Activation value itself is a measurement of neuron importance, and weights connected to the neuron get identical weight importance. One critical observation in prior experimental evaluations of existing continual learning methods is that the accuracy of each task significantly changes when the order of tasks is changed. As discussed in [20], proposing a continual learning method robust to the order of tasks is another critical aspect. Therefore, performance evaluation

Table 1. Performance disparity(%) between Fig. 2a and 2b on Split CIFAR 10. "Absolute task order" represents the sequence of tasks that a model learns. (Additional explanation is discussed in Sect. 3.)

Method	Absolute task order				
	T_1	T_2	T_3	T_4	T_5
SI [22]	11.9	5.9	2.4	**1.5**	9.48
MAS [2]	11.4	5.2	7.0	3.3	4.5
UCL [1]	**10.2**	3.1	2.0	5.0	4.7
OURS	11.4	**1.5**	**0.5**	1.7	**1.6**

with fixed task order does not coincide with the fundamental aim of continual learning where no dedicated order of tasks is given in reality. Figure 2 shows sample test results of state-of-the-art continual learning methods compared to our proposed method. As summarized in Table 1, classification accuracy of prior methods fluctuate as the order of tasks changes(from Fig. 2a to Fig. 2b).

In this work, we propose a regularization approach for continual learning assigning neuron importance by the measurement of average neuron activation. We balance neuron importance distribution among layers based on the average neuron activation divided by standard deviation, which is critical to performance consistency along the changes of task order. We assign calculated neuron importance to all weights of incoming edges connected to the neuron. A Neuron with high activation to the majority of instances is defined as an essential neuron. We freeze essential neurons by freezing the weights of all connected incoming edges (essential weights) during the learning of a new task so that our model remembers past tasks. We propose to evaluate the robustness to the order of tasks in a comprehensive manner in which we evaluate the average and standard deviation of classification accuracy with multiple sets of randomly shuffled tasks. Our approach remembers past tasks robustly compared to recent regularization methods [1, 2, 15, 22]. To measure performance fluctuation along the change of task order, we evaluate our method with numerous shuffled orders. We quantitatively evaluate our classification performance based on a measure of interference from past tasks on MNIST [6, 13], CIFAR10, CIFAR100 [12] and Tiny ImageNet [5] data sets.

2   Proposed Method

2.1   Neuron Importance by Average Neuron Activation

The proposed method extracts neuron importance based on the average activation value of all instances. And then the neuron importance is assigned to all weights of incoming edges connected to the neuron. In convolutional neural networks, activation value of a neuron corresponds to the average value of one feature map (i.e., global average pooling value). The average activation value of neuron corresponds to the average of global average pooling value. The average activation values at each layer are independently calculated but are considered together. In other words, the individual average activation values represent the importance of each neuron of a whole model. However, encoded features at each layer describe different aspects of an input image and, as a result, the average activation values at each layer should not be evaluated together. Therefore, the average activation value is not able to fully represent the characteristics of the essential neuron. Besides, in convolution neural networks, the absolute magnitude of average activation value (i.e., the average of global average pooling value) varies along the location of layer: in high-level feature maps, the portion of activated area decreases. Due to the difference in absolute average activation values across the layers, weights of earlier layers tend to be considered more essential. If the average activation value is used as neuron importance, networks will prefer to keep the weights of earlier layers. Instead, we propose to use layer-wise average activation divided by the respective standard deviation for neuron importance measurement. Compared to the average activation-based neuron importance, ours prevents earlier layers from getting excessive importance compared to other layers, which, in turn, prevents a network from vulnerable to changing the order of tasks in terms of forgetting past tasks.

Then, why this improves the continual learning performance regardless of task order? In prior works, more weights of lower layers tend to be frozen in earlier tasks that eliminate the chance of upcoming tasks to build new low-level feature sets. Only a new task that is fortunately able to rebuild higher-layer features based on the frozen lower layer weights from previous tasks could survive. On the other hand, ours keeps the balance of frozen weights in all layers securing more freedom of feature descriptions for new tasks in both lower and higher layers. Indeed, lower layer features such as edges are not class (task) dependent features. Therefore, excessively freezing lower layer features is not preferable in continual learning. Even though tasks change, a new task may find alternative low-level features that have high similarity with them of past tasks, as discussed in [11]. In order to encode such relation, we propose to use the average and standard deviation of neuron activation values at each layer. Our loss function is described as follows.

$$L_t = \tilde{L}_t + \alpha \sum_l \Omega_k^t (w_l^{t-1} - w_l^t)^2, \tag{1}$$

where $\tilde{L}_t$ is loss of current task (e.g., cross entropy loss), t is task index, l is weight index, and Ω_k^t indicates k^{th} neuron importance. α is a strength parameter to control the amount of weights consolidation. Neuron importance is defined as follows.

$$\Omega_k^t = \frac{\frac{1}{N_t} \sum_{i=1}^{N_t} f_k(x_i^{(t)})}{\sigma + \epsilon}, \tag{2}$$

$$\sigma = \sqrt{\frac{\sum_{i=1}^{N_t} \{f_k(x_i^{(t)}) - \frac{1}{N_t} \sum_{i=1}^{N_t} f_k(x_i^{(t)})\}^2}{N_t}},$$

where N_t is the number of instances, x is input, k is neuron index, $f_k(\cdot)$ is activation value (global average value, in the case of convolution neural network), and i is instance index. We introduce ϵ to prevent the numerator from being zero when the standard deviation becomes zero. Proposed method considers the variation of average activation value among instances and the differences of average activation value among different layers. It encourages freezing more weights of later layers than earlier layers which are more likely to describe given task-specific features.

Our experiments(Table 2 in Sect. 3.2) show that prior methods tend to forget past tasks in learning new tasks. In the prior methods, weights of later layers are more likely to change than weights of earlier layers during learning a new task. In general, if the essential weights of later layers of previous tasks change, the network forgets past tasks and hardly recovers previous task-specific features. On the other hand, even though weights of earlier layers of previous tasks change, there are other chances to recover general low-level features which are shared with following new tasks. Since our method puts relatively more constraints on the weights of task-specific features not to change than the prior methods, our method forgets past tasks less showing stable performance along the change in the order of tasks.

2.2   Weight Re-initialization for Better Plasticity

In continual learning, networks have to not only avoid catastrophic forgetting but also learn new tasks. According to the extent of difference in optimal classification feature space of different tasks, optimized feature space in the previous task might be significantly changed with a new task. In the learning of a new task, we can let the model start either from random weights or from optimized weights with previous tasks. Even though the optimized weights on previous tasks can be considered as a set of random weights for a new task, we avoid a situation where the optimized weights for one task work as a local optimal for another similar task that may hinder new training from obtaining new optimal weights through weight re-initialization. The situation can be explained with $\Omega_k(w_k^{t-1} - w_k^t)^2$ term in the loss function of our network. During the learning of a new task, the network is informed of past tasks by $\Omega_k(w_k^{t-1} - w_k^t)^2$ term which lets the network maintain essential weights of the past tasks assigning high Ω_k values. In other words, $\Omega_k(w_k^{t-1} - w_k^t)^2$ delivers the knowledge of previous tasks. Whatever the magnitude of Ω_k is, however, $\Omega_k(w_k^{t-1} - w_k^t)^2$ term is ignored if w_k^{t-1} almost equals to $w_k t$ already in the initial epoch of the training of a new task, which prevents the network from learning a new task. This situation is alleviated by weight re-initialization that allows the value of $\Omega_k(w_k^{t-1} - w_k^t)^2$ to be high enough regardless of the magnitude of Ω_k in the training of a new task. In this case, still the knowledge of previous tasks will be delivered by Ω_k and affect the training of a new task.

3 Experimental Evaluations

We perform experimental evaluations of our method compared to existing state-of-the-art methods for continual learning on several benchmark data sets; Split and permuted MNIST [6,13], and incrementally learning classes of CIFAR10, CIFAR100 [12] and Tiny ImageNet [5]. We set hyper-parameters of other existing approaches based on the description in [1]. We train all different tasks with a batch size of 256 and Adam [9] using the same learning rate (0.001). For the Split CIFAR tasks and Split Tiny ImageNet, as aforementioned, we perform the evaluation multiple times shuffling the order of tasks randomly to evaluate the robustness to task orders. We test with all 120, 200, and 50 random orders for Split CIFAR10, Split CIFAR10-100 and Split Tiny ImageNet respectively. To minimize statistical fluctuations of accuracy, each combination of task sequences is repeated three times.

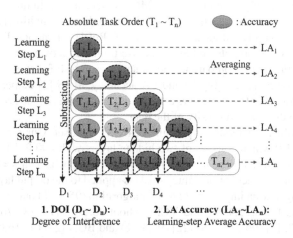

Fig. 3. Evaluation Metrics: DOI(Degree of Interference) and LA Accuracy of task. T, L and n stands for task, learning step, and the number of tasks respectively.

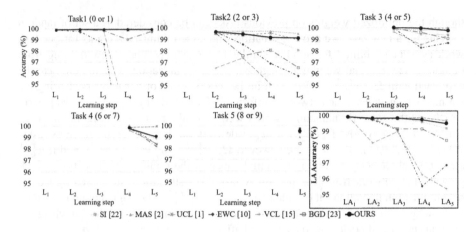

Fig. 4. Results on Split MNIST benchmark. VCL [15] indicates VCL (without coreset).

As described in Fig. 3, we define several evaluation metrics. "Absolute task order" indicates the sequence of tasks that a model learns. For instance, task 1 stands for the first task that a model learns no matter which classes comprise the task. "Learning step-wise average accuracy(LA Accuracy)" represents the accuracy of each learning step averaged through the whole tasks involved. (i.e., $LA_k = Average(L_k)$). "Degree of interference(DOI)" indicates the decreased extent of accuracy of each task after all learning steps are conducted. It is calculated by $(T_k, L_k) - (T_k, L_n)$. When we report the performance of randomly shuffled order experiment, we respectively average LA accuracy and DOI of randomly shuffled ordered test.

3.1 MNIST

We first evaluate our algorithm on a Split MNIST benchmark. In this experiment, two sequential classes compose each task (total 5 tasks). We use multi-headed and multi-layer perceptrons with two hidden layers with 400 ReLU activations. Each task has its output layer with two outputs and Softmax. We train our network for 40 epochs with $\alpha = 0.0045$. In Fig. 4, we compare the accuracy of each task for at every learning step (column-wise comparison in Fig. 3) and LA accuracy. MAS [2] outperforms all other baselines reaching 99.81% while ours achieves 99.7%. However, the accuracy is almost saturated due to the low complexity of the data.

We also evaluate methods on permuted MNIST data set. Our model used in this evaluation is MLP which consists of two hidden layers with 400 ReLUs each and one output layer with Softmax. The network is trained for 20 epochs with $\lambda = 0.005$. Also, to normalize the range of activation value, ReLU is applied to the output layer additionally when computing neuron importance Ω_k. Our algorithm (95.21%) outperforms MAS [2] (94.70%), EWC [10] (82.45%) and VCL(without coreset) [15] (89.76%) and on the other hand, UCL [1] (96.72%), SI [22] (96.39%) and BGD [23] (96.168%) show better results. However, most results on this data set achieve almost saturated accuracy.

3.2   Split CIFAR10

We test our method on a Split CIFAR10 benchmark. Two sequential classes compose each task (total 5 tasks). We use a multi-headed network with six convolution layers and two fully connected layers where the output layer is different for each task.

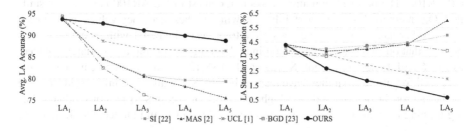

Fig. 5. Average LA Accuracy and its std. of Split CIFAR10 benchmark.

We train our network for 100 epochs with $\alpha = 0.7$. The order of 5 tasks that comprise CIFAR10 is randomly shuffled. As Fig. 5 describes, our method overall outperforms all other methods with large margins. Also, the standard deviation graph shows that our algorithm is more robust to the order of tasks. As Table 2 shows, proposed method shows better stability in the order of tasks and

Table 2. Average DOI (Degree of interference) and its std. (%) on Split CIFAR10. Note that proposed method forgets past tasks less regardless of the order of tasks.

| Method | Average DOI of absolute task order | | | |
	D_1	D_2	D_3	D_4
SI [22]	28.05(±11.7)	20.00(±7.4)	15.51(±8.2)	8.68(±5.8)
MAS [2]	33.59(±11.7)	27.37(±11.3)	19.15(±10.6)	11.45(±6.7)
UCL [1]	11.36(±5.8)	8.56(±3.6)	5.94(±3.0)	3.55(±6.5)
BGD [23]	39.06(±10.1)	34.83(±8.5)	29.19(±8.8)	19.71(±2.1)
OURS	**1.44(±1.1)**	**1.59(±1.2)**	**1.18(±0.8)**	**0.70(±0.7)**

also has a low degree of forgetting. In our method, average degraded degree of performance is lowest as 1.23%, whereas SI [22] is 18.06%, UCL [1] is 7.35%, MAS [2] is 22.89%, and BGD [23] is 30.7%.

Ablation Study. To verify the effect of weight re-initialization for the learning of new tasks, we compare performance of ours and UCL [1] with those without weight re-initialization. As Table 3 indicates, accuracy increases in both methods when weight re-initialization is applied. It suggests that weight re-initialization

Table 3. Performance difference(%) = (accuracy with weight re-initialization) – (accuracy without weight re-initialization). Note that the task order is fixed.

| Method | Task order | | | | |
	T_5	T_4	T_3	T_2	T_1
UCL [1]	0	−0.425	−0.9	4.93	6.38
OURS	−1	15.7	21.44	18.62	21.44

encourages better plasticity. Note that several weight importance based methods [2, 10, 22] cannot employ weight re-initialization since they consider the amount of weight changes in the methods.

3.3 Split CIFAR10-100

We evaluate our method on Split CIFAR10-100 benchmark where each task has 10 consecutive classes (total 11 tasks). We use the same multi-headed setup as in the case of Split CIFAR10. We train our network for 100 epochs with $\alpha = 0.5$. We fix task 1 as CIFAR10 due to the difference in the size of data set between CIFAR10 and CIFAR100. The order of remaining tasks that consist of CIFAR100 is randomly shuffled. Our method shows better stability showing the best accuracy values in old tasks. On the other hand, previous methods seem to prefer to be better with recent new tasks proving that our importance based continual learning is working appropriately. Indeed, as Fig. 6 and Table 4 represent, SI [22] and MAS [2] seem that they learn new tasks very well forgetting what they have learned before. Since all incoming weights are tied to the neuron in our method, the higher number of weights to be consolidated during training new tasks causes lower accuracy of final task. In practice, the decrease of

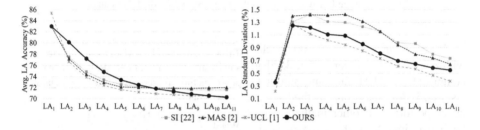

Fig. 6. Average LA Accuracy and its std. of CIFAR10-100 benchmark.

Table 4. Average DOI (Degree of interference) and its std. (%) on Split CIFAR10-100. Note that proposed method forgets past tasks less regardless of the order of tasks.

Method	Average DOI of absolute task order									
	D_1	D_2	D_3	D_4	D_5	D_6	D_7	D_8	D_9	D_{10}
SI [22]	5.85(±0.9)	7.37(±2.3)	6.58(±2.1)	5.87(±2.0)	5.57(±1.7)	5.09(±1.7)	4.45(±1.5)	3.97(±1.3)	3.28(±1.2)	2.17(±1.0)
MAS [2]	9.32(±1.4)	9.18(±2.8)	8.19(±2.2)	7.39(±2.1)	6.65(±1.9)	6.17(±1.9)	5.30(±1.6)	4.64(±1.4)	3.70(±1.2)	2.50(±1.0)
UCL [1]	3.74(±0.4)	**1.20(±1.2)**	1.31(±1.1)	0.98(±1.0)	**0.78(±0.9)**	**0.68(±0.8)**	0.64(±0.7)	0.59(±0.7)	**0.37(±0.5)**	0.27(±0.4)
OURS	**2.03(±0.4)**	2.08(±0.9)	**1.26(±0.9)**	**0.94(±0 8)**	0.84(±0.8)	0.77(±0.7)	**0.61(±0.7)**	**0.59(±0.6)**	0.47(±0.5)	**0.24(±0.4)**

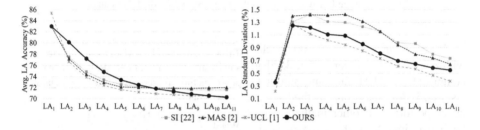

(a) LA accuracy of Split Cifar10   (b) LA accuracy of Split Cifar10-100

Fig. 7. The performance on Split CIFAR10 and CIFAR10-100 with doubled channel. Accuracy increases when we use a doubled channel network. Note that the task order is fixed.

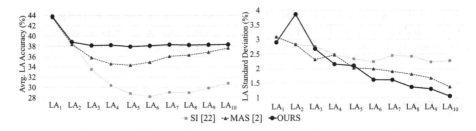

Fig. 8. Average LA Accuracy and its std. of Tiny ImageNet data set

Table 5. Average DOI (Degree of interference) and its std. (%) on Split Tiny ImageNet. Note that proposed method forgets past tasks less regardless of the order of tasks.

Method	Average DOI of absolute task order								
	D_1	D_2	D_3	D_4	D_5	D_6	D_7	D_8	D_9
SI [22]	27.35(±3.7)	29.02(±4.4)	27.09(±4.9)	22.67(±4.2)	20.24(±3.7)	18.25(±4.4)	14.86(±3.8)	11.70(±3.2)	7.78(±3.0)
MAS [2]	20.49(±4.2)	16.32(±3.6)	14.40(±4.1)	11.02(±3.6)	9.14(±3.6)	8.27(±2.7)	6.20(±2.3)	5.41(±2.3)	3.86(±1.8)
OURS	**11.47(±3.6)**	**6.75(±2.8)**	**5.78(±2.3)**	**3.98(±1.5)**	**3.16(±1.2)**	**2.76(±1.5)**	**2.15(±1.1)**	**1.70(±0.9)**	**0.98(±0.7)**

plasticity in our method can be addressed by using a larger network (e.g., the larger number of channels). We test the performance with a network of a doubled number of channels (256 to 512). Figure 7 shows that our network with doubled number of channels has improved accuracy keeping its stability and better plasticity. Table 4 shows that our method obtains lowest average degraded degree of performance 0.98% compared to SI [22], MAS [2], UCL [1] achieving 5.02%, 6.3%, 1.06% respectively. Also, the proposed method shows the lowest standard deviation of DOI, which indicates that our method is robust to the interference from various combinations of tasks.

3.4 Split Tiny ImageNet

We evaluate our method on Split Tiny ImageNet data set where each task has 20 consecutive classes (total 10 tasks). We use the same multi-headed setup as in the case of Split CIFAR10 and Split CIFAR10-100. We train our network for 100 epochs with $\alpha = 0.5$. The order of tasks is randomly shuffled. Only convolution neural networks based methods are tested for a fair comparison. In Fig. 8, our method outperforms all other methods with large margins. The standard deviation graph shows that our method algorithm shows the least performance disparity under the change in the order of tasks. Table 5 presents that our method acquires lowest average degraded degree of performance among SI [22], MAS [2] and ours, achieving 19.08%, 10.5%, and 4.3% respectively. Also, ours has the lowest standard deviation of DOI. This implies that our method is robust to the interference from various combinations of tasks.

4 Conclusion

We have proposed an activation importance-based continual learning method that consolidates important neurons of past tasks. Comprehensive evaluation has proved that the proposed method has implemented regularization-based continual learning achieving the fundamental aim of continual learning tasks not only balancing between stability and plasticity but also keeping robustness of the performance to the changes in the order of tasks.

References

1. Ahn, H., Cha, S., Lee, D., Moon, T.: Uncertainty-based continual learning with adaptive regularization. In: Advances in Neural Information Processing Systems, pp. 4392–4402 (2019)
2. Aljundi, R., Babiloni, F., Elhoseiny, M., Rohrbach, M., Tuytelaars, T.: Memory aware synapses: learning what (not) to forget. In: Proceedings of the European Conference on Computer Vision (ECCV), pp. 139–154 (2018)
3. Aljundi, R., Lin, M., Goujaud, B., Bengio, Y.: Gradient based sample selection for online continual learning. In: Advances in Neural Information Processing Systems, pp. 11816–11825 (2019)
4. Aljundi, R., Rohrbach, M., Tuytelaars, T.: Selfless sequential learning. arXiv preprint arXiv:1806.05421 (2018)
5. Deng, J., Dong, W., Socher, R., Li, L.J., Li, K., Fei-Fei, L.: ImageNet: a large-scale hierarchical image database, 2009. In: IEEE Conference on Computer Vision and Pattern Recognition (CVPR), pp. 248–255 (2020)
6. Goodfellow, I.J., Mirza, M., Xiao, D., Courville, A., Bengio, Y.: An empirical investigation of catastrophic forgetting in gradient-based neural networks. arXiv preprint arXiv:1312.6211 (2013)
7. Javed, K., White, M.: Meta-learning representations for continual learning. In: Advances in Neural Information Processing Systems, pp. 1820–1830 (2019)
8. Jung, S., Ahn, H., Cha, S., Moon, T.: Adaptive group sparse regularization for continual learning. arXiv preprint arXiv:2003.13726 (2020)
9. Kingma, D.P., Ba, J.: Adam: a method for stochastic optimization. arXiv preprint arXiv:1412.6980 (2014)
10. Kirkpatrick, J., et al.: Overcoming catastrophic forgetting in neural networks. Proc. Nat. Acad. Sci. **114**(13), 3521–3526 (2017)
11. Kornblith, S., Norouzi, M., Lee, H., Hinton, G.: Similarity of neural network representations revisited. In: International Conference on Machine Learning, pp. 3519–3529. PMLR (2019)
12. Krizhevsky, A., Hinton, G., et al.: Learning multiple layers of features from tiny images (2009)
13. LeCun, Y., Bottou, L., Bengio, Y., Haffner, P.: Gradient-based learning applied to document recognition. Proc. IEEE **86**(11), 2278–2324 (1998)
14. Li, Z., Hoiem, D.: Learning without forgetting. IEEE Trans. Pattern Anal. Mach. Intell. **40**(12), 2935–2947 (2017)
15. Nguyen, C.V., Li, Y., Bui, T.D., Turner, R.E.: Variational continual learning. arXiv preprint arXiv:1710.10628 (2017)
16. Riemer, M., et al.: Learning to learn without forgetting by maximizing transfer and minimizing interference. arXiv preprint arXiv:1810.11910 (2018)
17. Rusu, A.A., et al.: Progressive neural networks. arXiv preprint arXiv:1606.04671 (2016)
18. Razavian, A.S., Azizpour, H., Sullivan, J., Carlsson, S.: CNN features off-the-shelf: an astounding baseline for recognition. In: Proceedings of the IEEE Conference on Computer Vision and Pattern Recognition Workshops, pp. 806–813 (2014)
19. Thrun, S.: Is learning the n-th thing any easier than learning the first? In: Advances in Neural Information Processing Systems, pp. 640–646 (1996)

20. Yoon, J., Kim, S., Yang, E., Hwang, S.J.: Scalable and order-robust continual learning with additive parameter decomposition. arXiv preprint arXiv:1902.09432 (2019)
21. Yoon, J., Yang, E., Lee, J., Hwang, S.J.: Lifelong learning with dynamically expandable networks. arXiv preprint arXiv:1708.01547 (2017)
22. Zenke, F., Poole, B., Ganguli, S.: Continual learning through synaptic intelligence. Proc. Mach. Learn. Res. **70**, 3987 (2017)
23. Zeno, C., Golan, I., Hoffer, E., Soudry, D.: Task agnostic continual learning using online variational bayes. arXiv preprint arXiv:1803.10123 (2018)

AD-CGAN: Contrastive Generative Adversarial Network for Anomaly Detection

Laya Rafiee Sevyeri[✉] and Thomas Fevens

Gina Cody School of Engineering and Computer Science, Concordia University,
Montréal, QC, Canada
laya.rafiee@gmail.com, thomas.fevens@concordia.ca

Abstract. Anomaly detection (AD), a fundamental challenge in machine learning, aims to find samples that do not belong to the distribution of the training data. Among unsupervised anomaly detection models, models based on generative adversarial networks show promising results. These models mainly rely on the rich representations learned from the normal training data to find anomalies. However, their performance is bounded by the limitations of GANs, known as mode collapse, in learning complex training distribution. This work presents a new GAN-based anomaly detection model with a combination of contrastive learning to mitigate the negative effect of mode collapse in more complex distributions. Our unsupervised Anomaly Detection model based on Contrastive Generative Adversarial Network, AD-CGAN, contrasts a sample with local feature maps of itself instead of only contrasting the given sample with other instances as in conventional contrastive learning approaches. Contrastive loss in AD-CGAN helps the model learn more discriminative representations of normal samples. Furthermore, we consider a new normality score to target anomalous samples. The normality score is defined on the encoded representations of samples obtained from the model. Extensive experiments showed AD-CGAN outperforms its counterparts on multiple benchmarks with a significant improvement in ROC-AUC over the previously proposed reconstruction-based approaches.

Keywords: Anomaly detection · Contrastive learning · Generative adversarial networks

1 Introduction

Anomaly detection (AD), also known as out-of-distribution detection, has a long history in artificial intelligence. Anomaly detection refers to identifying those samples that do not come from the expected distribution. Supervised learning models address AD using classification approaches such as outlier exposure [12]. On the other hand, unsupervised learning approaches, such as

© The Author(s), under exclusive license to Springer Nature Switzerland AG 2022
S. Sclaroff et al. (Eds.): ICIAP 2022, LNCS 13231, pp. 322–334, 2022.
https://doi.org/10.1007/978-3-031-06427-2_27

reconstruction-based methods [24,32], mitigate the problem of limited labeled data and unknown anomalies. In these approaches, the model learns the distribution of the normal training data and then a reconstruction loss targets anomalies. AnoGAN [24] proposes using generative adversarial networks (GANs) to find anomalies in the medical domain. AnoGAN suffers from its lengthy inference procedure to find the inverse mapping of an image in a low-dimensional representation. Several studies tried to overcome the limitations of AnoGAN [23,30,31]. However, intrinsic problems of GANs, such as mode collapse [13], catastrophic forgetting [4,16], unstable training, and difficulty in convergence [21], limit the ability of these models to learn a suitable representation for the task of AD.

Lee *et al.* [20] showed that adding contrastive learning on the generator side in training GANs while maximizing mutual information on the discriminator side increases the quality of generated images by simultaneously mitigating mode collapse and catastrophic forgetting of the generator and the discriminator respectively. Contrastive learning [3,11] is a self-supervised approach that learns representations of the data in such a way that similar samples stay close to each other while dissimilar samples remain at a distance. Considering Lee *et al.* [20], we investigate the incorporation of a contrastive GAN for anomaly detection.

In this work, we propose a reconstruction-based Anomaly Detection approach using Contrastive Generative Adversarial Network (AD-CGAN). The proposed model contains three main sub-modules: a contrastive GAN, an autoencoder, and a second discriminator (different from the discriminator in GAN) on the latent representations. We train all modules simultaneously on the normal data to learn a discriminative representation for each image while keeping each image's local and global features as close as possible. The second discriminator trains on the hidden representations of two different reconstruction-based models, i.e., GAN and autoencoder, to provide more discriminative representations. We show that having a contrastive GAN while maximizing the mutual information between local and global features of an image provides more semantic and discriminative features for anomaly detection. Experimental results show that the representations obtained by the contrastive GAN in our anomaly detection model greatly increase the performance of reconstruction-based AD approaches. To the best of our knowledge, our work is the first to investigate using contrastive generative adversarial networks for anomaly detection. The codes of AD-CGAN and all the conducted experiments are available here.

2 Related Work

Anomaly detection or, in general, out-of-distribution detection approaches can be grouped according to the following paradigms.

Distributional-based approaches try to build a probabilistic model on the distribution of normal data. They expect that the anomalous samples receive a lower likelihood under the probabilistic model than the normal samples. Gaussian mixture models [22] and kernel density estimation (KDE) [18] from traditional models and RDA [32] and deep autoencoding Gaussian mixture model

(DAGMM) [33] from deep models are among these approaches. **Classification-based approaches** such as One-Class SVM (OC-SVM) [25] and support vector data description (SVDD) [26] use the idea of separating the normal data from the anomalous data based on their feature spaces. These approaches suffer from the insufficient and biased representations the feature learning methods can provide. One remedy for this issue is to use self-supervised learning methods. GEOM [9] and GOAD [2] are classification-based AD models that use surrogate tasks for anomaly detection. **Reconstruction-based approaches** rely on the idea that normal samples should receive smaller reconstruction loss rather than anomalous samples. Various loss and reconstruction basis functions are used in each of these approaches. K-means is used as an early basis reconstruction function [14] while An and Cho [1] proposed using deep neural networks as the basis functions. In the class of deep neural networks, generative models such as GANs [24] and autoencoder [32] are used to learn the reconstruction basis functions. Following the presentation of AnoGAN [24], several other studies used similar ideas with modifications on their basis functions and losses [6,23,30,31] to increase the performance of anomaly detection models based on GANs. One major issue in using generative models, especially GANs as the reconstruction basis function, is their difficulty to recover the entire data distribution (aka mode-collapse in GANs), leading to lower performance in comparison with classification-based approaches. Our model falls in the category of reconstruction-based approaches, combining adversarial training with contrastive learning to mitigate the challenges of reconstruction-based approaches.

3   Background

Unsupervised anomaly detection models only have access to the normal training data. Reconstruction-based models are unsupervised approaches that rely on the reconstruction loss of samples, where a high reconstruction loss implies an anomalous sample. Our model uses a GAN and an autoencoder as its reconstruction methods.

Generative adversarial network (GAN) [10] is a generative model that formulates the process of learning in a two-player minimax game between two learning components; i.e., generator and discriminator. One of the obstacles in using GANs for tasks such as anomaly detection is related to the catastrophic forgetting (neural network forgetting prior tasks while working on the current task) of the discriminator [4,16] which can negatively affect the AD performance. Another barrier is known as mode collapse where the generator only learns a small subset of modes in the training data. Recently self-supervised learning has gained attention in generative models such as GANs [4,27]. While these approaches try to mitigate the catastrophic forgetting, they do not diminish the mode collapse [28]. On the other hand, maximizing mutual information on the discriminator side and contrastive learning on the generator side seems a way to overcome these two issues simultaneously [20]. Therefore, we consider the idea of using a contrastive GAN in our AD model to detect anomalous samples with the purpose of increasing the performance of reconstruction-based models.

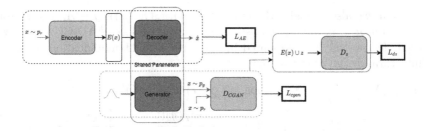

Fig. 1. Different components and losses of AD-CGAN are depicted.

4 Proposed Approach

In this work, we propose AD-CGAN, a reconstruction model based on Contrastive Generative Adversarial Network to find anomalies in images (see Fig. 1). In this approach, a contrastive GAN module learns to generate normal samples. An autoencoder module, which shares its decoder with the GAN's generator, learns to reconstruct normal samples from their latent representations. We also use a discriminator module on top of the autoencoder and the input random noise to the GAN as a regularizer. Since the model only trained on normal samples, we expect that during inference, it cannot reconstruct samples from any distribution other than training distribution. As a result, the dissimilarity between a given test sample and its corresponding generated output can be defined as our distance metric to target anomalous samples. Therefore, we define a normality score based on the reconstruction loss of the input sample during inference to find anomalous samples. In the following sections, we discuss all the modules and the normality score in more detail.

4.1 Contrastive GAN

Formal Definition: The training set $\mathcal{D}_{train} = \{x_1, x_2, ..., x_k \sim P_{ind}\}$ contains samples drawn from P_{ind}, normal distribution. To evaluate our model, we use a test set $\mathcal{D}_{test} = \{\bar{x}_1, \bar{x}_2, ..., \bar{x}_n \sim P_{ind} \cup P_{ood}\}$ including both normal and anomalous samples drawn from P_{ind} and the anomalous distribution (P_{ood}), respectively.

The contrastive GAN module which we refer to as $CGAN$ contains a generator G and a discriminator D_{cgan}. Training the $CGAN$ incorporates two losses: a contrastive loss L_{cgan} (See Eq. 1) and an adversarial loss L_{adv} (See Eq. 2).

In conventional contrastive learning, each image is contrasted with other samples, while in AD-CGAN, each image is contrasted with its own local feature maps to create positive/negative sets. Given an image $x \in X$, we consider the penultimate and ultimate representations of D_{cgan} as local ($C_\psi(x)$) and global ($E_\psi(x)$) features of x. We pass $E_\psi(x)$ through a dense layer ϕ_θ. Then, $\phi_\theta(E_\psi(x))$ and $C_\psi(x)$ go to the contrastive pairing phase to create positive/negative sets for the contrastive learning. For a given image x, the set of positive samples is the pair $(C_\psi^{(i)}(x), \phi_\theta(E_\psi(x)))$ for $i \in A = \{0, 1, ..., M^2 - 1\}$ of a $M \times M$ local feature map. Besides the local feature map of other images $x' \in X$ in the same mini-batch, we also consider the pairs $(C_\psi^{(j)}(x), \phi_\theta(E_\psi(x)))$ $j \in A$, $j \neq i$, from the same image x as the negative set. The contrastive loss of AD-CGAN, shown in Eq. 1, follows the loss presented in [20] with a slight modification to fit the architectural design of our model.

$$L_{cgan}(X) = -\mathbb{E}_{(x \in X)}\mathbb{E}_{(i \in A)}[\log\ p(C_\psi^{(i)}(x), E_\psi(x)|X)]$$

$$= -\mathbb{E}_{(x \in X)}\mathbb{E}_{(i \in A)}[\log \frac{exp(g_\theta(C_\psi^{(i)}(x), E_\psi(x)))}{\sum_{(x', i) \in X \times A} exp(g_\theta(C_\psi^{(i)}(x'), E_\psi(x)))}] \quad (1)$$

Here the function $g_\theta(C_\psi^{(i)}(x), E_\psi(x)) = C_\psi^{(i)}(x)^T \phi_\theta(E_\psi(x))$ maps the local/global features with K dimensions to a scalar score. For the adversarial loss, we used relativistic loss [15] shown in Eq. 2.

$$L_{adv_D} = -\mathbb{E}_{(x_r \in X_r,\ x_f \in X_f) \sim (\mathbb{P}, \mathbb{Q})}[\log(\sigma(C(x_r) - C(x_f)))]$$
$$L_{adv_G} = -\mathbb{E}_{(x_r \in X_r,\ x_f \in X_f) \sim (\mathbb{P}, \mathbb{Q})}[\log(\sigma(C(x_f) - C(x_r)))] \quad (2)$$

where L_{adv_G} and L_{adv_D} are the losses of the generator and the discriminator of the CGAN, σ is the sigmoid function, X_r and X_f represent sets of real and fake images respectively, $\mathbb{P}$ is the distribution of the real data, $\mathbb{Q}$ is the distribution of the fake data, and C is the critic.

In order to stabilize training, we constrained the discriminator D_{cgan} and the generator to learn only from the contrastive loss of real image and fake image features, respectively, as suggested in [20]. The final loss of the generator and the discriminator of our $CGAN$ is a combination of its adversarial and contrastive losses where α and β control the contribution of the contrastive loss:

$$L_G = L_{adv_G} + \alpha L_{cgan}(X_f)$$
$$L_{D_{cgan}} = L_{adv_D} + \beta L_{cgan}(X_r) \quad (3)$$

4.2   Autoencoder

The autoencoder AE trains with the mean squared error (MSE) reconstruction loss function, $L_{AE} = \|x - G(E(x))\|^2$, where $G(E(x))$ is the output of AE and G and E are the decoder (generator) and the encoder of AE. We use weights sharing for the decoder of AE and the generator of CGAN. In this way, we are using both training signals from GAN and AE to train the generator.

4.3   Latent Space Discriminator

We further apply a discriminator D_z on top of the encoded space of encoder, $E(x)$, and the random noise, z. The adversarial loss L_{dz} (Eq. 4) forces the encoder to encode images within the distribution of random noise. In this way, we decrease the instability of GAN by keeping its two input representations close to each other. This regularizer helps the model to better discriminate normal and anomalous samples (See Sect. 5.4).

$$L_{dz} = \mathbb{E}_{z \sim P_z}[\log D_z(z, z)] + \mathbb{E}_{z \sim P_z, x \in X}[1 - \log D_z(z, E(x))] \quad (4)$$

4.4   Normality Score

AD-CGAN relies on the reconstruction loss of each sample to find anomalies during inference. To see how far a generated image is from the actual test image, we present a normality score, a combination of multiple components. A well-trained AD-CGAN should only be able to generate samples belonging to the normal distribution seen during training. Hence, the normality score, which is defined based on the reconstruction loss, should be lower for normal samples and higher for anomalous samples. Our normality score contains two reconstruction losses: the generation reconstruction loss L_{Gr}, which involves the scores obtained from the trained $CGAN$, and the feature reconstruction loss L_{Fr}, which incorporates the scores obtained from latent representations of a given image. The normality score $NS(x)$ for a given image $x \in \mathcal{D}_{test}$ is defined as the summation of these two losses (Eq. 5).

$$NS(x) = L_{Gr} + L_{Fr} \tag{5}$$

where the generation and feature reconstruction losses are defined as

$$L_{Gr} = \lambda L_{GD}(x) + (1 - \lambda)L_{GR}(x), \ \ L_{Fr} = \rho L_{FE}(x) + (1 - \rho)L_{FL}(z) \tag{6}$$

Here, L_{Gr} includes discrimination loss, $L_{GD}(x) = \sum |f_D(x) - f_D(G(E(x)))|$ with intermediate representation of a given image x from D_{cgan} as $f_D(x)$, and residual loss, $L_{GR}(x) = \sum |x - G(E(x))|$, similar to the losses presented in [24]. It should be noted that $f_D(x)$ refers to the internal representation of image x obtained from the penultimate layer of the D_{cgan}. L_{Fr} contains encoded L_{FE} and latent L_{FL} feature reconstruction losses (Eq. 7).

$$
\begin{aligned}
L_{FE}(x) &= \sum |E(x) - E(G(E(x)))| \\
L_{FL}(z) &= \|D_z(E(G(z)), z) - D_z(E(x), E(G(z)))\|_1
\end{aligned}
\tag{7}
$$

where $E(x)$ is the encoded representation of x from the encoder E, z is the input random noise to the generator G of CGAN, and $G(z)$ is the output of G.

5   Experimental Results

We perform extensive experiments on several benchmark image datasets to evaluate our method. The detailed hyperparameters of AD-CGAN are shown in Table 1.

5.1   Datasets

We considered four benchmark datasets in our experiments: CIFAR-10 [17], FashionMNIST(fMNIST) [29], MNIST [19], and CatsVsDogs [8]. All of the datasets except CatsVsDogs include 10 classes. In order to evaluate AD-CGAN on AD tasks, we employ two different schemes. We introduce soft and hard anomaly detection experiments. In the soft experiments, we consider one-vs-all scheme. In

Table 1. AD-CGAN architecture and hyperparameters for the experiments on all datasets. Batch-size of 32 for all, $m = 2, 2, 3, 2$, $n = 4, 4, 4, 5$, $j = 4, 4, 6, 6$, and $k = 2, 2, 2, 2$ for MNIST, FashionMNIST, CIFAR10 and CatsVsDogs dataset respectively.

AD-CGAN architecture				AD-CGAN hyper-parameters	
Module	#Layers	Activation fn.	Dropout	Latent dimension	100 (except CatsVsDogs with 200)
$G(z)$	$m \times Conv2d$, $n \times Trans.Conv2d$	$PReLU$	×	Learning rate	Lr_{CGAN}: 3×10^{-4}, Lr_{AE}: 2×10^{-4}
$D_{cgan}(x)$	$j \times Conv2d$, $k \times Linear$	$LeakyReLU$	0.2	Optimizer	$Adam(\beta_1 = 0.5, \beta_2 = 0.999)$
$E(x)$	$5 \times Conv2d$	$LeakyReLU$	0.2	L_{cgan}	$\alpha = 0.3, \beta = 0.3$
$L_z(z, z)$	$3 \times Linear$	$LeakyReLU$	×	$NS(x)$	$\lambda = 0.1, \rho = 0.5$

this scheme, a dataset with C classes will lead to C different anomaly detection experiments. A given class c_{ind}, $1 \leq c_{ind} \leq C$, is considered as the *normal* class, while c_{ood} defines *anomalous* class of the rest of $C - 1$ classes. We introduced the hard scheme mainly to show how anomaly detection models based on GANs fail when the inlier class includes multiple distributions. In the hard AD scheme, we introduce all-vs-one scheme. Similar to the soft scheme, each dataset with C classes will lead to C different experiments. However, in contrast with the soft scheme, $1 \leq c_{ood} \leq C$ includes only a single class while c_{ind} contains the remaining $C - 1$ classes. Considering that CatsVsDogs has only two classes, each class was treated as *normal* in a separate experiment.

5.2   Baseline Methods

We compare the performance of our model with multiple AD models. DAGMM [33], OC-SVM [25], TIAE [5], ALAD [31], ADGAN [6], and AE-GAN [23] are the baselines where the last three models are based on GANs. DAGMM is an autoencoder-based model, which generates a low-dimensional representation of the training data, and leverages a Gaussian mixture model to perform density estimation on the low-dimensional representations. OC-SVM is a kernel-based method that typically uses an RBF kernel to learn a collection of closed sets in the input space. Samples that fall outside of these sets are assumed to be anomalous. TIAE uses a transformation invariant autoencoder with an additional training signal based on the most confident inlier samples to find anomalies. ALAD trains a modified bidirectional GAN [7] with multiple discriminators on normal samples and uses an L_1 reconstruction error on the feature space to find anomalies. AE-GAN trains on a mixed model of GAN and autoencoder and uses several scoring components to separate normal and anomalous samples. Unlike the former GAN-based approaches which benefit from a fast inference procedure, ADGAN uses gradient descent to find an inverse mapping of an image to a low-dimensional seed with a GAN trained on normal samples to generate a sample, which makes its inference very slow. ADGAN later uses an L_2 distance between the generated image and the original image to target anomalies.

In this work, we aim to address the difficulties of anomaly detection models based on GANs via introducing a contrastive GAN. Therefore, apart from the comparison with other anomaly detection models, we mainly focus on those

Table 2. ROC-AUC (%) comparison of AD models with *one-vs-all* scheme. The symbol [†] represents results reported from our implementations. All of the results from our implementations are averaged over three different runs.

Datasets	DAGMM	OC-CVM	ALAD	AE-GAN	ADGAN	TIAE	AD-CGAN
CIFAR-10	58.7[†]	62.0[†]	60.7	61.1	60.6	71.2 ± 1.44	**86.0 ± 0.04**
fMNIST	51.8[†]	92.8[†]	78.1 ± 0.12[†]	69.0 ± 0.16[†]	75.4[†]	86.8 ± 0.55	**93.9 ± 0.02**
MNIST	50.4[†]	91.7[†]	62.4 ± 0.09[†]	69.3	91.5	85.2 ± 0.81	**92.3 ± 0.03**
CatsVsDogs	50.6[†]	51.6[†]	53.4[†]	51.6[†]	49.0[†]	51.4[†]	**89.8 ± 0.04**

AD models which use GANs in their approach. For each of the experiments, if available, we reported their results from their original papers. For AD baselines based on GANs, we also ran their models on all of the datasets within the hard experiments. It should be noted that, due to the long inference process of ADGAN, we ran it only once using their implementation.

5.3 Results

The performance of AD-CGAN is summarized in Table 2. The Area Under the Curve (AUC) of the Receiver Operating Characteristics (ROC) measures the performance of a classifier under various threshold settings. In the context of this study, the ROC-AUC is a measurement of how well the classifier can distinguish between normal and anomalous samples. As illustrated in Table 2, AD-CGAN outperforms all the baselines in terms of ROC-AUC. The improvement is more notable on CatsVsDogs, with a large improvement for AD-CGAN, and CIFAR-10 by a minimum of 15% improvement on the soft scheme.

The detailed performance of each of the GAN-based models, in soft and hard schemes, for each of the classes of c_{ind}/c_{ood} is presented in Table 3. As the table shows, AD-CGAN surpasses all the baselines in each of the individual classes of c_{ind}/c_{ood} except c_1 on the MNIST dataset for both soft and hard schemes. We argue that AD-CGAN performs consistently in all classes of MNIST within both soft and hard schemes, while the performance of ADGAN in the hard scheme is not consistent across classes and their higher performance on c_1 could be related to the specific pattern of this class. This is in contrast with their competitive performance to AD-CGAN in the soft scheme in the MNIST dataset.

Our contrastive GAN without training on any pretext tasks was able to improve the current reconstruction-based anomaly detection models' performance by at least 7% improvement in several experiments. As expected, the performance in the hard scheme is lower compared with the soft scheme since the normal class contains multiple labels, each having a different distribution. This is more notable in FashionMNIST and MNIST datasets with around 7% drop in the performance. We argue that given the similar pattern in several labels of these two datasets, even AD-CGAN with its discriminative representations obtained by the contrastive loss may have difficulty in the hard scheme.

Table 3. ROC-AUC (%) comparison of GAN-based models on all four datasets with *one-vs-all* and *all-vs-one* schemes. In the *one-vs-all* scheme, the class number defines c_{ind}, while in *all-vs-one*, it refers to c_{ood}. The results are averaged over three different runs. $\lambda = 0.1$ and $\rho = 0.5$ are used for all the experiments. The symbol * represents results reported from the original paper. For simplicity, for each of the classes of CIFAR-10 and fMNIST, we use ordinal numbers instead of their label.

Datasets	Class	all-vs-one				one-vs-all			
		ALAD	ADGAN	AE-GAN	Ours	ALAD	ADGAN	AE-GAN	Ours
CIFAR-10	0	61.2 ± 0.02	44.3	63	**89.8** ± 0.12	67	62.7	67	**83.8** ± 0.04
	1	61.1 ± 0.02	39.6	63	**89.5** ± 0.13	46	54.6	49	**87.2** ± 0.03
	2	40.7 ± 0.00	58.2	60	**84.7** ± 0.06	64	56.1	63	**80.1** ± 0.10
	3	48.8 ± 0.01	44.7	54	**78.6** ± 0.09	63	59.5	56	**86.0** ± 0.07
	4	35.5 ± 0.01	66.1	35	**81.7** ± 0.14	66	58.6	73	**85.4** ± 0.04
	5	53.5 ± 0.02	44.5	52	**72.8** ± 0.01	53	62.8	52	**81.6** ± 0.02
	6	47.8 ± 0.01	61.5	60	**87.6** ± 0.06	78	60.4	72	**94.6** ± 0.03
	7	49.7 ± 0.01	47.4	51	**90.7** ± 0.03	52	62.3	63	**81.7** ± 0.04
	8	52.9 ± 0.03	45.7	54	**82.6** ± 0.02	75	70.2	68	**89.3** ± 0.06
	9	59.4 ± 0.01	31.3	63	**86.9** ± 0.02	43	59.1	48	**90.7** ± 0.04
	Average	51.1 ± 0.08	48.3	55.5*	**84.5** ± 0.05	60.7*	60.6*	61.1*	**86.0** ± 0.04
fMNIST	0	54.0 ± 0.02	48.4	45.3 ± 0.09	**89.5** ± 0.06	79.4 ± 0.02	74.1	74.4 ± 0.03	**93.3** ± 0.04
	1	68.2 ± 0.04	63.7	32.8 ± 0.12	**85.8** ± 0.03	94.1 ± 0.04	92.3	92.3 ± 0.01	**95.9** ± 0.04
	2	55.5 ± 0.03	40.4	57.9 ± 0.08	**90.4** ± 0.07	60.6 ± 0.09	71.1	67.7 ± 0.03	**94.0** ± 0.06
	3	47.9 ± 0.04	60.5	23.0 ± 0.05	**81.7** ± 0.10	79.5 ± 0.05	81.6	80.0 ± 0.03	**93.7** ± 0.03
	4	60.3 ± 0.13	47.8	34.9 ± 0.07	**81.4** ± 0.02	76.4 ± 0.06	73.6	82.5 ± 0.01	**93.1** ± 0.02
	5	22.2 ± 0.01	66.9	80.4 ± 0.02	**93.8** ± 0.04	85.5 ± 0.01	77.3	36.5 ± 0.06	**93.5** ± 0.02
	6	45.1 ± 0.01	34.5	52.1 ± 0.06	**92.1** ± 0.02	61.2 ± 0.08	70.0	55.1 ± 0.04	**91.7** ± 0.03
	7	44.1 ± 0.05	67.1	55.5 ± 0.09	**87.0** ± 0.09	94.9 ± 0.02	91.0	77.9 ± 0.07	**98.5** ± 0.01
	8	50.2 ± 0.09	54.1	76.0 ± 0.02	**83.8** ± 0.01	62.6 ± 0.03	50.3	49.9 ± 0.06	**91.6** ± 0.08
	9	60.7 ± 0.07	56.6	63.6 ± 0.09	**91.1** ± 0.06	86.5 ± 0.13	73.2	73.4 ± 0.13	**93.7** ± 0.07
	Average	50.8 ± 0.12	54.0	52.2 ± 0.18	**87.7** ± 0.04	78.1 ± 0.12	75.4	69.0 ± 0.16	**93.9** ± 0.02
MNIST	0	61.0 ± 0.05	42.7	73	**94.6** ± 0.07	74.7 ± 0.12	97.2	85	**97.1** ± 0.02
	1	87.1 ± 0.03	93.1	56	85.8 ± 0.11	69.8 ± 0.16	99.7	98	95.7 ± 0.02
	2	44.5 ± 0.04	39.7	61	**91.6** ± 0.05	50.4 ± 0.09	87.4	54	**92.1** ± 0.03
	3	47.7 ± 0.05	61.2	55	**86.6** ± 0.08	65.7 ± 0.05	84.8	69	**91.2** ± 0.03
	4	56.7 ± 0.05	70.2	49	**77.4** ± 0.04	63.6 ± 0.06	91.0	72	**95.5** ± 0.01
	5	50.1 ± 0.06	53.1	49	**82.8** ± 0.03	56.1 ± 0.04	**91.6**	54	87.8 ± 0.02
	6	51.8 ± 0.11	59.8	55	**86.9** ± 0.08	53.0 ± 0.08	**95.7**	60	88.6 ± 0.06
	7	56.4 ± 0.09	75.2	44	76.1 ± 0.02	49.6 ± 0.01	**93.7**	68	92.5 ± 0.01
	8	41.2 ± 0.08	58.5	59	**83.9** ± 0.02	75.3 ± 0.07	81.6	69	**87.2** ± 0.05
	9	42.4 ± 0.02	71.1	56	**84.5** ± 0.06	65.2 ± 0.09	92.4	64	**95.2** ± 0.02
	Average	53.9 ± 0.13	62.5	55.7*	**85.0** ± 0.05	62.4 ± 0.09	91.5*	69.3*	**92.3** ± 0.03
CatsVsDogs	Cats	–	–	–	–	52.6	53.1	51.7	**92.7** ± 0.03
	Dogs	–	–	–	–	54.1	44.9	52.1	**86.9** ± 0.05
	Average	–	–	–	–	53.4	49.0	51.6	**89.8** ± 0.04

5.4 Ablation Study

AD-CGAN is comprised of several training components as well as multiple normality score components. Each of the training components is critical in the models' performance. This can be inferred by comparing the performance of AD-CGAN with each of AE-GAN [23] and ALAD [31] where adding contrastive learning to GAN showed a notable performance gain on the anomaly detection performance. We also argue that the autoencoder is a key element in AD-CGAN where it removes the extensive and time-consuming inference procedure (as stated in the experiments in [23]). In order to have a better understanding of the effect of each of the components in the proposed normality score, we measure their effects in different anomaly detection settings (see Table 4).

Table 4. Ablation studies on MNIST and FashionMNIST given different normality score components of AD-CGAN. We used $\lambda = 0.1$ where the generation reconstruction loss had been used. We set $\rho = 1$ and $\rho = 0$ for AD-CGAN$_{LFE}$ and AD-CGAN$_{LFL}$, respectively. The ROC-AUC (%) results are averaged over three different runs.

	MNIST (%)		fMNIST (%)	
Model	all-vs-one	one-vs-all	all-vs-one	one-vs-all
$AD - CGAN_{LG}$	56.8 ± 0.13	72.3 ± 0.12	68.7 ± 0.11	80.8 ± 0.11
$AD - CGAN_{LFE}$	66.6 ± 0.12	84.4 ± 0.11	70.4 ± 0.13	86.1 ± 0.08
$AD - CGAN_{LFL}$	73.2 ± 0.09	84.9 ± 0.05	74.5 ± 0.08	87.0 ± 0.05
$AD - CGAN_{GF}$	$\mathbf{85.0} \pm 0.05$	$\mathbf{92.3} \pm 0.03$	$\mathbf{87.7} \pm 0.04$	$\mathbf{93.9} \pm 0.02$

Feature reconstruction loss is added to the normality score to measure how discriminative the latent representations of the two reconstruction models are. Several experiments on MNIST and FashionMNIST on soft and hard AD schemes showed that adding D_z leads to more discriminative latent representation, which affects the normality scores obtained by the feature reconstruction loss. We defined four distinct models of AD-CGAN based on the normality score components they have access to: AD-CGAN$_{LG}$ represents AD-CGAN with only generation reconstruction loss; AD-CGAN$_{LFL}$ with only latent feature reconstruction loss, L_{FL}; AD-CGAN$_{LFE}$ with only encoded feature reconstruction loss, L_{FE}; and AD-CGAN$_{GF}$ contains both L_{Gr} and L_{Fr} in its normality score. It should be mentioned that in each of these models, generation reconstruction loss is considered as part of the normality score. As the results reveal, removing the feature reconstruction loss (ignoring D_z) negatively affects the performance of AD-CGAN. The impact is more severe in the case of the all-vs-one (hard) scheme. On the other hand, AD-CGAN$_{LFE}$ that trains with D_z and encoded feature reconstruction loss, significantly improved AD-CGAN$_{LG}$ in both datasets. Similar behavior is observed on the results on AD-CGAN$_{LFL}$. However, it is important to note that in all the experiments, ignoring any training and/or normality score component results in lower performance. The best results are achieved when all the components with the right amount of contributions are considered, as it is shown in AD-CGAN$_{GF}$.

To further validate the effect of the contrastive loss, in another experiment, we found that applying contrastive loss to ADGAN [6] improves the ROC-AUC by 3% and 9% on CIFAR10 and FashionMNIST on all-vs-one, respectively.

6   Conclusion and Future Work

We presented a new reconstruction-based approach to tackle the problem of anomaly detection (AD) in images. The proposed approach adds contrastive learning to an anomaly detection model based on a generative adversarial network (GAN), AD-CGAN, to learn more discriminative and task agnostic features of normal data. AD-CGAN uses a normality score function including multiple components to further separate normal and anomalous samples. In this study, we considered two different AD schemes, soft and hard, to evaluate the performance of AD-CGAN. AD-CGAN was able to outperform all the previously reconstruction-based approaches on all four benchmark datasets within both soft and hard schemes. These results may open a new path for reconstruction-based anomaly detection models leading to more discriminative representations of normal data.

References

1. An, J., Cho, S.: Variational autoencoder based anomaly detection using reconstruction probability. Special Lecture IE **2**(1), 1–18 (2015)
2. Bergman, L., Hoshen, Y.: Classification-based anomaly detection for general data. In: International Conference on Learning Representations (2020)
3. Chen, T., Kornblith, S., Norouzi, M., Hinton, G.: A simple framework for contrastive learning of visual representations. In: International Conference on Machine Learning, pp. 1597–1607. PMLR (2020)
4. Chen, T., Zhai, X., Ritter, M., Lucic, M., Houlsby, N.: Self-supervised GANs via auxiliary rotation loss. In: Proceedings of the IEEE/CVF Conference on Computer Vision and Pattern Recognition, pp. 12154–12163 (2019)
5. Cheng, Z., Zhu, E., Wang, S., Zhang, P., Li, W.: Unsupervised outlier detection via transformation invariant autoencoder. IEEE Access **9**, 43991–44002 (2021)
6. Deecke, L., Vandermeulen, R., Ruff, L., Mandt, S., Kloft, M.: Anomaly detection with generative adversarial networks (2018). https://openreview.net/forum?id=S1EfylZ0Z
7. Donahue, J., Krähenbühl, P., Darrell, T.: Adversarial feature learning. In: 5th International Conference on Learning Representations, ICLR 2017, Toulon, France, 24–26 April 2017, Conference Track Proceedings. OpenReview.net (2017)
8. Elson, J., Douceur, J.R., Howell, J., Saul, J.: Asirra: a captcha that exploits interest-aligned manual image categorization. CCS **7**, 366–374 (2007)
9. Golan, I., El-Yaniv, R.: Deep anomaly detection using geometric transformations. In: Bengio, S., Wallach, H., Larochelle, H., Grauman, K., Cesa-Bianchi, N., Garnett, R. (eds.) Advances in Neural Information Processing Systems, vol. 31. Curran Associates, Inc. (2018)
10. Goodfellow, I., et al.: Generative adversarial nets. In: Proceedings of NIPS, pp. 2672–2680 (2014)
11. Hadsell, R., Chopra, S., LeCun, Y.: Dimensionality reduction by learning an invariant mapping. In: 2006 IEEE Computer Society Conference on Computer Vision and Pattern Recognition (CVPR 2006), vol. 2, pp. 1735–1742. IEEE (2006)
12. Hendrycks, D., Mazeika, M., Dietterich, T.: Deep anomaly detection with outlier exposure. In: International Conference on Learning Representations (2019)
13. Heusel, M., Ramsauer, H., Unterthiner, T., Nessler, B., Hochreiter, S.: Gans trained by a two time-scale update rule converge to a local nash equilibrium. Advances in neural information processing systems 30 (2017)

14. Jianliang, M., Haikun, S., Ling, B.: The application on intrusion detection based on k-means cluster algorithm. In: 2009 International Forum on Information Technology and Applications, vol. 1, pp. 150–152. IEEE (2009)

15. Jolicoeur-Martineau, A.: The relativistic discriminator: a key element missing from standard GAN. In: International Conference on Learning Representations (2019)

16. Kemker, R., McClure, M., Abitino, A., Hayes, T., Kanan, C.: Measuring catastrophic forgetting in neural networks. In: Proceedings of the AAAI Conference on Artificial Intelligence, vol. 32 (2018)

17. Krizhevsky, A.: Learning Multiple Layers of Features from Tiny Images. Master's thesis, Computer Science Department, University of Toronto (2009)

18. Latecki, L.J., Lazarevic, A., Pokrajac, D.: Outlier detection with kernel density functions. In: Perner, P. (ed.) MLDM 2007. LNCS (LNAI), vol. 4571, pp. 61–75. Springer, Heidelberg (2007). https://doi.org/10.1007/978-3-540-73499-4_6

19. LeCun, Y., Bottou, L., Bengio, Y., Haffner, P.: Gradient-based learning applied to document recognition. Proc. IEEE **86**(11), 2278–2324 (1998)

20. Lee, K.S., Tran, N.T., Cheung, N.M.: Infomax-GAN: improved adversarial image generation via information maximization and contrastive learning. In: Proceedings of the IEEE/CVF Winter Conference on Applications of Computer Vision, pp. 3942–3952 (2021)

21. Lucic, M., Kurach, K., Michalski, M., Gelly, S., Bousquet, O.: Are GANs created equal? A large-scale study. In: Proceedings of the NIPS, pp. 700–709 (2018)

22. Parzen, E.: On estimation of a probability density function and mode. Ann. Math. Stat. **33**(3), 1065–1076 (1962)

23. Rafiee, L., Fevens, T.: Unsupervised anomaly detection with a GAN augmented autoencoder. In: Farkaš, I., Masulli, P., Wermter, S. (eds.) ICANN 2020. LNCS, vol. 12396, pp. 479–490. Springer, Cham (2020). https://doi.org/10.1007/978-3-030-61609-0_38

24. Schlegl, T., Seeböck, P., Waldstein, S.M., Schmidt-Erfurth, U., Langs, G.: Unsupervised anomaly detection with generative adversarial networks to guide marker discovery. In: Niethammer, M., et al. (eds.) IPMI 2017. LNCS, vol. 10265, pp. 146–157. Springer, Cham (2017). https://doi.org/10.1007/978-3-319-59050-9_12

25. Schölkopf, B., Williamson, R.C., Smola, A.J., Shawe-Taylor, J., Platt, J.C., et al.: Support vector method for novelty detection. In: NIPS, vol. 12, pp. 582–588. Citeseer (1999)

26. Tax, D.M., Duin, R.P.: Support vector data description. Mach. Learn. **54**(1), 45–66 (2004)

27. Tran, N.T., Tran, V.H., Nguyen, B.N., Yang, L., Cheung, N.M.M.: Self-supervised GAN: analysis and improvement with multi-class minimax game. In: Wallach, H., Larochelle, H., Beygelzimer, A., d' Alché-Buc, F., Fox, E., Garnett, R. (eds.) Advances in Neural Information Processing Systems, vol. 32. Curran Associates, Inc. (2019)

28. Tran, N.T., Tran, V.H., Nguyen, B.N., Yang, L., Cheung, N.M.M.: Self-supervised GAN: analysis and improvement with multi-class minimax game. In: Wallach, H., Larochelle, H., Beygelzimer, A., d' Alché-Buc, F., Fox, E., Garnett, R. (eds.) Advances in Neural Information Processing Systems (2019)

29. Xiao, H., Rasul, K., Vollgraf, R.: Fashion-MNIST: a novel image dataset for benchmarking machine learning algorithms. arXiv preprint arXiv:1708.07747 (2017)

30. Zenati, H., Foo, C.S., Lecouat, B., Manek, G., Chandrasekhar, V.R.: Efficient GAN-based anomaly detection. arXiv preprint arXiv:1802.06222 (2018)

31. Zenati, H., Romain, M., Foo, C.S., Lecouat, B., Chandrasekhar, V.: Adversarially learned anomaly detection. In: 2018 IEEE International Conference on Data Mining (ICDM), pp. 727–736. IEEE (2018)
32. Zhou, C., Paffenroth, R.C.: Anomaly detection with robust deep autoencoders. In: Proceedings of the 23rd ACM SIGKDD International Conference on Knowledge Discovery and Data Mining, pp. 665–674. ACM (2017)
33. Zong, B., et al.: Deep autoencoding gaussian mixture model for unsupervised anomaly detection. In: International Conference on Learning Representations (2018)

Analyzing EEG Data with Machine and Deep Learning: A Benchmark

Danilo Avola[1], Marco Cascio[1], Luigi Cinque[1], Alessio Fagioli[1],
Gian Luca Foresti[2], Marco Raoul Marini[1], and Daniele Pannone[1(✉)]

[1] Sapienza University of Rome, Rome, Italy
{avola,cascio,cinque,fagioli,marini,pannone}@di.uniroma1.it
[2] Università degli Studi di Udine, Udine, Italy
gianluca.foresti@uniud.it

Abstract. Nowadays, machine and deep learning techniques are widely used in different areas, ranging from economics to biology. In general, these techniques can be used in two ways: trying to adapt well-known models and architectures to the available data, or designing custom architectures. In both cases, to speed up the research process, it is useful to know which type of models work best for a specific problem and/or data type. By focusing on EEG signal analysis, and for the first time in literature, in this paper a benchmark of machine and deep learning for EEG signal classification is proposed. For our experiments we used the four most widespread models, i.e., multilayer perceptron, convolutional neural network, long short-term memory, and gated recurrent unit, highlighting which one can be a good starting point for developing EEG classification models.

Keywords: Brain computer interfaces (BCI) · Electroencephalography (EEG) · Deep learning · Benchmark · Classification

1 Introduction

In the last years, machine learning (ML) and deep learning (DL) techniques have been used to overcome the limitation of classical image processing approaches in several fields [1,9]. The prominent results allowed researchers and industries to get outstanding outcomes in different tasks, such as environmental monitoring [5, 7,17], medical and rehabilitation [3,4,18], deception detection [6], and more [2]. In addition, ML and DL have also found application for cross-domain problems, such as trading, biology, and neuroscience. Despite these very different contexts, an ML/DL practitioner commonly follows a well-defined pipeline. The latter usually consists in: 1) choosing a model (or designing one from scratch), 2) setting its hyperparameters, 3) training the chosen model, and finally 4) evaluating it. Then, if the observed results are not convincing, the process is re-iterated from 2) and, in the worst case, even from 1). Hence, to avoid changing or redesigning the model, in order to speed up the process, it is useful to know a priori the class of models that works the best for a certain problem or typology of data.

S. Sclaroff et al. (Eds.): ICIAP 2022, LNCS 13231, pp. 335–345, 2022.
https://doi.org/10.1007/978-3-031-06427-2_28

For this reason, in this paper a benchmark on ML and DL models for EEG classification is proposed. The motivation of providing a benchmark for such a task is twofold. Firstly, EEG signals classification is increasingly used for different applications, such as security and smart prosthesis [19,21]. Secondly, to the best of our knowledge, there are no benchmarks available that highlight which class of models works best with EEG data. The comparisons have been performed by using four well-known ML and DL techniques, namely, the multilayer perceptron (MLP) [26], the convolutional neural network (CNN) [30], the long short-term memory (LSTM) [28], and the gated recurrent unit (GRU) [8]. These approaches have been tested on the EEG motor movement/imagery dataset [23], which contains EEG data of 109 volunteers performing different tasks.

The remainder of this paper is organized as follows. Section 2 reports the current state-of-the-art in EEG signal classification techniques. Section 3 provides a background on EEG signals and on the methods used for this benchmark. Section 4 presents the performed experiments on the chosen dataset, highlighting the best typology of models to perform EEG signals classification. Finally, Sect. 5 concludes the paper.

2 Related Work

In the last decades, several methods have been developed to analyze EEG signals and try to address different problems. Effective pipelines generally perform data pre-processing procedures, to reduce noise and to clean signals, a feature extraction step, to obtain a meaningful input representation, and an inference phase to correctly associate signals to the addressed task.

Concerning data pre-processing procedures, several techniques can be employed to remove artifacts from signals, such as filtering methods or wavelet transforms [12]. In [28], for instance, MI-BCI signals are improved via the fast Fourier transform (FFT) and stacked restricted Boltzmann machines (RBM). The refined signals can then be fed to a CNN to extract meaningful features, which are in turn analyzed and classified via an LSTM network. Differently, in [20], the authors explore various procedures, e.g., discrete wavelet or Fourier transforms, to polish the input signals. Subsequently, other techniques, i.e., PCA and fuzzification, are also applied as pre-processing to, respectively, reduce the input dimensionality and prepare the data for a fuzzy decision tree (FDT) classifier to detect epileptic seizures. While these approaches can be beneficial when analyzing EEG signals, other works concentrate on raw data by defining specific architectures. In [31], for example, an autoencoder is designed to reduce signal noise by exploiting brain activity of multiple persons. The XGBoost algorithm is then applied to these improved signals for MI classification.

Regarding the feature extraction step, it enables to capture diverse aspects from BCI signals and can be performed in many different ways. In [27], for example, an autoregressive discrete variational autoencoder (dVAE) is implemented to retain dynamic signal information. A Gaussian mixture hidden Markov model (GM-HMM) is then used to check for open or closed eyes. The authors of [14], instead, detect epileptic seizures using an SVM by means of multiscale radial basis functions (MRBF), and a modified particle swarm optimization (MPSO), that can simultaneously improve the EEG signals time-frequency extracted features associated to seizures. Differently, the authors of [15] employ a filter bank common spatial pattern (FB-CSP) in conjunction with a sliding window to extract spatial-frequency-sequential features that are used to classify MI by a gated recurrent unit (GRU) network.

In relation to the inference phase, it can be performed using various approaches such as machine or deep learning algorithms. In [16], for instance, a relevance vector machine (RVM) is used to predict speech from BCI signals represented through a Riemannian manifold covariance matrix. In [32], instead, a support vector machine (SVM) is employed to classify gender discrimination, alcholism and epilepsy from k-means clustered inter/intra BCI channels dependencies, in the latter. As for DL methods, the authors of [29] address the MI task exploiting the transfer learning paradigm, and fine-tune a pre-trained architecture using short time Fourier transformed BCI signals. The latter transform is also used in [25] to tackle the same MI task, although the authors introduce a custom model called CNN-SAE to automatically extract and combine time, frequency and location information from the input data. In [13], finally, a compact depth-wise and separable CNN, which is one of the chosen models used in this work benchmark, is designed to be effective on separate tasks such as event related potential, feedback error-related negativity, movement related cortical potential, as well as sensory motor rhythm classifications.

To conclude this overview, although many works tend to develop a specific methodology, others try to present a common ground by comparing existing approaches on complex BCI EEG signals analysis tasks. The authors of [11], for example, examine linear discriminant analysis (LDA), a MLP with a single hidden layer, and SVM performances on 5 different mental tasks, i.e., rest, math, letter, rotate, and count. In [10], instead, the best combination to detect epileptic seizures is searched among several wavelet transforms, i.e., Haar, Daubechies, Coiflets, and Biorthogonal, and either a SVM or probabilistic neural network classifier (PNN). Therefore, inspired by these works, and following their rationale, we present a benchmark with more advanced deep learning algorithms to address another task, i.e., the MI-BCI classification.

3   Background and Method

To provide a clearer overview concerning the context of the benchmark, this section provides the necessary background needed to understand EEG signals. In addition, a summary of ML and DL approaches used in the benchmark is provided.

3.1   Background on EEG Signals

The first step regarding the analysis of EEG data is acquiring the latter during the execution of a specific task, i.e. a trial, from the subject involved in the

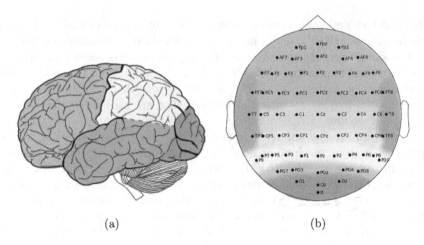

Fig. 1. The a) different areas of the human brain, and b) the corresponding 10–20 map system.

experiment. This acquisition is usually performed by placing electrodes, which capture electrical signals emitted by human neurons, on specific positions of the subject scalp, as shown in Fig. 1. These signals have an extremely low voltage (between $-70\,\mathrm{mV}$ and $+40\,\mathrm{mV}$), where some peaks could determine if an action occurred. Indeed, the brain activity is determined by sequences of such peaks. The entire EEG spectrum is comprised in the range $[1Hz, 50Hz]$ and can be divided in several frequency bandwidths, i.e., the rhythms. Such rhythm, indicating different brain activities, are defined as:

- δ waves: representing deep sleep and pathological states, in the range of $[0.5\,\mathrm{Hz}, 3\,\mathrm{Hz}]$;
- θ waves: indicating sleepiness and childhood memories, in the range of $[3\,\mathrm{Hz}, 7\,\mathrm{Hz}]$;
- α waves: representing adult relaxing state (rest status), in the range of $[8\,\mathrm{Hz}, 13\,\mathrm{Hz}]$;
- β waves: which identify the active state, involving calculus, focus, and muscular contraction, in the range of $[14\,\mathrm{Hz}, 30\,\mathrm{Hz}]$;
- γ waves: indicating tension or strong emotional states, over $30\,\mathrm{Hz}$.

Usually, the acquired signal presents different types of noise. Examples of noise recurrent in EEG data is the muscular activity (such as eye blinks and movements), external sources of electricity, electrodes movements or even sounds. To sanitize the data, some well-known techniques, such as Individual Component Analysis (ICA) [24], can be used. ICA decomposes the signals obtained from multiple electrodes during time in a sum of temporally independent (and fixed in terms of space) components. In particular, given an input matrix X of EEG signals, where the rows represent the electrodes and the columns the time, the algorithm finds a matrix W that decomposes and linearly smooths the data.

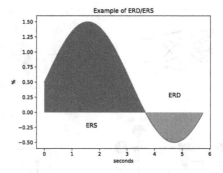

Fig. 2. Example of ERS and ERD. With an ERS, we have an increment of the bandwidth power (filled in blue), while with an ERD we have a decrement of the bandwidth power (filled in orange). (Color figure online)

The rows $U = WX$ represent the activation times of the ICA components, and have the form *components* $\times$ *time*. The columns of the inverse matrix W^{-1} provide the projection forces related to their respective components on each of the electrodes. In other words, these weights provide a topographical image of the scalp for each electrode. In this way, it is possible to locate brain areas involved in specific actions or feelings.

After data cleaning, it is easier to find the so-called *event related potential* (ERP), which is the response to an internal or external event, namely, a sensory, a cognitive, or a motor event. In addition, these events can induce temporary modulations in certain frequency bandwidths, e.g. the increment or decrement of the bandwidth oscillation power. These modulations are called *event related synchronization* (ERS) and *event related desynchronization* (ERD). Synchronization and desynchronization terms refer to the fact that the temporary modulations are due to an increment or a decrement of the synchronization of a neuron population. An example of ERD and ERS is shown in Fig. 2. Since the acquired EEG signal is a summary of the several brain activities, to get the signal regarding a specific stimulus, it is common practice to average the EEG signals obtained from the application of the same stimulus several times (e.g., 100 or more), allowing to remove disturbances not related to the stimulus itself.

Remember that EEG signals are continuously recorded during the experiments. Hence, the resulting data is organized as a 2D matrix having the form *electrodes* $\times$ *time*. To ease the analysis of EEG data in task-related experiments, usually the process that goes under the name of *epoching* is performed. The latter consists in re-arranging the 2D data in a 3D data matrix, having the form (*electrodes, time, events*).

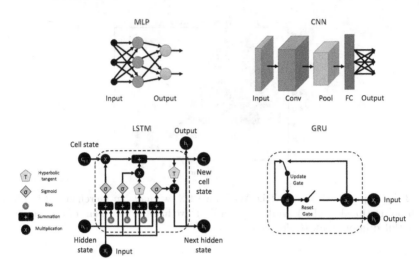

Fig. 3. Simplified examples of the used architectures. From top left to bottom right: multilayer perceptron (MLP), convolutional neural network (CNN), long short-term memory (LSTM) and the gate recurrent units (GRU). The depicted architectures represent the base structure of each model, meaning that only the main elements are present.

3.2 Summary on Used Machine and Deep Learning Approaches

For the proposed benchmark, four well-known approaches were tested, namely, MLP, CNN, LSTM, and GRU. These approaches were chosen since they are the most representative of their category: MLP for classical ML, a standard CNN for DL approaches that are not designed to work with temporal data, and LSTM and GRU for DL approaches designed to work with temporal data. As it is possible to see, two models that can correctly handle temporal data were chosen. This is due to the fact that the final accuracy of those models strongly depends on the input data, therefore there is no better model between them. In Fig. 3, examples of the chosen approaches are shown.

The first approach we have chosen is MLP, which belongs to the class of feedforward artificial neural networks, and is the most basic neural network approach that can be used in classification task. The simplest MLP consists of three layers, namely, an input layer, a hidden layer, and the output layer. In the proposed benchmark, the used MLP has 1 input layer, 3 hidden layers, and 1 output layer.

CNN, originally designed for image analysis, can be used for analyzing EEG without efforts. As the name suggests, CNN uses convolutions in order to extract features from the input data. Depending on the data type, CNN may use 1D, 2D, or even 3D convolutions. In our benchmark, we used 1D convolutions with kernels of size ($channels, 1$). This kernel size allowed analyzing contemporary all the channels at each time instant. The used CNN consists of 4 convolutional layers and 1 dense layer for classification.

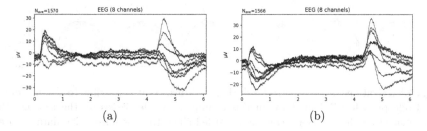

Fig. 4. Examples of signals corresponding to an event. In Figure a), the signal corresponding to the left hand is shown, while in b) the signal corresponding to the right hand is depicted.

LSTM, differently from MLP and CNN, has the purpose of handling temporal data. Belonging to the family of recurrent neural networks (RNN) [22], a single LSTM unit is composed of a cell, that contains values over time intervals, and several gates, namely, an input gate, an output gate, and a forget gate. The aim of these gates is to control the flow of the information into and out of the cell. This structure allows to overcome the vanishing gradient problem that affects the training of standard RNNs. The LSTM used in the proposed benchmark simply consists of 2 stacked LSTM units.

The last approach we are going to summarize is the GRU. As for LSTM, GRU belongs to RNNs family. In fact, GRU can be seen as an optimized version of the LSTM due to its different internal structure. In detail, GRU has gating units allowing to control the information flow, but it lacks the memory cell. The less complex structure makes GRU computationally more efficient with respect to LSTM. As for the latter, in our benchmark we used 2 stacked GRU units.

4   Experiments

In this section, the results of the benchmark, together with the used dataset, are presented. Concerning the data, the EEG signals were handled with MNE-Python, while all the tested methods were implemented in Pytorch. The machine used for training the model consisted in an AMD Ryzen 7 CPU, 16GB DDR4 of RAM, and an NVidia RTX 2080 GPU.

4.1   Dataset

For the experiments, the EEG-BCI Motor Imagery Dataset [23] has been used. It consists of more than 1500 records in European Data Format (EDF) files from 109 participants. The latter performed 14 runs while wearing a BCI2000 device. At each run, the subject could perform one of the following actions: open or close right or left hand, imagine opening or closing right or left hand, open or close both hands or feet, imagine opening or closing both hands or feet. The EDF files contain 64 EEG signals, with a frequency of 160 samples per second.

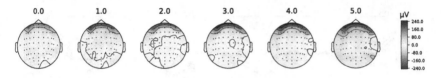

Fig. 5. Activations of the electrodes during the runs. As it is possible to see, not every electrode provide relevant data. In this example, the electrodes on the top, i.e. the ones placed on the forehead, can be discarded since they acquire noisy data such as eye blinks.

In the experiments, we have chosen arbitrarily to classify signals related to right or left hand rather than distinguish signals of hands from feet. Before training the several models, the data has been cleaned from noise, and the data related to subjects having id $(43, 88, 89, 92, 100, 104)$ was removed. This is due to the fact that in some EEG signals the sampling frequency was different, or there was too much noise in the data. The final dimension of the dataset consisted in 3136 events to be classified. These events are then splitted as follows: 1980 events for training, 216 events for validation, and 940 for testing. Each event is fed as input in the form of a 2D matrix having the form $[channels, samples]$.

4.2   Results

In Fig. 4, it is possible to see an example of signals related to the left (Fig. 4a) and right (Fig. 4b) hands. As it is possible to see, there are interesting patterns within the acquired signal. In detail, it is possible to observe that we have ERS in the intervals $[0, 1]$ and $[4.5, 5.4]$. Since these patterns are present in all the considered data, we choose to perform the experiments on different time intervals, which are the entire event, having length of $[0, 6]$, and the two ERS specified above. In Table 1, the results obtained with the different models are shown. As it is possible to see, the model that perform the best is the CNN, obtaining an accuracy of 90.4% on the interval $[0, 6]$, followed by MLP, with an accuracy of 85.2%. The worst results are obtained with the RNN models.

In detail, we have that GRU has obtained an accuracy of 76.9%, while the LSTM has obtained an accuracy of 57.9%, resulting as the worst model in our benchmark. Usually, RNNs performs better with respect to other methods when dealing with temporal or sequential data. This is amenable to the fact that the latter usually present some correlation among the several time instants. It is clear from the reported results that this aspect cannot be assumed in the case of brain signals. Instead, the several filters applied by the CNN seem to perform a better abstraction of the data, allowing to obtain the highest accuracy. This could be related to the fact that convolutions allow to better identify specific patterns, e.g. peaks, within the data.

During the experiments, we noticed that some channels provided more reliable data with respect to others. These channels are F7,F8, FT7, FT8, T9, TP7, TP8, and FC, as shown in the example in Fig. 5. Hence, we re-trained the cho-

Table 1. Accuracy values obtained with the different models on the different intervals. In detail, we have that Interval 1 = $[0, 1]$, Interval 2 = $[4.5, 5.4]$, and Interval 3 = $[0, 6]$.

	Interval 1	Interval 2	Interval 3
MLP	74.8%	84.6%	85.2%
CNN	78.8%	84.6%	90.4%
LSTM	65.4%	78.6%	57.9%
GRU	65.7%	85%	76.9%

Table 2. Accuracy values obtained by using only a subset of the channels. Also in this case, we have that Interval 1 = $[0, 1]$, Interval 2 = $[4.5, 5.4]$, and Interval 3 = $[0, 6]$.

	Interval 1	Interval 2	Interval 3
MLP	77.3%	85.5%	85.7%
CNN	76.3%	85.2%	88.8%
LSTM	60.3%	82.2%	65.2%
GRU	67.1%	84%	78.7%

sen models by using only the above-mentioned channels, obtaining the results shown in Table 2. In this case, there is a general accuracy improvement among the several models and intervals. Despite this, we have that the CNN still results the best performing model with an accuracy score of 88.8%. Again, MLP follows with an accuracy of 85.7%. Finally, we have GRU and LSTM with an accuracy of 78.7% and 65.2%, respectively.

5   Conclusions

In this paper, a benchmark of ML and DL models for EEG classification is presented. Four well-known models in ML and DL are compared, namely MLP, CNN, LSTM, and GRU. Extensive experiments have been performed on a dataset containing EEG data of 109 volunteers performing different tasks, highlighting that CNN is the best category of models for dealing with EEG data.

Acknowledgement. This work was supported by the MIUR under grant "Departments of Excellence 2018–2022" of the Sapienza University Computer Science Department and the ERC Starting Grant no. 802554 (SPECGEO).

References

1. Avola, D., Cinque, L., Foresti, G.L., Martinel, N., Pannone, D., Piciarelli, C.: Low-Level Feature Detectors and Descriptors for Smart Image and Video Analysis: A Comparative Study, pp. 7–29 (2018)

2. Avola, D., Bernardi, M., Cinque, L., Foresti, G.L., Massaroni, C.: Adaptive boot-strapping management by keypoint clustering for background initialization. Patt. Recogn. Lett. **100**, 110–116 (2017)

3. Avola, D., Cinque, L.: Encephalic NMR tumor diversification by textural inter-pretation. In: Proceedings of the 15th International Conference on Image Analysis and Processing (ICIAP), pp. 394–403 (2009)

4. Avola, D., Cinque, L., Fagioli, A., Filetti, S., Grani, G., Rodolà, E.: Multimodal feature fusion and knowledge-driven learning via experts consult for thyroid nodule classification. IEEE Trans. Circ. Syst. Video Technol., 1–8 (2021)

5. Avola, D., Cinque, L., Fagioli, A., Foresti, G.L., Massaroni, C., Pannone, D.: Feature-based SLAM algorithm for small scale UAV with Nadir view. In: Ricci, E., Rota Bulò, S., Snoek, C., Lanz, O., Messelodi, S., Sebe, N. (eds.) ICIAP 2019. LNCS, vol. 11752, pp. 457–467. Springer, Cham (2019). https://doi.org/10.1007/978-3-030-30645-8_42

6. Avola, D., Cinque, L., Foresti, G.L., Pannone, D.: Automatic deception detection in RGB videos using facial action units. In: Proceedings of the 13th International Conference on Distributed Smart Cameras, pp. 1–6 (2019)

7. Avola, D., Foresti, G.L., Cinque, L., Massaroni, C., Vitale, G., Lombardi, L.: A multipurpose autonomous robot for target recognition in unknown environments. In: Proceedings of the 14th International Conference on Industrial Informatics (INDIN), pp. 766–771 (2016)

8. Chen, J., Jiang, D., Zhang, Y.: A hierarchical bidirectional GRU model with atten-tion for EEG-based emotion classification. IEEE Access **7**, 118530–118540 (2019)

9. Dong, S., Wang, P., Abbas, K.: A survey on deep learning and its applications. Comput. Sci. Rev. **40**, 1–22 (2021)

10. Gandhi, T., Panigrahi, B.K., Anand, S.: A comparative study of wavelet families for EEG signal classification. Neurocomputing **74**(17), 3051–3057 (2011)

11. Garrett, D., Peterson, D.A., Anderson, C.W., Thaut, M.H.: Comparison of linear, nonlinear, and feature selection methods for EEG signal classification. IEEE Trans. Neural Syst. Rehabil. Eng. **11**(2), 141–144 (2003)

12. Jiang, X., Bian, G.B., Tian, Z.: Removal of artifacts from EEG signals: a review. Sensors **19**(5), 987 (2019)

13. Lawhern, V.J., Solon, A.J., Waytowich, N.R., Gordon, S.M., Hung, C.P., Lance, B.J.: EEGNet: a compact convolutional neural network for EEG-based brain-computer interfaces. J. Neural Eng. **15**(5), 056013 (2018)

14. Li, Y., Wang, X.D., Luo, M.L., Li, K., Yang, X.F., Guo, Q.: Epileptic seizure classification of EEGs using time-frequency analysis based multiscale radial basis functions. IEEE J. Biomed. Health Inform. **22**(2), 386–397 (2017)

15. Luo, T.J., Chao, F., et al.: Exploring spatial-frequency-sequential relationships for motor imagery classification with recurrent neural network. BMC Bioinform. **19**(1), 1–18 (2018)

16. Nguyen, C.H., Karavas, G.K., Artemiadis, P.: Inferring imagined speech using EEG signals: a new approach using Riemannian manifold features. J. Neural Eng. **15**(1), 016002 (2017)

17. Patrício, D.I., Rieder, R.: Computer vision and artificial intelligence in precision agriculture for grain crops: a systematic review. Comput. Electron. Agric. **153**, 69–81 (2018)

18. Petracca, A., et al.: A virtual ball task driven by forearm movements for neuro-rehabilitation. In: Proceedings of the International Conference on Virtual Rehabil-itation (ICVR), pp. 162–163 (2015)

19. Pham, T., Ma, W., Tran, D., Nguyen, P., Phung, D.: EEG-based user authentication in multilevel security systems. In: Advanced Data Mining and Applications, pp. 513–523 (2013)
20. Rabcan, J., Levashenko, V., Zaitseva, E., Kvassay, M.: Review of methods for EEG signal classification and development of new fuzzy classification-based approach. IEEE Access **8**, 189720–189734 (2020)
21. Ruhunage, I., Perera, C.J., Nisal, K., Subodha, J., Lalitharatne, T.D.: EMG signal controlled Transhumerai prosthetic with EEG-SSVEP based approach for hand open/close. In: IEEE International Conference on Systems, Man, and Cybernetics (SMC), pp. 3169–3174 (2017)
22. Rumelhart, D.E., Hinton, G.E., Williams, R.J.: Learning internal representations by error propagation. California Univ San Diego La Jolla Inst for Cognitive Science, Technical report (1985)
23. Schalk, G., McFarland, D., Hinterberger, T., Birbaumer, N., Wolpaw, J.: BCI 2000: a general-purpose brain-computer interface (BCI) system. IEEE Trans. Biomed. Eng. **51**(6), 1034–1043 (2004)
24. Stone, J.V.: Independent component analysis: an introduction. Trends Cogn. Sci. **6**(2), 59–64 (2002)
25. Tabar, Y.R., Halici, U.: A novel deep learning approach for classification of EEG motor imagery signals. J. Neural Eng. **14**(1), 016003 (2016)
26. Thimm, G., Fiesler, E.: High-order and multilayer perceptron initialization. IEEE Trans. Neural Netw. **8**(2), 349–359 (1997)
27. Wang, M., Abdelfattah, S., Moustafa, N., Hu, J.: Deep gaussian mixture-hidden Markov model for classification of EEG signals. IEEE Trans. Emerg. Top. Comput. Intell. **2**(4), 278–287 (2018)
28. Wang, P., Jiang, A., Liu, X., Shang, J., Zhang, L.: LSTM-based EEG classification in motor imagery tasks. IEEE Trans. Neural Syst. Rehabil. Eng. **26**(11), 2086–2095 (2018)
29. Xu, G., et al.: A deep transfer convolutional neural network framework for EEG signal classification. IEEE Access **7**, 112767–112776 (2019)
30. Yıldırım, Ö., Baloglu, U.B., Acharya, U.R.: A deep convolutional neural network model for automated identification of abnormal EEG signals. Neural Comput. Appl. **32**(20), 15857–15868 (2020)
31. Zhang, X., Yao, L., Zhang, D., Wang, X., Sheng, Q.Z., Gu, T.: Multi-person brain activity recognition via comprehensive EEG signal analysis. In: Proceedings of the EAI International Conference on Mobile and Ubiquitous Systems: Computing, Networking and Services (MobiQuitous), pp. 28–37 (2017)
32. Zhou, P.Y., Chan, K.C.: Fuzzy feature extraction for multichannel EEG classification. IEEE Trans. Cogn. Dev. Syst. **10**(2), 267–279 (2016)

A Two-Stage U-Net to Estimate the Cultivated Area of Plantations

Walysson Carlos dos Santos Oliveira$^{(\boxtimes)}$ [ID], Geraldo Braz Junior [ID],
Daniel Lima Gomes Junior, Anselmo Cardoso de Paiva [ID],
and Joao Dallyson Sousa de Almeida [ID]

Universidade Federal do Maranhão - UFMA, 322, São Luís, MA 65.086-110, Brazil
walysson.oliveira@discente.ufma.com.br,
{geraldo,paiva,jdallyson}@nca.ufma.br, daniellima@ifma.edu.br

Abstract. In order to reduce tax evasion in agribusiness, it is possible
to estimate the production of crops through the monitoring and analysis
of satellite images and compare with the values declared by the taxpayer.
For this, deep learning techniques can be applied to satellite images to
segment the cultivated area of plantations, and the segmented area can
be used to estimate crop yields. As an initial step, this work aims to
analyze the satellite images of plantations to estimate the cultivated
area of plantations using semantic segmentation. For this, we created a
dataset for planting areas data, and we proposed network architecture
for image segmentation, a two-stage U-net. The proposed methodology
returned average IoU results above 80% in both stages.

Keywords: Deep learning · Semantic segmentation · Agriculture
application

1 Introduction

Agriculture is responsible not only for a large part of the food items consumed
but also for a production chain that involves various segments of the economy.
In 2019, the sector represented 21.6% of the Brazilian GDP, according to the
Ministry of Agriculture, Livestock and Supply of Brazil [18]. Since agribusiness
is of paramount importance in the Brazilian economy, its participation in contributing to the functioning of the government through taxes and fees arising
from its activity is natural. However, agribusiness has a high rate of tax evasion
that occurs as a result of the current complexity of the Brazilian tax system, in
addition to the difficulty of inspection by the State, due to the high cost that
such inspection entails, making its execution unfeasible [4].

More and more attention has been paid to the application of deep learning techniques in agriculture [25]. Recent studies that propose to segment land
cover regions [15], vegetation areas [24] and plantation areas [12] using fully convolutional neural networks [11] for this purpose. In the context of agribusiness
taxation, the largest share of the tax burden is calculated on the production of

S. Sclaroff et al. (Eds.): ICIAP 2022, LNCS 13231, pp. 346–357, 2022.
https://doi.org/10.1007/978-3-031-06427-2_29

crops. In order to reduce tax evasion, it is possible to estimate the production of crops through the monitoring and analysis of satellite images and compare with the values declared by the taxpayer. For this, we can apply deep learning techniques in satellite images to segment the cultivated area of the plantations, and the segmented area can be multiplied by a factor (weight/area ratio), which depends on the crop planted, to estimate production of the crops. In this study, we will perform the segmentation of the planted area and, in future work, production estimation.

The present study uses deep learning techniques in satellite images to estimate the cultivated area of plantations using semantics with convolutional neural networks. As this application is still little explored in the agriculture segment, several contributions are included in this work. The main contributions stand out: (a) the construction of a dataset from satellite images, which could help other researchers, with the marking of eight classes identified in the land cover of planting regions (b) the proposed architecture of deep learning that could help in the automation of the inspection process of farms for tax purposes; and (c) the possibility of applying the proposed architecture to other problems in remote sensing images.

2 Related Works

To solve segmentation problems of satellite images and aerial images of plantations, the approaches found in the literature generally make use of deep learning techniques. These works focus on datasets created by companies, governments, universities, and researchers. The latter is often designed specifically for the particularities of the search. Some of these works are discussed in this section.

The Agriculture-Vision [8] dataset contains 21,061 aerial images of US farms captured throughout 2019. The dataset consists of six classes: cloud shadow, double-crop, crop failure, puddle water, water path, and weed block. Agriculture-Vision is a dataset for semantic segmentation of multiple classes, where these classes can overlap, for example, a puddle of water shaded by a cloud. A new fully convolutional neural network architecture called AgriSegNet [2] which uses attention gates and combines the result of image segmentation at different resolutions, was applied to this problem reaching 47.96% of average IoU. The winner of the Agriculture-Vision challenge used a U-net based architecture and combined a Residual DenseNet with Squeeze-and-Excitation (SE) [7] blocks. In addition, the winner network used expert networks with the same architecture but with fewer layers to segments less regular classes such as crop failure and water puddle. The result in average network IoU was 63.9%, obtaining the best average performance of the competition.

The DeepLabV3+ [6] convolutional neural network architecture was used to segment vegetation regions in satellite images into three classes: trees, shrubs, and grass [3]. In this approach, the Slovenia Dataset and Oregon Dataset datasets were used, and the network achieved 78.0% accuracy in the Slovenian dataset and 78.9% in the Oregon dataset.

A U-net-based architecture [17] was used to segment land cover areas in satellite images and classify them into different crop types [20]. The datasets used were the BigEarthNet Dataset [19] which presents 44 land cover classes, and the CORINE Dataset [5] which was developed by the European Environmental Agency and presents a variety of classes, including types of plantation crops. The architecture used was a modified U-net with a ResNet50 [10] in the encoder. The method had an accuracy of 77% for plantation areas and 86% for forest areas.

Although some datasets are already available in the literature for segmenting planting regions and land cover, they do not present the level of detail sought in this research and are unrelated to crops in Brazilian territory. Thus, a new dataset is built for this specific research and will be available to other researchers in this study. Most of the works used architectures based on U-net, and, in this study, these architectures will be used.

3   Methodology

In this section, we describe the proposed methodology to estimate the cultivated area of plantations using semantic segmentation with convolutional neural networks in satellite images. Figure 1 illustrates the proposed methodology.

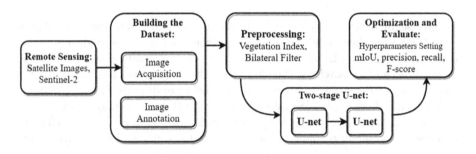

Fig. 1. Proposed methodology

3.1   Building the Dataset

From searches carried out in the literature, we found some image datasets for the segmentation of plantation areas. However, none of the datasets found presents the desired level of detail for the problem addressed, nor are they of Brazilian regions with its most common vegetation and plantations. Thus, we decided to create its dataset.

Image Acquisition. Image acquisition is made with the aid of the *Google Earth Engine* (GEE) cloud-based geospatial processing platform. GEE is a satellite imagery catalog and geospatial dataset that allows the user to view, manipulate, edit and create spatial data [13]. The satellite used is Sentinel-2, which has a spatial resolution of 10m in its four main bands and image updates every ten days [16].

The study region of this work is a continuous area of plantations of 78,590.5 km² of total area and 1,233.8 km of perimeter as illustrated in Fig. 2a). Point A, in Fig. 2a), is located at coordinates 10°47′19.41″S and 46°73′36.57″W and Point B at coordinates 14°32′85.88″S and 44°90′44.33″W. Images corresponding to this area are downloaded with the GEE API that allows you to define the region of interest, filter by date and cloud probability for images with better quality.

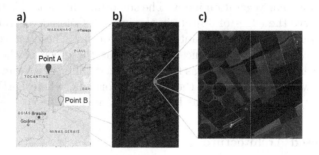

Fig. 2. a) Acquisition Area. b) Full image. c) 4 image blocks. (Color figure online)

The Sentinel 2 satellite has a variety of bands that capture different information from the Earth's surface. For the purpose of this work, four bands of interest are selected, which are bands B2 (Blue), B3 (Green), B4 (Red), and B8 (Near Infra-Red). The image acquisition region covers a large land area, so it is necessary to fragment the region into blocks of lower resolution. The selected area was fragmented in a tabular (or grid) form, composed of several blocks of 256 × 256 resolution images, as shown in Fig. 2c). The naming of the images will indicate their position in the image grid so that we can reassemble larger areas.

Image Annotation. After the acquisition steps, masks containing the annotations of each class in the images are created. We will use the annotations to carry out the training and evaluation of the networks. The annotation process is carried out with the aid of a tool for this purpose called *Labelme* [21].

The region of interest in this work is the area of cultivation of plantations. To define the classes, the plantation cultivation area is divided into three classes according to the plantation development stage: Plantation preparation area (class 1), which is the region of soil prepared to receive the planting, Young planting (class 2), which is the initial stage of the plantation that mixes green areas with still areas of soil and mature plantation (class 3) which are regions where the plantation is developed. It was thought important to note the paths that normally delimited the plantations and were named Plantation division lines (class 4). The fifth class represents areas of natural vegetation, green areas that are not cultivated, called Grassland or Forest Areas (class 5). The other classes are Areas of soil or rocks (class 6), Water (wetlands, lakes, rivers, etc.) (class 7), and Dwelling Areas (class 8), which are artificial houses or buildings in the farm regions. The definition of classes and annotation of areas was based on other datasets with similar intent: *DeepGlobe Land Cover* [9], *LandCoverNet* [1] and the *EOPatches Slovenia* [3].

3.2 Preprocessing

Two images pre-processing techniques are used, one being a smoothing filter and the other a color vegetation index. The smoothing filter used is the bilateral filter to preserve the edges of objects in the images. Colored vegetation indices are used in remote sensing of plantations and forests. These indices have the function of accentuating a specific color, such as the green of the plantation [22]. The vegetation index used was the Normalized Difference Vegetation Index (NDVI) [14] which is defined by the following formula, where RED is the red color channel, and NIR is the Near Infra-Red channel: $NDVI = \frac{NIR-RED}{NIR+RED}$.

3.3 Proposed Architecture

The proposed network architecture is composed of a combination of two U-net based [17] networks that will segment in two stages as illustrated in Fig. 3. In Stage 1, the first U-net receives satellite images with the RGB-NIR bands and is trained to learn how to segment the Plantation and Non-Plant regions. In Stage 2, the second U-net receives the RGB-NIR images without the regions classified as Non-Plant in Stage 1. This is because the Stage 2 input is the result of a filtering operation between the images and their predicted segmentation masks in Stage 1. Thus, Stage 2 serves to correct Stage 1 segmentation errors in the prediction of the No Plantation region and serves to focus on learning how to differentiate the Plantation sub-regions: Green Planting and Prepared Soil for the Planting.

The goal of the proposed architecture is to reduce the complexity of segmenting multiple classes considering that there are color similarities in the Grass or Forest (No Plantation) classes with the Young Planting and Mature Planting (Plantation) classes, the similarity between the sand or rock classes (No Plantation) with the Soil Prepared for Planting (Planting) class. With this, the idea is for Stage 1 to focus on typical characteristics of plantation areas such as plow lines, texture, and less focus and color characteristics.

3.4 Hyperparameter Optimization

Hyperparameters are parameters used to configure a Machine Learning model that cannot be estimated directly from network learning and must be configured before training a Machine Learning model, as they define the model architecture [23]. In the hyperparameter optimization step, the network is trained with a

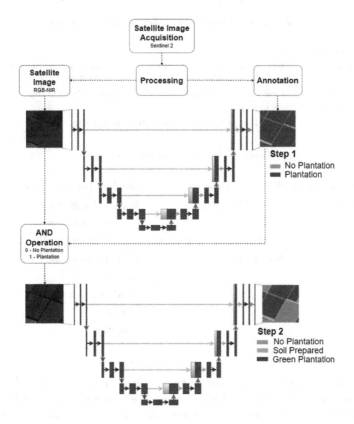

Fig. 3. Proposed architecture (Color figure online)

combination of hyperparameters that are tuned to optimize the results. In the experiments of this work, the best combination of hyperparameters that optimize the performance of the network is tested considering the sets in Table 1.

4   Experiments and Results

A dataset was created to segment plantation areas, where images were acquired from the *Sentinel-2* satellite and obtained through the *Google Earth* platform with the selection of the necessary bands and fragmentation of the region of interest in 9860 256 × 256 resolution image blocks. So far, 250 images have been tagged that have gone through the review and correction stages.

The *dataset* built contains 8 classes that can be used for a variety of types of applications. In the experiments of this article, the 8 classes are not used separately, but groupings of them, as the objective is to segment planting regions. The proposed network architecture performs a two-stage segmentation. In Table 2 are presented the classes of each one of these stages and their correspondence with the original classes of the dataset.

Table 1. Set of hyperparameters that is optimized

Hyperparameters	Test set
Backbone	VGG, ResNet, SeResNet, ResNext, SeResNext, SeNet, DenseNet, Inception, InceptionResNet, MobileNet and EfficientNet
Loss function	Jaccard, Dice, Categorical Focal, Categorical Cross Entropy
Batche size	2, 4, 6, 8, 10
Optimizer	Adam, Ftrl, Adagrad, Adamax, RMSprop, SGD, Nadam

Table 2. Definition of study classes.

Stage 1 classes	Stage 2 classes	Original Dataset Classes
Plantation	Green Plantation	Mature Plantation
		Young Plantation
	Soil Prepared for Planting	Soil Prepared for Planting
No Plantation	No Plantation	Path/Road in the Plantation
		Grass or Forest
		Sand or Rock
		Water
		Building

To obtain the best configuration of the architecture, tests were carried out on the hyperparameters listed in Table 1 using the hyperopt library following the set of hyperparameters presented in the methodology. 100 random combinations of hyperparameters were obtained from training for 250 epochs in the training set. A callback function was implemented to monitor the performance of the network and save the weights of the network at the time with the best mIoU in the validation set. At the end of training the 250 epochs of each combination, the weights of the best network were reloaded into the network and applied to the test set. Once the 1000 combinations were finished, the one with the best result in the test set was selected. The best combination is Backbone: EfficientNetB7, Batche Size: 8, Loss Function: Binary Focal Loss, and Optimizer: Adam. For network training, we used a hold-out validation where the dataset was shuffled and splitted into 50% for the training set, 20% for the validation set, and 30% for the test set. We used imagenet pre-trained weights and the tests each combination took an average of 3 h to complete, which implied 14 days of running the hyperparameter tests.

Segmentation is performed in two stages following the proposed architecture. The network of Stage 1 had as input four bands of satellite images, which are the RGB and NIR bands. The visual results from Stage 1 are shown in Fig. 4 with the segmentation of Plantation areas in dark green, indicating that the Plantation areas in the example images are well delimited and visually close to the real segmentation, but with noise in the segmentation.

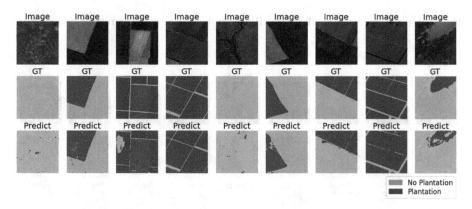

Fig. 4. On the first line, images in RGB. On the centerline, the actual segmentation mask for the Plantation and No Plantation classes. In the last line, the segmentation mask predicted in Stage 1. (Color figure online)

The numerical results of the segmentation of Stage 1 are presented in Table 3 and are consistent with the observations of the visual results. The average IoU had a result of 84.04%, which can be considered a median result for medical applications due to the criticality, but is a reasonably consistent result for an agricultural and satellite imagery application. Precision, Recall, and F-score results had results above 90%.

Table 3. The numerical result of metrics - Stage 1.

Class	IoU	Precision	Recall	F-score
Plantation	88.26%	94.83%	92.57%	93.67%
No Plantation	79.81%	86.92%	90.41%	88.57%
Mean	84.04%	90.87%	91.49%	91.12%

Figure 5 shows the qualitative results of Stage 2, which presents greater detail in the Plantation areas that were subdivided into regions of Green Plantation and Soil Prepared for Planting. It is possible to observe in Fig. 5 that the No Planting region had a very similar result to Stage 1, which was what was intended with the segmentation in two stages, keeping a good delimitation of the area and being very close to the real segmentation. Interestingly, we observed that the segmentation noises followed in Stage 1 were corrected in Stage 2 for the No Plantation class. This is a very positive fact and indicates the region predicted as No Planted in Stage 1 is also predicted as No Planted in Stage 2. However, the region predicted as Planted in Stage 1 is predicted as any of the three classes in Stage 2. On the other hand, the Plantation sub-areas: Prepared Soil and Green Plantation, did not have such a good delimitation of the areas, and it is clear that the noise from Stage 1 in the Plantation region is transmitted as an error to Stage 2.

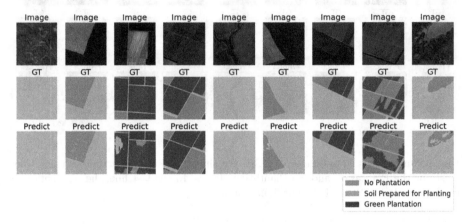

Fig. 5. On the first line, images in RGB. On the centerline, the actual segmentation mask for the No Plantation, Soil Prepared and Green Planting classes. In the last line, the segmentation mask predicted in Stage 2. (Color figure online)

The results from step 2 are presented in Table 4 and show an IoU result of the No Plantation class of 92.61% which improved compared to the result of 79.81% from step 1 and did well higher compared to the IoU values of 76.68% and 78.17% of the Green Plantation and Prepared Soil classes, respectively, which on average are lower than the result of 88.26% from step 1. Overall, the Mean IoU was 82.49%, and Precision, Recall, and F-score values were around 90%.

Table 4. The numerical result of metrics - Stage 2.

Class	IoU	Precision	Recall	F-score
Green Plantation	76.68%	83.41%	90.74%	86.77%
Soil Prepared	78.17%	93.07%	82.84%	87.65%
No Plantation	92.61%	92.90%	99.67%	96.15%
Mean	82.49%	89.79%	91.08%	90.19%

4.1 Comparison with a Single U-Net

In order to verify whether the proposed method actually has a better performance than segmentation with a single network, a single U-net was trained and the results are presented in Table 5.

The results of Table 5 for a single U-net show that the results of the proposed method, presented in Table 4 is superior in each of the measured metrics. As

Table 5. Numerical result of the segmentation of a single U-net.

Class	IoU	Precision	Recall	F-score
Green Plantation	73.00%	88.57%	80.69%	84.26%
Soil Prepared	74.33%	85.14%	85.35%	85.24%
No Plantation	76.44%	82.31%	91.02%	86.25%
Mean	74.59%	85.34%	85.69%	85.25%

expected, the most significant difference is found in the performance of the No Plantation class, since in the proposed method, there is an exclusive network to segment this region. Other studies in the Related Works section applied deep learning techniques in satellite imagery to segment vegetation regions and land cover. It was not possible to make a direct comparison (planting/non-planting segmentation area) with other works, but for a general comparison, the works cited in the Related Works section had the following results:

Table 6. Related works results

Author	Model	Dataset	Task	Result
[2]	AgriSegNet [2]	Agriculture-Vision [8]	Segmentation 6 plantation classes	47.96% mIoU
[7]	U-net based architecture [17]	Agriculture-Vision [8]	Segmentation 6 plantation classes	63.9% mIoU
[3]	DeepLabV3+ [6]	Slovenia Dataset	Segmentation trees, shrubs and grass	78.0% accuracy
[20]	U-net based architecture [17]	BigEarthNet [19] + CORINE [5]	Segmentation plantation areas	77% accuracy

5 Conclusion

In this work, a method for plantation area segmentation was presented, and as a contribution, the work also included the creation of a plantation area segmentation dataset. A two-stage segmentation architecture was proposed with two fully convolutional networks, U-net, in series in order to reduce the complexity of multiple class segmentation. We adjusted the network hyperparameters considering the U-net backbone, the optimizer, loss function, and batch size. The best combination of hyperparameters was obtained, and the proposed method results were superior to the technique of using only a single U-net.

The proposed methodology with the network with a two-stage segmentation architecture returned average IoU results above 80% in both stages. There was a significant improvement in the segmentation of the No Planting region from Stage 1, where the result was 79.81%, to Stage 2, where it rose to 92.61%, due to the removal of segmentation noise in this class in Stage 2. When compared with the results of a single U-net network, the architecture of the proposed method was better in all evaluated metrics. The main contributions achieved are the construction of a satellite image dataset that can help other researchers in similar studies; and the creation of a method for the segmentation of plantation areas that can be used as one of the initial Stages of an automated process of inspection of rural establishments for tax purposes.

As future work, the next Stage for creating an automated inspection process for rural establishments for tax purposes is to use the predicted segmentation area to estimate production in weight measurement units and compare with official data declared by rural producers for the tax administration. We can make various optimizations to the network architecture, such as using another base architecture in place of U-net, such as the DeepLab [6] architecture, or creating a new architecture. Perform a better comparison with related works by applying the proposed architecture to other datasets or using the architectures researched in the created dataset. Finally, it is still recommended to increase the number of images in the dataset and test other pre-processing possibilities.

Acknowledgements. The authors thank the National Council for Scientific and Technological Development (CNPq), This work was supported by the Coordenação de Aperfeiçoamento de Pessoal de Nível Superior (CAPES) - Finance Code 001 and the Foundation for Research and Scientific and Technological Development of the State of Maranhão (FAPEMA) for the financial support for the development of this work.

References

1. Alemohammad, H., Booth, K.: LandCoverNet: a global benchmark land cover classification training dataset (2020)
2. Anand, T., Sinha, S., Mandal, M., Chamola, V., Yu, F.R.: AgriSegNet: deep aerial semantic segmentation framework for IoT-assisted precision agriculture. IEEE Sens. J. **21**, 17581–17590 (2021)
3. Ayhan, B., Kwan, C.: Tree, shrub, and grass classification using only RGB images. Remote Sens. **12**(8), 1333 (2020). https://doi.org/10.3390/rs12081333. https://www.mdpi.com/2072-4292/12/8/1333
4. Brugnaro, R., Del Bel Filho, E., Bacha, C.J.C.: Avaliação da sonegação de impostos na agropecuária brasileira. Agric. São Paulo. SP **3**(50), 15–27 (2003)
5. Büttner, G., Feranec, J., Jaffrain, G., Mari, L., Maucha, G., Soukup, T.: The CORINE land cover 2000 project. EARSeL eProceedings **3**(3), 331–346 (2004). EARSeL Paris
6. Chen, L.C., Papandreou, G., Kokkinos, I., Murphy, K., Yuille, A.L.: DeepLab: semantic image segmentation with deep convolutional nets, Atrous convolution, and fully connected CRFs. IEEE Trans. Pattern Anal. Mach. Intell. **40**(4), 834–848 (2017)
7. Chiu, M.T., et al.: The 1st agriculture-vision challenge: methods and results. arXiv preprint arXiv:2004.09754 (2020)
8. Chiu, M.T., et al.: Agriculture-vision: a large aerial image database for agricultural pattern analysis. In: Proceedings of the IEEE/CVF Conference on Computer Vision and Pattern Recognition, pp. 2828–2838 (2020)

9. Demir, I., et al.: DeepGlobe 2018: a challenge to parse the earth through satellite images. In: The IEEE Conference on Computer Vision and Pattern Recognition (CVPR) Workshops, June 2018

10. He, K., Zhang, X., Ren, S., Sun, J.: Deep residual learning for image recognition. In: Proceedings of the IEEE Conference on Computer Vision and Pattern Recognition, pp. 770–778 (2016)

11. LeCun, Y., Bengio, Y., Hinton, G.: Deep learning. Nature **521**(7553), 436–444 (2015)

12. Rustowicz, R., Cheong, R., Wang, L., Ermon, S., Burke, M., Lobell, D.: Semantic segmentation of crop type in Africa: a novel dataset and analysis of deep learning methods. In: Proceedings of the IEEE/CVF Conference on Computer Vision and Pattern Recognition (CVPR) Workshops, June 2019

13. Mutanga, O., Kumar, L.: Google earth engine applications (2019)

14. Perez, A., Lopez, F., Benlloch, J., Christensen, S.: Colour and shape analysis techniques for weed detection in cereal fields. Comput. Electron. Agric. **25**(3), 197–212 (2000)

15. Rakhlin, A., Davydow, A., Nikolenko, S.: Land cover classification from satellite imagery with U-Net and Lovasz-softmax loss. In: Proceedings of the IEEE Conference on Computer Vision and Pattern Recognition (CVPR) Workshops, June 2018

16. Ribeiro, C.M.N.: Classificação do uso e cobertura do solo do estado de goiás empregando redes neurais artificiais (2019)

17. Ronneberger, O., Fischer, P., Brox, T.: U-Net: convolutional networks for biomedical image segmentation. In: Navab, N., Hornegger, J., Wells, W.M., Frangi, A.F. (eds.) MICCAI 2015. LNCS, vol. 9351, pp. 234–241. Springer, Cham (2015). https://doi.org/10.1007/978-3-319-24574-4_28

18. da Silva, M., Cesario, A.V., Cavalcanti, I.R.: Relevância do agronegócio para a economia brasileira atual. Apresentado em X ENCONTRO DE INICIAÇÃO À DOCÊNCIA, UNIVERSIDADE FEDERAL DA PARAÍBA. Recuperado de (2013). http://www.prac.ufpb.br/anais/IXEnex/iniciacao/documentos/anais/8.TRABALHO/8CCSADAMT01.pdf

19. Sumbul, G., Charfuelan, M., Demir, B., Markl, V.: BigEarthNet: a large-scale benchmark archive for remote sensing image understanding. In: IGARSS 2019–2019 IEEE International Geoscience and Remote Sensing Symposium, pp. 5901–5904. IEEE (2019)

20. Ulmas, P., Liiv, I.: Segmentation of satellite imagery using u-net models for land cover classification. arXiv preprint arXiv:2003.02899 (2020)

21. Wada, K.: labelme: Image Polygonal Annotation with Python. https://github.com/wkentaro/labelme (2016)

22. Woebbecke, D.M., Meyer, G.E., Von Bargen, K., Mortensen, D.A.: Color indices for weed identification under various soil, residue, and lighting conditions. Trans. ASAE **38**(1), 259–269 (1995)

23. Yang, L., Shami, A.: On hyperparameter optimization of machine learning algorithms: theory and practice. Neurocomputing **415**, 295–316 (2020)

24. Yang, M.D., Tseng, H.H., Hsu, Y.C., Tsai, H.P.: Semantic segmentation using deep learning with vegetation indices for rice lodging identification in multi-date UAV visible images. Remote Sens. **12**(4), 633 (2020)

25. Zhu, N., et al.: Deep learning for smart agriculture: concepts, tools, applications, and opportunities. Int. J. Agric. Biol. Eng. **11**(4), 32–44 (2018)

An Explainable Medical Imaging Framework for Modality Classifications Trained Using Small Datasets

Francesca Trenta[1]([✉])[ID], Sebastiano Battiato[1][ID], and Daniele Ravì[2,3,4][ID]

[1] IPLAB, University of Catania, Catania 95125, Italy
francesca.trenta@unict.it, battiato@dmi.unict.it
[2] University of Hertfordshire, Hatfield, UK
[3] University College London, London, UK
d.ravi@ucl.ac.uk
[4] Queen Square Analytics, London, UK

Abstract. With the huge expansion of artificial intelligence in medical imaging, many clinical warehouses, medical centres and research communities, have organized patients' data in well-structured datasets. These datasets are one of the key elements to train AI-enabled solutions. Additionally, the value of such datasets depends on the quality of the underlying data. To maintain the desired high-quality standard, these datasets are actively cleaned and continuously expanded. This labelling process is time-consuming and requires clinical expertise even when a simple classification task must be performed. Therefore, in this work, we propose to tackle this problem by developing a new pipeline for the modality classification of medical images. Our pipeline has the purpose to provide an initial step in organizing a large collection of data and grouping them by modality, thus reducing the involvement of costly human raters. In our experiments, we consider 4 popular deep neural networks as the core engine of the proposed system. The results show that when limited datasets are available simpler pre-trained networks achieved better results than more complex and sophisticated architectures. We demonstrate this by comparing the considered networks on the ADNI dataset and by exploiting explainable AI techniques that help us to understand our hypothesis. Still today, many medical imaging studies make use of limited datasets, therefore we believe that our contribution is particularly relevant to drive future developments of new medical imaging technologies when limited data are available.

Keywords: Modality classification · Quality control · Explainable artificial intelligence

1 Introduction

Medical imaging refers to a variety of different imaging techniques used to capture tissue information and used to facilitate the diagnosis, monitoring and

S. Sclaroff et al. (Eds.): ICIAP 2022, LNCS 13231, pp. 358–367, 2022.
https://doi.org/10.1007/978-3-031-06427-2_30

treatment of health conditions. The most popular techniques include ultra-sound, Magnetic Resonance Imaging (MRI), Positron emission tomography (PET), Computed tomography (CT), endoscopy, endomicroscopy, X-rays, hyper-spectral imaging, etc. Often, each of these techniques has many different modalities aimed to capture a specific property of the tissue (i.e. T1-weighted, T2-weighted, diffusion-weighted MRI, etc.). With the continued growth of these image modalities, the development of automatic tools for organizing medical images according to specific characteristics has become an urgent need. Indeed, performing these tasks without automation is time-consuming and expensive. In this regard, Machine Learning (ML) can automate image analysis more reliably. Traditional ML methods are based on the selection of hand-made features and require prior domain knowledge. More recently, the growing interest in Deep Learning (DL) technologies has led to the development of several approaches to accomplish these tasks in a much more efficient and data-driven way. In this work, we evaluated the performance obtained by some of the most popular DL frameworks developed for image classification. In particular, to reflect a real context in medical imaging, we train our system using a limited dataset and we draw conclusions about the effectiveness of each of the approaches on the considered modality classification task. In our system we use four deep models: ResNet-101 [10], VGGNet-16 [18], Normalizer-Free Networks (or NFNets) [2], and Vision Transformer [3]. First, we trained all the models from scratch on the considered dataset. Then, we performed the training by applying a Transfer Learning (TL) technique on each of these networks. Our results show that sophisticated networks struggle to perform well on these limited datasets, which is something confirmed also by other recent works [7,14]. Therefore we conclude that these advanced networks (e.g., Vision Transformer) work well in the context where millions of training data are available (i.e. ImageNet), but are not always a good solution in medical imaging where such large datasets are not accessible, and networks with fewer parameters seem to be a better choice.

To better understand why this happens, we provided an explainable AI analysis, which takes into account the number of parameters of each network and exploits the visual explanations of the activation functions obtained when the frameworks are trained on the proposed task.

Our work is intended to be an insight into the ability of the well-known neural networks to perform modality classification tasks for medical images.

We believe that our analysis is particularly important to drive future development for the classification of medical images or modalities aimed to reduce human workload and to help create an automatic tool to organize medical imaging datasets (i.e. during quality control).

2   Related Works

Several solutions have been proposed over the last few years to develop an effective system for modality classification of medical images [19]. In particular, in the literature, we found two main classes of approaches which are based on: (i) hand-crafted features and (ii) fully data-driven DL.

Approaches Based on Hand-Crafted Features. Several works proposed to use handcrafted features [9,19]. Khachane et al. [12], for example, developed a framework to extract a set of texture features in order to classify them using SVM, KNN and a fuzzy rule-based technique. In particular, these classifiers were combined to identify five types of medical modalities: CT-scan, X-ray, Ultrasound, MRI, and Microscopic images.

In [8], the authors investigated the modality classification by using both visual features (e.g., intensity histogram, and variance histogram) and textual features, such as the binary histogram of some predefined vocabulary words from image captions. Results proved that the fusion of both input data has led to good results, although the proposed approach struggles to classify accurately some modalities, such as CT and PET.

Approaches Based on Deep Learning. With the advance of DL, [19] proposed to compare some of the handcrafted features against DL networks trained with a multi-label strategy. The results show the superiority of these latest models which largely outperformed the other conventional methods. On a similar research line, Chiang et al. [5] proposed a modality classification technique using Convolutional Neural Networks (CNNs) designed to discriminate different modalities, including CT and MRI. In [11] the authors implemented a modality classification system consisting of a neural network that uses histogram and texture features for classifying images. Despite achieving remarkable results on the *CISMeF* database, the model often misclassifies MRI and CT scans.

Cheng et al. [4] developed instead a cascaded CNN to extract discriminative features from the ADNI dataset and aimed to distinguish MRI from PET images. The overall results indicate that the application of cascaded CNN further improves the classification capability of CNN. Yu et al. [22] proposed instead a DNN model to perform modality classification on the ImageCLEF dataset. Although this dataset contains 300,000 images, the authors only used 2,901 scans for training and 2,582 for testing. Results show that the proposed model achieved an accuracy of 70% confirming that the limited use of data can affect the resulting performance. Schouten et al. [16] investigated the performance of a CNN in both binary and multi-class problems taking into account the impact of the training data size. Specifically, the authors considered 200 images as training data and 25 as validation and test data. Results indicate good performance however, the authors observed some misclassifications due to very few training samples for specific classes. Experiments also suggest that increasing the training set improves class separation, leading to better performances. In [1], the authors also discussed the importance of training size. They demonstrated that increasing the number of training data improved the performance of their DL model. Specifically, they observed that the Dice score increased notably when more cases were included in the training data (from 0.858 at 160 cases to 0.867 at 320 cases). Finally, the authors of [20] investigated the performance of a CNN trained from scratch compared with a pre-trained one for the modality classification task. They presented consistent results which indicate that fine-tuned CNN should be preferred in the clinical domain for solving advanced and complex tasks. In

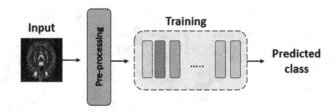

Fig. 1. The overall pipeline.

particular, they show that the deeply fine-tuned CNN models outperform CNN trained from scratch even when limited training data are available.

In conclusion one of the main challenges encountered in a clinical context concerns the lack of large-high-quality datasets. This is required since DL methods base their potentials on exploiting a massive amount of data [6] and large datasets have become a crucial point to building powerful data-driven solutions. However, the main barrier to collecting these quality data in this domain is the labelling process which is usually performed by human experts. This is time-consuming, subject to errors and suffers from inter-rater reliability typical of human experts. In the last few years, researchers have tried to find alternative solutions to deal with these problems including using federated learning, interactive reporting, and synoptic reporting [21], but still today these solutions cannot be applied easily.

3   Dataset

In this study, we used images from the ADNI study[1]. In particular, we considered scans from patients labelled as Cognitively Normal (CN), Mild Cognitive Impairment (MCI), and Alzheimer's disease (AD) and we selected the following image modalities: Diffusion Tensor Imaging (DTI), Functional Magnetic Resonance Imaging (fMRI), MRI T1, MRI T2 and, PET. To conduct our experiments, we subdivided the dataset into a training set of 8,500 slices (1,700 slices for each class) and a test set of 1,320 slices (300 images for each class, except the PET class which contains only 120 cases).

4   The Proposed Approach

Modality classification was performed by using four DL models: VGGNet-16, ResNet-101, NFNets model and Vision Transformer Net.

We performed experiments considering two configurations: (1) training the models from scratch, and (2) training the models with pre-trained weights from ImageNet through the TL technique. In Fig. 1, we summarize our pipeline.

---

[1] http://adni.loni.usc.edu/.

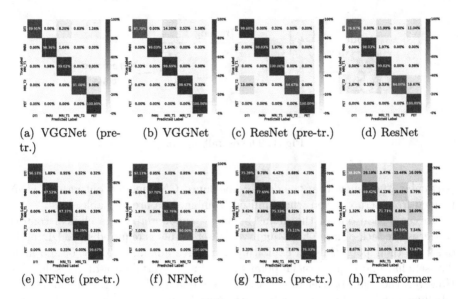

Fig. 2. Confusion matrices for our deep models. (a) Pre-trained VGGNet (b) VGGNet trained from scratch (c) Pre-trained ResNet (d) ResNet trained from scratch (e) Pre-trained NFNet (f) NFNet trained from scratch (g) Pre-trained Vision Transformer (h) Vision Transformer trained from scratch

4.1   Pre-processing

The pre-processing stage consists of the following step-by-step procedure applied to the raw image data. First of all, we remove 5% of intensities outliers from the entire MRI. Then, we extracted the central slices for each MRI which is intended to contain the most relevant information. Finally, each slice is resized to 224×224 pixels. This procedure creates a set of normalized slices and reduces irrelevant variations in the data. We discarded slices where we observed pre-processing failures.

4.2   Transfer Learning

In several applications, training a deep model from scratch (i.e., starting from randomly initialized weights) represents a challenging task, especially whether the dataset includes a small number of samples. To overcome this issue, the *transfer learning* strategy has been successfully proposed. Generally, transfer learning consists of using the weights of a network already trained on a larger dataset (e.g., ImageNet).

In our work, we used the transfer learning technique. To do so, we froze the models' parameters trained using ImageNet and added some new layers as follows: for the VGGNet model, we firstly implemented a linear transformation from high-dim features (4096) to low-dim features (2048) and then from 2048 to 512 features. All the hidden layers use ReLU as its activation function. Then,

Table 1. Table reporting the number of parameters, the training time and the inference time for used DL architectures. We report the speed computed over 100 iterations.

Method	No. parameters (M)	Training time (ms)	Inference time (ms)
ResNet-101	~44.5	1497×10^4	0.6×10^4
VGGNet-16	~138	1000×10^4	0.5×10^4
NFNet	~71	800×10^4	9×10^4
Vision Transformer	~86	472×10^4	0.4×10^4

the last feature vector is passed to the softmax layer to perform the final classification. With regard to ResNet101, the model consists of 101 layers. In our model, we froze the convolution layers from Layer 0 to Layer 4. Similarly to VGGNet-16, we defined a linear transformation from high-dim features (2048) to low-dim classes (5), followed by a Dropout layer and a ReLU layer. Finally, for both the network NFNet and Vision Transformer we just modified the last layer so that this has the same number of outputs as our classes (5).

4.3 Implementation Details

All experiments were conducted using PyTorch [15] over a cluster of GPU NVIDIA T4. A random horizontal flipping is also used for data augmentation of the training data. We applied Stochastic Gradient Descent (SGD) [13] optimizer. We trained all the models for 100 epochs using a batch size and the learning rate set to 16 and 0.0001, respectively.

4.4 Results and Discussion

Results show that the VGGNet and the ResNet networks outperform other neural networks in terms of accuracy. In particular, the pre-trained VGGNet-16 network (Fig. 2a) has been proved to be an outstanding performer, although it has more difficulty in identifying DTI and T2-MRI. Notable is that the overall accuracy achieved by the VGGNet model trained from scratch (Fig. 2b) is lower than that of its pre-trained counterpart (except for the T2 MRI). As shown in Fig. 2, the pre-trained ResNet-101 network (Fig. 2c) correctly classifies the fMRI, MRI_T1, and PET images, whereas it tends to identify the MRI_T2 images as DTI images. We can notice that both VGGNet and ResNet models achieved high accuracy also when they are trained from scratch, confirming their effectiveness in modality classification. A similar conclusion can be drawn for the ResNet model. The pre-trained model (Fig. 2c) achieved impressive results for all classes. Contrary to the pre-trained ResNet, the model trained from scratch tends to misclassify the class MRI_T2. Both models yielded a very high accuracy for the classification of the PET modality. With regard to NFNet network, it reached a high accuracy in both cases, however, the model trained from scratch fails to outperform the other networks. Specifically, results suggest that NFNet

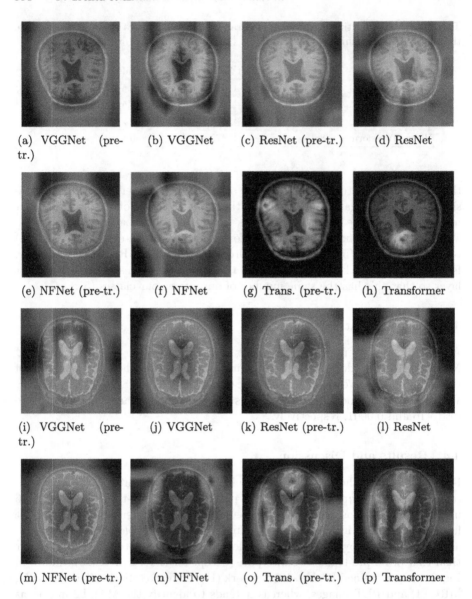

(a) VGGNet (pre-tr.)   (b) VGGNet   (c) ResNet (pre-tr.)   (d) ResNet

(e) NFNet (pre-tr.)   (f) NFNet   (g) Trans. (pre-tr.)   (h) Transformer

(i) VGGNet (pre-tr.)   (j) VGGNet   (k) ResNet (pre-tr.)   (l) ResNet

(m) NFNet (pre-tr.)   (n) NFNet   (o) Trans. (pre-tr.)   (p) Transformer

Fig. 3. The GRAD-CAM visualizations. (a) Pre-trained VGGNet, (b) VGGNet from scratch. (c) Pre-trained ResNet, (d) ResNet from scratch. (e) Pre-trained NFNets (f) NFNets from scratch (g) Pre-trained Vision Transformer (h) Vision Transformer from scratch (i) Pre-trained VGGNet, (j) VGGNet from scratch. (k) Pre-trained ResNet, (l) ResNet from scratch. (m) Pre-trained NFNets (n) NFNets from scratch (o) Pre-trained Vision Transformer (p) Vision Transformer from scratch

model is able to classify adequately T1 and PET scans and tends to misclassify DTI and T2 (Fig. 2). In Fig. 2, the confusion matrices related to NFNets are depicted. Regarding the Visual Transformer networks [3] results show that it achieved poor results in both cases (Fig. 2) yielding the worst performance. However, a general improvement can be observed when they are trained by applying a transfer learning technique. This performance could be due to the fact that the Vision Transformers do not work very well on limited datasets confirming that simpler approaches are more suitable for solving advanced tasks under challenging conditions. An exhaustive model visualization supports the comprehension of the model's predictive decisions and increases user trust in these deep models. Therefore, to further understand the behaviour of these networks and the obtained results we make use of GRAD-CAM [17] for deepening the gradient information of each deep model. Figure 3 reports the obtained GRAD-CAM visualizations. From our results, it is clear that VGGNet and ResNet models are class-discriminative and able to highlight the object of interest (i.e. the brain). More specifically, the GRAD-CAM algorithm confirms that the pre-trained VGGNet-16 network bases its classification on large regions of the brain whereas the no pre-trained counterpart focuses more on the cortex that is a very personalized feature (Fig. 3). However, we can note that pre-trained VGGNet model struggles to focus on MRI_T2 objects (Fig. 3i) as its counterpart trained from scratch (Fig. 3j). Although the ResNet-101 model is capable of locating the overall area of the brain, results indicate that it is not as accurate as the VGGNet model (Fig. 3c, d, k, l). Similarly, the NFNet network focuses more on the frontal lobes of the brain. With regard to the scans for the MRI_T2 class, the GRAD-CAM algorithm show that the entire brain area is used for image classification by the NFNet model (Fig. 3m). On the opposite side, the Vision Transformer network performs much worse than the other three methods since it bases its classification on the small local regions of the brain that are not sufficient to classify the modality (Fig. 3g, h, o, p).

To provide a further comparison between the deep models, we reported the training and inference time, and the complexity (measured in terms of the number of parameters) for each model. As expected more parameters the deep model has, the longer the training time is. However, the first observation from Table 1 is that the network structures including fully connected and convolutional layers (e.g., VGGNet and ResNet) cause the increase of training time. On the other hand, The NFNet and Vision transformer training proved to be faster than other models, despite their large number of parameters. The reason behind it is that NFNet networks don't make use of batch normalization avoiding in this way expensive computational power. Similarly, the Vision Transformer require less computational time than other models, since it consists of a full self-attention mechanism without involving a CNN structure.

5   Conclusions

In this work, we proposed an explainable AI framework that takes into account recent DL techniques aimed to perform modality classification. We designed a comparative study for these deep neural network models, showing their performances and outlining their strengths and drawbacks. More specifically, we compared convolutional-based networks (VGGNet and ResNet) with more advanced networks such as NFNet and Vision Transformer. Our results showed that the latest networks encountered several issues at classifying images when a limited dataset is used. On the other hand, deep networks such as ResNet are able to provide better results. Our analysis that is also based on the use of an explainable AI technique called GRAD-CAM, is particularly relevant to designing future deep learning models specifically for image modality classifications. In our future works, we aim to extend our pipeline and analysis to develop a more complex image retrieval system aimed to organize large datasets not only by image modalities but also by other characteristics including image quality (useful to perform quality control and remove samples with artefacts), type of organs, diseases status and even by the patient similarities.

References

1. Bardis, M., et al.: Deep learning with limited data: organ segmentation performance by U-Net. Electronics 9(8), 1199 (2020)
2. Brock, A., De, S., Smith, S.L., Simonyan, K.: High-performance large-scale image recognition without normalization. arXiv preprint arXiv:2102.06171 (2021)
3. Chen, X., Hsieh, C.J., Gong, B.: When vision transformers outperform ResNets without pretraining or strong data augmentations. arXiv preprint arXiv:2106.01548 (2021)
4. Cheng, D., Liu, M.: CNNs based multi-modality classification for ad diagnosis. In: 2017 10th International Congress on Image and Signal Processing, Biomedical Engineering and Informatics (CISP-BMEI), pp. 1–5. IEEE (2017)
5. Chiang, C.H., Weng, C.L., Chiu, H.W.: Automatic classification of medical image modality and anatomical location using convolutional neural network. PLoS ONE 16(6), e0253205 (2021)
6. Cho, J., Lee, K., Shin, E., Choy, G., Do, S.: How much data is needed to train a medical image deep learning system to achieve necessary high accuracy? arXiv preprint arXiv:1511.06348 (2015)
7. Dosovitskiy, A., et al.: An image is worth 16×16 words: transformers for image recognition at scale. arXiv preprint arXiv:2010.11929 (2020)
8. Han, X.H., Chen, Y.W.: Biomedical imaging modality classification using combined visual features and textual terms. Int. J. Biomed. Imaging 2011 (2011)
9. Hassan, M., Ali, S., Alquhayz, H., Safdar, K.: Developing intelligent medical image modality classification system using deep transfer learning and LDA. Sci. Rep. 10(1), 1–14 (2020)
10. He, K., Zhang, X., Ren, S., Sun, J.: Deep residual learning for image recognition. In: Proceedings of the IEEE Conference on Computer Vision and Pattern Recognition, pp. 770–778 (2016)
11. Kalpathy-Cramer, J., Hersh, W., et al.: Automatic image modality based classification and annotation to improve medical image retrieval. In: Medinfo 2007: Proceedings of the 12th World Congress on Health (Medical) Informatics; Building Sustainable Health Systems, p. 1334. IOS Press (2007)

12. Khachane, M.Y., Ramteke, R.J.: Modality based medical image classification. In: Shetty, N.R., Prasad, N.H., Nalini, N. (eds.) Emerging Research in Computing, Information, Communication and Applications, pp. 597–606. Springer, Singapore (2016). https://doi.org/10.1007/978-981-10-0287-8_55

13. Kiefer, J., Wolfowitz, J., et al.: Stochastic estimation of the maximum of a regression function. Ann. Math. Stat. **23**(3), 462–466 (1952)

14. Li, S., Chen, X., He, D., Hsieh, C.J.: Can vision transformers perform convolution? arXiv preprint arXiv:2111.01353 (2021)

15. Paszke, A., et al.: Automatic differentiation in PyTorch (2017)

16. Schouten, J.P., Matek, C., Jacobs, L.F., Buck, M.C., Bošnački, D., Marr, C.: Tens of images can suffice to train neural networks for malignant leukocyte detection. Sci. Rep. **11**(1), 1–8 (2021)

17. Selvaraju, R.R., Das, A., Vedantam, R., Cogswell, M., Parikh, D., Batra, D.: Grad-CAM: why did you say that? arXiv preprint arXiv:1611.07450 (2016)

18. Simonyan, K., Zisserman, A.: Very deep convolutional networks for large-scale image recognition. arXiv preprint arXiv:1409.1556 (2014)

19. Singh, S., Ho-Shon, K., Karimi, S., Hamey, L.: Modality classification and concept detection in medical images using deep transfer learning. In: 2018 International conference on image and vision computing New Zealand (IVCNZ), pp. 1–9. IEEE (2018)

20. Tajbakhsh, N., et al.: Convolutional neural networks for medical image analysis: full training or fine tuning? IEEE Trans. Med. Imaging **35**(5), 1299–1312 (2016)

21. Willemink, M.J., et al.: Preparing medical imaging data for machine learning. Radiology **295**(1), 4–15 (2020)

22. Yu, Y., et al.: Modality classification for medical images using multiple deep convolutional neural networks. J. Comput. Inf. Syst. **11**(15), 5403–5413 (2015)

Fusion of Periocular Deep Features in a Dual-Input CNN for Biometric Recognition

Andrea Abate⬤, Lucia Cimmino$^{(\boxtimes)}$⬤, Michele Nappi⬤, and Fabio Narducci⬤

Department of Computer Science, University of Salerno, 84084 Salerno, Italy
{abate,lcimmino,mnappi,fnarducci}@unisa.it

Abstract. Periocular recognition has attracted attention in recent times. The advent of the COVID-19 pandemic and the consequent obligation to wear facial masks made face recognition problematic due to the important occlusion of the lower part of the face. In this work, a dual-input Neural Network architecture is proposed. The structure is a Siamese-like model, with two identical parallel streams (called base models) that process the two inputs separately. The input is represented by RGB images of the right eye and the left eye belonging to the same subject. The outputs of the two base models are merged through a fusion layer. The aim is to investigate how deep feature aggregation affects periocular recognition. The experimentation is performed on the Masked Face Recognition Database (M²FRED) which includes videos of 46 participants with and without masks. Three different fusion layers are applied to understand which type of merging technique is most suitable for data aggregation. Experimental results show promising performance for almost all experimental configurations with a worst-case accuracy of 90% and a best-case accuracy of 97%.

Keywords: Periocular recognition · Biometric recognition · Siamese network

1 Introduction

In recent years, periocular recognition has attracted more and more attention. With the advent of the COVID-19 pandemic, several worldwide Countries introduced the obligation to wear face masks to protect people from infection. Facial recognition has become problematic due to the facial masks: the occlusion of the lower part of the face leads to the loss of more than half of the biometric information residing in the human face. As a result, the failure rate of the facial recognition systems is increased. Considering a typical facial recognition technology implemented on mobile devices, it is necessary to remove the facial mask to be correctly identified. A typical facial recognition system performs four main steps: face detection, face alignment, features extraction, and matching. Face detection and alignment phases are pre-processing phases that aim to segment

© The Author(s), under exclusive license to Springer Nature Switzerland AG 2022
S. Sclaroff et al. (Eds.): ICIAP 2022, LNCS 13231, pp. 368–378, 2022.
https://doi.org/10.1007/978-3-031-06427-2_31

and normalize (in a matter of size and pose, for example) the region of the image counting the face. Once the face has been detected and normalized, a series of distinctive features are extracted which allow the subjects to be effectively discriminated on the basis of specific geometric or photometric variations. The extracted characteristics are collected in a data structure (feature vector) capable of representing a specific subject. Finally, in the matching phase, the feature vector is compared with those present in the database and outputs the identity of the face. Recently, the extensive use of the face as a biometrics trait, also in several daily applications, made it necessary to adapt facial recognition systems to this new way of living. This adaptation can be realized by allowing standard facial recognition systems to identify subjects based on the periocular area, which is the only area not covered by the masks. As shown in Fig. 1 the periocular region refers to the area around the eyes, which includes features such as eyes, eyebrows, eyelashes, eye shape, eyelids, skin color, etc.

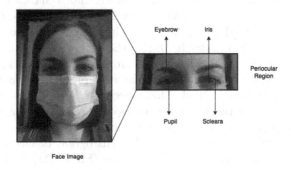

Fig. 1. The periocular region of a face.

In addition to being resilient to mask-induced occlusion, the periocular region is a reasonable trade-off between iris and face-based biometric verification systems, as the first one needs a high level of user collaboration. Obviously, the most important drawback of periocular recognition is that it is less accurate than recognition of the entire face. In fact, the accuracy of face recognition systems has reached very high levels, making it one of the most widely used biometrics in terms of accuracy and user acceptance.

In this work, a novel dual-input convolutional neural network structure is proposed, in order to efficiently recognize the subject using the solely periocular region. The structure is a Siamese-like network, with two identical parallel streams, called *base models*, that process the two inputs separately. The two inputs are represented by RGB images of the right eye and the left eye belonging to the same subject. The outputs of the two base models are merged through a fusion layer. Even if the two eyes come from the same image pose variation and irregular illumination can rise differences between the left and the right part of the periocular region, negatively affecting the identification process. For this reason, the experiments are conducted on an "in-the-wild" dataset, which is characterized by uncontrolled acquisition in different conditions of illumination and pose. The paper is organized as follows: in Sect. 2, related works are examined, Sect. 3 presents the proposed approach, and in Sect. 4 experimental settings and discussion is provided. Section 5 draws the conclusions of this work.

2   Related Works

A first attempt at the use of the periocular region as a biometric trait was presented by Park et al. in 2009 [9]. They achieve 77% of accuracy applying global and local matching. Using 1136 gallery periocular images acquired from 568 different participants, one year later, the authors in [8] improved the accuracy to 87.32% through the combination of three different matching. In soft biometric categorization, the periocular region is very important. This biometric trait's strength is that it takes almost no user cooperation, which makes the periocular area suitable for security, surveillance, and scenarios where faces are partially occluded. As a part of soft biometrics, over the years, the periocular region has also been used in combination with other biometric characteristics to improve the recognition rate. In [10,16] the periocular region was employed to improve face recognition, other works propose a combination with the iris trait [4,13]. A crucial part of the periocular recognition systems is represented by the feature extraction method. This aspect has been extensively studied in the literature. There are two types of existing features: global features, obtained from the entire periocular image, and local features, which are extracted from a group of discrete points (key points) [1]. Histogram of Oriented Gradients (HOG) [12], Scale-Invariant Feature Transform (SIFT) [11], and Local Binary Patterns (LBP) [6] are the most commonly adopted feature extraction techniques. These techniques can be used individually or in combination, [3,5,17]. The advent of Deep Learning and the development of efficient and sophisticated Neural Networks architectures has improved the performances of periocular biometric systems, especially through identifying optimal features. The Convolutional Neural Networks (CNNs), which are particularly suitable for image classification, are employed in the majority of these techniques. Zhao and Kumar [18] presented a Semantic-assisted Convolutional Neural Network (SCNN). The idea was to add an extra CNN branch to an existing CNN. The additional branch is trained with semantic information such as ethnicity and gender. In [2] Alonso-Fernandez et al. propose a multi-algorithmic approach with a fusion scheme based on linear logistic regression. An RGB-OCLBCP dual-stream CNN for periocular recognition in the wild is proposed in [15]. The model accepts an RGB image of the eye and an Orthogonal Combination-Local Binary Coded Pattern (OCLBCP) descriptor. The features are aggregated through the use of two distinct late-fusion layers. In this direction another research that aims to improve periocular recognition in the wild context is presented in [7] by Tiong et al. The authors extend the Label Smoothing Regularization (LSR) technique in convolutional neural networks cross-entropy loss, by implementing L2SR by considering learned to smoothen prediction distribution instead of predefined

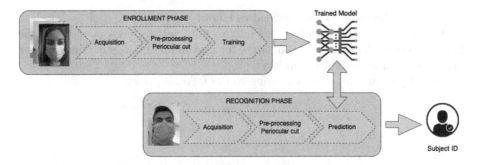

Fig. 2. The workflow of the proposed approach.

uniform distribution. They demonstrate that L2SR outperforms LSR improving the class generalization. A multi-modal facial biometrics recognition system that fuses face with periocular biometric, using dual-stream convolutional neural networks is presented in [14]. This network, in particular, is made up of two progressive components that use different fusion techniques to combine RGB data and texture descriptors for multi-modal facial biometrics. These last novel approaches have encouraged the investigation of the performances achievable using a dual-input CNN model that works on two RGB input images addressed in this work. The images are referred to the left and the right eyes of the same periocular region.

3   Proposed Approach

In this paper, a dual-input CNN that works on periocular RGB of the left eye and right eye images is proposed. An overall view of the workflow of the proposed approach is depicted in Fig. 2. It is a Siamese-like CNN model, consisting of two parallel artificial neural networks, each of which can learn the hidden representation of an input vector. Both neural networks are feed-forward models that use error back-propagation during training. They work in parallel and compare their outputs at the end, often using a distance function (i.e., Euclidean distance, Cosine distance, etc.). The output of a Siamese network is the distance between the projected representations of the two input vectors in order to express their semantic similarity or dissimilarity. Therefore, the most common application of Siamese networks is to determine the differences between two sets of characteristics, how much these characteristics are similar.

In this study, information aggregation is explored: the objective is to determine the impact of the combination of the features on periocular recognition. This aggregation is represented by the type of fusion layer employed in the model, in this case, three different fusion layers are considered: *Maximum*, *Average*, and *Concatenation* layer.

Formally, fusion layers are defined as follow:

Given $A = Flatten_1$ and $B = Flatten_2$ both feature vectors of size k the fusion layers are feature vectors of size k for maximum and average fusion layers and $2k$ for concatenation fusion layer.

Such layers are obtained according to the following rules:

$$f_{max}(i) = max[A(i), B(i)], i = 1, \cdots, k \tag{1}$$

$$f_{avg}(i) = avg[A(i), B(i)], i = 1, \cdots, k \tag{2}$$

$$f_{concat}(i) = concat[A(i), B(i)], i = 1, \cdots, 2k \tag{3}$$

As reported in Table 1, the proposed model consists of two identical CNN base models, both composed of four Convolutional layers (Conv), three Max-pooling 2D layers (MaxPool), and a final Flatten layer to prepare the output for the fusion. The two outputs of the convolutional process, meaning the two convolutional maps representing the deep features, are combined through the Fusion layer, which can be one of the functions listed above (Eqs. 1, 2 and 3); at the end three Dense layers performs the true classification task by which determining the identity of the subjects.

Table 1. Network architecture and layers configuration.

Network layers		Configurations
		M^{2}FRED
$Input_1$	$Input_2$	(None, 80, 80, 3)
$Conv1_1$	$Conv1_2$	64@80 × 80; $kernel$: 3 × 3
$MaxPool1_1$	$MaxPool1_2$	@78 × 78; $size$: 2 × 2
$Conv2_1$	$Conv2_2$	32@39 × 398; $kernel$: 3 × 3
$MaxPool2_1$	$MaxPool2_2$	@37 × 37; $size$: 2 × 2
$Conv3_1$	$Conv3_2$	64@80 × 80; $kernel$: 3 × 3
$Conv4_1$	$Conv4_2$	64@16 × 16; $kernel$: 3 × 3
$MaxPool3_1$	$MaxPool3_2$	@14 × 14; $size$: 2 × 2
$Flatten_1$	$Flatten_2$	$(None, 3136)$
$Fusion$		$(None, 3136)_{max}$
		$(None, 3136)_{avg}$
		$(None, 6272)_{con}$
$Dense$		$(None, 1000)$
$Dense$		$(None, 500)$
$Dense$		$(None, 100)$
$Dense$		$(None, 46)$

4   Experimentation

An overview of the dataset used, experimental findings and configuration is provided in this section. All testing sessions share the same experimental settings: a batch size of 64 and 20 epochs for the training process. The data was divided in a typical 70:30 ratio.

4.1   Dataset

The experiments are carried out on $\mathbf{M^2FRED}$ dataset. As briefly mentioned in Sect. 1, M^2FRED is an *in-the-wild* dataset characterized by non-controlled acquisitions made by users using personal mobile devices. This makes the dataset even more challenging, being also cross-platform.

The Mobile Masked Face REcognition Database is a frontal face multimodal database including digital videos of 46 people. The subjects involved in the acquisition process were asked to self-record their faces using a personal mobile device according to a protocol defining precise acquisition guidelines. According to the procedure, each person is recorded in different and time-scheduled sessions over a period of time that is not limited to a single day. The acquisitions were divided into 4 sessions, each of which was held on a different day, with interleaved indoor and outdoor acquisitions, in order to have different environmental conditions. The acquired subjects are recorded both wearing and not wearing the facial mask, which is a distinctive feature of this dataset.

Fig. 3. Examples of acquisitions with and without facial masks from M^2FRED dataset.

4.2   Data Pre-processing

Prior to being input into the neural model, the data from the dataset in Sect. 4.1 have been pre-processed. Considering the composition of M^2FRED, which is a facial dataset, a procedure for cutting the periocular area was applied. Since it

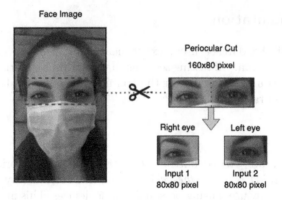

Fig. 4. Data pre-processing: from the facial image the periocular region is extracted and then split into solely left and right eyes.

includes video acquisitions, a collection of ten frames were sampled from the sixteen videos belonging to each subject, then 30 frames were randomly selected to compose the final image dataset. The entire periocular area was cut out from the face, resulting in an image of size 160×80. Following, there was applied a split into either the right and left eye. The final images which represent the input of the two base models have a size of 80×80 pixels. In Fig. 4 an example of pre-processed data is shown.

4.3   Results and Discussion

Table 2 presents the overall results achieved in this work, based on the experimental settings and applied model described in previous sections. Accuracy, precision, recall, and F1-score metrics are collected for each fusion layer used in the model. The number of samples and the learning rate is the same for every experimental session, in order to have a fair comparison on the result set. The values of the model parameters are reported in Table 2. The number of epochs was chosen empirically, applying early stopping technique during the training phase. To better highlight the motivations behind the adoption of a fusion model the results relating to the application of the basic model on the single left eye and right eye, are reported in Table 2. The performances achieved by applying the base model to the left and right eye are substantially the same, with an accuracy, recall and F1-score of 87% and a precision of 89% for both left and right eye. A **minimum difference can be seen in** Fig. 5b in which the left eye shows a slightly AUC score of 0.89 with respect to the right eye which achieve an AUC score of 0.88. Also, the Equal Error Rate values for basic models are homogeneous and congruent with the ROC curves results. It can be observed an EER of 0.14 for the right eye (Fig. 7b) and EER of 0.13 for left eye (Fig. 7a).

Focusing on the experiments involving the fusion layers, it can be observed that the Concatenation fusion layer (see Eq. 3) is the best performing, with an accuracy of 97%, an AUC score of 0.97 (see Fig. 5a) and an EER of 0.08 (see Fig. 6c). The Average fusion layer (see Eq. 2) behaves identically to the Concatenation layer, showing just a little decrease of 2% for all the metrics (accuracy, precision, recall, and F1-score equal to 0.95). The ROC curve referred to the Average fusion layer experiment is depicted in Fig. 5a. The worst performing fusion layer is the Maximum layer (see Eq. 1) with an accuracy of 90% and, a precision, recall, and F1-score of 92%. In this case, there is a more substantial drop in performance compared to the previous scenario (Average VS Concatenation). A similar decrease can be observed also in the ROC curve and AUC score shown in Fig. 5a. Although the ROC curves of the fusion model exhibit a general improvement over the basic models, the EER associated with the application of the maximum and average fusion layers is slightly higher, at 0.15 (Fig. 6a) and 0.14 (Fig. 6b), respectively. It is evident at this point, that the results produced by using the fusion model on both eyes are superior to the results obtained by using the single base model on either the left or right eye. The fusion appears to be an improvement in all cases, increasing performance by 10% in the best scenario and 3% in the worst case. Periocular recognition is surely positively affected by the fusion model.

Table 2. Experimental results and comparative analysis of the base model on separate eyes and by adding different fusion layers.

	Metrics				Model parameters				
	Accuracy	Precision	Recall	F1-score	N_samples	Epochs	lr	Fuse layer	Eye
M²FRED	0.87	0.89	0.87	0.86	30 × 46	20	$1.00e^{-04}$	None	Right
	0.87	0.89	0.87	0.86	30 × 46	20	$1.00e^{-04}$	None	Left
	0.90	0.92	0.92	0.92	30 × 46	20	$1.00e^{-04}$	Maximum	Both
	0.95	0.95	0.95	0.95	30 × 46	20	$1.00e^{-04}$	Average	Both
	0.97	0.97	0.97	0.97	30 × 46	20	$1.00e^{-04}$	Concatenation	Both

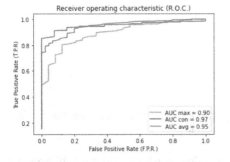
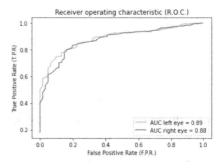

(a) Application of three different fuse layers.   (b) No fusion layer applied.

Fig. 5. ROC curves and AUC scores by using different fusion layers (left) and the ROC curves of the base models without fusion on right/left eyes only (right).

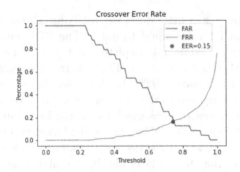

(a) EER maximum fusion layer applied.

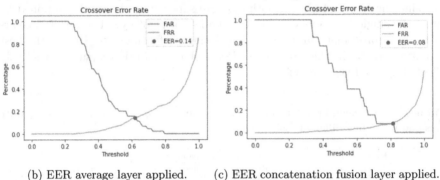

(b) EER average layer applied.   (c) EER concatenation fusion layer applied.

Fig. 6. EER of the fusion model with different fusion layers applied.

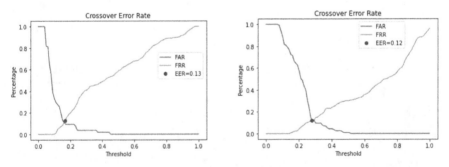

(a) EER left eye with no fusion layer applied.   (b) EER right eye with no fusion layer applied.

Fig. 7. EER plot of the base models without fusion on left (a) and right (b) eyes only.

5   Conclusions

This work proposed a dual-input CNN, which takes RGB images of the right eye and the left eye of the same periocular region as input, for periocular recognition in an uncontrolled environment. The features maps from the processing of the two base models, were merged by a fusion layer. Experiments were performed on M^2FRED dataset, using three different fusion layers with the same overall architecture in order to evaluate the most suitable fusion layer for the information aggregation improving the overall recognition performance. By comparing the results achieved by the model on eyes separately and the ones obtained from the fusion of the deep feature, it can be observed that the fusion layer contributes significantly on the recognition performance, by adding up to 10% of recognition accuracy in the best conditions. Such a best level of performance was achieved by the Concatenation layers, which reached an accuracy of 97% of recognition. These promising findings, also show how the proposed model can effectively recognize people, by first processing the two eye images separately and then merging the extracted features. In future works a benchmark operation including other datasets can be applied, in order to test the behaviour of this kind of network in different acquisition conditions and with a higher number of enrolled subjects to recognize. In this direction, the problem of head pose estimation may play a crucial role to adequate the neural model during the training process to possible different acquisition conditions as well as considering the extra occlusions coming from wearing eyeglasses of makeup.

References

1. Alonso-Fernandez, F., Bigun, J.: A survey on periocular biometrics research. Pattern Recogn. Lett. **82**, 92–105 (2016). https://doi.org/10.1016/j.patrec.2015.08.026

2. Alonso-Fernandez, F., et al.: Cross-sensor periocular biometrics for partial face recognition in a global pandemic: comparative benchmark and novel multialgorithmic approach. arXiv preprint arXiv:1902.08123 (2019)

3. Castrillón-Santana, M., Lorenzo-Navarro, J., Ramón-Balmaseda, E.: On using periocular biometric for gender classification in the wild. Pattern Recogn. Lett. **82**, 181–189 (2016). https://doi.org/10.1016/j.patrec.2015.09.014. an insight on eye biometrics

4. Eskandari, M., Toygar, Ö.: Selection of optimized features and weights on face-iris fusion using distance images. Comput. Vis. Image Underst. **137**, 63–75 (2015). https://doi.org/10.1016/j.cviu.2015.02.011

5. Joshi, A., Gangwar, A., Sharma, R., Singh, A., Saquib, Z.: Periocular recognition based on Gabor and Parzen PNN. In: 2014 IEEE International Conference on Image Processing (ICIP), pp. 4977–4981 (2014). https://doi.org/10.1109/ICIP.2014.7026008

6. Joshi, A., Gangwar, A.K., Saquib, Z.: Person recognition based on fusion of iris and periocular biometrics. In: 2012 12th International Conference on Hybrid Intelligent Systems (HIS), pp. 57–62 (2012). https://doi.org/10.1109/HIS.2012.6421309

7. Jung, Y.G., Park, J., Tiong, L.C.O., Teoh, A.B.J.: Periocular recognition in the wild with learned label smoothing regularization. In: Twelfth International Conference on Digital Image Processing (ICDIP 2020), vol. 11519, p. 115190T. International Society for Optics and Photonics (2020)
8. Park, U., Jillela, R.R., Ross, A., Jain, A.K.: Periocular biometrics in the visible spectrum. IEEE Trans. Inf. Forensics Secur. **6**(1), 96–106 (2011). https://doi.org/10.1109/TIFS.2010.2096810
9. Park, U., Ross, A., Jain, A.K.: Periocular biometrics in the visible spectrum: a feasibility study. In: 2009 IEEE 3rd International Conference on Biometrics: Theory, Applications, and Systems, pp. 1–6 (2009). https://doi.org/10.1109/BTAS.2009.5339068
10. Raja, K.B., Raghavendra, R., Stokkenes, M., Busch, C.: Fusion of face and periocular information for improved authentication on smartphones. In: 2015 18th International Conference on Information Fusion (Fusion), pp. 2115–2120 (2015)
11. Raja, K.B., Raghavendra, R., Stokkenes, M., Busch, C.: Multi-modal authentication system for smartphones using face, iris and periocular. In: 2015 International Conference on Biometrics (ICB), pp. 143–150 (2015). https://doi.org/10.1109/ICB.2015.7139044
12. Reddy, N., Derakhshani, R.: Emotion detection using periocular region: a cross-dataset study. In: 2020 International Joint Conference on Neural Networks (IJCNN), pp. 1–6 (2020). https://doi.org/10.1109/IJCNN48605.2020.9207542
13. Santos, G., Hoyle, E.: A fusion approach to unconstrained iris recognition. Pattern Recogn. Lett. **33**(8), 984–990 (2012). https://doi.org/10.1016/j.patrec.2011.08.017. https://www.sciencedirect.com/science/article/pii/S0167865511002686. noisy Iris Challenge Evaluation II - Recognition of Visible Wavelength Iris Images Captured At-a-distance and On-the-move
14. Tiong, L.C.O., Kim, S.T., Ro, Y.M.: Multimodal facial biometrics recognition: dual-stream convolutional neural networks with multi-feature fusion layers. Image Vis. Comput. **102**, 103977 (2020). https://doi.org/10.1016/j.imavis.2020.103977
15. Tiong, L.C.O., Lee, Y., Teoh, A.B.J.: Periocular recognition in the wild: implementation of RGB-OCLBCP dual-stream CNN. Appl. Sci. **9**(13) (2019). https://doi.org/10.3390/app9132709
16. Woodard, D.L., Pundlik, S., Miller, P., Jillela, R., Ross, A.: On the fusion of periocular and iris biometrics in non-ideal imagery. In: 2010 20th International Conference on Pattern Recognition, pp. 201–204 (2010). https://doi.org/10.1109/ICPR.2010.58
17. Xu, J., Cha, M., Heyman, J.L., Venugopalan, S., Abiantun, R., Savvides, M.: Robust local binary pattern feature sets for periocular biometric identification. In: 2010 Fourth IEEE International Conference on Biometrics: Theory, Applications and Systems (BTAS), pp. 1–8 (2010). https://doi.org/10.1109/BTAS.2010.5634504
18. Zhao, Z., Kumar, A.: Accurate periocular recognition under less constrained environment using semantics-assisted convolutional neural network. IEEE Trans. Inf. Forensics Secur. **12**(5), 1017–1030 (2017). https://doi.org/10.1109/TIFS.2016.2636093

Improve Convolutional Neural Network Pruning by Maximizing Filter Variety

Nathan Hubens[1,2]([✉]), Matei Mancas[1], Bernard Gosselin[1], Marius Preda[2], and Titus Zaharia[2]

[1] ISIA, Faculty of Engineering of Mons, UMONS, Mons, Belgium
{nathan.hubens,matei.mancas,bernard.gosselin}@umons.ac.be,
[2] ARTEMIS, Telecom SudParis, IP Paris, Paris, France
{nathan.hubens,marius.preda,titus.zaharia}@telecom-sudparis.eu

Abstract. Neural network pruning is a widely used strategy for reducing model storage and computing requirements. It allows to lower the complexity of the network by introducing sparsity in the weights. Because taking advantage of sparse matrices is still challenging, pruning is often performed in a structured way, *i.e.* removing entire convolution filters in the case of ConvNets, according to a chosen pruning criteria. Common pruning criteria, such as l_1-norm or movement, usually do not consider the individual utility of filters, which may lead to: (1) the removal of filters exhibiting rare, thus important and discriminative behaviour, and (2) the retaining of filters with redundant information. In this paper, we present a technique solving those two issues, and which can be appended to any pruning criteria. This technique ensures that the criteria of selection focuses on redundant filters, while retaining the rare ones, thus maximizing the variety of remaining filters. The experimental results, carried out on different datasets (CIFAR-10, CIFAR-100 and CALTECH-101) and using different architectures (VGG-16 and ResNet-18) demonstrate that it is possible to achieve similar sparsity levels while maintaining a higher performance when appending our filter selection technique to pruning criteria. Moreover, we assess the quality of the found sparse subnetworks by applying the Lottery Ticket Hypothesis and find that the addition of our method allows to discover better performing tickets in most cases.

Keywords: Neural network pruning · Neural network interpretation

1 Introduction

Convolutional Neural Networks have been applied to a wide variety of computer vision tasks and have exhibited state-of-the-art results in most of them. Their recent success was partially due to an increase in their complexity and depth

---

N. Hubens—This research has been conducted in the context of a joint-PhD between the two institutions.

at the expense of increasing the needs of parameter storage and computation. This drawback makes it challenging for deep neural networks to be used for applications with limitations in terms of memory and/or processing time, such as embedded systems or real-time applications.

Recent studies have exhibited a particular characteristic of neural networks, called the Lottery Ticket Hypothesis [1]. This hypothesis suggests that, in regular neural network architectures, there exists a subnetwork that can be trained to the same level of performance as the original one, as long as it starts from the same original conditions. This implies that an important reason why complex architectures are successful nowadays is because, by possessing many parameters, they have more chance to contain such a "winning ticket". At the same time, once a winning ticket has been discovered, all the other parameters can be removed without affecting the model's performance. The technique consisting of removing unnecessary parameters and inducing sparsity in a neural network is called *neural network pruning*.

To prune a neural network, the most commonly used criteria of selection is the magnitude pruning, also called l_1 pruning, *i.e.* removing parameters having the lowest absolute value. Even though it allows to reach non-trivial sparsity levels, magnitude pruning still presents two main shortcomings. First, while being very efficient when the network's weights have been randomly initialized, magnitude pruning has shown limitations in the transfer-learning regime, *i.e.* when the network's weights come from pre-training on a larger dataset. Indeed, in the transfer learning regime, final weight values are mostly predetermined by the original model [2]. Thus, high magnitude weights that were useful for the pre-training task are not necessarily useful for the new end task. This was recently solved by the movement pruning criteria [2], which removes weights whose value moves towards zero. Second, magnitude pruning does not explicitly seek to remove redundant parameters and to maximize the variety of filters that the network contains after pruning. Methods attempting to increase the variety of filters by grouping them by similar functionality have shown promising results [3].

In this work we propose to slightly alter the usual pruning process and more particularly the filter selection mechanism by introducing a clustering method, ensuring that parameters exhibiting similar/redundant behaviours can be pruned while those with unique behaviour are being kept. The contributions of this paper are the following:

- Propose a simple and intuitive algorithm grouping redundant features and ensuring that filters extracting unique features are not removed in the pruning process. This algorithm may be appended to any filter selection criteria.
- Empirically show that the proposed method improves storage and computation costs over recently proposed techniques across several benchmark models and datasets without affecting the network performance.
- Show that using the proposed method, we are able to discover better subnetworks, following the Lottery Ticket Hypothesis.

2   Related Work

Pruning techniques can differ in many aspects. The three main axis of differentiation are detailed in this section.

Granularity. When pruning a neural network, it is first required to define the granularity, *i.e.* the structure according to which parameters will be removed. Granularities of pruning are usually categorized into two groups: *unstructured* pruning, *i.e.* when individual weights are evaluated and removed [4,5]. This leads to sparse weight matrices, often difficult to optimize in terms of computation and speed efficiency. For those reasons, *structured* pruning was introduced. This kind of pruning takes care of removing blocks of weights, which can take the form of vectors, kernels or even filters [6–8]. While most of structured pruning granularities still leaves sparse weight matrices, as the weights are removed in blocks, it makes it easier for dedicated hardware or libraries to take advantage of the removal of those weights. Filter pruning is a particular case as sparse filters can just simply be removed from the network architecture, thus keeping the network dense, and not requiring any specialized sparse computation library to obtain computation speed-up and storage reduction.

Criteria. In order to know which weights, or group of weights, will be pruned, we need to evaluate their importance. For that purpose, we define a criteria, ranking each parameter, and remove those that have the lowest score. Early work made use of second-order approximation of the loss surface to select the parameters to remove [4,5]. Due to the important computation overhead that such a technique introduces, other criteria have emerged, such as l_1-norm, l_2-norm or first-order Taylor expansion. Other training techniques enforcing sparse structures such as l_0 regularization [9] or variational dropout [10] have also been introduced. To this day, the criteria of selection which is the most commonly used due to its simplicity yet providing good and generalizable results across many datasets and architectures is the magnitude pruning [11], based on l_1-norm, the weights with the lowest absolute value assumed to be the least important, as they will produce the weakest activations and thus, participate the least to the output of the neural network. Moreover, the criteria may be used to compare weights belonging to a common layer, *i.e.* local pruning, thus leading to a structure with layers of equivalent sparsities. The criteria may also be applied to the whole network, comparing weights from all layers, *i.e.* global pruning, and leading to a network with layers of different sparsity levels. While global pruning possesses more computation overhead since it potentially compares millions of parameters at each pruning step, it usually provides better results as it allows for more freedom of parameter selection.

Scheduling. The pruning scheduling defines how pruning is integrated in the training process. Early pruning methods cared about removing redundant weights after the network has been trained to convergence, and performing the pruning in a single-step, which is nowadays called *one-shot pruning* [6]. It was soon discovered that *iterative pruning*, performing the pruning in several steps, alternating pruning and fine-tuning of the network to allow it to recover from the lost performance, helped to reach more extreme sparsity levels and to let the network to more easily recover the pruning of many weights [12,13]. Recently, research about the Lottery Ticket Hypothesis have shown that the optimal pruned network could be discovered from the very initial state of a neural network, *i.e.* before any training has occurred [1]. While it is still very difficult to uncover such an optimal network right from initialization, it inspired many research to prune the network earlier in the training process, *i.e.* not starting from a first pre-training phase, but also to propose more complex scheduling functions which integrates pruning early in the training process [14,15].

In our work, we focus on a particular structured pruning approach, filter pruning. More particularly, our contribution concerns the criteria of selection of filters. We propose a slight alteration that can be inserted into the iterative pruning process, allowing to use any state-of-the-art criteria, but helping them to select redundant filters to remove, while preserving unusual, thus potentially discriminative filters.

3   Proposed Methodology

Performing filter pruning following an iterative schedule usually consists of a three-step method, represented in Fig. 1: (1) train the network to convergence, (2) prune a portion of the convolution filters, according to a chosen criterion and, (3) fine-tune the model to recover from the lost performance. Steps (2) and (3) are then repeated, alternating pruning and fine-tuning until the desired sparsity is reached. We propose to add an additional step between (1) and (2). Indeed, before selecting the weights to remove according to a chosen criteria, we first would like to cluster filters exhibiting similar behaviour and to perform pruning in each cluster separately, and consequently only on redundant filters. By doing so, pruning will only retain independent filters while also retaining filters which have uncommon behaviours, thus maximizing the variety of remaining filters.

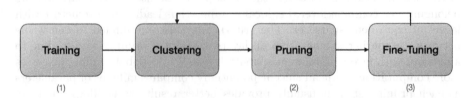

Fig. 1. The proposed pruning pipeline. We introduce a fourth step in the common iterative pruning process, aiming to cluster convolution filters by similar functionality. By then performing pruning in each cluster, we ensure that we remove redundant filters, while preserving the rare ones that could have been removed otherwise.

To learn about the functionality of each filter, we make use of a neural network interpretation technique named *Activation Maximization* [16]. This method uses the gradient ascent optimization technique, starting from a random noise image, to modify each pixel of the image in order to maximize the activation of a particular convolution filter. In other words, the synthesized image is the image of features that excites the most a selected filter and thus, the feature it is the most sensitive to when processing a natural image.

For each convolutional filter in a layer of the network, we can synthesize its corresponding "signature" image based on Activation Maximization. The goal is then to perform K-Means clustering of those images, effectively grouping similar images together while also keeping unique ones into their dedicated group. To effectively reduce dimensionality and facilitate the task of K-Means clustering, we first feed our images to the convolutional part of an AlexNet model [17] pretrained on ImageNet [18], encoding those into a feature vector, which will serve as input data to the clustering algorithm (Fig. 2).

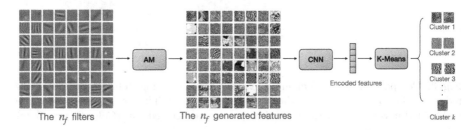

Fig. 2. Representation of the clustering process. We first extract each filter of a given layer, then generate the corresponding feature images with Activation Maximization technique. Those synthesized images are then encoded to a lower dimension by a pretrained ConvNet, and clustered with K-Means. This technique allows to group filters sensitive to similar features together.

Once each feature image has been clustered, we can then apply the pruning process, selecting remaining filters according to a chosen criteria but, this time not by comparing all the filters in the layer, but by comparing filters whose feature images are located in the same group. By doing so in each group, we will then only preserve the single best representative filter of each feature. The number of clusters k is thus chosen as a compression parameter, depending on the desired sparsity. By setting a high number of clusters k, this creates more groups and thus removes less filters.

When comparing common criteria before and after the addition of our clustering method, we observe that a greater variety of filters are retained. As an example, Fig. 3 represents features extracted from the filters of the first layer of a simple ConvNet, AlexNet [17], by using the Activation Maximization technique. Three clusters of similar features have been highlighted in color, and the corresponding remaining features are shown for each pruning technique, with removed one being greyed out. We can observe that, by clustering similar features, we ensure that: (1) redundant features are removed and (2) rare features are being kept, which is not the case when using magnitude and movement pruning alone, where some features belonging to a same cluster are still present.

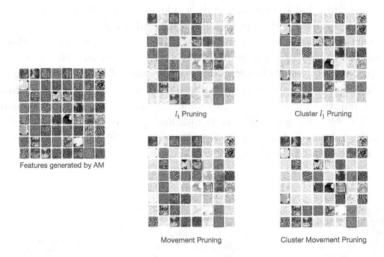

Fig. 3. Comparison of the remaining features after applying different pruning techniques until a sparsity of 50% in the first layer of AlexNet. Three dominant clusters are highlighted in color. Features removed by pruning are greyed out. (Color figure online)

4   Comparison to Common Criteria

In this section, we evaluate the effects of adding the extra clustering step presented in Sect. 3 in the iterative pruning pipeline and apply the pruning criteria in each group separately.

Datasets and Architectures. For our experiments, the datasets have been chosen to be varied in terms of image resolution and number of classes. In particular, we evaluate our methods on the three following datasets: (1) CIFAR-10 [19], composed of RGB images distributed in 10 classes, and of resolution 32×32. (2) CIFAR-100 [19], composed of RGB training images distributed in 100 classes and of resolution 32×32 and (3) Caltech-101 [20], composed of pictures of objects, distributed in 101 classes and of resolution 300×200. Those datasets are then tested on two types of popular convolutional network architectures: VGG-16 [21] and ResNet-18 [22]. In particular, we use a modified version of VGG-16 which consists of 13 convolutional layers and 2 fully-connected layers, with each convolutional layer being followed by a batch normalization layer [23]. ResNet-18 belongs to a family of ConvNets using residual connections [22], and contains 17 convolutional layers and a single fully-connected layer.

Training Procedure. The networks we use for our experiments are initialized from pre-trained weights, *i.e.* networks were previously trained on ImageNet and we reuse their weights. Images of our dataset are first resized to 224 × 224 and are augmented by using horizontal flips, rotations, image warping and random cropping, then aggregated in batches of size 64. We train each model using the 1cycle learning rate method [24], where the training starts with a learning rate warmup until a nominal value, then gradually decay until the end of the training.

Pruning Method. We use the 4-step schedule presented in Fig. 1. We propose to iterate over each layer and remove a specific amount of filters in order to reach the desired sparsity. As each layer will eventually have the same sparsity level, it can be considered as a form of local pruning. The first step is performed for 15 epochs, at a learning rate of $1e-3$. Our clustering method then takes care of grouping the filters of a target layer into k groups, $k = s \times n_f$, s being the desired sparsity in percent and n_f the amount of filters in that layer. After performing the pruning of filters and only keeping a single filter from each cluster, we fine-tune our model for 3 epochs, with a learning rate of $3e-4$, to allow the network to recover from the loss of its parameters. This process is performed iteratively for each layer in the network. We evaluate the benefits of our extra step according to two pruning criteria: (1) l_1-norm of the filters, *i.e.* remove the filters that possess the lowest norm, computed for each filter by $\sum_i |w_i|_t$ and (2) movement pruning, *i.e.* only keep the filters whose magnitude has increased the most during training, computed by $\sum_i |w_i|_t - |w_i|_0$, with $|w_i|_t$, the weights values at training step t and $|w_i|_0$, the weights values at initialization.

Frameworks and Hardware. These experiments are conducted using the PyTorch [25] and fastai [26] libraries for the implementation of the training loop, fasterai [27] and Lucent [28] for the implementation of the clustering and pruning methods and using a 12 GB Nvidia GeForce GTX 1080 Ti GPU for computation.

Results. The experiments, conducted on VGG-16 (Table 1) and ResNet-18 (Table 2), show that, for almost all sparsity levels, dataset and criteria tested, it is beneficial to the pruning process to add the proposed clustering step. Indeed, for the same sparsity level, accuracy increases up to 5% can be observed, which also means that the same networks could be pruned to a higher sparsity level without witnessing performance degradation.

Table 1. Results of applying different pruning criteria on VGG-16. The benefit of applying our clustering method before selecting the filters to remove translates to a higher accuracy for most sparsity levels and datasets. Values in bold are the best when comparing a single criteria with and without the clustering process. Accuracies and standard deviation over 3 runs are reported.

		l_1	Cluster l_1	Movement	Cluster movement
CIFAR-10					
Sparsity	60%	90.89 ± 0.17	**92.39 ± 0.07**	91.55 ± 0.18	**92.45 ± 0.26**
	70%	89.47 ± 0.09	**90.91 ± 0.13**	90.35 ± 0.12	**90.89 ± 0.10**
	80%	84.95 ± 0.08	**87.04 ± 0.23**	86.50 ± 0.19	**87.48 ± 0.10**
CIFAR-100					
Sparsity	60%	54.94 ± 0.21	**58.72 ± 0.30**	54.81 ± 0.34	**57.29 ± 0.71**
	70%	47.56 ± 0.94	**51.67 ± 0.61**	47.59 ± 0.29	**51.25 ± 0.41**
	80%	35.67 ± 0.90	**39.30 ± 0.65**	36.82 ± 1.42	**42.30 ± 0.94**
Caltech-101					
Sparsity	60%	86.63 ± 0.39	**87.44 ± 0.28**	86.94 ± 0.28	**87.40 ± 0.51**
	70%	84.87 ± 0.41	**86.01 ± 0.79**	82.89 ± 0.19	**84.43 ± 0.25**
	80%	**79.18 ± 0.63**	79.03 ± 0.72	76.00 ± 0.59	**78.75 ± 0.63**

Table 2. Results of applying different pruning criteria on ResNet-18. The benefit of applying our clustering method before selecting the filters to remove translates to a higher accuracy for most sparsity levels and datasets. Values in bold are the best when comparing a single criteria with and without the clustering process. Accuracies and standard deviation over 3 runs are reported.

		l_1	Cluster l_1	Movement	Cluster movement
CIFAR-10					
Sparsity	60%	93.32 ± 0.11	**93.76 ± 0.18**	92.73 ± 0.16	**93.57 ± 0.10**
	70%	92.17 ± 0.11	**92.20 ± 0.11**	90.79 ± 0.14	**92.35 ± 0.03**
	80%	89.58 ± 0.17	**90.13 ± 0.09**	87.04 ± 0.24	**89.53 ± 0.23**
CIFAR-100					
Sparsity	60%	71.65 ± 0.22	**72.61 ± 0.41**	70.95 ± 0.13	**72.27 ± 0.37**
	70%	67.18 ± 0.17	**68.46 ± 0.25**	66.19 ± 0.15	**68.44 ± 0.21**
	80%	59.14 ± 0.13	**60.32 ± 0.30**	58.50 ± 0.41	**59.86 ± 0.23**
Caltech-101					
Sparsity	60%	92.63 ± 0.05	**93.00 ± 0.16**	90.77 ± 0.18	**91.81 ± 0.26**
	70%	88.75 ± 0.42	**89.48 ± 0.30**	85.32 ± 0.27	**87.75 ± 0.24**
	80%	79.89 ± 0.10	**80.31 ± 0.35**	75.05 ± 0.56	**76.29 ± 0.39**

5   Application to Lottery Ticket Hypothesis

In order to not only compare the relative performance of the addition of our proposed clustering technique, we also would like to compare the quality of the remaining subnetworks, obtained after pruning. Such an analysis may be performed using the Lottery Ticket Hypothesis.

Finding Lottery Tickets. The Lottery Ticket Hypothesis states that: in each network, there exists a subnetwork that, trained in isolation, and from the same initial weight values, is able to achieve comparable performance in a comparable training time as the whole network. The authors proposed to find the said winning-tickets through an iterative process, repeatedly pruning a portion of remaining weights according to their l_1-norm, then resetting them to their initial value, *i.e.* the value before any training has occurred. Despite having shown promising results on small datasets, and for simple architectures, this method has shown difficulties to generalize to higher complexity use-cases [1]. To overcome this limitation, the same authors propose a slight weakening of the hypothesis. Instead of reinitializing the weights to their original value, they should be reinitialized to a value of an earlier step in the training process. This new hypothesis is called Lottery Ticket Hypothesis with Rewinding, and the sparse subnetworks are then called matching tickets instead of winning tickets [29].

Comparison of Tickets. We compare the performance of discovered tickets using the same pruning criteria, datasets and training procedure as described in Sect. 4, and for the ResNet-18 architecture. As our experiment concerns large datasets and complex architecture, we propose to study the effect of our pruning technique on the Lottery Ticket Hypothesis with Rewinding. To uncover the tickets, we adopt the same methodology as presented in Sect. 3, but reinitializing the weights after each pruning step to the value they had after the initial training step, *i.e.* to their value after Step (1) in Fig. 1. The operation is performed for each criteria evaluated in the Sect. 4 and for sparsity levels ranging from 10% to 80%. After extracting the subnetwork, we train it for 15 epochs and then compare the versions obtained with and without the addition of our variety enforcing clustering method.

Results. From this experiment, whose results are reported in Figs. 4, we can observe that in most cases, the addition of the clustering method prior to the criteria selection helps to find a better performing ticket, thus validating the quality of the pruned network. While the addition of a clustering technique before applying the pruning criteria seems profitable in most cases, it benefits movement pruning the most. Indeed, increases up to 2% in accuracy may be observed in the case of l_1 pruning, and up to 5% in the case of movement pruning (Figs. 5 and 6).

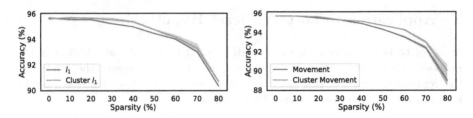

Fig. 4. Results of the Lottery Ticket Hypothesis with Rewind test for different sparsities, performed with ResNet-18 on CIFAR-10.

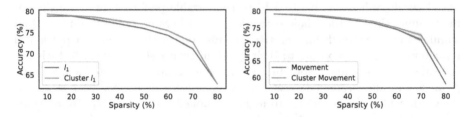

Fig. 5. Results of the Lottery Ticket Hypothesis with Rewind test for different sparsities, performed with ResNet-18 on CIFAR-100.

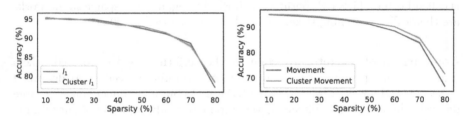

Fig. 6. Results of the Lottery Ticket Hypothesis with Rewind test for different sparsities, performed with ResNet-18 on CALTECH-101.

6   Conclusion

In this work, we propose a novel pruning method, introducing a clustering process before applying the pruning criteria. This clustering process, based on an interpretation technique called Activation Maximization, groups filters sensitive to similar features in the input image. By then applying the pruning criteria to each feature group, we ensure that pruning is applied on redundant filters, and that rare filters, which may be alone in their group, are retained. Experiments have shown that our method leads to better results than classical methods on both VGG-16 and ResNet-18 architectures and for CIFAR-10, CIFAR-100 and CALTECH-101 datasets. Those results demonstrate that one should avoid pruning rare or unique filters and that keeping a wide filter variability is crucial to achieving both a higher pruning rate and a lower accuracy loss. Moreover, by performing Lottery Ticket Hypothesis with Rewinding tests, we have demonstrated that the subnetworks discovered after pruning were of better quality, as they were able to reach higher performance in the same training time.

References

1. Frankle, J., Carbin, M.: The lottery ticket hypothesis: finding sparse, trainable neural networks. In: International Conference on Learning Representations (ICLR) (2019)
2. Sanh, V., Wolf, T., Rush, A.M.: Movement pruning: adaptive sparsity by fine-tuning. In: Advances in Neural Information Processing Systems (NeurIPS) (2020)
3. Qin, Z., Fuxun, Y., Liu, C., Chen, X.: Functionality-oriented convolutional filter pruning. In: British Machine Vision Conference (BMVC) (2019)
4. LeCun, Y., Denker, J.S., Solla, S.A.: Optimal brain damage. In: Advances in Neural Information Processing Systems (NeurIPS) (1990)
5. Hassibi, B., Stork, G.D., Wolff, G.: Optimal brain surgeon and general network pruning. In: International Conference on Neural Networks (ICANN) (1993)
6. Li, H., Kadav, A., Durdanovic, I., Samet, H., Graf, H.: Pruning filters for efficient ConvNets. In: International Conference on Learning Representations (ICLR) (2017)
7. He, Y., Zhang, X., Sun, J.: Channel pruning for accelerating very deep neural networks. In: International Conference on Computer Vision (ICCV) (2017)
8. Hubens, N., et al.: An experimental study of the impact of pre-training on the pruning of a convolutional neural network. In: International Conference on Applications of Intelligent Systems (APPIS) (2020)
9. Tartaglione, E., Lepsøy, S., Fiandrotti, A., Francini, G.: Learning sparse neural networks via sensitivity-driven regularization. In: Advances in Neural Information Processing Systems (NeurIPS) (2018)
10. Molchanov, D., Ashukha, A., Vetrov, D.: Variational dropout sparsifies deep neural networks. In: International Conference on Machine Learning (ICML) (2017)
11. Gale, T., Elsen, E., Hooker, S.: The state of sparsity in deep neural networks. In: The International Conference on Machine Learning (ICML) (2019)
12. Han, S., Pool, J., Tran, J., Dally, W.: Learning both weights and connections for efficient neural networks. In: International Conference on Neural Information Processing Systems (ICPS) (2015)
13. Molchanov, P., Tyree, S., Karras, T., Aila, T., Kautz, J.: Pruning convolutional neural networks for resource efficient inference. In: International Conference on Learning Representations (ICLR) (2017)
14. Zhu, M., Suyog, G.: To prune, or not to prune: exploring the efficacy of pruning for model compression. In: International Conference on Learning Representations (ICLR) (2018)
15. Hubens, N.: One-cycle pruning: pruning ConvNets with tight training budget. arXiv: abs/2107.02086 (2021)
16. Erhan, D., Bengio, Y., Courville, A., Vincent, P.: Visualizing higher-layer features of a deep network. University of Montreal, vol. 1341, p. 3 (2009)
17. Krizhevsky, A., Sutskever, I., Hinton, G.: ImageNet classification with deep convolutional neural networks. In: Advances in Neural Information Processing Systems (NeurIPS) (2012)

18. Deng, J., Dong, W., Socher, R., Li, L., Li, K., Fei-Fei, L.: ImageNet: a large-scale hierarchical image database. In: International Conference on Computer Vision and Pattern Recognition (CVPR) (2009)
19. Krizhevsky, A., Hinton, G.: Learning multiple layers of features from tiny images. University of Toronto (2009)
20. Li, F., Fergus, R., Perona, P.: Learning generative visual models from few training examples: an incremental Bayesian approach tested on 101 object categories. In: International Conference on Computer Vision and Pattern Recognition Workshop (CVPR) (2004)
21. Simonyan, K., Zisserman, A.: Very deep convolutional networks for large-scale image recognition. In: International Conference on Learning Representations (ICLR) (2015)
22. He, K., Zhang, X., Ren, S., Sun, J.: Deep residual learning for image recognition. In: Conference on Computer Vision and Pattern Recognition (CVPR) (2015)
23. Liu, Z., Sun, M., Zhou, T., Huang, G., Darrell, T.: Rethinking the value of network pruning. In: International Conference on Learning Representations (ICLR) (2019)
24. Smith, L., Topin, N.: Super-convergence: very fast training of neural networks using large learning rates. In: SPIE Artificial Intelligence and Machine Learning for Multi-Domain Operations Applications (2019)
25. Paszke, A., et al.: PyTorch: an imperative style, high-performance deep learning library. In: Advances in Neural Information Processing Systems (NeurIPS) (2019)
26. Howard, J., Gugger, S.: fastai: a layered API for deep learning. In: MDPI Information (2020)
27. Hubens, N.: FasterAI: a library to make smaller and faster neural networks (2020). https://github.com/nathanhubens/fasterai
28. Swee Kiat, L.: Lucent (2019). https://github.com/greentfrapp/lucent
29. Frankle, J., Dziugaite, K., Roy, D., Carbin, M.: Linear mode connectivity and the lottery ticket hypothesis. In: International Conference on Machine Learning (ICML) (2020)

Improving Autoencoder Training Performance for Hyperspectral Unmixing with Network Reinitialisation

Kamil Książek[1,2(✉)] , Przemysław Głomb[1] , Michał Romaszewski[1] ,
Michał Cholewa[1] , Bartosz Grabowski[1] , and Krisztián Búza[3]

[1] Institute of Theoretical and Applied Informatics, Polish Academy of Sciences,
44-100 Gliwice, Poland
{kksiazek,przemg,mromaszewski,mcholewa,bgrabowski}@iitis.pl
[2] Department of Data Sciences and Engineering, Silesian University of Technology,
44-100 Gliwice, Poland
kamil.ksiazek@polsl.pl
[3] Biointelligence Group, Department of Mathematics-Informatics,
Sapientia Hungarian University of Transylvania, 540485 Târgu Mureș, Romania
buza@biointelligence.hu

Abstract. Neural networks, in particular autoencoders, are one of the most promising solutions for unmixing hyperspectral data, i.e. reconstructing the spectra of observed substances (endmembers) and their relative mixing fractions (abundances), which is needed for effective hyperspectral analysis and classification. However, as we show in this paper, the training of autoencoders for unmixing is highly dependent on weights initialisation; some sets of weights lead to degenerate or low-performance solutions, introducing negative bias in the expected performance. In this work, we experimentally investigate autoencoders stability as well as network reinitialisation methods based on coefficients of neurons' dead activations. We demonstrate that the proposed techniques have a positive effect on autoencoder training in terms of reconstruction, abundances and endmembers errors.

Keywords: Autoencoders · Hyperspectral unmixing · Training stability · Network reinitialisation

1 Introduction

Hyperspectral imaging (HSI) combines reflectance spectroscopy with image processing – image pixels contain information about hundreds of spectral bands that can characterise chemical composition and properties of visible objects. In HSI the spectra of pixels are often a mixture of different substances [14], as the sensor captures light

Supplementary Information The online version contains supplementary material available at https://doi.org/10.1007/978-3-031-06427-2_33

The original version of this chapter was revised: Table 2 has been updated. The correction to this chapter is available at https://doi.org/10.1007/978-3-031-06427-2_66.

S. Sclaroff et al. (Eds.): ICIAP 2022, LNCS 13231, pp. 391–403, 2022.
https://doi.org/10.1007/978-3-031-06427-2_33

reflected from nearby objects or aggregated from several sources due to low spatial resolution. The task of hyperspectral unmixing (HU) is to reconstruct the original spectra of observed substances, called endmembers, and their fractional mixture coefficients, called abundances. On the one hand, HU facilitates further data analysis and improves classification results [10]. On the other hand, correlations between pixels and huge data volume resulting from the fact that every pixel can be treated as an example in a high-dimensional feature space, make neural networks particularly suitable models for HU.

Although a number of machine learning algorithms for HU based on statistical and geometric principles have been developed [4], for the aforementioned reasons, deep learning models seem to be the most promising solution, with autoencoders (AE) emerging as an architecture of choice. Most prominent examples include: a deep AE network [28] which is a sequence of stacked AE followed by a variational AE, generative and encoder models trained using pure pixel information [6], EndNet architecture [21] or deep convolutional autoencoders [24].

One of key elements of training a neural model is a proper initialisation of weights, which prevents the phenomenon of vanishing or exploding gradients. Several weight initialisation methods have been introduced, e.g. [9, 12]. However, while those methods are derived from examining gradient flow principles, they are usually applied without verification of the quality of initial weights.

In this work we study the problem of AE training failures, resulting from bad initialisation weights, focusing on the problem of HU. We also present network reinitialisation methods which can alleviate a problem of bad weights and improve the network performance. In particular, we present the following contributions:

1. We have experimentally verified the presence of failed trainings of autoencoders in HU scenarios. We have investigated this effect through $n = 100000$ individual autoencoder training sessions across a diverse range of variables, and found that this effect persists across all studied variants of autoencoder architectures, datasets, weight initialisation methods, loss function types, and hyperparameter choices.
2. To the best of our knowledge, this is the first detailed study of such failures in a standard autoencoder training scenario on a real-world hyperspectral dataset.
3. Based on our results, we use statistical analysis with the Kruskal-Wallis H-test to empirically confirm the thesis that a specific autoencoder initialisation affects the final data reconstruction error of a trained model.
4. To resolve this issue, we propose network reinitialisation methods based on dead activations' coefficients. We show that for networks with ReLU activation function these approaches can both mitigate the impact of bad weights initialisations as well as unfavourable weight values encountered during the training.

1.1 Related Work

An overview of the HU methods can be found in [4]. The approaches range from simple pure pixel algorithms e.g. Pixel Purity Index (PPI) [5] or N-FINDR [29] to more complex ones e.g. SISAL [3] which work with non-pure pixels and noisy data.

An autoencoder (AE) is a neural network that through hidden layers compresses an input into a lower-dimensional (latent) space and reconstructs the original data. Reduction of the input dimensionality makes the AE well-suited for HU, thus they are often used as a base for HU algorithms. In [23], authors analysed fully connected AE-based architectures for blind unmixing in an unsupervised setting. The use of AE for unmixing in a nonlinear case was a focus of [30], where authors showed how in certain situation the linearity assumption will not hold. The approach via convolutional AE was tested in [22] and [26]. However, both approaches use spectral and spatial features of data at the same time.

The general problem of random weight initialisation leading to inadequate results was observed in [13,15] and it was mitigated by the use of a stacked Restricted Boltzmann Machines (RBMs) to determine the initial weights for AE networks. In [9], a weight initialisation scheme was proposed that maintains activation and back-propagated gradients variance as one moves up or down the network.

The authors of [17] discuss the convergence of back-propagation, using several heuristics to support NN construction. In [25] authors propose to mitigate the instability of the unmixing algorithm with multiple initialisations. In [28] a scheme is proposed, where the first layer of a neural network determines the initialisation parameters for the unmixing engine, similarly to [11] where a cascade model is proposed. In [21] the End-Net algorithm is paired with VCA (Vertex Component Analysis) filter or FCLS (Fully Constrained Least Squares) for initialisation. The FCLS is also used in [6]. Results of the first run of the network may be used to re-initialise it in order to improve results [20].

We point out that naive initialisation of the weights may lead to dead neurons. The problem of dead neurons is that certain neurons output 0 regardless of the input. This makes them impossible to train using gradient-based optimisation methods. In [19], the authors provide a theoretical analysis regarding death of neurons with ReLU activation function. They prove that ReLU network will eventually die as its depth goes to infinity. To alleviate the problem of dying neurons, they propose a new weight initialisation procedure. In [27], authors show that it is possible to increase the network depth with guaranteed probability of living weights initialisation, as long as the network width increases accordingly. They also propose a sign flipping scheme to make sure the ratio of living data points in a k-layer network is at least 2^{-k}.

Various attempts are made to improve the optimization of weights. In [1], the authors study the impact of different reinitialisation methods on generalisation using multiple convolutional neural networks (CNNs). In [2] Autoinit, an algorithm is proposed for dynamic scaling of weights of neural networks. Its potential is demonstrated on various architectures, including CNNs or residual networks.

2 Performance Investigation of Autoencoders in a Hyperspectral Unmixing Problem

2.1 Linear Spectral Mixing

In this work we use the Linear Mixing Model (LMM) of the pixel spectra, i.e. a B-band pixel $\boldsymbol{x} = \begin{bmatrix} x_1, ..., x_B \end{bmatrix}^\top$ is written as a linear combination of E endmembers with the addition of a noise vector, i.e. $\boldsymbol{x} = \sum_{j=1}^{E} a_j \cdot \boldsymbol{w}_j + \boldsymbol{n}$, where $\boldsymbol{a} = \begin{bmatrix} a_1, ..., a_E \end{bmatrix}^\top$ is a vector of abundances, $\boldsymbol{W} = \begin{bmatrix} \boldsymbol{w}_1, ..., \boldsymbol{w}_E \end{bmatrix}$ is a matrix of endmembers, $\boldsymbol{W} \in \mathbb{R}^{B \times E}$ and

$n = [n_1, ..., n_B]^\top$ is a noise vector. To preserve the physical properties of abundances, it is necessary to ensure that the nonnegativity constraint is fulfilled and the sum of all fractional abundances equals one. It means that $\forall j \in \{1, ..., E\}$ $a_j \geq 0$, $\sum_{j=1}^{E} a_j = 1$.

For M pixels in the image $X \in \mathbb{R}^{B \times M}$ the corresponding abundance matrix is denoted as $A = [a_1, ..., a_M]$, $A \in \mathbb{R}^{E \times M}$ and $N = [n_1, ..., n_M]$, $N \in \mathbb{R}^{B \times M}$ is the noise matrix, where $n_i \in \mathbb{R}^{B \times 1}$ is a noise vector, $i \in \{1, ..., M\}$. Accordingly, the LMM Equation can be rewritten as $X = WA + N$. The aim of the HU process is to estimate the endmembers matrix W and the abundances matrix A which provide an estimate of the pure spectra of substances and their fractions present in different pixels of the image.

2.2   Architectures of Autoencoders for Hyperspectral Unmixing

We focus on the architecture presented in [23] (see Fig. 1). Its encoder part consists of multiple linear layers that transform the input data into a latent space, according to the pattern: $\text{Enc}(X_b) : \mathbb{R}^{B \times b_s} \rightarrow \mathbb{R}^{E \times b_s}$, where $X_b \in \mathbb{R}^{B \times b_s}$ is a batch of input data, b_s is a batch size. Then, in the decoder part, input spectra are reconstructed based on a latent representation: $\text{Dec}(\text{Enc}(X_b)) : \mathbb{R}^{E \times b_s} \rightarrow \mathbb{R}^{B \times b_s}$.

We denote the reconstructed image by $\hat{X} = \text{Dec}(\text{Enc}(X))$. The goal of the autoencoder is to minimise difference between X and $\hat{X}$. The design of this architecture allows it to be used for HU; as a decoder has only one layer, its neurons' activations on the last encoder layer can be treated as abundance vectors while weights connecting the encoder part with the output of the autoencoder can be considered as endmembers.

The authors of [23] studied several architectures with different number of layers and activation functions. We focus on the one that achieved one of the highest efficiency, i.e. a version with sigmoid activation function. This architecture consists of four linear layers in the encoder part, having $9E$, $6E$, $3E$ and E neurons, respectively. After that, a batch normalisation (BN) layer is applied. Then, a dynamical soft thresholding (ST) is used which can be written as $x_{ST}^i = \max(0, x_{BN}^i - \alpha)$, where $i \in \{1, ..., b_s\}$, $X_{ST} = [x_{ST}^1, ..., x_{ST}^{b_s}]$, $X_{BN} = [x_{BN}^1, ..., x_{BN}^{b_s}]$ are matrices with all batch pixels after soft thresholding or batch normalisation, respectively; 0 is a zero vector and α is a vector of trainable parameters. Then, to ensure that the sum to one constraint for abundances is met, each pixel vector is normalised, i.e. $\forall i \in \{1, ..., b_s\}$ $x_{norm}^i = x_{ST}^i / \left(\sum_{k=1}^{E} x_{ST}^{i,k} \right)$, where x_{norm}^i is the i-th batch vector after normalisation, $x_{ST}^{i,k}$ is the k-th coordinate of the i-th vector. Gaussian Dropout (GD) is applied as the last encoder layer, but only during training. Finally, the single decoder layer reconstructs the signal from the latent to the input space. We denote this architecture as the *original*.

To investigate the underlying dependencies of AEs for HU we have also prepared a simplified version of the described architecture, denoted *basic*, where an encoder has two linear layers in which ReLU activation function is used. The number of neurons in the first hidden layer is a hyperparameter and is equal to $n_1 E$, where E is the number of endmembers, while the second hidden layer ending the encoder part has E neurons.

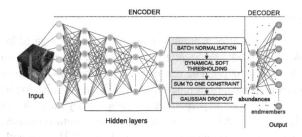

Fig. 1. The pipeline of the autoencoder architecture from [23], denoted as *original*.

There are no BN, ST or GD layers. A normalisation layer is left to ensure that the sum of fractional abundances per each pixel is equal to one. A decoder part has one linear layer.

2.3 Performance Evaluation

Datasets. We used two well-known datasets for HU: Samson and Jasper Ridge [31].

Samson is an image with dimensions $95 \times 95 \times 156$, spectral range of $401-889$ nm and spectral resolution ~ 3.13 nm. Pixels spectra are a mixture of three endmembers: water, trees and soil.

Jasper Ridge is an image with dimensions $100 \times 100 \times 198$. Originally, there were 224 bands covering the range of $380-2500$ nm but bands $1-3$, $108-112$, $154-166$ and $220-224$ were removed due to disturbances, as did authors in [21,31]. Endmembers represent: trees, water, soil and road.

Experiment. The goal of the experiment is to confirm the impact of weight initialisation on the final reconstruction error of the AE. For each of $N = 50$ randomly initialised AEs, we have performed $k = 50$ separate training sessions for each AE initialisation with the same dataset and hyperparameter values, resulting in $n_a = 2500$ trained models. This has been repeated across ten different hyperparameters sets (see Table 1) and four methods of weight initialisation, i.e. He [12] and Glorot [9] with normal or uniform distribution, bringing the total number of models to $n_b = 100000$.

Each model is evaluated as follows. Let D be an HSI image and let V_{GT} be a set of correct endmembers for image D. Let also D_{GT} be ground truth for fractional abundances, i.e. correct abundances for endmembers V_{GT}. The AE $A_{i,j}, i = 1, \ldots, N, j = 1, \ldots, k$ is trained with a set of vectors from dataset D. The endmembers $V_{i,j}$ and abundances vectors are then extracted; the endmembers set $V_{i,j}$ is matched to V_{GT}. The match used is the permutation of endmembers with the smallest distance to V_{GT}. Abundances vectors are compared to ground truth image D_{GT} to calculate error $E_{i,j}^a$ in terms of RMSE, like in [28], while endmembers are compared to their counterparts from V_{GT} and the value of $E_{i,j}^e$ is calculated according to the SAD function. For a given experiment mean endmembers error is calculated as follows: $E^e = \sum_{i=1}^{N} \sum_{j=1}^{k} \frac{E_{i,j}^e}{Nk}$. Mean abundances error E^a is calculated analogously.

2.4   Parameters

We have explored the effect of weight initialisation with a range of hyperparameters (see Table 1). We have used: two architectures (*original* and *basic*), two datasets, two loss functions (MSE and SAD). The rest of hyperparameters (e.g. learning rate, batch size, GD parameter etc.) have been tuned using RayTune optimisation library [18]. For comparison, we also included hyperparameters indicated by the authors of the *original* architecture [23]. In all experiments we have used the Adam optimiser. Additionally, we have used four initialisation algorithms from [9] and [12], each with uniform[1] and normal distribution. The source code necessary for the replication of weights initialisation experiments is publicly available in our GitHub repository: https://github.com/iitis/AutoencoderTestingEnvironment.

Table 1. The list of performed experiments with all hyperparameters used. An encoder means the number of neurons on the first hidden layer and it concerns only the *basic* architecture. Gaussian Dropout (GD) is applied only for the *original* architecture.

Experiment ID	Architecture	Hyperparameters origin	Loss	Dataset	Encoder	Batch size	Learning rate	GD
1	*original*	RayTune	MSE	Samson	–	100	0.01	0
2	*original*	RayTune	SAD	Samson	–	100	0.01	0
3	*original*	Article	SAD	Samson	–	20	0.01	0.1
4	*basic*	RayTune	MSE	Samson	10E	4	0.0001	–
5	*basic*	RayTune	SAD	Samson	20E	4	0.0001	–
6	*original*	RayTune	MSE	Jasper Ridge	–	100	0.01	0
7	*original*	RayTune	SAD	Jasper Ridge	–	100	0.01	0
8	*original*	Article	MSE	Jasper Ridge	–	5	0.01	0.1
9	*original*	Article	SAD	Jasper Ridge	–	5	0.01	0.1
10	*basic*	RayTune	MSE	Jasper Ridge	10E	20	0.001	–

2.5   Statistical Verification

To confirm a non-uniform behaviour of weights (the existence of initialisations leading to worse models after the network training than other initialisations) we have used the Kruskal-Wallis H-test for a one-way analysis of variance [16]. The test is performed as follows: our models are treated as N different populations where every population corresponds to a single set of initial weights and samples correspond to error estimates of consecutive training runs. The hypotheses of an H-test are as follows:

H_0: All population means are equal, i.e. $\mu_{A_1} = \mu_{A_2} = ... = \mu_{A_N}$.
H_1: At least one population has a statistically significantly different mean than the others.

Since the rejection of the null hypothesis does not give an answer as to which population differs from others, a post-hoc analysis is performed using the Conover-Iman test [7,8] which can be used if and only if the null hypothesis of the Kruskal-Wallis H-test is rejected. By performing the pairwise comparison for all population pairs, we can conclude which differences between populations are statistically significant.

---

[1] With [12] method using uniform initialisation we have used the version from the PyTorch library, which differs from the paper with: 1) biases are not initialised to 0; 2) bounds of a uniform distribution are constant and not dependent on the number of connections.

2.6 Network Reinitialisation Methods

To minimise the impact of bad neural network weights we propose statistics that allow us to detect and alleviate this phenomenon.

Let $\mathfrak{N}$ be a $n-$layer autoencoder network for which $\mathbf{c} = [c_0, c_1, c_2, ..., c_n]$ determines the number of neurons in subsequent layers. A number c_0 denotes the input size while for $i \in \{1, 2, ..., n\}$, c_i is the number of neurons in the $i-$th network layer. We assume that ReLU activation function is used in all hidden layers. Let us also define a matrix of neurons' activation values of the $i-$th layer of $\mathfrak{N}$, $\mathbf{G}_i(\boldsymbol{X_b}) = [\mathbf{g_1}, ..., \mathbf{g_{b_s}}]$, where $\boldsymbol{X_b} \in \mathbb{R}^{B \times b_s}$ is a batch of network input data and $i \in \{1, 2, ..., n\}$. Columns of the matrix $\mathbf{G}_i(\boldsymbol{X})$ store vectors of neurons' activations values for consecutive input data points, i.e. $\mathbf{g_j} = [g_{j,1}, g_{j,2}, ..., g_{j,c_i}]^\top$, where $j \in \{1, ..., b_s\}$.

The $q-$th neuron of the $k-$th layer is called *dead for a given input data* $\mathbf{x_i}$, $\mathbf{x_i} \neq \mathbf{0}$, where $\mathbf{x_i}$ is an element of $\boldsymbol{X_b}$, if $g_{j,q} = 0$ for $\mathbf{x_i}$ as the input of the network, $q \in \{1, ..., c_k\}$ and $k \in \{1, ..., n\}$. During a given training iteration we have $c_k \cdot b_s$ activations values for the $k-$th network layer. We introduce a dead activations' coefficient for the $k-$th network layer, d_{dead}^k:

$$d_{dead}^k = \frac{\mathcal{N}_0^k}{c_k \cdot b_s} \in [0, 1], \tag{1}$$

where $\mathcal{N}_0^k$ is a number of zero activations for all neurons of the $k-$th layer. We also calculate a dead activations' coefficient for the $q-$th neuron of the $k-$th layer, $d_{dead}^{k,q}$:

$$d_{dead}^{k,q} = \frac{\mathcal{N}_0^{k,q}}{b_s} \in [0, 1], \tag{2}$$

where $\mathcal{N}_0^{k,q}$ is a number of zero activations for the $q-$th neuron of the $k-$th network layer. During preliminary research using *basic* architecture, it was found that there is a significant correlation between the number of dead activations for the second encoder layer in a selected run of the trained model and mean reconstruction error. In such AE architectures for HU the last encoder layer has only few neurons, according to the number of endmembers in a given dataset (e.g. 3 for Samson and 4 for Jasper Ridge) so this layer is a bottleneck. We investigated models generated during Experiments 4 and 10 (according to Table 1), i.e. using MSE loss function and two HSI datasets. Results of Spearmank's rank correlation coefficient for all weight initialisation methods ranged between 0.76 and 0.89. Based on the above observations we have proposed three network reinitialisation methods dependent on the number of dead activations for the second encoder layer (d_{dead}^2) or the number of dead activations for consecutive neurons of the second encoder layer, i.e. ($d_{dead}^{2,1}, ..., d_{dead}^{2,c_2}$):

- *whole network reinitialisation*: if $d_{dead}^2 > t$ then all model's weights are randomly generated according to the given weights initialisation approach;
- *single layer reinitialisation*: if $d_{dead}^2 > t$ then all weights of the second encoder layer are reinitialised;
- *partial reinitialisation of a single layer*: weights of the second encoder layer connected with neurons which exceeded the dead activations' thresholds are reinitialised.

During each iteration of the network training one of the above conditions is verified, depending on the selected method. In our experiments, for all the methods used, we decided to set the threshold value t at 0.6.

Table 2. Results for different weight initialisation methods: He [12] and Glorot [9] with normal/uniform distribution. For each experiment a H-test statistic and a logarithm of p-value are presented. The significance level α is equal to 0.05. The column 'ph' corresponds to the ratio of p-values $< \alpha$ in post-hoc analysis. The bold font indicates the experiment with p-value $\geq \alpha$. In the case of very small p-values, logarithmic p-values are denoted '$-$ inf'.

Init.	He normal (KHN)			He uniform (KHU)			Glorot normal (XGN)			Glorot uniform (XGU)		
Exp. ID	H-stat	log p-val	ph	H-stat	log p-val	ph	H-stat	log p-val	ph	H-stat	log p-val	ph
1	200.9	-45.07	0.33	243.2	-61.76	0.42	78.8	-5.41	0.12	70.4	-3.72	0.10
2	891.2	-355.38	0.70	443.2	-147.76	0.56	625.9	-231.04	0.64	660.3	-246.96	0.66
3	763.8	-295.33	0.67	267.8	-71.82	0.41	239.1	-60.11	0.39	180.9	-37.49	0.33
4	2185.8	$-$ inf	0.88	2025.3	$-$ inf	0.72	1997.6	$-$ inf	0.75	2141.5	$-$ inf	0.76
5	1954.3	$-$ inf	0.88	2093.4	$-$ inf	0.90	1840.8	$-$ inf	0.87	1777.2	$-$ inf	0.85
6	134.0	-20.95	0.24	75.8	-4.79	0.11	76.3	-4.90	0.12	98.8	-10.32	0.16
7	903.8	-361.36	0.71	761.1	-294.07	0.67	953.8	-385.10	0.70	871.0	-345.85	0.69
8	77.3	-5.09	0.12	75.4	-4.71	0.11	78.4	-5.33	0.12	93.3	-8.88	0.16
9	69.2	-3.50	0.10	**66.2**	**-2.98**	–	74.2	-4.46	0.11	**47.9**	**-0.65**	–
10	1767.3	$-$ inf	0.85	2041.9	$-$ inf	0.82	1344.6	-572.45	0.70	1155.1	-481.29	0.73

3   Results

Results of the experimental evaluation of the non-uniform behaviour of weights are presented in Table 2 as Kruskal-Wallis H-test statistics and the corresponding p-values for the significance level $\alpha = 0.05$. In all but one experiment H_0 was rejected due to p-values lower than 0.05. An alternative hypothesis H_1 states that at least one initialisation of initial weights resulted in a value of the reconstruction error that was significantly different from the rest. This confirms that the reconstruction error of the trained network depends on weight initialisation.

To compare individual initialisations, a post-hoc analysis with the Conover-Iman test was performed. Example results are presented in Fig. 2 as heat-maps where each cell represents the statistical significance of difference between RMSE values of a pair of experiments. *NS* denotes a statistically insignificant difference. An analysis of these matrices reveals that for some cases, e.g. Experiment 1, there exists a subset of outlying initialisations while, e.g. for Experiment 4 the majority of initialisation pairs are significantly different. A summary of post-hoc analysis is presented in Table 2. Values in 'ph' columns are ratios of statistically significant ($p < 0.05$) differences between individual experiments according to the Conover-Iman test. Results indicate that using KHN initialisation slightly increased the number of outlying initialisations.

3.1   Autoencoder Training Improvement

We performed experiments validating the efficiency of network reinitialisation methods described in Sect. 2.6. We used 200 sets of initial weights from Experiment 4 (50

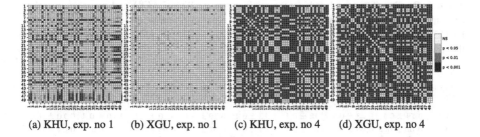

(a) KHU, exp. no 1   (b) XGU, exp. no 1   (c) KHU, exp. no 4   (d) XGU, exp. no 4

Fig. 2. Results of the post-hoc Conover-Iman test for pairs of initialisations.

per each initialisation approach like XGN, XGU, KHN and KHU). Each model was trained once for each training improvement method. Table 3 presents mean results of the described experiments with the corresponding. We denoted as *baseline* initial training results, without the use of reinitialisation techniques. Furthermore, to check whether the improvement is statistically significant, the Wilcoxon signed-rank tests between results for baseline models and for models with the application of reinitialisation methods were performed. Conducted experiments clearly indicate that proposed reinitialisation techniques significantly outperform the baseline training approach. In the case of Glorot initialisations *whole network* method achieved lowest RMSE and abundances error while for He algorithm *partial reinitialisation* was on average the most efficient.

Table 3. Results of network reinitialisation methods. The consecutive rows present mean results with standard deviation for each network setup. The filled square (■) means that the improvement between a given reinitialisation method and baseline is statistically significant, in terms of the Wilcoxon signed-rank test.

Init. method	Reinitialisation method	RMSE	Abundances error	Endmembers error
XGN	Baseline	0.036 ± 0.04	0.349 ± 0.04	0.744 ± 0.14
	Partial reinitialisation	0.089 ± 0.03	0.413 ± 0.08	$\mathbf{0.451 \pm 0.24}$ ■
	Single layer	0.019 ± 0.03 ■	0.310 ± 0.06 ■	0.777 ± 0.19
	Whole network	$\mathbf{0.007 \pm 0.00}$ ■	$\mathbf{0.276 \pm 0.03}$ ■	0.687 ± 0.11 ■
XGU	Baseline	0.052 ± 0.05	0.344 ± 0.05	0.723 ± 0.12
	Partial reinitialisation	0.089 ± 0.03	0.407 ± 0.08	$\mathbf{0.375 \pm 0.03}$ ■
	Single layer	0.020 ± 0.04 ■	0.307 ± 0.05 ■	0.707 ± 0.16
	Whole network	$\mathbf{0.007 \pm 0.00}$ ■	$\mathbf{0.270 \pm 0.03}$ ■	0.695 ± 0.11
KHN	Baseline	0.047 ± 0.04	0.365 ± 0.03	1.176 ± 0.13
	Partial reinitialisation	$\mathbf{0.024 \pm 0.01}$ ■	$\mathbf{0.330 \pm 0.03}$ ■	$\mathbf{1.100 \pm 0.13}$ ■
	Single layer	0.028 ± 0.01 ■	0.355 ± 0.04	1.153 ± 0.11
	Whole network	0.027 ± 0.01 ■	0.355 ± 0.04	1.134 ± 0.13 ■
KHU	Baseline	0.053 ± 0.05	0.357 ± 0.03	0.758 ± 0.19
	Partial reinitialisation	$\mathbf{0.007 \pm 0.00}$ ■	$\mathbf{0.308 \pm 0.04}$ ■	0.733 ± 0.12
	Single layer	0.014 ± 0.01 ■	0.347 ± 0.03	0.732 ± 0.15
	Whole network	0.008 ± 0.01 ■	0.316 ± 0.04 ■	$\mathbf{0.663 \pm 0.17}$ ■

Table 4. Results of mean abundances error in terms of RMSE (E^a) and mean endmembers error in terms of SAD (E^e) with standard deviations for different weight initialisation methods: He [12] and Glorot [9] with normal/uniform distribution. Bold font indicates the experiment with the lowest value of error.

Init.	He normal (KHN)		He uniform (KHU)		Glorot normal (XGN)		Glorot uniform (XGU)	
Exp. ID	Abundances	Endmembers	Abundances	Endmembers	Abundances	Endmembers	Abundances	Endmembers
1	0.34 ± 0.0	0.84 ± 0.2	0.36 ± 0.0	0.68 ± 0.2	$\mathbf{0.32 \pm 0.0}$	$\mathbf{0.36 \pm 0.3}$	0.33 ± 0.1	0.43 ± 0.3
2	0.19 ± 0.1	0.23 ± 0.1	0.13 ± 0.1	0.13 ± 0.1	0.10 ± 0.1	0.10 ± 0.1	$\mathbf{0.10 \pm 0.1}$	$\mathbf{0.10 \pm 0.1}$
3	$\mathbf{0.06 \pm 0.1}$	0.05 ± 0.1	0.07 ± 0.1	0.04 ± 0.1	0.07 ± 0.1	0.05 ± 0.1	0.07 ± 0.1	$\mathbf{0.04 \pm 0.1}$
4	0.36 ± 0.0	1.17 ± 0.1	0.36 ± 0.0	0.76 ± 0.2	0.35 ± 0.0	0.75 ± 0.1	$\mathbf{0.34 \pm 0.0}$	$\mathbf{0.73 \pm 0.1}$
5	0.35 ± 0.0	0.95 ± 0.1	0.35 ± 0.0	0.97 ± 0.2	$\mathbf{0.34 \pm 0.0}$	0.88 ± 0.2	0.34 ± 0.0	$\mathbf{0.87 \pm 0.1}$
6	0.29 ± 0.0	0.80 ± 0.1	0.24 ± 0.1	0.58 ± 0.2	0.22 ± 0.0	0.51 ± 0.2	$\mathbf{0.22 \pm 0.0}$	$\mathbf{0.50 \pm 0.2}$
7	0.20 ± 0.0	0.42 ± 0.1	0.17 ± 0.0	0.31 ± 0.1	$\mathbf{0.16 \pm 0.0}$	$\mathbf{0.28 \pm 0.1}$	0.16 ± 0.0	0.28 ± 0.1
8	0.29 ± 0.1	0.51 ± 0.4	$\mathbf{0.29 \pm 0.1}$	0.49 ± 0.4	0.29 ± 0.1	0.48 ± 0.3	0.29 ± 0.1	$\mathbf{0.48 \pm 0.4}$
9	0.27 ± 0.1	0.30 ± 0.1	0.27 ± 0.1	0.29 ± 0.1	0.27 ± 0.1	0.28 ± 0.1	$\mathbf{0.26 \pm 0.1}$	$\mathbf{0.28 \pm 0.1}$
10	0.30 ± 0.0	1.04 ± 0.1	$\mathbf{0.27 \pm 0.0}$	$\mathbf{0.89 \pm 0.1}$	0.29 ± 0.0	0.89 ± 0.1	0.28 ± 0.0	0.89 ± 0.1

3.2   Discussion

We observed that optimisation using MSE as a loss function also minimises the reconstruction error in terms of SAD function, to some extent. This relationship is particularly evident in the case of *basic* architecture. Indeed, if the value of MSE is close to 0, then also the SAD has to be close to 0. However, the opposite is not true, because optimisation using SAD function does not reduce error in the sense of MSE. Moreover, in most cases, after training with SAD function, all or almost all reconstructed points are outside of the simplex designated by endmembers. This phenomenon occurs because SAD function is scale invariant which means that only the spectral angle between input and output points is minimised. It is not necessarily related to the reduction of Euclidean distance between spectra. Despite this, abundances error in terms of RMSE and endmembers SAD error are comparable or even lower than when training with the MSE function which can be seen in the detailed results in Table 4.

Regarding the loss function, unmixing results for corresponding pairs of experiments, i.e. experiments with the same architecture and on the same dataset but with different loss functions are comparable. For pairs 1–2, 6–7 and 8–9, and for all weight initialisation methods, both abundances and endmembers errors were slightly smaller when the autoencoder was trained using the SAD loss, compared to the MSE. The lowest average error values are usually achieved for XGU initialisation. Furthermore, for all experiments, the weight initialisation method led to a lower mean endmembers error than KHN initialisation. Overall, Glorot initialisation technique seems to be on average better than He methods, when considering abundances/endmembers error values.

For some models with ReLU activation function, due to a large number of dead activations, a problem with the gradient flow during backpropagation steps emerged. This is especially important for bottleneck autoencoder architectures for HU. Presented reinitialisation methods which limit the number of dead activations can alleviate this problem allowing the signal to flow.

4   Conclusions

We have explored this phenomenon for the case of HU using autoencoders. We have observed cases of the vanishing gradient in the first AE layers and confirmed that initialisation has a crucial impact on the final AE performance. A weak initialisation leads to high reconstruction or endmembers errors, despite proper values of hyperparameters (e.g. selected by RayTune [18]). The problem was observed under a range of hyperparameter values, datasets, architectures, and initialisation methods. The phenomenon was confirmed by statistical verification, based on a large set of training experiments. Finally, we have presented three AE improvement methods based on a reinitialisation of all or some network weights. Our results may be applicable beyond the HU problem, into all domains where similar AE architectures are used. In the future, we would like to look into the impact of reinitialisation techniques on the performance of multilayer perceptrons in classification or regression tasks.

Acknowledgements. K.K. acknowledges funding from the European Union through the European Social Fund (grant POWR.03.02.00-00-I029). B.G. acknowledges funding from the budget funds for science in the years 2018-2022, as a scientific project "Application of transfer learning methods in the problem of hyperspectral images classification using convolutional neural networks" under the "Diamond Grant" program, no. DI2017 013847.

References

1. Alabdulmohsin, I., Maennel, H., Keysers, D.: The impact of reinitialization on generalization in convolutional neural networks (2021)
2. Bingham, G., Miikkulainen, R.: AutoInit: analytic signal-preserving weight initialization for neural networks (2021)
3. Bioucas-Dias, J.M.: A variable splitting augmented Lagrangian approach to linear spectral unmixing. In: 2009 First Workshop on Hyperspectral Image and Signal Processing: Evolution in Remote Sensing, pp. 1–4 (2009). https://doi.org/10.1109/WHISPERS.2009.5289072
4. Hyperspectral unmixing overview: geometrical, statistical, and sparse regression-based approaches. IEEE J. Sel. Top. Appl. Earth Obs. Remote Sens. **5**(2), 354–379 (2012). https://doi.org/10.1109/JSTARS.2012.2194696
5. Boardman, J., Kruse, F.A., Green, R.: Mapping target signatures via partial unmixing of AVIRIS data. In: Summaries of the Fifth Annual JPL Airborne Earth Science Workshop. Volume 1: AVIRIS Workshop (1995)
6. Borsoi, R.A., Imbiriba, T., Bermudez, J.C.M.: Deep generative endmember modeling: an application to unsupervised spectral unmixing. IEEE Trans. Comput. Imaging **6**, 374–384 (2020). https://doi.org/10.1109/TCI.2019.2948726
7. Conover, W.J.: Practical Nonparametric Statistics, vol. 350, 3rd edn. Wiley, Hoboken (1998)
8. Conover, W.J., Iman, R.L.: On multiple-comparisons procedures (1979)
9. Glorot, X., Bengio, Y.: Understanding the difficulty of training deep feedforward neural networks. In: Proceedings of AISTATS (2010), vol. 9, pp. 249–256 (2010)
10. Guo, A.J., Zhu, F.: Improving deep hyperspectral image classification performance with spectral unmixing. Signal Process. **183**, 107949 (2021). https://doi.org/10.1016/j.sigpro.2020.107949
11. Guo, R., Wang, W., Qi, H.: Hyperspectral image unmixing using autoencoder cascade. In: 2015 7th Workshop on Hyperspectral Image and Signal Processing: Evolution in Remote Sensing (WHISPERS), pp. 1–4 (2015). https://doi.org/10.1109/WHISPERS.2015.8075378

12. He, K., Zhang, X., Ren, S., Sun, J.: Delving deep into rectifiers: surpassing human-level performance on ImageNet classification. In: Proceedings of ICCV, pp. 1026–1034 (2015). https://doi.org/10.1109/ICCV.2015.123

13. Hinton, G.E., Salakhutdinov, R.R.: Reducing the dimensionality of data with neural networks. Science **313**(5786), 504–507 (2006). https://doi.org/10.1126/science.1127647

14. Keshava, N., Mustard, J.F.: Spectral unmixing. IEEE Signal Process. Mag. **19**(1), 44–57 (2002). https://doi.org/10.1109/79.974727

15. Krizhevsky, A., Hinton, G.E.: Using very deep autoencoders for content-based image retrieval. In: ESANN (2011)

16. Kruskal, W.H., Wallis, W.A.: Use of ranks in one-criterion variance analysis. J. Am. Stat. Assoc. **47**(260), 583–621 (1952). https://doi.org/10.2307/2280779

17. LeCun, Y.A., Bottou, L., Orr, G.B., Müller, K.-R.: Efficient BackProp. In: Montavon, G., Orr, G.B., Müller, K.-R. (eds.) Neural Networks: Tricks of the Trade. LNCS, vol. 7700, pp. 9–48. Springer, Heidelberg (2012). https://doi.org/10.1007/978-3-642-35289-8_3

18. Liaw, R., Liang, E., Nishihara, R., Moritz, P., Gonzalez, J.E., Stoica, I.: Tune: a research platform for distributed model selection and training (2018)

19. Lu, L., Shin, Y., Su, Y., Em Karniadakis, G.: Dying ReLU and initialization: theory and numerical examples. Commun. Comput. Phys. **28**(5), 1671–1706 (2020). https://doi.org/10.4208/cicp.OA-2020-0165

20. Lv, J., Shao, X., Xing, J., Cheng, C., Zhou, X.: A deep regression architecture with two-stage re-initialization for high performance facial landmark detection. In: Proceedings of CVPR 2017, pp. 3691–3700 (2017). https://doi.org/10.1109/CVPR.2017.393

21. Ozkan, S., Kaya, B., Akar, G.B.: EndNet: sparse autoencoder network for endmember extraction and hyperspectral unmixing. IEEE Trans. Geosci. Remote Sens. **57**(1), 482–496 (2019). https://doi.org/10.1109/TGRS.2018.2856929

22. Palsson, B., Ulfarsson, M.O., Sveinsson, J.R.: Convolutional autoencoder for spatial-spectral hyperspectral unmixing. In: Proceedings of IGARSS 2019, pp. 357–360 (2019). https://doi.org/10.1109/IGARSS.2019.8900297

23. Palsson, B., Sigurdsson, J., Sveinsson, J.R., Ulfarsson, M.O.: Hyperspectral unmixing using a neural network autoencoder. IEEE Access **6**, 25646–25656 (2018). https://doi.org/10.1109/ACCESS.2018.2818280

24. Palsson, B., Ulfarsson, M.O., Sveinsson, J.R.: Convolutional autoencoder for spectral-spatial hyperspectral unmixing. IEEE Trans. Geosci. Remote Sens. **59**(1), 535–549 (2021). https://doi.org/10.1109/IGARSS.2019.8900297

25. Plaza, A., Chang, C.: Impact of initialization on design of endmember extraction algorithms. IEEE Trans. Geosci. Remote Sens. **44**(11), 3397–3407 (2006). https://doi.org/10.1109/TGRS.2006.879538

26. Ranasinghe, Y., et al.: Convolutional autoencoder for blind hyperspectral image unmixing (2020)

27. Rister, B., Rubin, D.L.: Probabilistic bounds on neuron death in deep rectifier networks (2021)

28. Su, Y., Li, J., Plaza, A., Marinoni, A., Gamba, P., Chakravortty, S.: DAEN: deep autoencoder networks for hyperspectral unmixing. IEEE Trans. Geosci. Remote Sens. **57**(7), 4309–4321 (2019). https://doi.org/10.1109/TGRS.2018.2890633

29. Winter, M.E.: N-FINDR: an algorithm for fast autonomous spectral end-member determination in hyperspectral data. In: Descour, M.R., Shen, S.S. (eds.) Imaging Spectrometry V, vol. 3753, pp. 266–275. International Society for Optics and Photonics, SPIE (1999). https://doi.org/10.1117/12.366289

30. Zhao, M., Wang, M., Chen, J., Rahardja, S.: Hyperspectral unmixing via deep autoencoder networks for a generalized linear-mixture/nonlinear-fluctuation model (2019)
31. Zhu, F.: Hyperspectral unmixing: ground truth labeling, datasets, benchmark performances and survey (2017)

Cluster Centers Provide Good First Labels for Object Detection

Gertjan J. Burghouts$^{(\boxtimes)}$ ⓘ, Maarten Kruithof, Wyke Huizinga,
and Klamer Schutte

TNO, The Hague, The Netherlands
gertjan.burghouts@tno.nl
https://gertjanburghouts.github.io

Abstract. Learning object detection models with a few labels, is possible due to ingenious few-shot techniques, and due to clever selection of images to be labeled. Few-shot techniques work with as few as 1 to 10 randomized labels per object class. We are curious if performance of randomized label selection can be improved by selecting 1 to 10 labels per object class in a non-random manner. Several active learning techniques have been proposed to select object labels, but all started with a minimum of several tens of labels. We explore an effective and simple label selection strategy, for the case of 1 to 10 labels per object class. First, the full unlabeled dataset is clustered into N clusters, where N is the desired number of labels. Clustering is based on k-means on embedding vectors from a state-of-the-art pretrained image classification model (SimCLR v2). The image closest to the center is selected to be labeled. It is effective: on Pascal VOC we validate that it improves over randomized selection over 25%, with large improvements especially when having 1 label per object class. We have several benefits to report on this simple strategy: it is easy to implement, it is effective, and it is relevant in practice where one often starts with a dataset without any labels.

Keywords: Label selection · Few labels · Object detection · Clustering

1 Introduction

Learning good object detection models with few labels, is essential for practical applications where one needs to have a model in place quickly and annotation resources are scarce. Our interest is in as few labeled samples as 1 label per object class. Progress has been realized by adapting image classification strategies to object detection, mainly few-shot techniques and active learning.

Active learning for object detection is challenging compared to image classification, since there are many possible candidate image regions that may contain an object. One of the first approaches (without deep learning) was to cluster images and then select those that reduce uncertainty most by minimizing entropy as the criterion [10]. For object detection based on deep learning, other criteria were proposed, e.g., to select uncertain images based on a model's confidences. In [15], boxes of intermediate confidences are considered, because these are uncertain, where selection is based on large variation of confidences across image scales. A similar approach is taken in [2,4,8], where the score of the image is obtained after aggregation of confidence divergence scores locally (spatial patches, feature cells or pixels). Another approach was taken in [20], by formulating the image selection as a multiple instance learning problem. Uncertainty is estimated from images which are modeled as bag of objects instances. These earlier works typically selected several tens [4,15] to hundreds or more labels [2]. Our interest, however, is the range of only 1 to 10 labels per object class.

Few-shot object detection deals with the problem of learning good object models with few labels. Several methods have been proposed to deal with few labeled images for the novel classes. Approaches vary from feature transfer and advanced reweighting by meta-learning [11,18,19], and proper pretraining and careful two-stage finetuning [17]. Currently the best performance on the Pascal VOC object detection benchmark [7] is achieved by the careful two-stage finetuning by [17]. These few-shot techniques work with as few as 1 to 10 randomized labels per object class. We are curious if performance of randomized label selection can be improved by selecting 1 to 10 labels per object class in a non-random manner. To that end, we adopt the state-of-the-art technique and validation from [17] and we only change the way that labels are selected. We validate whether labels can be selected more effectively than random.

We explore a very simple yet effective label selection strategy, for the case of 1 to 10 labels per object class. It addresses two criteria that are well-known in active learning: diversity [3] and representative power [16]. First, the full dataset (without any labels) is clustered into N clusters, where N is the desired number of labels. Clustering is based on k-means [13] on embedding vectors from a state-of-the-art pretrained image classification model (SimCLR v2 [6]) optionally followed by a dimension reduction [14]. The image closest to the center is selected to be labeled. Diversity and representativeness are well-known criteria in active learning [3,16], which we adopt. The found clusters will be spread by design, which facilitates diversity. For each cluster, we select the most representative sample for that cluster.

Our simple strategy is effective: on the Pascal VOC benchmark, it improves randomized selection over 25%. Large improvements are established especially when having 1 label per object class. At all amounts of labels per object class, our method improves over random selection. We report this simple strategy, because: it is easy to implement, it is effective, and it is relevant in practice when one often starts with a dataset without any labels.

Section 2 discusses related work. In Sect. 3, we explain our method. Section 4 shows experimental results and shares the findings, followed by a conclusion in Sect. 5.

2   Related Work

2.1   Few-Shot Object Detection

Various methods have been proposed to deal with few labeled bounding boxes for the novel classes. Meta-learning approaches include feature reweighting for single-stage YOLOv2 detection [11] and two-stage Faster R-CNN [19]. A meta-learner takes a few labeled images of the object classes and their bounding boxes. It learns to reweight the vector representation of the image to maximally represent each object category. The meta-learning approach in [18] learns to generate model parameters from base classes, using a large amount of labeled bounding boxes. The rationale is that this generation model is able to learn novel classes more efficiently. Recently, a simpler approach including proper pretraining and careful two-stage finetuning [17] outperformed the meta-learning approaches. First a base model was learned from the base classes for which many labeled object boxes are available. Then a model was finetuned to adapt to novel classes, by retraining only the last layer on a balanced set of few labels from base and novel classes. Currently the best performance on the Pascal VOC object detection benchmark is achieved by careful two-stage finetuning. Therefore, we adopt this approach of two-stage finetuning.

2.2   Label Selection for Object Detection

We consider the problem where for novel object classes only a few labels per class can be selected. Diversity [3] and representativeness [16] are well-known label selection criteria. Earlier works have studied label selection for object detection. Clustering has proven an effective means divide the space of unlabeled images, followed by a method to select the clusters of which an image will to be labeled [10]. The criterion for this selection is based on reducing uncertainty, by sequentially minimizing the expected entropy. For object detection based on deep learning, recently other label selection methods have been proposed, based on the internal representations and/or outputs of the model. One proposed criterion is to label the images of which the margin between the highest confidence and the second highest confidence is low. In [15], boxes of intermediate confidences are considered, as these are uncertain. Largest margins of confidence are determined across image scales, by comparing the highest confidence to the second highest confidence from another auxiliary layer. Other auxiliary layers encode the image at a different spatial scale. A large margin between close spatial scales indicates uncertainty. A similar approach is taken in [4] the score of the image is obtained after aggregation of the confidences at all spatial cells in the image, because objects are contained in regions within the image. This approach is shown to be beneficial for ≥50 labels per object class. Pixel level distributions of object class confidences are considered in [2,8], where the goal is to look for images which locally have a large divergence of confidence distributions, as a measure of uncertainty. In [8] the entropy of such distributions is computed per pixel and accumulated into an image score. In [2], the divergence score is computed in patches. An image score is obtained by pooling across patches. This approach is demonstrated to be effective for ≥ 500 labeled images. Multiple instance learning has been studied for object label selection [20]. The Active Object Detection (MI-AOD) selects the most informative images based on predicting instance-level uncertainty of unlabeled images. Unlabeled images are modeled as instance bags. Uncertainty is estimated by re-weighting instances in a multiple instance learning (MIL) fashion. The model was validated with a starting point of 5% of the dataset being labeled.

The methods described above have proven their effectiveness at several tens or hundreds of labeled images. In contrast, we are interested in object detection with only 1 to 10 labels per object class.

3   Method

The rationale is to cluster the embeddings of the images in such a way, that the clusters are diverse, and one representative image is selected per cluster. When allowed to have N labeled images per class, with C classes, then all images are clustered into $N * C$ clusters and one image is selected per cluster. The selected $N * C$ images will be labeled. Our label selection method consists of several consecutive components, which are explained in the next subsections.

3.1   Image Representation

Images are represented by an embedding vector. For this purpose, we consider the state-of-the-art embedding by the SimCLR v2 model [6]. This embedding has been shown to generalize well to new vision tasks, in particular when few labels are available. SimCLR v2 is an improvement of SimCLR [5]. SimCLR is based on contrastive learning. It learns representations by maximizing agreement between differently augmented views of the same data example via a contrastive loss in the latent space. SimCLR v2 has improvements over SimCLR: larger ResNet models, increased capacity of the non-linear projection head by making it deeper, careful finetuning of particular layers, and a memory mechanism to stabilize weights by a moving average during training. Given its generalization with few labels, we consider SimCLR v2 suitable for representing images for label selection. For comparison, as a reference, we will also validate our method using Resnet50 embeddings [9]. Both the SimVLR v2 and the Resnet50 models used are pretrained on ImageNet1K, providing knowledge on discriminating image structure, which we expect to be useful for low amounts of labels.

3.2   Clustering to Select Labels

The state-of-the-art method [17] balances the number of labels from the base and novel classes. Despite the many labels for the base classes, this class balancing is important for the optimization of the model parameters [17]. Hence it is useful to also carefully select the few labels not only for the novel classes, but also for the base classes.

The selection will be based on clustering the dataset's images, as it will facilitate diversity via the clusters which are spread out in embedding space. Representativeness of the selected images is taken into account by selecting a representative image per cluster. The clustering will be performed on the vectors resulting from embedded the images by SimCLR v2 [6]. We expect that the clustering may be improved by reducing the dimensionality of the embeddings (4096 dimensions), as distances in high-dimensional spaces may be problematic [1] and taking into account local structures in the data may be beneficial [14]. For this purpose, we leverage UMAP [14] which has proven useful to learn useful dimension reductions (typically ≤ 30) on the target dataset by taking into account the local manifold structure of the target dataset.

In the resulting vector space, we apply K-means clustering [13]. When the label budget is to select N labeled images per class, with C classes, then $N * C$ clusters are determined. Per cluster the representative image is selected by taking the image with its vector closest to the cluster center. For $N * C$ clusters, selecting one image per cluster, yields a selection of $N * C$ images to be labeled.

4   Experiments

4.1   Setup

For few-shot object detection on Pascal VOC [7], recently a new standard was proposed by [17] to perform 3 randomized splits of 15 base classes and 5 novel classes. The images are taken from the Pascal VOC 2007 and 2012 sets. For the 15 base classes, all labels are available. For the 5 novel classes, only a few labels are available. We follow this setup. For increasing amounts of labels (1, 2, 3, 5, 10 per class), the result is shown by the average mAP (mean average precision) across the 3 splits.

4.2   Implementation

Our label selection method is applicable to any object detection model that requires a (possibly small) set of images to learn the model. For experiments in this paper, we used the Retinanet object detection model [12], which is one of the standard models today. It performs well due to its balancing of object classes versus the large background of image regions that contain no object of interest (focal loss). It is a single-stage model, which we found easier to train. We use this implementation[1] with all standard parameters, which has a straightforward training procedure with build-in image and box augmentations.

---

[1] https://github.com/fizyr/keras-retinanet.

4.3   Results

The original setup [17] is to select an amount of labeled objects per class. Our setup is a bit different, as we select the actual images to be labeled per class. Therefore we report the amount of labeled images per class. As a baseline, we consider random selection of labeled images per class. That is, for a class, we extract the list of all images that contain that class, and select randomly from that list. As a reference, we report the state-of-the-art performance [17]. Please note that they consider an amount of labeled objects instead of labeled images, and therefore their results are not directly comparable.

We show the performance for two types of label selection:

– Label selection per class. We take all images that contain that class as a starting point for our method. We select N labels for each of the C classes. The purpose of this setup is to provide a reference to compare with the state-of-the-art where methods also select an amount of labels per class. Practical use of performing label selection per class is arguable, as it requires that all images to be assigned to a certain class, which not often is the case. However, note that similar problems exists with the commonly used per-class random selection, which also requires observation of quite some data to make sure all classes have sufficient randomly selected samples.
– Label selection from the full dataset. We take all images as a starting point for our method. We select $N * C$ images at once, where N is the amount of labels per object class, and C the number of classes. This is a realistic setup, because projects often start with an unlabeled dataset, without any prior knowledge on which images contain which object.

The results are summarized in Table 1. Best setting per clustering variant is indicated by underlining. Best result for 1 label per object class is indicated in bold. Our findings are listed in the next subsection.

4.4   Findings

Our findings are:

1. For random selection, a selection of N labels per class works much better (0.36) than random on the full dataset (0.30). A selection per class guarantees that all classes are labeled, and probably more diverse labels. Especially with 1 label/class, the risk of not selecting a class from the full dataset is large, leading to a low performance (0.17) compared to per-class selection (0.32).
2. Selecting labels by clustering works better than random selection. In particular, the result at 1 label/class by clustering on the full dataset, using the SimCLR embedding, stands out (0.39) in comparison to random selection on the full dataset (0.17). Also, this is better than state-of-the-art and the common scheme to select randomly per class.

Table 1. mAP results on Pascal VOC for the variants of our method and state-of-the-art, for various amounts of labeled images per class (we report the average across the three randomized splits of 15 base classes vs. 5 novel classes).

Method			Labeled images per object class					
			1	2	3	5	10	Average
Random selection per class								
Wang et al., ICML, 2020 [17]*			0.31	0.32	0.41	0.47	0.48	0.40
Baseline: random images per class			0.32	0.36	0.41	0.48	0.51	0.42
Label selection per class								
Random			0.32	0.36	0.41	0.48	0.51	0.42
Clustering	Resnet50		0.33	0.41	0.46	0.47	0.50	0.43
Clustering	Resnet50	UMAP	0.33	0.41	0.45	<u>0.52</u>	0.52	0.45
Clustering	SimCLR		<u>0.35</u>	<u>0.44</u>	0.45	0.49	0.52	0.45
Clustering	SimCLR	UMAP	0.34	0.43	<u>0.47</u>	0.50	<u>0.52</u>	<u>0.45</u>
Label selection on full dataset								
Random			0.17	0.33	0.39	0.42	0.47	0.36
Clustering	Resnet50		0.30	0.37	0.43	0.48	0.49	0.41
Clustering	Resnet50	UMAP	0.28	0.37	0.36	0.46	0.47	0.39
Clustering	SimCLR		**0.39**	<u>0.41</u>	0.42	0.46	0.49	<u>0.43</u>
Clustering	SimCLR	UMAP	0.27	0.38	<u>0.46</u>	<u>0.46</u>	<u>0.49</u>	0.41

*) [17] reports amount of labeled objects (not labeled images) per class. Therefore we repeated experiments for labeled images per class.

3. Interestingly, the result by clustering on the full dataset, using the SimCLR embedding, with 1 label/class (0.39) is better than the clustering per class (0.35). Apparently, the clustering is able to select more diverse labels than per class (this is to be expected) at no cost of missing classes in the labeling (which is a common problem with random selection, which resulted in a low performance of 0.17).

4. The success of the clustering on the full dataset is beneficial for everyday object detection applications, as these projects often start with a dataset without any labels. The clustering per class requires prior knowledge about which images contain which object. Clustering on the full dataset does not require any such prior knowledge.

5. At several labels/class, clustering on the full dataset no longer outperforms clustering or random selection per class (e.g. at 5 labels/class 0.46 versus 0.50 and 0.48).

6. Interestingly, for ≥ 3 labels/class, it is better to use dimension reduction by UMAP. We hypothesize that at more labels/class, the local structures in the embedding space become more important. UMAP helps to make better separations between the local structures of the target dataset.

7. For ≥ 5 labels per object class, all sampling strategies yield very similar performances.

Fig. 1. Examples of selected images for {sofa, motorbike, bicycle, cat, bus}, for one labeled image per class on average, using our method on SimCLR v2.

8. Generally, in our label selection approach, SimCLR v2 embeddings lead to higher performance compared to ResNet50 embeddings. The best choice across the various amounts of labels/class, is clustering on the full dataset using the SimCLR v2 embedding.

4.5 Illustrations of Selected Images

Figure 1 shows examples of selected images for one split, with 5 novel classes: {sofa, motorbike, bicycle, cat, bus}. These selected images are a result of our method when labeling 1 image per class on average, using SimCLR v2 embeddings. The selected images are very representative: the object is displayed in very common backgrounds, and the objects are well visible.

5 Conclusion

We have found that random selection of object labels can be improved when having only 1 to 10 labels per object class by a simple strategy. The strategy is to cluster the full dataset and label the image that is closest to each cluster center. It is effective and easy to implement. Moreover, selecting only 1 to 10 labels per object class, is the relevant case in practice where one often starts with a dataset without any labels. Earlier works have studied the selection of several tens of labels per object class, but that is not always possible when there are few images of the object or resources or time-to-deployment is limited. That is why we think that the simple strategy and improved performances as presented in this paper are relevant.

References

1. Aggarwal, C.C., Hinneburg, A., Keim, D.A.: On the surprising behavior of distance metrics in high dimensional space. In: Van den Bussche, J., Vianu, V. (eds.) ICDT 2001. LNCS, vol. 1973, pp. 420–434. Springer, Heidelberg (2001). https://doi.org/10.1007/3-540-44503-X_27

2. Aghdam, H.H., Gonzalez-Garcia, A., Weijer, J.v.d., López, A.M.: Active learning for deep detection neural networks. In: Proceedings of the IEEE/CVF International Conference on Computer Vision, pp. 3672–3680 (2019)

3. Ash, J.T., Zhang, C., Krishnamurthy, A., Langford, J., Agarwal, A.: Deep batch active learning by diverse, uncertain gradient lower bounds. In: International Conference on Learning Representations (2020)

4. Brust, C.A., Käding, C., Denzler, J.: Active learning for deep object detection. In: Computer Vision Theory and Applications (VISAPP) (2019)

5. Chen, T., Kornblith, S., Norouzi, M., Hinton, G.: A simple framework for contrastive learning of visual representations. In: Proceedings of International Conference on Machine Learning (2020)

6. Chen, T., Kornblith, S., Swersky, K., Norouzi, M., Hinton, G.: Big self-supervised models are strong semi-supervised learners. In: Proceedings of NeurIPS (2020)

7. Everingham, M., Van Gool, L., Williams, C.K., Winn, J., Zisserman, A.: The PASCAL visual object classes challenge 2007 (VOC2007) results (2007)

8. Haussmann, E., et al.: Scalable active learning for object detection. In: 2020 IEEE Intelligent Vehicles Symposium (IV), pp. 1430–1435. IEEE (2020)

9. He, K., Zhang, X., Ren, S., Sun, J.: Deep residual learning for image recognition. In: Proceedings of the IEEE Conference on Computer Vision and Pattern Recognition, pp. 770–778 (2016)

10. Holub, A., Perona, P., Burl, M.C.: Entropy-based active learning for object recognition. In: 2008 IEEE Computer Society Conference on Computer Vision and Pattern Recognition Workshops, pp. 1–8. IEEE (2008)

11. Kang, B., Liu, Z., Wang, X., Yu, F., Feng, J., Darrell, T.: Few-shot object detection via feature reweighting. In: Proceedings of the IEEE/CVF International Conference on Computer Vision, pp. 8420–8429 (2019)

12. Lin, T.Y., Goyal, P., Girshick, R., He, K., Dollár, P.: Focal loss for dense object detection. In: Proceedings of the IEEE International Conference on Computer Vision, pp. 2980–2988 (2017)

13. Lloyd, S.P.: Least squares quantization in PCM. IEEE Trans. Inf. Theory **28**(2), 129–137 (1982)

14. McInnes, L., Healy, J.: UMAP: uniform manifold approximation and projection for dimension reduction. arXiv e-prints 1802.03426 (2018)

15. Roy, S., Unmesh, A., Namboodiri, V.P.: Deep active learning for object detection. In: BMVC, p. 91 (2018)

16. Sinha, S., Ebrahimi, S., Darrell, T.: Variational adversarial active learning. In: Proceedings of the IEEE/CVF International Conference on Computer Vision, pp. 5972–5981 (2019)

17. Wang, X., Huang, T.E., Darrell, T., Gonzalez, J.E., Yu, F.: Frustratingly simple few-shot object detection. In: International Conference on Machine Learning (2020)

18. Wang, Y.X., Ramanan, D., Hebert, M.: Meta-learning to detect rare objects. In: Proceedings of the IEEE/CVF International Conference on Computer Vision, pp. 9925–9934 (2019)

19. Yan, X., Chen, Z., Xu, A., Wang, X., Liang, X., Lin, L.: Meta R-CNN: towards general solver for instance-level low-shot learning. In: Proceedings of the IEEE/CVF International Conference on Computer Vision, pp. 9577–9586 (2019)
20. Yuan, T., et al.: Multiple instance active learning for object detection. In: Proceedings of the IEEE/CVF Conference on Computer Vision and Pattern Recognition, pp. 5330–5339 (2021)

Unsupervised Detection of Dynamic Hand Gestures from Leap Motion Data

Andrea D'Eusanio[1], Stefano Pini[1], Guido Borghi[2], Alessandro Simoni[1(✉)],
and Roberto Vezzani[1]

[1] Department of Engineering "Enzo Ferrari" (DIEF),
University of Modena and Reggio Emilia, 41125 Modena, Italy
{andrea.deusanio,s.pini,alessandro.simoni,roberto.vezzani}@unimore.it
[2] Dipartimento di Informatica Scienza e Ingegneria (DISI), University of Bologna,
47521 Cesena, Italy
guido.borghi@unibo.it

Abstract. The effective and reliable detection and classification of dynamic hand gestures is a key element for building Natural User Interfaces, systems that allow the users to interact using free movements of their body instead of traditional mechanical tools. However, methods that temporally segment and classify dynamic gestures usually rely on a great amount of labeled data, including annotations regarding the class and the temporal segmentation of each gesture. In this paper, we propose an unsupervised approach to train a Transformer-based architecture that learns to detect dynamic hand gestures in a continuous temporal sequence. The input data is represented by the 3D position of the hand joints, along with their speed and acceleration, collected through a Leap Motion device. Experimental results show a promising accuracy on both the detection and the classification task and that only limited computational power is required, confirming that the proposed method can be applied in real-world applications.

Keywords: Dynamic hand gestures recognition · Unsupervised gesture detection · Transformer-based architecture · Leap motion

1 Introduction

Nowadays, *Natural User Interfaces* (NUIs) [26], *i.e.* interfaces in which the interaction relies on free movements of the user body instead of the adoption of mechanical tools (such as mouse and keyboard), represent a powerful solution in the *Human Computer Interaction* (HCI) field to build intuitive and user-friendly applications. Dynamic hand gestures are one of the most-used ways to interact [31], along with voice commands [1,4] and gaze [18,20]. In this context, the growing interest in dynamic hand gestures has been supported by the recent introduction of affordable devices that are capable of acquiring both 2D and 3D data. Moreover, some devices can also provide additional semantic information,

S. Sclaroff et al. (Eds.): ICIAP 2022, LNCS 13231, pp. 414–424, 2022.
https://doi.org/10.1007/978-3-031-06427-2_35

such as hand keypoints [35] or skeleton joints of the human body [34], with high accuracy and real time performance. In this work, we assume to acquire information on the user's hand using the *Leap Motion Controller* device [21], an infrared stereo camera capable of estimating the 3D positions of the hand joints in real time through its proprietary software tools.

Unfortunately, NUIs are still limited in real world applications, due to the lack of effective and robust methods that correctly and quickly detect and classify the gestures. In addition, most of the existing methods, especially the deep learning-based ones, require a large amount of labeled training data, in which the class and the temporal boundaries of each gesture have to be annotated. However, the annotation of the temporal segmentation of each gesture is a time-consuming and error-prone procedure, in particular in case of long sequences. Moreover, the use of specific datasets collected for a given use case or application requires the user to label new data if the method is applied in a different scenario or if a new gesture is added.

To address these issues, in this paper we propose a *Transformer*-based [37] model and a specific training approach that, in an unsupervised manner, allow the network to learn to detect the presence of dynamic hand gestures within an input sequence. Indeed, during the training phase, we assume to have access to a large set of single dynamic gestures (*i.e.* a set of sequences, each containing only one gesture) and their gesture class. In this scenario, we propose to exploit the *Connectionist Temporal Classification* (CTC) loss [13]: using this objective, a neural network can learn to temporally segment (*i.e.* detect) an element without explicit segmentation labels, while requiring only sequences of multiple gestures and the associated list of gesture classes. Thus, we apply this loss to "synthetic" gesture sequences, generated by combining several single gestures, and we show that this training approach leads to a learned model that is capable of successfully segmenting and classifying dynamic hand gestures. Moreover, during the testing procedure, we assume to have a continuous data stream that can contain none, one, or multiple dynamic gestures. We show that, even in this challenging case, the network trained with "synthetic" sequences successfully segments and classifies the gestures.

2 Related Work

In the literature, several approaches have been proposed to tackle the detection of hand gestures. Low-level motion parameters such as acceleration, velocity, and trajectory curvature [17] or, in general, body activity [16] can be used to detect the gesture boundaries. In addition, methods based on *Continuous-time Dynamic Programming* (CDP) [28], *Dynamic Time Warping* (DTW) [9], *Hidden Markov Models* (HMMs) [6,36] and *Conditional Random Fields* (CRF) [32] have been presented. Predicted likelihood scores are compared with a given threshold to detect the gesture boundaries even though, in terms of generalization capabilities, defining a fixed threshold is hard. Indeed, some methods [22,38] propose to compute an adaptive threshold at inference.

A great variety of methods tackles the classification of detected hand gestures, either static or dynamic. The task is commonly addressed through the use of machine learning-based methods, such as HOG and SVMs [12,33], HMMs [2, 22], and neural networks. The latter can be split, based on the used model, in recurrent networks, such as RNNs [15,23], LSTMs [5,40], and CNNs (2D [7,11] or 3D [27,42]). Moreover, recent works propose the combination of multiple features extracted by CNNs [25] or GNNs [41]. Recently, Vaswani *et al.* [37] proposed the Transformer model, an effective self-attention mechanism that has rapidly replaced recurrent methods in many natural language processing and computer vision tasks, including gesture classification [10].

As mentioned, the input of the framework is represented by the 3D hand joints provided by the Leap Motion Controller. In order to improve the usability on application scenarios, many efforts have been conducted by researchers in order to accurately detect 3D hand joints [24] even from single RGB [3] and infrared images [29] and depth maps [39].

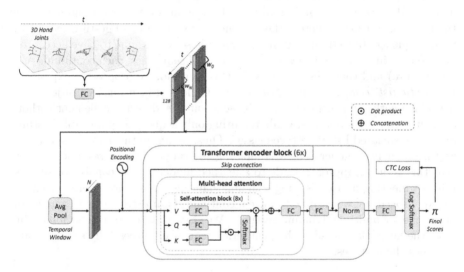

Fig. 1. Overview of the proposed method. A sequence of 3D hand joint features, provided by the *Leap Motion*, are fed into a fully connected (FC) layer and temporally pooled. Then, a Transformer-based network, based on self-attention blocks (see Sect. 3.1) and optimized with the CTC loss, (see Sect. 3.2), learns to detect and classify dynamic gestures.

3   Proposed Method

In this section, we present the architectural details of our method and the proposed training procedure that solves the detection and classification tasks in an unsupervised and supervised manner, respectively.

3.1 Network Architecture

The proposed gesture detection and classification architecture is defined by a Transformer-based model [37] which enables the temporal analysis of the input data. A schematic overview is shown in Fig. 1.

The input consists of a series of M feature vectors v_j representing the dynamic hand gesture at time j. The first module of the framework is represented by a fully connected layer with ReLu activation function that remaps each input vector into a 128-d feature. These features are then subdivided along the time dimension into a set of N temporal windows $W_i = \{v_j \,|\, (i-1)\frac{M}{N} < j \leq i\frac{M}{N}\}$ and passed through an average pooling layer that extracts an embedding e of aggregated features. This preliminary feature mapping operations can be defined as:

$$e = \mathrm{AvgPool}(\mathrm{FC}(v)) \tag{1}$$

The temporal analysis of the embedding e is then performed by a Transformer-based network which is composed of 6 consecutive encoder blocks E. Each block E contains a set of 8 self-attention blocks followed by two fully connected layers, one with a ReLu activation function and the other one with a normalization layer. An encoder is defined by:

$$E_i(e) = \mathrm{Norm}(e + \mathrm{MultiHead}(e)) \tag{2}$$

where MultiHead represents a multi-head attention block with 8 self-attention blocks. In details, a single self-attention operator is represented as:

$$\mathrm{Att}(e) = \mathrm{softmax}\left(\frac{Q\,K}{\sqrt{d_k}}\right) V \tag{3}$$

where Q, K, and V are linear projections of e into a 32-dimensional feature space, $d_k = 32$ is the scaling factor corresponding to the size of K. The multi-head attention module is a combination of multiple attention operators:

$$\mathrm{MultiHead} = (\mathrm{Att}_1 \oplus \ldots \oplus \mathrm{Att}_8)W^O \tag{4}$$

where $\oplus$ is the concatenation operator and W^O is a linear projection from and to a 128-d feature space.

The final part of our network is composed of a linear classifier that predicts $N + 1$ classes (corresponding to N different hand gestures and a "no gesture" label). The output features are then passed through a log-softmax that generates the final scores π.

3.2 Proposed Training Procedure

Our training procedure is built around the *Connectionist Temporal Classification* (CTC) loss function [13] and the generation of synthetic gesture sequences during training.

Synthetic Sequence Generation. The CTC loss can be used to learn an additional "None" class in an unsupervised manner. However, it requires sequences of multiple gestures split by a "no gesture" action, but obtaining this kind of data is hard. Indeed, these sequences are more complex to collect compared to single-gesture clips and the gestures must not be always performed in the same order. In addition, their annotation is more expensive, due to the need for a temporal segmentation of each gesture. To address these issues, we propose an alternative approach to construct synthetic gesture sequences from single gestures and show that they can be successfully used for training the proposed model. In details, we randomly combine single-gesture annotated sequences in longer sequences composed of multiple gestures and create a ground truth vector as a list of annotations of the single gestures. Then, without the need for any temporal annotation other than the ordered list of gesture classes, we train our network with the CTC loss.

Connectionist Temporal Classification. We employ the CTC loss function [13] to optimize the neural model during the training procedure. In particular, we adopt this loss for learning to segment and label gesture sequences from unsegmented data streams. In details, the model is forced by the CTC loss to predict one of the ground truth gesture classes or an additional label "no gesture" for each input window W_i. The result is a 1-d vector or path π, which maps the input to a sequence of class labels. Moreover, a function $\beta(\pi) = y$ removes the "no gesture" labels and collapses the sequentially-repeated class labels in single instances. For example, giving an input sequence of 7 frames and an output path such as $\pi = [-, 3, 3, -, -, 2, -, 2]$ (where $-$ is the "no gesture" class), the decoded output is $y = [3, 2, 2]$.

Gesture Detection and Classification. Given the predicted path π, we consider each switch from a "no gesture" label to any other gesture class as a detection of a new gesture. In a similar way, the switch from a gesture class to another gesture label or to the "no gesture" one is used to identify the end of the current gesture. That is, we use the class change in the prediction from/to the "no gesture" status or from/to another gesture as the beginning and the end of a gesture. Given that the model predicts a gesture class for each temporal window W_i, the gesture classification is simply given by this prediction. It is worth noting that, in this way, both the detection and the classification of the gestures are computed by the same model in a single pass.

4   Experimental Results

In this section, we present the experimental setting in terms of exploited data and model results. Finally, we analyze the performance of the proposed approach.

Table 1. Experimental results for the hand gesture detection task, obtained on the Briareo dataset [27].

Model	Detection metrics			
	Jaccard Index (%)	FPR (%)	Δ Start (s)	Δ End (s)
LSTM [14]	22.93	64.44	0.42 ± 0.47	0.08 ± 0.43
GRU [8]	42.30	45.50	0.51 ± 0.80	0.29 ± 0.77
Ours	**53.42**	**39.25**	$\mathbf{0.40 \pm 0.72}$	$\mathbf{-0.06 \pm 0.69}$

4.1 Experimental Setting

As input of the proposed model, we chose to use high-level 3D data, giving the extraction of this information as granted. Indeed, many sensor SDKs and existing neural networks are capable of computing high-level hand features in real time. In details, we use the location and rotation of the 3D hand joints retrieved by the Leap Motion SDK. In addition, we compute the speed and the acceleration of each joint using the joint locations in the previous time steps. An input gesture v_j can be defined as multiple hand joint features g^i:

$$g^i = \left([x, y, z], [\alpha, \beta, \gamma], [\mathbf{s}_x, \mathbf{s}_y, \mathbf{s}_z], [\mathbf{a}_x, \mathbf{a}_y, \mathbf{a}_z] \right) \tag{5}$$

where $[x, y, z]$ are the 3D coordinates of the hand joint i, $[\alpha, \beta, \gamma]$ are its rotation as Euler angles, $[\mathbf{s}_x, \mathbf{s}_y, \mathbf{s}_z]$ and $[\mathbf{a}_x, \mathbf{a}_y, \mathbf{a}_z]$ are respectively the speed and acceleration vectors computed with regard to the previous two frames. Since the Leap Motion device collects 16 hand joints, each joint information g^i is concatenated to the others obtaining a 192-d feature vector v_j.

During training, we optimize the network parameters using the *Adam* [19] optimizer with learning rate 10^{-4}, weight decay 10^{-4} and dropout with probability $p = 10^{-1}$ within the Transformer block. The model is developed using the PyTorch [30] framework. The code will be published online[1].

Given that we are not bound to predefined multiple-gesture sequences, we test using different numbers of gestures within the synthetic sequences. Similarly, we test multiple values of N by fixing the number of time steps within each window W_i. Thanks to the designed architecture, we do not have to set a fixed number of time steps M per each gesture sequence. In other words, we directly give the gesture sequence to the network regardless of its length. We report the results of our experiments in Sect. 4.3.

4.2 Dataset

We train and evaluate our method on a publicly released dataset, namely *Briareo*, following official train and test splits. Briareo [27] is a hand gesture dataset

---

[1] https://aimagelab.ing.unimore.it/go/unsupervised-gesture-segmentation.

Table 2. Experimental results, split between detection and classification metrics, obtained on the Briareo dataset [27].

Model	Classification metrics		
	F1 score (%)	Accuracy (%)	Recall (%)
Manganaro *et al.* [27]	–	94.40	–
LSTM [14]	89.87	98.98	82.29
GRU [8]	92.36	99.13	86.46
Ours	**95.99**	**99.52**	**92.71**

recorded in a realistic car simulator using multiple devices including the Leap Motion sensor, placed in the tunnel console looking upwards. Briareo was conceived for the automotive context, in which the infrared capabilities of the acquisition devices can be used to develop light-invariant vision-based solutions. Gestures from 12 classes are performed by 40 different subjects (33 males and 7 females). Each gesture is performed by each subject 3 times and the dataset contains an additional recording containing all the gestures in sequence. While the single-gesture sequences are used for training, the all-gestures sequence is used for testing. In order to evaluate the performance of the proposed approach, we create a small validation set sampling from the training data and manually annotate the all-gestures sequences.

4.3   Results

To evaluate our method, we select a set of metrics for both the detection and the classification task. In particular, we use the *Jaccard Index* as main metric for the detection, expressed as the intersection over union between the predicted path π and the ground truth path π^{GT}. This metric rewards a correct detection if there is at least one overlapping frame between the predicted and the ground truth gesture. All frames that are detected as gesture but do not have a correspondence in the ground truth are considered False Positives. In addition, we assess the temporal delay of the predictions as two time intervals Δ, one at the start and one at the end of the gesture, between the predicted and the target gesture. To evaluate the classification task, we use the F1 score, the classification accuracy, and the recall metric.

We report the results on Briareo both for the detection (Table 1) and the classification (Table 2) tasks. We compare the proposed Transformer-based method with two methods, based on LSTM [14] and GRU [8], and literature competitors. As shown in the left part of the tables, our method, trained on synthetic gesture sequences, obtains promising results on the detection metrics and outperforms the recurrent-based approaches. In the classification task, all the networks reach similar results, but our method is still able to outperform other approaches. Indeed, the transformer module has the ability to elaborate long sequences without recurrent structures, which is beneficial for the overall performance.

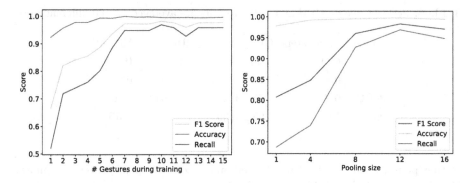

Fig. 2. F1 score, accuracy, and recall changing the number of gestures in the synthetic training sequences (left) and varying the number of time steps (which are average pooled) within each window W_i (right).

4.4 Ablation Study

In this section, we assess the performance of our method using different hyper-parameter values.

In Fig. 2 (left), we evaluate how the number of gestures that compose the training sequence impacts the classification accuracy during the test phase (when the number is fixed to 12). The experiment shows that the network achieves the best classification metrics when using 7 or more gestures and that results are very similar for higher values.

Similarly, we evaluate how the temporal pooling affects the model and report the results in the right part of Fig. 2. While there is a huge gap between the smallest pooling sizes (*i.e.* 1, 4) and the highest ones (*i.e.* 8, 12, 16), there is not a substantial difference within the highest group, showing that the pooling operation is beneficial to the classification accuracy.

4.5 Performance Analysis

We test the computational load of our architecture in terms of number of parameters and the required GPU VRAM in the testing phase. The performance analysis is conducted on a workstation equipped with a *Intel Core i7-7700K* and a *NVidia GeForce GTX 1080Ti*. Results are reported in Table 3. As shown, our method obtains comparable results on both GPU and CPU during inference.

4.6 Limitations

In this section, we analyze the difficulties encountered using the framework. First, we note that the training procedure based on the CTC loss function is unstable: the loss value has high variance – from low to high values and vice versa – even within consecutive training steps. Second, due to the unsupervised nature of the proposed method, we obtain a relatively high false positive rate during the

Table 3. Performance on GPU and CPU. Results are averaged on 100 different runs.

Model	Parameters (M)	GPU (ms)	CPU (ms)	VRAM (GB)
LSTM	0.9	**1.67 ± 0.23**	7.12 ± 0.64	0.60
GRU	**0.7**	1.75 ± 0.31	6.58 ± 0.66	0.60
Ours	1.0	4.65 ± 0.39	**4.72 ± 1.13**	**0.59**

testing phase, as shown in Table 1: that may require a post-processing phase. Finally, as reported in Table 3, the VRAM required at inference is low, while a great amount may be required for the training of the framework, depending on the length of the sequences and the pooling size W_i (see Sect. 3.1).

5   Conclusions and Future Work

In this paper, we propose a method to detect and classify dynamic hand gestures. The model, based on the Transformer architecture, is trained in an unsupervised manner for the temporal segmentation task and in a traditional supervised setting for the classification one. The CTC loss is exploited to learn the temporal segmentation without explicit labels. Experimental results, obtained on the Briareo dataset, reveal that the proposed method achieves satisfying accuracy scores with limited computational load. As future work, aiming at implementing the system in a real world Human-Car Interaction framework, we will test the model on embedded boards with a dedicated GPU and limited computational power, such as the *NVidia TX2* or the *NVidia Xavier* platforms.

References

1. Bala, A., Kumar, A., Birla, N.: Voice command recognition system based on MFCC and DTW. Int. J. Eng. Sci. Technol. **2**(12), 7335–7342 (2010)
2. Borghi, G., Vezzani, R., Cucchiara, R.: Fast gesture recognition with multiple stream discrete HMMs on 3D skeletons. In: 2016 23rd International Conference on Pattern Recognition (ICPR), pp. 997–1002. IEEE (2016)
3. Boukhayma, A., Bem, R.d., Torr, P.H.: 3D hand shape and pose from images in the wild. In: Proceedings of the IEEE/CVF Conference on Computer Vision and Pattern Recognition, pp. 10843–10852 (2019)
4. Brenon, A., Portet, F., Vacher, M.: Preliminary study of adaptive decision-making system for vocal command in smart home. In: 2016 12th International Conference on Intelligent Environments (IE), pp. 218–221. IEEE (2016)
5. Caputo, F.M., et al.: Online gesture recognition. In: Eurographics Workshop on 3D Object Retrieval. The Eurographics Association (2019)
6. Chen, F.S., Fu, C.M., Huang, C.L.: Hand gesture recognition using a real-time tracking method and hidden Markov models. Image Vis. Comput. **21**(8), 745–758 (2003)

7. Cheng, W., Sun, Y., Li, G., Jiang, G., Liu, H.: Jointly network: a network based on CNN and RBM for gesture recognition. Neural Comput. Appl. **31**(1), 309–323 (2019)
8. Cho, K., et al.: Learning phrase representations using RNN encoder-decoder for statistical machine translation. arXiv preprint arXiv:1406.1078 (2014)
9. Darrell, T.J., Essa, I.A., Pentland, A.P.: Task-specific gesture analysis in real-time using interpolated views. IEEE Trans. Pattern Anal. Mach. Intell. **18**(12), 1236–1242 (1996)
10. D'Eusanio, A., Simoni, A., Pini, S., Borghi, G., Vezzani, R., Cucchiara, R.: A transformer-based network for dynamic hand gesture recognition. In: International Conference on 3D Vision (2020)
11. D'Eusanio, A., Simoni, A., Pini, S., Borghi, G., Vezzani, R., Cucchiara, R.: Multi-modal hand gesture classification for the human-car interaction. Informatics **7**, 31 (2020). Multidisciplinary Digital Publishing Institute
12. Feng, K.p., Yuan, F.: Static hand gesture recognition based on hog characters and support vector machines. In: 2013 2nd International Symposium on Instrumentation and Measurement, Sensor Network and Automation (IMSNA), pp. 936–938. IEEE (2013)
13. Graves, A., Fernández, S., Gomez, F., Schmidhuber, J.: Connectionist temporal classification: labelling unsegmented sequence data with recurrent neural networks. In: Proceedings of the 23rd International Conference on Machine Learning, pp. 369–376 (2006)
14. Hochreiter, S., Schmidhuber, J.: Long short-term memory. Neural Comput. **9**(8), 1735–1780 (1997)
15. Hu, Y., Wong, Y., Wei, W., Du, Y., Kankanhalli, M., Geng, W.: A novel attention-based hybrid CNN-RNN architecture for sEMG-based gesture recognition. PLoS ONE **13**(10), e0206049 (2018)
16. Kahol, K., Tripathi, P., Panchanathan, S.: Automated gesture segmentation from dance sequences. In: Sixth IEEE International Conference on Automatic Face and Gesture Recognition 2004, Proceedings, pp. 883–888. IEEE (2004)
17. Kang, H., Lee, C.W., Jung, K.: Recognition-based gesture spotting in video games. Pattern Recogn. Lett. **25**(15), 1701–1714 (2004)
18. Ki, J., Kwon, Y.M.: 3D gaze estimation and interaction. In: 2008 3DTV Conference: The True Vision-Capture, Transmission and Display of 3D Video, pp. 373–376. IEEE (2008)
19. Kingma, D.P., Ba, J.: Adam: a method for stochastic optimization. arXiv preprint arXiv:1412.6980 (2014)
20. Lander, C., Gehring, S., Krüger, A., Boring, S., Bulling, A.: GazeProjector: accurate gaze estimation and seamless gaze interaction across multiple displays. In: Proceedings of the 28th Annual ACM Symposium on User Interface Software & Technology, pp. 395–404 (2015)
21. Leap Motion. https://www.ultraleap.com/product/leap-motion-controller. Accessed 09 Dec 2021
22. Lee, H.K., Kim, J.H.: An HMM-based threshold model approach for gesture recognition. IEEE Trans. Pattern Anal. Mach. Intell. **21**(10), 961–973, e0206049 (1999)
23. Lefebvre, G., Berlemont, S., Mamalet, F., Garcia, C.: BLSTM-RNN based 3D gesture classification. In: Mladenov, V., Koprinkova-Hristova, P., Palm, G., Villa, A.E.P., Appollini, B., Kasabov, N. (eds.) ICANN 2013. LNCS, vol. 8131, pp. 381–388. Springer, Heidelberg (2013). https://doi.org/10.1007/978-3-642-40728-4_48
24. Li, R., Liu, Z., Tan, J.: A survey on 3D hand pose estimation: cameras, methods, and datasets. Pattern Recogn. **93**, 251–272 (2019)

25. Liu, J., Liu, Y., Wang, Y., Prinet, V., Xiang, S., Pan, C.: Decoupled representation learning for skeleton-based gesture recognition. In: Proceedings of the IEEE/CVF Conference on Computer Vision and Pattern Recognition, pp. 5751–5760 (2020)

26. Liu, W.: Natural user interface-next mainstream product user interface. In: 2010 IEEE 11th International Conference on Computer-Aided Industrial Design & Conceptual Design, vol. 1, pp. 203–205. IEEE (2010)

27. Manganaro, F., Pini, S., Borghi, G., Vezzani, R., Cucchiara, R.: Hand gestures for the human-car interaction: the Briareo dataset. In: Ricci, E., Rota Bulò, S., Snoek, C., Lanz, O., Messelodi, S., Sebe, N. (eds.) ICIAP 2019. LNCS, vol. 11752, pp. 560–571. Springer, Cham (2019). https://doi.org/10.1007/978-3-030-30645-8_51

28. Oka, R.: Spotting method for classification of real world data. Comput. J. **41**(8), 559–565 (1998)

29. Park, G., Kim, T.K., Woo, W.: 3D hand pose estimation with a single infrared camera via domain transfer learning. In: 2020 IEEE International Symposium on Mixed and Augmented Reality (ISMAR), pp. 588–599. IEEE (2020)

30. Paszke, A., et al.: Automatic differentiation in PyTorch. In: Neural Information Processing Systems Workshops (2017)

31. Pavlovic, V.I., Sharma, R., Huang, T.S.: Visual interpretation of hand gestures for human-computer interaction: a review. IEEE Trans. Pattern Anal. Mach. Intell. **19**(7), 677–695 (1997)

32. Quattoni, A., Wang, S., Morency, L.P., Collins, M., Darrell, T.: Hidden conditional random fields. IEEE Trans. Pattern Anal. Mach. Intell. **29**(10), 1848–1852 (2007)

33. Ren, Y., Gu, C.: Hand gesture recognition based on hog characters and SVM. Bull. Sci. Technol. **2**, 011 (2011)

34. Shotton, J., et al.: Real-time human pose recognition in parts from single depth images. In: CVPR 2011, pp. 1297–1304. IEEE (2011)

35. Simon, T., Joo, H., Matthews, I., Sheikh, Y.: Hand keypoint detection in single images using multiview bootstrapping. In: Proceedings of the IEEE Conference on Computer Vision and Pattern Recognition, pp. 1145–1153 (2017)

36. Starner, T., Weaver, J., Pentland, A.: Real-time American sign language recognition using desk and wearable computer based video. IEEE Trans. Pattern Anal. Mach. Intell. **20**(12), 1371–1375 (1998)

37. Vaswani, A., et al.: Attention is all you need. In: Advances in Neural Information Processing Systems (NIPS), pp. 5998–6008 (2017)

38. Yang, H.D., Park, A.Y., Lee, S.W.: Robust spotting of key gestures from whole body motion sequence. In: 7th International Conference on Automatic Face and Gesture Recognition (FGR06), pp. 231–236. IEEE (2006)

39. Yuan, S., et al.: Depth-based 3D hand pose estimation: From current achievements to future goals. In: Proceedings of the IEEE Conference on Computer Vision and Pattern Recognition, pp. 2636–2645 (2018)

40. Zhang, L., Zhu, G., Mei, L., Shen, P., Shah, S.A.A., Bennamoun, M.: Attention in convolutional LSTM for gesture recognition. In: Advances in Neural Information Processing Systems, pp. 1953–1962 (2018)

41. Zhang, W., Lin, Z., Cheng, J., Ma, C., Deng, X., Wang, H.: STA-GCN: two-stream graph convolutional network with spatial–temporal attention for hand gesture recognition. Vis. Comput. **36**(10), 2433–2444 (2020). https://doi.org/10.1007/s00371-020-01955-w

42. Zhu, G., Zhang, L., Shen, P., Song, J.: Multimodal gesture recognition using 3-D convolution and convolutional LSTM. IEEE Access **5**, 4517–4524 (2017)

SCAF: Skip-Connections in Auto-encoder for Face Alignment with Few Annotated Data

Martin Dornier[1,2(✉)], Philippe-Henri Gosselin[1], Christian Raymond[2],
Yann Ricquebourg[2], and Bertrand Coüasnon[2]

[1] InterDigital, Cesson-Sévigné, France
{martin.dornier,philippehenri.gosselin}@interdigital.com
[2] Univ Rennes, CNRS, IRISA, Rennes, France
{christian.raymond,yann.ricquebourg,bertrand.couasnon}@irisa.fr

Abstract. Supervised face alignment methods need large amounts of training data to achieve good performance in terms of accuracy and generalization. However face alignment datasets rarely exceed a few thousand samples making these methods prone to overfitting on the specific training dataset. Semi-supervised methods like TS^3 or 3FabRec have emerged to alleviate this issue by using labeled and unlabeled data during the training. In this paper we propose Skip-Connections in Auto-encoder for Face alignment (SCAF), we build on 3FabRec by adding skip-connections between the encoder and the decoder. These skip-connections lead to better landmark predictions, especially on challenging examples. We also apply for the first time active learning to the face alignment task and introduce a new acquisition function, the Negative Neighborhood Magnitude, specially designed to assess the quality of heatmaps. These two proposals show their effectiveness on several face alignment datasets when training with limited data.

Keywords: Face alignment · Semi-supervised training · Active learning

1 Introduction

Face alignment (also called facial landmark detection) aims to localize a set of pre-defined facial anatomical keypoints such as the corners of the mouth, the boundaries of the eyes or the tip of the nose [16,24,27]. Many applications rely on this task, for example, facial expression recognition or face swapping.

Although the rise of deep learning methods significantly improved the performances, the algorithms are still limited by the amount of labeled data available for training. Semi-supervised methods [1,8,9,12,13,22,23,32] have emerged in the field of face alignment to alleviate the lack of labeled data. In this work, we follow this principle, we try to train face alignment models with as little labeled data as possible. To do so, we build on 3Fabrec [1], this semi-supervised method achieves impressing performance in face alignment even with very limited training data. However, because of its relatively simple architecture, its performance degrades significantly on challenging datasets such as WFLW [27]. We try in this work to alleviate this issue. Our contribution is twofold:

- We enhance 3FabRec architecture with skip-connections between its encoder and decoder during the supervised training. This addition significantly improves its performances on both 300-W [24] and WFLW [27] datasets.
- We successfully apply active learning to face alignment and introduce a new acquisition function, the Negative Neighborhood Magnitude, improving even further the performance of our method when training with limited data.

The rest of the paper is organized as follows. Section 2 sums up the existing work on face alignment, in particular with limited data, and introduces the active learning procedure. In Sect. 3, we present our proposed methods to address face alignment with limited data. The results of these methods are shown in Sect. 4. Finally, we conclude this paper in Sect. 5.

2   Related Work

In our context, face alignment methods can be divided into two families: supervised and semi-supervised methods.

2.1   Supervised Face Alignment

Before the deep learning development in the computer vision domain, face alignment algorithms usually relied on parametric models, such as active shape model [6] or active appearance model [19], or on cascade regression [5,28,29]. Nowadays, almost all methods are based on artificial neural networks. Among recent approaches, while some methods still try to regress directly the landmark coordinates [10], most of them are now based on heatmap regression [3,7,17,21,26,27]. In this latter case, the network outputs a probabilistic heatmap for each landmark and the landmark coordinates are computed from it, usually with the best local maximum. Wu et al. [27] use facial boundaries heatmaps instead of landmark heatmaps making the algorithm more robust to large poses and occlusions. To take into account occlusions Kumar et al. [17] model the uncertainty and visibility of landmarks as a mixture of random variables while Zhu et al. [31] add in the model weights based on occlusion probability.

2.2   Semi-supervised Face Alignment

Annotating facial landmarks is time-consuming and can be difficult on faces with large pose or occlusions. For this reason, face alignment datasets rarely exceed a few thousand annotated faces. Semi-supervised methods try to alleviate the lack of annotated training data by incorporating non annotated, or weakly annotated, data into the learning process. Zhu et al. [32] augment the training dataset with synthetics faces generated from a 3D face model. Similarly, Qian et al. [22] generate images with different styles from an input pose image. Honari et al. [13] impose the equivariance of landmark predictions over multiple transformations of a face image. To deal with the large variance of different images styles Dong et al. [8] transforms images into style-aggregated images, and Robinson et al. [23] generate fake landmark heatmaps from unlabeled images using a Generative Adversarial Network [12]. Dong et al. [9] train a teacher to assess the quality of student predicted landmarks, the best samples are added, along with real data, to the next training set for retraining the student detectors. Finally, Browatzki et al. [1] propose 3FabRec that we will detail in the next section.

2.3 3FabRec

In 3FabRec [1], first, an auto-encoder is trained on a large number of unlabeled face images. During this unsupervised training, the hidden representation of the auto-encoder learns implicit knowledge about face features (shape, skin color, gender...) in order to reconstruct the image. The massive amounts of images used for training make this representation robust to a large diversity of faces.

After the unsupervised training, the auto-encoder is modified to perform face alignment, the decoder (also called generator) weights are frozen and convolutional layers called Interleaved Transfer Layers (ITLs) are added between its layers to take advantage of its generative power to also generate landmark heatmaps. The model is then trained on labeled face alignment datasets.

This method achieves impressive results even with few labeled data.

2.4 Active Learning

In the academic field, authors usually sample randomly labeled data from the full annotated training set to demonstrate the effectiveness of their method with limited training data. However, in real-world applications, at first, no labeled data is available and one must decide which samples to annotate.

Active learning aims to select the best samples to annotate to get the best possible model. It is particularly useful when annotation is time-consuming such as facial landmark annotations. It follows an iterative procedure: from an unlabeled dataset $\mathcal{U}_N$, an initial set $\mathcal{L}^0$ is annotated, the model is trained on this labeled set and all the remaining unlabeled samples are ranked using an *acquisition function*, the K best samples are annotated and added to $\mathcal{L}^0$ giving a new labeled set $\mathcal{L}^1$. The model is then trained from scratch on this new labeled set and this procedure is repeated until the annotation budget has been exhausted.

The acquisition functions can be divided into two approaches even though some combine both [15]. The first one is based on "uncertainty sampling" [11,30], meaning the acquisition function will try to select the samples where the model is the least confident, the acquisition function acts as a proxy of the training loss which is not available. The second approach is based on "diversity sampling" [25], the acquisition function tries to find samples that represent the diversity existing in the unlabeled dataset, it is particularly suited for classification tasks where having a class-balanced training dataset is crucial. To the best of our knowledge, before this work active learning had never been applied to face alignment.

3   Methods

3.1   SCAF: Adding Skip-Connections to 3FabRec

To detect precisely a facial landmark, spatial information must be kept. However, in common convolutional networks such as the auto-encoder of 3FabRec [1], the spatial dimensions of the features maps are progressively reduced in the encoder as global information emerges leading to a compact representation of the face image. This representation contains strong semantic information useful to reconstruct the whole face but may lack local details crucial to detect precisely the landmarks. To address this issue, many recent architectures for face alignment [3,9,21,26,27] use the Hourglass architecture where "skip-connections" are added between the encoder and decoder. These skip-connections between the two parts of the network preserve spatial information at multiple resolutions, the decoder can combine these different resolutions to generate better heatmaps.

Following this principle, we propose SCAF which stands for Skip-Connections in Auto-encoder for Face alignment, we enhance the 3FabRec architecture with skip-connections between the encoder and the ITLs. Thus, the input of an ITL is the element-wise sum of the output of the previous ResNet layer of the generator and the output of the corresponding encoder layer (the one with the same spatial dimensions). Before the sum, the output of the encoder layer is transformed by a set of convolutions called "bottleneck block" as it is done in Hourglass architectures. The full architecture can be seen in Fig. 1.

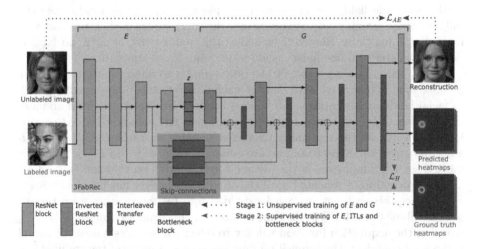

Fig. 1. Our network architecture and the 2-stage training pipeline. We add skip-connections between the encoder and the generator of 3FabRec.

We also noticed that splitting the supervised training into two steps: (1) Training only the ITLs, (2) Finetuning the ITLs and the encoder, is not necessary and we obtained better results training directly both. As in 3FabRec, we use as loss the $L2$ distance between the predicted ($\tilde{H}$) and ground truth heatmaps H.

3.2 Active Learning for Face Alignment

When training with very few examples, we optionally use active learning to select the best examples. We introduce a *new* acquisition function called Negative Neighborhood Magnitude (NNM) based on uncertainty sampling applied to the landmark heatmaps. When the model is not confident about its predictions, we noticed that the magnitude of the heatmaps near the predicted landmark is lower than when it is confident (see Fig. 2). Thus, to compute the NNM, for each predicted heatmap $\tilde{H}$, we compute the sum of the heatmap pixels in a square window W_i of size s around the predicted landmark position $\tilde{l}_i$, then we sum all heatmaps and take the negative so that the NNM behaves the same way as entropy, the less confident the model is, the greater NNM is.

$$NNM(\tilde{H}) = -\sum_{i=1}^{L} \sum_{u,v \in W_i} \tilde{H}_i(u,v) \tag{1}$$

After each model training, we rank the unlabeled samples. Some datasets contain very hard images where the most face is occluded which, thus, have a large NNM but are not useful for annotation. So, when we select images to label, we discard a percentage of the images with the largest NNM (see Sect. 4.6).

Fig. 2. Original image with ground truth (green dots) and predicted (blue dots) landmarks, ground truth heatmaps and predicted heatmaps for two images from WFLW. (Color figure online)

4 Experiments

4.1 Datasets

Unsupervised Training Datasets. We used a combination of two datasets for the unsupervised training:

AffectNet [20]: dataset created to capture a wide range of facial emotions. It contains 748K images.

CelebA [18]: dataset with 202K images of celebrities.

Our final dataset for unsupervised training contains about 950K images. The authors of 3FabRec [1] used as dataset a combination of 228K images from AffectNet [20] and 1.8M images from the VGGFace2 dataset [4] yielding a total of 2.1M images. However, due to copyright issues, VGGFace2 [4] is no longer available, so we could not use it for our experiments.

Supervised Training Datasets. We trained and evaluated our supervised models on two facial landmark datasets.

300-W [24]: combination of several facial landmark datasets re-annotated with 68 landmarks. Following the usual splits [1,9], our training set contains 3148 images, the *full* test set contains 689 images and is split into a *common* test set of 554 images and a *challenging* test set of 135 images.

WFLW [27]: dataset containing 7500 training images and 2500 testing images annotated with 98 landmarks. Many faces are heavily occluded or blurred making it a challenging dataset.

4.2   Experimental Settings

Unsupervised Training. Apart from the training datasets (see Sect. 4.1), we follow the same procedure as in 3FabRec [1], we use the same network (without ITLs and skip-connections) and the same hyperparameters.

Supervised Training. Using the ground truth bounding boxes, we crop and resize the images to 256×256 pixels. To generate the ground truth heatmaps, we use Gaussian kernels with $\sigma = 7$. Each ITL layer is a 3×3 convolutional layer and for the bottleneck blocks we use the hierarchical, parallel, and multi-scale block of [2]. The modified auto-encoder generates landmark heatmaps of size 128×128 pixels by skipping the last generator layer (the authors of 3FabRec [1] showed that the higher generator layers contain mostly decorrelated local appearance information). We use the same data augmentations as in 3FabRec [1], we also use Adam [14] to optimize the ITL and bottleneck layers, their learning rate is set to 0.001 while the encoder's one remains to 2×10^{-5}, the Adam's β_1 is reset to the default value of 0.9. Unlike 3FabRec [1], we train directly the three modules in parallel without any ITL-only-training stage before.

Active Learning. The initial labeled $\mathcal{L}^0$ set always contains 10 random samples from the unlabeled set $\mathcal{U}_N$, the number K of added samples after each training depends on the final training set size. For 300-W we used $K = 60, 30, 10$ for a final training size of 315 (10% of dataset), 158 (5%), 50 respectively. For WFLW, we used $K = 100, 75, 40, 20, 10$ for a final training size of 750 (10% of dataset), 375 (5%) 200, 100, 50 respectively.

Evaluation. To evaluate our models, we use the Normalized Mean Error (NME) with the distance between outer eye-corners as "inter-ocular" normalization.

4.3   Unsupervised Training Results

We trained the auto-encoder with the same training parameters as in 3FabRec
[1] except for the datasets (see Sect. 4.1) but because we had less than half of the
number of images used in 3FabRec [1] for our unsupervised training, we obtained
worse results on the supervised training. Fortunately, the authors of 3FabRec [1]
provide the source code and pre-trained weights for the auto-encoder at https://
github.com/browatbn2/3FabRec, so we decided to use these weights to focus on
the supervised training and get fair comparisons with their results.

4.4   Qualitative Results

During the supervised training of SCAF, the reconstruction error increases
because details non-necessary for landmark detection such as gender or skin color
fade away. Only the shape of the face remains but sometimes some reconstructed
facial parts do not even match with the predicted landmarks. For example (see
Fig. 3), the mouth is always reconstructed as close even if it is open in the original
face, but the landmark predictions align with the original mouth.

Fig. 3. Comparison of the reconstructions and landmark predictions. Top row shows
some original images from the WFLW Full test set and their ground truth landmarks
(green dots). Bottom row shows the reconstructed images with SCAF. Predicted landmarks are displayed in blue along with the ground truth landmarks in green. (Color
figure online)

4.5   Comparison with State-of-the-art

Comparison with Fully Supervised Methods. Table 1 compares our methods with fully supervised methods on 300-W [24] and WFLW [27] when training
on the full training set. We re-trained 3FabRec [1] from the provided unsupervised weights available at https://github.com/browatbn2/3FabRec to get a fair
comparison between the original network and our modified architecture. When

training on 300-W, our implementation of 3FabRec gets worse results than the ones reported in the paper of 3FabRec [1] however SCAF improves our results, especially on the Challenging test set. For WFLW, this time, our implementation of 3Fabrec obtain NME results slightly better than the ones reported in the 3FabRec paper [1]. The addition of the skip-connections improves again the NME from 5.58 to 5.50. Recent fully supervised methods beat our approach when trained on the full training set of WFLW or 300-W but the point of our semi-supervised method is to keep good performance even when training with limited data as we will see in the next paragraph.

Training with Limited Data. For 300-W, our implementation of 3FabRec gets better results than the ones reported in 3FabRec paper [1] when training on reduced training size. Apart from the training size of 50, SCAF outperforms our implementation of 3FabRec for any training dataset size. If we also apply active learning, the NME is reduced on the Challenging test but increased on the Common test set meaning that the model is more robust to challenging examples but a bit less precise for common examples. Results are reported in Table 2.

Table 3 compares our models to other semi-supervised methods on WFLW. Firstly, when training 3FabRec network on WFLW, skipping the ITL-only training step significantly improves the NME, especially when training with very limited data. Apart from the training size of 50, SCAF consistently outperforms 3FabRec when training on full or limited training data. When combined with Active learning, its performance is improved even further, especially with the training size of 50 where it beats our implementation of 3FabRec this time.

Table 1. Normalized mean error (%) on 300-W on the Common, Challenging and Full test sets and on WFLW Full test set

300-W				WFLW	
Method	Com.	Chall.	Full	Method	Full
SDM [28]	5.57	15.40	7.52	SDM [28]	10.29
SAN [8]	3.34	6.60	3.98	SAN [8]	5.22
LAB [27]	2.98	5.19	3.49	LAB [27]	5.27
ODN [31]	3.56	6.67	4.17	SA [22]	4.39
SA [22]	3.21	6.49	3.86	AWing [26]	4.36
TS^3 [9]	2.91	5.90	3.49	LUVLi [17]	4.37
AWing [26]	2.72	4.52	3.07	3FabRec [1]	5.62
LUVLi [17]	2.76	5.16	3.23	3FabRec (Our impl.)	5.58
3FabRec [1]	3.36	5.74	3.82	SCAF (Ours)	5.50
3FabRec (Our impl.)	3.54	5.93	4.01		
SCAF (Ours)	3.48	5.83	3.95		

Table 2. Normalized mean error (%) with reduced training sets on 300-W on the Common, Challenging and Full test sets (first, second and third columns respectively for each training set size). AL stands for active learning.

300-W dataset															
Method	Training set size														
	100%			20%			10%			5%			50 (1.5%)		
RCN+ [13]	3.00	**4.98**	**3.46**	–	**6.12**	**4.15**	–	6.63	4.47	–	9.95	5.11	–	–	–
SA [22]	3.21	6.49	3.86	3.85	–	–	4.27	-	–	6.32	–	–	–	–	–
TS³ [9]	**2.91**	5.90	3.49	4.31	7.97	5.03	4.67	9.26	5.64	–	–	–	–	–	–
3FabRec [1]	3.36	5.74	3.82	3.76	6.53	4.31	3.88	6.88	4.47	4.22	6.95	4.75	4.55	7.39	5.10
3FabRec (Our impl.)	3.54	5.93	4.01	3.79	6.33	4.29	3.93	6.70	4.47	4.10	6.86	4.64	**4.27**	7.23	4.85
SCAF (Ours)	3.48	5.89	3.95	**3.66**	6.23	4.17	**3.87**	6.60	**4.40**	**3.93**	6.84	**4.50**	4.33	7.60	4.97
SCAF+AL (Ours)	–	–	–	–	–	–	3.99	**6.49**	4.48	4.19	**6.78**	4.70	4.29	**6.93**	**4.81**

4.6   Ablation Studies

Comparison of Acquisition Functions. We tried three different acquisition functions on the 3FabRec [1] and SCAF architectures. Two of them are based on uncertainty sampling: our proposed Negative Neighborhood Magnitude (NNM) and the mean of the spatial entropy of the heatmaps. The last function is based on diversity sampling, it is the K-center-greedy algorithm used in [25].

Table 3. Normalized mean error (%) with reduced training sets on WFLW Full test set. AL stands for active learning.

WFLW dataset					
Method	Training set size				
	100%	20%	10%	5%	50
SA [22]	**4.39**	**6.00**	7.20	–	–
3FabRec [1]	5.62	6.51	6.73	7.68	8.39
3FabRec (Our impl.)	5.58	6.23	6.42	6.84	7.74
SCAF (Ours)	5.50	6.07	6.28	6.72	8.06
SCAF+AL (Ours)	–	–	**6.24**	**6.59**	**7.60**

Table 4 reports the NME on WFLW for these acquisition functions. Apart from the final training size of 50 on the 3FabRec network, the K-center-greedy function improves consistently the results over random sampling. When using NNM or Entropy to select the samples among *all* the unlabeled samples, when the final training size is small (≤ 200), the results are worse (or barely superior) than sampling at random. However, if we discard the top 10% ranked samples when selecting the samples, then we improve the NME and strongly outperform the random and K-center-greedy samplings. These methods are referred as $NNM_{10\%}$ and $Entropy_{10\%}$ in Table 4. This shows that very hard samples in the

Table 4. Normalized mean error (%) on WFLW Full test set for different active learning methods and different training set sizes (5% = 375 examples and 10% = 750 examples), for our implementation of 3FabRec and SCAF.

WFLW dataset										
Method	3FabRec (Our impl.)					SCAF				
	Final training set size					Final training set size				
	50	100	200	5%	10%	50	100	200	5%	10%
Random	7.74	7.44	7.04	6.84	6.42	8.06	7.40	6.88	6.72	6.28
NNM	8.27	7.57	7.15	6.77	6.36	8.04	7.44	7.01	6.63	6.22
Entropy	8.17	7.53	7.06	6.71	6.32	7.95	7.44	7.02	6.61	6.22
$NNM_{10\%}$	**7.63**	7.20	6.82	**6.62**	**6.31**	7.60	6.99	**6.72**	**6.59**	6.24
$Entropy_{10\%}$	7.71	**7.12**	**6.83**	**6.62**	6.34	**7.53**	**6.96**	6.73	6.62	**6.22**
K-center-greedy	7.85	7.36	6.95	6.65	6.32	7.74	7.18	6.82	6.61	6.28

WFLW training dataset should not be added to the training set because they are outliers and won't help the model to generalize to unseen data. However, as the final training set size increases, the benefit of discarding the worst examples tends to disappear; because more samples are added, the proportion of outliers decreases, and "normal" challenging samples are added to the training set.

Figure 4 shows the top-5 ranked samples according to the NNM after training SCAF on 10 random samples for the WFLW training set. The top row displays the top-5 samples among *all* unlabeled samples, the five images are clearly outliers: blue color, distorted face for the second image, non-human face for the last image and won't help much the model to generalize to unseen data after training. The bottom row displays the top-5 images after discarding the top-10% images from the unlabeled dataset. These five images are still challenging (low resolution, occlusion, baby face) but closer to "normal" images, adding them to the training set should improve the model predictions.

Entropy and NNM have close results in terms of NME. However, in the case of Entropy, the whole heatmaps must be normalized before computing the entropy on the whole heatmaps too, whereas the computation of the NNM only requires summing heatmaps values of small windows. In our experiments, with an Intel Core i7-9850H CPU, computing the Entropy took on average 0.042 s whereas computing the NNM only took 0.012 s. Thus, the NNM is 3.5 times faster to compute than the Entropy while achieving comparable results.

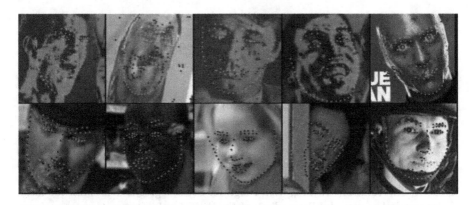

Fig. 4. Top-5 ranked images for NNM after training the model on 10 random samples. Ground truth landmarks are displayed with green dots while blue ones are the predicted landmarks. Top row shows the top-5 ranked images among all the unlabeled samples while bottom row displays the top-5 ranked images after removing the top-10% images. (Color figure online)

5   Conclusion

In this paper, we addressed the problem of training face alignment models with limited labeled data. To achieve this goal, we improved 3FabRec [1] architecture by adding skip-connections between the encoder and decoder during the supervised training. This makes the network predict more accurately the facial landmarks heatmaps, especially for challenging examples where the hidden representation fails to capture all the specificities of the image. We also applied active learning to the face alignment task to improve even further the performance with limited training data and showed its effectiveness by introducing a new acquisition function for heatmaps called Negative Neighborhood Magnitude. This function achieves similar performance to spatial entropy in terms of NME while being much faster to compute.

Acknowledgement. This work was granted access to the HPC resources of IDRIS under the allocation 2021-AD011012376 made by GENCI.

References

1. Browatzki, B., Wallraven, C.: 3fabRec: fast few-shot face alignment by reconstruction. In: Proceedings of the IEEE/CVF Conference on Computer Vision and Pattern Recognition, pp. 6110–6120 (2020)
2. Bulat, A., Tzimiropoulos, G.: Binarized convolutional landmark localizers for human pose estimation and face alignment with limited resources. In: Proceedings of the IEEE International Conference on Computer Vision, pp. 3706–3714 (2017)

3. Bulat, A., Tzimiropoulos, G.: How far are we from solving the 2D & 3D face alignment problem? (and a dataset of 230,000 3D facial landmarks). In: Proceedings of the IEEE International Conference on Computer Vision, pp. 1021–1030 (2017)
4. Cao, Q., Shen, L., Xie, W., Parkhi, O.M., Zisserman, A.: Vggface2: a dataset for recognising faces across pose and age (2018)
5. Cao, X., Wei, Y., Wen, F., Sun, J.: Face alignment by explicit shape regression. Int. J. Comput. Vis. **107**(2), 177–190 (2014)
6. Cootes, T.F., Taylor, C.J., Cooper, D.H., Graham, J.: Active shape models-their training and application. Comput. Vis. Image Understanding **61**(1), 38–59 (1995)
7. Dapogny, A., Bailly, K., Cord, M.: Decafa: deep convolutional cascade for face alignment in the wild. In: Proceedings of the IEEE/CVF International Conference on Computer Vision, pp. 6893–6901 (2019)
8. Dong, X., Yan, Y., Ouyang, W., Yang, Y.: Style aggregated network for facial landmark detection. In: Proceedings of the IEEE Conference on Computer Vision and Pattern Recognition, pp. 379–388 (2018)
9. Dong, X., Yang, Y.: Teacher supervises students how to learn from partially labeled images for facial landmark detection. In: Proceedings of the IEEE/CVF International Conference on Computer Vision, pp. 783–792 (2019)
10. Feng, Z.H., Kittler, J., Awais, M., Huber, P., Wu, X.J.: Wing loss for robust facial landmark localisation with convolutional neural networks. In: Proceedings of the IEEE Conference on Computer Vision and Pattern Recognition, pp. 2235–2245 (2018)
11. Gal, Y., Islam, R., Ghahramani, Z.: Deep Bayesian active learning with image data. In: International Conference on Machine Learning, pp. 1183–1192. PMLR (2017)
12. Goodfellow, I., et al.: Generative adversarial nets. In: Ghahramani, Z., Welling, M., Cortes, C., Lawrence, N.D., Weinberger, K.Q. (eds.) Advances in Neural Information Processing Systems 27, pp. 2672–2680. Curran Associates, Inc. (2014)
13. Honari, S., Molchanov, P., Tyree, S., Vincent, P., Pal, C., Kautz, J.: Improving landmark localization with semi-supervised learning. In: Proceedings of the IEEE Conference on Computer Vision and Pattern Recognition, pp. 1546–1555 (2018)
14. Kingma, D.P., Ba, J.: Adam: a method for stochastic optimization. arXiv preprint arXiv:1412.6980 (2014)
15. Kirsch, A., Van Amersfoort, J., Gal, Y.: Batchbald: efficient and diverse batch acquisition for deep Bayesian active learning. Adv. Neural. Inf. Process. Syst. **32**, 7026–7037 (2019)
16. Koestinger, M., Wohlhart, P., Roth, P.M., Bischof, H.: Annotated facial landmarks in the wild: a large-scale, real-world database for facial landmark localization. In: 2011 IEEE International Conference on Computer Vision Workshops (ICCV Workshops), pp. 2144–2151. IEEE (2011)
17. Kumar, A., et al.: LUVLi face alignment: estimating landmarks' location, uncertainty, and visibility likelihood. In: Proceedings of the IEEE/CVF Conference on Computer Vision and Pattern Recognition, pp. 8236–8246 (2020)
18. Liu, Z., Luo, P., Wang, X., Tang, X.: Deep learning face attributes in the wild. In: Proceedings of International Conference on Computer Vision (ICCV), December 2015
19. Matthews, I., Baker, S.: Active appearance models revisited. Int. J. Comput. Vis. **60**(2), 135–164 (2004)
20. Mollahosseini, A., Hasani, B., Mahoor, M.H.: Affectnet: a database for facial expression, valence, and arousal computing in the wild. IEEE Trans. Affective Comput. **10**(1), 18–31 (2017)

21. Newell, A., Yang, K., Deng, J.: Stacked hourglass networks for human pose estimation. In: Leibe, B., Matas, J., Sebe, N., Welling, M. (eds.) ECCV 2016. LNCS, vol. 9912, pp. 483–499. Springer, Cham (2016). https://doi.org/10.1007/978-3-319-46484-8_29

22. Qian, S., Sun, K., Wu, W., Qian, C., Jia, J.: Aggregation via separation: boosting facial landmark detector with semi-supervised style translation. In: Proceedings of the IEEE/CVF International Conference on Computer Vision, pp. 10153–10163 (2019)

23. Robinson, J.P., Li, Y., Zhang, N., Fu, Y., Tulyakov, S.: Laplace landmark localization. In: Proceedings of the IEEE/CVF International Conference on Computer Vision, pp. 10103–10112 (2019)

24. Sagonas, C., Tzimiropoulos, G., Zafeiriou, S., Pantic, M.: 300 faces in-the-wild challenge: the first facial landmark localization challenge. In: Proceedings of the IEEE International Conference on Computer Vision Workshops, pp. 397–403 (2013)

25. Sener, O., Savarese, S.: Active learning for convolutional neural networks: a core-set approach. arXiv preprint arXiv:1708.00489 (2017)

26. Wang, X., Bo, L., Fuxin, L.: Adaptive wing loss for robust face alignment via heatmap regression. In: Proceedings of the IEEE International Conference on Computer Vision, pp. 6971–6981 (2019)

27. Wu, W., Qian, C., Yang, S., Wang, Q., Cai, Y., Zhou, Q.: Look at boundary: a boundary-aware face alignment algorithm. In: Proceedings of the IEEE Conference on Computer Vision and Pattern Recognition, pp. 2129–2138 (2018)

28. Xiong, X., De la Torre, F.: Supervised descent method and its applications to face alignment. In: Proceedings of the IEEE Conference on Computer Vision and Pattern Recognition, pp. 532–539 (2013)

29. Yan, J., Lei, Z., Yi, D., Li, S.: Learn to combine multiple hypotheses for accurate face alignment. In: Proceedings of the IEEE International Conference on Computer Vision Workshops, pp. 392–396 (2013)

30. Yoo, D., Kweon, I.S.: Learning loss for active learning. In: Proceedings of the IEEE/CVF Conference on Computer Vision and Pattern Recognition, pp. 93–102 (2019)

31. Zhu, M., Shi, D., Zheng, M., Sadiq, M.: Robust facial landmark detection via occlusion-adaptive deep networks. In: Proceedings of the IEEE/CVF Conference on Computer Vision and Pattern Recognition, pp. 3486–3496 (2019)

32. Zhu, X., Lei, Z., Liu, X., Shi, H., Li, S.Z.: Face alignment across large poses: a 3D solution. In: Proceedings of the IEEE Conference on Computer Vision and Pattern Recognition, pp. 146–155 (2016)

Full Motion Focus: Convolutional Module for Improved Left Ventricle Segmentation Over 4D MRI

Daniel M. Lima[1,2]([⊠])(iD), Catharine V. Graves[2](iD), Marco A. Gutierrez[2](iD), Bruno Brandoli[3](iD), and Jose F. Rodrigues Jr.[1](iD)

[1] Institute of Mathematics and Computer Science, Universidade de Sao Paulo, Sao Carlos, SP, Brazil
danielm@usp.br
[2] Laboratorio de Informatica Biomedica, Instituto do Coracao, Hospital das Clinicas, Faculdade de Medicina, Universidade de Sao Paulo, Sao Paulo, SP, Brazil
[3] Dalhousie University, B3H 4R2 Halifax, NS, Canada

Abstract. Magnetic Resonance Imaging (MRI) is a widely known medical imaging technique used to assess the heart function. Over Cardiac MRI (CMR) images, Deep Learning (DL) models perform several tasks with good efficacy, such as segmentation, estimation, and detection of diseases. Such models can produce even better results when their input is a Region of Interest (RoI), that is, a segment of the image with more analytical potential for diagnosis. Accordingly, we describe *Full Motion Focus (FMF)*, an image processing technique sensitive to the heart motion in a 4D MRI sequence (video) whose principle is to combine static and dynamic image features with a Radial Basis Function (RBF) to highlight the RoI found in the motion field. We experimented *FMF* with the U-Net convolutional DL architecture over three CMR datasets in the task of Left Ventricle segmentation; we achieved a rate of detection (Recall score) of 99.7% concerning the RoIs, improved the U-Net segmentation (mean Dice score) by 1.7 ($p < .001$), and improved the overall training speed by 2.5 times (+150%).

Keywords: Cardiac MRI · Motion · Deep learning · Localization · Segmentation

1 Introduction

Magnetic resonance imaging (MRI) is a medical imaging technique used to capture volumetric image sequences of internal soft tissues, such as cardiac muscles. In comparison to X-ray imaging (XR) and computed tomography (CT), MRI provides images with improved structural details via finer spatial resolutions. Cardiac MRI (CMR) focuses on the heart, allowing trained cardiologists to measure heart parameters, for example, the mass of the cardiac muscle (myocardium mass), volumes of blood cavities (atrial and ventricular volumes) and volume of

S. Sclaroff et al. (Eds.): ICIAP 2022, LNCS 13231, pp. 438–450, 2022.
https://doi.org/10.1007/978-3-031-06427-2_37

blood pumped per heartbeat (ejection fraction) [9]. Those parameters are used to assess how healthy is the heart, by recognizing early conditions and signs before the onset of infarcts and other complications.

Due to the complexity of CMR images, comparably complex techniques are required to produce detailed analyses. One of these techniques is deep learning (DL). Many of the tasks related to the cardiac function analysis have benefited from DL methods—for example segmentation of structures [1], estimation of heart parameters [17], and detection of diseases [7]. For even better results, research in DL has pointed out that models based on convolutional neural networks (CNN) had higher efficacy when provided with regions of interest (RoI) either explicitly or implicitly [17]. The RoI proposal is a preprocessing step whose goal is to identify the most prominent regions of an image for discovering clinically relevant artifacts.

The explicit RoI proposal approaches usually follow a combination of methods, for example: (a) pipelining a segmentation and a regression network; (b) preprocessing the input with a region proposal algorithm [5] or with a CNN [16]; or (c) by using manual cropping [18]. The implicit RoI detectors are added to the DL network abstracted as additional operators and variables; e.g. multiscale Inception [14] and attention [15] modules, which benefit from the RoI to down-weight less-informative neurons and inputs inside the network. Inception modules weight convolutions of different sizes, while attention modules assign a weight to each feature channel. This additional neural information processing guides which input features or channels shall have more weight.

In this paper, we develop a module that highlights regions in the image sequence by analyzing the motion field using a radial basis function (RBF) [8]. In our experiments, we analyze our method by using the RBF for cropping the input before having it processed by a pretrained segmentation U-Net convolutional network [10]. Our methodology is an innovation in the task of region proposal for CMR analysis; we demonstrate results that justify the use and further investigation of the employed principles. We named it after its working mechanism as Full Motion Focus (*FMF*).

2   Theory and Related Work

2.1   Cardiac MRI

MRI is the most precise medical imaging technique for examination of the heart structures, it records heart images along a complete heartbeat cycle [2,12]. In practice, the magnetization signal is triggered by a reference pulse, then captured several times for noise reduction and, finally, reconstructed by inverse Fast Fourier Transform. The resulting image is usually visualized in slices along a positional axis: the long axis has a frontal or lateral view of the heart, and the short axis aligns to a cross-sectional plane.

The short-axis view is split in three regions: the base or basal region near the top where blood vessels connect to the heart (slice B); the middle or medial in between (slice M); and the apex or apical region at the bottom tip of the heart (slice A)—refer to Fig. 1. The normal human heart has four chambers: right atrium (RA), right ventricle (RV), left atrium (LA), and left ventricle (LV). The atria receive blood and the ventricles pump it out of the heart. Even though all chambers are important, the LV is of special interest because it is the cardiac muscle that does the "heavy lifting" of pumping oxygenated blood from the lungs to the whole body. In the short-axis view, the LV appears as a ring shape, whose thickness and internal volume measurements are essential to estimate the myocardium mass and ejection fraction, respectively.

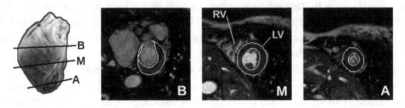

Fig. 1. Picture of heart, and examples of CMR cross-sections in the Base (B), Middle (M) and Apex (A) regions, with LV boundaries. Adapted from Saber *et al.* [11].

2.2 Computer Methods for CMR Analysis and LV Segmentation

Computer methods for functional analysis of CMR images, as reviewed by Peng *et al.* [9], were organized in three ways: image-driven, model-driven, and by direct estimation. Further subdivisions of the LV segmentation spans five groups: (1) image processing methods such as thresholding, morphology operators, and region-growing; (2) pixel/voxel-based classification by Gaussian mixture models, neural networks, k-means, k-NN, or SVM; (3) active contours (snakes), deformable models, level sets, and motion tracking; (4) PCA or ICA with strong priors from anatomic heart models; and (5) direct estimation by, e.g., latent discriminant analysis combined with SVM. This work refers to category 2 as it employs a pixel-based classification by a neural network.

Recent approaches for LV segmentation use CNNs (such as U-Net) experimented over many datasets and methodology combinations. U-Net [10] is a general segmentation model which combines a tower of downscaled-then-upscaled deep representations. For CMR images, U-Net displayed good results (89% Dice) when trained from min-cut priors [4]. A U-Net architecture with residual blocks and optical flow information achieved 89%, 95% and 85% Dice in the base, middle, and apex regions of the heart, respectively, in the work of Yan *et al.* [19]. In the work of Wu *et al.* [16], the authors combine a custom CNN for region proposal with a U-Net segmentation to achieve 95% Dice. Overall, the combination of region proposal to U-Net had good results in particular datasets, but still has room for improvement when the evaluation generalizes across multiple datasets. Different from former works, our methodology, *FMF*, uses RBFs to propose RoIs that will aid a CNN processing in the task of LV segmentation, which we demonstrate with experimental U-net improvements in multiple datasets.

3   Materials and Methods

FMF starts with a 4D image input $\mathbf{x} = I(t, x, y, z)$, that is, a sequence of volumes (frames) each one with a time t coordinate. Initially, we normalize $\mathbf{x}$ to produce sequence $\mathbf{x}^*$ with frames in a format more adequate for Neural Network processing—see Sect. 3.1. From $\mathbf{x}^*$, we extract static visual features to produce $\mathbf{x}_s$, and motion estimation to produce $\mathbf{x}_t$, detailed in Sect. 3.2. Next, we apply two sets of weights: w_s are weights related to visual features, and w_t are weights related to the motion (time). After weighting, both features are combined in tensor $\mathbf{v}$. Then, as presented in Sect. 3.3, we compute the center voxel μ_v defined by the weighted sum of all the voxels' coordinates. The segmentation map $\mathbf{y_S}$ is produced by applying a threshold to $\mathbf{v}$ and extracting the bounds of non-zero voxels. Afterwards, we compute the scale σ_v given by the standard deviation of all the voxels' distances from center μ_v. At this point, refer to Sect. 3.4, we can apply a radial basis function at center μ_v with radius σ_v, computing $\mathbf{y_L}$, then we scale the region defined by $\mathbf{y_L}$ to the CNN input shape. In Sect. 3.6, we explain the neural network processing, its parameters, and training.

3.1   Intensity Normalization

We normalize the image intensities between 0 and 1 by subtracting the minimum value and dividing by the range of values, using an small constant ϵ to avoid division-by-zero:

$$\mathbf{x}^* = \frac{\mathbf{x} - \min(\mathbf{x}) + \epsilon}{\max(\mathbf{x}) - \min(\mathbf{x}) + \epsilon} \tag{1}$$

3.2   Visual Features and Motion Estimation

For visual features extraction, we consider the statistical mean and standard deviation obtained with $3 \times 3 \times 3$ kernels. The mean image I_μ is computed using the convolution operation with mean kernel M, defined as:

$$M = \frac{1}{1 \cdot 3 \cdot 3 \cdot 3} \cdot J_{1,3,3,3} \tag{2}$$

where $J_{1,3,3,3}$ is a $1 \times 3 \times 3 \times 3$ matrix-of-ones. That is, M is just the arithmetic mean of a 3^3 matrix, in a convolution form. With kernel M, we compute I_μ as:

$$I_\mu = I * M \tag{3}$$

In turn, the standard deviation image I_σ is computed by taking the differences between image I and mean image I_μ, then squaring the differences element-wise (Hadamard power); after that, we apply a convolution operation with the mean kernel M, before taking the Hadamard square-root, as follows:

$$I_\sigma = [(I - I_\mu)^{(2)} * M]^{(1/2)} \tag{4}$$

Accordingly, the mean image I_μ and the standard deviation image I_σ define $\mathbf{x}_s = \{I_\mu, I_\sigma\}$. Notice that we express the Hadamard powers using the definitions and notations defined in the work of Fallat and Johnson [3].

Motion is estimated by function $E(I)$—in Eq. 5, given by the root-mean-squared differences of intensity along the time coordinate, where T is the time interval (or number of frames) in image I, and S_t is the Sobel kernel w.r.t. time, instead of the default S_x and S_y spatial Sobel kernels. This function is related to the magnitude of the optical flow vector field [6] in each voxel, as follows:

$$\mathbf{x}_t = E(I) = \sqrt{\frac{1}{T} \int_{t=0}^{T} \left(\frac{\partial I}{\partial t}\right)^2 dt} \approx \sqrt{\frac{1}{T} \sum_{t=0}^{T-1} [I(t) * S_t]^{(2)}} \tag{5}$$

Figure 2 illustrates a sequence of six CMR frames spaced by $1/6$ of the cardiac cycle (upper row in the figure), and the respective absolute derivatives in each point along the time dimension (lower row) as computed by Eq. 5. Figure 3a shows the total motion estimate in the sequence, while Fig. 3b presents the cumulative energy histogram. By combining both static and motion features $\mathbf{x}_s$ and $\mathbf{x}_t$, we obtain:

$$\mathbf{v} = w_s \mathbf{x}_s + w_t \mathbf{x}_t \tag{6}$$

The weights w_s and w_t are initialized to 0.1 and 0.9 respectively for $\mathbf{x}_s$ and $\mathbf{x}_t$, i.e., although we emphasize the motion features, we also include the static visual features, which will address the problematic cases when the heart has limited motility, which might be the case for heart complications.

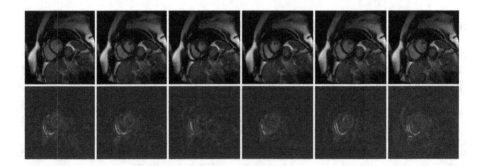

Fig. 2. Images in the medial slice, showing six frames (upper) and absolute time derivatives (lower). This figure only shows medial slices, but *FMF* is defined for volumes.

3.3   Center and Scale Computation

The center of energy μ_v, formalized by Eq. 7, is the voxel defined by the energy-weighted sum of all the voxels' coordinates $\mathbf{r}$ in dimensions x, y, z; that is, every

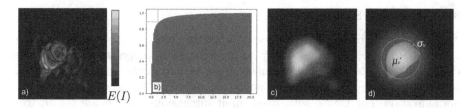

Fig. 3. a) Motion estimate $E(I)$ in whole cardiac cycle. b) Cumulative histogram's shoulder lies near threshold $p = 0.9$. c) $E(I) > 90\%$ of Figure a, convolved with Gaussian($\sigma = 5$). d) The RBF is fitted to this region with center μ_v and radius $\propto \sigma_v$.

voxel i has $\mathbf{r}_i = \langle x_i, y_i, z_i \rangle \mid 0 \leq x_i < \text{width}, 0 \leq y_i < \text{height}, 0 \leq z_i < \text{slices}, 1 \leq i \leq N$, for N is the number of voxels in the image, as follows:

$$\mu_v = \langle \bar{x}, \bar{y}, \bar{z} \rangle = \frac{\sum_{i=1}^{N} \mathbf{v}_i \cdot \mathbf{r}_i}{\sum \mathbf{v}} \tag{7}$$

The segmentation $\mathbf{y_S}$ is given by the thresholded image from the previous step, see Fig. 3c. The scale estimate σ_v is the cube root of the volume (in voxels) of the thresholded image considering the values above the 90% percentile:

$$(\mathbf{y_S})_i = \mathbb{1}[\mathbf{v}_i > Q(0.9, \mathbf{v})] \tag{8}$$

$$\sigma_v = \frac{3}{\|\mathbf{r}_{\max}\|} \sqrt[3]{\sum_{i=1}^{N} (\mathbf{y_S})_i} \tag{9}$$

where the indicator function $\mathbb{1}[c]$ returns 1 if c is true, or 0 otherwise; and $Q(p, v)$ is the quantile function, returning the maximum of the lowest p (%) values in v.

3.4   Segmentation and Localization Focus

The segmentation focus $\mathbf{y_S}$ is derived from thresholded $\mathbf{v}$ (Eq. 8), and the localization $\mathbf{y_L}$ is found by fitting an RBF to $\mathbf{y_S}$. The radius d_i is the distance from the center μ_v to each voxel i. The Euclidean distance (L_2-norm $\| \cdot \|$) was used:

$$d_i = \left\| \frac{\mathbf{r}_i - \mu_v}{\mathbf{r}_{\max}} \right\| = \sqrt{\left(\frac{x_i - \bar{x}}{x_{\max}}\right)^2 + \left(\frac{y_i - \bar{y}}{y_{\max}}\right)^2 + \left(\frac{z_i - \bar{z}}{z_{\max}}\right)^2} \tag{10}$$

The chosen RBF is a Gaussian ϕ of the voxels' distances:

$$(\mathbf{y_L})_i = \phi(i) = \exp\left[-(d_i/\sigma_v)^2\right] \tag{11}$$

Both outputs $\mathbf{y_S}$ and $\mathbf{y_L}$ are illustrated in Fig. 3d. According to the framework explained so far, we can design a focus for many different objects in the images by changing the functions for image feature extraction, motion estimation, center, scale, and RBF. It is also possible to detect multiple objects or objects of complex shapes by a mixture of RBF models.

3.5 Crop and Scaling

This step performs the final preparation of the image so that it will fit the CNN input shape; this is necessary because the MRI images may have diverse dimensions. Besides, the RBF estimation is executed in the original resolution, so the proposed region must be cropped then adjusted, followed by an intensity normalization. For CNNs with fixed input dimensions, we rescale the images using bicubic spline interpolation. For CNNs with variable input shape, we only adjust the image proportions as requested by the model, e.g. the U-Net we use has five max-pooling layers with a down-scaling factor of 2, which means the input dimensions should be multiples of 2^5.

Table 1. Overview of the CMR datasets. n = number of patients; Sxy = spatial resolution (pixel spacing) in the axial plane (mm/pixel); Sz = slice resolution (mm/pixel).

Dataset	n	width	height	slices	frames	Sxy	Sz
LVSC	100	138–512	138–512	8–24	18–35	0.68–2.14	6–10
ACDC	100	154–428	154–512	6–18	12–35	0.70–1.92	5–10
M&Ms	320	196–548	192–512	6–20	18–36	0.68–1.82	5–10

3.6 Segmentation CNN

In this section, we employ method *FMF* in combination with the 2D U-Net CNN [10], a consolidated technique widely tested for image processing—refer to Sect. 2.2; the methodology, though, is suitable to any other CNN. The *FMF* pipeline executes along the whole cardiac volume and, after cropping, the RoI pass to the segmentation CNN, which in this case is the 2D U-Net. The CNN was trained with outputs obtained with method *FMF* during 30 epochs using an adaptive momentum optimizer, initial learning rate $\eta = 0.001$, Nesterov β_1 = 0.9, L_∞-decay β_2 = 0.999, and loss function binary cross-entropy plus the Sørensen-Dice coefficient (DSC):

$$\mathcal{L}(y,p) = -[y \cdot \log(p) + (1-y) \cdot \log(1-p)] + DSC(y,p) \qquad (12)$$

where DSC is a performance metric defined in Sect. 4.1, Eq. 14 .

3.7 Datasets

To evaluate method *FMF*, we used three CMR datasets specified across the short-axis orientation. The metadata in the datasets include: a) binary masks for the LV and RV; b) physiological parameters such as myocardium mass, ventricle area, volume, ejection fraction, thickness, and dimensions of structures; c) image acquisition parameters such as spatial resolution (mm), temporal resolution (frames per cardiac cycle) and slice gaps (mm). Not all of the datasets encompass the same information; following, we provide more details, with a summary presented in Table 1.

- **LVSC** - Cardiac Atlas Project (CAP) 2011 LV Segmentation Challenge [13]; we used the 100 patients in the training set with LV masks on all frames;
- **ACDC** - MICCAI 2017 Automated Cardiac Diagnosis Challenge [1]; it was created from clinical data, including sequences of 100 patients with RV and LV masks in two frames, from the University Hospital of Dijon (France) over 6 years;
- **M&Ms** - MICCAI 2020 Multi-Centre, Multi-Vendor & Multi-Disease Cardiac Image Segmentation Challenge [11]; this database was collected from six hospitals in Spain, Canada, and Germany using several MRI scanners (Siemens, GE, Philips, and Canon), it includes RV and LV masks in two frames—we used 320 patients for which the labels are publicly available.

4   Experiments and Results

We evaluated our methodology by analyzing the selected datasets and comparing the results of two CNNs: one base U-Net CNN on the raw images without the *FMF* RoI proposal, and another U-Net CNN on the images processed by *FMF*. After running *FMF* and cropping the RoI, the cropped frames were passed to the U-Net. The myocardium labels were obtained by subtracting epicardium and endocardium masks. The datasets were individually split with 75% for training and 25% for validation of models; for testing, we performed an all-versus-all scheme, that is, fitting the CNNs to the training set of one dataset and testing on the validation set of another dataset.

The experiments ran in a computer with Intel i7-7700k CPU and NVIDIA Titan-Xp GPU. Our software was written in Python, scipy, tensorflow, and matplotlib. The steps were: load an image; compute features; fit RBF; crop and resize both the input image and the label mask; Base CNN prediction on the original image and *FMF*-CNN prediction on the cropped image; then, evaluate the predicted masks against the labels using the following metrics.

4.1   Metrics

We compared three indices: (1) Recall, the proportion of the labels that was preserved in the *FMF* region proposals; it is also known as Sensitivity or True Positive Rate—TPR as defined by Eq. 13, which aims to verify if the bounding boxes cover the labels entirely; (2) the Sørensen-Dice coefficient—DSC as defined by Eq. 14, which is equivalent to the F1-score (average of Precision and Recall) that refers to the segmentation output when not-using vs using method *FMF*; (3) speedup, the ratio of time taken by the CNN when not-using vs using method *FMF*—it is defined as $(t_{\text{base}}/t_{\text{ours}})$.

$$TPR(y, \hat{y}) = \frac{|y \cap \hat{y}|}{|y|} = \frac{TP}{TP + FN} \tag{13}$$

$$DSC(y, \hat{y}) = \frac{2|y \cap \hat{y}|}{|y| + |\hat{y}|} = \frac{2TP}{2TP + FP + FN} \tag{14}$$

Recall ranges from 0, when none of the marked voxels are detected; to 1 when 100% of the marked voxels are detected. DSC ranges from 0, for no intersection; to 1 for a perfect match between y and $\hat{y}$. Speedup is a positive ratio with value $=1$ when times are equal; >1 when *FMF* is faster; and <1 when *FMF* is slower.

4.2   Results

Table 2 presents the results of the *FMF* RoI proposal as Recall, and the results of the *FMF*-CNN segmentation as Dice score. The Recall, that is, the ability to identify the regions of interest, was nearly perfect for all the datasets. Concerning the Dice score, the U-Net trained on dataset M&Ms (the largest dataset) was significantly improved when using the *FMF* RoI proposal, with a mean performance increase of $+7.2$ (percent points) considering all the datasets.

Table 2. Performance results. In the first (left-most) column, the name of the dataset and the Recall metric. In the remaining columns, the DSC (Dice score) for the Base CNN segmentation and for the segmentation obtained after the *FMF* RoI proposal.

Train	Test	Base	FMF	ΔDSC	Train	Test	Base	FMF	ΔDSC
M&Ms 99.75% recall	LVSC	62.4%	70.9%	+8.5		LVSC	50.1%	44.8%	−5.7
	ACDC	60.8%	67.0%	+6.2	ACDC	ACDC	74.6%	69.9%	−4.7
	M&Ms	81.3%	82.2%	+0.9	98.84% recall	M&Ms	67.6%	47.4%	−20.2
	all	65.4%	72.6%	+7.2		all	56.6%	48.3%	−8.3
LVSC 99.75% recall	LVSC	74.3%	70.8%	−3.5		LVSC	76.5%	77.5%	+1.0
	ACDC	75.7%	79.7%	+4.0	all	ACDC	85.9%	86.8%	+0.9
	M&Ms	67.0%	69.5%	+2.5	99.69% recall	M&Ms	77.2%	83.0%	+5.8
	all	74.5%	72.7%	−1.8		all	77.7%	79.4%	+1.7

Table 3. Contingency matrix considering all voxels in subsets *all-train* and *all-test*, refer to Table 2. A corrected McNemar's chi-squared test comparing CNN without *FMF* (Base) versus with *FMF* asserted significantly different predictions, with $p < .001$.

	FMF (T)	FMF (F)	
Base (T)	6,041,403	2,543,222	$\chi^2 = 885305.1$
Base (F)	818,157	709,767,778	$p < 10^{-302}$

Table 4. Mean training time for CNN without *FMF* (Base) and with *FMF*, concerning 30 epochs. In all the cases, *FMF* accelerated the training speed, as the CNN processes less data.

Dataset	ACDC	M&Ms	LVSC	all
Base	21.0s	93.2s	254.7s	427.2s
FMF	11.6s	37.6s	95.1s	170.3s
Speed-up	1.81x	2.48x	2.68x	2.51x

With dataset LVSC used for training, we observed a lower performance when validating over itself (-3.5), but an improvement when validating with ACDC and M&Ms (+4.0 and +2.5, respectively), which corresponds to the overall mean performance of +0.3. With ACDC used for training, we did not observe performance improvements, with a mean decrease of 8.3% when considering all the datasets—it is noticeable that the Recall of RoIs had the smallest performance for this particular dataset. Finally, when using all the datasets for training, we observed a mean performance increase of +2.35 concerning all the datasets.

We applied the corrected McNemar's chi-squared test from package statsmodels 0.12.2, whose results asserted that the Base predictions and predictions after *FMF* were significantly different, with $p < .001$ (see Table 3). From a practical perspective, we also compared the training speed, which is paramount to larger experiments such as hyper-parameter search. Table 4 shows the training time of the networks without and with method *FMF*; in all the cases our method improved the training speed by 150%.

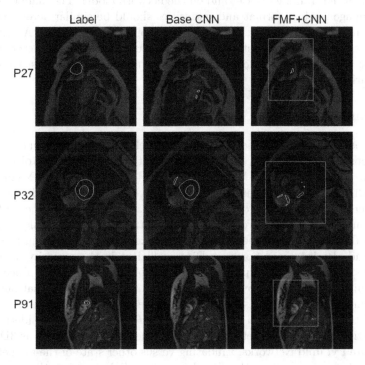

Fig. 4. Slices with DSC<0.2 from patients P27, 32 and 91. Columns are labels, predictions by base CNN, and predictions by *FMF* (rectangles) plus CNN (contours).

5   Discussion

Based on our results, the main achievement of method *FMF* is the ability to identify and crop the RoIs with a very small error (Recall = 99.69%). Furthermore, by cropping the RoIs before the CNN training, we observed that the training process was remarkably accelerated, increasing the training speed by 150%. When considering metric DSC, we observed significant improvements when considering the entire bundle of experiments, as presented in Table 2.

For the majority of the datasets, all but ACDC used for training, we observed significant improvements. This inefficacy with ACDC is possibly related to imaging artifacts and standard operating procedures used to label the ACDC images (refer to Fig. 4). In general, *FMF* is able to automatically focus on the correct region and guide the segmentation CNN (Fig. 4, P27). However, in some cases, the CNN lost track of the LV when fed with a cropped RoI (P32). In a few cases, the CNN is unable to detect anything, even when fed with a centered RoI (P91). This is a known problem when segmenting the cardiac apex. These findings should be investigated in future works.

Overall, method *FMF* demonstrated significant improvements in the task of RoI detection; from the results, it became evident that the performance depends on the training dataset and on the network model. The main characteristic is image quality; format and intensities should be similar across datasets, and labeling standard operating procedures should be compatible. An extended preprocessing has the potential to overcome such issues across heterogeneous datasets; possibly, a dynamically adjusted module would be insensitive to data noise/variation.

6   Conclusions

In this paper, we proposed method *FMF*, a novel approach based on convolution operations and on the use of a radial basis function to detect the RoI in cardiac magnetic resonance images. We validated *FMF* with a U-Net CNN comparing our results to those of the canonical U-Net and of the *FMF*-CNN in three public reference datasets. According to our results, *FMF* was able to crop 99.69% (Recall metric) of the RoI voxels in all the datasets, being suitable to preprocess the data for CNN segmentation. *FMF* accelerated the training process by 150%, and also increased Sørensen-Dice coefficient in the majority of our test cases ($p < .001$). Further improvement is possible by extended preprocessing of the training datasets, and by the use of more advanced CNNs that support a fine-tuning to deal with the specificities of more challenging datasets.

As future work, we intend to expand the *FMF* methodology considering more possibilities: (1) use other networks in the segmentation step, such as 3D U-Net or Feature Pyramid Networks, validating versus other state-of-the-art networks; (2) embed this method as the first layer of a CNN, so that the parameters are adjusted automatically; (3) experiment with more datasets, preprocessing, and augmentation methods; (4) test different feature extractors for the static visual features, such as gray-level co-occurrence matrices and Gabor filters; other motion estimation methods, and radial basis functions.

Acknowledgement. This research was financed by Brazilian agencies Fundacao de Amparo a Pesquisa do Estado de Sao Paulo (2018/11424-0, 2016/17078-0); Coordenacao de Aperfeicoamento de Pessoal de Nivel Superior (Finance Code 001); and Conselho Nacional de Desenvolvimento Cientifico e Tecnologico (406550/2018-2).

References

1. Bernard, O., et al.: Deep learning techniques for automatic MRI cardiac multi-structures segmentation and diagnosis: is the problem solved? IEEE Trans. Med. Imaging **37**(11), 2514–2525 (2018). https://doi.org/10.1109/TMI.2018.2837502

2. Earls, J.P., Ho, V.B., Foo, T.K., Castillo, E., Flamm, S.D.: Cardiac MRI: recent progress and continued challenges. J. Magn. Reson. Imaging **16**(2), 111–127 (2002). https://doi.org/10.1002/jmri.10154

3. Fallat, S.M., Johnson, C.R.: Hadamard powers and totally positive matrices. Linear Algebra Appl. **423**(2–3), 420–427 (2007). https://doi.org/10.1016/j.laa.2007.01.012

4. Guo, F., Ng, M., Wright, G.: Cardiac MRI left ventricle segmentation and quantification: a framework combining U-Net and continuous max-flow. In: Pop, M., Sermesant, M., Zhao, J., Li, S., McLeod, K., Young, A., Rhode, K., Mansi, T. (eds.) STACOM 2018. LNCS, vol. 11395, pp. 450–458. Springer, Cham (2019). https://doi.org/10.1007/978-3-030-12029-0_48

5. He, K., Zhang, X., Ren, S., Sun, J.: Spatial pyramid pooling in deep convolutional networks for visual recognition. IEEE Trans. Pattern Anal. Mach. Intell. **37**(9), 1904–1916 (2015). https://doi.org/10.1109/TPAMI.2015.238982

6. Horn, B.K., Schunck, B.G.: Determining optical flow. Artif. Intell. **17**(1–3), 185–203 (1981)

7. Khened, M., Alex, V., Krishnamurthi, G.: Densely connected fully convolutional network for short-axis cardiac cine MR image segmentation and heart diagnosis using random forest. In: Pop, M., et al. (eds.) STACOM 2017. LNCS, vol. 10663, pp. 140–151. Springer, Cham (2018). https://doi.org/10.1007/978-3-319-75541-0_15

8. Majdisova, Z., Skala, V.: Radial basis function approximations: comparison and applications. Appl. Math. Model. **51**, 728–743 (2017). https://doi.org/10.1016/j.apm.2017.07.033

9. Peng, P., Lekadir, K., Gooya, A., Shao, L., Petersen, S.E., Frangi, A.F.: A review of heart chamber segmentation for structural and functional analysis using cardiac magnetic resonance imaging. Magn. Reson. Mater. Phys., Biol. Med. **29**(2), 155–195 (2016). https://doi.org/10.1007/s10334-015-0521-4

10. Ronneberger, O., Fischer, P., Brox, T.: U-Net: convolutional networks for biomedical image segmentation. In: Navab, N., Hornegger, J., Wells, W.M., Frangi, A.F. (eds.) MICCAI 2015. LNCS, vol. 9351, pp. 234–241. Springer, Cham (2015). https://doi.org/10.1007/978-3-319-24574-4_28

11. Saber, M., Abdelrauof, D., Elattar, M.: Multi-center, multi-vendor, and multi-disease cardiac image segmentation using scale-independent multi-gate UNET. In: Puyol Anton, E., et al. (eds.) STACOM 2020. LNCS, vol. 12592, pp. 259–268. Springer, Cham (2021). https://doi.org/10.1007/978-3-030-68107-4_26

12. Seraphim, A., Knott, K.D., Augusto, J., Bhuva, A.N., Manisty, C., Moon, J.C.: Quantitative cardiac MRI. J. Magn. Reson. Imaging **51**(3), 693–711 (2020). https://doi.org/10.1002/jmri.26789

13. Suinesiaputra, A., et al.: A collaborative resource to build consensus for automated left ventricular segmentation of cardiac MR images. Med. Image Anal. **18**(1), 50–62 (2014). https://doi.org/10.1016/j.media.2013.09.001

14. Szegedy, C., et al.: Going deeper with convolutions. In: CVPR 2015, pp. 1–9. IEEE (2015). https://doi.org/10.1109/CVPR.2015.7298594
15. Vaswani, A., et al.: Attention is all you need. In: Guyon, I., et al. (eds.) NIPS 2017. Advances in Neural Information Processing Systems, vol. 30, pp. 5998–6008. Curran Associates Inc, Red Hook (2017)
16. Wu, B., Fang, Y., Lai, X.: Left ventricle automatic segmentation in cardiac MRI using a combined CNN and U-net approach. Comput. Med. Imaging Graph. **82**, 101719 (2020). https://doi.org/10.1016/j.compmedimag.2020.101719
17. Xue, W., Brahm, G., Pandey, S., Leung, S., Li, S.: Full left ventricle quantification via deep multitask relationships learning. Med. Image Anal. **43**, 54–65 (2018). https://doi.org/10.1016/j.media.2017.09.005
18. Xue, W., Lum, A., Mercado, A., Landis, M., Warrington, J., Li, S.: Full quantification of left ventricle via deep multitask learning network respecting intra- and inter-task relatedness. In: Descoteaux, M., Maier-Hein, L., Franz, A., Jannin, P., Collins, D.L., Duchesne, S. (eds.) MICCAI 2017. LNCS, vol. 10435, pp. 276–284. Springer, Cham (2017). https://doi.org/10.1007/978-3-319-66179-7_32
19. Yan, W., Wang, Y., Li, Z., van der Geest, R.J., Tao, Q.: Left ventricle segmentation via optical-flow-net from short-axis cine MRI: preserving the temporal coherence of cardiac motion. In: Frangi, A.F., Schnabel, J.A., Davatzikos, C., Alberola-López, C., Fichtinger, G. (eds.) MICCAI 2018. LNCS, vol. 11073, pp. 613–621. Springer, Cham (2018). https://doi.org/10.1007/978-3-030-00937-3_70

Super-Resolution of Solar Active Region Patches Using Generative Adversarial Networks

Rasha Alshehhi[⊠]

New York University, Abu Dhabi, United Arab Emirates
ra130@nyu.edu

Abstract. Monitoring solar active region patches from Helioseismic and Magnetic Imager (HMI) instruments is essential for space weather forecasting. However, recovering small bipolar details in HMI patches requires additional pre-processing steps to obtain better quality. This work uses a generative adversarial network, with transposed convolution and super-pixel convolution up-sampling layers, to generate the higher quality of HMI patches. It trains and validates the network based on binary cross-entropy, mean absolute error and multi-scale dice-coefficient functions. It illustrates the performance of the generative method in two image types (magnetogram and continuum intensity patches) from two instruments (SDO/HMI and SOT/NET). It also compares its performance with state-of-the-art methods. The results demonstrate that the generative method produces high-quality images by increasing polarity contrast and retrieving smaller structures.

Keywords: Space weather · Solar active region patches · Generative adversarial network · Multi-scale dice-coefficient

1 Introduction

Solar active regions are the sources of energetic phenomena such as solar flares and coronal mass ejections (CME). These phenomena cause electromagnetic and particle radiation to interfere with telecommunications and power transmissions on Earth and pose significant hazards to astronauts and spacecraft [16]. Therefore, understanding active regions in solar patches is essential to understand their influence on space weather on Earth. Astronomers use Space-weather HMI Active Region Patches (SHARPs) [4], which are small patches of the HMI solar image to analyze solar phenomena related to solar active regions. However, the image quality of SHARPs is not sufficient for further image processing (e.g., detecting and classifying flares). Therefore, astronomers need to use image pre-processing techniques (e.g., image enhancement or noise reduction) or new relatively expensive instruments to have higher quality images or better resolution and measurement frequency. On the other hand, noise, saturation levels, spectral inversion techniques and instruments introduce inhomogeneity. Therefore, astronomers need an efficient and effective sup-resolution method to enhance SHARPs. They acquire SHARPs from Helioseismic and Magnetic Imager (HMI) instruments [20]. The HMI provides full-disk images (called HMI images) of the solar photosphere with a cadence

© The Author(s), under exclusive license to Springer Nature Switzerland AG 2022
S. Sclaroff et al. (Eds.): ICIAP 2022, LNCS 13231, pp. 451–462, 2022.
https://doi.org/10.1007/978-3-031-06427-2_38

better than one every minute to study solar oscillations and understand the interior structure of the Sun. These observations are the best astronomical observations because of the absence of atmosphere and good astronomical imaging. However, the trade-off between full-disk observations and spatial resolution makes HMI images insufficient to detect, classify, and track small-scale structures of interest (active regions), particularly with the higher bipolar contrast. Therefore, enhancement of image quality of SHARPs helps to identify the characteristics of small-scale structures in the HMI images. It also improves the estimation of physical characteristics of small-scale structures such as total unsigned magnetic flux [7]. Nowadays, convolutional neural networks (CNNs), especially generative adversarial networks (GANs) [9] show remarkable success in various applications such as image-to-image translation, image-to-paint translation, harmonization, sup-resolution, etc. They capture the internal distribution of images. However, highly diverse data with a wide range is still a significant challenge. This work concentrates on super-resolution. Sup-resolution is a method of enhancing one or many degraded or corrupted images by adding high-frequency content and removing the degradation because of image processing of astronomical telescopes [10]. It uses the generative network to improve solar polarity contrast and retrieve small-structure with higher quality. The proposed method can be summarized as follows. First, it uses two transposed convolution blocks and super-pixel convolution blocks as up-sampling blocks in the generative network. Second, it uses binary cross-entropy, mean absolute error and multi-scale dice-coefficient as adversarial and content loss functions. Third, it is evaluated in two Sun datasets from diverse instruments: SDO/HMI and SOT/NET images (described in Sect. 4.1). This paper is organized as follows. Section 2 describes the previous works related to super-resolution techniques. Section 3 illustrates the used super-resolution method. Section 4 presents experimental settings and results in various datasets. Section 5 summarizes and provides an outlook for the future.

2   Related Work

There are different methods designed for super-resolution tasks. The classical techniques used image pre-processes such as interpolation, which does not recover the high-frequency information and leads to image blurring [8]. The reconstruction method is another classical method that retrieves high-frequency lost details; however, it does not reconstruct the texture features. The current works use convolutional networks to capture spatial regularities in complex imagery, categorized into two types according to the number of training samples: multiple-images super-resolution (MISR) and single-image super-resolution (SISR). The MISR employs multiple low-resolution images of the same scene acquired from the same or different sensors to construct a high-resolution image such as SRGAN [13] and EDSR [14]. The SISR employs a single image to reconstruct a high-resolution image from it such as DIP [24], ZSSR [22], SinGAN [21], SRResCGAN [25] and SRResCycGAN [26]. In astrophysics, Baso and Ramos [7] applied a convolutional network for simultaneous deconvolution and super-resolution low-resolution SHARP images based on higher-resolution images. They used DeepVel [19] and trained in synthetic data obtained from stimulated SHARP images and validated in actual SHARP images. Their approach required ground-truth, which is usually difficult in astrophysics and therefore, they train and validate their method in simulated datasets. Their approach was employed in various SHARP images (magnetogram and continuum intensity patches, described in Sect. 4.1) from the HMI instrument only. Rahman et al. [18] proposed an unsupervised convolutional network based on residual attention and progress generative networks to improve 4×4 degraded image into a higher-resolution image. Their method was applied for certain SHARP images (magnetogram images, described in Sect. 4.1) captured by both HMI and SOT instruments [2]. This work uses the generative network based on transposed convolution and super-pixel convolution blocks (up-sampling layers). It trains the generative network based on binary cross-entropy function (adversarial loss), mean absolute error and multi-scale dice-coefficient functions (reconstruction loss). The up-sampling blocks and multi-scale loss function are applied to enhance polarity contrast and retrieve smaller structures from various SHARP images (magnetogram and continuum intensity patches from both HMI and SOT instruments [2]).

3 Method

We use original input SHARP images and its degraded version to test the validity of the proposed method to enhance image quality. Therefore, to generate a higher-resolution image $\hat{Y}$ from the input SHARP image Y, this work degrades Y into lower-resolution image X by fixed scale.

3.1 Network Architecture

The proposed network consists of two network branches. The first network branch is a generator to produce sup-resolved image $\hat{Y}$ from degraded image X. The second network branch is a discriminator to distinguish between the super-resolved image $\hat{Y}$ and the original image Y.

Generator: This work proposes a generator architecture, inspired by super-FAN architecture [6] (Fig. 1), to generate the enhanced SHARP image $\hat{Y}$ from the degraded SHARP image X. The super-FAN consists of 12 and then 3 local residual convolution blocks (each with convolution, batch normalization and ReLU activation followed by another convolution, batch normalization and element-wise sum layers). The local residual convolution blocks are followed by two deconvolution blocks (each with normalization, ReLU activation and deconvolution layers), and then by two convolution blocks (each with convolution and ReLU activation layers). This work uses 2 transposed-convolution blocks, instead of the first deconvolution block, after the first 6 local residual convolution blocks and after the second 6 local residual convolution blocks. These blocks are followed by three local residual convolution blocks with the sub-pixel convolution block, followed by one convolution block. The motivation behind using two transposed convolution blocks and super-pixel convolution block is they maintain spatial integrity and apply only pixel-shuffle blocks at the high-resolution (as in [13]) or apply one deconvolution block after 12 and 3 local residual convolution blocks (as in [6]) is insufficient for the generation sharper details of input images. In addition, we found using short skip-connection after every 6 blocks (unlike

SRGAN [13] and Super-FAN [6]) helps to fuse low-level, middle-level and high-level features in the training process and has better gradient flow. The PReLU activation function is also used instead of ReLU activation function (similar to SRGAN [13]) because it improves the overall performance. The size of the degraded input image X is 16×16 of input image Y 128×128 and the size of the higher-resolution output image is 512×512. Similar to [6], 3×3 kernel is used in local residual convolution blocks, 2×2 kernel in transposed convolution blocks, and 2 kernel in the sup-pixel convolution block and 1×1 in the convolution block.

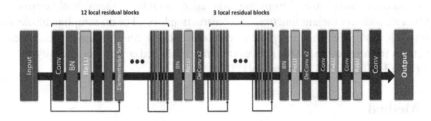

Fig. 1. An overview of the Super-FAN architecture.

Discriminator: This work uses the patchGAN [12] based on the ResNet architecture to differentiate between the enhanced SHARP images $\hat{Y}$ and the original SHARP image Y of same size. The ResNet architecture consists of 10 local residual convolution blocks followed by a 1×1 convolution block. This work uses a patch size equals 64×64.

3.2 Training

The loss function is a combination of binary cross-entropy as adversarial loss and multiscale dice-coefficient and mean absolute error functions as reconstruction loss.

Adversarial Loss: This work employs a binary cross-entropy to train the generative model G to fool a discriminator D that is trained to differentiate between enhanced SHARP image $\hat{Y}$ from SHARP image Y based on:

$$
L_{adv}(G_{\theta_G}, D_{\theta_D}) = min_{\theta_G} max_{\theta_D}
$$
$$
\mathbb{E}_{y \sim \mathbb{P}_{HR}} \left[log D_{\theta_D}(y) \right] + \mathbb{E}_{x \sim \mathbb{P}_{LR}} \left[log(1 - D_{\theta_G}(G_{\theta_G}(x))) \right], \tag{1}
$$

where $\mathbb{P}_{HR}$ is data distribution of the SHARP image Y and $\mathbb{P}_{LR}$ is data distribution of the degraded SHARP image X. G and D are generator and discriminator parameterized by θ_G and θ_D, respectively.

Reconstruction Loss: To retrieve details of images, this work uses convolutional feature maps to capture small-scale to large-scale structures; from pixel-based to region-based. The discriminator consists of N_L convolution layers. Each convolutional layer L consists of N_M feature maps. The dice-coefficient function of patch i and patch j

at same feature map M of the same convolution layer L of both original image Y and high-resolution output image $\hat{Y}$ are presented as follows:

$$E(Y, \hat{Y}) = \sum_{i,j} \frac{2Y_i * \tilde{Y}_j}{Y_i + \tilde{Y}_j} \tag{2}$$

$$L_D(Y, \hat{Y}) = \sum_{s=0}^{N_s} \sum_{l=0}^{N_L} \sum_{m=0}^{N_M} W_L W_M E_{(s,l,m)}(Y, \hat{Y}), \tag{3}$$

where W_L and W_M are weight factors assigned to each convolutional layer and each feature map (W_L and W_M are assigned randomly between 0–1). N^s is the number of input degraded SHARP images X, The reconstruction loss is a combination of the previous multi-scale dice-coefficient function L_D and mean absolute error (MAE) L_{MAE} function.

$$L_{MAE} = \sum_{s=0}^{N_s} |\hat{Y} - Y|, \tag{4}$$

$$L_{rec}(Y, \hat{Y}) = L_D(Y, \hat{Y}) + \gamma L_{MAE}, \tag{5}$$

where γ is a weight factor of the function L_{MAE}. The final loss function is a combination of the reconstruction and binary cross-entropy functions.

$$L_{total} = \alpha L_{rec} + \beta L_{adv}, \tag{6}$$

where α and β are the weighting factors for both loss functions, respectively. L_{rec} is equal to reconstruction loss and L_{adv} is equal to adversarial loss.

4 Performance

4.1 Experimental Settings

Data: This work uses Space-weather HMI Active Region Patches (SHARPs) from Helioseismic and Magnetic Imager (HMI) of Solar Dynamics Observatory (SDO) satellite [23]. The SHARPs are sub-regions of the full-disk HMI images with a 12-minute cadence with spatial resolution 0.1 arcsec and sampling of 0.5 arcsec per pixel. The SHARPs are patches of various sizes and the horizontal dimension is different from the vertical dimension (dimensionless). The SHARP images are readily accessible from https://sdo.gsfc.nasa.gov/. Each SHARP has four maps. This work focuses on Line-of-sight (LOS) magnetic field B_{los}, called magnetogram, and continuum intensity l_c. The magnetogram measures line-of-sight magnetic-field strength data (outflow: white positive regions and inflow: black negative regions) and the continuum measures intensity on the Sun surface. Astronomers use these images to generate physical parameters that help to understand the Sun and consequently Space-weather [4]. This work validates the proposed network in another dataset; from Narrowband Filter Imager (NFI). The NFI images are generated by a Solar Optical Telescope (SOT) instrument on-board the

Hinode satellite. The NFI images are readily accessible from https://sot.lmsal.com/data/ sot/. It measures the magnetic field and continuum intensity with spatial resolution 0.3 arcsec and sampling of 0.16 arcsec per pixel. It is aligned to the HMI magnetogram based on visual inspection. Its pixel resolution is set to be $1/4$ of the HMI by using the cubic interpolation. All measurements of the NFI are more likely to be more accurate compared to the HMI [2]. For simplicity, the SDO/HMI data is referred to as the first dataset and the SOT/NFI data is referred to as the second dataset.

Training Details: This work trains, validates and evaluates the proposed model in random SDO/HMI samples from 2014 January to December 2020, where samples are not repeated in all three datasets (10,000 training, 2,000 validation and 5,000 testing samples). The input image is degraded using bicubic interpolation. It is normalized by subtracting the mean from images and dividing by standard deviation. This work trains the model with an Adam optimization function by setting momentum parameters $B_1 = 0.5$ and $B_2 = 0.999$. The mini-batch size is assigned as 64. The learning rate is initialized as 0.0001. The network is trained for 5000 epochs. The weight factors α and β are assigned as 0.6 and 0.4. The networks are implemented with Nvidia Tesla V100 GPU-32 GB.

Evaluation Metrics: This work evaluates the proposed method by comparison the enhanced SHARP output with outputs of the state-of-the-art methods: MISR methods such as EDRS [14] and SRGAN [13] and SISR methods such as ZSSR [22], DIP [24] and SinGAN [21]. It quantitatively compares the proposed method with the MISR methods. However, it compares the performance of the proposed method with SISR methods using different scores. It uses mean opinion score (MOS) [17], distortion root mean square error (RMSE) [3], structure similarity (SSIM) [11] and peak signal-to-noise ratio (PSNR) [5]. The MOC is calculated based on the evaluation of THE enhanced SHARP images from twenty astrophysical experts, where one means bad quality and five means excellent quality. The RMSE is calculated based on the difference between the enhanced and the original SHARP images, where lower is better. The SSIM measures the difference depending on the mean, variance and covariance of both images, where higher is better. The PSNR measures the peak ratio of signal over noise; where higher is better.

4.2 Experimental Results in the SDO/HMI Dataset

Figure 2 shows an example of applying the network in the whole SHARP image, observed on February 8, 2018. The upper rows show SDO/HMI magnetogram images, while the lower row shows SDO/HMI continuum images. The leftmost column is the original SHARP image and the rightmost column is the enhanced SHARP images with 5-scale dimensions of the original SHARP images. The proposed network improves polarity contrast between pixels of a strong magnetic field (white pixels, max = 1500 kG) and weak magnetic field (black pixels, min = -1500 KG) and remaining pixels (gray pixels). On the other hand, it retrieves small-scale structures in continuum images, particularly with high intensity ($lc > 4000\ kDN/s$).

Comparison with Multiple-image Super-resolution Methods: Figure 3 selects a part from one of SHARP images observed on July 07, 2014 and compares the super-resolution output with outputs of MISR networks. The first row shows an original

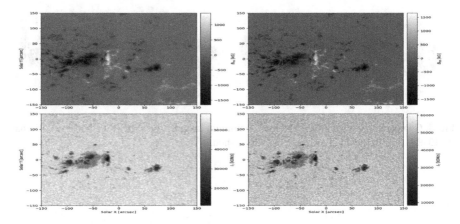

Fig. 2. Comparison between original and enhanced SDO/HMI images. The upper images are magnetogram images and the lower images are continuum images, respectively.

image and an output image of the network EDRS [14]. The second row presents the results of the network SRGAN [13] and the proposed method. The results show that the proposed method mainly reconstructs small-scale magnetic structures much better than previous methods. It has better performance in alleviating blurring artifacts, especially when compared with original low-resolution images. It also has the sharpest polarity inversion line. Figure 4 compares continuum outputs of EDSR, SRGAN and the proposed method. Here, it is more notable that proposed methods improve low-contrast in all granule structures, in particular lines between very-high intensity value and very-low intensity value. Figure 5 shows the spatial Fourier power spectrum of the original image, the super-resolved output of the input image and the super-resolved output of the smoothed image with Gaussian filter ($kernel = 9 \times 9$). The power spectrum of the super-resolved image is closer to the original image, in particular for more minor frequencies. However, it is slightly higher with higher frequencies. The power spectrum of the super-resolved output of a smoothed image is higher compared to one of the original images. The super-resolved of magnetogram outputs is more stable compared to continuum outputs, particularly with minor frequencies.

| (a) | (b) | (c) | (d) |

Fig. 3. Comparison between the enhanced magnetogram images from MISR methods: (a) original image, (b) EDRS, (c) SRGAN and (d) the proposed method.

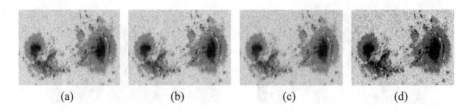

Fig. 4. Comparison between the enhanced continuum images from MISR methods: (a) original, (b) EDRS, (c) SRGAN and (d) the proposed method.

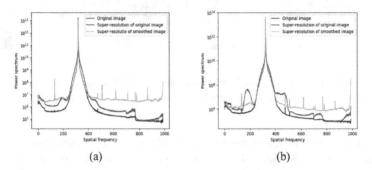

Fig. 5. Comparison between enhanced SDO/HMI magnetogram images (a) and enhanced SDO/HMI continuum images (b): original image (blue), super-resolution output of original image (orange), super-resolution output of original smoothed image (green).

Comparison with Single-image Super-resolution Methods: Table 1 compares between unsupervised single image super-resolution methods: ZSSR [22], DIP [24], SinGAN [21] and the proposed method. The comparison is based on metrics MOS, RMSE and SSIM. Remarkably, the average RMSE of the proposed method is the lowest and the average MOC and SSIM are the highest compared to all methods.

Comparison with the State-of-the-art Methods: Table 2 shows a comparison between the proposed super-resolution method with the most recent methods in enhancing SHARP patches of SDO/HMI magnetogram images [18]. The comparison is based on the pixel to pixel correlation coefficient (P2P-CC), PSNR, RMSE and SSIM. The proposed method achieves significantly higher pixel-correlation (P2P-CC $>$ 9%) and lower errors (RMSE $>$ 7). However, the previous method is slightly better than the proposed method in image quality with lower noise (PSNR $>$ 2) and higher similarity (SSIM $>$ 2%). The PSNR is small because we use a weighted average function of convolutional feature maps as reconstruction loss. Applying other functions or assigned other weights could help to reduce the noise in the output images.

The Impact of Up-sampling Blocks and Loss Functions: Table 3 validates the performance of the generative networks using different up-sampling blocks: shuffle-pixel (as in [13]), deconvolution (as in [6]) and transposed convolution based on RSNR, RMSE and SSIM metrics in the output SHARP images. It also evaluates the performance using different reconstruction loss functions using only the dice-coefficient function of the last

Table 1. Comparison with SISR methods in SDO/HMI images.

Method	MOS	RMSE	SSIM
ZSSR [22]	3.3	14.5 ± 1.21	0.89 ± 0.33
DIP [24]	3.7	14.8 ± 2.23	0.90 ± 0.31
SinGAN [21]	4.2	16.2 ± 1.5	0.89 ± 0.23
The proposed method	4.5	13.2 ± 1.2	0.91 ± 0.12

Table 2. Comparison between Rahman and the proposed methods in SDO/HMI magnetogram images.

Method	Rahman et al. [18]	The proposed method
P2P-CC	0.87 ± 0.4	$\mathbf{0.96 \pm 0.2}$
PSNR	$\mathbf{23 \pm 1.21}$	19 ± 1.41
RMSE	20.65 ± 2.3	$\mathbf{13.2 \pm 1.12}$
SSIM	$\mathbf{0.93 \pm 0.32}$	0.91 ± 0.12

layer, the multi-scale dice-coefficient function of all layers with and without absolute error (MAE) function. Using various up-sampling blocks in the network architecture (e.g., super-pixel convolution followed by transposed convolution blocks or transposed convolution followed by super-pixel convolution blocks) improves the quality of output SHARP images with PSNR > 2, RMSE > 4 and SSIM $> 3\%$. On the other hand, a combination of multi-scale dice-coefficient function L_D and mean absolute error function L_{MAE} improves the quality of the SHARP images with low error (RMSE $= 13.3 \pm 1.2$), low noise (PSNR $= 19.1 \pm 1.41$) and a high degree of similar structures (SSIM $= 0.91 \pm 0.12$).

An example of super-resolution application: Super-resolution of HMI patches is an important pre-processing stage in various applications such as detection of solar flares and predicting front-side and far-side magnetogram images [1] and consequently help astronomers to access non-visible regions more than a week ahead before they rotate to face the Earth [15,16]. The classical method uses normalization with 98% and linear stretching as a pre-processing stage before training, validation and testing the generative model to produce a whole HMI magnetogram image. This work uses the proposed super-resolution method as a pre-processing stage, instead of normalization and stretching. Table 4 compares between physical parameters of generated magnetogram images after applying normalization and stretching [1] and after applying the proposed super-resolution method as pre-processing stages. It measures physical properties such as total unsigned of magnetic flux (TUMF) correlation coefficient (CC), relative error (R1) of the total unsigned magnetic flux (TUMF) [4] and normalized mean square error (R2) of the magnetic field (MF) [4] and PSNR. The proposed super-resolution method demonstrates better estimation of physical metrics (higher correction coefficient and lower error) and higher image quality.

Table 3. The impact of up-sampling blocks and loss functions in SDO/HMI magnetogram images.

Up-sampling block	PSNR	RMSE	SSIM
3 Pixel-shuffle blocks	13.2 ± 3.35	22.2 ± 3.40	0.82 ± 0.13
3 Deconvolution blocks	13.3 ± 3.12	20.2 ± 2.50	0.85 ± 0.22
3 Transposed convolution blocks	15.4 ± 2.13	21.2 ± 1.25	0.83 ± 0.17
3 Super-Pixel convolution blocks	15.1 ± 2.30	19.2 ± 2.20	0.86 ± 0.14
2 Super-Pixel + 1 Transposed convolution blocks	17.3 ± 1.14	15.2 ± 2.15	0.89 ± 0.21
2 Transposed + 1 Super-Pixel convolution blocks	**19.1 ± 1.41**	**13.2 ± 1.12**	**0.91 ± 0.12**
Loss function	PSNR	RMSE	SSIM
Single-scale dice	13.1 ± 2.14	18.1 ± 2.50	0.81 ± 0.27
Multi-scale dice	14.3 ± 1.26	16.7 ± 2.13	0.85 ± 0.12
Absolute error	15.9 ± 1.71	19.3 ± 1.11	0.82 ± 0.24
Single-scale dice + absolute error	17.3 ± 1.12	15.2 ± 2.4	0.87 ± 0.17
Multi-scale dice + absolute error	**19.1 ± 1.41**	**13.2 ± 1.2**	**0.91 ± 0.12**

Table 4. Comparison between physical parameters using Alshehhi and the proposed method in the whole SDO/HMI magnetogram images.

Method	Alshehhi [1]	The proposed method
TUMF-CC	0.75	**0.80**
TUMF-R1	0.70 ± 0.019	**0.60 ± 0.021**
MF-R2	0.45 ± 0.005	**0.35 ± 0.015**
PSNR	16.0 ± 1.21	**19.1 ± 1.41**

4.3   Experimental Results in the SOT/NFI Dataset

This section evaluates the performance of the proposed super-resolution method in the SOT/NFI images. As it is observable in Fig. 6, the SOT/NFI images have better resolution than the SDO/HMI images. The absolute values of the magnetic field strength (Fig. 6(a)) in the SOT/NFI images are higher than those from SDO/HMI images. Applying the proposed method increases polarity sharpness better (as shown in Fig. 6(b)). Moreover, the super-resolved continuum intensity image (as shown in Fig. 6(d)) clearly shows detailed solar granule structures. The improvement in all images can be interpreted because the proposed method uses the multi-scale dice-coefficient function and transposed convolution blocks to capture smaller structures in the input images.

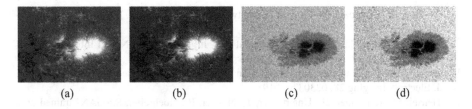

(a)   (b)   (c)   (d)

Fig. 6. Comparison between the original and enhanced SOT/NFI image: (a) magnetogram image (b) enhanced magnetogram image (RMSE = 11.2 ± 1.11), (c) continuum intensity, (d) enhanced continuum intensity (RMSE = 10.2 ± 1.23).

5 Conclusion

This work uses a generative network to enhance the solar magnetogram and continuum intensity patches. It uses transposed convolution and super-pixel convolution blocks to generate better higher-resolution solar images. It uses a combination of multi-scale dice-coefficient, mean absolute error and binary cross-entropy functions to capture content and adversarial losses. This method is evaluated in two datasets: SDO/HMI and SOT/NFI. It is compared with state-of-the-art methods. All methods can generate good super-resolution versions from original images, but the proposed method is more consistent. It produces sup-resolved images with higher quality and low errors. It also reconstructs polarity contrast (e.g., the contrast between white/positive and black/negative regions) more efficiently in magnetogram images. It also captures small-scale to middle-scale details (e.g., spatial and texture) in continuum intensity images. In the future, this method may be used to enhance full-disk HMI images and other Sun observations for better understanding solar physical mechanisms.

References

1. Alshehhi, R.: Deep regression for imaging solar magnetograms using pyramid generative adversarial networks. In: IEEE Conference on Computer Vision and Pattern Recognition (2020)
2. Bamba, Y., Kusano, K., Imada, S., Iida, Y.: Comparison between Hinode/SOT and SDO/HMI, AIA data for the study of the solar flare trigger process. Astronomical Soc. Japan **66**, S16 (2014)
3. Blau, Y., Michaeli, T.: The perception-distortion tradeoff. In: IEEE Conference on Computer Vision and Pattern Recognition, pp. 6228–6237 (2018)
4. Bobra, M.G., et al.: The Helioseismic and Magnetic Imager (HMI) Vector Magnetic Field Pipeline: SHARPs - Space-Weather HMI active region patches. Solar Phys. **289**(9), 3549–3578 (2014)
5. Borji, A.: Pros and Cons of GAN evaluation measures. Comput. Vis. Image Understanding **179**, 41–65 (2019)
6. Bulat, A., Tzimiropoulos, G.: Super-FAN: integrated facial landmark localization and super-resolution of real-world low resolution faces in arbitrary poses with GANs. In: IEEE Conference on Computer Vision and Pattern Recognition (2017)
7. Díaz Baso, C.J., Asensio Ramos, A.: Enhancing SDO/HMI images using deep learning. Astronomy Astrophys. **614**, A5 (2018)

8. Duchon, C.E.: Lanczos filtering in one and two dimensions. J. Appl. Meteorol. **18**(8), 1016–1022 (1979)
9. Goodfellow, I, et al.: Generative adversarial nets. In: Advances in Neural Information Processing Systems 27, pp. 2672–2680 (2014)
10. Guo, R., Shi, X., Wang, Z.: Super-resolution from unregistered aliased astronomical images. J. Electron. Imaging **28**, 023032 (2019)
11. Heusel, M., Ramsauer, H., Unterthiner, T., Nessler, B., Hochreiter, S.: GANs trained by a two time-scale update rule converge to a local nash equilibrium. In: Advances in Neural Information Processing Systems 30, pp. 6626–6637 (2017)
12. Isola, P., Zhu, J., Zhou, T., Efros, A.A.: Image-to-image translation with conditional adversarial networks. In: IEEE Conference on Computer Vision and Pattern Recognition, pp. 5967–5976 (2017)
13. Ledig, C., et al.: Photo-realistic single image super-resolution using a generative adversarial network. In: IEEE Conference on Computer Vision and Pattern Recognition, pp. 105–114 (2017)
14. Lim, B., Son, S., Kim, H., Nah, S., Lee, K.M.: Enhanced deep residual networks for single image super-resolution. In: IEEE Conference on Computer Vision and Pattern Recognition (2017)
15. Lindsey, C., Braun, D.C.: Seismic images of the far side of the sun. Science **287**(5459), 1799–1801 (2000)
16. Lindsey, C., Braun, D.: Seismic imaging of the sun's far hemisphere and its applications in space weather forecasting. Space Weather Int. J. Res. Appl. **15**(6), 761–781 (2017)
17. Moorthy, A.K., Bovik, A.C.: Blind image quality assessment: from natural scene statistics to perceptual quality. IEEE Trans. Image Process. **20**(12), 3350–3364 (2011)
18. Rahman, S., Moon, Y.J., Park, E., Siddique, A., Cho, I.H., Lim, D.: Super-resolution of SDO/HMI magnetograms using novel deep learning methods. Astrophys. J. **897**(2), L32 (2020)
19. Ramos, A.A., Requerey, I.S., Vitas, N.: DeepVel: Deep Learning for the Estimation of Horizontal Velocities at the Solar Surface. CoRR (2017)
20. Schou, J., et al.: Design and ground calibration of the Helioseismic and Magnetic Imager (HMI) instrument on the Solar Dynamics Observatory (SDO). Solar Phys. **275**, 229–259 (2012)
21. Shaham, T.R., Dekel, T., Michaeli, T.: SinGAN: learning a generative model from a single natural image. In: IEEE International Conference on Computer Vision (2019)
22. Shocher, A., Cohen, N., Irani, M.: Zero-shot super-resolution using deep internal learning. In: IEEE Conference on Computer Vision and Pattern Recognition (2018)
23. Smirnova, V., Riehokainen, A., Solov'ev, A., Kallunki, J., Zhiltsov, A., Ryzhov, V.: Long quasi-periodic oscillations of sunspots and nearby magnetic structures. Astronomy Astrophys. **552**, A23 (2013)
24. Ulyanov, D., Vedaldi, A., Lempitsky, V.: Deep image prior. In: IEEE Conference on Computer Vision and Pattern Recognition (2018)
25. Umer, R.M., Foresti, G.L., Micheloni, C.: Deep generative adversarial residual convolutional networks for real-world super-resolution. In: IEEE Conference on Computer Vision and Pattern Recognition (2020)
26. Umer, R.M., Micheloni, C.: Deep cyclic generative adversarial residual convolutional networks for real image super-resolution. In: Bartoli, A., Fusiello, A. (eds.) ECCV 2020. LNCS, vol. 12537, pp. 484–498. Springer, Cham (2020). https://doi.org/10.1007/978-3-030-67070-2_29

Avoiding Shortcuts in Unpaired Image-to-Image Translation

Tomaso Fontanini$^{(\boxtimes)}$ (ID), Filippo Botti, Massimo Bertozzi(ID),
and Andrea Prati(ID)

IMP Lab, Department of Engineering and Architecture,
University of Parma, Parma, Italy
{tomaso.fontanini,massimo.bertozzi,andrea.prati}@unipr.it,
filippo.botti2@studenti.unipr.it

Abstract. Image-to-image translation is a very popular task in deep learning. In particular, one of the most effective and popular approach to solve it, when a paired dataset of examples is not available, is to use a cycle consistency loss. This means forcing an inverse mapping in order to reverse the output of the network back to the source domain and reduce the space of all the possible mappings. Nevertheless, the network could learn to take shortcuts and softly apply the target domain in order to make the reverse translation easier therefore producing unsatisfactory results. For this reason, in this paper an additional constraint is introduced during the training phase of an unpaired image-to-image translation network; this forces the model to have the same attention both when applying the target domains and when reversing the translation. This approach has been tested on different datasets showing a consistent improvement over the generated results.

Keywords: Generative adversarial network · Attention generation

1 Introduction

Unpaired image-to-image translation aims at finding a mapping from an input image belonging to a source domain to an output image belonging to a target domain when a paired dataset of samples is not available. The preferred architecture for approaching this task is a generative adversarial network (GAN), where the generator is trained to apply the target domain to an input image and the discriminator is trained to distinguish if an image is real or generated. However, if no additional constraint is imposed during training, the generator could not only simply apply the target domain but also alter the overall shape/identity of the input in order to more easily fool the discriminator.

For this reason, CycleGAN [27] introduced a cycle consistency loss by adding an additional generator trained to learn an inverse mapping from the target domain back to the source domain. This solution allows to apply the target domains to an image without changing its overall content. Nevertheless, a

S. Sclaroff et al. (Eds.): ICIAP 2022, LNCS 13231, pp. 463–475, 2022.
https://doi.org/10.1007/978-3-031-06427-2_39

significant drawback introduced by the cycle consistency loss is that the generator could take a shortcut applying the target domain just enough to fool the discriminator, but also just as softly to make the reverse translation easier. A practical demonstration of this can be seen in Fig. 1, where the resulting images (second column) show a negligible transformation and therefore retain lots of information from the original domain (first column). This effect eases the reverse translation, but does not represent an *optimal* result, i.e. a result image which is indistinguishable from an image belonging to the target domain.

Fig. 1. Some samples that demonstrate the shortcuts introduced by the cycle consistency loss in CycleGAN for horse → zebra, orange → apple and apple → orange.

To solve this limitation, inspiration has been drawn by the recent applications of attention as valuable information to be used during the training of CNNs. The concept of attention in CNNs was first introduced by Zeiler and Fergus [23] as a way to visualize regions in the images that are important for the network when taking a certain decision or performing a certain task. Recently, attention was not only used as a mean of visual explanation of CNNs, but also actively during training. For example, attention was transferred from a teacher to a student model in order to improve the classification performance of the student in [22], attention maps were used effectively for semantic segmentation in [15] and, lastly, Dhar *et al.* [5] introduced an attention distillation loss for incremental learning that allowed to preserve the information about base classes when adding new ones, without storing any of their data. In addition to that, Liu *et al.* [16] showed that attention map can be generated effectively even in generative models like Variational AutoEncoders (VAE).

In this paper, we propose to actively use attention maps during the training of a CycleGAN. In particular, the intuition is that the attention obtained when translating an input image to a target domain and the attention obtained when translating the output image back to the source domain should be the same, because the network needs to focus on the same area of the network with the same intensity in both cases. This allowed to prevent the generator from taking shortcuts when applying the target domain, then resulting in images with a much higher quality.

To sum up, the main contributions of this paper are the following:

– A system that utilizes attention maps during the training of an unpaired image-to-image translation network allowing to limit the introduction of shortcuts caused by the cycle consistency loss. This improves the generated results without the need of any additional module;

- A quantitative and qualitative evaluation over common unpaired image-to-image translation datasets.

2 Related Work

Conditional GANs for Image-to-image Translation. Generative Adversarial Networks (GANs) were first introduced by Goodfellow *et al.* in [7] and since then they become very common in lots of deep learning applications. In addition to that, conditional GANs (cGANs) [19] allowed to achieve control over the generated samples by feeding the GAN model with additional information like labels [2] or text [24]. When both the input and the output of the generator are images, this is often referred to as *cGANs for image-to-image translation*. Initially, a great success was achieved using paired datasets of images [11], but very often it is not possible to have a ground truth when applying a target domain to an image.

For this reason, DiscoGAN [12] and, more notably, CycleGAN [27] were introduced. CycleGAN, which will be described in detail later in this paper, works by simultaneously training two generators and two discriminators. One generator is trained to produce images belonging to a domain Y starting from a domain X, while the other generator is trained in the opposite way. Also, CycleGAN uses a cycle consistency loss to force the output of a generator to be reversed back to its original domain when fed to the other generator, allowing to maintain the shape of the original input intact during the translation. Since then, the idea of introducing a cycle consistency loss for unpaired image-to-image translation has become very popular [4,9,10,14]. In addition to that, recently other authors noted that cycle loss can limit the efficacy of the translation task. Particularly, [20] relies on a council of networks that collaborates between each other and [25] proposes an adversarial-consistency loss for image-to-image translation. Yet, in our paper, cycle consistency loss is maintained and results are improved without the need of designing a completely-new architecture.

Attention Maps Generation. A very active research field consists in understanding how neural networks perform their tasks or take their decisions. Some preliminary results in this direction were achieved by [17] and [23]. After that, CAM (Combined Attention Model) was introduced by Zhou *et al.* [26], but was limited by the fact that was applicable only to some types of CNNs. More general and effective methods are represented by GradCAM [21] and GradCAM++ [3]. They both are *gradient-based* methods that use the gradient (generated by the classification output of the network in a specific layer L) to produce the attention maps.

The concept of attention can be also partially exploited in unpaired image-to-image translation. In particular, Mejjati *et al.* [18] added an attention-guided generator to the CycleGAN architecture and used the attention as a way to separate foreground and background in order to apply the target domain to the former and not the to latter. Finally, Emami *et al.* [6] calculated attention in the CycleGAN discriminator and multiplied it with input image to guide the generation. Both these approaches use the attention maps more as a mask than actively during the training as we propose. They are effective when a clear separation between foreground and background is present in the training samples, which is not always the case. For this reason, they are not designed to solve the cycle consistency limitation mentioned before.

3 Proposed Approach

Given two different domains X and Y, where $\{x_i\}_{i=1}^N$ are images belonging to X and $\{y_j\}_{j=1}^M$ are images belonging to Y, our model follows the CycleGAN formulation and therefore it is composed by two different generators G and F that learn the mapping $X \rightarrow Y$ and $Y \rightarrow X$, respectively. In addition to that, the model is also composed by two discriminators D_X and D_Y that learn to distinguish between real samples $\{x\}$ or $\{y\}$ and translated samples $\{G(x)\}$ or $\{F(y)\}$.

The baseline model is trained using an *adversarial loss* and a *cycle consistency loss*. The first one forces the generated samples to match the distribution of the target domain X or Y, while the second one allows to reverse the translation and avoids the generated samples to diverge from the input samples' shape. Nevertheless, cycle consistency loss introduces a drawback when the network tries to translate a domain X to a different domain Y or viceversa. Indeed, the cycle could prevent the generator to apply consistently the target domain in order to ease the reverse translation.

The main objective of this paper is to solve this drawback. This was achieved introducing a new loss term that forces the attention in the latent space of the two generators G and F to be the same when applying and reversing the translation, avoiding to lower the intensity of focus of the network during the cycle and denying the introduction of shortcuts when applying the target domain.

3.1 Attention Generation

Our method to calculate attention maps draws inspiration from GradCAM [21], but it introduces some modifications. In particular, since both D_X and D_Y follow the PatchGAN architecture [11], we compute the gradient by backpropagating the mean of the discriminator output $d = \frac{1}{P}\sum_p D(x)$ to the latent space of the corresponding generator and, more specifically, to the last layer of the last residual block $\mathbf{L}$. Global Average Pooling is then applied to the gradient to obtain the weight w_k:

$$w_k = \frac{1}{R}\sum_i\sum_j\left(\frac{\partial d}{\partial \mathbf{L}_k^{ij}}\right) \tag{1}$$

where $\mathbf{L}_k$ is the kth feature map with dimensions $w \times h$ of the layer $\mathbf{L}$ and $R = w \times h$. After this step, w_k is multiplied with the feature maps of the layer $\mathbf{L}_k$ obtaining the attention map $\mathbf{A}^d$:

$$\mathbf{A}^d = ReLU(\sum_k w_k \mathbf{L}_k) \tag{2}$$

where $ReLU$ is the rectified linear unit function.

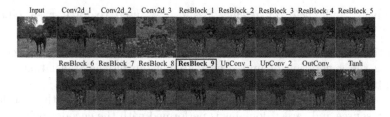

Fig. 2. Overview of attention maps extracted from the generator network at different layers. The proposed system uses the attention from the last residual block (in bold).

Finally, Fig. 2 shows all the attention maps generated from each layer of the generator network starting from an input image. Looking at the different attention maps, the one extracted from the last residual block is the one that precisely highlights the region where the domain needs to be applied. For this reason, the attention extracted from the last residual block is the one used in the loss that will be presented in the next section.

3.2 Network Architecture

Underlying Architecture. The proposed architecture is based on CycleGAN and therefore it is composed by two generators and two discriminators. More in detail, the generators are both composed by an encoder, a decoder and 9 residual blocks in between, while the discriminators follows the PatchGAN architecture introduced by Isola *et al.* in [11]. Indeed, an overview of the system is presented in Fig. 3.

During training, *adversarial loss* $\mathcal{L}_{adv}$ is used to push the generators G and F to produce realistic results belonging to the target domains X and Y,

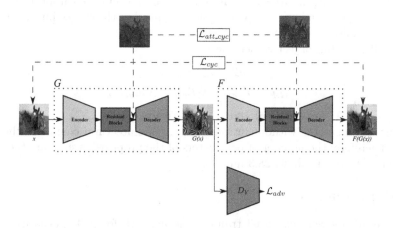

Fig. 3. Overview of the proposed system when translating an image from source domain X to target domain Y.

respectively, and *cycle consistency loss* $\mathcal{L}_{cyc}$ is used to force $x \rightarrow G(x) \rightarrow F(G(x)) \approx x$ and $y \rightarrow F(y) \rightarrow G(F(y)) \approx y$.

Attention Consistency Loss. The cycle consistency loss is very effective in avoiding the generation of undesired mapping during the translation, but, as stated before, it can also introduce shortcuts during training.

More in detail, in the generator's latent space the majority of the translation between source and target domains is performed with the encoder being mainly responsible of reducing the input dimension by encoding the image information, and the decoder that allows to go back to the original input shape. Nevertheless, during the decoding phase, the network can learn to reduce the intensity with which the domain is applied in order to facilitate the job of the inverse mapping generator. For example, an horse that was only half turned into a zebra would probably be considered a zebra by the corresponding discriminator, but it would be much easier to reverse it back to its original domain.

For this reason, we introduced a new term called *attention consistency loss* $\mathcal{L}_{att\_cyc}$ to improve the domain translation task. More specifically, the objective is to push the network to maintain the same attention over the whole translation cycle. Therefore, having $(x, F(y))$ and $(y, G(x))$ as input of G and F, respectively, the loss is:

$$\mathcal{L}_{att\_cyc} = \|\mathbf{A}^{D_Y}(x) - \mathbf{A}^{D_X}(G(x))\|_2 + \tag{3}$$
$$\|\mathbf{A}^{D_X}(y) - \mathbf{A}^{D_Y}(F(y))\|_2$$

where $\mathbf{A}^{D_Y}(x)$ and $\mathbf{A}^{D_Y}(F(y))$ are the attentions generated from the last residual block of G using the gradient obtained backpropagating from D_Y, while, similarly, $\mathbf{A}^{D_X}(y)$ and $\mathbf{A}^{D_X}(G(x))$ are the attentions generated from F backpropagating from D_X.

Imposing this new constraint during training helps the network to avoid any shortcut when applying the target domain, since the decoder will not dilute anymore the translation to ease the work of the cycle consistency loss term. The reason is that, in order to maintain the same level of attention in the two generators, the domain needs to be strongly applied to the source image without any compromise.

Finally, the full objective becomes:

$$\mathcal{L}_{D_X, D_Y} = \mathcal{L}_{adv} \tag{4}$$
$$\mathcal{L}_{G,F} = \mathcal{L}_{adv} + \lambda_1 \mathcal{L}_{cyc} + \lambda_2 \mathcal{L}_{att\_cyc} \tag{5}$$

where λ_1 and λ_2 are set to 10 and 1, respectively.

Our system works without introducing any architectural change to the network. Therefore, the number of parameters of the proposed system are the same of Cyclegan, that is about 28.3 million.

4   Experiments

All experiments were executed training the network for 200 epochs using the Adam optimizer [13] with a learning rate of 0.0002. A qualitative and quantitative evaluation will also be performed. In particular, the latter has been done using both the FID (Frechet Inception Distance) score [8] and the KID (Kernel Inception Distance) score [1].

4.1   Datasets

A subset of datasets used by CycleGAN have been selected for the experiments: *horse2zebra* (939 horse images and 1177 zebra images), *orange2apple* (996 apple images, and 1020 orange images), *photo2map* (1096 maps and 1096 aerial photos) and, finally, *monet2photo* (1074 Monet paintings and 6853 pictures).

Among all these, the last two datasets contain images where the translation process can not take advantage of a strict separation between foreground and background proving that our architecture will be effective also in these cases.

4.2   Results When a Foreground/background Separation Is Present

The objective of this set of experiments is to prove that introducing the attention consistency loss in the training has a positive effect over the results. In particular, it should solve the limitation of the cycle consistency loss that tends to maintain lots of features from the source domain in the translated image. In order to validate this claim, qualitative and quantitative results on the first two datasets will be presented.

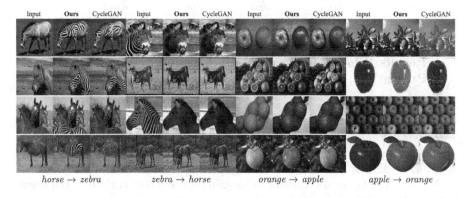

Fig. 4. Samples generated by our model wrt those generated by CycleGAN.

A comparison between samples generated by our model and sample generated by CycleGAN can be seen in Fig. 4. It is clear how results obtained with the aid of attention consistency loss are qualitatively superior to the ones obtained using a vanilla version of CycleGAN. In particular, CycleGAN is able to translate the domain somehow correctly, but the application is not consistent over the image. This is particularly evident in the *horse* → *zebra* domain transfer task where the stripes do not cover completely the original horse shape and most of the original color is still visible. On the other hand, when using the attention consistency loss, the stripes are applied much more strongly and precisely over the animal body and almost no trace of the source domain is left. In addition to that, when doing the opposite transformation (which is much harder), the proposed system allows to remove stripes from the zebras more firmly and the overall translation is more convincing. On the other side, considering *orange → apple* and *apple → orange* translations, CycleGAN sometimes tends to left the original fruit almost unchanged or to apply a color filter over the whole image during the domain transfer, while after the application of the attention consistency loss the original fruit is not recognizable anymore in the image and no filter is applied. This leads to a sharper result.

Finally, a full quantitative evaluation has been carried out and the results are reported in Table 1. Our method outperforms CycleGAN in all the different domains, in terms of both FID and KID, proving that the introduction of the attention consistency loss is beneficial for the network.

Table 1. Quantitative results of our method compared with CycleGAN.

	FID ↓			
	Horse → Zebra	*Zebra → Horse*	*Apple → Orange*	*Orange → Apple*
CycleGAN [27]	33.66	64.57	105.15	81.05
Ours	**27.94**	**61.54**	**103.89**	**75.79**
	KID ↓			
CycleGAN [27]	0.013	0.026	0.058	0.042
Ours	**0.009**	**0.024**	**0.054**	**0.040**

4.3   Results When a Foreground/Background Separation Is Not Present

After proving the effectiveness of the attention consistency loss over domains where a clear separation between background and foreground was possible, we tested our system on datasets where there is no such separation.

Firstly, we experimented with style transfer and trained the network to produce images similar to a Monet painting starting from pictures, and vice versa. In this case, the translation needs to happen on the whole surface of the image and therefore methods like [18] would not be applicable.

Qualitative exemplar results of this task are shown in Fig. 5a. Indeed, it can be observed how, when translating from panting to photo, CycleGAN tends to maintain visible the brush strokes affecting the realism of the produced results, while our method produces more colorful and plausible results. In the opposite case, CycleGAN struggles with color variety, whereas our method is able to transfer the painter style much better. Finally, the improvement of our method wrt CycleGAN is confirmed by quantitative results (Table 1), where both FID and KID scores are lower (and therefore better) when applying attention consistency loss.

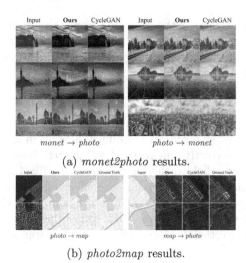

(a) *monet2photo* results.

(b) *photo2map* results.

Fig. 5. Some results with no foreground/background separation.

To further demonstrate the superiority of the proposed approach, we also experimented with map image translated in aerial view, and vice versa. Qualitative results for this experiment are shown in Fig. 5b. Indeed, our method is much better and precise than CycleGAN in reproducing water and highways on the maps images. Furthermore, when translating from maps to photos, it produces more realistic results than CycleGAN, especially for water and trees areas. These results were also validated by FID and KID values, reported in Table 2. CycleGAN has a slightly better FID score only in the case of the *photo* → *map* translation, but the corresponding KID score, a more reliable quality estimator, shows the effectiveness of our method.

4.4 Ablation Study

As mentioned before, the final setting of our architecture considers $\lambda_1 = 10$, $\lambda_2 = 1$ (see Eq. 5) and the generation of attention from the last residual block. This final setting has been obtained through an ablation study (performed using the *horse2zebra* dataset). Table 3 reports the results achieved with this study, where last line (row #6) corresponds to the final setting.

A first interesting experiment is to change the value of λ_2, while the value λ_1 has not been changed to be compliant with the original choice of CycleGAN. Increasing λ_2 to 10 (row #1) or decreasing it to 0.1 (row #2) do not bring to better results in terms of FID (results with KID are very similar in general). We also tried to apply the attention consistency loss $\mathcal{L}_{att\_cyc}$ only when translating the domain from X to Y and not when translating from Y to X (row #3) and only to F in the $X \to Y$ case and only to G in the $Y \to X$ case (row #4) in order to impose the loss only on the first generator in each cycle. Finally, the loss was calculated extracting the attention from the last four residual blocks of the two generators instead of using only the last one (row #5). All these experiments lead us to the final setting mentioned above and reported in row #6, which achieves the best results.

4.5 Drawbacks

We have proved that when CycleGAN produces a result in which the translation is applied softly but correctly, our method will greatly boost the quality of the generated results. Nevertheless, there are some cases, like the ones presented in Fig. 6, where our attention transfer has the effect of enhancing the failure of CycleGAN in applying the target domain. For example, in the *horse → zebra* translation, if CycleGAN paints some stripes over the background our method could amplify it. Nevertheless, these effects rarely happen and only in some extreme cases.

Table 2. FID and KID results comparison between CycleGAN and the proposed method on *monet2photo* and *photo2map*.

	FID ↓			
	Monet → Photo	*Photo → Monet*	*Map → Photo*	*Photo → Map*
CycleGAN [27]	144.18	145.55	70.70	**63.61**
Ours	**141.90**	**140.69**	**55.55**	64.70
	KID ↓			
CycleGAN [27]	0.022	0.012	0.026	0.033
Ours	**0.019**	**0.011**	**0.013**	**0.025**

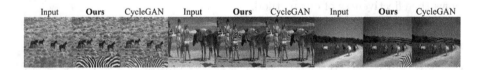

Input   **Ours**   CycleGAN   Input   **Ours**   CycleGAN   Input   **Ours**   CycleGAN

Fig. 6. Translation errors, already present in the CycleGAN output, that were enhanced by the attention consistency loss.

This limitation could be solved by combining the proposed method with ones like [18] where the translation is applied after separating foreground and background using the attention generated from an additional network. This is out of the scope of this paper and it is applicable only in the datasets like the ones in Sect. 4.2.

Table 3. Ablation study for different applications of the attention consistency loss.

		FID ↓	
		$Horse \rightarrow Zebra$	$Zebra \rightarrow Horse$
1	$\lambda_2 = 10$	30.66	62.69
2	$\lambda_2 = 0.1$	31.90	66.01
3	$\mathcal{L}_{att\_cyc}$ only $X \rightarrow Y$	30.60	65.32
4	$\mathcal{L}_{att\_cyc}$ single gen	32.74	62.71
5	$\mathcal{L}_{att\_cyc}$ 4 res blocks	33.18	61.77
6	**Ours**	**27.94**	**61.54**

5 Conclusions and Future Works

The objective of this paper was to cope with an important drawback of the cycle consistency loss used in unpaired image-to-image translation. In particular, this loss has the side effect of encouraging shortcuts when translating an image from a source to a target domain. The proposed solution exploits the attention maps extracted from the two generators of the network by introducing a new loss term called *attention consistency loss*. This loss forced the two generators to have the same attention in order to maintain the focus of the network high during the whole cycle.

Eventually, we proved the efficacy of the method by performing several experiments showing both qualitative and quantitative results, and in two main scenarios: the foreground and background clearly separated in the image, and scenarios where this separation is not present (typical cases of style transfer).

Future works will consist in testing the proposed loss to other architectures like [18] and also expand the use of attention maps to different tasks other than unpaired image-to-image translation.

Acknowledgments. This research has financially been supported by the Programme "FIL-Quota Incentivante" of University of Parma and co-sponsored by Fondazione Cariparma.

References

1. Bińkowski, M., Sutherland, D.J., Arbel, M., Gretton, A.: Demystifying mmd Gans. arXiv preprint arXiv:1801.01401 (2018)
2. Brock, A., Donahue, J., Simonyan, K.: Large scale GAN training for high fidelity natural image synthesis. arXiv preprint arXiv:1809.11096 (2018)
3. Chattopadhay, A., Sarkar, A., Howlader, P., Balasubramanian, V.N.: Grad-cam++: generalized gradient-based visual explanations for deep convolutional networks. In: 2018 IEEE Winter Conference on Applications of Computer Vision (WACV), pp. 839–847. IEEE (2018)

4. Choi, Y., Choi, M., Kim, M., Ha, J.W., Kim, S., Choo, J.: Stargan: unified generative adversarial networks for multi-domain image-to-image translation. In: Proceedings of the IEEE Conference on Computer Vision and Pattern Recognition, pp. 8789–8797 (2018)
5. Dhar, P., Singh, R.V., Peng, K.C., Wu, Z., Chellappa, R.: Learning without memorizing. In: Proceedings of the IEEE/CVF Conference on Computer Vision and Pattern Recognition (CVPR), June 2019
6. Emami, H., Aliabadi, M.M., Dong, M., Chinnam, R.B.: Spa-GAN: spatial attention GAN for image-to-image translation. IEEE Trans. Multimedia **23**, 391–401 (2020)
7. Goodfellow, I.J., et al.: Generative adversarial networks. arXiv preprint arXiv:1406.2661 (2014)
8. Heusel, M., Ramsauer, H., Unterthiner, T., Nessler, B., Hochreiter, S.: GANs trained by a two time-scale update rule converge to a local nash equilibrium. arXiv preprint arXiv:1706.08500 (2017)
9. Hoffman, J., et al.: Cycada: cycle-consistent adversarial domain adaptation. In: International Conference on Machine Learning, pp. 1989–1998. PMLR (2018)
10. Huang, X., Liu, M.Y., Belongie, S., Kautz, J.: Multimodal unsupervised image-to-image translation. In: Proceedings of the European conference on computer vision (ECCV), pp. 172–189 (2018)
11. Isola, P., Zhu, J.Y., Zhou, T., Efros, A.A.: Image-to-image translation with conditional adversarial networks. In: Proceedings of the IEEE Conference on Computer Vision and Pattern Recognition, pp. 1125–1134 (2017)
12. Kim, T., Cha, M., Kim, H., Lee, J.K., Kim, J.: Learning to discover cross-domain relations with generative adversarial networks. In: International Conference on Machine Learning, pp. 1857–1865. PMLR (2017)
13. Kingma, D.P., Ba, J.: Adam: a method for stochastic optimization. arXiv preprint arXiv:1412.6980 (2014)
14. Lee, H.Y., Tseng, H.Y., Huang, J.B., Singh, M., Yang, M.H.: Diverse image-to-image translation via disentangled representations. In: Proceedings of the European Conference on Computer Vision (ECCV), pp. 35–51 (2018)
15. Li, K., Wu, Z., Peng, K.C., Ernst, J., Fu, Y.: Tell me where to look: Guided attention inference network. In: Proceedings of the IEEE Conference on Computer Vision and Pattern Recognition (CVPR), June 2018
16. Liu, W., Li, R., Zheng, M., Karanam, S., Wu, Z., Bhanu, B., Radke, R.J., Camps, O.: Towards visually explaining variational autoencoders. In: Proceedings of the IEEE/CVF Conference on Computer Vision and Pattern Recognition, pp. 8642–8651 (2020)
17. Mahendran, A., Vedaldi, A.: Understanding deep image representations by inverting them. In: Proceedings of the IEEE Conference on Computer Vision and Pattern Recognition, pp. 5188–5196 (2015)
18. Mejjati, Y.A., Richardt, C., Tompkin, J., Cosker, D., Kim, K.I.: Unsupervised attention-guided image to image translation. arXiv preprint arXiv:1806.02311 (2018)
19. Mirza, M., Osindero, S.: Conditional generative adversarial nets. arXiv preprint arXiv:1411.1784 (2014)
20. Nizan, O., Tal, A.: Breaking the cycle-colleagues are all you need. In: Proceedings of the IEEE/CVF Conference on Computer Vision and Pattern Recognition, pp. 7860–7869 (2020)

21. Selvaraju, R.R., Cogswell, M., Das, A., Vedantam, R., Parikh, D., Batra, D.: Grad-cam: visual explanations from deep networks via gradient-based localization. In: Proceedings of the IEEE International Conference on Computer Vision, pp. 618–626 (2017)

22. Zagoruyko, S., Komodakis, N.: Paying more attention to attention: improving the performance of convolutional neural networks via attention transfer. arXiv preprint arXiv:1612.03928 (2016)

23. Zeiler, M.D., Fergus, R.: Visualizing and understanding convolutional networks. In: Fleet, D., Pajdla, T., Schiele, B., Tuytelaars, T. (eds.) ECCV 2014. LNCS, vol. 8689, pp. 818–833. Springer, Cham (2014). https://doi.org/10.1007/978-3-319-10590-1_53

24. Zhang, H., Xu, T., Li, H., Zhang, S., Wang, X., Huang, X., Metaxas, D.N.: Stack-GAN++: realistic image synthesis with stacked generative adversarial networks. IEEE Trans. Pattern Anal. Mach. Intell. **41**(8), 1947–1962 (2018)

25. Zhao, Y., Wu, R., Dong, H.: Unpaired image-to-image translation using adversarial consistency loss. In: Vedaldi, A., Bischof, H., Brox, T., Frahm, J.-M. (eds.) ECCV 2020. LNCS, vol. 12354, pp. 800–815. Springer, Cham (2020). https://doi.org/10.1007/978-3-030-58545-7_46

26. Zhou, B., Khosla, A., Lapedriza, A., Oliva, A., Torralba, A.: Learning deep features for discriminative localization. In: Proceedings of the IEEE Conference on Computer Vision and Pattern Recognition, pp. 2921–2929 (2016)

27. Zhu, J.Y., Park, T., Isola, P., Efros, A.A.: Unpaired image-to-image translation using cycle-consistent adversarial networks. In: Proceedings of the IEEE International Conference on Computer Vision, pp. 2223–2232 (2017)

Towards Efficient and Data Agnostic Image Classification Training Pipeline for Embedded Systems

Kirill Prokofiev[1] and Vladislav Sovrasov[2]($\boxtimes$)

[1] Intel, Higher School of Economics, Nizhny Novgorod, Russia
kirill.prokofiev@intel.com
[2] Intel, Nizhny Novgorod State University, Nizhny Novgorod, Russia
vladislav.sovrasov@intel.com

Abstract. Nowadays deep learning-based methods have achieved a remarkable progress at the image classification task among a wide range of commonly used datasets (ImageNet, CIFAR, SVHN, Caltech 101, SUN397, etc.). SOTA performance on each of the mentioned datasets is obtained by careful tuning of the model architecture and training tricks according to the properties of the target data. Although this approach allows setting academic records, it is unrealistic that an average data scientist would have enough resources to build a sophisticated training pipeline for every image classification task he meets in practice. This work is focusing on reviewing the latest augmentation and regularization methods for the image classification and exploring ways to automatically choose some of the most important hyperparameters: total number of epochs, initial learning rate value and it's schedule. Having a training procedure equipped with a lightweight modern CNN architecture (like MobileNetV3 or EfficientNet), sufficient level of regularization and adaptive to data learning rate schedule, we can achieve a reasonable performance on a variety of downstream image classification tasks without manual tuning of parameters to each particular task. Resulting models are computationally efficient and can be deployed to CPU using the OpenVINO™ toolkit. Source code is available as a part of the OpenVINO™ Training Extensions (https://github.com/openvinotoolkit/training_extensions).

Keywords: Image classification · Deep learning · Lightweight models

1 Introduction

Throughout the past decade deep learning-based image classification methods have made a great progress increasing their performance by 45% [40] from the AlexNet [20] level on well-known ImageNet benchmark [7]. Although SOTA results are obtained by a fine-grained adaptation of all the training pipeline components to the target task, in practice a data engineer could not have enough

S. Sclaroff et al. (Eds.): ICIAP 2022, LNCS 13231, pp. 476–488, 2022.
https://doi.org/10.1007/978-3-031-06427-2_40

resources to do it. Typically to solve an image classification task we need to make the following decisions:

- Choose a model architecture;
- Build data augmentation pipeline;
- Choose optimization method and it's parameters (learning rate schedule, length of the training);
- Apply some extra regularization methods in case of overfitting;
- Apply additional techniques to handle hard classes imbalance or high level of label noise if needed.

Wrong decisions on each step can hurt the resulting classification model performance or bring misalignments with the requirements to computational complexity, for instance. At the same time, techniques and models which are successful on one dataset may not be beneficial on others, i.e. they are not dataset-agnostic. From the practical perspective, although reaching the ultimate performance is very demanding task, one can still think about some fail-safe training configuration that would allow obtaining moderate results on a wide range of middle-sized image classification datasets with minimum effort to further tuning. In this work we aim to propose such a configuration for a couple of modern computationally effective architectures: EfficientNet-B0 [34] and MobileNetV3 [14] family. To reach the goal we added the adaptability to the optimization process of scheduling the weights, designed a robust initial learning rate estimation heuristic and curated a set of regularization techniques that suit well to the considered model architectures.

In brief, the key contributions of this paper can be summarized as follows:

- Designed an optimization controlling policy that includes an optimal initial learning rate estimator and a modified version of the ReduceLROnPlateau [2] scheduler equipped with an early stopping procedure;
- Proposed a way of applying Deep Mutual Learning [44] to reduce overconfidence of model predictions;
- For each of the considered models curated a suitable bundle of data augmentation and regularization methods and validated it on various middle-sized downstream image classification datasets.

2   Related Work

Optimal Learning Rate Estimation and Scheduling. The problem of initial learning rate setting can be seen as a general hyperparameter optimization and, thus, a variety of general methods can be directly applied [3,21]. From the other side, several simple heuristics were designed to directly tackle it [1,24]. The last approach is more lightweight because it doesn't imply dependencies on any hyperparameters estimation frameworks, but at the same time, it is not very reliable since assumptions that these algorithms are based on, could not strictly hold in a wide range of real-world tasks. Considering a fine-grained grid search as the most robust method, we believe that a combination of a hyperparameters tuning framework and properly designed trials execution process will be a good trade-off between robustness and accuracy. When the initial learning rate is chosen, the schedule and stopping criteria define the final result of the training. Typically researchers set reasonable amount of training epochs in advance, especially if they focus on a single dataset, but to save computational resources and avoid overfitting it is beneficial to stop training early [31,42]. Early stopping can break logic of popular schedulers like cosine or 1cycle [33], especially if too large number of total epochs was set initially, so in this work we use drops to decrease learning rate in combination with initial linear warm-up.

Data Augmentation. Recently a wide range of effective data augmentation methods has been proposed [11,25]. Augmix [13] allows combining several simple augmentations (like random crop, color perturbation, rotation, flip) into a pipeline with adjustable applying policy. That allows to use Augmix as a replacement for classic sequential pipeline of transformations. In case of high capacity models, additional mixing-sample augmentations like fmix [11] can be applied on the top of Augmix output to further diversify the input data.

Regularization. To achieve higher classification accuracy, a training pipeline should keep a balance between fitting ability and regularization. We have tested a lot of approaches acting on different directions: continuous dropout [32], mutual learning [44], label smoothing and confidence penalty [29], no bias decay [12], and found a suitable combination for each of considered architecture type. Complexity of data augmentation, batch size and learning rate values can also be viewed as regularization factors. Taking this into account, transferring regularization parameters between different datasets and architectures should be done with care.

Optimization. Currently even SOTA approaches in classification still use SGD [30,39] for finetuning on small target datasets, while use stronger adaptive optimizers [23,38] for initial pre-training with huge amount of samples. We are aiming to finetune already pre-trained models and choose the SGD-based Sharpness-Aware Minimizer (SAM) [10] as a default optimizer. SAM, like SGD, preserves the mentioned fitting-regularization tradeoff, it's authors claim that SAM also provides robustness to label noise. This property is useful if we aim to build a reliable training pipeline.

3   Method

In this section, we describe the overall training pipeline from models architectures to training tricks.

3.1 Models

We chose MobileNetV3 [14] and EfficientNet [34] as base architectures for performing image classification. Namely, we conducted all the experiments on MobileNetV3 small 1x, large 0.75x, large 1x and EfficientNet-B0. The chosen models form a strong performance accuracy trade-off in the range from 0.06 to 0.4 GFLOPs, which is enough in most of edge-oriented applications.

3.2 Training Tricks

Data Augmentation. Properly chosen data augmentation pipeline can boost classification accuracy on a variety of downstream tasks. At the same time, optimal augmentations are different for different datasets. Thus, any hand-crafted pipeline would be suboptimal and considering this, our goal is to find a pipeline that maintains fitting-regularization tradeoff for MobileNetV3 and EfficientNet. After conducting experiments with modern techniques [11,13,37,43], we found AugMix [13] with a pre-trained on ImageNet policy is the most beneficial for our setup. When we add MixUp [43], CutMix [37] or FMix [11] to the pipeline we observe a performance drop compared to pure AugMix. This indicates that Aug-Mix provides a better fitting-regularization tradeoff for the chosen lightweight models, while MixUp-like augmentations are too hard for them.

Optimization. Conventionally, SGD with momentum is widely used for fine-tuning in downstream classification tasks [30,39]. An extension to SGD, Sharpness Aware Minimization (SAM) [10], allows to achieve higher results in the fine-tuning than SGD, while requiring two forward-backward passes of the model per training iteration. We operate with lightweight models, so additional cost of SAM is not critical. We also tried AdamW [16], but it performed slightly worse than SGD.

Additionally, we employ no bias decay (turning off weight decay for biases in all layers) [12] for better generalization.

Optimal Learning Rate Estimation. Since SGD performance on a given dataset is highly correlated with the initial learning rate magnitude, we have to incorporate estimation of this parameter into our training pipeline. Straightforward approach is to use grid or random search within a given range, but we were focusing on less time consuming methods. Fast-ai's heuristic [15] can generate learning rate proposal after performing several training iterations, but in our experiments it tends to output too high values which destroy ImageNet initialization even if warmup strategy is applied. To overcome this problem, we propose to finetune the model with a pre-defined small learning rate for one epoch on the target data and then run the original fast-ai algorithm. In this case the model would react to fast increase of the learning rate more smoothly and the algorithm will select a value which is quite close to one located with the grid search (see Fig. 1). Fast-ai's heuristic with pre-training is almost as lightweight as the original one:

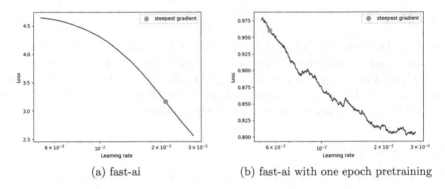

(a) fast-ai   (b) fast-ai with one epoch pretraining

Fig. 1. Without pretraining the estimated learning rate is too large (≈ 0.02). Whereas with one epoch pretraining the estimated learning rate is more suitable for the CIFAR-100 dataset (≈ 0.0056).

for instance if the training is scheduled for 100 epochs, one extra epoch for the initial finetuning will introduce only 1% of additional overhead.

For fine-grained learning rate selections we use Tree-structured Parzen Estimator (TPE) from the Optuna [3] framework. We set optimization criterion as top-1 accuracy on the validation subset after training on the target data for several epochs. We use median trial pruning criterion which breaks a trial if the best intermediate result is worse than the median of intermediate results of the previous trials at the same step. Pruning heuristic also could be used for grid search, but TPE also generates locations of the next trials based on previous trials locations instead of random choice or using a grid, which allows us to set a lower total trails limit compared to the grid search. We set TPE as a default initial learning estimator in our pipeline, though fast-ai with pre-training could be used in case of limited resources.

Learning Rate Scheduling and Early Stopping. Complexity of the input dataset can drastically vary and thus the number of epochs that are sufficient for training also varies. For efficient training we have to adapt to different data. To achieve this, we need a flexible learning rate schedule and reliable early stopping criterion. Popular SOTA schedulers like cosine [22] and 1cycle [33] require a pre-defined number of epochs and the shape of the produced learning rate curve depends on it. If we, for an instance, predefine 200 training epochs and early stopping criterion returns a stop flag on the 20th epoch, the training will be stopped at an unstable state of the model, because convergence hasn't been reached yet. If we take into account this drawback and prohibit early stopping criterion returning a stop flag till the half of the training pass, this will cause too long training, but the model will converge.

To overcome these problems we propose to use a modified version of ReduceLROnPlateau [2] scheduler denoted hereafter as ReduceLROnPlateauV2. In ReduceLROnPlateauV2 we force learning rate decay if 75% of the maximum training length had been reached, but learning rate drop have not been performed yet. This helps model to converge when average training loss, which we use as a criterion for learning rate drop, is unstable. Also we incorporated early stopping into ReduceLROnPlateauV2: if the learning rate was decayed to a pre-defined minimal value and the best top-1 score on validation subset hadn't been improving for a predefined number of epochs, we stop the training.

Additionally we use 5 epoch linear learning rate warmup and, following no bias decay practice, we increase learning rate for biases by a factor of 2.

Mutual Learning. To further boost the performance of classification models we tried to apply deep mutual learning (DML) [44]. This technique implies mutual learning of a collection of simple student models. As a result, performance of each student individually is supposed to be increased. We directly apply this technique to pairs of identical models and didn't obtain substantial accuracy gain. This framework could also be used in a different manner: if one of the students has a stronger regularization and trains slower than the others, it may prevent others from overfitting and softly transfer properties of it's regularized distribution to faster students. We applied the second setup to pairs of models with the same architecture, but different loss functions. Fast student is trained with the cross-entropy loss while slow one is trained with the AM-Softmax [8] loss with scale $s = 1$ and margin $m = 0$. Such settings of the angular loss make output distribution of the slower model less confident and it also pushes the faster model to output a smoother distribution while strong discriminative properties formed by AM-Softmax are transferred to the faster model as well. In the described scheme the overall training losses are defined as follows:

$$L_{fast}(x) = L_{CE}(p_1(x), y) + D_{KL}(p_2(x)||p_1(x))$$
$$L_{slow}(x) = L_{AMS}(p_2(x), y) + D_{KL}(p_1(x)||p_2(x))$$

where (x, y) is a training sample and it's label, $D_{KL}(p||q)$ is Kullback Leibler divergence between discrete distributions p and q, $p_1(x)$ – distribution estimated by the fast model on the sample x, $p_2(x)$ – distribution estimated by the slow model on the sample x. Results of applying this approach are demonstrated on the Fig. 2.

Drawback of the DML is significant increase of the training time and memory footprint. Taking this into account, we decide to include this technique only to the training strategy for MobileNetV3 family, besides for EfficientNet-B0 it is not so beneficial. Instead, we train EfficientNet-B0 with AM-Softmax directly. In that case $s = \max(\sqrt{2} \cdot \log(C - 1), 3)$ (where C is the number of classes) and margin $m = 0.35$. For each MobileNetV3 model we employ MobileNetV3-large as a slower student.

4   Experiments

This section describes the evaluation process, metrics, datasets and reports the evaluation results. For a given model family we use the same training techniques and parameters across all the considered datasets; this allows us to validate how well the proposed training pipeline can work on variable data.

4.1 Datasets

To validate the proposed approach we choose 11 widespread classification datasets, that are listed in Table 1.

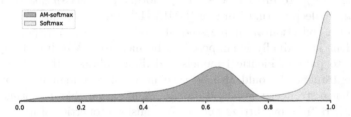

Fig. 2. Distributions of the most probable class confidence produced by the fast model on validation data. If slow student is trained with AM-Softmax, fast student does not tend to return overconfident predictions. Experiment is conducted with a pair of MobileNetV3 small models on Cars Dataset [18] containing 196 classes.

Table 1. Image classification datasets which were used for training.

Dataset	Number of classes	Number of images	
		Train	Validation
Oxford-IIIT Pets [28]	37	3680	3369
Describable Textures (DTD)* [6]	47	4826	814
Oxford 102 Flowers* [27]	102	6614	1575
Caltech 101* [9]	101	6941	1736
Cars Dataset [18]	196	8144	8041
Birdsnap* [4]	500	47386	2443
CIFAR-100 [19]	100	50000	10000
Fashion-MNIST [35]	10	60000	10000
SVHN [26]	10	73257	26032
Food-101 [5]	101	75750	25250
SUN397* [36]	397	92440	16314

* for experiments on these datasets we do custom random splits.

4.2 Evaluation Protocol

To evaluate image classification models besides commonly used *top-k* accuracy metric, we also calculate mean average precision (mAP) in the same sense as it is considered in person re-identification field [45]. We would denote a set of C classes as a gallery and each of N samples in the evaluation set as a query. Then, in terms of the retrieval task, we have to rank the gallery by similarity with the query, compute the average precision of this query, and then collect the mAP. In terms of the classification task, the similarity is the predicted probability of a class, and, thus, computing average precision is straightforward:

$$
mAP = \frac{\sum_{q=1}^{Q} AP(q)}{Q} = \frac{\sum_{i=1}^{N} AP(i)}{N} = \frac{1}{N} \sum_{i=1}^{N} \frac{1}{K_i},
$$

where K_i is the index of the probability of ground truth class in the sorted set of predicted class probabilities for a sample. $AP(i)$ varies from 1 (if the model prediction is always correct) to $\frac{1}{C}$ (if the model always assigns the lowest probability to the true class). mAP can be viewed as an aggregation of all C possible *top-k* scores and reveals the ranking ability of the evaluated classification models.

After single dataset evaluation metrics are defined, we also have to define a way to compare different strategies for a given model taking into account results on a set of datasets. In this work, we will follow [41] and consider metrics averaging across datasets as a measure of the quality of a training strategy for a given model. This approach doesn't imply any weighting procedure, so each dataset equally contributes to the final metric.

4.3 Results

We trained all the models starting from the ImageNet-pretrained weights. We consider the following training setup as a baseline: training length is 200 epochs; learning rate is annealed to zero with cosine schedule; optimization is performed with SGD with momentum; training images are augmented with random flip, random color jittering, random crop and random rotation; dropout with $p = 0.2$ is applied to the classifier layer; learning rate is set as an average of optimal learning rates obtained by grid search over all the 11 considered datasets for each model individually.

Table 2. Results of our adaptive training pipeline against baseline.

Model	AVG *top-1*, %	AVG *top-5*, %	AVG *mAP*	AVG *epochs*
MobileNetV3-small baseline	82.24	95.29	82.35	200
Our MobileNetV3-small	85.00	96.15	88.01	86
MobileNetV3-large-0.75x baseline	85.28	96.30	87.03	200
Our MobileNetV3-large-0.75x	87.60	96.97	91.14	83
MobileNetV3-large baseline	85.79	96.47	87.99	200
Our MobileNetV3-large	88.26	97.34	91.79	81
EfficientNet-B0 baseline	86.47	96.77	89.27	200
Our EfficientNet-B0	89.13	97.79	92.75	82

For our adaptive pipeline we set the maximum training length to 200 epochs, maximum amount of trials in TPE to 15 (each trial takes 6 epochs or less); learning rate search range for EfficientNet-B0 is $[0.001 - 0.01]$, while for MobileNetV3 the range is $[0.005 - 0.03]$; Augmix with a pre-trained on ImageNet policy, random flip, random crop and random rotation are employed for data augmentation; early stopping and learning rate decay patiences are equal to 5 epochs; ρ in SAM is 0.05. Weight decay is always set to $5 \cdot 10^{-4}$. Input resolution for all versions of the considered models is 224×224, no test time augmentation is applied.

The final results are presented for all our models in the Table 2. Our adaptive training strategy clearly outperforms baseline by 2–3% AVG top-1 and 3–5% AVG mAP. Also it reduces average number of epochs required for training more than twice: from 200 to 86 or less.

Top-1 scores by datasets are presented in the Table 3 as well as comparison with other solutions. Our MobileNetV3-small with adaptive training outperforms both our baseline and one of the best publicly available repositories with advanced training tricks for MobileNetV3-small (it includes label smoothing [29],

Table 3. Detailed comparison with other methods. Top-1 metric is presented.

Model	CIFAR-100*	DTD*	Food-101*	SUN397	SVHN*	Birdsnap	Caltech101	Cars*	Fashion-MNIST*	Flowers*	Pets*
Our MNV3-small baseline	79.57	68.58	74.71	57.38	96.02	77.12	90.41	87.24	95.13	93.89	84.65
Our MNV3-small	83.49	72.45	79.59	62.83	97.29	79.04	92.94	91.33	95.74	94.86	85.55
MNV3-small+	82.43	68.13	69.45	61.13	96.49	78.75	91.74	86.66	95.48	92.81	85.41
Our MNV3-large-0.75x baseline	81.05	72.28	80.97	61.97	96.13	81.39	92.87	91.18	95.02	95.36	89.86
Our MNV3-large-0.75x	85.36	74.50	85.53	67.02	97.54	83.21	95.23	93.75	95.83	96.22	90.51
Our MNV3-large baseline	81.70	73.53	81.57	62.65	96.18	82.21	93.46	91.29	95.31	95.69	90.13
Our MNV3-large	86.24	76.49	85.77	67.96	97.57	84.1	95.42	93.76	96.17	96.63	91.21
Our EffNet-B0 baseline	84.84	74.81	83.75	64.43	96.88	80.03	94.54	90.46	95.80	95.26	91.74
Our EffNet-B0	**86.52**	**77.18**	86.06	72.10	**97.82**	83.33	95.63	93.77	96.28	97.10	92.00
VTAB ResNet-50 [41]	84.00	76.8	–	–	97.40	–	–	–	–	**97.40**	**92.6**
Inception v4 [17]	87.5	78.1	90.00	–	–	–	–	93.30	–	98.50	94.50
EffNet-B7+SAM [10]	92.56	–	92.98	–	–	–	–	94.82	–	99.37	96.03

* for these datasets we do splits as defined by the authors, for others we use custom splits

+Implementation is taken from https://github.com/ShowLo/MobileNetV3, learning rate is set the same as for our baseline.

Mixup [43], no bias decay [12], EMA decay and learning rate warmup). At the same time, two non-adaptive strategies for MobileNetV3-small perform on par. EfficientNet-B0 with adaptive strategy demonstrates results similar to ResNet-50 from VTAB [41], which is trained with heavyweight search over hyperparameters (learning rate, schedule, optimizers, batch size, train preprocessing functions, evaluation pre-processing, and weight decay). The results of heavyweight SOTA models are also presented in the Table 3 for the reference.

4.4 Ablation Study

We conduct an ablation study by removing each single component of our training pipeline, while others are enabled (see Table 4). For experiments we use all the data except the SUN397 for MobilenetV3 and subset of 6 datasets for EfficientNet-B0 (CIFAR-100, DTD, Flowers, Cars, Pets, Caltech101). Each of the training tricks has roughly equal gain ($< 2\%$ top-1), excepting AM-Softmax related ones that add 2–3% of mAP to the result.

Table 4. Impact of each training trick to MobileNetV3-large and EfficientNet-B0 results.

Configuration	MobileNetV3-large		EfficientNet-B0	
	AVG $top\text{-}1$	AVG mAP	AVG $top\text{-}1$	AVG mAP
Final solution	**90.63**	**94.17**	**90.63**	**94.14**
w/o SAM	89.74	93.34	90.44	94.14
w/o Mutual learning	89.58	90.71	–	–
w/o AugMix	90.14	93.75	90.41	93.88
w/o NBD	90.16	93.49	90.43	94.14
w/o AM-Softmax	–	–	90.51	92.14
w/o adaptive learning rate strategy	89.97	93.38	90.16	93.83

5   Conclusion

In this work, we presented adaptive training strategies for several lightweight image classification models that can perform better on a wide range of downstream datasets in finetuning from ImageNet scenario than conventional training pipelines if there are no resources available for heavyweight models and extensive hyperparameters optimization. These strategies are featured with optimal learning rate estimation and early stopping criterion, allowing them to adapt to the input datasets to some extent. The ability to adapt is shown by conducting experiments on 11 diverse image classification datasets.

References

1. fast.ai: Learning rate finder. https://fastai1.fast.ai/callbacks.lr_finder.html#Learning-Rate-Finder

2. pytorch: Reducelronplateau scheduler. https://pytorch.org/docs/stable/generated/torch.optim.lr_scheduler.ReduceLROnPlateau.html#torch.optim.lr_scheduler.ReduceLROnPlateau

3. Akiba, T., Sano, S., Yanase, T., Ohta, T., Koyama, M.: Optuna: a next-generation hyperparameter optimization framework. In: Proceedings of the 25rd ACM SIGKDD International Conference on Knowledge Discovery and Data Mining (2019)

4. Berg, T., Liu, J., Lee, S.W., Alexander, M.L., Jacobs, D.W., Belhumeur, P.N.: Birdsnap: large-scale fine-grained visual categorization of birds. In: 2014 IEEE Conference on Computer Vision and Pattern Recognition, pp. 2019–2026 (2014). https://doi.org/10.1109/CVPR.2014.259

5. Bossard, L., Guillaumin, M., Van Gool, L.: Food-101 – mining discriminative components with random forests. In: Fleet, D., Pajdla, T., Schiele, B., Tuytelaars, T. (eds.) ECCV 2014, Part VI. LNCS, vol. 8694, pp. 446–461. Springer, Cham (2014). https://doi.org/10.1007/978-3-319-10599-4_29

6. Cimpoi, M., Maji, S., Kokkinos, I., Mohamed, S., Vedaldi, A.: Describing textures in the wild. In: Proceedings of the IEEE Conference on Computer Vision and Pattern Recognition (CVPR) (2014)

7. Deng, J., Dong, W., Socher, R., Li, L.J., Li, K., Fei-Fei, L.: ImageNet: a large-scale hierarchical image database. In: CVPR09 (2009)

8. Deng, J., Guo, J., Zafeiriou, S.: Arcface: additive angular margin loss for deep face recognition. ArXiv (2018)

9. Fei-Fei, L., Fergus, R., Perona, P.: Learning generative visual models from few training examples: an incremental Bayesian approach tested on 101 object categories. In: 2004 Conference on Computer Vision and Pattern Recognition Workshop, p. 178 (2004)

10. Foret, P., Kleiner, A., Mobahi, H., Neyshabur, B.: Sharpness-aware minimization for efficiently improving generalization. ArXiv abs/2010.01412 (2020)

11. Harris, E., Marcu, A., Painter, M., Niranjan, M., Prügel-Bennett, A., Hare, J.S.: Fmix: enhancing mixed sample data augmentation. arXiv: Learning (2020)

12. He, T., Zhang, Z., Zhang, H., Zhang, Z., Xie, J., Li, M.: Bag of tricks for image classification with convolutional neural networks. In: 2019 IEEE/CVF Conference on Computer Vision and Pattern Recognition (CVPR), pp. 558–567 (2019)

13. Hendrycks, D., Mu, N., Cubuk, E.D., Zoph, B., Gilmer, J., Lakshminarayanan, B.: AugMix: a simple data processing method to improve robustness and uncertainty. In: Proceedings of the International Conference on Learning Representations (ICLR) (2020)

14. Howard, A.G., et al.: Searching for mobileNetV3. In: 2019 IEEE/CVF International Conference on Computer Vision (ICCV), pp. 1314–1324 (2019)

15. Jeremy Howard, S.G.: Deep Learning for Coders with fastai and PyTorch. O'Reilly Media, Inc., USA (2020)

16. Kingma, D.P., Ba, J.: Adam: a method for stochastic optimization. CoRR abs/1412.6980 (2015)

17. Kornblith, S., Shlens, J., Le, Q.V.: Do better ImageNet models transfer better? In: 2019 IEEE/CVF Conference on Computer Vision and Pattern Recognition (CVPR), pp. 2656–2666 (2019)

18. Krause, J., Stark, M., Deng, J., Fei-Fei, L.: 3D object representations for fine-grained categorization. In: 4th International IEEE Workshop on 3D Representation and Recognition (3dRR-13), Sydney, Australia (2013)

19. Krizhevsky, A.: Learning multiple layers of features from tiny images (2009)

20. Krizhevsky, A., Sutskever, I., Hinton, G.E.: ImageNet classification with deep convolutional neural networks. In: Pereira, F., Burges, C.J.C., Bottou, L., Weinberger, K.Q. (eds.) Advances in Neural Information Processing Systems, vol. 25, pp. 1097–1105. Curran Associates, Inc. (2012). http://papers.nips.cc/paper/4824-imagenet-classification-with-deep-convolutional-neural-networks.pdf

21. Liaw, R., Liang, E., Nishihara, R., Moritz, P., Gonzalez, J.E., Stoica, I.: Tune: a research platform for distributed model selection and training. arXiv preprint arXiv:1807.05118 (2018)

22. Loshchilov, I., Hutter, F.: Sgdr: stochastic gradient descent with warm restarts. arXiv: Learning (2017)

23. Loshchilov, I., Hutter, F.: Decoupled weight decay regularization. In: ICLR (2019)

24. Mukherjee, K., Khare, A., Verma, A.: A simple dynamic learning rate tuning algorithm for automated training of DNNs. ArXiv abs/1910.11605 (2019)

25. Naveed, H.: Survey: Image mixing and deleting for data augmentation. ArXiv abs/2106.07085 (2021)

26. Netzer, Y., Wang, T., Coates, A., Bissacco, A., Wu, B., Ng, A.: Reading digits in natural images with unsupervised feature learning (2011)

27. Nilsback, M.E., Zisserman, A.: Automated flower classification over a large number of classes. In: Indian Conference on Computer Vision, Graphics and Image Processing (December 2008)

28. Parkhi, O.M., Vedaldi, A., Zisserman, A., Jawahar, C.V.: Cats and dogs. In: IEEE Conference on Computer Vision and Pattern Recognition (2012)

29. Pereyra, G., Tucker, G., Chorowski, J., Kaiser, L., Hinton, G.E.: Regularizing neural networks by penalizing confident output distributions. ArXiv abs/1701.06548 (2017)

30. Pham, H.H., Xie, Q., Dai, Z., Le, Q.V.: Meta pseudo labels. ArXiv abs/2003.10580 (2020)

31. Prechelt, L.: Early stopping - but when? (March 2000)

32. Shen, X., Tian, X., Liu, T., Xu, F., Tao, D.: Continuous dropout. IEEE Trans. Neural Netw. Learn. Syst. **29**, 3926–3937 (2018)

33. Smith, L.N., Topin, N.: Super-convergence: very fast training of neural networks using large learning rates. In: Defense + Commercial Sensing (2019)

34. Tan, M., Le, Q.V.: Efficientnet: rethinking model scaling for convolutional neural networks. ArXiv abs/1905.11946 (2019)

35. Xiao, H., Rasul, K., Vollgraf, R.: Fashion-MNIST: a novel image dataset for benchmarking machine learning algorithms (2017)

36. xiong Xiao, J., Hays, J., Ehinger, K.A., Oliva, A., Torralba, A.: Sun database: large-scale scene recognition from abbey to zoo. In: 2010 IEEE Computer Society Conference on Computer Vision and Pattern Recognition, pp. 3485–3492 (2010)

37. Yun, S., Han, D., Oh, S.J., Chun, S., Choe, J., Yoo, Y.: CutMix: regularization strategy to train strong classifiers with localizable features. In: 2019 IEEE/CVF International Conference on Computer Vision (ICCV), pp. 6022–6031 (2019)

38. Zeiler, M.D.: Adadelta: an adaptive learning rate method. ArXiv abs/1212.5701 (2012)

39. Zhai, X., Kolesnikov, A., Houlsby, N., Beyer, L.: Scaling vision transformers. ArXiv abs/2106.04560 (2021)

40. Zhai, X., Kolesnikov, A., Houlsby, N., Beyer, L.: Scaling vision transformers (2021)

41. Zhai, X., et al.: The visual task adaptation benchmark. ArXiv abs/1910.04867 (2019)

42. Zhang, C., Bengio, S., Hardt, M., Recht, B., Vinyals, O.: Understanding deep learning requires rethinking generalization. ArXiv abs/1611.03530 (2017)

43. Zhang, H., Cissé, M., Dauphin, Y., Lopez-Paz, D.: MixUp: beyond empirical risk minimization. ArXiv abs/1710.09412 (2018)
44. Zhang, Y., Xiang, T., Hospedales, T.M., Lu, H.: Deep mutual learning. In: 2018 IEEE/CVF Conference on Computer Vision and Pattern Recognition, pp. 4320–4328 (2018)
45. Zheng, L., Shen, L., Tian, L., Wang, S., Wang, J., Tian, Q.: Scalable person re-identification: a benchmark. In: 2015 IEEE International Conference on Computer Vision (ICCV), pp. 1116–1124 (2015)

Medicinal Boxes Recognition on a Deep Transfer Learning Augmented Reality Mobile Application

Danilo Avola[1], Luigi Cinque[1], Alessio Fagioli[1], Gian Luca Foresti[2], Marco Raoul Marini[1], Alessio Mecca[1], and Daniele Pannone[1(✉)]

[1] Sapienza University, Rome, Italy
{avola,cinque,fagioli,marini,mecca,pannone}@di.uniroma1.it
[2] Udine University, Udine, Italy
foresti@di.uniroma1.it

Abstract. Taking medicines is a fundamental aspect to cure illnesses. However, studies have shown that it can be hard for patients to remember the correct posology. More aggravating, a wrong dosage generally causes the disease to worsen. Although, all relevant instructions for a medicine are summarized in the corresponding patient information leaflet, the latter is generally difficult to navigate and understand. To address this problem and help patients with their medication, in this paper we introduce an augmented reality mobile application that can present to the user important details on the framed medicine. In particular, the app implements an inference engine based on a deep neural network, i.e., a densenet, fine-tuned to recognize a medicinal from its package. Subsequently, relevant information, such as posology or a simplified leaflet, is overlaid on the camera feed to help a patient when taking a medicine. Extensive experiments to select the best hyperparameters were performed on a dataset specifically collected to address this task; ultimately obtaining up to 91.30% accuracy as well as real-time capabilities.

Keywords: Convolutional neural network · Deep learning · Augmented reality

1 Introduction

In the last decade, machine learning and deep learning algorithms have become widespread tools to address many computer vision-based problems, including medical imaging analysis [12], person re-identification [7,11], environment monitoring [8,14,15], emotion recognition [10], handwriting validation [3,4], background modeling [2], and video synthesis [5]. Although effective, these methods usually require a large amount of training data which, however, is not always available. To solve this issue, transfer learning approaches are being exploited to retain previous knowledge from different tasks, and reduce the amount of

S. Sclaroff et al. (Eds.): ICIAP 2022, LNCS 13231, pp. 489–499, 2022.
https://doi.org/10.1007/978-3-031-06427-2_41

data required to address the new one [6,31]. A paradigm that is proving particularly relevant to the medical field, where sensitive data is generally employed. As a matter of fact, automatic procedures for detection, classification, and analysis are already being developed [9,24]. What is more, in concert with these diagnosis-related procedures, mobile applications are also being explored to support patients in their daily routine [21,23,28] since, in general, dealing with the medical aspects of life is a challenging task for both medical operators and sick persons. In particular, operators have to manage several, often elder, people with different pathologies throughout the day. To help this category, applications noticing anomalous patterns can provide a huge healthcare boost, as shown in [30], where bed falls are detected even at nighttime to safeguard patients. Regarding the latter, instead, they tend to either forget when to take a medicine or its required dosage, as reported by [22]. Although it is possible to look up these instructions on the internet or the patient information leaflet (PIL), it can be difficult to find the correct information; suggesting that there is a technological gap that can be filled.

On a different yet related note, hardware advances are enabling mobile devices to execute ever increasingly complex deep learning algorithms, leading to many performance improvements in different fields such as mobile biometrics security through face [18,26] and fingerprint [17,25] recognition. Moreover, the improved hardware, in conjunction with optimized libraries, allows augmented reality (AR) techniques to be run smoothly on most mobile phones. In general, AR applications add a semitransparent layer on top of the camera feed to provide the user with more information with respect to what is being framed by the device. This augmented content can then be decided by analyzing the scene, by exploiting markers, or by using GPS coordinates [27]. What is more, this technology can be applied in heterogeneous application fields such as rehabilitation [13,16], art [1,19], or teaching [29], and it is used to make the application more engaging and compelling to the user.

In this paper, to both leverage hardware advances and help people to take a correct medicinal dosage, we present a mobile AR application that recognizes different kinds of medicines from their package, and provides the user with useful information such as its posology or simplified leaflet. In particular, starting from the smartphone camera feed, video frames are sent to an inference engine for classification. Specifically, the engine is based on a deep neural network exploiting the transfer learning paradigm to overcome the limited amount of training data. Finally, AR is employed to present information on the recognized medicine, and enable real-time interactions by the users. To ensure this time constraint, extensive experiments were performed on a dataset specifically collected for this task, so that the best performing model could be used inside the application.

The rest of this paper is organized as follows. Section 2 introduces the AR mobile application and inference engine used to classify medicinal boxes. Section 3 summarizes information on the collected dataset, relevant implementation details, as well as the obtained results. Finally, Sect. 4 draws some conclusions on the presented work.

2   Background and Method

In this section, we present the proposed AR pipeline used to classify medicinal packages, and provide details on the application interface as well as the underlying inference engine exploiting the transfer learning paradigm.

2.1   Mobile App AR Interface

To provide patients with extra information on a given medicine in real-time, we devised an iOS AR application that rapidly shows the suggested posology as well as more detailed PIL guidelines when prompted. The mobile application was implemented following the common model-view-controller (MVC) design pattern, summarized in Fig. 1, where different software components are organized into one of the MVC roles. Specifically, application logic and operations, such as the inference engine, are associated to the model; interface aspects, including rendered AR elements properties, are linked to the view; while the controller is responsible for the data flow between model and view components, allowing for a responsive and well-organized application. By following this pattern, it is possible to develop highly re-usable software and it is easier to extend pre-existing libraries. In particular, we exploited the Apple ARKit framework to implement AR capabilities. In detail, the application controller component sends video frames captured through the mobile device camera to the inference engine and, whenever a medicinal package is recognized, the view component is updated in real-time so that the RGB camera feed is overlaid with meaningful 2D text information, e.g., posology or PIL. Furthermore, the latter are correctly bounded to the corresponding object through the aforementioned framework functionalities; effectively presenting relevant data of a particular medicine to the user. The intuitive AR interface showing either posology or simplified PIL information is reported in Fig. 2.

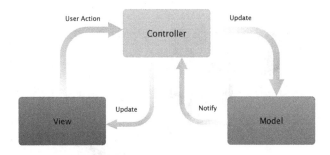

Fig. 1. MVC pattern design scheme.

Fig. 2. In-app AR user interface screenshots. On the left, the base overlay showing quick details on the recognized medicine posology. On the right, additional PIL extracted information are instead displayed.

2.2   Inference Engine

To discriminate between medicinal packages and show the correct information inside the AR interface, the application requires a component to perform inference. The latter was implemented as a deep neural network based on transfer learning and imported as the application inference engine via the iOS CoreML framework. In particular, an ImageNet pretrained densenet [20] was selected due to its generally high classification performances as well as its internal architecture. In detail, this network leverages a dense connectivity by introducing links from any layer to any subsequent one, as shown in Fig. 3. Formally, given the

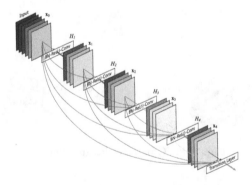

Fig. 3. Densenet dense block scheme. Image courtesy of [20].

l-th layer, its input x_l will be composed by the feature maps of all preceding layers, as follows:

$$x_l = H_l([x_0, x_1, \ldots, x_{l-1}]),\tag{1}$$

where $[x_0, x_1, \ldots, x_{l-1}]$ represents the concatenation operation of all previous feature maps; while $H_l(\cdot)$ is a composite function applying batch normalization, a rectified linear unit (ReLU) activation function and a 1×1 convolution acting as a bottleneck to consolidate and limit the output size of a given layer. While these dense connections are effective tools to feed-forward information inside a network, they must have the same size in order to be concatenated through Eq. (1). However, this aspect collides with the essential downsampling procedure of a CNN. Therefore, to address this issue, the authors organized their architecture into dense blocks. What is more, transition layers were designed to reduce the feature map size between these blocks. In particular, each transition layer contains a batch normalization operation, a 1×1 convolution, and a 2×2 average pooling layer. Moreover, a compression hyperparameter ϕ is also applied on transition layers to increase the architecture compactness by reducing the next block input features maps m to $\lfloor \phi m \rfloor$. Furthermore, notice that m directly depends on another hyperparameter k, called growth rate, that indicates the number of feature maps (i.e., filters) to be produced per layer. Intuitively, if H_l produces k feature maps, the l-th layer will have $k_0 + k * (l - 1)$ maps as input, where k_0 corresponds to the input image channels. Thus, by modifying k, it is possible to substantially change the number of network parameters. An overview of a densenet with dense blocks and transition layers is shown in Fig. 4.

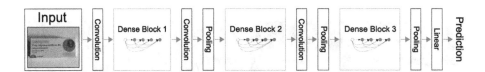

Fig. 4. Densenet architecture showing dense blocks interleaved by transition layers.

To apply the transfer learning paradigm and retain previous knowledge computed on the ImageNet dataset, all model weights are frozen apart from the linear layer used as a classifier component. The latter is then modified by changing its size to handle the right number of classes. Subsequently, the classifier is trained on the medicinal package classification task using the densenet extracted feature maps via the classical backpropagation algorithm and cross-entropy loss, thus exploiting prior knowledge via the frozen layers.

Finally, since the model receives frames from the camera feed in real-time, a λ threshold is applied on the confidence scores produced by the densenet to discard all classifications below it, to avoid showing the user information on uncertain recognitions. Indeed, by using this strategy, possible misclassification of objects with similar shapes, e.g., a generic box, are ignored by the application and only medicinal packages will be enhanced through the proposed AR interface.

3   Experiments

In this section, we first introduce the dataset used to test the mobile AR application, which was specifically collected to address the medicinal box classification task. Subsequently, implementation details for all of the described technologies are reported. Finally, a discussion on summarized performances obtained by the inference engine is presented.

3.1   Dataset

To correctly assess the presented model, 978 images for 63 distinct medicinal packages were collected from different sources such as Google and Bing images, as well as offline, directly through the phone camera. In addition, to reinforce the dataset difficulty, we ensured that among the 63 categories there were several distinct boxes presenting various similarities, e.g., shape or color scheme. Moreover, different lighting conditions, as well as distance and angles from the camera, were captured to further increase data variability. To train and test the model, the collection was divided using a stratified 10-fold cross-validation procedure using 80%/20% splits for the train and test sets, respectively. Notice that this approach retains the requested number of samples per class, thus ensuring there are enough samples for both training and test phases. Moreover, to further improve the model abstraction capabilities, a data augmentation strategy was devised by means of random horizontal flips and random rotations $\theta \in [-15, 15]$ degrees. Samples from the collected dataset are shown in Fig. 5.

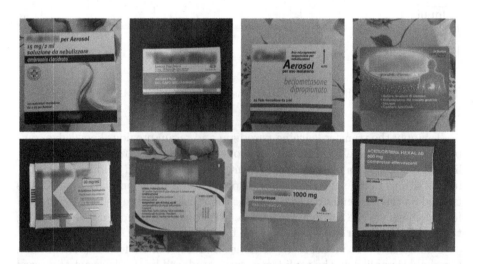

Fig. 5. Medicinal boxes samples from the collected dataset.

3.2   Implementation Details

The presented mobile application was developed on an iOS mobile phone, i.e., iPhone 12 Pro. This device implements ARKit 5.0 and CoreML 4.1 frameworks, which enable, respectively, AR capabilities and neural network translation as well as its execution. This phone was selected due to both its ability to run the latest frameworks versions, as well as its hardware specifications, which allowed for a smooth real-time AR experience.

Concerning the transfer learning procedure, before implementing the application inference engine, the densenet was developed using the PyTorch framework and torchvision library, which contains an ImageNet pre-trained densenet version. In particular, we selected the densenet-121 model and modified its last linear layer to have 63 nodes, to match the dataset classes. Moreover, the default growth rate $k = 32$ and compression $\phi = 0.5$ were set for the architecture. A $\lambda = 0.85$ was also found to be a good confidence threshold to avoid uncertain classification showing in the AR interface. Regarding the training hyperparameters, the model was fine-tuned on the collected dataset for 100 epochs, using the SGD algorithm, with a learning rate lr set to 0.1, decreased by a factor of 10 at epochs 40, and 80, a weight decay of 5e−4, and a Nesterov momentum of 0.9.

Finally, both training and experiments were performed using a 6-Core Intel i7 2.60 GHz CPU with 32 GB RAM and single GPU, i.e., a GeForce GTX 1070 with 8 GB of dedicated RAM.

3.3   Performance Evaluation

In order to choose the best model parameters, ablation studies were performed on hyperparameters k and ϕ. Common classification metrics, i.e., accuracy, precision, recall, and f1-score, were used to assess each configuration that was tested using the aforementioned 10-fold cross-validation procedure.

Concerning the growth rate parameter k, the obtained results are summarized in Table 1. As shown, by increasing the number of filters generated by each convolutional layer, performances improve across all metrics, reaching up to an average 91.35% f1-score. More interestingly, using a low k number does not allow to represent and learn the input data distribution, resulting in consistent 30% gaps. For higher k values, instead, there are diminished performance returns. As a matter of fact, there is roughly a 0.02% difference between models using $k = 32$ and $k = 64$, indicating that enough information is already captured by the former value. What is more, the number of parameters and memory used at inference time increase superlinearly with respect to the growth rate k. Thus, since there are negligible performance gains, it is important to choose the correct k value to reduce the mobile device computational burden during inference; resulting in $k = 32$ as the better choice for the presented work.

Regarding the compression rate ϕ, results for various values are summarized in Table 2. As can be seen, reducing or increasing the extracted feature maps on transition layers via ϕ, has relatively little impact on the number of parameters, memory consumption, as well as all metrics. Nevertheless, using the default

Table 1. Ablation study on growth rate hyperparameter k. Results correspond to the 10-fold cross-validation scores average. For all networks, ϕ was set to 0.5.

Model	k	Params	Memory	Accuracy	Precision	Recall	F1-score
densenet-121	4	0.2 M	0.9 MB	61.13%	64.40%	61.13%	61.89%
densenet-121	8	0.6 M	2.4 MB	79.90%	80.74%	79.90%	80.09%
densenet-121	16	1.9 M	7.7 MB	89.07%	89.42%	89.07%	89.14%
densenet-121	32	7.0 M	28.4 MB	91.30%	91.51%	91.30%	91.33%
densenet-121	64	27.3 M	109.7 MB	91.32%	91.55%	91.32%	91.35%

0.5 value allows to achieve the best performances since first, it avoids losing too much information from the preceding dense block with respect to a lower 0.1 size; and second, it improves the model abstraction capabilities without incurring in possible overfitting scenarios, as opposed to the $\phi = 1.0$ case where performances begin to deteriorate.

Table 2. Ablation study on compression hyperparameter ϕ. Results correspond to the 10-fold cross-validation scores average. For all networks, k was set to 32.

Model	ϕ	Params	Memory	Accuracy	Precision	Recall	F1-score
densenet-121	0.1	6.5 M	26.2 MB	90.00%	90.28%	90.00%	90.06%
densenet-121	0.5	7.0 M	28.4 MB	91.30%	91.51%	91.30%	91.33%
densenet-121	1.0	7.7 M	31.2 MB	91.26%	91.51%	91.26%	91.31%

Although the ablation studies present interesting information, a better system overview can be provided by a confusion matrix, which is reported in Fig. 6. As expected, most predictions lie in the matrix diagonal, indicating that the model can correctly recognize all classes with a high accuracy. Furthermore, for a given box, misclassifications tend to concentrate on specific medicines, suggesting that there are similarities between those packages. As a matter of fact, this outcome can be confirmed by observing Fig. 7, which shows boxes misclassified by the presented densenet. As can be observed, the reported packages have similar sizes and colors but can have different product names, active substance weights or molecules. This indicates that while the system uses the entire box to recognize a medicine, it is still not able to accurately discriminate text; suggesting that the model can be further improved by refining its ability to classify written content, a task left as future work.

Finally, the densenet model with $k = 32$ and $\phi = 0.5$ was employed as inference engine inside the mobile application to enable stable real-time capabilities over a 30FPS video stream. Notice that the selected device achieved similar performances even with the more demanding $k = 64$ model, however since the AR interface showed several, albeit infrequent, frame drops, the less complex model was still the preferred choice to reduce the device burden.

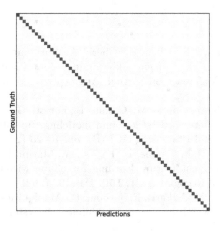

Fig. 6. Confusion matrix of a densenet-121 model with $k = 32$ and $\phi = 0.5$.

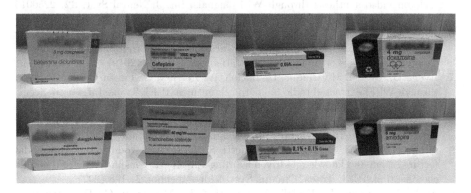

Fig. 7. Examples of misclassified medicine boxes.

4   Conclusions

In this paper we presented an AR mobile application that can classify medicinal boxes and show the user useful information, such as the posology or PIL, to help them using medicinals in a proper way. Extensive experiments were performed on a dataset specifically collected to address this task, and the best model configuration was selected to achieve real-time execution on the mobile device. Specifically, the chosen architecture, i.e., a densenet, obtains significant performances by exploiting the transfer learning paradigm, reaching up to 91.30% accuracy on 63 classes; highlighting the effectiveness of the proposed methodology.

Acknowledgment. This work was supported by the MIUR under grant "Departments of Excellence 2018–2022" of the Sapienza University Computer Science Department and the ERC Starting Grant no. 802554 (SPECGEO).

References

1. Aitamurto, T., Boin, J.B., Chen, K., Cherif, A., Shridhar, S.: The impact of augmented reality on art engagement: liking, impression of learning, and distraction. In: International Conference on Virtual, Augmented and Mixed Reality, pp. 153–171 (2018)
2. Avola, D., Bernardi, M., Cascio, M., Cinque, L., Foresti, G.L., Massaroni, C.: A new descriptor for keypoint-based background modeling. In: International Conference on Image Analysis and Processing (ICIAP), pp. 15–25 (2019)
3. Avola, D., Bigdello, M.J., Cinque, L., Fagioli, A., Marini, M.R.: R-signet: reduced space writer-independent feature learning for offline writer-dependent signature verification. Pattern Recognit. Lett. **150**, 189–196 (2021)
4. Avola, D., Caschera, M.C., Ferri, F., Grifoni, P.: Ambiguities in sketch-based interfaces. In: Annual Hawaii International Conference on System Sciences (HICSS), pp. 290b–290b (2007)
5. Avola, D., Cascio, M., Cinque, L., Fagioli, A., Foresti, G.L.: Human silhouette and skeleton video synthesis through Wi-Fi signals. Int. J. Neural Syst. **32**, 2250015 (2022)
6. Avola, D., Cascio, M., Cinque, L., Fagioli, A., Foresti, G.L., Massaroni, C.: Master and rookie networks for person re-identification. In: International Conference on Computer Analysis of Images and Patterns, pp. 470–479 (2019)
7. Avola, D., Cascio, M., Cinque, L., Fagioli, A., Petrioli, C.: Person re-identification through Wi-Fi extracted radio biometric signatures. IEEE Trans. Inf. Foren. Secur. Early Access **17**, 1145–1158 (2022)
8. Avola, D., et al.: Low-altitude aerial video surveillance via one-class SVM anomaly detection from textural features in UAV images. Information **13**(1), 2 (2021)
9. Avola, D., Cinque, L., Fagioli, A., Filetti, S., Grani, G., Rodolà, E.: Multimodal feature fusion and knowledge-driven learning via experts consult for thyroid nodule classification. IEEE Trans. Circuits Syst. Video Technol. **1**, 1–8 (2021)
10. Avola, D., Cinque, L., Fagioli, A., Foresti, G.L., Massaroni, C.: Deep temporal analysis for non-acted body affect recognition. IEEE Trans. Affect. Comput. **1**, 1–12 (2020)
11. Avola, D., Cinque, L., Fagioli, A., Foresti, G.L., Pannone, D., Piciarelli, C.: Bodyprint-a meta-feature based LSTM hashing model for person re-identification. Sensors **20**(18), 5365 (2020)
12. Avola, D., Cinque, L., Fagioli, A., Foresti, G., Mecca, A.: Ultrasound medical imaging techniques: a survey. ACM Comput. Surv. (CSUR) **54**(3), 1–38 (2021)
13. Avola, D., Cinque, L., Foresti, G.L., Marini, M.R.: An interactive and low-cost full body rehabilitation framework based on 3D immersive serious games. J. Biomed. Inform. **89**, 81–100 (2019)
14. Avola, D., Foresti, G.L., Cinque, L., Massaroni, C., Vitale, G., Lombardi, L.: A multipurpose autonomous robot for target recognition in unknown environments. In: International Conference on Industrial Informatics (INDIN), pp. 766–771 (2016)
15. Avola, D., Foresti, G.L., Martinel, N., Micheloni, C., Pannone, D., Piciarelli, C.: Real-time incremental and geo-referenced mosaicking by small-scale UAVs. In: International Conference on Image Analysis and Processing (ICIAP), pp. 694–705 (2017)
16. Avola, D., Spezialetti, M., Placidi, G.: Design of an efficient framework for fast prototyping of customized human-computer interfaces and virtual environments for rehabilitation. Comput. Methods Programs Biomed. **110**(3), 490–502 (2013)

17. Baldini, G., Steri, G.: A survey of techniques for the identification of mobile phones using the physical fingerprints of the built-in components. IEEE Commun. Surv. Tutor. **19**(3), 1761–1789 (2017)

18. Freire-Obregón, D., Narducci, F., Barra, S., Castrillón-Santana, M.: Deep learning for source camera identification on mobile devices. Pattern Recognit. Lett. **126**, 86–91 (2019)

19. He, Z., Wu, L., Li, X.R.: When art meets tech: the role of augmented reality in enhancing museum experiences and purchase intentions. Tour. Manag. **68**, 127–139 (2018)

20. Huang, G., Liu, Z., Van Der Maaten, L., Weinberger, K.Q.: Densely connected convolutional networks. In: IEEE Conference on Computer Vision and Pattern Recognition (CVPR), pp. 4700–4708 (2017)

21. Konig, A., et al.: Use of speech analyses within a mobile application for the assessment of cognitive impairment in elderly people. Curr. Alzheimer Res. **15**(2), 120–129 (2018)

22. Mayo-Gamble, T.L., Mouton, C.: Examining the association between health literacy and medication adherence among older adults. Health Commun. **33**(9), 1124–1130 (2018)

23. Petersen, M., Hempler, N.F.: Development and testing of a mobile application to support diabetes self-management for people with newly diagnosed type 2 diabetes: a design thinking case study. BMC Med. Inform. Decis. Mak. **17**(1), 1–10 (2017)

24. Petracca, A., et al.: A virtual ball task driven by forearm movements for neurorehabilitation. In: International Conference on Virtual Rehabilitation (ICVR), pp. 162–163 (2015)

25. Rahmawati, E., et al.: Digital signature on file using biometric fingerprint with fingerprint sensor on smartphone. In: International Electronics Symposium on Engineering Technology and Applications (IES-ETA), pp. 234–238 (2017)

26. Ríos-Sánchez, B., Costa-da Silva, D., Martín-Yuste, N., Sánchez-Ávila, C.: Deep learning for face recognition on mobile devices. IET Biom. **9**(3), 109–117 (2020)

27. de Souza Cardoso, L.F., Mariano, F.C.M.Q., Zorzal, E.R.: A survey of industrial augmented reality. Comput. Ind. Eng. **139**, 106159 (2020)

28. Tun, S.Y.Y., Madanian, S., Mirza, F.: Internet of Things (IoT) applications for elderly care: a reflective review. Aging Clin. Exp. Res. **33**(4), 855–867 (2021)

29. Turkan, Y., Radkowski, R., Karabulut-Ilgu, A., Behzadan, A.H., Chen, A.: Mobile augmented reality for teaching structural analysis. Adv. Eng. Inform. **34**, 90–100 (2017)

30. Zhao, F., Cao, Z., Xiao, Y., Mao, J., Yuan, J.: Real-time detection of fall from bed using a single depth camera. IEEE Trans. Autom. Sci. Eng. **16**(3), 1018–1032 (2018)

31. Zhuang, F., et al.: A comprehensive survey on transfer learning. Proc. IEEE **109**(1), 43–76 (2020)

Consistency Regularization for Unsupervised Domain Adaptation in Semantic Segmentation

Sebastian Scherer[✉], Stephan Brehm, and Rainer Lienhart

Machine Learning and Computer Vision Lab, University of Augsburg,
Augsburg, Germany
{sebastian1.scherer,stephan.brehm,rainer.lienhart}@uni-a.de

Abstract. Unsupervised domain adaptation is a promising technique for computer vision tasks, especially when annotating large amounts of data is very costly and time-consuming, as in semantic segmentation. Here it is attractive to train neural networks on simulated data and fit them to real data on which the models are to be used. In this paper, we propose a consistency regularization method for domain adaptation in semantic segmentation that combines pseudo-labels and strong perturbations. We analyse the impact of two simple perturbations, dropout and image mixing, and show how they contribute enormously to the final performance. Experiments and ablation studies demonstrate that our simple approach achieves strong results on relevant synthetic-to-real domain adaptation benchmarks.

Keywords: Domain adaptation · Semi-supervised learning · Unsupervised learning · Semantic segmentation · Synthetic data

1 Introduction

Semantic segmentation has accomplished amazing performance on annotated data and has become one of the most important tasks in computer vision. However, labelling data for semantic segmentation requires assigning a class label to each pixel in an image, which is an extremely tedious and expensive task. For example, annotating a single image of the Cityscapes dataset [6], which consists of images of urban scenes, takes up to 90 min [6]. As a result, datasets for semantic segmentation of urban scenes are generally much smaller than datasets for image classification. Synthetic images from computer games are a powerful alternative to real images because they can be labelled automatically, since the geometric 3D scene and the objects it contains, that are projected into the image, are known. This results in high-resolution datasets with precise object boundaries that are inexpensive to obtain and offer almost infinite possibilities for the automatic creation of synthetic data. The problem here is, however, that computer simulations are not perfectly realistic. In general, convolutional

© The Author(s), under exclusive license to Springer Nature Switzerland AG 2022
S. Sclaroff et al. (Eds.): ICIAP 2022, LNCS 13231, pp. 500–511, 2022.
https://doi.org/10.1007/978-3-031-06427-2_42

neural networks (CNNs) learn features only from the domain on which they were trained. For this reason, CNNs trained on synthetic data tend to perform poorly on real images, even when the synthetic data consists of significantly more images. For example, in our experiments we noticed that as few as 30 images from the Cityscapes dataset were enough to get similar performance to 24000 images from a dataset that consists of images from the GTA5 game [16].

Unsupervised domain adaptation (UDA) tries to bridge this domain gap. It aims to transfer the knowledge of a label-rich source domain to an unlabelled target domain with similar class information. When synthetic data is used as source domain, the problem of *synthetic-to-real* domain adaptation arises. Recently, researchers utilised self-training (ST) techniques for UDA [1,5,15,25,30], that allows the usage of images from the target domain directly for the training of the segmentation network via pseudo-labels. They do so by adding a new loss term to the training objective that encourage the model to make consistent prediction of images from the target domain under different perturbations of the image. These approaches achieve great results and represent the current state of the art. In such a framework, the used perturbations are the key for the success of those approaches [9]. Current approaches however use perturbations on the image level, that are sometimes not realistic, such as heavy noise [30], Fourier Mixing [15,28] or Style Mixing [15,20]. Even though these perturbations are non-realistic, they improve the general performance when applied. We argue that this improvement also comes from the fact, that the perturbations lead to a much worse prediction, making the pseudo labels, which may not be perfectly correct as well, still better and therefore leading to an improvement of the model. In our experiments, we found that perturbations that do not degrade the prediction of the model do not improve the UDA task, while perturbations that degrade the prediction of the model do. This comes close to the bootstrapping idea in its idiomatic meaning, which refers to a self-starting process that is supposed to continue or improve itself without external input. Based on this situation, the question arises as to how the prediction of the model can be deteriorated the most. While perturbations on the image level are effective, we found that perturbation within the network itself are powerful as well. Surprisingly, we found that a simple baseline model, which uses heavily dropout as perturbation, achieves strong results on current benchmarks for UDA. Combined with a perturbation on the image, we achieve state-of-the-art results. As image perturbation, we utilise a recent perturbation that has originally been proposed for the image classification task: the CowMask [8] image mixing method. It mixes two images and their predictions by a network with a specific mask looking similar to the typical black and white skin pattern of a cow. We argue that this perturbation is perfect for segmentation tasks, as it simulates the occlusion of objects and introduces an additional segmentation task.

2 Related Work

UDA for semantic segmentation has been extensively studied in the last years. Adversarial training was the previously dominant approach applied either on the input space, the feature space or the output space [23] of a segmentation network. Popular input space adaptation techniques try to change the style of the source domain by performing image-to-image translation, for example by making synthetic images look more realistic [2,5,11,19]. The biggest disadvantage of adversarial training is the unstable training behaviour. Recently, a new line of methods introduced semi-supervised learning (SSL) techniques for UDA and showed remarkable results. SSL aims to include unlabelled data alongside labelled data in the training of a neural network. These approaches are either based on consistency regularization [5,15,30] or self-training [31,32]. Self-training aims to generate pseudo-labels for the unlabelled data and fine-tune the model on them iteratively [13,27]. Zou et al. [31] and Li et al. [14] applied the pseudo-labelling approach for UDA task and achieved strong improvements. The key idea of consistency regularization is that the predictions of a model should be invariant under different perturbations. These approaches usually adopt a teacher - student framework, where the teacher model is an exponential moving average (EMA) of the student model. The teacher model transfers the learned knowledge to the student, who is additionally influenced by perturbations that are normally applied to the input image. In comparison to self-training, these approaches are typically not trained iteratively, but end-to-end. Choi et al. [5] combined this approach with a GAN-based augmentation module for image translation. Zhou et al. [30] further incorporated an additional uncertainty module that tries to approximate the uncertainty of the predictions to filter uncertain pixel predictions from the loss calculation. To do so, they perform several forward passes of an image with different Gaussian Noise applied and calculate the pixel-wise entropy based on those predictions. Melas-Kyriazi and Manrai [15] proposed a similar approach with different perturbations on the image, namely simple data augmentation, CutMix [29], style consistency and Fourier consistency. Compared to the other work, they did not apply the exponential moving average of the trained model and use the prediction of the model itself as guidance. Our work is mostly related to this line of research. Building on PixMatch [15], we show that simply by applying dropout as perturbation for the student model, we are able to improve the general results. Combined with the recent image-mixing technique using CowMask [8], we achieve strong results with a very simple and easy to implement approach. Specially, we show that the used perturbations are the key for this kind of approaches.

3   Methodology

Let $\mathcal{S}, \mathcal{T}$ be the source and target domain and let $X_{\mathcal{S}}$, $X_{\mathcal{T}}$ be sets of images from each domain, respectively. We denote $\mathbf{x}_s \in X_{\mathcal{S}}$ and $\mathbf{x}_t \in X_{\mathcal{T}}$ as data samples from the source and target domain. At the source domain we have access to N labelled segmentation masks, i.e., $X_{\mathcal{S}} = \{(\mathbf{x}_s^i, \mathbf{y}_s^i)\}_{i=1}^{N}$. We denote $\mathbf{y}$ as ground truth annotation from the dataset and $\hat{\mathbf{y}}$ as pseudo-label. The target domain has no labelled samples and shares C categories of the source domain. Our task is to train a segmentation network that performs well on the target domain.

This problem formulation can further be extended to multiple source domains $X_S^1, X_S^2, ..., X_S^K$ or target domains $X_T^1, X_T^2, ..., X_T^M$, with K and M the number of source or target domains, respectively.

Figure 1 shows an overview of our proposed architecture. The overall objective function for the segmentation training is defined by a supervised part based on images from the source domain as well as a self-supervised part based on images from the target domain. The self-supervised part employs consistency regularization and pseudo-labels on the target domain. Similar to Tarvainen et al. [22], we make use of two networks of identical architecture: a student network F_S and a teacher network F_T. The predictions of the teacher network are used to produce pseudo-labels for images of the target domain, which are subsequently used to train the student network. In this framework, the teacher network is simply the exponential-moving average (EMA) of the student model and will not receive any gradient-based parameter updates. The overall objective $\mathcal{L}^{F_S}$ is defined as follows:

$$\mathcal{L}^{F_S} = \mathcal{L}_S^{F_S} + \lambda_T \mathcal{L}_T^{F_S}, \tag{1}$$

where λ_T is a trade-off parameter. $\mathcal{L}_S^{F_S}$ indicates the softmax cross-entropy objective $\mathcal{L}_{ce}$ between the prediction of the student network $F_S(\mathbf{x}_s)$ for an image of the source domain $\mathbf{x}_s \in X_S$ and its pixel-level annotation map $\mathbf{y}_s$, i.e., $\mathcal{L}_S^{F_S} = \mathcal{L}_{ce}(F_S(\mathbf{x}_s), \mathbf{y}_s)$.

Given an image from the target domain $\mathbf{x}_t \in X_T$ and its perturbed version $\bar{\mathbf{x}}_t$, we feed the image through the teacher network F_T to obtain the soft pseudo-label $\hat{\mathbf{y}}_t^{soft} = F_T(\mathbf{x}_t)$. We can get a hard pseudo-label by calculating the argmax at the class dimension, i.e. $\hat{\mathbf{y}}_t^{hard} = argmax(\hat{\mathbf{y}}_t^{soft})$. Both soft and hard pseudo labels may now be used as targets for the student network. By incorporating $\hat{\mathbf{y}}_t^{soft}$, we can calculate a loss by applying the Mean Squared Error (MSE):

$$\mathcal{L}_T^{F_S} = \mathcal{L}_{MSE}(F_S(\bar{\mathbf{x}}_t), \hat{\mathbf{y}}_t^{soft}). \tag{2}$$

By incorporating $\hat{\mathbf{y}}_t^{hard}$, we can calculate a loss by applying the standard cross entropy loss:

$$\mathcal{L}_T^{F_S} = \mathcal{L}_{ce}(F_S(\bar{\mathbf{x}}_t), \hat{\mathbf{y}}_t^{hard}). \tag{3}$$

Comparing soft and hard pseudo-labels, soft pseudo-labels are generally more robust against noisy labels (false classifications), while hard pseudo-labels bring an additional learning effect as the highest activation is reinforced across classes. We will investigate both in our ablation study.

Perturbations. As stated in [9], the success of SSL techniques based on consistency regularization depends on the used perturbations. These are used only at the forward pass of the student model when calculating the self-supervised loss on images from the target domain. We experiment with two different perturbations, one at the input image and one within the model itself.

An easy way to perform a perturbation on the student side is by using dropout layers. The role of dropout is to improve generalization performance by preventing the model from overfitting. It forces the network to learn more

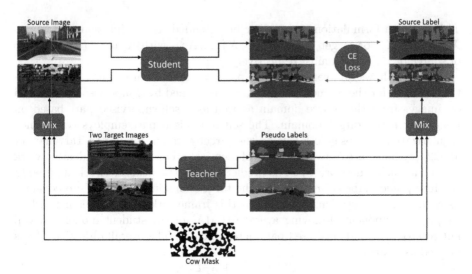

Fig. 1. Illustration of the proposed solution.

robust features that can deal with many random subsets of neurons. In addition, dropout usually also leads to a deterioration of the prediction and thus to an increase of the error, since, in a sense, only a part of the original network is used. As we work with CNNs, we will make usage of SpatialDropout as proposed by Tompson et al. [24]. Given a feature tensor from a convolutional layer of size *height* × *width* × *depth*, where *depth* is the amount of filters of the layer, the dropout is applied at some dimensions of the depth across the entire feature map. It therefore simulates that certain filters had no activation. The dropout within the network has the positive effect, that the resulting sub-network of the original network will be trained on clean images, with lets the network learn particular features of that domain. Other perturbations such as heavy random noise or Style change [15,20] applications let the network learn on images that will not occur in reality.

Following [8], we further use the Cutmix [29] augmentation using a Cow-Mask [8] as perturbation. In this augmentation, individual parts of the image will be replaced by another image. These augmentations do not occur in reality, but they serve as an additional perturbation by suppressing areas and introducing additional object boundaries at the edge of the applied mask. The replacement of certain areas within the image further results in an occlusion of the objects. To generate a single sample for the student network, we take two images from the target domain $\mathbf{x}_t^1$ and $\mathbf{x}_t^2$. Both images will be fed to the teacher network F_T to obtain the pseudo labels $\hat{\mathbf{y}}_t^1$ and $\hat{\mathbf{y}}_t^2$. We then calculate the image and target for the training of the student network by mixing the images and the pseudo labels according to the generated mask M by:

$$\bar{\mathbf{x}}_t = M \odot \mathbf{x}_t^1 + (1 - M) \odot \mathbf{x}_t^2,$$
$$\hat{y}_t = M \odot \hat{\mathbf{y}}_t^1 + (1 - M) \odot \hat{\mathbf{y}}_t^2,$$

$$(4)$$

where $M \in \{0,1\}^{W \times H}$ denotes a binary mask and $\odot$ is element-wise multiplication. For the calculation of a random CowMask M, please refer to [8].

4 Experiments and Results

In this section, we detail the experiments that we conducted in order to show the benefit and performance of our proposed approach.

4.1 Datasets

We use five datasets in our experiments. The Cityscapes dataset [6] contains images from real-world urban scenes, split into 2975 images for training and 500 for validation. The GTA5 dataset [16] and the SYNTHIA dataset [17] contain 24966 and 9400 synthetic images with pixel wise annotations, respectively. The annotations of both datasets are compatible with Cityscapes. Both synthetic datasets serve as source domain, while the Cityscapes dataset serves as target domain. This results in two popular *synthetic-to-real* domain adaptation scenarios: GTA5 to Cityscapes (GTA → CS) and SYNTHIA to Cityscapes (SYN → CS). We further experiment with a third synthetic dataset Synscapes [26], which contains 25000 photo-realistic images. Besides *synthetic-to-real* domain adaptation, we also evaluate our approach at the CS → ACDC benchmark. The ACDC [18] dataset contains images of four common adverse visual conditions: fog, nighttime, rain and snow. Images from the Cityscapes dataset are taken at normal weather conditions and at daytime. This domain adaptation attempts to improve the segmentation model at different visual conditions as they occur in the labelled dataset. As it considers four different visual conditions, it can be seen as a multi-target domain adaptation problem. Each visual condition contains of 400 training images, 100 validation images and 500 test images.

4.2 Implementation Details

For a fair comparison to earlier works, we adopt the VGG16 [21] and the ResNet101 [10] backbone pre-trained on the ImageNet dataset [7]. Following Deeplab-V2 [3], we incorporate Atrous Spatial Pyramid Pooling (ASPP) as the decoder and then use bilinear upsampling to get the segmentation output. We use color jittering as augmentation on images from the source domain with the same settings as used in [4]. For the dropout perturbation, we place a dropout layer before each pooling or strided convolutional layer for simple reproducibility and re-implementation. If not stated otherwise, we use a dropout rate of 0.3 and an EMA value of 0.999. For the ResNet101, the Batch Normalization layers are frozen during training. We set λ_{real} to 50 when using soft pseudo-labels, otherwise to 1. All experiments were conducted on a single NVIDIA V100 GPU with 16 GB of VRAM. We perform 25.000 training steps where no loss is calculated on the target domain as warm-up phase. Afterwards, each mini-batch consist of an image from the source and target domain respectively. We train our models with early stopping.

Table 1. Ablation study of our proposed perturbations within the consistency regularization framework using the VGG16 network as backbone. 'SL' and 'HL' stands for soft and hard pseudo-labels, respectively. 'CM' stands for the image-mixing technique using a CowMask.

Method	$mIoU^{19}$
Baseline	35.23
SL	40.57
SL + Dropout	49.03
SL + CM	51.29
SL + Dropout + CM	49.81
HL	40.05
HL + Dropout	48.0
HL + CM	52.26
HL + Dropout + CM	**53.4**

(a) GTA5 → Cityscapes

Method	$mIoU^{16}$	$mIoU^{13}$
Baseline	31.0	35.9
SL	32.05	36.86
SL + Dropout	37.6	42.90
SL + CM	38.74	43.85
SL + Dropout + CM	41.47	47.34
HL	33.17	38.01
HL + Dropout	38.12	43.53
HL + CM	41.82	47.50
HL + Dropout + CM	**44.74**	**51.21**

(b) SYNTHIA → Cityscapes

4.3 Ablation Study

In this section, we study the effectiveness of each component in our approach and investigate how they contribute to the final performance on both benchmarks when using the VGG16 as backbone. Table 1 compares the use of soft and hard pseudo-labels (SL vs. HL) as well as the perturbations alone and in combination. Comparing the results for soft and hard pseudo-labels, we can identify only minor differences. However, when both perturbations are used, hard pseudo-labels lead to better results. This is presumably because when both perturbations are used, the model improves and so do the pseudo labels, giving the additional learning effect of the pseudo labels a better impact. Comparing the approach without any perturbation to the baseline, we can observe an improvement of 5% on the GTA → CS and 1% on the SYN → CS benchmark. Using only dropout as perturbations improves the results for 8 − 9% on the GTA → CS and 5% on the SYN → CS benchmark. The same applies for the CowMask image mixing perturbation, which improves the result even more. Thus, the use of dropout or image mixing as perturbations leads to a stronger improvement than the use of pseudo-labels in general. Combining both perturbations gives a slight improvement in comparison to both perturbations alone. In general for the GTA → CS benchmark, we improve the result by 17% compared to the baseline, while 12% come alone from the used perturbations. This shows that the used perturbations are indeed the key of the success of such methods.

We also study the effect of different dropout and EMA score rates in Fig. 2. For the dropout experiment, the image mixing perturbation is not used, but it is for the EMA experiment. It can be seen, that different EMA values hardly have an effect on the final performance. Higher values achieve a slightly better result. It should also be noted that higher EMA values lead to more stable

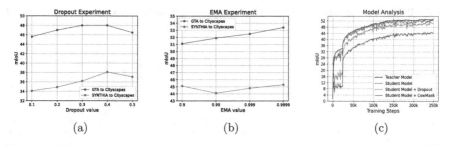

Fig. 2. (a), (b) Ablation study for different dropout and EMA decay values for the GTA → CS and the SYNTHIA → CS benchmark. (c) Performance of the teacher model, the student model and the student model with the proposed perturbations during training. The experiments were performed with a VGG16 as backbone. Best viewed in colour.

training and the results are much better reproducible. For the dropout values, the performance increases up to 0.4, then decreases again slightly. This may be due to the fact that at very high dropout values too much information is lost.

To explain the effect of perturbations on the final performance, we conducted an experiment where we continuously evaluate the teacher model and the student model with and without our proposed perturbations. The result is shown in Fig. 2c. It can be seen, that the teacher model is always better as the student model impaired by perturbations. That means, that the feedback the student gets from the teacher during training is always on average better as its own predictions. As it receives a positive feedback, the model may be able to continuously improve itself and learn better decision boundaries.

4.4 Comparisons to State-of-the-Art

For a fair comparison, we compare our method on both benchmarks with similar methods from the last two years that are primary based on consistency regularization for SSL in any kind. Note that not all published methods report results for the VGG16 and the ResNet101 backbone. We report results using hard pseudo-labels and both perturbations. The results are shown in Table 2 and Table 3. For the VGG backbone, we compare our method with two approaches that combine SSL with additional image-to-image translation [5,30]. For the ResNet101 backbone, we compare our method with four state-of-the art methods that utilize SSL [1,12,15,25]. We can see that our much simpler approach achieves a substantial improvement on both benchmarks and achieves the best results for the GTA → CS benchmark with both backbones. At the SYN → CS benchmark we achieve the second-best results, while SAD [1] performs best. However, we believe that this comes mainly from their additional class balanced training and importance sampling, where they increase the sample frequency of certain classes during training, as they reported a drop from 49.9% to 44.5% at the GTA → CS benchmark when not using it. Specially importance sampling of rare classes is a typical approach of semantic segmentation applications in general and not

Table 2. Results on the GTA5 → Cityscapes benchmark. We compare our method using the VGG16 (A) and the ResNet-101 (B) backbone.

GTA5 → Cityscapes

	Model	road	side.	buil.	wall	fence	pole	t-light	t-sign	vege.	terr.	sky	pers.	rider	car	truck	bus	train	motor	bike	mIoU
Baseline	A	88.0	39.8	79.3	27.5	7.4	26.0	25.9	11.4	81.3	24.6	67.0	52.0	12.8	81.5	20.8	12.2	0.0	8.3	0.7	35.2
Choi et al. [5]	A	90.2	51.5	81.1	15.0	10.7	37.5	35.2	28.9	84.1	32.7	75.9	62.7	19.9	82.6	22.9	28.3	0.0	23.0	25.4	42.5
Zhou et al. [30]	A	95.1	66.5	84.7	35.1	19.8	31.2	35.0	32.1	86.2	43.4	82.5	61.0	25.1	87.1	35.3	46.1	0.0	24.6	17.5	47.8
SAD [1] (w/o CBT-IS-FL)	A	88.1	41.0	85.7	30.8	30.6	33.1	37.0	22.9	86.6	36.8	90.7	67.1	27.1	86.8	34.4	30.4	8.5	7.5	0.0	44.5
SAD [1]	A	90.0	53.1	86.2	33.8	32.7	38.2	46.0	40.3	84.2	26.4	88.4	65.8	28.0	85.6	40.6	52.9	17.3	13.7	23.8	49.9
Ours	A	93.6	58.7	88.4	41.3	40.6	33.9	47.4	59.5	85.0	37.4	86.0	57.7	33.9	86.7	38.7	53.5	24.9	42.7	4.1	**53.4**
Baseline	B	89.1	41.0	81.9	31.0	5.3	28.7	27.9	14.4	82.3	28.9	84.9	51.7	12.5	81.5	21.6	15.9	0.0	5.1	0.2	37.0
MLSL [12]	B	89.0	45.2	78.2	22.9	27.3	37.4	46.1	43.8	82.9	18.6	61.2	60.4	26.7	85.4	35.9	44.9	36.4	37.2	49.3	49.0
PixMatch [15]	B	91.6	51.2	84.7	37.3	29.1	24.6	31.3	37.2	86.5	44.3	85.3	62.8	22.6	87.6	38.9	52.3	0.65	37.2	50.0	50.3
DACS [25]	B	89.9	39.7	87.9	30.7	39.5	38.5	46.4	52.8	88.0	44.0	88.8	67.2	35.8	84.5	45.7	50.2	0.0	27.3	34.0	52.1
SAD [1]	B	90.4	53.9	86.6	42.4	27.3	45.1	48.5	42.7	87.4	40.1	86.1	67.5	29.7	88.5	49.1	54.6	9.8	26.6	45.3	53.8
Ours	B	95.1	65.1	88.3	46.6	28.2	36.5	44.6	49.0	86.9	42.0	89.0	64.1	30.9	89.7	53.0	64.6	0.0	25.4	49.0	**55.2**

specific to UDA. Compared to PixMatch [15] that is mostly related to our work, we observe a substantial improvement on both benchmarks.

We further evaluate our proposed solution for multi-source or multi-target domain adaptation problems. At these experiments, we simply merge the different domains or datasets into one. Table 4a shows that the performance increases when we simply combine different synthetic datasets. Combining GTA5 and SYNTHIA, we can improve the performance from 55.2% using GTA5 only to 59.9%. Including Synscapes as well, we can achieve a performance of 63.3%. This shows that using multiple different synthetic datasets has a positive effect on the performance, and that by exploring and combining more advanced synthetic datasets, we could achieve similar results to a model trained fully supervised on Cityscapes. Table 4b shows the result for the CS → ACDC benchmark, where our proposed method is able to improve the performance on different visual conditions. It should also be mentioned that we did not observe any deterioration in the performance of the adapted model at the Cityscapes validation set. In this context, the *Source-only* model is trained fully supervised on Cityscapes, while the *Oracle* model used the labelled training set of Cityscapes and the ACDC dataset as supervision. Compared to the *Source-only* model, we can improve the performance relative to the *Oracle* model by nearly 55%. Since the ACDC dataset contains only 400 images per visual condition at the training set, we believe that a larger unlabelled dataset can further improve the results, making the annotation process unnecessary in this case. The ability of SSL to use unlabelled data is not exploited in this experiment, as both our UDA trained model as well as the *Oracle* see the same images during training.

Table 3. Results on the SYNTHIA → Cityscapes benchmark. We compare our method using the VGG16 (A) and the ResNet-101 (B) backbone.

SYNTHIA → Cityscapes

	Model	road	side.	buil.	wall*	fence*	pole*	t-light	t-sign	vege.	sky	pers.	rider	car	bus	motor	bike	mIoU[16]	mIoU[13]
Baseline	A	44.8	19.6	64.7	3.1	0.1	26.0	4.8	12.7	75.9	75.5	46.4	12.3	65.1	15.7	10.0	18.5	31.0	35.9
Choi et al. [5]	A	90.1	48.6	80.7	2.2	0.2	27.2	3.2	14.3	82.1	78.4	54.4	16.4	82.5	12.3	1.7	21.8	38.5	46.6
Zhou et al. [30]	A	93.1	53.2	81.1	2.6	0.6	29.1	7.8	15.7	81.7	81.6	53.6	20.1	82.7	22.9	7.7	31.3	41.5	48.6
SAD [1]	A	77.9	38.6	83.5	15.8	1.5	38.2	41.3	27.9	80.8	83.0	64.3	21.2	78.3	38.5	32.6	62.1	**49.1**	**56.2**
Ours	A	65.0	25.6	81.9	19.1	0.0	31.1	1.3	40.9	79.1	82.4	61.5	27.9	86.5	61.4	14.6	37.7	44.7	51.2
Baseline	B	52.5	20.6	72.8	3.0	0.0	27.6	0.0	6.8	78.8	78.7	42.7	15.8	67.7	18.5	8.6	18.6	32.1	37.2
MLSL [12]	B	59.2	30.2	68.5	22.9	1.0	36.2	32.7	28.3	86.2	75.4	68.6	27.7	82.7	26.3	24.3	52.7	45.2	51.0
PixMatch [15]	B	92.5	54.6	79.8	4.8	0.1	24.1	22.8	17.8	79.4	76.5	60.8	24.7	85.7	33.5	26.4	54.4	46.1	54.5
DACS [25]	B	80.6	25.1	81.9	21.5	2.9	37.2	22.7	24.0	83.7	90.8	67.6	38.3	82.9	38.9	28.5	47.6	48.3	54.8
SAD [1]	B	89.3	47.2	85.5	26.5	1.3	43.0	45.5	32.0	87.1	89.3	63.6	25.4	86.9	35.6	30.4	53.0	**52.6**	**59.3**
Ours	B	89.0	53.6	85.0	23.7	3.2	34.4	6.0	41.3	82.2	80.6	54.1	39.0	86.2	68.1	25.6	45.1	51.1	58.1

Table 4. Results for multi-source and multi-target domain adaptation.

	Sources	mIoU[19]
Single Source DA	G	55.2
	S	44.3
	C	55.4
Multi Source DA	G+S	59.9
	G+S+C	63.3

(a) Results on the Cityscapes validation set combining different source domains. G: GTA5, S: SYNTHIA, C:Synscapes.

Method	mIoU[19]
Source-only	50.6
Ours	59.1
Oracle	66.3

(b) Results on the ACDC test set for the Cityscapes → ACDC benchmark using the ResNet101 backbone.

5   Conclusion

In this work, we investigated the problem of unsupervised domain adaptation for semantic segmentation. To address this problem, we presented the use of an approach for consistency regularization combined with perturbations on the input image as well as the model itself. Through a comprehensive series of ablation studies, we have sought to understand which aspects of this approach are most important to the final performance of the model. We were able to show that the type of perturbation is the key to success. Even a simple perturbation such as dropout is able to improve the performance of the model by a large margin. Combined with an image mixing method, the approach is able to achieve state-of-the-art results. Future work may explore the combination of other existing perturbation functions.

References

1. Araslanov, N., Roth, S.: Self-supervised augmentation consistency for adapting semantic segmentation. In: Proceedings of the IEEE/CVF Conference on Computer Vision and Pattern Recognition, pp. 15384–15394 (2021)

2. Brehm, S., Scherer, S., Lienhart, R.: Semantically consistent image-to-image translation for unsupervised domain adaptation. In: 2022 International Conference on Agents and Artificial Intelligence (2022)

3. Chen, L.C., Papandreou, G., Kokkinos, I., Murphy, K., Yuille, A.L.: DeepLab: semantic image segmentation with deep convolutional nets, Atrous convolution, and fully connected CRFs. IEEE Trans. Pattern Anal. Mach. Intell. **40**(4), 834–848 (2017)

4. Chen, T., Kornblith, S., Norouzi, M., Hinton, G.: A simple framework for contrastive learning of visual representations. In: International Conference on Machine Learning, pp. 1597–1607. PMLR (2020)

5. Choi, J., Kim, T., Kim, C.: Self-ensembling with GAN-based data augmentation for domain adaptation in semantic segmentation. In: Proceedings of the IEEE/CVF International Conference on Computer Vision, pp. 6830–6840 (2019)

6. Cordts, M., et al.: The cityscapes dataset for semantic urban scene understanding. In: Proceedings of the IEEE Conference on Computer Vision and Pattern Recognition, pp. 3213–3223 (2016)

7. Deng, J., Dong, W., Socher, R., Li, L.J., Li, K., Fei-Fei, L.: ImageNet: a large-scale hierarchical image database. In: 2009 IEEE Conference on Computer Vision and Pattern Recognition, pp. 248–255. IEEE (2009)

8. French, G., Oliver, A., Salimans, T.: Milking CowMask for semi-supervised image classification. arXiv preprint arXiv:2003.12022 (2020)

9. French, G., Laine, S., Aila, T., Mackiewicz, M., Finlayson, G.: Semi-supervised semantic segmentation needs strong, varied perturbations. In: British Machine Vision Conference (2020)

10. He, K., Zhang, X., Ren, S., Sun, J.: Deep residual learning for image recognition. In: Proceedings of the IEEE Conference on Computer Vision and Pattern Recognition, pp. 770–778 (2016)

11. Hoffman, J., et al.: Cycada: cycle-consistent adversarial domain adaptation. In: International Conference on Machine Learning, pp. 1989–1998. PMLR (2018)

12. Iqbal, J., Ali, M.: MLSL: multi-level self-supervised learning for domain adaptation with spatially independent and semantically consistent labeling. In: Proceedings of the IEEE/CVF Winter Conference on Applications of Computer Vision, pp. 1864–1873 (2020)

13. Lee, D.H.: Pseudo-label: the simple and efficient semi-supervised learning method for deep neural networks. In: ICML 2013 Workshop: Challenges in Representation Learning (WREPL) (July 2013)

14. Li, Y., Yuan, L., Vasconcelos, N.: Bidirectional learning for domain adaptation of semantic segmentation. In: Proceedings of the IEEE/CVF Conference on Computer Vision and Pattern Recognition, pp. 6936–6945 (2019)

15. Melas-Kyriazi, L., Manrai, A.K.: Pixmatch: unsupervised domain adaptation via pixelwise consistency training. In: Proceedings of the IEEE/CVF Conference on Computer Vision and Pattern Recognition, pp. 12435–12445 (2021)

16. Richter, S.R., Vineet, V., Roth, S., Koltun, V.: Playing for data: ground truth from computer games. In: Leibe, B., Matas, J., Sebe, N., Welling, M. (eds.) ECCV 2016, Part II. LNCS, vol. 9906, pp. 102–118. Springer, Cham (2016). https://doi.org/10.1007/978-3-319-46475-6_7

17. Ros, G., Sellart, L., Materzynska, J., Vazquez, D., Lopez, A.M.: The Synthia dataset: a large collection of synthetic images for semantic segmentation of urban scenes. In: Proceedings of the IEEE Conference on Computer Vision and Pattern Recognition, pp. 3234–3243 (2016)

18. Sakaridis, C., Dai, D., Van Gool, L.: ACDC: the adverse conditions dataset with correspondences for semantic driving scene understanding. In: Proceedings of the IEEE/CVF International Conference on Computer Vision, pp. 10765–10775 (2021)
19. Scherer, S., Schön, R., Ludwig, K., Lienhart, R.: Unsupervised domain extension for nighttime semantic segmentation in urban scenes. In: 2021 International Conference on Deep Learning Theory and Applications (2021)
20. Sheng, L., Lin, Z., Shao, J., Wang, X.: Avatar-net: multi-scale zero-shot style transfer by feature decoration. In: Proceedings of the IEEE Conference on Computer Vision and Pattern Recognition, pp. 8242–8250 (2018)
21. Simonyan, K., Zisserman, A.: Very deep convolutional networks for large-scale image recognition. arXiv preprint arXiv:1409.1556 (2014)
22. Tarvainen, A., Valpola, H.: Mean teachers are better role models: weight-averaged consistency targets improve semi-supervised deep learning results. arXiv preprint arXiv:1703.01780 (2017)
23. Toldo, M., Maracani, A., Michieli, U., Zanuttigh, P.: Unsupervised domain adaptation in semantic segmentation: a review. Technologies **8**(2), 35 (2020)
24. Tompson, J., Goroshin, R., Jain, A., LeCun, Y., Bregler, C.: Efficient object localization using convolutional networks. In: Proceedings of the IEEE Conference on Computer Vision and Pattern Recognition, pp. 648–656 (2015)
25. Tranheden, W., Olsson, V., Pinto, J., Svensson, L.: DACS: domain adaptation via cross-domain mixed sampling. In: Proceedings of the IEEE/CVF Winter Conference on Applications of Computer Vision, pp. 1379–1389 (2021)
26. Wrenninge, M., Unger, J.: Synscapes: a photorealistic synthetic dataset for street scene parsing. arXiv preprint arXiv:1810.08705 (2018)
27. Xie, Q., Hovy, E.H., Luong, M., Le, Q.V.: Self-training with noisy student improves ImageNet classification. CoRR abs/1911.04252 (2019). http://arxiv.org/abs/1911.04252
28. Yang, Y., Soatto, S.: FDA: Fourier domain adaptation for semantic segmentation. In: Proceedings of the IEEE/CVF Conference on Computer Vision and Pattern Recognition, pp. 4085–4095 (2020)
29. Yun, S., Han, D., Oh, S.J., Chun, S., Choe, J., Yoo, Y.: CutMix: regularization strategy to train strong classifiers with localizable features. In: Proceedings of the IEEE/CVF International Conference on Computer Vision, pp. 6023–6032 (2019)
30. Zhou, Q., Feng, Z., Cheng, G., Tan, X., Shi, J., Ma, L.: Uncertainty-aware consistency regularization for cross-domain semantic segmentation. arXiv preprint arXiv:2004.08878 (2020)
31. Zou, Y., Yu, Z., Vijaya Kumar, B.V.K., Wang, J.: Unsupervised domain adaptation for semantic segmentation via class-balanced self-training. In: Ferrari, V., Hebert, M., Sminchisescu, C., Weiss, Y. (eds.) ECCV 2018, Part III. LNCS, vol. 11207, pp. 297–313. Springer, Cham (2018). https://doi.org/10.1007/978-3-030-01219-9_18
32. Zou, Y., Yu, Z., Liu, X., Kumar, B., Wang, J.: Confidence regularized self-training. In: Proceedings of the IEEE/CVF International Conference on Computer Vision, pp. 5982–5991 (2019)

Towards an Efficient Facial Image Compression with Neural Networks

Maria Ausilia Napoli Spatafora[(✉)][ID], Alessandro Ortis[ID],
and Sebastiano Battiato[ID]

Department of Mathematics and Computer Science, University of Catania,
Catania, Italy
maria.napolispatafora@phd.unict.it, {ortis,battiato}@dmi.unict.it

Abstract. Digital images are more and more part of everyday life. Efficient compression methods are needed to reduce the disk-space usage for their storage and the bandwidth for their transmission while keeping the resolution and the visual quality of the reconstructed images as close to the original images as possible. Not all images have the same importance. The facial images are being extensively used in many applications (e.g., law enforcement, social networks) and require high efficient facial image compression schemes in order to not compromise face recognition and identification (e.g., for surveillance and security scenarios). For this reason, we propose a promising approach that consists of a custom loss that combines the two tasks of image compression and face recognition. The results show that our method compresses efficiently face images guaranteeing high perceptive quality and face verification accuracy.

Keywords: Convolutional autoencoder · Face images compression · Custom loss function

1 Introduction

The use of images is increasing in many different applications. Digital images require huge amounts of space for storage and large bandwidth for transmissions. Image compression is a solution and so it is a fundamental problem in computer vision and image processing, which finds several applications. Image compression techniques consist in two tasks [23]:

- coding: it reduces the number of bits required to represent an image by taking advantage of these redundancies.
- decoding or decompression: it is the inverse process; it is applied to the compressed data to get the reconstructed image.

It is very crucial to reduce the size of disk-space used as well as reduce the amount of internet bandwidth used for the transmission of images while keeping the resolution and the visual quality of the reconstructed image as close to the original image as possible. The signal-processing-oriented compression has

© The Author(s), under exclusive license to Springer Nature Switzerland AG 2022
S. Sclaroff et al. (Eds.): ICIAP 2022, LNCS 13231, pp. 512–523, 2022.
https://doi.org/10.1007/978-3-031-06427-2_43

a long history studded with algorithms that are still widely employed today (e.g., JPG). Nevertheless, the attainable performances have already achieved the maximum. For this reason, it is taking new approaches to image processing such as the Neural Networks due to the proliferation of specific hardware (e.g., GPUs) and the meaningful results in other computer vision fields. The increasing interest in image compression with Neural Networks is proven by international workshops and challenges on this subject and by human and financial efforts of lead companies in this area. Such is the case of NVIDIA Corporation that has recently developed a compression algorithm for video conferencing based on Neural Networks that is able to outperform competing methods [29]. In addition, the potential in this area has been demonstrating by profitable literature. During the year, autoencoders [1,24], Recurrent Neural Networks [25,26] and Generative Adversarial Networks [21] achieve good and promising performance.

Particularly, facial images are being extensively used and stored in large databases by social networks, web services, or various organizations such as states, law enforcement, schools, universities, and private companies. Thus, efficient storage of such images is very useful. For this reason, face-specific image compression schemes are developed. This motivated us to explore the possibility to build new architectures for images specifically addressed for compression of face images. Facial image compression can be regarded as a special application of general image compression but this task has the need of high efficient facial image compression algorithms (e.g., in surveillance system) in order to not compromise face recognition and identification that have been developing rapidly in recent years.

This drove us to propose a new approach for the compression of facial images through a custom loss function that joins efficiently the image compression need with the face recognition task.

The rest of the paper is organised as follow: in Sect. 2, we present related works on image compression with neural networks; Sect. 3 describes the proposed method; Sect. 4 reports experimental settings and results; Sect. 5 reports conclusions and future works.

2 Related Works

2.1 Image Compression

Image compression is an important research topic in the field of signal processing to achieve efficient image storage and transmission. During the past decades, many *ad hoc* algorithms and heuristics [13,14,19,28,32] have been proposed and some of them became common standards such as JPG.

Recently, computer vision is applied in many research fields showing high capabilities including image compression in order to obtain perceptual and disk-space performance better than those of previous approaches. Different neural approaches have been adopted such as Autoencoder [1,24], Recurrent Neural Network [25,26] and Generative Adversarial Network [21].

Autoencoder architectures have been adopted also for purposes that bring forward the compression image. Autoencoder has been employed to reduce the dimensionality of images [10], to convert images to compressed binary codes for retrieval [15] and to extract compact visual representations for several purposes [27]. More recently, the authors in [1] and [24] both exploited methods based on convolutional autoencoder architectures; Ballè et al. [1] used Generalized Divise Normalization (GDN) for joint nonlinearity and approximated the quantized values with additive uniform noise for entropy rate loss; while Theis et al. [24] employed a smooth approximation of the derivative of the rounding function and upper-bound the discrete entropy rate loss.

Works in [25] and [26] explored various transformations for binary features extraction based on different types of RNNs that adopt stochastic quantization, and finally, the binary representations are entropy-coded. Toderici et al. [25] proposed a RNN to compress 32×32 images; while, the work in [26] further introduced a set of full-resolution compression methods for progressive encoding and decoding of images.

Finally, GAN architectures [9] have been applied in the field of image compression, too. Santurkar et al. [21] described the concept of generative compression that is the compression of data using generative models; this approach produces accurate and visually pleasing reconstructions at much deep compression levels for image data.

2.2 Facial Image Compression

Although previous approaches have been widely used and designed for general purposes, facial images are specific objects and play a very important role in technology and science, military, medical and other industries. Facial images differ from other natural images and have specific prior knowledge as presented by [16] that analyzes statistically gradient features of facial images. The gradient features of facial images are compared with those of other natural images showing that they are different in symmetry and geometry structure. For this reason, there are also some face-specific image compression schemes such as works [3,7,20] that design dictionary-based coding schemes on this specific type of image. Liu et al. [17] proposed a PCA method to realize facial image compression extracting common features from this type of image in order to reduce redundancy to achieve compression.

In recent years, some approaches based on deep learning have been proposed to represent and compress facial images. Hu et al. [11] proposed a content-aware facial image compression method to compress facial images at low bit rates using Convolutional Autoencoder. Chen et al. [4] proposed a GAN architecture that is able to automatically optimize codec configuration according to the minimization of semantic distortion. Finally, Bian et al. [2] developed a Convolutional Autoencoder combined with a standard lossless image code PNG. This framework is followed by the measurement of performance for facial recognition on compressed images.

Differently from previous works, we propose a new approach for facial image compression that is able to simultaneously save enough space and ensure good performance in face recognition task. The method is summarised in the next section.

3   Proposed Method

In this section, we describe the developed convolutional autoencoder architecture for face image compression and the workflow of the algorithm.

The proposed architecture is made by three parts: the encoder (*Enc*), the decoder (*Dec*), and the binarizer (*Bin*). The overall pipeline of the process is:

1. *Enc* generates a compact representation of the input image for the encoding: it takes RGB PNG image of any size as input and produces a compressed representation in **xz**[1] format exploiting LZMA algorithm [8];
2. *Bin* is used for quantifying the output of *Enc*;
3. *Dec* is the mirror symmetry of *Enc* and reconstructs an image in PNG format from its binary representation.

The input of the binarizer - marked as x - is the output feature vector from the encoder part and consists of continuous floating-point numbers. The output of the binarizer - marked as $Bin(x)$ - is the final compressed data after the quantization step provided by the binarizer, and it is composed of discrete binary values. The binarizer follows the rule defined in Eq. (1), where i is the i-th element of encoding output vector and of P that is a vector x-like filled by values drawn from a uniform probability.

$$Bin(x[i]) = \begin{cases} -1, \text{if } \frac{1-x[i]}{2} <= P[i] \\ +1, \qquad \texttt{otherwise} \end{cases} \qquad (1)$$

The Fig. 1 shows the structure of the neural network employed. For the encoding process, input images pass through two convolutional structures containing a convolutional layer, followed by batch normalization and Rectified Linear Unit (ReLU) activation function each. The first convolutional structure contains 32 filters of size 8×8 (with stride of 4 and padding of 2) while the second contains 64 filters of size 2×2 (with stride of 2). After these two convolutional structures, the processed input image goes through the binarizer that consists of a single convolutional layer with 128 filters with a size of 1×1 followed by the hyperbolic tangent (tanh function). This processed data vector, finally, passes through the function detailed in Eq. (1). At this point, the network produces the compressed archive in **xz** format.

For the decoding process, input data pass through two transposed convolutional structures containing transposed convolutional layer, batch normalization,

---

[1] This acronym is a file format for LZMA archives.

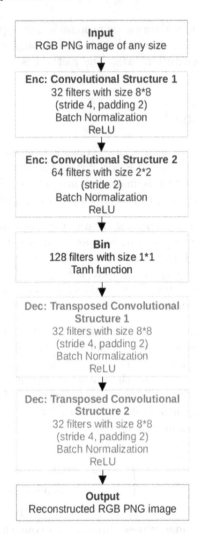

Fig. 1. Structure of the employed neural network.

and ReLU activation function. The first transposed convolutional structure contains 32 filters of size 8×8 (with stride of 4 and padding of 2) while the second contains 3 filters of size 2×2. The output is the image in PNG format.

We choose to use batch normalization after convolution and transposed convolution. While we adopt the ReLU activation function.

The architecture becomes specific for facial image compression with a custom loss function whose goal is to bind the task of image compression and the task of face recognition. The custom loss is a linear combination of two losses: the one works on pixels space and the other works on features loss

$$\text{custom loss} = a \cdot \text{pixel loss} + b \cdot \text{feature loss} \qquad (2)$$

Here, a and b are the weights assigned to the two parts of the loss. The pixel loss aims to obtain a reconstructed image more and more similar to the original. While, the feature loss has the goal of achieving a face-embedding of reconstructed image more and more near to that of the original. We select FaceNet [22] as the face-embedding. It is a neural-network-based tool[2] that maps face to a compact Euclidean space. Such space amplifies distances of faces from distinct people, while reduce distances of faces from the same person. This model is pre-trained with triplet loss and center loss.

4   Experimental Settings and Results

In this section, we will introduce the datasets and the conducted experiments and comparisons for facial compression. The results are encouraging and promising.

The training of the proposed convolutional autoencoder happens in two phases:

1. *Image Compression*: the above network is first trained on dataset provided by the "3rd Workshop and Challenge on Learned Image Compression" [5] that has taken place during the "CVPR 2020" [6] (the Fig. 2 shows some images from this dataset). The organizers of this workshop and challenge, provide a set of images which represent a very realistic view of the types of images that are widely available nowadays to ensure realism. This dataset contains 568 images. For the purpose of the training, we adopt the L1 loss, the Multi Scale Structural Similarity (MS-SSIM) as learning metric (it is a convenient way to incorporate image details at different resolutions [30] and it is suggested by the challenge [5]) and the following hyper-parameters: dynamic learning rate 0.0001; momentum 0.99. Finally, we use Adam optimizer.
2. *Face Image Compression*: it is used the model pre-trained in the previous phase. The employed dataset is CelebA aligned [18] converted to PNG. CelebA contains 10177 number of identities and 202599 number of facial images. We eliminated the faces that cannot be detected by MTCNN [31]. Here, the combined loss (Eq. 2) is employed together with MTCNN-FaceNet pipeline for the features-space loss in the following way:
   - MTCNN provides the bounding-boxes of the face in the groundtruth image;
   - these bounding-boxes are used to extract faces in the groundtruth image and in the output image of the network;
   - the patches corresponding to the faces are fed to FaceNet that computes two face-embeddings;
   - the face-embeddings are compared with the features-space loss.

   For this loss, we adopt L2 Loss for both parts and the weights $a = b = 1$. The other hyperparameters are static learning rate 0.003, momentum 0.99, weight decay 0.001 and we use Adam as optimizer.

Fig. 2. Some images from dataset provided by the "3rd Workshop and Challenge on Learned Image Compression" [5]

After the training of our autoencoder, we test it on another dataset containing only faces. For this purpose, we choose the Labeled Faces in the Wild (LFW) [12]. This benchmark dataset contains 13233 face images of 5749 people with a picture size of 112×96.

---

[2] It is available here https://github.com/timesler/facenet-pytorch.

Table 1. Results of the image compression task with our proposed approach about LFW dataset.

	Average	Maximum	Minimum
bpp	0.761	0.864	0.507
Compression ratio	5.892	8.065	3.336
PSNR	30.905	35.064	28.671
MS-SSIM	0.989	0.998	0.857

In order to properly evaluate the validity of the method, we use on the one hand compression ratio, bit-per-pixels (bpp), Peak signal-to-noise ratio (PSNR), and MS-SSIM as metrics for the image compression task, and on the other average cosine similarity (1 indicates two vectors with the same orientation; -1 indicates two vectors diametrically opposed) and face verification through 1-NN with the cosine distance as metric for the face recognition task. We compare our results in the face recognition task with the JPG to have a point of reference. Moreover, we compare our method injected with the custom loss (Eq. 2) with the same method without the custom loss. All these results are reported in the Tables 1, 2 respectively for the image compression task and for the face recognition task with the performances of JPG at different compression quality levels. The results prove the validity of the proposed loss that improves our performance enough to reach JPG's performance. Actually, the custom loss function decreases the number of faces not detected in reconstructed images and increases both the average cosine similarity and the face verification accuracy. It should be noted that JPG is a little bit better than our method with custom loss only for the face verification accuracy. The Fig. 3 shows some faces compressed with our method and JPG (quality factor 80) proving that our method with the custom loss preserves good perceptive quality of the images. These results releave interesting insights in the research topic of the facial image compression with neural networks.

Table 2. Results of the face recognition task after compression and comparison with the approach without the custom loss and JPG about LFW dataset.

	Compressed dataset (MB)	Average cosine similarity	Face verification accuracy	Faces not detected
Our method without custom loss	34	0.422	0.306	68
Our method with custom loss	40	0.801	0.944	1
JPG (quality factor 70)	35	0.768	0.986	2
JPG (quality factor 80)	41	0.780	0.986	2
JPG (quality factor 90)	56	0.793	0.990	2

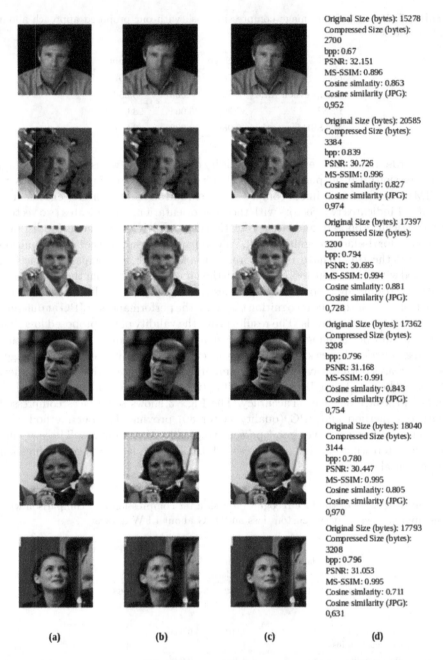

Original Size (bytes): 15278
Compressed Size (bytes): 2700
bpp: 0.67
PSNR: 32.151
MS-SSIM: 0.896
Cosine simlarity: 0.863
Cosine similarity (JPG): 0,952

Original Size (bytes): 20585
Compressed Size (bytes): 3384
bpp: 0.839
PSNR: 30.726
MS-SSIM: 0.996
Cosine simlarity: 0.827
Cosine similarity (JPG): 0,974

Original Size (bytes): 17397
Compressed Size (bytes): 3200
bpp: 0.794
PSNR: 30.695
MS-SSIM: 0.994
Cosine simlarity: 0.881
Cosine similarity (JPG): 0,728

Original Size (bytes): 17362
Compressed Size (bytes): 3208
bpp: 0.796
PSNR: 31.168
MS-SSIM: 0.991
Cosine simlarity: 0.843
Cosine similarity (JPG): 0,754

Original Size (bytes): 18040
Compressed Size (bytes): 3144
bpp: 0.780
PSNR: 30.447
MS-SSIM: 0.995
Cosine simlarity: 0.805
Cosine similarity (JPG): 0,970

Original Size (bytes): 17793
Compressed Size (bytes): 3208
bpp: 0.796
PSNR: 31.053
MS-SSIM: 0.995
Cosine simlarity: 0.711
Cosine similarity (JPG): 0,631

(a)   (b)   (c)   (d)

Fig. 3. (a) Original image; (b) Compressed and reconstructed image by our convolutional autoencoder; (c) Image in JPG with quality factor 80; (c) Metrics.

5   Conclusions

In this paper, we explore the problem of image compression for a specific type of image such as those containing only faces. Facial image compression is a significant topic because surveillance and security scenarios need efficient compression algorithms to not compromise facial recognition and identification. We develop simplistic convolutional autoencoder architecture that is trained first on images of various genres and next on facial images after the injection of a custom loss that is able to guarantee the face recognition and face verification. Experimental results validate our proposed method achieved a state-of-the-art performance compared to the coding standard of JPG from the point of view of objective quality metrics and face verification task.

The experiments reported in this paper represent a first investigation attempt on this research topic, which outcomes revealed insights suggesting that there is room for improvements in this research direction. Moreover, there are not many studies in literature for facial image compression and those existing do not have a unique touchstone that makes a fair comparison difficult with the state-of-the-art for the face image compression field. For this reason, we are planning to build a benchmark dataset of facial images with the indication of different quality factors of some algorithms (e.g., JPG), and the face recognition performance with some face embedding algorithms (e.g., FaceNet). Future works will be also devoted to the improvement of the compression ratio preserving the face recognition performance in order to outperform JPG. This could be done with a more complex neural network architecture characterized by residual blocks and entropy coding that have already been applied in the image compression topic.

References

1. Ballé, J., Laparra, V., Simoncelli, E.: End-to-end optimized image compression. In: Proceedings of the International Conference on Learning Representation (ICLR) (November 2017)
2. Bian, N., Liang, F., Fu, H., Lei, B.: A deep image compression framework for face recognition. In: 2019 2nd China Symposium on Cognitive Computing and Hybrid Intelligence (CCHI), pp. 99–104 (2019). https://doi.org/10.1109/CCHI.2019.8901914
3. Bryt, O., Elad, M.: Compression of facial images using the K-SVD algorithm. J. Vis. Commun. Image Represent. **19**(4), 270–282 (2008). https://doi.org/10.1016/j.jvcir.2008.03.001
4. Chen, Z., He, T.: Learning based facial image compression with semantic fidelity metric. Neurocomputing **338**, 16–25 (2019). https://doi.org/10.1016/j.neucom.2019.01.086
5. CLIC: CLIC, Workshop and Challenge on Learned Image Compression. http://www.compression.cc/. Accessed 26 Apr 2021
6. CVPR: CVPR 2020. http://cvpr2020.thecvf.com/. Accessed 26 Apr 2021
7. Elad, M., Goldenberg, R., Kimmel, R.: Low bit-rate compression of facial images. IEEE Trans. Image Process. **16**, 2379–2383 (2007)

8. Epiphany, J.L., Danasingh, A.A.: Hardware implementation of LZMA data compression algorithm. Int. J. Appl. Inf. Syst. **5**, 52–56 (2013)
9. Goodfellow, I., et al.: Generative adversarial networks. In: Proceedings of the 27th International Conference on Neural Information Processing Systems (ICONIP), vol. 63, pp. 2672–2680 (2014)
10. Hinton, G.E., Salakhutdinov, R.: Reducing the dimensionality of data with neural networks. Science **313**, 504–507 (2006)
11. Hu, S., Duan, Y., Tao, X., Liu, Y., Zhang, X., Lu, J.: Content-aware facial image compression with deep learning method. In: 2020 International Conference on Wireless Communications and Signal Processing (WCSP), pp. 516–521 (2020). https://doi.org/10.1109/WCSP49889.2020.9299680
12. Huang, G.B., Ramesh, M., Berg, T., Learned-Miller, E.: Labeled faces in the wild: a database for studying face recognition in unconstrained environments. Tech. rep., University of Massachusetts, Amherst (2007)
13. Hurtik, P., Perfilieva, I.: A hybrid image compression algorithm based on jpeg and fuzzy transform. In: 2017 IEEE International Conference on Fuzzy Systems (FUZZ-IEEE), pp. 1–6 (2017). https://doi.org/10.1109/FUZZ-IEEE.2017.8015614
14. Jassim, F., Qassim, H.: Five modulus method for image compression. Signal Image Process.: Int. J. **3**, 19–28 (2012). https://doi.org/10.5121/sipij.2012.3502
15. Krizhevsky, A., Hinton, G.: Using very deep autoencoders for content-based image retrieval. In: Proceedings of the 19th European Symposium on Artificial Neural Networks (ESANN) (2011)
16. Liu, X., Gan, Z., Liu, F.: Hierarchical subspace regression for compressed face image restoration. In: 2018 10th International Conference on Wireless Communications and Signal Processing (WCSP), pp. 1–6 (2018). https://doi.org/10.1109/WCSP.2018.8555682
17. Liu, Y., Kau, L.: Scalable face image compression based on principal component analysis and arithmetic coding. In: 2017 IEEE International Conference on Consumer Electronics - Taiwan (ICCE-TW), pp. 265–266 (2017). https://doi.org/10.1109/ICCE-China.2017.7991097
18. Liu, Z., Luo, P., Wang, X., Tang, X.: Deep learning face attributes in the wild. In: 2015 IEEE International Conference on Computer Vision (ICCV), pp. 3730–3738 (2015). https://doi.org/10.1109/ICCV.2015.425
19. Rabbani, M., Joshi, R.: An overview of the JPEG 2000 still image compression standard. Signal Process.: Image Commun. **17**, 3–48 (2002)
20. Ram, I., Cohen, I., Elad, M.: Facial image compression using patch-ordering-based adaptive wavelet transform. IEEE Signal Process. Lett. **21**(10), 1270–1274 (2014). https://doi.org/10.1109/LSP.2014.2332276
21. Santurkar, S., Budden, D., Shavit, N.: Generative compression. In: Proceedings of the Picture Coding Symposium (PCS) (2017)
22. Schroff, F., Kalenichenko, D., Philbin, J.: FaceNet: a unified embedding for face recognition and clustering. In: 2015 IEEE Conference on Computer Vision and Pattern Recognition (CVPR), pp. 815–823 (2015). https://doi.org/10.1109/CVPR.2015.7298682
23. Subramanya, A.: Image compression technique. IEEE Potentials **20**(1), 19–23 (2001). https://doi.org/10.1109/45.913206
24. Theis, L., Shi, W., Cunningham, A., Huszár, F.: Lossy image compression with compressive autoencoders. In: Proceedings of the International Conference on Learning Representation (ICLR) (2017)

25. Toderici, G., et al.: Variable rate image compression with recurrent neural networks. In: Proceedings of the International Conference on Learning Representation (ICLR) (2016)
26. Toderici, G., et al.: Full resolution image compression with recurrent neural networks. In: Proceedings of the Computer Vision and Pattern Recognition (CVPR), pp. 5435–5443 (2017). https://doi.org/10.1109/CVPR.2017.577
27. Vincent, P., Larochelle, H., Bengio, Y., Manzagol, P.: Extracting and composing robust features with denoising autoencoders. In: Proceedings of the 25th International Conference on Machine Learning, pp. 1096–1103 (2008). https://doi.org/10.1145/1390156.1390294
28. Wallace, G.K.: The JPEG still picture compression standard. IEEE Trans. Consum. Electron. **38**(1), xviii–xxxiv (1992). https://doi.org/10.1109/30.125072
29. Wang, T., Mallya, A., Liu, M.: One-shot free-view neural talking-head synthesis for video conferencing. In: Proceedings of the IEEE Conference on Computer Vision and Pattern Recognition (2021)
30. Wang, Z., Simoncelli, E., Bovik, A.: Multiscale structural similarity for image quality assessment. In: Proceedings of the Thirty-Seventh Asilomar Conference on Signals, Systems and Computers, vol. 2, pp. 1398–1402 (2003). https://doi.org/10.1109/ACSSC.2003.1292216
31. Zhang, K., Zhang, Z., Li, Z., Qiao, Y.: Joint face detection and alignment using multitask cascaded convolutional networks. IEEE Signal Process. Lett. **23**(10), 1499–1503 (2016). https://doi.org/10.1109/LSP.2016.2603342
32. Zhang, Y., Cai, Z., Xiong, G.: A new image compression algorithm based on non-uniform partition and u-system. IEEE Trans. Multimed. **23**, 1069–1082 (2021). https://doi.org/10.1109/TMM.2020.2992940

Avalanche RL: A Continual Reinforcement Learning Library

Nicoló Lucchesi[✉], Antonio Carta, Vincenzo Lomonaco, and Davide Bacciu

Deparment of Computer Science, University of Pisa, Pisa, Italy
nicolo.lucchesi@gmail.com, antonio.carta@di.unipi.it,
{vincenzo.lomonaco,bacciu}@unipi.it

Abstract. Continual Reinforcement Learning (CRL) is a challenging setting where an agent learns to interact with an environment that is constantly changing over time (the stream of *experiences*). In this paper, we describe `Avalanche RL`, a library for Continual Reinforcement Learning which allows users to easily train agents on a continuous stream of tasks. `Avalanche RL` is based on PyTorch [23] and supports any OpenAI Gym [4] environment. Its design is based on Avalanche [16], one of the most popular continual learning libraries, which allow us to reuse a large number of continual learning strategies and improve the interaction between reinforcement learning and continual learning researchers. Additionally, we propose Continual Habitat-Lab, a novel benchmark and a high-level library which enables the usage of the photorealistic simulator Habitat-Sim [28] for CRL research. Overall, `Avalanche RL` attempts to unify under a common framework continual reinforcement learning applications, which we hope will foster the growth of the field.

Keywords: Continual learning · Reinforcement learning · Reproducibility

1 Introduction

Recent advances in data-driven algorithms, the so-called Deep Learning revolution, has shown the possibility for AI algorithms to achieve unprecedented performances on a narrow set of specific tasks. On the contrary, humans are able to quickly learn new tasks and generalize to novel scenarios. Continual Learning (CL) in the same way seeks to develop data-driven algorithms able to incrementally learn behaviors from a stream of data. Reinforcement Learning (RL) is yet another Machine Learning paradigm which formulates the learning process as a sequence of interactions between an agent and the environment. The agent must learn off of this interaction how to achieve a goal of a particular task by taking actions in the environment while receiving a (scalar) reward. Continual Reinforcement Learning (CRL) combines the non-stationarity assumption of a stream of data with the RL setting, having an agent learn multiple tasks in sequence.

S. Sclaroff et al. (Eds.): ICIAP 2022, LNCS 13231, pp. 524–535, 2022.
https://doi.org/10.1007/978-3-031-06427-2_44

While still in its early stages, CRL has seen a rising interest in publications in recent years (according to Dimensions [10] data). To support this growth, we focus on benchmarks and tools, introducing AvalancheRL: we extend Avalanche [16], the staple framework for Continual or Lifelong Learning, to support Reinforcement Learning in order to seamlessly train agents on a continuous stream tasks.

Existing RL libraries [6,21,24,25] do not focus on lifelong applications and force users to write custom code to develop continual solutions. Avalanche gives us re-usability by providing pre-implemented CL strategies as well as code structure when experimenting with them, but lacked support altogether when coming to RL. Related CRL projects instead either focus on providing a specific benchmark [33] or combine multiple frameworks results [22], limiting the overall flexibility and methods customization options.

`Avalanche RL` attempts to address both problems aiming to offer a malleable framework encompassing a variety of RL algorithms with fine-grained control over their internals, leveraging pre-existing CL techniques to learn efficiently from the interaction with multiple environments. In particular, we support any environment exposing the OpenAI Gym `gym.Env` interface.

The availability of compelling benchmarks has always lead the progress of data-driven algorithms [5,13,14], therefore our second effort is aimed at providing a challenging dataset for realistic Continual Reinforcement Learning.

Habitat-Lab allows an embodied agent to roam a photorealistic (typically indoor) scene in the attempt of solving a particular task; unfortunately, it does not offer support for the continual scenario. Therefore, we developed Continual Habitat-Lab, a high-level library enabling the usage of Habitat-Sim [28] for CRL, allowing the creation of sequences of tasks while integrating with `Avalanche RL`.

We first outline the design principles that guided the development of `Avalanche RL` (Sect. 2), describe its structure (Fig. 1) and go over the main features of the framework with code examples (Sect. 3). We then introduce Continual Habitat-Lab and describe its integration with `Avalanche RL` (Sect. 4).

All the source code of the work hereby presented is publicly available on GitHub for both `Avalanche RL`[1] and Continual Habitat-Lab[2].

2 Design Principles

`Avalanche RL` is built as an extension of Avalanche [16], and it retains the same design principles and a similar API. The target users are practitioners and researchers, and therefore the library must be simple, allowing to setup an experiment with a few lines of code, as well as highly customizable. As a result, `Avalanche RL` provides high-level APIs with ready-to-use components, as well as low-level features that allow heavy customization of existing implementations by leveraging an exhaustive callback system (Sect. 3.2).

---

[1] https://github.com/continualAI/avalanche-rl.
[2] https://github.com/NickLucche/continual-habitat-lab.

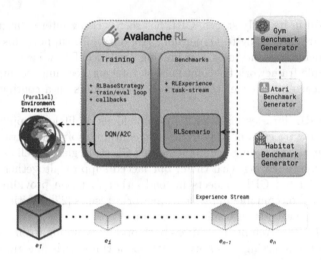

Fig. 1. Avalanche RL core-functionalities overview. The Benchmarks module capabilities, providing access to a stream of environments, are addressed in Sect. 3.1. Data is obtained through (parallel, Sect. 3.3) interaction with the stream and it is consumed by the algorithm in the learning process, as motivated in Sect. 3.2. Streams can be easily created through benchmark generators (right-hand side).

Avalanche RL codebase is comprises 5 main modules: **Benchmarks, Training, Evaluation, Models,** and **Logging.** We give a brief overview of them in the remainder of this section, but we refer the reader to [16] for more details about the general architecture of Avalanche.

Benchmarks maintains a uniform API for data handling, generating a *stream* of data from one or more datasets, conveniently divided into temporal *experiences*; this is the core abstraction over the *task stream* formalism which is distinctive of CL and it is accessible through a **Scenario** object. In order to create benchmarks more easily, this module provides *benchmark generators* which allow one to specify particular configurations through a simple API.

Training provides all the necessary utilities concerning model training. It includes simple and efficient ways of implementing new *strategies* as well as a set pre-implemented CL baselines and state-of-the-art algorithms. A **Strategy** abstracts a general learning algorithm implementing a training and an evaluation loop while consuming experiences from a *benchmark*. Continual behaviors can be added when needed through **Plugins**: they operate latching on the callback system defined by Strategies and are designed in such a modular way so that they can be easily composed to provide hybrid behaviors.

Evaluation provides all the utilities and metrics that can help evaluate a CL algorithm. Here we can find *pluggable* metric monitors such as (Train/Test/Batch)

Accuracy, RAM, CPU and GPU usage, all designed with the same modularity principles in mind.

Models contains several model architectures and pre-trained models that can be used for continual learning experiments (similar to `torchvision.models`), from simple customizable networks to implementation of state-of-the-art models.

Logging includes advanced logging and plotting features with the purpose of visualizing the metrics of the Evaluation module, such as highly readable output, file and `TensorBoard` support.

2.1 Notation

We adopt the well renowned notation from [32] for Reinforcement Learning related formulations while we make use of the formalization introduced in [15] regarding Continual Learning.

In particular, we refer to the RL problem as consisting of a tuple of five elements commonly denoted as $<\mathcal{S}, \mathcal{A}, \mathcal{R}, \mathcal{P}, \gamma>$ in the MDP formulation, where $\mathcal{S}$ and $\mathcal{A}$ are sets of **states** and **actions**, respectively. $\mathcal{R}$ or $r()$ is the **reward function**, with $r(s, a, s')$ being the expected immediate reward for transition from state $s \in \mathcal{S}$ to $s' \in \mathcal{S}$ under action $a \in \mathcal{A}$. $\mathcal{P}$ or $p()$ is the **transition function** defining the dynamics of the environment, with $p(s', r|s, a)$ denoting the probability of transitioning from s into s' with scalar reward r under a. Finally, γ represents the discount factor which weights the importance of immediate and future rewards.

An agent follows a policy π, which maps states to action probabilities. In Deep RL, learned policies are parameterized function (such as a neural network) which we indicate with π_θ.

We refer to a *Dataset* as a collection of samples $\{x_i\}_i^N$, optionally with labels $\{<x_i, y_i>\}_i^N$ in the case of supervised learning. We then denote a general task to be solved by some agent with τ and define the data relative to that task with D_τ.

3   Avalanche RL

CRL applications in `Avalanche RL` are implemented by modeling the interaction between core components: the task-stream abstraction (i.e., the continuously changing environment) and the RL strategy (i.e., the agent and its learning algorithm).

`Avalanche RL` implements these two components in the **Benchmarks** and **Training** module, respectively. In the remainder of this section, we describe the environment and the implementation of its continual shift in Sect. 3.1. Then, in Sect. 3.2, we describe the implementation of RL algorithms and their integration in the Training module. Section 3.3 and 3.4 highlight some important implementation details and useful features offered by the framework, such as the automatic parallelization of the RL environment.

3.1   Benchmarks: Stream of Environments

Most continual learning frameworks [15] assume that the stream of data is made of static datasets of a fixed size. Instead, in CRL problems the stream consists of different environments, and samples are obtained through the interaction between the agent and the environment.

To support streams of environments, `Avalanche RL` defines a stream $S = \{e_1, e_2, ..\}$ as a sequence of experiences e_i, where each experience provides access to an environment with which the agent can interact to generate state transitions (samples) online. Over time, this means that the agent learns by interacting with a stream of environments $\{\mathcal{E}_1, \mathcal{E}_2, ..\}$, as in Fig. 1. In the source code, `RLExperience` is the class which defines the CRL experience.

Using this **task-stream abstraction**, it is easy to define CRL benchmarks as a set of parallel streams of environments. Notice that each experience may be a small shift, such as a change in the background, as well as a completely different tasks, such as a different game. Different tasks may provide a task label which can be used by the agent to distinguish among them. The `RLScenario` is the class responsible for the CRL benchmark's definition, and it can be thought as a container of streams.

RL Environments implement a common interface, which is the one of OpenAI Gym environments. This common interface allows to abstract away the interaction with the environment, decoupling the data generation process from the data sampling and freeing the user from the hassle of manually re-writing the data-fetching loop.

New CRL benchmarks can be easily created using the `gym_benchmark_generator`, which allows to define an `RLScenario` by providing any sequence of Gym environments (including custom ones). We can see an example in Fig. 2, in which we instantiate an `RLScenario` handling a stream of tasks which gives access to two randomly sampled environments.

Note that unlike static datasets, the environment can be used to produce an endless amount of data. Therefore, the interaction with the experience must be explicitly limited by some number of steps or episodes rather than epochs, which we can express during the creation of a *Strategy* as in Sect. 3.2.

As the Atari game suite [3] has become the main benchmark for RL algorithms in recent years, we also provide a tailored `atari_benchmark_generator` (Fig. 2) which takes care of adding common pre-processing techniques (e.g. frame stacking) as Gym Wrappers around each environment. This allows to minimize the time in between experiments as one can easily reproduce setups such as the one in [12] (sampling random Atari games to learn in sequence) by simply specifying a few arguments when creating a scenario. The benchmark interface also promotes the pattern of environment *wrapping*, which is Gym's intended way of organizing data processing methods to favor reproducibility. Reproducibility of experiments in particular is of great importance to `Avalanche` and one of the main reasons that drove us to propose an end-to-end framework for CRL.

```
1 # Scenario with 4 experiences alternating 2 random
    environments.
2 simple_scenario =
    gym_benchmark_generator(n_random_envs=2,
        n_parallel_envs=8, n_experiences=4)
3
4 # Scenario alternating 2 Atari games
5 atari_scenario = atari_benchmark_generator(
    ['BreakoutNoFrameskip-v4', 'PongNoFrameskip-v4'],
    frame_stacking=True,normalize_observations=True,
    # You can specify additional custom wrappers.
6   extra_wrappers=[ReducedActionSpaceWrapper],
    # Evaluate on both games during every experience.
7   eval_envs=['BreakoutNoFrameskip-v4',
    'PongNoFrameskip-v4'])
```

```
1 # Model
2 model = ActorCriticMLP(num_inputs=4, num_actions=2)
3 # CRL Benchmark Creation
4 scenario = gym_benchmark_generator(...)
5 # Prepare for training & testing
6 optimizer = Adam(model.parameters(), lr=1e-4)
7
8 # RL strategy
9 strategy = A2CStrategy(model, optimizer,
    per_experience_steps=10000,
    max_steps_per_rollout=5,
10   eval_every=1000, eval_episodes=10)
11 # train loop and final evaluation
12 strategy.train(scenario.train_stream)
    strategy.eval(scenario.test_stream)
```

(a) Benchmark creation   (b) Minimal training setup

Fig. 2. Example of `Avalanche RL` usage. (a) defines a task stream alternating two randomly sampled environments for 4 experiences. `n_parallel_envs` specifies the number of parallel actors (Sect. 3.3). The second scenario instead creates a stream of 2 Atari games with pre-processing attached. (b) puts everything together, instantiating a pre-implemented model (Sect. 3.4) and creating an "A2C agent" which is trained on the stream of games. The agent will perform 10000 *Update* steps per-experience while gathering 5 data samples at every *Rollout* step (Sect. 3.2). Evaluation will take place with the specified parameters.

3.2   Training: Reinforcement Learning Strategies

`Avalanche RL` provides several learning algorithms (listed at the end of this section) which have been implemented to be highly modular and easily customizable. The framework offers full-access to their internals in order to provide fine-grained customization options and specific support for continual learning techniques.

There are two main patterns to adapt a learning algorithm: subclassing and **Plugins** (as introduced in Sect. 2). In particular, `Avalanche RL` implements most continual learning strategies as Plugins. The modularity of the implementation allows to combine many RL strategies with popular CL strategies, such as replay buffers, regularization methods, and so on. As far as we are aware, `Avalanche` is the only library that allows the seamless composition of different learning algorithms. RL strategies inherit from `RLBaseStrategy`, a class which provides a common "skeleton" for both on and off-policy algorithms and abstracts many of the most repetitive patterns, including environment interaction, tracking metrics, CPU-GPU tensor relocation and more. `RLBaseStrategy` also provides callbacks which can be used by plugins.

Inspired by the open-source framework `stable-baseline3` [25] (sb3), RL strategies are divided into two main steps: **rollout** collection and **update**. Unlike sb3, we grouped both *on* and *off-policy* algorithms under this simple workflow.

The rollout stage abstracts the *data gathering* process which iterates the following steps:

1. $a_t \sim \pi_\theta(s_t)$: `sample_rollout_action`, to be implemented by the specific algorithm, returns the action to perform during a rollout step.
2. play action and observe next state and reward: s', r, *done, info*=`env.step`(a_t) referring to `Gym` interface.
3. store state transition in some data structure: **Step**. Store multiple Steps in a **Rollout**. These data structures are optimized for insertion speed and lazily delay "re-shaping" operations until they are needed by the update phase.
4. test rollout terminal condition, number of steps or episodes to run.

The **update** step is instead entirely delegated to the specifics of the algorithm: it boils down to implementing a method which has access to the rollouts collected at the previous stage and must define and compute a loss function which is then used to execute the actual parameters update. To enable the user with fine-grained control over the strategy workflow, we added callbacks which are executed just before and after the two stages.

At a higher level, the workflow we described happens within a single experience. To learn from a stream, the process is repeated for each experience in the stream, a behavior which is implemented by the `RLBaseStrategy`.

To summarize, one can implement a RL algorithm by sub-classing `RLBaseStrategy` and implementing the `sample_rollout_action` and **update** step. For example, A2C can be implemented in less than 30 lines of code[3]. Alternatively, customization of any algorithm is always possible by implementing a plugin, which allows to "inject" additional behavior, or by subclassing any of the available strategies. All the algorithm implementations expose their internals through class attributes, so one can for instance access the loss externally (e.g. from plugins) simply with `strategy.loss`.

Along with the release of our framework we provide an implementation of A2C and DQN [18], including popular "variants" with target network [19] and DoubleDQN [9].

3.3   Parallel Actors Interaction: VectorizedEnv

Since the data gathering supports any environment exposing the `Gym` interface, we are also able to automatically parallelize the agent-environment interaction in a transparent way to the user. This common practice [6,17,25] relies on using multiple *Actors*, each owning a local copy of the environment in which they perform actions, while synchronizing updates on a shared network; varying the amount of local resources available to each worker we can obtain different *degrees* of *asynchronicity* [7], allowing to scale computations on multiple CPUs.

To implement this behavior we leveraged `Ray` [20], a framework for parallel and distributed computing with a focus on AI applications. `Ray` abstracts away the parallel (and distributed) execution of code, sharing data between master

---

[3] https://github.com/ContinualAI/avalanche-rl/blob/master/avalanche_rl/training/ strategies/actor_critic.py.

and workers by serializing `numpy` arrays, which, in the case of execution on a single machine, are written once to shared memory in read-only mode and only referred to by actors.

This feature is opaque to the user, as it happens entirely inside a `VectorizedEnv`: this component wraps a single `Gym` environment and exposes the same interface, while under the hood it instantiates a pool of actors and handles results gathering and synchronization, acting as master. The API of our implementation was inspired by the work of `sb3`, although we opted to use `Ray` as a backend instead of Python's `multiprocessing` library due to distributed setting support.

`RLBaseStrategy` takes care of wrapping any environment with a `VectorizedEnv`, so the user can exploit parallel execution by simply specifying the number of workers/environment replicas, as shown in Fig. 2.

3.4   Additional Features

To complement the features we described in the previous sections we also implemented a series of utility components which one expects from a serviceable framework. Most of the changes listed in this section are not as important when taken singularly but as a whole they contribute significantly to `Avalanche RL` functionalities and as such they are hereby reported.

- **Models** from *seminal* papers such as [9,17–19] have been re-implemented in Pytorch and are available in the *Models* module.
- Evaluation Metrics: `RLBaseStrategy` automatically records gathered rewards and episode lengths during training, smoothing scalars with a window average by default. Additionally, one can record any significant value (e.g. loss, ϵ-greedy's ϵ) with minimal effort thanks to improved metrics builders.
- *Continual* Control Environments: classic control environments provided by `Gym` have been wrapped in order to expose hard-coded parameters (e.g. gravity, force..) which can now be modified to obtain varying conditions. This is useful for rapidly testing out algorithms on well renowned problems.
- Extended available **Plugins**, including EWC [12] and a ReplayMemory-based one inspired by works from [27] and [11].
- Miscellaneous tools such as environment wrappers for easily re-mapping actions keys (useful when learning multiple games with a single network) or reducing the action set and an additional *logger* with improved readability. `Avalanche RL` is compatible with `Avalanche` logging methods, such as Tensorboard [1].

4   Continual Habitat Lab

`Continual-Habitat-Lab` (CHL) is a high-level library for FAIR's simulator Habitat [28]: inspired by Habitat-Lab, we created a library with the goal of adding support for continual learning. CHL defines the abstraction layer needed to work with a stream of tasks $\{\tau_1, \tau_2..\}$, the core of CL systems.

We designed the library to be a shallow wrapper on top of Habitat-Sim functionalities and API while "steering" its intended usage toward learning applications, enforcing the data generation process to be carried out through online interaction and dropping the need for a pre-computed Dataset of positions altogether.

We also revisited the concept of Task to make it simpler and yet give it more control over the environment: while the next-state transition function $p(s'|s, a)$ is implemented by the dynamics of the simulator (Habitat-Sim), we bundled the reward function r into the task definition. To define a Task one must hence define a reward function $r(s, a, s') \rightarrow r$, a goal test function $g(s) \rightarrow \{T, F\}$ and an action space $\mathcal{A}$ as defined by Gym.

As Task is meant to be the main component through which the user can inject logic and behavior to be learned by the agent, we give direct access to the simulator at specific times through callbacks (e.g. to change environment condition, lighting, add objects..).

In order to natively support CRL a **TaskIterator** is assigned to the handling of the stream of tasks, hiding away the logic behind task sampling and duration while giving access to the current active task to be used by the environment.

We leveraged the multitude of 3D scenes datasets compatible with Habitat-Sim with the goal of specifying changing environment conditions, a most important feature to CL. To do so, we bundled the functionalities regarding scene switch in a sole component named **SceneManager**. It provides utilities for loading and switching scenes with a few configurable behaviors: scene swapping can happen on task change or after a number of episodes or actions is reached, even amid a running episode, maintaining current agent configuration and avoiding any expensive simulator re-instantions.

To offer a easily configurable system we re-designed the configuration system from scratch basing it on the popular OmegaConf library for Python: apart from providing a unified configuration entry-point which can be created programmatically or from a yaml file, the system dynamically maps Task and Actions parameters to configuration options. This allows the user to change experiments conditions by changing class arguments directly from the configuration file.

Continual Habitat Lab is integrated with Avalanche RL through a specialized benchmark generator (habitat_benchmark_generator) that takes care of *synchronizing* the stream of tasks defined in the CHL configuration with the one served to a Strategy. It does so by defining an *experience* each time a task or scene is changed, while serving the same object reference to the Habitat-Sim environment.

5   Conclusion and Future Work

In this paper, we have presented two novel libraries for Continual Reinforcement Learning: Avalanche RL and Continual Habitat Lab. We believe that these libraries can be helpful for the CRL community by extending and adapting work from the Continual Learning community on supervised and unsupervised continual learning (`Avalanche`) while also integrating a realistic simulator (`Habitat-Sim`) to benchmark CRL algorithms on complex embodied real-life scenarios.

In particular, `Avalanche RL` allows users to easily train and evaluate agents on a continual stream of tasks defined as a sequence of any Gym Environment. It is based on implementing a simple API upon the interaction of RL algorithms and task-streams, while offering a fine-grained control over their internals.

Through `Avalanche` researchers can exploit and extend the large amount of work done by the Continual Learning community while benefiting from the integration of highly modular and easily extensible RL algorithms. The library implements a large set of highly desirable features, such as parallel environment interaction, and provides implementations for popular baselines such as EWC [12], including benchmarks, learning strategies and architectures, all of which can be easily instantiated with a single line of code.

`Avalanche RL` can improve code reusability, ease-of-use, modularity and reproducibility of experiments, and we strongly believe that the whole CRL community would benefit from a collective effort such as `Avalanche RL` as a tool to speed-up the research in the field.

Having the goal of providing a shared and collaborative open-source codebase for CRL applications, `Avalanche RL` is constantly looking to add and refine functionalities. In the short term, we plan to implement a broader range of state-of-the art RL algorithms, including (but not limited to) PPO [30], TRPO [29] and SAC [8]. Additionally, we are also looking to increment the number of CL strategies such as *pseudo-rehersal* [2,26].

We are aiming to keep on expanding the supported simulators targeting a wider range of applications, from robotics to games engines [31] to widen the CRL benchmarks suite. Finally, we are expecting to merge `Avalanche RL` into `Avalanche`, striving to provide a single end-to-end framework for all continual learning applications.

References

1. Abadi, M., Agarwal, A., Barham, P., Brevdo, E., Chen, Z., Citro, C., et al.: TensorFlow: large-scale machine learning on heterogeneous systems (2015). https://www.tensorflow.org/
2. Atkinson, C., McCane, B., Szymanski, L., Robins, A.V.: Pseudo-rehearsal: achieving deep reinforcement learning without catastrophic forgetting. CoRR abs/1812.02464 (2018). http://arxiv.org/abs/1812.02464
3. Bellemare, M.G., Naddaf, Y., Veness, J., Bowling, M.: The arcade learning environment: an evaluation platform for general agents. J. Artif. Intell. Res. **47**, 253–279 (2012). arxiv:1207.4708
4. Brockman, G., Cheung, V., Pettersson, L., Schneider, J., Schulman, J., Tang, J., et al.: OpenAI gym. CoRR abs/1606.01540 (2016). http://arxiv.org/abs/1606.01540
5. Deng, J., Dong, W., Socher, R., Li, L.J., Li, K., Fei-Fei, L.: ImageNet: a large-scale hierarchical image database. In: 2009 IEEE Conference on Computer Vision and Pattern Recognition, pp. 248–255. IEEE (2009)

6. Denoyer, L., la Fuente, A.D., Duong, S., Gaya, J., Kamienny, P., Thompson, D.H.: Salina: sequential learning of agents. CoRR abs/2110.07910 (2021). https://arxiv.org/abs/2110.07910

7. Espeholt, L., Soyer, H., Munos, R., Simonyan, K., Mnih, V., Ward, T., et al.: Impala: scalable distributed deep-RL with importance weighted actor-learner architectures (2018)

8. Haarnoja, T., Zhou, A., Abbeel, P., Levine, S.: Soft actor-critic: off-policy maximum entropy deep reinforcement learning with a stochastic actor. CoRR abs/1801.01290 (2018). http://arxiv.org/abs/1801.01290

9. van Hasselt, H., Guez, A., Silver, D.: Deep reinforcement learning with double q-learning. CoRR abs/1509.06461 (2015). http://arxiv.org/abs/1509.06461

10. Hook, D.W., Porter, S.J., Herzog, C.: Dimensions: building context for search and evaluation. Front. Res. Metr. Anal. **3**, 23 (2018)

11. Isele, D., Cosgun, A.: Selective experience replay for lifelong learning. CoRR abs/1802.10269 (2018). http://arxiv.org/abs/1802.10269

12. Kirkpatrick, J., Pascanu, R., Rabinowitz, N.C., Veness, J., Desjardins, G., Rusu, A.A., et al.: Overcoming catastrophic forgetting in neural networks. CoRR abs/1612.00796 (2016). http://arxiv.org/abs/1612.00796

13. Krizhevsky, A., Nair, V., Hinton, G.: Cifar-10 (Canadian institute for advanced research). http://www.cs.toronto.edu/~kriz/cifar.html

14. LeCun, Y., Cortes, C.: MNIST handwritten digit database (2010). http://yann.lecun.com/exdb/mnist/

15. Lesort, T., Lomonaco, V., Stoian, A., Maltoni, D., Filliat, D., Díaz-Rodríguez, N.: Continual learning for robotics: Definition, framework, learning strategies, opportunities and challenges. Inf. Fusion **58**, 52–68 (2020)

16. Lomonaco, V., Pellegrini, L., Cossu, A., Carta, A., Graffieti, G., Hayes, T.L., et al.: Avalanche: an end-to-end library for continual learning (2021)

17. Mnih, V., et al.: Asynchronous methods for deep reinforcement learning. CoRR abs/1602.01783 (2016). http://arxiv.org/abs/1602.01783

18. Mnih, V., et al.: Playing Atari with deep reinforcement learning (2013)

19. Mnih, V., Kavukcuoglu, K., Silver, D., Rusu, A.A., Veness, J., Bellemare, M.G., et al.: Human-level control through deep reinforcement learning. Nature **518**(7540), 529–533 (2015). http://dx.doi.org/10.1038/nature14236

20. Moritz, P., et al.: Ray: a distributed framework for emerging AI applications (2018)

21. Moritz, P., Nishihara, R., Wang, S., Tumanov, A., Liaw, R., Liang, E., et al.: Ray: a distributed framework for emerging AI applications. CoRR abs/1712.05889 (2017). http://arxiv.org/abs/1712.05889

22. Normandin, F., et al.: Sequoia: a software framework to unify continual learning research. CoRR abs/2108.01005 (2021). https://arxiv.org/abs/2108.01005

23. Paszke, A., Gross, S., Massa, F., Lerer, A., Bradbury, J., Chanan, G., et al.: Pytorch: an imperative style, high-performance deep learning library. In: Wallach, H., Larochelle, H., Beygelzimer, A., d' Alché-Buc, F., Fox, E., Garnett, R. (eds.) Advances in Neural Information Processing Systems, vol. 32, pp. 8024–8035. Curran Associates, Inc. (2019). http://papers.neurips.cc/paper/9015-pytorch-an-imperative-style-high-performance-deep-learning-library.pdf

24. Plappert, M.: keras-rl (2016). https://github.com/keras-rl/keras-rl

25. Raffin, A., Hill, A., Ernestus, M., Gleave, A., Kanervisto, A., Dormann, N.: Stable baselines3 (2019). https://github.com/DLR-RM/stable-baselines3

26. Robins, A.V.: Catastrophic forgetting, rehearsal and pseudo rehearsal. Connect. Sci. **7**, 123–146 (1995)

27. Rolnick, D., Ahuja, A., Schwarz, J., Lillicrap, T.P., Wayne, G.: Experience replay for continual learning. CoRR abs/1811.11682 (2018). http://arxiv.org/abs/1811.11682

28. Savva, M., Kadian, A., Maksymets, O., Zhao, Y., Wijmans, E., Jain, B., et al.: Habitat: a platform for embodied AI research (2019)

29. Schulman, J., Levine, S., Moritz, P., Jordan, M.I., Abbeel, P.: Trust region policy optimization. CoRR abs/1502.05477 (2015). http://arxiv.org/abs/1502.05477

30. Schulman, J., Wolski, F., Dhariwal, P., Radford, A., Klimov, O.: Proximal policy optimization algorithms (2017)

31. Schwarz, J., Altman, D., Dudzik, A., Vinyals, O., Teh, Y.W., Pascanu, R.: Towards a natural benchmark for continual learning (2018). https://marcpickett.com/cl2018/CL-2018_paper_48.pdf

32. Sutton, R.S., Barto, A.G.: Reinforcement Learning: An Introduction, 2nd edn. The MIT Press, Cambridge (2018). http://incompleteideas.net/book/the-book-2nd.html

33. Wolczyk, M., Zajac, M., Pascanu, R., Kucinski, L., Milos, P.: Continual world: a robotic benchmark for continual reinforcement learning. CoRR abs/2105.10919 (2021). https://arxiv.org/abs/2105.10919

CVGAN: Image Generation with Capsule Vector-VAE

Rita Pucci$^{(\boxtimes)}$ ⓘ, Christian Micheloni ⓘ, Gian Luca Foresti ⓘ,
and Niki Martinel ⓘ

University of Udine, Udine, Italy
{rita.pucci,christain.micheloni,gialuca.foresti,niki.martinel}@uniud.it

Abstract. In unsupervised learning, the extraction of a representational learning space is an open challenge in machine learning. Important contributions in this field are: the Variational Auto-Encoder (VAE), on a continuous latent representation, and the Vector Quantized - VAE (VQ-VAE), on a discrete latent representation. VQ-VAE is a discrete latent variable model that has been demonstrated to learn nontrivial features representations of images in unsupervised learning. It is a viable alternative to the continuous latent variable models, VAE. However, training deep discrete variable models is challenging, due to the inherent non-differentiability of the discretization operation. In this paper, we propose Capsule Vector - VAE(CV-VAE), a new model based on VQ-VAE architecture where the discrete bottleneck represented by the quantization code-book is replaced with a capsules layer. We demonstrate that the capsules can be successfully applied for the clusterization procedure reintroducing the differentiability of the bottleneck in the model. The capsule layer clusters the encoder outputs considering the agreement among capsules. The CV-VAE is trained within Generative Adversarial Paradigm (GAN), CVGAN in short. Our model is shown to perform on par with the original VQGAN, VAE in GAN. CVGAN obtains images with higher quality after few epochs of training. We present results on ImageNet, COCOStuff, and FFHQ datasets, and we compared the obtained images with results with VQGAN . The interpretability of the training process for the latent representation is significantly increased maintaining the structured bottleneck idea. This has practical benefits, for instance, in unsupervised representation learning, where a large number of capsules may lead to the disentanglement of latent representations.

Keywords: VAE · Capsules · VQ-VAE · VQGAN · GAN · Computer vision

1 Introduction

In machine learning, one of the hallmarks is the automatic extraction of useful features from images. At the state of the art (SOTA), models reach a better generalisation by filtering their inputs to retain only the information that is relevant to the task at hand [26]. Such a behaviour is obtained with bottleneck

© The Author(s), under exclusive license to Springer Nature Switzerland AG 2022
S. Sclaroff et al. (Eds.): ICIAP 2022, LNCS 13231, pp. 536–547, 2022.
https://doi.org/10.1007/978-3-031-06427-2_45

layers, which indirectly reduce the amount of information allowed through by limiting feature dimensionality [28,30]. In Variational Auto-Encoder (VAE) the amount of information is directly constrained, and a continuous stochastic bottleneck represents each sample with a probability distribution [12]. The Vector-Quantized Variational Auto-Encoder (VQ-VAE) [17] encodes features as discrete latent variables. The usage of discrete latent variables offers a direct control over the information content of the learned representation. VQ-VAE demonstrates to achieve results competitive with the SOTA obtaining a good property of features separation [4,27]. However, training deep discrete latent representation models is challenging due to the lack of a derivative function in the bottleneck of the model. A strategy is to assume that the system is stochastic and optimise the mean output of discrete random units [10]. Alternative strategies improve performances by altering the training dynamics methods [13] or applying a straight-through gradient estimator, which copies the gradients from the decoder to the encoder [5,17]. VQ-VAE is a deterministic discrete variable model which operates by solving an approximate on-line k-means clustering in which the gradient is approximated using the straight through estimator [1]. VQ-VAE is recently used as the backbone of many proposed works for image generation such as with transformers in VQGAN [5] and with Contrastive Language-Image Pre-training in VQGAN-CLIP [5,22]. We propose CV-VAE , a new implementation of the bottleneck in VAE, that is a trade-off between a fully differentiable model and the idea of a discrete latent representation. CV-VAE introduces a new continuous bottleneck based on capsules layer, which represents each samples as the clustered probability distribution of features. We demonstrate the potentiality of this method by replacing the vector quantization in VQ-VAE with the capsule vectors. We remove the code-book and we use the capsule layer. This layer is trained by the routing-by-agreement (RbA) method [25] to learn capsules vectors and obtain the dynamic representation of the features extracted from the input. The proposed bottleneck is differentiable and it is trained within the encoder and decoder of the VAE model in Generative Adversarial paradigm to generate images, in such scenario CV-VAE is referred as CVGAN. At the SOTA, the main application of capsules layer is on image classification task [3,18,19,25]. Other works on this task, stress out structure to improve the performances [9,15,16,29]. We focus on another application of capsule layers which is on the image manipulation task. The representative features extracted by the capsule layer focus on the probabilistic distribution at entity-level [20,21] to reconstruct missing information of the input image. We think that this ability provides the model with a bottleneck able to extract latent variables that are representative for the reconstruction of the image in the decoding phase of the model. We discuss the clusterization property of capsule layer and we propose a proper alternative of such a bottleneck for the VAE based on capsule layer. Then, we replace the optimisation of the mean output of discrete random units (VQ-VAE) with the routine by agreement strategy to identify the significant clusters of features for the representation of the input. Lastly, we evaluate the results obtained with CVGAN comparing it with the VQGAN at the same conditions trained on ImageNet, the values provided for the VQGAN have been reproduced. We establish that the results obtained with CVGAN are competitive with the one of VQGAN, and that the capsule vectors are a valid trade-off between the differentiability function of the continuous bottle-necks and the clustered definition of probability distribution. This concept is close to the discrete code-book definition. In the following sections we introduce the proposed CVGAN, and the clusterization for representative learning. *Our contribution is a new model CVGAN which embraces the VAE idea and the capsule layer exploiting the RbA procedure as clusterization paradigm. This is obtained with a newly proposed application of capsule layer in a new bottleneck for latent space generation. The bottleneck is completely differentiable still maintaining the clusterization idea and the learning proves to be more interpretable. At SOTA, this is the first attempt of applying capsules for the generation of latent space and we think that the clusterization offered by the capsules layer is a valid candidate for the bottleneck implementation. A detailed discussion on this point will be presented in Sect. 3.2.*

2   Background

In this work, we present a new way of implementing Variational Auto-Encoder bottleneck [12,23] with continuous latent variables [14] based on capsule layers [25]. The proposed bottleneck learns to cluster useful features from the input image. We summarise VAE , VQ-VAE and VQGAN, capsules, and GANs ideas.

2.1   Variational Auto-Encoder (VAE)

VAE is a deep convolutional neural network algorithm introduced in [12], it belongs to the families of probabilistic graphical models and variational Bayesian methods. The VAE models consists of an encoder network and a decoder network. The encoder network parameterises a posterior distribution $q(z|\mathbf{X})$ of latent random variables z given the input data $\mathbf{X}$, and a prior distribution $p(z)$. The encoder must learn an efficient compression of the data into this lower-dimensional space, this is typically referred to as a 'bottleneck'. The decoder network receives the representation z and outputs the probability distribution of data, so it is denoted by a distribution $p(\mathbf{X}|z)$. The posteriors and priors in VAE are assumed normally distributed with diagonal covariance.

2.2   Vector Quantized Variational Auto-Encoders (VQ-VAE)

The VQ-VAE models are VAE models in which the bottleneck is implemented with a quantization function based on latent embedding vectors [17]. The latent embedding space is defined with $\mathbf{e} \in \mathbb{R}^{K \times D}$ where K is the size of the discrete latent space (i.e., K-way categories for the quantization), and D is the dimensionality of each latent embedding vector $\mathbf{e}^i \in \mathbb{R}^D$, $i \in 1, 2, ..., K$ that compose the code-book, as shown in Fig. 1. An input $\mathbf{X}$ is passed through an encoder

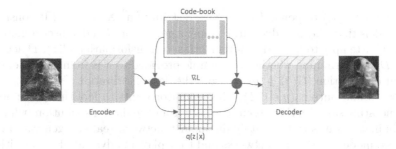

Fig. 1. The VQ-VAE [5,17]. The output of the encoder $z^{e(\mathbf{X})}$ is quantized based on a code-book randomly initialised. The application of a discrete quantization makes the connections with $q(.)$ not differentiable. By the straight through estimator, the gradient $\triangledown_z L$ will push the encoder to change its output without taking into consideration the quantization step.

producing output $z^{e(\mathbf{X})}$. The discrete latent variables z is calculated by a nearest neighbour look-up using the code-book. The input to the decoder is the corresponding embedding vector $\mathbf{e}^k$. The pipeline is implemented as a regular autoencoder with a particular non-linearity that maps the latents to 1-of-K codebook. The complete set of parameters for the model is the union of parameters of the encoder, decoder, and the embedding space $\mathbf{e}$.

2.3 Capsules Layer

Capsules capture the presence of an entity in an image [25] by clustering the features extracted. A capsule is a set of neurons; they collectively produce an activation vector with one element from each neuron to hold that neuron's instantiation value. In a hierarchical structure of capsule layers, the RbA procedure sends the activation vector at the lower layer to all the capsules at a higher layer. The activation vector of each capsule of the higher layer makes predictions for the parent capsule. The routing mechanism compares the higher layer prediction with the activation vector of the parents-capsules. The activation vectors which match are clusterised to provide the final vector that represents the probability of the entities to be present. Capsules and RbA are promising for a dynamic clusterization of the features. Such a solution is still not explored at the SOTA in VAE and we think that can be a good trade-off between the VAE and VQ-VAE. Rather than having a static bottleneck made by a fixed code-book (VQ-VAE), we propose to remove the quantization procedure and use a capsules layer to cluster the representative variables to compute the features distribution. This modification makes the training procedure completely differentiable while maintaining the clustering idea of the feature representation.

2.4 Generative Adversarial Network

The Generative Adversarial Network [7] is a framework for generative modeling of data through learning a transformation from points belonging to a prior distribution ($z \sim p_z$) to points from a data distribution ($\mathbf{X} \sim p_{data}$). It consists of two models that play an adversarial game: a generator G and a discriminator D. While G attempts to learn the aforementioned transformation $G(z)$, D acts as a critic $D(\cdot)$ determining whether the sample provided to it is from the generator's output distribution ($G(z) \sim p_G$) or from the data distribution ($\mathbf{X} \sim p_{data}$), thus giving a scalar output ($y \in 0, 1$). The generator wants to fool the discriminator by generating samples that resemble those from the data distribution, while the discriminator wants to accurately distinguish between real and generated data. The two models are neural networks and they play an adversarial game with the objective:

$$min_G max_D V(D, G) = E_{\mathbf{X} \sim p_{data}(\mathbf{X})}[log D(\mathbf{X})] + E_{z \sim p_z(z)}[log(1 - D(G(z)))] \quad (1)$$

3   Capsules Vector-Variational Auto-Encoders (CV–VAE)

3.1   Capsule Layer

Let $\mathbf{u}_i \in \mathbb{R}^{d_u}$ be an output of a capsule i at layer L, and j the index at layer $L + 1$, the prediction is calculated as:

$$\hat{\mathbf{u}}_{i|j} = \mathbf{W}_{ij} \mathbf{u}_i \quad (2)$$

where the $\mathbf{W}_{ij} \in \mathbb{R}^{d_u \times d_{\hat{u}}}$ is a weighted matrix applied to compute the affine transformation that given an activation vector $\mathbf{u}_i$ provides a prediction vector $\hat{\mathbf{u}}_{i|j}$. Then the procedure identifies the importance of capsule i at lower layer for capsule j at higher layer, by the coupling coefficients c_{ij}. The coupling coefficients c_{ij} are computed by applying the soft-max function over b_{ij}

$$c_{ij} = \frac{exp(b_{ij})}{\sum_k exp(b_{ik})} \quad (3)$$

where b_{ij} is log probability of capsule i being coupled with capsule j. The b_{ij} variable is initialised at 0, then it is updated α times, at each iteration of the procedure. The input vector to parent capsule j is the weighted sum of the probability vectors at capsule i multiplied by the coupling coefficient:

$$\mathbf{s}_j = \sum_i c_{ij} \hat{\mathbf{u}}_{j|i} \quad (4)$$

The capsule layer implements the clustering procedure by computing the agreement, as shown in Fig. 2. The output vectors of the capsule layer represent the probability of an object of been present in the given input or not. These vectors can exceed value one, depending on the output, so to make the output vector represents a probability, a non linear squashing function is used to restrict the vector length to 1, where $\mathbf{s}_j$ is input to capsule j and $\mathbf{v}_j$ is the output.

$$\mathbf{v}_j = \frac{||\mathbf{s}_j||^2}{1 + ||\mathbf{s}_j||^2} \frac{\mathbf{s}_j}{||\mathbf{s}_j||} \quad (5)$$

Finally b_{ij} are updated by computing the inner product of $\mathbf{v}_j$ and $\hat{\mathbf{u}}_{j|i}$. If two vectors agree, the product would be larger leading to longer vector length.

3.2   Clustering of Representative Learning

Clustering implies partitioning data into meaningful groups such that items are similar within each group and dissimilar across different groups. Capsules with RbA implement a clusterization procedure. In Eq. 3, c_{ij} should be large for (i, j) such that $\hat{\mathbf{u}}_{i,j}$ is similar to others $\hat{\mathbf{u}}_{k,j}$, with $k \neq i$ and small otherwise. The coupling coefficient c_{ij} is responsible of the clustering phase in Fig. 2. The $\hat{\mathbf{u}}_{i|j}$, which denote the agreement at layer L, are depicted by blue circles, they are empowered by c_{ij} in the clusterization phase to obtain $\mathbf{v}_j$. By squashing $\mathbf{s}_j$, we interpret $\mathbf{v}_j$ as the probability that capsule i is grouped to cluster j, Fig. 2: capsule $\mathbf{v}_{i,j}$ should be learnt as representative for the input of the capsule layer. We think that the clusterization obtained by the capsules layer is a trade-off between the random latent variables learned in VAE and the discrete latent representation obtained with VQ-VAE . In contrast with VQ-VAE, in CV-VAE we have the differentiability of the entire structure with the RbA procedure and similar to the VQ-VAE , we introduce the clusterization property of the latent variables which reduces the latent space generated by the model.

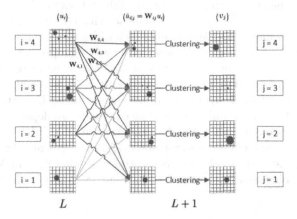

Fig. 2. Visual representation of the RbA paradigm: the output of the capsules at layer L, $\mathbf{u}_i$, is forwarded to the capsules at the upper layer $L + 1$. An affine transformation is applied to obtain $\hat{\mathbf{u}}_i$. Then the features are clustered in $\mathbf{v}_j$ based on the agreement among capsules.

3.3 CV-VAE Architecture

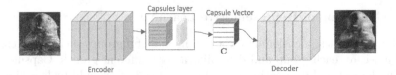

Fig. 3. The proposed CV-VAE: The model consists of an encoder network E and a decoder network G connected through an newly proposed bottleneck based on the capsule layer which implement the clustering of features.

In Fig. 3, we present the new model CV-VAE. We learn a convolutional model consisting of an encoder E and a decoder G, such that they learn to represent images with latent variables from a learned representation made by capsule vectors. We follow the implementation of E and G proposed in VQGAN. The E network consists of six blocks with $\{ResNet - Attention - Convolution\}$. E receives as input an image $\mathbf{Y} \in \mathbb{R}^{c_{input} \times h_{input} \times w_{input}}$ and extracts an hidden/latent representation $\mathbf{X} \in \mathbb{R}^{c_z \times h_z \times w_z}$ defined in the latent space z. $\mathbf{X}$ is the input of the proposed capsule layer. The capsule layer has β capsules implemented with convolutional layers that compute β different views, $\mathbf{U}$, of the input $\mathbf{X}$, Fig. 4. The $\mathbf{U} \in \mathbb{R}^{\beta \times d_u \times h \times w}$ consists of the activity vectors, each of which with d_u digits, and computed by the neurons which are in the capsules at layer L, and h and w are the resolution of each view. These vectors represent with length and orientation the presence and properties of entities in the image. Following the RbA procedure described in Sect. 3.1, the capsule layer is introduced in the bottleneck of the structure. It is able to cluster the extracted $\mathbf{U}$ features, to remove the unclustered features, and maintain the ones that are depicted as present by the agreement among the capsules. The RbA provides $\mathbf{V} \in \mathbb{R}^{\beta \times d_a \times h \times w}$, where d_a is the number of digits of each vector obtained. We aggregate the features in $\mathbf{V}$ by computing the $L_1 norm$ of each vector, this operation flattens $\mathbf{V}$ to $\mathbb{R}^{\beta \times h \times w}$. Finally, a convolutional layer maps the flattened $\mathbf{V}$ into $\mathbf{C} \in \mathbb{R}^{c_C \times h_C \times w_C}$ that

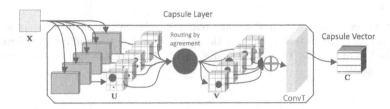

Fig. 4. Proposed bottleneck of CV-VAE: it consists of a capsule layer and a convolution transpose layer. The capsules extract $\mathbf{U}$ features which are clusterised by the RbA procedure, to obtain $\mathbf{V}$. We aggregate the matrices and upsample by the convolutional transposed layer.

is the input of the decoder network. The decoder network G is composed by six blocks, each of which is composed of $\{ResNet - Attention - Upsampling\}$, this network generates an output image $\hat{\mathbf{Y}} \in \mathbb{R}^{c_{input} \times h_{input} \times w_{input}}$.

3.4 Loss Function

We apply an adversarial training procedure with a patch-based discriminator D [8] that aims to differentiate between real and reconstructed images:

$$\mathcal{L}_{GAN}(E, G, CL, D) = [logD(\mathbf{Y}) + log(1 - D(\hat{\mathbf{Y}}))] \tag{6}$$

where the CL denotes the capsule layer of CVGAN. The $\hat{\mathbf{Y}}$ denotes the output obtained by G while $\mathbf{Y}$ is the ground-truth image.

4 Experiments

Implementation Details. Following SOTA methods [5], we considered the ImageNet dataset [24]. CVGAN is trained on the training split suggested for the ImageNet dataset, the 1.3M images (with no labels). The input images are resized to 256×256.

The input of the network is $\mathbf{Y} \in \mathbb{R}^{3 \times 256 \times 256}$, with $batch = 6$. The network E outputs $z(\mathbf{X}) \in (R)^{256 \times 16 \times 16}$, which is the input of the capsules layer. Capsule layer consists of $\beta = 32$ capsules and outputs $\mathbf{U} \in \mathbb{R}^{32 \times 16 \times 9 \times 9}$ through convolutional layers. The $[32 \times 9 \times 9]$ tensors, which are extracted by the capsules, describe the point of views of the capsules and each position has $d_u = 16$ digits. In RbA , we set $\alpha = 3$ for the $c_{i,j}$ update. The output of RbA is $\mathbf{V} \in \mathbb{R}^{32 \times 64 \times 9 \times 9}$, where each vector obtained by clusterisation has $d_a = 64$ digits. The transposed convolution layer in the bottleneck outputs $\mathbf{C} \in \mathbb{R}^{256 \times 16 \times 16}$. Finally G generates the output matrix $\hat{\mathbf{Y}} \in \mathbb{R}^{3 \times 256 \times 256}$. The CVGAN models $X = 3 \times 256 \times 256$ images by compressing them to $z = 256 \times 16 \times 16$ latent space via a capsule layer $p(\mathbf{X}|z)$. So it is obtained a reduction of a factor of $\frac{3 \times 256 \times 256}{256 \times 16 \times 16} = 3$. We evaluate the CVGAN on the ImageNet (validation split), COCOStuff [2], and FFHQ [11] datasets.

Results. We present the results obtained with the model trained for 800K steps on ImageNet. In Fig. 5, we compare the results obtained with CVGAN and with VQGAN. In this paper, with VQGAN we consider the first phase presented by the authors in [5]. For each group of images, the ground-truth images are presented for reference. It is worth noting that the E and G networks for CVGAN and VQGAN are the same. The two models are trained in GAN against a D Patch-Discriminator. Results obtained with CVGAN appear with better border and colour quality than VQGAN. The colours are vibrant and consistent with the ground-truth. We observe that both models have some problems in reconstructing fine details such as the architectural structure, we think this is due to the lack of quality of these details. We want to focus on the details obtained with CVGAN, shown in Fig. 6. In the figure, we show three images depicting three different details being reconstructed. In the detail of the dog's face, we note CVGAN's

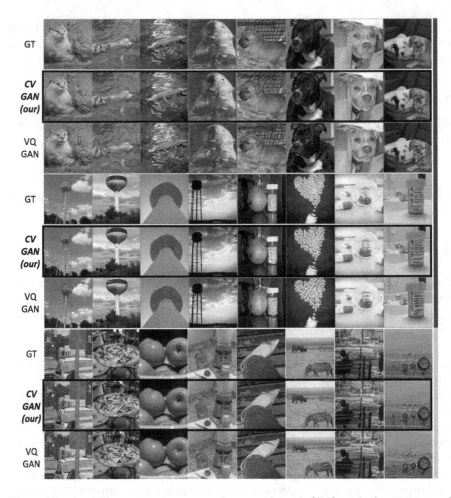

Fig. 5. Results: We compare images at the ground-truth (GT) with the CVGAN, and the VQGAN reconstructions. On the right side: the dark line denotes results on ImageNet (test split), and the light blue line is for COCOStuff results. Orange rectangles highlight our results (Color figure online).

ability to reconstruct the smallest details of the face. The capsules have been shown to be able to maintain relationships between the parts with the whole. In this case, the parts are the dog's eyes, nose and mouth and the whole is the dog's face. This is visible in the facial symmetry reconstructed in Fig. 6a.

In Fig. 6b, the peel of the pear reconstructed by CVGAN appears to be smooth and without any filling pattern, which is instead visible in the VQGAN reconstruction. Finally, in Fig. 6c we observe that CVGAN provides a greater quality of the architectural details of the structure in the image, the edges are smoother and clearer without any blurring than the image of VQGAN. Table 1 shows the results for comparing the Fréchet Inception Distance [6] (FID↓) score for the

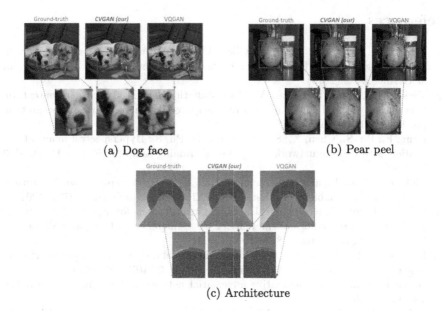

(a) Dog face   (b) Pear peel

(c) Architecture

Fig. 6. In this Figure, we compare details of the reconstructed images from ImageNet (test split) with CVGAN and VQGAN.

Table 1. FID↓ on images of ImageNet (test split), COCOStuff [2], and FFHQ [11]

	ImageNet	COCOStuff	FFHQ
CVGAN (our)	**10.30392**	**13.76736**	**6.7976**
VQGAN	10.31322	13.15105	5.892715

synthesis of semantic images. The CVGAN results are on par with VQGAN, which makes CVGAN a promising model furthermore it is completely differentiable.

5   Conclusions

In this article, we have introduced CVGAN, a new model that combines VAE with capsule layers to achieve a fully differentiable architecture and grouped latent variables. CVGAN captures important relationships between data characteristics in a completely unsupervised way. Furthermore, CVGAN performs image reconstructions that have a higher quality than VQGAN on ImageNet data under similar conditions. We believe that this is the first latent variable model based on the capsule layer and that the clustering offered by RbA has proved to be a promising application on the VAE model family.

References

1. Bengio, Y., Léonard, N., Courville, A.: Estimating or propagating gradients through stochastic neurons for conditional computation. arXiv preprint arXiv:1308.3432 (2013)
2. Caesar, H., Uijlings, J., Ferrari, V.: Coco-stuff: thing and stuff classes in context. In: Proceedings of the IEEE Conference on Computer Vision and Pattern Recognition, pp. 1209–1218 (2018)
3. Deng, F., Pu, S., Chen, X., Shi, Y., Yuan, T., Pu, S.: Hyperspectral image classification with capsule network using limited training samples. Sensors **18**(9), 3153 (2018)
4. Eloff, R., et al.: Unsupervised acoustic unit discovery for speech synthesis using discrete latent-variable neural networks. arXiv preprint arXiv:1904.07556 (2019)
5. Esser, P., Rombach, R., Ommer, B.: Taming transformers for high-resolution image synthesis. In: Proceedings of the IEEE/CVF Conference on Computer Vision and Pattern Recognition, pp. 12873–12883 (2021)
6. Fréchet, M.: Sur la distance de deux lois de probabilité. Comptes Rendus Hebdomadaires des Seances de L Academie des Sciences **244**(6), 689–692 (1957)
7. Goodfellow, I., et al.: Generative adversarial networks. Commun. ACM **63**(11), 139–144 (2020)
8. Isola, P., Zhu, J.Y., Zhou, T., Efros, A.A.: Image-to-image translation with conditional adversarial networks. In: Proceedings of the IEEE Conference on Computer Vision and Pattern Recognition, pp. 1125–1134 (2017)
9. Jaiswal, A., AbdAlmageed, W., Wu, Y., Natarajan, P.: Capsulegan: generative adversarial capsule network. In: Proceedings of the European Conference on Computer Vision (ECCV) Workshops (September 2018)
10. Jang, E., Gu, S., Poole, B.: Categorical reparameterization with gumbel-softmax. arXiv preprint arXiv:1611.01144 (2016)
11. Karras, T., Laine, S., Aila, T.: A style-based generator architecture for generative adversarial networks. In: Proceedings of the IEEE/CVF Conference on Computer Vision and Pattern Recognition, pp. 4401–4410 (2019)
12. Kingma, D.P., Welling, M.: Auto-encoding variational Bayes. arXiv preprint arXiv:1312.6114 (2013)
13. Łańcucki, A., et al.: Robust training of vector quantized bottleneck models. In: 2020 International Joint Conference on Neural Networks (IJCNN), pp. 1–7. IEEE (2020)
14. Mnih, A., Gregor, K.: Neural variational inference and learning in belief networks. In: International Conference on Machine Learning, pp. 1791–1799. PMLR (2014)
15. Mukhometzianov, R., Carrillo, J.: Capsnet comparative performance evaluation for image classification. arXiv preprint arXiv:1805.11195 (2018)
16. Nair, P., Doshi, R., Keselj, S.: Pushing the limits of capsule networks. Technical note (2018)
17. Oord, A.v.d., Vinyals, O., Kavukcuoglu, K.: Neural discrete representation learning. arXiv preprint arXiv:1711.00937 (2017)
18. Pucci, R., Micheloni, C., Foresti, G.L., Martinel, N.: Deep interactive encoding with capsule networks for image classification. Multimed. Tools Appl. **79**(43), 32243–32258 (2020)
19. Pucci, R., Micheloni, C., Foresti, G.L., Martinel, N.: Fixed simplex coordinates for angular margin loss in capsnet. In: 2020 25th International Conference on Pattern Recognition (ICPR), pp. 3042–3049. IEEE (2021)

20. Pucci, R., Micheloni, C., Foresti, G.L., Martinel, N.: Pro-ccaps: progressively teaching colourisation to capsules. In: Proceedings of the IEEE/CVF Winter Conference on Applications of Computer Vision, pp. 2271–2279 (2022)

21. Pucci, R., Micheloni, C., Martinel, N.: Collaborative image and object level features for image colourisation. In: Proceedings of the IEEE/CVF Conference on Computer Vision and Pattern Recognition, pp. 2160–2169 (2021)

22. Radford, A., Sutskever, I., Kim, J.W., Krueger, G., Agarwal, S.: Clip: connecting text and images (2021)

23. Rezende, D.J., Mohamed, S., Wierstra, D.: Stochastic backpropagation and approximate inference in deep generative models. In: International Conference on Machine Learning, pp. 1278–1286. PMLR (2014)

24. Russakovsky, O., et al.: Imagenet large scale visual recognition challenge. Int. J. Comput. Vis. **115**(3), 211–252 (2015)

25. Sabour, S., Frosst, N., Hinton, G.E.: Dynamic routing between capsules. In: Advances in Neural Information Processing Systems, pp. 3856–3866 (2017)

26. Tishby, N., Pereira, F.C., Bialek, W.: The information bottleneck method. arXiv preprint physics/0004057 (2000)

27. Tjandra, A., Sisman, B., Zhang, M., Sakti, S., Li, H., Nakamura, S.: Vqvae unsupervised unit discovery and multi-scale code2spec inverter for zerospeech challenge 2019. arXiv preprint arXiv:1905.11449 (2019)

28. Veselỳ, K., Karafiát, M., Grézl, F., Janda, M., Egorova, E.: The language-independent bottleneck features. In: 2012 IEEE Spoken Language Technology Workshop (SLT), pp. 336–341. IEEE (2012)

29. Xiang, C., Zhang, L., Tang, Y., Zou, W., Xu, C.: Ms-capsnet: a novel multi-scale capsule network. IEEE Signal Process. Lett. **25**(12), 1850–1854 (2018)

30. Yu, D., Seltzer, M.L.: Improved bottleneck features using pretrained deep neural networks. In: Twelfth Annual Conference of the International Speech Communication Association (2011)

Self-Adaptive Logit Balancing for Deep Learning Robustness in Computer Vision

Jiefei Wei[1(✉)], Qinggang Meng[1], and Luyan Yao[2]

[1] Loughborough University, Epinal Way, Loughborough LE11 3TU, UK
{J.Wei,Q.Meng}@lboro.ac.uk
[2] University of Nottingham, University Park, Nottingham NG7 2RD, UK
luyan.yao@nottingham.ac.uk

Abstract. With wide applications of machine learning algorithms, machine learning security has become a significant issue. The vulnerability to adversarial perturbations exists in most machine learning algorithms, including cutting-edge deep neural networks. The standard adversarial perturbation defence techniques with adversarial training need to generate adversarial examples during the training process, which require high computational costs. This paper proposed a novel defence method using self-adaptive logit balancing and Gaussian noise boost training. This method can improve the robustness of deep neural networks without high computational cost and achieve competitive results compared with the adversarial training methods. Meanwhile, this defence method enables deep learning systems to have proactive and reactive defence during the operation. A sub-classifier is trained to determine whether the system is under attack and detect attack algorithms via the patterns of the Log-Softmax values. It can achieve high accuracy for detecting clean inputs and adversarial examples created by seven attack methods.

Keywords: Machine learning security · Adversarial robustness · Adversarial examples · Deep neural networks

1 Introduction

The vulnerability to adversarial perturbations is a significant risk of machine learning algorithms, including deep neural networks (DNNs) [26]. The weakness under adversarial perturbation attacks proves that DNNs are challenged to discover a model with all the desired underlying functions based on the current learning rules and datasets [12]. In an image classification problem, adversarial perturbations are the tiny changes or modifications to the original inputs, which are often imperceptible to the sense of humans; however, they can lead machine learning algorithms to misclassify the modified inputs [12,14]. The inputs modified by adversarial perturbations are also known as adversarial examples [12,14].

Adversarial robustness is essential for deep learning applications, especially in safety-critical areas, such as self-driving vehicles and facial recognition authorisation. The robustness of DNNs can be improved in two directions. The proactive defence is to train DNNs models that are strong when taking clean data and adversarial examples. The reactive defence allows the deep learning systems can detect adversarial attacks and reject them. This paper aims to resolve the issue of adversarial robustness in both directions. The contributions of this paper are summarised as the followings:

- This paper proposed a novel regularisation algorithm that drives the Log-Softmax values of incorrect classes to distribute close to their mean value. This algorithm can significantly improve the adversarial robustness of deep neural networks.
- The proposed defence method can defend the popular attack methods with less computational cost than adversarial training. Moreover, it achieves competitive results compared with advanced adversarial training algorithms, such as TRADES [29] and adversarial weight perturbation (AWP) [27].
- The DNNs trained by the proposed algorithm can have strong reactive defence via a sub-classifier. It can determine whether the DNNs are misleading and detect the attack algorithms based on the Log-Softmax value patterns.

2   Related Work

Since discovering adversarial perturbations, many adversarial attack methods, such as fast gradient sign method (FGSM) [12], projected gradient descent (PGD) [19] and automatic parameter-free attacks [8], have been developed to test the robustness of DNNs. Meanwhile, adversarial defence approaches can reduce this machine learning security risk. However, design and evaluation defences against adversarial perturbations have proven to be extremely difficult [4].

Adversarial training is a popular and intuitive defence method that can improve adversarial robustness by introducing adversarial examples into the training process, for example, adversarial training for free [24], TRADES [29] and Madry [19]. Generative adversarial networks (GANs) [11] can also be used for adversarial defence, for instance, APE-GAN [15], which can improve the robustness via adversarial perturbation elimination. The approaches described above are highly time-consuming and computational power costly; for instance, the TRADES is typically more than ten times slower than natural training [29].

The regularisation methods, for instance, the logit pairing [16] can also improve adversarial robustness without getting access to extra data and high computational cost. The key idea of clean logit paring is learning the model to predict logits of small magnitude and receive a penalty for overconfidence [16]. However, the paring mechanism does not care whether the randomly paired training examples are appropriate and fragile under well-designed attacks [9].

Meanwhile, other methods are trying to improve adversarial robustness, including improving the quality and quantity of training datasets, such as label smoothing [10], leveraging unlabelled data [5], and using data augmentation [22]. Label smoothing is initially proposed by Szegedy et al. to improve the performance of DNNs, which uses smoothed uniform label vectors in place of one-hot label vectors [25]. Furthermore, it has been proved that a model trained with label smoothing can invalidate some simple gradient-based attacks [10]. However, signing uniform label vectors to varying data instances is not good enough for general robustness and fails under solid attacks [23].

The reactive defence methods have been proposed to detect adversarial examples from logits or other feature patterns and achieved high accuracy [1,20]. Nevertheless, these detection methods are evaluated on very few attack algorithms and can be fooled by adaptive attacks [13].

3   Algorithm Design for Adversarial Defence

3.1   Preliminary

In this paper, the adversarial defence problem is simplified to train an image classifier to have high test accuracy on clean data and adversarial examples. An N-layers DNN is a mapping $f : \mathbb{R}^n \to \mathbb{R}^m$ that takes $x \in \mathbb{R}^n$ as an input and produces $y \in \mathbb{R}^m$ as an output. It can be expressed as:

$$f(x) = F^N(F^{N-1} \dots (F^1(x))) = y \tag{1}$$

When using a DNN to solve a K-class image classification problem, let S denote the output of the SoftMax layer, such that:

$$S = SoftMax(f(x)) = \{s_1, s_2, \dots, s_K\} \tag{2}$$

where $0 \le s_i \le 1$ and $s_1 + s_2 + \dots + s_K = 1$. The DNN will assign the label of input x according to:

$$predict = agrmax_i S(f(x)) \tag{3}$$

It is necessary to have a threat model for evaluating the defences' performance. Without a threat model, defence proposals are likely not to be falsifiable or trivially falsifiable [4]. For any clean input x and a similarity metric $\mathcal{D}$, x' is a valid adversarial example if x' is misclassified and $\mathcal{D}(x, x') \le \epsilon$, where ϵ is perturbation limit. The white-box adversarial attacks can access complete knowledge of deep neural networks and the inner workings of the defence. Meanwhile, black-box attacks, transfer attacks and adaptive attacks are also deployed to evaluate the proposed defence method. The proposed defence method should be robust against non-targeted attacks at different L_p norms.

3.2   Rethink of Vulnerability

Training Data. One-hot labelling [30] is a standard method to organise the datasets for machine learning because it is efficient and straightforward. Under one-hot labelling, every instance of a dataset is labelled as one group bits, among which only the ground truth is encoded as 1, the single high, and all others as 0. However, the one-hot label may create noise and biases because there is no sufficient information within every single data instance to support that instance is 100% true to be in the labelled class. Under the ideal conditions, every instance in the dataset should be labelled individually according to its feature quality.

Learning Rules. It is intuitive to utilise zero-one loss [21] when training a machine learning classifier with a one-hot labelled dataset. However, the problem of adversarial defence is challenging because the zero-one loss is NP-hard to optimise, and the surrogate loss has too little theoretical guarantee on the tightness [29]. Meanwhile, the current learning rules, such as backpropagation with cross-entropy loss, force the DNNs to give all target labels 100% confidence. Therefore, the DNNs are easily overconfident under the fixed training dataset and under-fitted to solve the general problem [12,26]. The DNNs cannot capture all meaningful features from the limited database and assign correct weights to every feature. This issue is hard to be resolved by a deeper or more complex network structure [12].

3.3   Self-adaptive Logit Balancing

Zero-Cross-Entropy Loss. We proposed a zero-cross-entropy loss to solve the overconfident issue by making the loss equal to 0 when the prediction is equal to its target label. Instead of forcing DNNs to have cross-entropy loss equal to 0 for every training example towards its true label, the zero-cross-entropy loss only encourages DNNs to believe that the training examples are most likely to be in their labelled classes (give the biggest Softmax to the target label among all labels).

Furthermore, the zero-cross-entropy loss introduces a self-adaptive early stop mechanism to punish the overconfidence, which will help DNNs to learn more appropriate weights for all extracted features. During training progress, the instances with apparent features will be more likely to achieve 0 loss (correct classification) at the earlier training stage. When the DNNs are trained with mini-batches, the loss of clear instances will keep decreasing until all examples in their mini-batch reach 0 loss. Therefore, with the zero-cross-entropy loss, the instances with more features related to their corresponded classes will have lower losses than the poor-quality examples.

Logit Balancing Loss. The Log-Softmax values of incorrect classes are not regularised when the costs are calculated, only relying on the target label. Most of the untargeted white-box attacks are trying to decrease the confidence of the target label. When the confidence of the target label has decreased, the sum of confidence of all other labels will increase because the sum of all confidences has to be 1. Suppose the additional confidence reallocated from target label to incorrect labels can be distributed evenly. In that case, the model is hard to attack until the confidence of the correct label drop to $1/K$ (K *is the number of classes*).

Therefore, squeezing the distribution of Log-Softmax values of all incorrect classes to their mean could increase the adversarial robustness. Then a novel logit balancing loss:

$$Loss_{LB} = \beta \times SD\left((-log(S(f(x_i)) \setminus s_t)\right) \tag{4}$$

Algorithm 1: Self-Adaptive Logit Balancing Training.

Data: Logit balancing weight β, batch size m, network architecture
  parametrised by θ, Randomly initialise network f_θ.

Result: Robust network f_θ.

begin

    Read mini-batch $B = (x_1, x_2, \ldots, x_m)$ from training set, and its
  corresponding target label $T = (t_1, t_2, \ldots, t_m)$

    repeat

        for $i = 1$ **to** m **do**

            $GN\_x_i = x_i + \sigma \times GaussianNoise$

            if $argmax_{index} S(f(GN\_x_i)) = t_i$ **then**

                $loss_i = \beta \times SD\left((-log(S(f(GN\_x_i)) \setminus s_t)\right) \times s_t$

            else

                $loss_i = L_{CE}(f_\theta(GN\_x_i), t_i) = -log(s_t)$

        Update: $\theta \leftarrow \theta - \sum_{i=1}^{m} \nabla[loss_i] \div m$

    until *training is converged.*

---

is introduced to control the distribution of the Log-Softmax value of all incorrect labels, where β is the logit balancing weight and SD stands for standard deviation. Decreasing the standard deviation of the incorrect labels' Log-Softmax values has two effects. The first one is making the distribution of the incorrect labels' logits closer to their mean then improving the adversarial robustness of DNNs. The second effect is that decreasing the mean of the Log-Softmax values of all classes will increase the loss value of the target label. Therefore, heavy and fixed logit balancing weight will lead DNNs to give low confidence to correct labels. Finally, the logit balancing will lead the model to provide the same confidence for every class. For example, a ten-classes data will have 0.1 SoftMax confidence for all classes. Therefore, a self-adaptive regularisation parameter is introduced:

$$Loss_{LB} = \beta \times SD\left((-log(S(f(x_i)) \setminus s_t)\right) \times s_t \tag{5}$$

where s_t is the SoftMax value of the target label. So, the logit balancing loss of each training example will be changed based on its confidence in the target label - the higher confidence of the correct label, the more attenuation of logit balancing loss.

3.4 Gaussian Noise Boost Training with Logit Balancing

Gaussian noise boost training is utilised to solve the limited training dataset's data distribution problem [22]. Instead of training the DNNs on the clean images, Gaussian noise boost training will add different random Gaussian noise to each training example for every training epoch.

As described in Algorithm 1, the proposed adversarial defence algorithm is designed by integrating Gaussian noise boost training and self-adaptive logit balancing. The proposed defence algorithm suggests minimising Eq. 5 to maximise the robust accuracy under adversarial attacks. The self-adaptive logit balancing mechanism is activated if the DNN can classify the current input correctly, thereby avoiding the negative effect of logit balancing and maximising the overall accuracy.

4   Experimental Results and Evaluations

4.1   Experiment Setup

MNIST Setup. The CNN architecture in TRADES [29] is used for a fair comparison, which has four convolutional layers, followed by three fully-connected layers. We set logit balancing weights $\beta = 0.16$, Gaussian noise weight $\sigma = 0.4$.

CIFAR-10 Setup. The models are trained with WideResNet-34-10 architecture [28], which is the same as in TRADE and AWP. We set logit balancing weight $\beta = 0.02$ and Gaussian noise weight $\sigma = 24/255$.

Attack Algorithms. The models are tested by various attacks under L_∞ and L_2 norms. In this paper, Torchattacks [17] is used to perform all the tests. Torchattacks is a PyTorch library containing the most popular adversarial attacks to generate adversarial examples for verifying the robustness of deep learning models [17]. The PGD and CW are proposed by Madry Lab MIT [19]. The APGD and APGDT are designed by combining Expectation Over Transformation and PGD [3,18]. The fast adaptive boundary attack (FAB) has an intuitive geometric meaning, yields quickly high-quality results and is robust to the phenomenon of gradient masking [7]. The square attack is a query-efficient black-box adversarial attack via random search to find adversarial examples [2]. The perturbation limits for testing DNNs on MNIST data are 0.1 and 0.3. The DNNs on CIFAT-10 data are tested with perturbation $\epsilon = 8/255$.

Adversarial Training Methods for Comparison. Two defence methods are investigated for a detailed comparison with the proposed defence method because they share the same model architecture and use the same benchmark datasets. TRADES is one of the advanced adversarial training algorithms, the first-place winner of the NeurIPS 2018 Adversarial Vision Challenge (Robust Model Track) [29]. AWP is a regularisation method that regularises the weight loss landscape of adversarial training, forming a double-perturbation mechanism that injects the worst-case input and weight perturbations [27]. The TRADES model on MNIST data is adversarial trained with perturbation $\epsilon = 0.3$. The TRADES and AWP models on CIFAR-10 data are adversarial trained with perturbation $\epsilon = 8/255$.

Table 1. Robust accuracy on MNIST data under various attacks.

Perturbation	Defence model	PGD	APGDT	Square	FAB	PGD-L_2	CW-L_2
$\epsilon = 0.1$	TRADES	98.88	98.82	98.81	98.86	97.19	97.24
	LB	96.81	94.01	94.49	94.58	95.60	96.72
$\epsilon = 0.3$	TRADES	96.33	93.58	92.93	94.93	82.02	96.51
	LB	56.41	0.6	0.5	18.2	31.88	89.68

Table 2. Robust accuracy on CIFAR-10 data under various attacks.

Defence model	Natural accuracy	PGD	APGD	APGDT	Square	FAB	PGD-L_2	CW-L_2
Logit balancing	85.27	74.67	48.18	3.97	40.06	53.80	73.30	74.48
	85.05	**77.82**	**61.58**	**32.64**	**73.47**	**80.25**	**79.06**	**83.53**
TRADES ($1/\lambda = 6$)	84.17	56.61	54.81	55.07	58.91	54.87	28.49	11.28
	74.53	**53.34**	**61.61**	**61.22**	**67.07**	**67.50**	**52.49**	**64.10**
AWP-TRADES	85.36	60.13	58.85	56.20	60.40	56.84	33.98	22.19
	76.95	**56.91**	**64.05**	**63.44**	**68.68**	**68.54**	**50.80**	**67.57**

4.2 Robustness Under Various Attacks

After exploring the logit balancing models with various logit balancing weights and Gaussian noise weights, it is found that increasing the σ of Gaussian noise can improve the overall robustness. However, heavy Gaussian noise will cause the natural accuracy to be impaired. For example, the natural accuracy of logit balancing model on CIFAR-10 data will drop to 74% from 85% when rising $\sigma = 24/255$ to $\sigma = 64/255$. Furthermore, large logit balancing weights will give the deep learning models only gradient masking without actual robustness.

As shown in Table 1, on MNIST data, the logit balancing (LB) model has the overall adversarial robustness that is close to the performance of TRADES when setting perturbation limit $\epsilon = 0.1$. However, when increasing the perturbation limit to $\epsilon = 0.3$, the overall robustness of the logit balancing model is challenged. On CIFAR-10 data (see Table 2), the logit balancing model has better general robustness than TRADES and AWP, except for the strong white-box targeted attack - APGDT.

Furthermore, it is found that adding Gaussian noise to test inputs can eliminate the effects of adversarial perturbations and improve robustness further. By adding Gaussian noise masks to adversarial inputs, the adversarial robustness of the logit balancing model has been improved significantly against all tested attack algorithms. The results of Gaussian noise masked inputs are shown in Table 2 in bold text.

The robust accuracy of APGDT increased from 3.97% to 32.64%. The robust accuracy of AGPD reached 61.58%, which is almost equal to the accuracy of TRADES (61.61%). The robust accuracy under Square and FAB attacks increased from 40.06% and 53.80% to 73.47% and 80.25%, which outperformed the TRADES and AWP with accuracy around 68%. Furthermore, the proposed defence method has even more advantages than TRADES and AWP for defending L_2 norm attacks with or without Gaussian noise masking. Gaussian noise-masking can also slightly improve the robustness of the TRADES and AWP

Table 3. Robustness under transfer attacks on CIFAR-10 data.

Target model	Source model	PGD	APGD	APGDT	SQUARE	FAB	PGD-L_2	CW-L_2
Logit balancing	Natural model	79.31	80.66	83.14	84.09	84.73	80.45	85.15
	TRADES model	59.60	72.03	74.42	82.63	80.08	65.39	78.86
	AWP-TRADES	61.64	74.11	74.03	82.13	79.83	65.77	79.43

model against adversarial attacks. However, Gaussian noise masking will impede the natural accuracy of TRADES and AWP models by about 10%.

4.3 Results Under Transfer Attacks

Adversarial examples can transfer between models, which means the adversarial examples generated against one model with a specific architecture will be more likely to be adversarial against other models [12]. Adversarial transfer attacks are the adversarial examples generated from source models to attack the target model. Furthermore, such transfer attacks are successful at circumventing defences that are utilising gradient masking [4].

As shown in Table 3, the logit balancing model is attacked by three source models. The natural model is trained with clean data and without defences. The transfer attacks from the natural model barely can not affect the targeted model. It proves that the robustness of the logit balancing model does not come from gradient masking. The targeted model can ensure at least 60% robust accuracy under the transfer attacks from TRADES and AWP models.

4.4 Adaptive Attack and Attack Convergence

Testing the defence via various non-adaptive attacks is necessary but not sufficient [4]. Therefore, it is essential to apply adaptive adversaries that are designed adapted to the proposed defence's details and try to invalidate the robustness claims. The adaptive attack method for logit balancing is proposed by replacing the loss function of regular gradient-based attacks with logit balancing loss - Eq. 4. For exploring the limitations of the proposed defence method, it is necessary to adjust attack parameters until the attack is converged. As illustrated in Table 4, the logit balancing model can achieve 55% robust accuracy under adaptive and converged attacks for L_∞ and L_2 distance.

Table 4. Adaptive attacks and attack convergence on CIFAR-10 data.

Defence model	Number of iterations	PGD-L_∞	Adaptive PGD-L_∞	PGD-L_2	Adaptive PGD-L_2
Logit balancing	20	74.67	57.43	74.12	57.56
	160	73.65	54.77	73.77	55.65

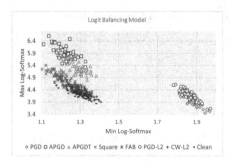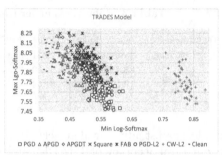

Fig. 1. The scatter plot of the min and max Log-Softmax values under various attacks on CIFAR-10 data.

4.5 Attack Detection

The prior works rarely studied the attack detection topic and proactive defence together. When applying machine learning algorithms in safety-critical environments, such as autonomous robotics and biometric authorisation, while improving the proactive robustness, it is important to identify whether an attack is in progress. The proposed algorithm has more advantages than TRADES for adversarial example detection.

As illustrated in Fig. 1, for the logit balancing model, the average values of the minimum and the maximum Log-Softmax values of each mini-batch (batch size equals to 100) of test data under seven adversarial attack methods are separately distributed. It can intuitively tell whether the logit balancing model is under attack and identify the attack methods by observing the abnormal Log-Softmax values. However, it is difficult to know what kind of attack methods the TRADES model is attacked by investigating their Log-Softmax values. This attack detection feature allows engineers and researchers to design solutions for different attacks in advance to ensure the deep learning systems with logit balancing can stop safely and promptly when attacks are happening.

Table 5 is the confusion matrix of a logistic regression [6] classifier that is trained to identify adversarial attacks feeding to the logit balancing model on CIFAR-10 data. The overall accuracy is 77.25%. Except for the adversarial attacks from the CW and FAB methods, all other attacks can be identified with at least 84% accuracy and not be confused with clean images. The clean images can be classified correctly at 72% and tend to confuse the CW attacks when they are misclassified. As shown in Table 2, after Gaussian noise masking, the robust accuracy of the logit balancing model under the CW attacks is 83.53%, which is very close to its natural accuracy of 85.05%.

Table 5. Confusion matrix of adversarial attacks detection.

Classified as →	a	b	c	d	e	f	g	h
a = APGD	44 (92%)	4	0	0	0	0	0	0
b = APGDT	4	46 (92%)	0	0	0	0	0	0
c = CW-L2	0	0	16 (32%)	15	0	0	0	19
d = FAB	0	0	13	34 (68%)	0	0	2	1
e = PGD	0	0	0	0	44 (88%)	6	0	0
f = PGD-L2	0	0	0	0	8	42 (84%)	0	0
g = Square	0	0	0	5	0	0	45 (90%)	0
h = Clean	0	0	12	2	0	0	0	36 (72%)

5   Conclusion

In this paper, the issues of DNNs robustness against human-designed adversarial perturbations are studied. The proposed defence method focuses on improving the robustness of DNNs against adversarial attacks via self-adaptive logit balancing and Gaussian noise boost training. The experimental results of this paper have been evaluated carefully and rigorously, and the results outperformed adversarial training defence approaches under non-targeted attacks. In the future, pre-trained DNNs with logit balancing can be used as a data augmentation tool, which can improve the label quality by replacing the one-hot label vectors with the SoftMax vectors produced by the pre-trained logit balancing model. Meanwhile, the proposed detection approach can achieve high accuracy and precise adversarial attack recognition without complex computation. However, this paper has not verified the robustness of the adversarial example detection method. The robustness of adversarial example detection can also improve the safety and trustworthiness of DNNs, which is worth studying in the future.

References

1. Aigrain, J., Detyniecki, M.: Detecting adversarial examples and other misclassifications in neural networks by introspection. arXiv preprint arXiv:1905.09186 (2019). 10.48550/arXiv. 1905.09186
2. Andriushchenko, M., Croce, F., Flammarion, N., Hein, M.: Square attack: a query-efficient black-box adversarial attack via random search. In: Vedaldi, A., Bischof, H., Brox, T., Frahm, J.-M. (eds.) ECCV 2020. LNCS, vol. 12368, pp. 484–501. Springer, Cham (2020). https://doi.org/10.1007/978-3-030-58592-1_29
3. Athalye, A., Engstrom, L., Ilyas, A., Kwok, K.: Synthesizing robust adversarial examples. In: International conference on machine learning, pp. 284–293. PMLR (2018). https://doi.org/10.48550/arXiv.1707.07397
4. Carlini, N., et al.: On evaluating adversarial robustness. arXiv preprint arXiv:1902.06705 (2019). 10.48550/arXiv. 1902.06705
5. Carmon, Y., Raghunathan, A., Schmidt, L., Liang, P., Duchi, J.C.: Unlabeled data improves adversarial robustness. arXiv preprint arXiv:1905.13736 (2019). 10.48550/arXiv. 1905.13736
6. le Cessie, S., van Houwelingen, J.: Ridge estimators in logistic regression. Appl. Stat. **41**(1), 191–201 (1992)

7. Croce, F., Hein, M.: Minimally distorted adversarial examples with a fast adaptive boundary attack. In: International Conference on Machine Learning, pp. 2196–2205. PMLR (2020). 10.48550/arXiv. 1907.02044

8. Croce, F., Hein, M.: Reliable evaluation of adversarial robustness with an ensemble of diverse parameter-free attacks. In: International Conference on Machine Learning, pp. 2206–2216. PMLR (2020). 10.48550/arXiv. 2003.01690

9. Engstrom, L., Ilyas, A., Athalye, A.: Evaluating and understanding the robustness of adversarial logit pairing. arXiv preprint arXiv:1807.10272 (2018). 10.48550/arXiv. 1807.10272

10. Fu, C., Chen, H., Ruan, N., Jia, W.: Label smoothing and adversarial robustness. arXiv preprint arXiv:2009.08233 (2020). 10.48550/arXiv. 2009.08233

11. Goodfellow, I.J., et al.: Generative adversarial networks. arXiv preprint arXiv:1406.2661 (2014). 10.48550/arXiv. 1406.2661

12. Goodfellow, I.J., Shlens, J., Szegedy, C.: Explaining and harnessing adversarial examples. arXiv preprint arXiv:1412.6572 (2014). 10.48550/arXiv. 1412.6572

13. Hosseini, H., Kannan, S., Poovendran, R.: Are odds really odd? bypassing statistical detection of adversarial examples. arXiv preprint arXiv:1907.12138 (2019). 10.48550/arXiv. 1907.12138

14. Huang, X., Kwiatkowska, M., Wang, S., Wu, M.: Safety verification of deep neural networks. In: Majumdar, R., Kunčak, V. (eds.) CAV 2017. LNCS, vol. 10426, pp. 3–29. Springer, Cham (2017). https://doi.org/10.1007/978-3-319-63387-9_1

15. Jin, G., Shen, S., Zhang, D., Dai, F., Zhang, Y.: APE-GAN: Adversarial perturbation elimination with gan. In: ICASSP 2019–2019 IEEE International Conference on Acoustics, Speech and Signal Processing (ICASSP), pp. 3842–3846. IEEE (2019). https://doi.org/10.1109/ICASSP.2019.8683044

16. Kannan, H., Kurakin, A., Goodfellow, I.: Adversarial logit pairing. arXiv preprint arXiv:1803.06373 (2018). 10.48550/arXiv. 1803.06373

17. Kim, H.: Torchattacks: A pytorch repository for adversarial attacks. arXiv preprint arXiv:2010.01950 (2020). 10.48550/arXiv. 2010.01950

18. Liu, X., Li, Y., Wu, C., Hsieh, C.J.: Adv-BNN: Improved adversarial defense through robust bayesian neural network. arXiv preprint arXiv:1810.01279 (2018). 10.48550/arXiv. 1810.01279

19. Madry, A., Makelov, A., Schmidt, L., Tsipras, D., Vladu, A.: Towards deep learning models resistant to adversarial attacks. arXiv preprint arXiv:1706.06083 (2017). 10.48550/arXiv. 1706.06083

20. Roth, K., Kilcher, Y., Hofmann, T.: The odds are odd: a statistical test for detecting adversarial examples. In: International Conference on Machine Learning, pp. 5498–5507. PMLR (2019)

21. Sammut, C., Webb, G.I.: Encyclopedia of Machine Learning. Springer Science & Business Media (2011)

22. Schmidt, L., Santurkar, S., Tsipras, D., Talwar, K., Madry, A.: Adversarially robust generalization requires more data. arXiv preprint arXiv:1804.11285 (2018). 10.48550/arXiv. 1804.11285

23. Shafahi, A., Ghiasi, A., Najibi, M., Huang, F., Dickerson, J.P., Goldstein, T.: Batch-wise logit-similarity: generalizing logit-squeezing and label-smoothing. In: European Conference on Computer Vision. British Machine Vision Conference (2019)

24. Shafahi, A., Najibi, M., Ghiasi, A., Xu, Z., Dickerson, J., Studer, C., Davis, L.S., Taylor, G., Goldstein, T.: Adversarial training for free! arXiv preprint arXiv:1904.12843 (2019). 10.48550/arXiv. 1904.12843

25. Szegedy, C., Vanhoucke, V., Ioffe, S., Shlens, J., Wojna, Z.: Rethinking the inception architecture for computer vision. In: Proceedings of the IEEE Conference on Computer Vision and Pattern Recognition, pp. 2818–2826 (2016). https://doi.org/10.1109/CVPR.2016.308
26. Szegedy, C., et al.: Intriguing properties of neural networks. arXiv preprint arXiv:1312.6199 (2013). 10.48550/arXiv. 1312.6199
27. Wu, D., Xia, S.T., Wang, Y.: Adversarial weight perturbation helps robust generalization. Adv. Neural Inf. Process. Syst. **33**, 2958–2969 (2020)
28. Zagoruyko, S., Komodakis, N.: Wide residual networks. arXiv preprint arXiv:1605.07146 (2016). 10.48550/arXiv. 1605.07146
29. Zhang, H., Yu, Y., Jiao, J., Xing, E., El Ghaoui, L., Jordan, M.: Theoretically principled trade-off between robustness and accuracy. In: International Conference on Machine Learning, pp. 7472–7482. PMLR (2019)
30. Zheng, A., Casari, A.: Feature Engineering for Machine Learning: Principles and Techniques for Data Scientists. O'Reilly Media, Inc. (2018)

Don't Wait Until the Accident Happens: Few-Shot Classification Framework for Car Accident Inspection in a Real World

Kyung Ho Park[✉] and Hyunhee Chung

Data Intelligence Group, SOCAR, Seoul, Republic of Korea
{kp,esther}@socar.kr

Abstract. Car accident inspection is a binary classification task to recognize whether a given car image includes a damaged surface or not. While the prior studies utilized various computer vision algorithms under the fully supervised, high data availability regime, these studies bear several limits for application in the real world. First, acquiring a large amount of car accident images is challenging due to their scarcity. Second, the supervised classifier would fail to recognize a sample not seen a priori. To improve the aforementioned drawbacks, we propose a few-shot classification framework for the accident inspection task and illustrate several takeaways to the practitioners. First, we designed a few-shot classification framework and validated our approach precisely identifies the accident, although the practitioner has a few accident images. Second, we analyzed the fine-grained discriminative characteristics between normal and accident images; thus, fine-grained feature extractor architecture is adequate for our accident inspection task. Third, we scrutinized optimal image resizing strategy varies along with the feature extractor architecture; therefore, we recommend that practitioners be cautious in handling real world car images. Lastly, we analyzed a larger number of acquired accident images that are advantageous in a few-shot classification. Based on these contributions, we highly expect further studies can realize the benefits of automated car part recognition in the real world shortly.

Keywords: Car accident inspection · Few-shot learning · Fine-grained classification

1 Introduction

Recent developments in computer vision and deep neural networks have empowered various applications in the intelligent transportation industry, such as traffic surveillance [3], traffic law enforcement [6,13], and especially in automating

---

K. H. Park and H. Chung—Equal Contribution.

© The Author(s), under exclusive license to Springer Nature Switzerland AG 2022
S. Sclaroff et al. (Eds.): ICIAP 2022, LNCS 13231, pp. 560–571, 2022.
https://doi.org/10.1007/978-3-031-06427-2_47

post-accident procedures [17,19]. When a car accident happens, the traditional procedures necessitate a visit of the human inspector to the scene of an accident, and it creates a burden of cost to the investigating entities. With the widespread of smartphones, investigating entities such as car insurance companies and car-sharing companies started to let their customers take pictures of their damaged cars and send them to the company. To filter accident images necessary to the inspection, the inspectors classify whether a given image includes the damaged car surface or not. As the aforementioned classification requires a particular resource consumption, academia and industry started developing a classifier with computer vision algorithms. Most approaches utilized deep neural networks such as Convolutional Neural Networks (CNN) to identify whether a given image is an accident or a normal image [17,19]. The prior studies acquired a large amount of finely-labeled datasets, established a classifier under the supervised learning paradigm, and achieved a promising binary classification performance.

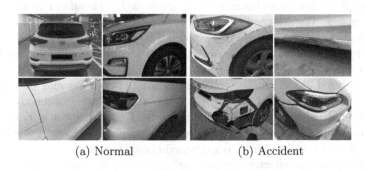

(a) Normal   (b) Accident

Fig. 1. Examples of normal and accident samples in the real world. The surface of the cars in the normal class does not have any defects in the image. On the other hand, there exist various defects (i.e., scratch, dent, severe crash separation of car part) at the surface of the cars in the accident class. We presumed that the practitioner cannot acquire a large labeled dataset of accident images under the supervised learning regime.

However, we figured out that the prior approaches have several drawbacks when applied to large-scale real-world settings. First, due to the scarcity of accident images, acquiring a labeled dataset requires a cost burden to the practitioners. The practitioner should wait until a particular amount of accident images are accumulated from their customers, and they should manually annotate every image. Second, the past approaches have a risk of failing at an accident image not seen during the training phase. Referring to the several real-world samples of the normal images and accident images, the normal samples at Fig. 1 (a) have a similar pattern with a clean car surface. On the other hand, accident samples at Fig. 1 (b) bear various damages such as scratches, dents, separation of car parts, severe crashes, and the mixture of the damages mentioned above. In a real world settings with emerging novel types of damage at the accident image, a classifier under the supervised learning paradigm would challenge to expand its capability to learn from the limited new samples.

To improve the drawbacks of past studies, our study proposes a few-shot learning framework for an efficient car accident inspection. The few-shot learning aims to automatically adapt the neural networks such that they work precisely on samples from classes not seen at the training phase, given a few labeled samples for the new class. In the classification setting, the few-shot classification model first trains a model with a large number of the labeled dataset that can be easily acquired. Then, it aims to establish a method that adapts to a novel classification task at the test phase where a small number of labeled samples are available at each class [1]. While the prior studies under the fully supervised, high data availability regimes, our few-shot learning framework only requires a small number of accident images. We trained the deep neural networks with easily accessible *car model dataset*, and designed the networks to adapt to the car accident images (which is novel to the trained classes) only with a few samples.

Throughout the study, the key contributions are as follows.

- We propose a few-shot classification framework that classifies normal and accident images. We designed the framework with two effective few-shot classification approaches **Siamese network** and **Relation network**, and validated our approach with **Relation network** can accomplish precise classification performance.
- We analyzed discriminative characteristics between normal and accident images are fine-grained. Along with the analogy, we examined a feature extractor architecture utilized in fine-grained classification is more advantageous than simple CNN and conventional ResNet architecture, which was conventionally utilized in coarse-grained classification.
- We scrutinized an effective image resizing strategy to deal with real world car accident images. We experimentally showed that the optimal image resizing strategy varies along with the feature extractor architectures; thus, practitioners should be cautious when handling car images in the real world.
- We showed more samples at the support set elevate the classification performance. Based on the analogy, we highly recommend the practitioners to acquire as many samples (car accident images in our problem setting) as possible to establish support sets.

2   Related Works

2.1   Car Accident Inspection

An early study [9] on car accident inspection utilized an image processing algorithm [9] acquired a 3D CAD model of the car and matched a given accident image with the 3D model to identify the damage. This approach leveraged image processing algorithms to figure out the correspondence between the 2D image and 3D model and accomplished a promising detection performance. Recently, an advance of deep learning empowered numerous car accident inspection methods with deep neural networks [17] acquired annotated accident images at 7 damaged car parts and 1 normal image and cast the car accident inspection task as an 8-class classification problem [19] accumulated 32 classes of damaged car parts to cover various accident types. Based on the annotated dataset, they trained a binary classifier that discriminates accident images from normal images. While the aforementioned studies achieved a promising classification performance, there exists a limit of the supervised learning paradigm. As they commonly utilize finely-labeled datasets for training the classifier, the practitioner should bear the cost of data collection and annotation. As the supervised classifier only knows the pattern of images included in the training set, these approaches have a risk of failing at classification given an image not seen a priori. In a nutshell, these limits of prior car accident inspection approaches have motivated our study to reduce the resource consumption at data acquisition and robustness to the unseen samples.

2.2 Few-Shot Classification

As an early approach to the few-shot learning, [10] proposed a Siamese networks that utilize a shared feature extractor to produce representation vectors for both the support and query images. Given a query image, Siamese networks identify the most similar class in the support set by selecting the smallest $\ell 1$ distance between the support and query images. As an improved study from the Siamese networks, [21] suggested Relation networks that extended the Siamese networks by parameterizing and learning the distance measuring module with Multi-Layered Perceptrons (MLP). As an extended study from the Siamese networks, Matching networks [22] utilizes cosine similarities, and ProtoNet [20] employed squared Euclidean distance. While the aforementioned approaches froze the trained feature extractors during the test phase, several works designed a framework to adapt the feature extractor to the few samples in the unseen support sets during the test phase. Model-Agonistic Meta Learning (MAML) [7] and its extended studies [14, 16, 18] designed a feature extractor to learn a set of meta-parameters which enables an adaptation into the unseen support classes. Although adapting feature extractors during the test phase accomplished precise performance, it has a limit that retraining inherently creates a burden of computation. We analyzed this burden of computation could create a burden to the practitioners in a real world setting; thus, our study selected Siamese networks and Relation networks, which are representative approaches under the paradigm of frozen feature extractor.

3 Few-Shot Classification Framework

3.1 Problem Setting

First, the few-shot classification utilizes three datasets: a training set, a support set, and a query set. The training set is a set of data used to train the few-shot classification model, and we utilize it during the training phase. The support set and the query set are utilized during the test phase. The support set and query set consist of the unseen classes during the training phase; thus, their labels space differs from the training set. There exist few samples at each class of the support set, while the query set includes a particular amount of labeled samples, which is manually labeled for model validation. If the support set includes K samples for each of N classes, we conventionally define the few-shot classification problem as N-way K-shot setting. Under scarce data regimes of car accident inspection, the training set for the few-shot classification should be easily acquirable large dataset. In this study, we collected car model images from 6 different car models for the training set, and we denoted this training set as *car model dataset*. As car insurance companies or car-sharing companies conventionally have many car images of various models, we analyzed car models dataset can be easily established. For the support set and query set, we retrieved normal and accident images and denoted this dataset as *accident inspection dataset*; thus, the test phase is consists of 2 classes.

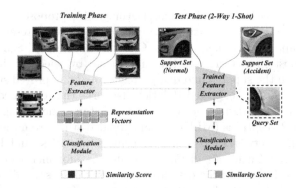

Fig. 2. Illustration of our few-shot classification framework in the car accident inspection setting. The training phase utilizes 6-way *car models dataset* and the test phase employs 2-way *accident inspection dataset*.

3.2   Few-Shot Classification

In this study, we employed two few-shot learning methods of **Siamese network** [10] and **Relation network** [21]. The difference between the two approaches is the classification module. Given a pair of representation vectors from the support sample and query sample, the **Siamese network** measures a $\ell 1$ distance between representation vectors and regress the final similarity score. On the other hand, **Relation network** concatenates representation vectors from the support sample and query sample and provides to the MLP model to finally estimate the similarity score. As numerous studies on few-shot classification noted, the similarity score metric is a significant factor in elevating the classification accuracy [1]. Thus, we analyzed comparing the classification performance between **Siamese network** and **Relation network** would provide a meaningful take-away towards candidate practitioners. We illustrated an overall structure of our framework in Fig. 2.

3.3   Feature Extractor

We analyzed simple feature extractor architecture proposed in the original studies of [10, 21] can be improved in the accident inspection problem setting for the following reasons. First, real world car images are much sophisticated than the public benchmark datasets. In these datasets (i.e., Omniglot [12], CIFAR-10 [11]), there exists a single object at the image of low resolution and size; thus, simple feature extractor architecture could have effectively scrutinized the pattern of each class. However, understanding the real world car images with various sizes and resolution scales would be challenging to the simple CNN architecture. Second, we analyzed the discriminative patterns between the normal class and accident class is fine-grained. Referring to Fig. 3, we compared sample images from the normal class and the accident class. While they have a similar shape and color, but the accident sample differs from the normal sample as it has scratches on its surface. We acknowledged that a simple CNN architecture would challenge learning the aforementioned fine-grained discriminative patterns; therefore, the feature extractor architecture has a room for improvement. Following the notion that wide and deep neural network architectures are more advantageous in learning sophisticated patterns [15], our study employed three types of feature extractor architectures as below.

- **Simple CNN**: A CNN architecture proposed in the studies of both **Siamese network** and **Relation network**. It consists of several convolution layers followed by the activation layer and pooling layer. Refer [10, 21] for a detailed implementation details.
- **ResNet** [8]: A wide and deep neural networks proposed in a ILSVRC challenge [4]. As conventional computer vision studies and applications in a real-world employ the extended versions of ResNet, we employed ResNet-50 architecture in the study.
- **Progressive Multi-Granularity Network (PMG)** [5]: A CNN architecture proposed in a fine-grained classification task. The PMG network is known to be advantageous in learning fine-grained discriminative patterns at each class.

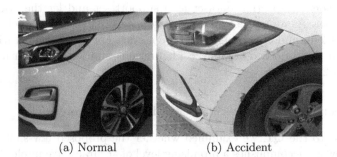

(a) Normal   (b) Accident

Fig. 3. Examples of normal and accident which look similar to each other, implying fine-grained difference between them. While these samples have a similar shape and color, the accident image at (b) differs from the normal image at (a) as a scratched surface exists at the accident image.

4   Experiments

4.1   Research Questions

Before the experiments, we defined three research questions illustrating key take-aways that our study aims to examine. The questions are described below.

- **Q1. Is our few-shot classification framework effective?** As a foremost takeaway, we examined whether the proposed few-shot classification framework can precisely discriminate accident images from normal images.
- **Q2. What is an optimal resizing strategy for real world car images?** While the images at the public benchmark datasets have the same size and resolution, car images from the real world do not; thus, we analyzed unifying the shape of real world images without much loss of information is important to the practitioners.
- **Q3. Is large K advantageous in 2-way K-shot classification?** Lastly, we scrutinized the relationship between the classification performance and the number of samples at each class of the support set, which is K.

4.2   Experiment Setup

Throughout the experiments, we acquired two labeled real world datasets: *1) car model dataset* for the training phase and *2) accident inspection dataset* for the test phase. Both datasets were provided by (Anonymous company name for blind submission), which is the largest car-sharing company in the (Anonymous country name for blind submission). *car model dataset* is a set of car images with 6 classes of car models, and each class includes 1000 images. *accident inspection dataset* consists if 2 classes of images: normal images and accident images. The dataset includes 2947 images at the normal class and 1222 samples at the accident class. Note that *car inspection dataset* are labeled by professional car accident inspectors at (Anonymous company name for blind submission) for the purpose of academic research. Under the setting of 2-way K-shot classification with episode training [20, 22], we randomly selected K samples at each class of the car accident inspection dataset as a support set and let the other images as a query set. As an evaluation metric, we measured the average classification accuracy at 200 randomly generated test episodes. For an implementation detail, we followed the proposed training configurations and loss function in previous studies [10, 21].

4.3   Effectiveness of Few-Shot Classification Framework

In response to the **Q1**, we examined whether the proposed few-shot classification framework accomplishes a promising level of accuracy. We implemented two methods of few-shot classification (**Siamese network** and **Relation network**) and one supervised classifier as a baseline noted as **Supervised**. The supervised classifier is a binary classifier trained with a few support set only. Under the scarce data availability regime, the practitioner can train a supervised classifier with a few samples at the support set, but the performance is not usually satisfactory [2,21]. We established a supervised classifier as a baseline of our approach; thus, we expect the few-shot classification framework would achieve better classification accuracy than the supervised classifier. We compared each framework's performance along with three different feature extractor architectures (**Simple CNN**, **ResNet**, and **PMG**) to determine the most effective one. Based on the aforementioned setups, experiment results are shown in Table 1.

Table 1. Few-shot classification accuracy on the combinations of approaches and the feature extractor architecture. Our few-shot classification framework based on **Relation network** accomplished the best accuracy among classification methods, and **PMG** contributed to the best accuracy among feature extractor archictectures.

Feature Extractor	Classification method		
	Supervised	Siamese network	Relation network
Simple CNN	0.4945	0.5450	**0.7151**
ResNet	0.5321	0.6100	**0.7405**
PMG	0.5054	0.5600	**0.7779**
Average	0.5106	0.5650	**0.7445**

Throughout the experiment results in Table 1, we figured out two key takeaways. First, we discovered our few-shot classification framework with Relation network accomplished a promising classification performance while the baseline (Supervised) and our approach with the Siamese network achieved an accuracy of around 50%. We analyze this supremacy derives from the parameterizable module at the few-shot classification method. While the framework with **Siamese network** measures the similarity between a support sample and a query sample based on $\ell 1$-distance, the Relation network employs MLP layers. As $\ell 1$-ddistance is non-parameterizable, the Siamese network could not adapt its similarity measuring module along with the dataset type. On the other hand, the Relation network can adapt its similarity measuring module as it is consists of learnable layers. We analyzed this adaptability at the similarity measuring module, which enabled our few-shot framework to adapt to real world car images, contributing to better classification accuracy.

Second, we discovered that fine-grained neural network architecture is more advantageous in understanding real world car images. Following the experiment result in Table 1, our few-shot classification framework with **PMG**-based feature extractor accomplished the best performance compared to the others. We presumed discriminative characteristics between accident and normal images were fine-grained. As shown in Fig. 3, we analyzed coarse-grained characteristics (i.e., shape, color) would not be enough to recognize accident images from normal images. As the performance of the few-shot classification framework with Relation network showed the best accuracy than others, we resulted in the discriminative characteristics between the accident and normal images are fine-grained rather than coarse-grained.

4.4   Optimal Image Resizing Strategy

In response to the **Q2**, we scrutinized an optimal image resizing strategy which sustains discriminative characteristics between the normal and the accident image. We utilized the framework with the **Relation network** following its superior performance shown in Sect. 4.3. We employed three conventional image resizing strategies (**CenterCrop, Interpolation**, and a serial operation of **Interpolation** after **CenterCrop**) and compared the classification performance along with different feature extractor architectures. As shown in Fig. 4, the **CenterCrop** option primarily focuses on the car object while the **Interpolation** option illustrates the object with backgrounds. The **Interpolation&CenterCrop** option describes a middle-level close-up between the **CenterCrop** and the **Interpolation**. The experiment results are shown in Table 2.

Table 2. Few-shot classification accuracy along with the combinations of resizing strategies and feature extractor architecture. As the accuracy varies along with the resizing strategy and feature extractor architecture, we recommend the practitioners be cautious in dealing with real world images.

Feature extractor	Resizing Strategy		
	CenterCrop	Interpolation & CenterCrop	Interpolation
Simple CNN	0.6126	0.6614	**0.7151**
ResNet-50	**0.7730**	0.7258	0.7405
PMG	0.4945	0.7455	**0.7779**
Average	0.6917	0.7109	0.7445

Original Image   Interpolation   Interpolation and CenterCrop   CenterCrop

Fig. 4. The illustration of resized images after **Interpolation, Interpolation & CenterCrop**, and **CenterCrop**

An interesting result is a similar performance between the **CenterCrop** option at the **ResNet** architecture and the **Interpolation** option at **PMG** architecture. We analyzed this result derived from the characteristics of the information that exists at the resized image. When we apply Interpolation to real world images, the resized image does not lose much information at the original image, such as background and the car's shape. On the other hand, the CenterCrop primarily focuses on the area at the center of the image; thus, the contextual information is disappeared. As fine-grained feature extractor (PMG) scrutinizes an image from the coarse to the fine level, contextual information is critical to learn the coarse-to-fine representations of the image. As the Interpolation option sustains many contextual details of the real world image, it contributes to the **PMG**-based approach. The **CenterCrop** option contrariwise hinders the **PMG** architecture from learning the pattern. On the other hand, **ResNet** scrutinizes well when a given image includes coarse-level characteristics; thus, the **ResNet** architecture is benefited from the **CenterCrop** option as it eliminates the background, which emphasizes the damaged part. In a nutshell, we discovered an optimal resizing strategy varies along with the feature extractor architecture. We resulted in the practitioners should precisely consider the resizing strategy along with their few-shot classification approach.

4.5   Impact of Large K

In response to the **Q3**, we examined whether a large number of samples at the support set contributes to better accuracy. Following the superior performance shown in Sect. 4.3 and 4.4, we utilized the Relation network with PMG feature extractors architecture and the Interpolation image resizing strategy. Under the setting of 2-way K-shot classification, we increased the level of K to compare the accuracy. Following the experiment results in Table 3, we resulted in the larger K size concretely contribute to the better classification accuracy; thus, the large number of samples is advantageous to the performance. In a nut shell, given our few-shot classification framework, the practitioners can easily enhance the classification performance by adding more samples to the support set.

Table 3. Few shot classification accuracy along with the size of K. As the larger K contributes to the escalation of accuracy, we recommend the practitioners acquire as many samples as possible for the support set.

2-way K-shot	Number of K			
	1-shot	3-shot	5-shot	7-shot
	0.5974	0.7509	0.7779	0.7981

5   Conclusion

In this study, we proposed a few-shot classification framework to establish an accident inspection system in the real world with the following contributions. First, we designed a few-shot classification framework and examined the **Relation network**-based approach is superior performance due to its parameterizable similarity measuring approach. Second, we analyzed the feature extractor with fine-grained neural networks contributes to escalating the classification accuracy. Third, we scrutinized that an optimal image resizing strategy is different along with the feature extractor architecture; thus, the practitioner should be cautious at resizing real world images into the fixed size. Lastly, as we examined the larger number of samples at the support set contributes to better accuracy, the practitioners are encouraged to acquire many samples to the support set as possible. Still, our framework shall be improved to elevate the classification performance, and it should be validated with more various scenarios. We highly expect our study becomes a concrete baseline to the studies of automating car accident inspection, and the extended studies can benefit society shortly.

References

1. Bateni, P., Goyal, R., Masrani, V., Wood, F., Sigal, L.: Improved few-shot visual classification. In: Proceedings of the IEEE/CVF Conference on Computer Vision and Pattern Recognition, pp. 14493–14502 (2020)
2. Chen, W.Y., Liu, Y.C., Kira, Z., Wang, Y.C.F., Huang, J.B.: A closer look at few-shot classification. arXiv preprint arXiv:1904.04232 (2019)
3. Datondji, S.R.E., Dupuis, Y., Subirats, P., Vasseur, P.: A survey of vision-based traffic monitoring of road intersections. IEEE Trans. Intell. Transp. Syst. **17**(10), 2681–2698 (2016)
4. Deng, J., Dong, W., Socher, R., Li, L.J., Li, K., Fei-Fei, L.: Imagenet: a large-scale hierarchical image database. In: 2009 IEEE Conference on Computer Vision and Pattern Recognition, pp. 248–255. IEEE (2009)
5. Du, R., et al.: Fine-grained visual classification via progressive multi-granularity training of jigsaw patches. In: Vedaldi, A., Bischof, H., Brox, T., Frahm, J.-M. (eds.) ECCV 2020. LNCS, vol. 12365, pp. 153–168. Springer, Cham (2020). https://doi.org/10.1007/978-3-030-58565-5_10
6. Fang, J., Zhou, Y., Yu, Y., Du, S.: Fine-grained vehicle model recognition using a coarse-to-fine convolutional neural network architecture. IEEE Trans. Intell. Transp. Syst. **18**(7), 1782–1792 (2016)
7. Finn, C., Abbeel, P., Levine, S.: Model-agnostic meta-learning for fast adaptation of deep networks. In: International Conference on Machine Learning, pp. 1126–1135. PMLR (2017)
8. He, K., Zhang, X., Ren, S., Sun, J.: Deep residual learning for image recognition. In: Proceedings of the IEEE Conference on Computer Vision and Pattern Recognition, pp. 770–778 (2016)
9. Jayawardena, S., et al.: Image based automatic vehicle damage detection (2013)
10. Koch, G., Zemel, R., Salakhutdinov, R., et al.: Siamese neural networks for one-shot image recognition. In: ICML Deep Learning Workshop, vol. 2. Lille (2015)

11. Krizhevsky, A., Hinton, G., et al.: Learning multiple layers of features from tiny images (2009)
12. Lake, B.M., Salakhutdinov, R., Tenenbaum, J.B.: Human-level concept learning through probabilistic program induction. Science **350**(6266), 1332–1338 (2015)
13. Lu, L., Huang, H.: A hierarchical scheme for vehicle make and model recognition from frontal images of vehicles. IEEE Trans. Intell. Transp. Syst. **20**(5), 1774–1786 (2018)
14. Mishra, N., Rohaninejad, M., Chen, X., Abbeel, P.: Meta-learning with temporal convolutions. **2**(7), 23 (2017). arXiv preprint arXiv:1707.03141
15. Nguyen, T., Raghu, M., Kornblith, S.: Do wide and deep networks learn the same things? uncovering how neural network representations vary with width and depth. arXiv preprint arXiv:2010.15327 (2020)
16. Nichol, A., Achiam, J., Schulman, J.: On first-order meta-learning algorithms. arXiv preprint arXiv:1803.02999 (2018)
17. Patil, K., Kulkarni, M., Sriraman, A., Karande, S.: Deep learning based car damage classification. In: 2017 16th IEEE International Conference on Machine Learning and Applications (ICMLA), pp. 50–54. IEEE (2017)
18. Ravi, S., Larochelle, H.: Optimization as a model for few-shot learning (2016)
19. Singh, R., Ayyar, M.P., Pavan, T.V.S., Gosain, S., Shah, R.R.: Automating car insurance claims using deep learning techniques. In: 2019 IEEE Fifth International Conference on Multimedia Big Data (BigMM), pp. 199–207. IEEE (2019)
20. Snell, J., Swersky, K., Zemel, R.S.: Prototypical networks for few-shot learning. arXiv preprint arXiv:1703.05175 (2017)
21. Sung, F., Yang, Y., Zhang, L., Xiang, T., Torr, P.H., Hospedales, T.M.: Learning to compare: relation network for few-shot learning. In: Proceedings of the IEEE Conference on Computer Vision and Pattern Recognition, pp. 1199–1208 (2018)
22. Vinyals, O., Blundell, C., Lillicrap, T., Wierstra, D., et al.: Matching networks for one shot learning. Adv. Neural Inf. Process. Syst. **29**, 3630–3638 (2016)

Robust Object Detection with Multi-input Multi-output Faster R-CNN

Sebastian Cygert(✉) and Andrzej Czyżewski

Faculty of Electronics, Telecommunication and Informatics,
Multimedia Systems Department, Gdańsk University of Technology,
Gdansk, Poland
sebcyg@multimed.org

Abstract. Recent years have seen impressive progress in visual recognition on many benchmarks, however, generalization to the out-of-distribution setting remains a significant challenge. A state-of-the-art method for robust visual recognition is model ensembling. However, recently it was shown that similarly competitive results could be achieved with a much smaller cost, by using multi-input multi-output architecture (MIMO).

In this work, a generalization of the MIMO approach is applied to the task of object detection using the general-purpose Faster R-CNN model. It was shown that using the MIMO framework allows building strong feature representation and obtains very competitive accuracy when using just two input/output pairs. Furthermore, it adds just 0.5% additional model parameters and increases the inference time by 15.9% when compared to the standard Faster R-CNN. It also works comparably to or outperforms the Deep Ensemble approach in terms of model accuracy, robustness to out-of-distribution setting, and uncertainty calibration when the same number of predictions is used. This work opens up avenues for applying the MIMO approach in other high-level tasks such as semantic segmentation and depth estimation.

Keywords: CNN · Robustness · Detection · Uncertainty · Ensembling

1 Introduction

Convolutional Neural Networks (CNNs) have recently become a standard method for image processing as they achieve excellent results on many benchmarks. Despite their impressive performance, the current machine learning techniques lack robustness when presented with an image that does not follow the

A. Czyżewski—This work was supported in part by the Polish National Centre for Research and Development (NCBR) through the European Regional Development Fund entitled: INFOLIGHT Cloud-Based Lighting System for Smart Cities under Grant POIR.04.01.04/2019.

training dataset distribution (out-of-domain data). It was shown that the current models are vulnerable to noisy input [17,33], novel weather conditions [22], and background changes [32], which creates safety considerations for models deployed to the real world, e.g., autonomous driving or medical applications. To improve visual recognition models' robustness it was proposed to use data augmentations that change objects' appearance, for example by using style-transfer [7,11] data augmentation or color distortions [4,8].

Additionally, the current models are often overconfident in their predictions [12], and the problem becomes more evident with out-of-domain data [8,24]. Sampling based-methods were shown to obtain very good results in terms of accuracy, out-of-domain robustness, and improving model predictive uncertainty [8,21,24]. The gold standard is the ensemble approach [14], which involves combining the output of several, diverse models. Several methods were developed to reduce the high computational cost, such as test-time dropout [10] or batch ensemble [30]. Another competitive approach is the m-heads model [19], which can be viewed as an ensemble with parameter sharing in the first layers of the network. Those methods, however, do not always match ensembling accuracy, as the success of sampling-based methods lies in the diversity of the predictions, which is a challenging problem [1,18].

Recently, the literature on obtaining many predictions from one model using a single inference step has increased. These methods were inspired by the compression methods, which show that it is possible to remove even 90% of the weights, without affecting the final model accuracy [9,13]. Therefore, instead of compressing the model, it should be possible to fit more than one subnetwork within the main network. For example, [5,31] use a single model in the multi-task setting, and the latter approach retrieves a subnetwork (from the main model) to efficiently solve the target task. Another method uses the multi-input multi-output (MIMO) approach, where a single model makes multiple predictions simultaneously. MIMO was shown to only slightly increase the computational cost while matching the accuracy of model ensembling and was showcased on the image classification task. Yet, whether the MIMO approach would work in a multi-task setting such as object detection, particularly when regressing objects' localizations is unknown. In this work, the MIMO method is adapted for object detection tasks and further evaluated. To summarize, the contributions of this work are as follows:

- The multi-input multi-output model was adapted to the object detection task and the architectural changes and implementation details are presented,
- It was shown that such a model acting as a strong regularizer brings significant improvements in terms of in-domain and out-of-domain accuracy, and in classification calibration, by adding only a small computational cost at inference time,
- The robustness of the MIMO approach robustness is presented by comparing its results to different Deep Ensemble approaches, which it outperforms (unless a larger number of models are used for the Deep Ensemble) or matches in accuracy.

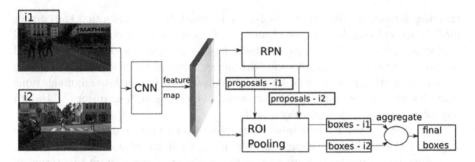

Fig. 1. Architecture of the proposed MIMO Faster R-CNN. Both images are sampled independently during training, and each subchannel in the network is responsible for predicting boxes in the corresponding image. During testing, both inputs to the network are the same, and the final results are obtained by running aggregation on both channel results.

2   Method

The standard Faster R-CNN model consists of two main modules [26]: first, region proposals are generated using a Region Proposal Network (RPN), and in the second stage, the region proposals are classified and refined using a Region of Interest Pooling network. The feature map is obtained using some standard CNN architecture. The RPN predicts a set of candidate boxes (anchors), for which it predicts the probability of the anchor being an object and its coordinates. In the final stage, each region proposal is processed by classification layers and regression layers. The first outputs a logit vector $z \in \mathbf{R}^K$ over all K classes, per anchor. The second outputs bounding-box regression offsets, which are used to refine the initial bounding boxes. Finally, a softmax function is applied $p = softmax(z)$, which results in a list of predicted class probabilities. The whole model is trained using a multi-task loss function, consisting of the classification part (cross-entropy loss) and the regression part of the bounding boxes localizations (L1 smooth loss).

The architecture overview of the proposed multi-input multi-output Faster R-CNN model is presented in the Fig. 1. To adapt the Faster R-CNN model into the MIMO framework following changes were applied:

- Multiply the number of input channels by M (ensemble size),
- Region Proposal Network now outputs M sets of region proposals (each per input image),
- The ROI Pooling layer independently processes M set of proposals and the outputs need to be aggregated at test-time. This can be done using the standard non-maximum suppression (NMS) method or more advanced methods, such as Weighted Boxes Fusion [28].

Note that the feature map (output from the convolutional backbone) is of the same size as before, however, now it contains information about M images, without forcing any explicit structure on how to share the information from different images. Using that shared feature map RPN returns M independent sets of region proposals, which requires changing the RPN loss function to:

$$L(\{\hat{p_{im}}\}, \{\hat{t_{im}}\}) = \frac{1}{M} \sum_{m=1}^{M} (\frac{1}{N_{cls}} \sum_{i} L_{cls}(\hat{p_{im}}, p_{im}) + \frac{\lambda}{N_{box}} \sum_{i} p_i smooth_{L1}(\hat{t_{im}} - t_{im})) \quad (1)$$

where i is the anchor index, and m is the index of input/output pair. $\hat{t_{im}}$ are predicted parametrized bounding boxes, $\hat{p_{im}}$ are predicted probabilities of the anchor being an object, and t_{im}, p_{im} are the ground-truth counterparts. The equation is normalized by N_{cls} - mini-batch size, and N_{box} - number of anchors. L_{cls} is simply the log loss over two classes (object vs. not object). Note that when $M = 1$, this refers to the standard RPN loss in the Faster R-CNN. Similarly, the loss for the ROI layer, now becomes a sum over M input/output pairs.

During training, each input is being sampled independently. However, during testing, the input is repeated M times so that M possibly different outputs are obtained for **the same input image**. It was empirically shown in the task of image classification that each of the M outputs provides good accuracy on its own and that the results are diverse enough, which allows them to be efficiently combined. In practice, $M = 2$ is often used. For more complex tasks, the network capacity does not allow processing a larger number of images in parallel. This agrees with the literature on model compression which shows that usually modest compression rates (up to 50%) are achievable, when using structured pruning (removing whole filters) [13,29].

After M sets of results are obtained, they need to be efficiently combined together. A standard approach in object detection to reduce redundant boxes is the NMS algorithm, which clusters together detections with high overlap, and keeps only detections with high confidence. Such a procedure might be non-optimal when combining predictions from different models. Recently, the Weighted Boxes Fusion (WBF) [28] method was proposed. It efficiently combines different predictions by updating the final bounding box coordinates by using the confidence-weighted average of coordinates forming a cluster. Similarly, the final confidence score is also an average of all boxes forming a cluster. In this work, both aggregation methods (NMS and WBF) are evaluated.

3 Experiments

3.1 Experimental Setting

Datasets. For the evaluation, Cityscapes [6], Berkeley Deep Drive (BDD) [34] and MS-COCO [20] datasets were used. Also a corrupted version [17] of the Cityscapes dataset was used. It included a number of synthetically generated distortions types grouped into four main categories: noise, blur, weather corruptions, and digital noise. Each corruption has five levels of intensity, and for simplicity, the distortions were applied using a medium intensity level. Usage of this benchmark is a popular way to measure models' robustness in the o.o.d. setting [7,15,17,22–25,33].

Implementation Details. For Cityscapes experiments, the models were trained for 64 epochs, using SGD optimizer, with an initial learning rate of 0.01 and a learning rate step reduction by a factor of 10 at epoch 48, similar as in [22]. All models were trained on a single GPU (Tesla V100) using a batch size of 6. During the training standard vision-based augmentations are applied: horizontal flipping and random resize. All of the models were pretrained on ImageNet [27]. For BDD and COCO datasets, the training lasted for 12 epochs, with learning rate reductions at epochs 8 and 11. Results are reported on the held-out validation sets. MMDetection library was used [3].

The color jittering data augmentation was applied using the Albumentations library [2] with default parameters and the following transformations: random changes in brightness, contrast, saturation and hue. In addition, style-transfer data augmentation was also used for some models, to improve the diversity of the ensemble approach. A standard procedure was applied, as in [11,22]: as the source of the style images, Kaggle's Painter By Numbers dataset was used, and during training, a stylized image was sampled with probability $p = 0.5$, otherwise, the original image was used.

Uncertainty Estimation. The Expected Calibration Error (ECE) is one of the ways to compute classification calibration [12]. Based on the confidence of the prediction, they are partitioned into M bins, and the ECE is computed as the weighted average of the calibration error within each bin:

$$ECE(c) = \sum_{m=1}^{M} \frac{|B_m|}{n} |acc(B_m) - conf(B_m)| \qquad (2)$$

with B_m being the set of indices of the samples for which the prediction confidence falls into the mth interval. A lower score means better calibration.

3.2   MIMO Faster R-CNN

Standard MIMO architecture struggles with fitting more subnetworks, especially on more challenging datasets. For example, in [15] the authors found that when training a ResNet-50 [16] classifier on the ImageNet dataset with $M = 2$, the model performed worse than a baseline. It was hypothesized that this happens when the main network does not have sufficient capacity to correctly classify two independent images at once. To improve that, the authors proposed relaxing independence between the inputs and added another hyperparameter p,

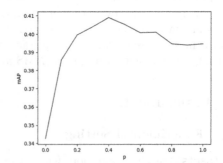

Fig. 2. Model accuracy (mAP) on Cityscapes dataset as a function of probability p that the same images are sampled, when the model is trained with $M = 2$ input/output pairs.

which defines the probability that the networks use the same data during training. Namely when $p = 0$, both images are sampled independently, and when $p = 1$, the training images are the same. As a result, in our first experiments on the Cityscapes dataset, a model with $M = 2$ input/output pairs was trained, and the p parameter was varied to see how it affected the final model performance (Fig. 2).

At $p = 0$ the inputs were fully independent, however, the final performance is limited by the network capacity. As the p grew, the subnetworks used the same image during training (with p probability), which allowed some of the features to be shared, which improved the performance. The performance peaked at $p = 0.4$ and then is slightly decreased. It is a similar result to that described in [15], when using ResNet-50 for ImageNet classification task. As a result, further experiments were performed using $M = 2$, and $p = 0.4$.

Further, the results are compared with the standard Faster R-CNN model (*baseline*) and Deep Ensemble approach (also consisting of $M = 2$ models) (Table 1). First, the MIMO Faster R-CNN outperformed a single model, improving the mAP score from 0.386 to 0.409. It also slightly outperformed the Deep Ensemble model. Importantly, the MIMO model brought only a slight increase in the parameters compared to the standard Faster R-CNN (from 41.38M to 41.4M). In terms of inference time (as measured on a Tesla V-100 GPU), it has increased by 15.9% (from 88ms to 102 ms per image). Note, that when applying the MIMO framework to the image classification task the increase in inference time was very small (around 1%) [15]. For the task of object detection, a larger increase in the processing time is attributed to M-times larger number of proposal regions being processed and the additional aggregation method. However, the processing time was still significantly smaller when compared to the Deep Ensemble method.

It is important to note that starting the training with ImageNet weights was crucial for the Cityscapes dataset (for all models). Additionally, when training the MIMO Faster R-CNN, one must also copy the ImageNet to the new filters (in the first channel). For the Deep Ensemble approach, the WBF aggregation method provided better results, yielding an improvement in the mAP score of 0.01, over the NMS approach. Overall, the WBF method performed the same or better than the NMS method, and all of the results were achieved using WBF aggregation. We also conducted experiments on the naive m-heads architecture [19], in which the backbone was shared and the RPN and ROI nets were doubled. However, such an approach resulted in poor performance (mAP score of 0.378) and a larger increase in number of parameters (55.9M). Such poor performance might be a result of the non-optimal structure of the proposed m-heads approach, and as such other variants could be explored.

3.3 Robustness and Uncertainty

In this section, further experiments are described which focus on robustness and uncertainty estimation. First, it can be noted that the accuracy in the o.o.d.

Table 1. Accuracy and computational cost of different methods.

Model	mAP	num. params	inf. time
Baseline	0.386	41.384M	0.088
MIMO (M=2)	0.409	41.397M	0.102
Deep ensemble (M=2)	0.406	82.768M	0.176

Table 2. Models' accuracy and calibration using different models and augmentation methods. Last two columns present results for corrupted Cityscapes. CJ stands for the color jittering augmentation and DE for the Deep Ensemble.

Model	mAP	ECE	c-mAP	c-ECE
Baseline	0.386	0.066	0.106	0.113
CJ	0.388	0.064	0.124	0.115
MIMO ($M = 2$)	**0.409**	0.045	**0.172**	0.075
MIMO ($M = 2$) + CJ	0.408	**0.04**	**0.172**	**0.071**
DE ($M = 2$, baseline)	0.406	0.068	0.116	0.124
DE ($M = 2$, CJ)	0.408	0.062	0.134	0.112
DE (MIMO, $M=2$)	**0.426**	0.05	0.184	0.087
DE (MIMO+CJ, $M=2$)	0.425	**0.046**	**0.186**	**0.068**
DE (baseline, $M=5$)	0.417	0.078	0.122	0.129
DE (CJ + style, $M=5$)	0.421	0.075	0.139	0.114

setting is severely impacted (Table 2), as it was previously shown in the literature. For the baseline model, the accuracy on the corrupted version of the dataset was equal to 0.106. The accuracy was significantly improved when using the MIMO approach (0.172), outperforming Deep Ensemble by a large margin (0.116).

Further, the impact of adding color jittering data augmentation was measured. As expected, it improved the accuracy of the baseline model in the o.o.d. setting. On the other hand, the MIMO approach did not result in significant changes, except for slightly improving model calibration. Deep Ensemble also benefited from the added data augmentation, but it clearly lacks the robustness of the MIMO approach (e.g., 0.134 mAP score in the o.o.d. testing, compared to the 0.172 of the MIMO approach). When using color jittering, no significant changes were observed when measuring the impact of the p value on the final accuracy (as in the Fig. 2), and as a result, $p = 0.4$ was further used.

It can also be noticed that the MIMO approach provided the best classification calibration results above of the tested models. The ECE score on the clean dataset equaled 0.045 (compared to 0.066 of the baseline model) and was further reduced to 0.042 when color jitter data augmentation was used. The ECE in the o.o.d. setting was again the lowest out of the evaluated methods, also significantly outperforming the Deep Ensemble approach.

Table 3. Accuracy on BDD Dataset when training on daytime images. $M = 2$ was used.

Model	BDD-day	BDD-night
Baseline	0.293	0.233
CJ	0.293	0.237
MIMO $(p = 0.7)$	0.301	0.244
MIMO + CJ $(p = 0.7)$	0.3	0.241
DE (baseline)	0.3	0.24
DE (CJ)	**0.302**	**0.246**

A potential critique of the evaluated Deep Ensemble approach is that a very small ensemble size was used and that the models' diversity is limited. As such, the ensemble method was also tested when 5 models were used, which was shown in the literature to provide good results already [24]. Additionally, one of the ensembles consisted of models of which some used color jittering data augmentation, and some used style-transfer data augmentation, to improve ensemble diversity. That setting allowed the Deep Ensemble approach to obtain very competitive results (Table 2, bottom part). The mAP increased to 0.421 and 0.139 on the clean and corrupted versions of the Cityscapes datasets, respectively (note that the accuracy in the o.o.d. is still worse than when using the MIMO approach). It was also checked whether using a MIMO Faster R-CNN models ensemble could further improve the results. When combining the outputs from two MIMO Faster R-CNN models, an impressive mAP of 0.426 was obtained on the clean dataset, and 0.184 on the corrupted version, which is an improvement over the Deep Ensemble approach consisting of 5 models. The usage of color jittering has improved the model calibration. Overall, these results further confirm the robustness of the MIMO Faster R-CNN model. Sample detections are presented in the Fig. 4.

It is also interesting to look at the accuracy of the MIMO Faster R-CNN when using only one output. In such a scenario, model accuracy equals 0.405, which is a 0.004 drop compared to the full MIMO approach, but it is still a significant improvement over the baseline (0.386). This shows that the MIMO framework acts as a strong regularizer during training, which leads to strong feature representations.

The proposed method was also evaluated on the **BDD dataset** (Table 3). The model was trained using daytime images only and evaluated on daytime and nighttime images in this setting, the same as in [22]. Overall, compared to the standard training, the MIMO approach improved the accuracy on the clean daytime images from 0.293 to 0.301 and on nighttime images (o.o.d. test) from 0.233 to 0.244. The probability p of sampling the same images during training also had to be further increased to observe improvements when using the MIMO model. This might be because a BDD is more challenging and includes a larger and more diverse set of images than the Cityscapes dataset. In that setting, the results of the MIMO approach are very similar to those obtained by Deep Ensemble. Looking at the single model accuracy within the MIMO method, it was found that it achieved almost the same accuracy (0.3 mAP value) as the full model. However, since the BDD dataset is very challenging, and the probability p of sampling the same images had to be increased, the outputs from single channels are no longer diverse, limiting the accuracy of the MIMO framework for this dataset. We experimented with a larger backbone (ResNet-101), but it provided a similar increase over the standard model.

When evaluating the model on the COCO dataset no gains in accuracy were observed. Given the hypothesis that significant gains of the MIMO approach come from the regularization property, it should work better when using only the fraction of the training dataset. In fact, such an observation was made and it was shown that the MIMO approach was in particular useful in the low data regime (Fig. 3). Each model was trained for the same number of steps. MIMO framework was, in particular, effective when less than 50% of the

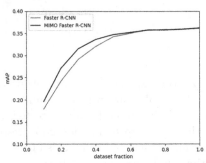

Fig. 3. Accuracy of the standard and MIMO-based Faster R-CNN on the COCO dataset when using only a fraction of the training dataset.

training dataset was used, e.g. when using 30% of the data, using the MIMO approach improved the accuracy from 0.291 to 0.315 of the mAP score.

Fig. 4. Detection results for the baseline and MIMO Faster R-CNN on different distortion types (motion blur, snow effect, Gaussian noise in the consecutive columns). Note, smaller confidence values for the MIMO model (i.e., 1st column). MIMO model performs on par or better than the standard model, however the corruptions vulnerability remains challenging (3rd column). Best viewed in digital format.

3.4 Discussion

The experiments showed that the MIMO approach could bring significant accuracy improvements compared to the standard training when using just $M = 2$ input/output pairs. Further, using just one output from the MIMO approach brings a significant gain in the model accuracy. The MIMO approach turned out to be especially effective when using a fraction of the original dataset on the COCO dataset. Given those observations, we conjecture that training the model in multi-input multi-output works as a very strong regularizer, which allows the model to build a robust feature representation, and therefore even when using a small M, the model can work very well. This finding adds more context to the original MIMO paper [15], which attributed its great performance mainly to ensembling diverse predictions. A similar observation was made in the literature for structured pruning, which showed that model compression could actually increase its performance (at modest compression rates) [29].

It was also interesting to note that using specific data augmentation (for example, style transfer, color jittering) was not essential for the MIMO model to significantly improve the accuracy in the out-of-distribution setting. Again, this might indicate that using such texture-invariant data augmentation is unnecessary for the model to increase its robustness when its build representation is strongly regularized. This can be viewed as a complementary finding to a recent work [23], which shows that the increased shape bias (using the aforementioned data augmentations) does not necessarily improve model robustness.

A potential drawback of the MIMO approach is that when the task or dataset is especially challenging, it requires increasing the probability p of sampling the same images during training. This reduces the diversity between model outputs and diminishes the potential gains of having multiple outputs. The presented results could be further improved. For example, it was shown that using batch repetition during training for the MIMO framework has improved the results [15,25], however, this came at the cost of significantly increased training time. Also, no specific optimization of the hyperparameters for MIMO was performed.

4 Conclusions

This work showed that using a multi-input multi-output approach can be generalized to object detection on real-world datasets. The MIMO Faster R-CNN model presented very competitive results in terms of model accuracy, uncertainty calibration and-out-of distribution robustness when using only $M = 2$ input/output pairs. The model adds only 0.5% of model parameters and increases the inference time by 15.9%. Similar accuracy can also be obtained when using the Deep Ensemble approach, but on the Cityscapes dataset, it required using a larger number of models (and a significantly higher computational cost). The MIMO approach works as a regularizer during training, which significantly increases the accuracy of a single subnetwork compared to the standard training. A current limitation of the MIMO framework is that when the target dataset or task is challenging, the probability p of sampling the same image during training must be increased, limiting the diversity of the MIMO outputs. The authors believe that this work opens up many directions for further research, including applying the approach to other high-level tasks such as semantic segmentation.

References

1. Ashukha, A., Lyzhov, A., Molchanov, D., Vetrov, D.P.: Pitfalls of in-domain uncertainty estimation and ensembling in deep learning. In: 8th International Conference on Learning Representations. ICLR (2020)
2. Buslaev, A., Iglovikov, V.I., Khvedchenya, E., Parinov, A., Druzhinin, M., Kalinin, A.A.: Albumentations: fast and flexible image augmentations. Information 11(2), 125 (2020)
3. Chen, K., et al.: MMDetection: Open mmlab detection toolbox and benchmark. preprint arXiv:1906.07155 (2019)
4. Chen, T., Kornblith, S., Norouzi, M., Hinton, G.E.: A simple framework for contrastive learning of visual representations. In: Proceedings of the 37th International Conference on Machine Learning. ICML (2020)
5. Cheung, B., Terekhov, A., Chen, Y., Agrawal, P., Olshausen, B.A.: Superposition of many models into one. Ann. Conf. Neural Inf. Process. Syst. 32 (2019)
6. Cordts, M., et al.: The cityscapes dataset for semantic urban scene understanding. In: Proceedings of the IEEE Conference on Computer Vision and Pattern Recognition, pp. 3213–3223 (2016)
7. Cygert, S., Czyżewski, A.: Toward robust pedestrian detection with data augmentation. IEEE Access 8, 136674–136683 (2020)
8. Cygert, S., Wróblewski, B., Woźniak, K., Słowiński, R., Czyżewski, A.: Closer look at the uncertainty estimation in semantic segmentation under distributional shift. In: 2021 International Joint Conference on Neural Networks (IJCNN) (2021)
9. Frankle, J., Carbin, M.: The lottery ticket hypothesis: finding sparse, trainable neural networks. In: 7th International Conference on Learning Representations. ICLR (2019)
10. Gal, Y., Ghahramani, Z.: Dropout as a Bayesian approximation: representing model uncertainty in deep learning. In: Proceedings of the 33nd International Conference on Machine Learning. ICML, vol. 48, pp. 1050–1059 (2016)
11. Geirhos, R., et al.: Imagenet-trained cnns are biased towards texture; increasing shape bias improves accuracy and robustness. In: 7th International Conference on Learning Representations. ICLR (2019)
12. Guo, C., Pleiss, G., Sun, Y., Weinberger, K.Q.: On calibration of modern neural networks. In: Proceedings of the 34th International Conference on Machine Learning-Volume, vol. 70, pp. 1321–1330 (2017)
13. Han, S., Pool, J., Tran, J., Dally, W.: Learning both weights and connections for efficient neural network. In: Advances in Neural Information Processing Systems, vol. 28 (2015)
14. Hansen, L.K., Salamon, P.: Neural network ensembles. IEEE Trans. Pattern Anal. Mach. Intell. 12(10), 993–1001 (1990)
15. Havasi, M., et al.: Training independent subnetworks for robust prediction. In: International Conference on Learning Representations. ICLR (2021)
16. He, K., Zhang, X., Ren, S., Sun, J.: Deep residual learning for image recognition. In: 2016 IEEE Conference on Computer Vision and Pattern Recognition. https://doi.org/10.1109/CVPR.2016.90

17. Hendrycks, D., Dieterich, T.G.: Benchmarking neural network robustness to common corruptions and perturbations. In: 7th International Conference on Learning Representations. ICLR (2019)

18. Lakshminarayanan, B., Pritzel, A., Blundell, C.: Simple and scalable predictive uncertainty estimation using deep ensembles. Adv. Neural Inf. Process. Syst. (NeurIPS) **30** (2017)

19. Lee, S., Purushwalkam, S., Cogswell, M., Crandall, D.J., Batra, D.: Why M heads are better than one: Training a diverse ensemble of deep networks. preprint arXiv:1511.06314 (2015)

20. Lin, T..Y.., et al.: Microsoft COCO: common objects in context. In: Fleet, David, Pajdla, Tomas, Schiele, Bernt, Tuytelaars, Tinne (eds.) ECCV 2014. LNCS, vol. 8693, pp. 740–755. Springer, Cham (2014). https://doi.org/10.1007/978-3-319-10602-1_48

21. Mehrtash, A., Wells, W.M., Tempany, C.M., Abolmaesumi, P., Kapur, T.: Confidence calibration and predictive uncertainty estimation for deep medical image segmentation. IEEE Trans. Med. Imaging **39**(12), 3868-3878 (2020)

22. Michaelis, C., et al.: Benchmarking robustness in object detection: autonomous driving when winter is coming. In: Machine Learning for Autonomous Driving Workshop, NeurIPS (2019)

23. Mummadi, C.K., Subramaniam, R., Hutmacher, R., Vitay, J., Fischer, V., Metzen, J.H.: Does enhanced shape bias improve neural network robustness to common corruptions? In: Proceedings of the 38th International Conference on Machine Learning. ICML (2021)

24. Ovadia, Y., et al.: Can you trust your model's uncertainty? evaluating predictive uncertainty under dataset shift. Adv. Neural Inf. Process. Syst. (2019)

25. Ramé, A., Sun, R., Cord, M.: Mixmo: Mixing multiple inputs for multiple outputs via deep subnetworks. preprint arXiv:2103.06132

26. Ren, S., He, K., Girshick, R., Sun, J.: Faster r-cnn: towards real-time object detection with region proposal networks. IEEE Trans. Pattern Anal. Mach. Intell. **28** (2016)

27. Russakovsky, O., et al.: ImageNet large scale visual recognition challenge. Int. J. Comput. Vis. (IJCV) **115**(3), 211-252 (2015)

28. Solovyev, R.A., Wang, W., Gabruseva, T.: Weighted boxes fusion: ensembling boxes from different object detection models. Image Vis. Comput. **107**, 104117 (2021)

29. Wen, W., Wu, C., Wang, Y., Chen, Y., Li, H.: Learning structured sparsity in deep neural networks. Adv. Neural Inf. Process. Syst. **29** (2016)

30. Wen, Y., Tran, D., Ba, J.: Batchensemble: an alternative approach to efficient ensemble and lifelong learning. In: 8th International Conference on Learning Representations. ICLR (2020)

31. Wortsman, M., et al.: Supermasks in superposition. In: Annual Conference on Neural Information Processing Systems 2020, NeurIPS 2020 (2020)

32. Xiao, K.Y., Engstrom, L., Ilyas, A., Madry, A.: Noise or signal: the role of image backgrounds in object recognition. In: 9th International Conference on Learning Representations. ICLR 2021 (2021)

33. Yin, D., Lopes, R.G., Shlens, J., Cubuk, E.D., Gilmer, J.: A fourier perspective on model robustness in computer vision. Adv. Neural Inf. Process. Syst. **32** (2019)

34. Yu, F., et al.: BDD100K: a diverse driving dataset for heterogeneous multitask learning. In: 2020 IEEE/CVF Conference on Computer Vision and Pattern Recognition. IEEE (2020)

Multiple Input Branches Shift Graph Convolutional Network with DropEdge for Skeleton-Based Action Recognition

Yan Liu, Yuelin Deng, Jinping Su[✉], Ruonan Wang, and Chi Li

Taikang Insurance Group Co., Ltd., Taikang Innovation Center, No. 21-1 Science Park Road, Changping District, Beijing, People's Republic of China
{liuyan146,dengyl29,sujp05,wangrn07,lichi01}@taikanglife.com

Abstract. Graph Convolutional Networks (GCNs) achieve remarkable success in the skeleton-based action recognition tasks. However, the recent state-of-the-art (SOTA) methods for this task usually have a large model size and too heavy computational complexity. In this work, we propose an early fused model, Multiple Input Branches Shift Graph Convolutional Network with DropEdge (MIBSD-GCN). First, to reduce the complexity of the multi-stream model, we introduce a lightweight Shift Graph Convolutional Network (Shift-GCN) block. It is embedded into an early fused architecture, Multiple Input Branches (MIB), which can enrich input features and suppresses the model redundancy. Then, a novel spherical coordinate representation is added as one of the input branches to enhance the recognition effect. Finally, we design the Shift Graph Convolutional Network with DropEdge (SD-GCN) to prevent over-fitting and over-smoothing, while maintain the model accuracy. Extensive experiments on two large-scale datasets, NTU RGB+D 60 and NTU RGB+D 120, show that the proposed model outperforms previous SOTA methods. We achieve 96.6% accuracy on the Cross-view benchmark of the NTU RGB+D 60, while being 3.4–16.5 times fewer FLOPs than other SOTA models.

Keywords: Skeleton-based action recognition · Multiple Input Branches · Shift Graph Convolutional Network · DropEdge · Spherical coordinate

1 Introduction

Skeleton-based action recognition plays an essential role in video understanding and human-computer interaction, owing to its robustness against dynamic circumstances and ability to work in complicated backgrounds.

Since Spatial Temporal Graph Convolutional Network (ST-GCN) [1] was first proposed to model the skeleton data with Graph Convolutional Network (GCN), a growing number of studies for skeleton-based action recognition based on GCNs have been reported. To learn discriminative and rich features from skeleton sequences, current state-of-the-art (SOTA) models are often extremely sophisticated and over-parameterized.

© The Author(s), under exclusive license to Springer Nature Switzerland AG 2022
S. Sclaroff et al. (Eds.): ICIAP 2022, LNCS 13231, pp. 584–596, 2022.
https://doi.org/10.1007/978-3-031-06427-2_49

These models generally contain a multi-stream architecture with numerous model parameters, resulting in a complicated training procedure and high computational cost. To tackle this problem, some efforts have been made in recent research from the perspectives of the GCN block innovation [2] or architecture improvement [3]; nevertheless, there are still a large number of redundant calculations. Therefore, our goal is to develop a faster, more powerful, and efficient method. In this paper, an early fused model, Multiple Input Branches Shift Graph Convolutional Network with DropEdge (MIBSD-GCN), is proposed. The main contributions of this study can be summarized as follows:

(1) An early fused Multiple Input Branches (MIB) architecture is designed to take inputs from four individual spatial-temporal feature sequences, including joint in Cartesian coordinate, joint in spherical coordinate, velocity and bone. In particular, the novel spherical coordinate joint is added as one of the input branches, to improve the discernibility of motion features.
(2) The Shift Graph Convolutional Network (Shift-GCN) block, which is composed of shift graph operations and lightweight point-wise convolutions, is embedded into the early fused MIB architecture to reduce the computational cost of the unit block and the architecture simultaneously.
(3) To avoid over-fitting and over-smoothing, an improved Shift Graph Convolutional Network with DropEdge (SD-GCN) is proposed, and the work includes defining a DropEdge operation in Shift-GCN and designing two DropEdge strategies.

2   Related Work

Early skeleton-based action recognition methods view human skeleton as a set of independent features that are aggregated by hand-crafted [4, 5] or by learning [6–9], and they model the spatial and temporal joint correlations. Subsequently, traditional deep-learning-based methods manually construct the skeleton as a sequence joint-coordinate vectors [7, 10–15] and feed it into RNNs or CNNs to generate prediction. In recent years, CNN is generalized to GCNs of arbitrary structures and successfully applied to image classification and other tasks. In 2018, Yan et al. proposed ST-GCN [1], which applied GCN to skeleton-based action recognition method for the first time. This algorithm models the dynamic skeleton-based on the time series representation of human joint positions, and extends the graph convolution into a spatial temporal convolution network to capture the spatial-temporal relationship. Later, many researchers put forward further improvements based on ST-GCN. Some of these works introduce incremental modules to enhance the expressiveness and network capacity [16–19]. Others adopt multi-stream fusion strategy [3, 13, 20, 21] which can learn more discriminative and richer features from skeleton sequences, while making the model more expressive. For instance, Shi et al. present a multi-stream attention-enhanced adaptive graph convolutional network (MS-AAGCN) [21], it extracts the joint, bone and their motion information and integrates them in a unified multi-stream framework. This approach increases the flexibility and generalization capacity of the model, thus it becomes commonly utilized.

Generally, most GCN-based models capture the dependency between joints of the human body structure on the spatial dimension and use the regular convolution on the time dimension for action recognition, meanwhile the multi-stream architecture is also widely adept. There are two common weaknesses in these models: (1) Higher computational cost is caused by the GCN operation or multi-stream architecture. (2) The accepting fields of both spatial and temporal graphs are pre-defined.

Some improvements are made to solve the above problems. Recently, a novel lightweight Shift-GCN [2] block is proposed instead of using heavy regular graph convolutions. Its non-local spatial shift graph operation is computationally efficient and achieves powerful performance. However, it is still a late fused architecture which leads to large model size and great computational cost. Part-wise Attention Residual Graph Convolutional Network (PA-ResGCN) [3] has advantages in this respect. It is designed as an early fused MIB architecture, which greatly reduces the Floating-point Operations (FLOPs) of the multi-stream model. Given the above, we hope to design a model that integrates their advantages and improves them.

3 Methods

3.1 Model Architecture

Taking into consideration the overall model optimization, we utilize PA-ResGCN and Shift-GCN as our baseline methods, and make three improvements: (1) The joint in spherical coordinate is added as one of the input branches for the feature expansion. (2) The lightweight Shift-GCN block is integrated into an early fused MIB architecture instead of the ResGCN block. (3) A DropEdge operation in Shift-GCN and two DropEdge strategies are proposed to prevent over-fitting and over-smoothing.

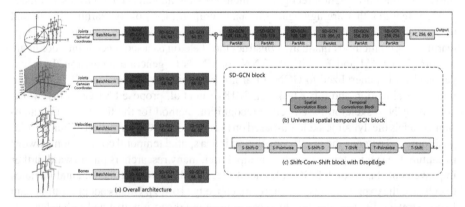

Fig. 1. The overview of MIBSD-GCN ("-D" denotes the DropEdge operation)

Figure 1 shows the overview of the proposed Multiple Input Branches Shift Graph Convolutional Network with DropEdge (MIBSD-GCN). After the original skeleton data pass the data pre-processing module introduced in Sect. 3.2, four input branches are obtained: joint in Cartesian coordinate, joint in spherical coordinate, velocity, and bone. Then, we develop the MIB architecture that merges the four input branches in the early stage of our model. The input branches consist of a batchnorm layer, an initial SD-GCN block and two SD-GCN blocks for feature extraction. After the input branches, a concatenation operation is employed to fuse the feature maps and then send them into the main stream, which is constructed by six SD-GCN blocks. At the same time, the Part-wise Attention (PartAtt) block is applied to highlight the importance of different body parts over the whole action sequence in the main stream [3]. Finally, the output feature map of the main stream is globally averaged to a feature vector, and a Fully Connected (FC) layer is used to determine the final action class. Notably, all the SD-GCN blocks are Shift-Conv-Shift mode blocks, as shown in Fig. 1(c), and the DropEdge operation is adopted, which is described in detail in Sect. 3.4.

3.2 Data Processing

The multi-stream fusion architecture is an effective way to augment the input data and enrich the input features. Based on the data composing and processing of the work [3], we add a new data representation, namely spherical coordinate, besides other three input branches. Different from common features, we discover spherical coordinate is also expressive and discriminative, which can more clearly show the changing process of body movements, and better improve the accuracy of action recognition. Therefore, we use spherical coordinate as another spatial representation of human skeleton data, and ensure that it is consistent with other representation methods for the feature description of the same posture [22, 23]. After data processing, we feed these four branches data (including eight data formats in total) into the early fused MIB structure in parallel. Next, we mainly introduce the transformation process from Cartesian coordinates to spherical coordinates.

Concretely, we map 3D skeleton joints to a spherical coordinate system. Suppose the 3D coordinates of a human skeleton joint v_{ti} is (x, y, z). The transformation relation between Cartesian coordinate system (x, y, z) and corresponding spherical coordinate system (r, θ, φ) is shown as Eq. (1):

$$\begin{cases} r = \sqrt{x^2 + y^2 + z^2} \\ \theta = \arccos \frac{z}{r} \\ \varphi = \arctan \frac{y}{x} \end{cases} \tag{1}$$

where r is the distance from the origin O to the joint v_{ti}, and θ is the angle between the line connecting the origin O and the joint v_{ti} to the positive Z-axis, denoted as the polar angle. The azimuth φ is the angle between the projection line of the connecting line of two points on the XY-plane to the positive X-axis. In this calculation we should make sure that θ is in the range $(-\frac{\pi}{2}, \frac{\pi}{2}]$ and φ is in the range $(-\pi, \pi]$.

3.3 Shift Graph Convolutional Network in Multiple Input Branches

Multi-stream approaches are divided into early fused and late fused architectures according to different fusion stages, and most of them applied in skeleton-based action recognition are late fused architectures. Although fusing at the score layer results in outstanding accuracy, the model size and computational cost are greatly increased, and many parameters in multi-stream models are redundant. An early fused MIB architecture is constructed in the work [3]. The results of different fusion stages show that fusing after the second stage is the best model to balance accuracy and complexity. This structure not only retains the rich input features, but also significantly suppresses the model complexity and redundancy, making the training procedure easier to converge.

Different from the structure amelioration in the MIB architecture, Shift-GCN [2] makes the model lighter from the angle of the GCN block improvement. Instead of using heavy regular graph convolutions, the Shift-GCN block is composed of novel shift graph operations and lightweight point-wise convolutions. Where, the adaptive non-local spatial shift graph operation makes the receptive field of each node cover the full skeleton graph in spatial dimension, and the adaptive temporal shift operation provides more flexible receptive fields for temporal graph. What's more, the ingenious combinations in the Shift-Conv-Shift mode further improve Shift-GCN and make it outperform the mainstream methods with strong performance and high efficiency.

In order to minimize the complexity and maintain the high accuracy, we replace each original ResGCN block in the MIB architecture by the more lightweight Shift-GCN block. It is expected that the integration of two lightweight and excellent methods leads to more effective comprehensive performance, which is verified in our experiments in Sect. 4.3.

3.4   Shift Graph Convolutional Network with DropEdge

In this subsection, SD-GCN is proposed to improve the effect of Shift-GCN embedded in the MIB framework. In the non-local spatial shift graph convolution, all nodes are connected to each other. As the number of edges increases in the early fused MIB architecture, the relationship between the links becomes more complex, causing over-fitting and over-smoothing to become more severe. The work [24] proves that DropEdge is a good solution to this problem. It can be used to generate various randomly deformed copies of the original graph, which better prevents over-fitting. Moreover, DropEdge is treated as a message-passing reducer in GCNs. By removing certain edges, the connections between nodes become sparser, preventing over-smoothing to some extent. Therefore, we introduce DropEdge to our model.

DropEdge Operation in Shift Graph Convolutional Network. A DropEdge operation in Shift-GCN is defined for the first time in this work. The original non-local shift graph operation is illustrated in Fig. 2(a). Given a spatial skeleton feature map $F \in \mathbf{R}^{N \times C}$, where N is the joint size and C is the channel size, the shift distance of the i-th channel is $i \bmod N$. The shifted-out channels are used to fill the corresponding empty spaces. The shift operation of node 1 is shown as an example in Fig. 2(a). After the non-local shift, the feature map looks like a spiral, which allows every node to obtain information from all other nodes [2]. Considering that the non-local shift graph operation is a directional operation, we regard the entire non-local graph as a directed graph and all edges as directed edges. When we plan to delete an edge from one node to another, we only need to stop the shift operation from the shifted-out node to the shifted-in node and keep the shifted-in channel at its original value on the feature map instead of being filled with the value of the shifted-out channel. An example of the DropEdge operation in the non-local shift graph operation is illustrated in Fig. 2(b). Supposing that the edges from node 3 to node 1 and node 6 to node 1 are dropped, the original values of node 1 in the corresponding shifted-in channels are retained on the feature map without shift operation. It is worth mentioning that the DropEdge operation only works in the non-local spatial shift graph convolution during training.

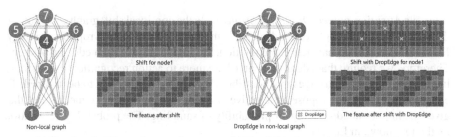

(a) Non-local shift graph operation   (b) DropEdge in non-local shift graph operation

Fig. 2. The process of the dropedge operation in shift graph convolutional network

Two Different DropEdge Strategies. In this subsection, we propose two different DropEdge strategies for Shift-GCN.

Random DropEdge Strategy. Consistent with an ordinary GCN, we use the adjacency matrix $\mathbf{A}_{shift}$ to represent the shift relationship in the directed graph. $\mathbf{A}_{shift}$ is an $N \times N$ matrix whose elements $\mathbf{A}_{shift} = \{a_{ij} | i = 1, ..., N, j = 1, ..., N\}$ indicate the shift relationship between the corresponding nodes. Here, v_i denotes the shift-in node, and v_j denotes the shift-out node. If there is a shift relationship from v_j to v_i, that is, the edge from v_j to v_i remains connected, it is expressed as $v_j \rightarrow v_i$. Thus, the elements in matrix $\mathbf{A}_{shift}$ can be expressed as Eq. (2):

$$a_{ij} = \begin{cases} 1, & if \ v_j \rightarrow v_i \\ 0, & else \end{cases} \quad (2)$$

All the elements in $\mathbf{A}_{shift}$ are 1 in the non-local shift graph. When the random DropEdge strategy works, the non-local shift graph randomly drops out a certain rate of edges. Formally, it enforces $N \times N \times (1 - p)$ non-zero elements of the adjacency matrix $\mathbf{A}_{shift}$ to be zero, where $N \times N$ is the total number of edges, and p is the kept rate. Following the expression in [24], $\mathbf{A}_{shift-drop}$ denotes the resulting adjacency matrix, and $\mathbf{A}_{drop}$ is a sparse matrix representing the dropped edges. Accordingly, the $\mathbf{A}_{shift-drop}$ relation with $\mathbf{A}_{shift}$ becomes:

$$\mathbf{A}_{shift-drop} = \mathbf{A}_{shift} - \mathbf{A}_{drop} \quad (3)$$

$\mathbf{A}_{shift-drop}$ determines whether the shift operation is executed between the corresponding nodes in the non-local shift graph convolution. If the element a_{ij} in $\mathbf{A}_{shift-drop}$

is 1, the normal shift operation is performed from v_j to v_i. Otherwise, the shifted-in channels corresponding to $v_j \rightarrow v_i$ on the feature map maintain the original values without shift operation. Specifically, we prefer to obtain $\mathbf{A}_{shift-drop}$ by independently computing for each l-th layer expressed as $\mathbf{A}_{shift-drop}{}^l$, instead of using the same $\mathbf{A}_{shift-drop}$ in a training batch. Such layer-wise DropEdge mode brings in more randomness and deformations of the original data and leads to higher performance than the batch-wise DropEdge mode [24].

Adaptive Attention-guided DropEdge Strategy. In the non-local shift graph convolution, the probability of all edges being dropped is the same with random DropEdge strategy. However, in actual human action recognition, the importance of each edge is different. Therefore, we require that edges that contribute more to action recognition should more likely to be retained, and edges that contribute less should more likely to be dropped. To enhance the effect, we propose an adaptive attention-driven DropEdge mechanism that gives the important edges a higher probability of sampling. The probability calculation method is shown in Eq. (4):

$$\mathbf{P}^l = d \cdot \tanh(k \cdot \mathbf{M_A}^l) + (1 - d) \tag{4}$$

$\mathbf{P}^l$ and $\mathbf{M_A}^l$ have the same size as $\mathbf{A}_{shift-drop}{}^l$. $\mathbf{P}^l$ is a matrix indicating the retained probability of the edges. $\mathbf{M_A}^l$ is a learnable and standardized attention mask. To assess the distribution of the attention area, it is often implicitly assumed that the absolute value of an activation is an indication of the importance of one unit [25]. We follow this assumption and generate $\mathbf{M_A}^l$ by averaging the absolute values across the channels corresponding to the edges. k is a scale factor, and d is a drop parameter for the probabilistic interval control. Through the constraint of this formula, the probability of the edges being retained is controlled in the interval $[1 - 2 * d, 1]$. In implementation, $\mathbf{A}_{shift-drop}{}^l$ is obtained through sample edges with the Bernoulli distribution with probability matrix $\mathbf{P}^l$.

4   Experimental Results

In this section, we introduce two large-scale datasets, NTU RGB+D 60 and NTU RGB+D 120, as well as implementation details of our experiments. In addition, we conduct ablation studies to evaluate the performance of each model component. Finally, we present the experimental results on these two datasets.

4.1   Datasets

NTU RGB+D 60. This dataset consists of a total of 56,880 3D human action videos collected simultaneously by three Kinect V2 multi-modal sensors. 40 subjects are invited to demonstrate different actions, including 60 action categories in total. In these action videos, each action sample corresponds to an action category, and there are four data patterns. In this work, we use a time series of 3D coordinates of multiple skeleton joints.

There are at most two human skeletons in each frame, each skeleton marks the coordinate positions of 25 human joints. Besides, the dataset is divided into two benchmarks for evaluation: Cross-subject (X-sub) and Cross-view (X-view). X-sub divides 40 subjects into a training set and a test set on average, while X-view divides the training set and the test set according to different cameras.

NTU RGB+D 120. This dataset is expanded in many aspects on the basis of NTU RGB+D 60. Namely, the number of subjects is increased to 106, the action category is supplemented with 60 additional categories, and 57,600 samples are added to the video. Similarly, The X-sub benchmark takes the videos shot by 53 volunteers as the training set and the rest as the validation set. X-setup selects the samples with even setup IDs as the training data and the samples with odd setup IDs as the testing data.

4.2 Experiment Settings

In our experiments, the maximum number of training epochs is set to 140. The stochastic gradient descent (SGD) with a Nesterov momentum of 0.9 and a weight decay of 0.0001 is employed to tune the parameters. The initial learning rate is set to 0.1 and decays with a cosine schedule. For NTU RGB+D 60 and NTU RGB+D 120, the batch size is 32.

4.3 Ablation Study

The Contribution of Spherical Coordinate Feature. An important improvement of our method is the addition of the spherical coordinate feature to the MIB framework. Here, we compare the accuracy of 3-stream (without spherical coordinate features) and 4-stream (with spherical coordinate features) MIB architectures. The ResGCN and Shift-GCN blocks are used as unit blocks of the architecture, respectively. The comparisons show that the spherical coordinate feature added to the input branches play a positive role in the final result, as observed in Table 1 and Table 2.

Table 1. Comparison between different MIB architectures using the ResGCN block on NTU RGB +D 60 X-view in accuracy (%).

MIB architecture	Unit block	Accuracy
3-stream	ResGCN	96.0
4-stream	ResGCN	96.2

Table 2. Comparison between different MIB architectures using the Shift-GCN block on NTU RGB+D 60 X-view in accuracy (%).

MIB architecture	Unit block	Accuracy
3-stream	Shift-GCN	95.8
4-stream	Shift-GCN	96.0

The Effect of Shift-GCN Embedded in the MIB Architecture. It is verified that integrating the Shift-GCN block into the MIB framework improves the overall performance. It can be seen from Table 3 and Table 4 that after the Shift-GCN block is embedded into the MIB architectures with 3-stream and 4-stream, the accuracy is close to that of the ResGCN block, while the FLOPs are reduced by more than 6 times.

Table 3. Comparison between different unit blocks in the 3-stream MIB architecture on NTU RGB +D 60 X-view in FLOPs (G) and accuracy(%).

MIB architecture	Unit block	FLOPs	Accuracy
3-stream	ResGCN	18.52	96.0
3-stream	Shift-GCN	2.89	95.8

Table 4. Comparison between different unit blocks in the 4-stream MIB architecture on NTU RGB+D 60 X-view in FLOPs (G) and accuracy(%).

MIB architecture	Unit block	FLOPs	Accuracy
4-stream	ResGCN	20.32	96.2
4-stream	Shift-GCN	2.96	96.0

The Contribution of SD-GCN. To prove the effectiveness of SD-GCN introduced in Sec. 3.4, respectively, we adopt different unit blocks and DropEdge strategies in the MIB architecture with 4-stream. The results of the ablation study are presented in Table 5 and Table 6. In summary, the best SD-GCN block gains 0.6% improvement than the baseline Shift-GCN block on NTU RGB+D 60 X-view. In addtion, the results also imply that the adaptive attention-guided DropEdge strategy performs better than the random DropEdge strategy, and that the layer-wise mode is preferred to the batch-wise mode. Based on the aforementioned results, we use the same best strategy and parameters for experiments on other datasets, as shown in Sect. 4.4.

4.4 Comparison with the State-of-the-Art

In this subsection, we compare the FLOPs and accuracy of the proposed MIBSD-GCN and other SOTA methods using two datasets: NTU-RGB+D 60 and NTU-RGB+D 120.

Table 5. Comparison between different random drop rates on NTU RGB+D 60 X-view in accuracy (%).

MIB architecture	Unit block	Random drop rate	Drop mode	Accuracy
4-stream	Shift-GCN	–	–	96.0
4-stream	SD-GCN	0.1	Batch-wise	96.1
4-stream	SD-GCN	0.2	Batch-wise	96.4
4-stream	SD-GCN	0.3	Batch-wise	96.3
4-stream	SD-GCN	0.1	Layer-wise	96.2
4-stream	SD-GCN	0.2	Layer-wise	96.5
4-stream	SD-GCN	0.3	Layer-wise	96.3

Table 6. Comparison between different attention-guided drop parameters on NTU RGB+D 60 X-view in accuracy (%).

MIB architecture	Unit block	Attention-guided drop parameter d	Drop mode	Accuracy
4-stream	Shift-GCN	–	–	96.0
4-stream	SD-GCN	0.1	Layer-wise	96.5
4-stream	SD-GCN	0.2	Layer-wise	96.6
4-stream	SD-GCN	0.3	Layer wise	96.4

As shown in Table 7, these results demonstrate the effectiveness and competitiveness of the proposed method. We consider that the combination of the MIB architecture and the lightweight Shift-GCN block significantly reduces the complexity of our model, while the expansion of input features and the proposed SD-GCN block make important contributions to the high accuracy.

For a clear illustration, Fig. 3 shows the FLOPs v.s. accuracy performance of our MIBSD-GCN and other SOTA methods, where the proposed MIBSD-GCN exceeds the SOTA methods in balancing accuracy and complexity, with high accuracy and 3.4–16.5 times less computation cost.

Table 7. Comparison with the SOTA methods on NTU RGB+D 60 and 120 datasets in FLOPs (G) and accuracy (%).

Models	NTU RGB+D 60			NTU RGB+D 120	
	FLOPs	Accuracy		Accuracy	
		X-view	X-sub	X-setup	X-sub
ST-GCN [1]	16.32	88.3	81.5	73.2	70.7
SR-TSL [27]	4.20	92.4	84.8	79.9	74.1
AS-GCN [17]	26.76	94.2	86.8	78.5	77.9
AGC-LSTM [18]	54.40	95.0	89.2	–	–
2s-AGCN [13]	37.32	95.1	88.5	84.2	82.5
MS-AAGCN [21]	–	96.2	90.0	–	–
DGNN [20]	126.80	96.1	89.9	–	–
MS-G3D [28]	48.88	96.2	91.5	88.4	86.9
PA-ResGCN-B19 [3]	18.52	96.0	90.9	88.3	87.3
4s-Shift-GCN [2]	10.00	96.5	90.7	87.6	85.9
DC-GCN+ADG [29]	25.72	96.6	90.8	88.1	86.5
EfficientGCN-B4 [30]	15.24	95.7	91.7	89.1	88.3
MIBSD-GCN (ours)	2.96	96.6	91.1	88.0	86.4

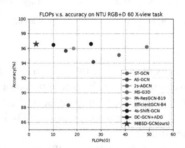

Fig. 3. FLOPs v.s. accuracy on NTU RGB+D 60 X-view task.

5   Conclusion

In this paper, we propose an efficient and strong model, Multiple Input Branches Shift Graph Convolutional Network with DropEdge (MIBSD-GCN) for skeleton-based action recognition. Different from other multi-stream frameworks, the proposed method combines a lightweight Shift-GCN unit block with an early fused MIB architecture, to reduce the computational complexity of the model. Furthermore, we design a feature representation in the spherical coordinate to enhance the recognition effect, and propose a DropEdge operation in Shift-GCN and two DropEdge strategies which ensure high model accuracy. On the challenging datasets, NTU RGB+D 60 and NTU RGB+D 120, the performance of our MIBSD-GCN model is equal to the SOTA methods, while the complexity is 3.4–16.5 times lower than that of them.

References

1. Yan, S., Xiong, Y., Lin, D.: Spatial temporal graph convolutional networks for skeleton-based action recognition. In: 32nd AAAI Conference on Artificial Intelligence, pp. 7444–7452. AAAI, New Orleans, LA (2018)
2. Cheng, K., Zhang, Y., He, X., Chen, W., Cheng, J., Lu, H.: Skeleton-based action recognition with shift graph convolutional network. In: 33rd IEEE/CVF Conference on Computer Vision and Pattern Recognition, pp. 183–192. IEEE, ELECTR NETWORK (2020)
3. Song, Y.F., Zhang, Z., Shan, C., Wang, L.: Stronger, faster and more explainable: a graph convolutional baseline for skeleton-based action recognition. In: 28th ACM International Conference on Multimedia, pp. 1625–1633. ACM, Seattle, US (2020)
4. Vemulapalli, R., Arrate, F., Chellappa, R.: Human action recognition by representing 3d skeletons as points in a lie group. In: 27th IEEE Conference on Computer Vision and Pattern Recognition, pp. 588–595. IEEE, Columbus, OH (2014)
5. Wang, J., Liu, Z., Wu, Y., Yuan, J.: Mining actionlet ensemble for action recognition with depth cameras. In: 25th IEEE Conference on Computer Vision and Pattern Recognition, pp. 1290–1297. IEEE, Providence, RI (2012)
6. Shahroudy, A., Liu, J., Ng, T.T., Wang, G.: Ntu rgb+d: A large scale dataset for 3D human activity analysis. In: 29th IEEE Conference on Computer Vision and Pattern Recognition, pp. 1010–1019. IEEE, Seattle, WA (2016)
7. Du, Y., Wang, W., Wang, L.: Hierarchical recurrent neural network for skeleton based action recognition. In: 28th IEEE Conference on Computer Vision and Pattern Recognition, pp. 1110–1118. IEEE, Boston, MA (2015)
8. Li, C., Xie, C., Zhang, B., Han, J., Zhen, X., Chen, J.: Memory attention networks for skeleton-based action recognition. IEEE Trans. Neural Netw. Learn. Systems (2021)
9. Zhang, P., Lan, C., Xing, J., Zeng, W., Xue, J., Zheng, N.: View adaptive recurrent neural networks for high performance human action recognition from skeleton data. In: 16th IEEE International Conference on Computer Vision, pp. 2136–2145. IEEE, Venice, ITALY (2017)
10. Liu, J., Wang, G., Duan, L.Y., Abdiyeva, K., Kot, A.C.: Skeleton-based human action recognition with global context-aware attention LSTM networks. IEEE Trans. Image Process. **27**(4), 1586–1599 (2017)
11. Li, R., Wang, S., Zhu, F., Huang, J.: Adaptive graph convolutional neural networks. In: 32nd AAAI Conference on Artificial Intelligence, pp. 3546–3553. AAAI, New Orleans, LA (2018)
12. Luan, S., Zhao, M., Chang, X.W., Precup, D.: Break the ceiling: Stronger multi-scale deep graph convolutional networks. arXiv preprint arXiv:1906.02174 (2019)
13. Shi, L., Zhang, Y., Cheng, J., Lu, H.: Two-stream adaptive graph convolutional networks for skeleton-based action recognition. In: 32nd IEEE/CVF Conference on Computer Vision and Pattern Recognition, pp. 12018–12027. IEEE Comp Soc, Long Beach, CA (2019)
14. Li, C., Cui, Z., Zheng, W., Xu, C., Yang, J.: Spatio-temporal graph convolution for skeleton based action recognition. In: 32nd AAAI Conference on Artificial Intelligence, pp. 3482–3489. AAAI, New Orleans, LA (2018)
15. Li, C., Zhong, Q., Xie, D., Pu, S.: Co-occurrence feature learning from skeleton data for action recognition and detection with hierarchical aggregation. In: 27th International Joint Conference on Artificial Intelligence, pp. 786–792. IJCAI, Stockholm, Sweden (2018)
16. Li, B., Li, X., Zhang, Z., Wu, F.: Spatio-temporal graph routing for skeleton-based action recognition. In: 33rd AAAI Conference on Artificial Intelligence, pp. 8561–8568. AAAI, Honolulu, HI (2019)

17. Li, M., Chen, S., Chen, X., Zhang, Y., Wang, Y., Tian, Q.: Actional-structural graph convolutional networks for skeleton-based action recognition. In: 32nd IEEE/CVF Conference on Computer Vision and Pattern Recognition, pp. 3590–3598. IEEE Comp Soc, Long Beach, CA (2019)

18. Si, C., Chen, W., Wang, W., Wang, L., Tan, T.: An attention enhanced graph convolutional LSTM network for skeleton-based action recognition. In: 32nd IEEE/CVF Conference on Computer Vision and Pattern Recognition, pp. 1227–1236. IEEE Comp Soc, Long Beach, CA (2019)

19. Wen, Y.H., Gao, L., Fu, H., Zhang, F. L., Xia, S.: Graph CNNs with motif and variable temporal block for skeleton-based action recognition. In: 33rd AAAI Conference on Artificial Intelligence, pp. 8989–8996. AAAI, Honolulu, HI (2019)

20. Shi, L., Zhang, Y., Cheng, J., Lu, H.: Skeleton-based action recognition with directed graph neural networks. In: 32nd IEEE/CVF Conference on Computer Vision and Pattern Recognition, pp. 7912–7921. IEEE Comp Soc, Long Beach, CA (2019)

21. Shi, L., Zhang, Y., Cheng, J., Lu, H.: Skeleton-based action recognition with multi-stream adaptive graph convolutional networks. IEEE Trans. Image Process. **29**, 9532–9545 (2020)

22. Xia, L., Chen, C.C., Aggarwal, J.K.: View invariant human action recognition using histograms of 3D joints. In: 25th IEEE Computer Society Conference on Computer Vision and Pattern Recognition Workshops, pp. 20–27. IEEE, Providence, RI (2012)

23. Yang, X., Tian, Y.: Effective 3d action recognition using eigenjoints. J. Vis. Commun. Image Represent. **25**(1), 2–11 (2014)

24. Rong, Y., Huang, W., Xu, T., Huang, J.: Dropedge: Towards deep graph convolutional networks on node classification. arXiv preprint arXiv:1907.10903 (2019)

25. Zagoruyko, S., Komodakis, N.: Paying more attention to attention: Improving the performance of convolutional neural networks via attention transfer. arXiv preprint arXiv:1612.03928 (2016)

26. Liu, J., Shahroudy, A., Perez, M., Wang, G., Duan, L.Y., Kot, A.C.: Ntu rgb+d 120: a large-scale benchmark for 3D human activity understanding. IEEE Trans. Pattern Anal. Mach. Intell. \textbf{42}(10), 2684–2701 (2019)

27. Si, C., Jing, Y., Wang, W., Wang, L., Tan, T.: Skeleton-based action recognition with spatial reasoning and temporal stack learning. In: 15th European Conference on Computer Vision, pp. 103–118. Springer-verlag Berlin, Munich, Germany (2018)

28. Liu, Z., Zhang, H., Chen, Z., Wang, Z., Ouyang, W.: Disentangling and unifying graph convolutions for skeleton-based action recognition. In: 33rd IEEE/CVF Conference on Computer Vision and Pattern Recognition, pp. 143–152. IEEE, ELECTR NETWORK (2020)

29. Cheng, K., Zhang, Y., Cao, C., Shi, L., Cheng, J., Lu, H.: Decoupling GCN with dropgraph module for skeleton-based action recognition. In: 16th European Conference on Computer Vision, pp. 536–553. Springer International Publishing, Glasgow, UK (2020)

30. Song, Y. F., Zhang, Z., Shan, C., Wang, L.: Constructing stronger and faster baselines for skeleton-based action recognition. arXiv preprint arXiv:2106.15125 (2021)

Contrastive Supervised Distillation for Continual Representation Learning

Tommaso Barletti[✉] (ID), Niccoló Biondi[✉] (ID), Federico Pernici(ID),
Matteo Bruni(ID), and Alberto Del Bimbo(ID)

Media Integration and Communication Center (MICC), Dipartimento di Ingegneria
dell'Informazione, Università degli Studi di Firenze, Florence, Italy
{tommaso.barletti,niccolo.biondi,federico.pernici,matteo.bruni,
alberto.bimbo}@unifi.it

Abstract. In this paper, we propose a novel training procedure for
the continual representation learning problem in which a neural network model is sequentially learned to alleviate catastrophic forgetting in visual search tasks. Our method, called *Contrastive Supervised Distillation* (CSD), reduces feature forgetting while learning discriminative features. This is achieved by leveraging labels information in a distillation setting in which the student model is con-
trastively learned from the teacher model. Extensive experiments show
that CSD performs favorably in mitigating catastrophic forgetting
by outperforming current state-of-the-art methods. Our results also
provide further evidence that feature forgetting evaluated in visual
retrieval tasks is not as catastrophic as in classification tasks. Code at:
https://github.com/NiccoBiondi/ContrastiveSupervisedDistillation.

Keywords: Representation learning · Continual learning · Image
retrieval · Visual search · Contrastive learning · Distillation

1 Introduction

Deep Convolutional Neural Networks (DCNNs) have significantly advanced the
field of visual search or visual retrieval by learning powerful feature representations from data [1–3]. Current methods predominantly focus on learning feature
representations from static datasets in which all the images are available during
training [4–6]. This operative condition is restrictive in real-world applications
since new data are constantly emerging and repeatedly training DCNN models on
both old and new images is time-consuming. Static datasets, typically stored on
private servers, are also increasingly problematic because of the societal impact
associated with privacy and ethical issues of modern AI systems [7,8].

T. Barletti and N. Biondi—Contributed equally.

© The Author(s), under exclusive license to Springer Nature Switzerland AG 2022
S. Sclaroff et al. (Eds.): ICIAP 2022, LNCS 13231, pp. 597–609, 2022.
https://doi.org/10.1007/978-3-031-06427-2_50

These problems may be significantly reduced in incremental learning scenarios as the computation is distributed over time and training data are not
required to be stored on servers. The challenge of learning feature representation
in incremental scenarios has to do with the inherent problem of catastrophic forgetting, namely the loss of previously learned knowledge when new knowledge
is assimilated [9,10]. Methods for alleviating catastrophic forgetting has been
largely developed in the classification setting, in which catastrophic forgetting
is typically observed by a clear reduction in classification accuracy [11–15]. The
fundamental differences with respect to learning internal feature representation
for visual search tasks are: (1) evaluation metrics do not use classification accuracy (2) visual search data have typically a finer granularity with respect to
categorical data and (3) no classes are required to be specifically learned. These
differences might suggest different origins of the two catastrophic forgetting phenomena. In this regard, some recent works provide some evidence showing the
importance of the specific task when evaluating the catastrophic forgetting of the
learned representations [16–19]. In particular, the empirical evidence presented
in [16] suggests that feature forgetting is not as catastrophic as classification
forgetting. We argue that such evidence is relevant in visual search tasks and
that it can be exploited with techniques that learn incrementally without storing
past samples in a memory buffer [20].

According to this, in this paper, we propose a new distillation method for the
continual representation learning task, in which the search performance degradation caused by feature forgetting is jointly mitigated while learning discriminative features. This is achieved by aligning current and previous features of the
same classes while simultaneously pushing away features of different classes. We
follow the basic working principle of contrastive loss [21], used in self-supervised
learning, to effectively leverage label information in a distillation-based training
procedure in which we replace anchor features with the feature of the teacher
model.

Our contributions can be summarized as follows:

1. we address the problem of continual representation learning proposing a novel
   method that leverages label information in a contrastive distillation learning
   setup. We call our method Contrastive Supervised Distillation (CSD).
2. experimental results on different benchmark datasets show that our CSD
   training procedure achieves state-of-the-art performance.
3. Our results confirm that feature forgetting in visual retrieval using fine-
   grained datasets is not as catastrophic as in classification.

2   Related Works

Continual Learning (CL). CL has been largely developed in the classification setting, where methods have been broadly categorized based on exemplar
[22–25] and regularization [20,26–28]. Only recently, continual learning for feature representation is receiving increasing attention and few works pertinent to
the regularization-based category has been proposed [17–19]. The work in [17]

proposed an unsupervised alignment loss between old and new feature distributions according to the Mean Maximum Discrepancy (MMD) distance [29]. The work [19] uses both the previous model and estimated features to compute a semantic correlation between representations during multiple model updates. The estimated features are used to reproduce the behaviour of older models that are no more available. Finally, [18] addresses the problem of lifelong person re-identification in which the previously acquired knowledge is represented as similarity graphs and it is transferred on the current data through graphs convolutions. While these methods use labels only to learn new tasks, our method leverages labels information to both learn incoming tasks and for distillation.

Reducing feature forgetting with feature distillation is also related to the recent backward compatible representation learning in which newly learned models can be deployed without the need to re-index the existing gallery images [30–32]. This may have an impact on privacy as also the gallery images are not required to be stored on servers. Finally, the absence of the cost re-indexing is advantageous in streaming learning scenarios as [33,34].

Contrastive Learning. Contrastive learning has been proposed in [35] for metric learning and then it is demonstrated to be effective in unsupervised/self-supervised representation learning [21,36,37]. All these works focus on obtaining discriminative representations that can be transferred to downstream tasks by fine-tuning. In particular, this is achieved as, in the feature space, each image and its augmented samples (the positive samples) are grouped together while the others (the negative samples) are pushed away. However, [38] observed that, given an input image, samples of the same class are considered as negative and, consequently, pushed apart from it. We follow a similar argument which considers as positive also these images.

3   Problem Statement

In the continual representation learning problem, a model $\mathrm{M}(\,\cdot\,; \theta, \mathbf{W})$ is sequentially trained for T tasks on a dataset $\mathcal{D} = \{(\mathbf{x}_i, y_i, t_i) \mid i = 1, 2, \ldots, N\}$, where $\mathbf{x}_i$ is an image of a class $y_i \in \{1, 2, \ldots, L\}$, N is the number of images, and $t_i \in \{1, 2, \ldots, T\}$ is the task index associated to each image. In particular, for each task k, M is trained on the subset $\mathcal{T}_k = \mathcal{D}|_{t_i = k} = \{(\mathbf{x}_i, y_i, t_i) \mid t_i = k\}$ which represents the k-th training-set that is composed by L_k classes. Each training-set has different classes and images with respect to the others and only $\mathcal{T}_k$ is available to train the model M (memory-free).

At training time of task k, in response to a mini-batch $\mathcal{B} = \{(\mathbf{x}_i, y_i, t_i)\}_{i=1}^{|\mathcal{B}|}$ of $\mathcal{T}_k$, the model M extracts the feature vectors and output logits for each image in the batch, i.e., $\mathrm{M}(\mathbf{x}_i) = C(\phi(\mathbf{x}_i))$, where $\phi(\cdot, \theta)$ is the representation model which extracts the feature vector $f_i = \phi(\mathbf{x}_i)$ and C is the classifier, which projects the feature vector f_i in an output vector $z_i = C(f_i)$. At the end of the training phase, M is used to index a gallery-set $\mathcal{G} = \{(\mathbf{x}_g, y_g) \mid g = 1, 2, \ldots, N_g\}$ according to the extracted feature vectors $\{(f_g, y_g)\}_{g=1}^{N_g}$.

At test time, a query-set $\mathcal{Q} = \{\mathbf{x}_q \mid q = 1, 2, \ldots, N_q\}$ is processed by the representation model $\phi(\cdot, \theta)$ in order to obtain the set of feature vectors $\{f_q\}_{q=1}^{N_q}$. According to cosine distance function d, the nearest sample in the gallery-set $\mathcal{G}$ is retrieved for each query sample f_q, i.e.,

$$f^* = \underset{g=1,2,\ldots,N_g}{\arg\min} \; d(f_g, f_q), \tag{1}$$

4 Method

To mitigate the effect of catastrophic forgetting while acquiring novel knowledge from incoming data, we propose a training procedure that follows the *teacher-student* framework, where the teacher is the model before the update and the student is the model that is updated. The teacher is leveraged during the train of the student to preserve the old knowledge as old data is not available.

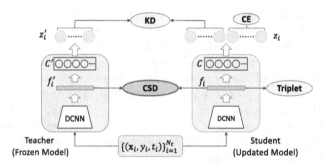

Fig. 1. Proposed method is based on the teacher-student framework. During the training of the student, CE and triplet losses are minimized to learn the new task data, are KD and CSD are used to preserve the old knowledge using the teacher (not trainable).

With reference to Fig. 1, at each task k, the student is trained on the training-set $\mathcal{T}_k = \{(\mathbf{x}_i, y_i, t_i) \mid t_i = k\}$ and the teacher is set as frozen, i.e., not undergoing learning. The loss function that is minimized during the training of the student is the following:

$$\mathcal{L} = \mathcal{L}_{plasticity} + \mathcal{L}_{stability} \tag{2}$$

where $\mathcal{L}_{stability} = 0$ during the training of the model on the first task. In the following, the components of the plasticity and stability loss are analyzed in detail. In particular, we adopt the following notation. Given a mini-batch $\mathcal{B}$ of training data, both the student and the teacher networks produce a set of feature vectors and classifier outputs in response to training images $\mathbf{x}_i \in \mathcal{B}$. We refer to as $\{f_i\}$, $\{z_i\}$ for the feature vectors and classifier outputs of the student, respectively, with $\{f_i'\}$, $\{z_i'\}$ for the teacher ones, and with $|\mathcal{B}|$ to the number of elements in the mini-batch.

4.1   Plasticity Loss

Following [17], during the training of the updated model, the plasticity loss is defined as follows:

$$\mathcal{L}_{plasticity} = \mathcal{L}_{\text{CE}} + \mathcal{L}_{\text{triplet}} \tag{3}$$

with

$$\mathcal{L}_{\text{CE}} = \frac{1}{|\mathcal{B}|} \sum_{i=1}^{|\mathcal{B}|} y_i \log\left(\frac{\exp\left(z_i\right)}{\sum_{j=1}^{|\mathcal{B}|} \exp\left(z_j\right)} \right) \tag{4}$$

$$\mathcal{L}_{\text{triplet}} = \max(\|f_i - f_p\|_2^2 - \|f_i - f_n\|_2^2). \tag{5}$$

$\mathcal{L}_{\text{CE}}$ and $\mathcal{L}_{\text{triplet}}$ are the cross-entropy loss and the triplet loss, respectively. The plasticity loss of Eq. 3 is optimized during the training of the model and it is used in order to learn the novel tasks.

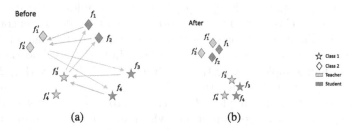

Fig. 2. Proposed CSD loss. *(a)* The features of four samples of two classes are firstly mapped in the feature space by the teacher (blue) and the student (orange). *(b)* With CSD samples belonging to the same class (same symbol) are clustered together and separated from the others. (Color figure online)

4.2   Stability Loss

The stability loss preserves the previously acquired knowledge in order to limit the catastrophic forgetting effect, that is typically performed using the teacher model for distillation. The stability loss we propose is formulated as follows:

$$\mathcal{L}_{stability} = \lambda_{\text{KD}} \, \mathcal{L}_{\text{KD}} + \lambda_{\text{CSD}} \, \mathcal{L}_{\text{CSD}} \tag{6}$$

where λ_{KD} and λ_{CSD} are two weights factors that balance the two loss components, namely Knowledge Distillation (KD) and the proposed Contrastive Supervised Distillation (CSD). In our experimental results, we set both λ_{KD} and λ_{CSD} to 1. An evaluation of different values is reported in the ablation studies of Sect. 6.

Knowledge Distillation. KD [39] minimizes the log-likelihood between the classifier outputs of the student and the soft labels produced by the teacher, instead of the ground-truth labels (y_i) used in the standard cross-entropy loss. This encourages the outputs of the updated model to approximate the outputs produced by the previous one. KD is defined as follows:

$$\mathcal{L}_{KD} = \frac{1}{|\mathcal{B}|} \sum_{i=1}^{|\mathcal{B}|} \frac{\exp(z_i')}{\sum_{j=1}^{|\mathcal{B}|} \exp(z_j')} \log\left(\frac{\exp(z_i)}{\sum_{j=1}^{|\mathcal{B}|} \exp(z_j)} \right) \tag{7}$$

Contrastive Supervised Distillation. We propose a new distillation loss, i.e., the Contrastive Supervised Distillation (CSD) that aligns current and previous feature models of the same classes while simultaneously pushing away features of different classes. This is achieved at training time imposing the following loss penalty:

$$\mathcal{L}_{CSD} = -\frac{1}{|\mathcal{B}|} \sum_{i=1}^{|\mathcal{B}|} \frac{1}{|\mathcal{P}(i)|} \sum_{p \in \mathcal{P}(i)} \log\left(\frac{\exp(f_i' \cdot f_p)}{\sum_{\substack{a=1 \\ a \neq i}}^{|\mathcal{B}|} \exp(f_i' \cdot f_a)} \right) \tag{8}$$

where $\mathcal{P}(i) = \{(x_p, y_p, t_p) \in \mathcal{B} \mid y_p = y_i\}$ is a set of samples in the batch which belong to the same class of $\mathbf{x}_i$, i.e., the *positive* samples. Equation 8 encourage for each class, the alignment of the student representations to the ones of the same class of the teacher model, which acts as anchors. In Fig. 2, we show the effect of CSD loss on four samples $\{(\mathbf{x}_i, y_i)\}_{i=1}^{4}$ with $y_i \in \{1, 2\}$. Initially (Fig. 2(a)) the feature vectors extracted by the student f_i (orange samples) are separated from the teacher ones f_i' (blue samples). CSD clusters together features of the same class moving the student representations, which are trainable, towards the fixed ones of the teacher while pushing apart features belonging to different classes. For the sake of simplicity, this effect is shown just for f_1' and f_3'. Indeed, f_1 and f_2 become closer to f_1', while f_3 and f_4 are spaced apart with respect to f_1' as they are of class 2. The same effect is visible also for f_3' which attracts f_3 and f_4 and push away f_1 and f_2 as shown in Fig. 2(b).

CSD imposes a penalty on feature samples considering not only the overall distribution of features of the teacher model with respect to the student one, but it also clusters together samples of the same class separating from the clusters of the other classes. Our method differs from KD as the loss function is computed directly on the features and not on the classifier outputs resulting in more discriminative representations. CSD also considers all the samples of each class as positive samples that are aligned with the same anchor of the teacher and not pairs (teacher-student) of samples as in [40].

5   Experimental Results

We perform our experimental evaluation on CIFAR-100 [41] and two fine-grained datasets, namely CUB-200 [42] and Stanford Dogs [43]. The CIFAR-100 dataset consist of 60000 32×32 images in 100 classes. The CUB-200 dataset contains 11788 224×224 images of 200 bird species. Stanford Dogs includes over 22000 224×224 annotated images of dogs belonging to 120 species.

The continual representation learning task is evaluated following two strategies. In CIFAR-100, we evenly split the dataset into T training-set, where the model is trained sequentially. The experiments are evaluated with $T = 2, 5, 10$. In CUB-200 and Stanford Dogs, following [44,45], we use half of the data to pre-train a model and split the remaining data into T training-set. CUB-200 is evaluated with $T = 1, 4, 10$ while Stanford Dogs with $T = 1$.

Implementation Details. We adopt ResNet32 [46][1] as representation model architecture on CIFAR-100 with 64-dimension feature space. We trained the model for 800 epochs for each task using Adam optimizer with a learning rate of $1 \cdot 10^{-3}$ for the initial task and $1 \cdot 10^{-5}$ for the others. Random crop and horizontal flip are used as image augmentation. Following [19], we adopt pretrained Google Inception [47] as representation model architecture on CUB-200 and Stanford Dogs with 512-dimension feature space. We trained the model for 2300 epochs for each task using with Adam optimizer with a learning rate of $1 \cdot 10^{-5}$ for the convolutional layers and $1 \cdot 10^{-6}$ for the classifier. Random crop and horizontal flip are used as image augmentation. We adopt RECALL@K [44,48] as performance metric using each image in the test-set as query and the others as gallery.

Table 1. Evaluation on CIFAR-100 of CSD and compared methods.

METHOD	RECALL@1 (1–50)	RECALL@1 (51–100)	RECALL@1 Average
Initial model	67.6	21.7	44.7
Fine-Tuning	37.4	**64.1**	50.8
LwF [20]	64.0	59.4	61.7
MMD loss [17]	61.8	60.9	61.4
CSD (Ours)	**65.1**	61.6	**63.4**
Joint training	70.5	71.9	71.2

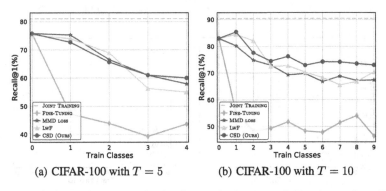

(a) CIFAR-100 with $T = 5$ (b) CIFAR-100 with $T = 10$

Fig. 3. Evolution of RECALL@1 on the first task as new tasks are learned on CIFAR-100. Comparison between our method (CSD) and compared methods.

---

[1] https://github.com/arthurdouillard/incremental_learning.pytorch

5.1   Evaluation on CIFAR-100

We compare our method on CIFAR-100 dataset with the Fine-Tuning baseline, LwF [20], and [17] denoted as MMD loss. As an upper bound reference, we report the Joint Training performance obtained using all the CIFAR-100 data to train the model.

We report in Table 1 the scores obtained with $T = 2$. In the first row, we show the Initial Model results, i.e., the model trained on the first half of data from CIFAR-100. Our approach achieves the highest recall when evaluated on the initial task and the highest recall on the second task between methods trying to preserve old knowledge, being second only to Fine-Tuning that focuses only on learning new data. This results in our method achieving the highest average recall value with an improvement of −2% RECALL@1 with respect to LwF and MMD loss and −10.4% with respect to the Fine-Tuning baseline. The gap between all the continual representation learning methods and Joint Training is significant (∼8%). This underlines the challenges of CIFAR-100 in a continual learning scenario since there is a noticeable difference in the appearance between images of different classes causing a higher feature forgetting.

Figure 3(a) and Fig. 3(b) report the evolution of RECALL@1 on the initial task as new tasks are learned with $T = 5$ and $T = 10$, respectively. In both

Table 2. Evaluation on Stanford Dogs and CUB-200 of CSD and compared methods.

METHOD	STANFORD DOGS			CUB-200		
	RECALL@1 (1-60)	RECALL@1 (61-120)	RECALL@1 Average	RECALL@1 (1-100)	RECALL@1 (101-200)	RECALL@1 Average
Initial model	81.3	69.3	75.3	79.2	46.9	63.1
Fine-Tuning	74.0	**83.7**	78.8	70.2	75.1	72.7
MMD loss [17]	79.5	83.4	81.4	77.0	74.1	75.6
Feat. Est. [19]	79.9	83.5	81.7	77.7	75.0	76.4
CSD (Ours)	**80.9**	83.5	**82.2**	**78.6**	**78.3**	**78.5**
Joint training	80.4	83.1	81.7	78.2	79.2	78.7

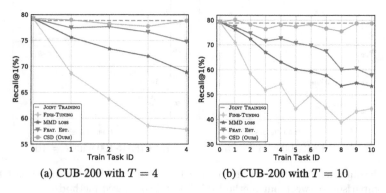

(a) CUB-200 with $T = 4$ (b) CUB-200 with $T = 10$

Fig. 4. Evolution of RECALL@1 on the first task as new tasks are learned on CUB-200. Comparison between our method (CSD) and compared methods.

experiments, our approach does not always report the highest scores, but it achieves the most stable trend obtaining the best result as the training end. This confirms that our approach is effective also when the model is updated multiple times.

5.2   Evaluation on Fine-grained Datasets

We compare our method on CUB-200 and Stanford Dogs datasets with the Fine-Tuning baseline, MMD loss [17], and [19] denoted as Feature Estimation. As an upper bound reference, we report the Joint Training performance obtained using all the data to train the model.

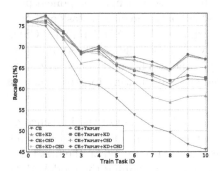

Fig. 5. Ablation on loss component on CUB-200 with $T = 10$. "+" represents the combination of components.

Table 3. Ablation on the weight factors for KD and CSD in Eq. 6 on CUB-200 with $T = 1$.

λ_{KD}	λ_{CSD}	RECALL@1 (1-100)	RECALL@1 (101-200)	RECALL@1 Average
0.1	0.1	78.24	76.82	77.53
0.1	1	79.19	77.50	78.35
0.1	10	78.56	76.07	77.32
1	0.1	79.32	73.82	76.57
1	1	78.62	**78.34**	**78.48**
1	10	79.12	75.32	77.22
10	0.1	78.35	76.76	77.56
10	1	**79.53**	76.93	78.23
10	10	78.90	75.53	77.22

We report in Table 2 the scores obtained with $T = 1$ on the fine-grained datasets. On Stanford Dogs, our approach achieves the highest recall when evaluated on the initial task and comparable result with other methods on the final task with a gap of only 0.2% with respect to Fine-Tuning that focus only on learning new data. This results in our method achieving the highest average recall value with an improvement of 0.5% RECALL@1 concerning Feature Estimation, 0.8% for MMD loss, and 3.4% for Fine-Tuning. On the more challenging CUB-200 dataset, we obtain the best RECALL@1 on both the initial and the final task outperforming the compared methods. Our method achieves the highest average recall value with an improvement of 2.1% RECALL@1 with respect to Feature Estimation, 2.9% for MMD loss, and 5.8% for Fine-Tuning. Differently from CIFAR-100, on fine-grained datasets, there is a lower dataset shift between different tasks leading to a higher performance closer to the Joint Training upper bound due to lower feature forgetting.

We report in Fig. 4(a) and Fig. 4(b) the challenging cases of CUB-200 with $T = 4$ and $T = 10$, respectively. These experiments show, consistently with Table 2, how our approach outperforms state-of-the-art methods. In particular, with $T = 10$ (Fig. 4(b)), our method preserves the performance obtained on the initial task during every update. CSD largely improves over the state-of-the-art methods by almost 20%–25% with respect to [19] and [17] achieving similar performance to the Joint Training upper bound. By leveraging labels information for distillation during model updates, CSD provides better performance and favorably mitigates the catastrophic forgetting of the representation compared to other methods that do not make use of this information.

6   Ablation Study

Loss Components. In Fig. 5, we explore the benefits given by the components of the loss in Eq. 2 (i.e., CE, triplet, KD, and CSD) and their combinations in terms of RECALL@1 on CUB-200 with $T = 10$. To observe single component performance, we analyze the trend of RECALL@1 on both the current task and previous ones evaluated jointly. When CSD is used, (i.e., CE+CSD, CE+KD+CSD, CE+triplet+CSD, CE+triplet+KD+CSD), we achieve higher RECALL@1 and maintain a more stable trend with respect to others. This underlines how CSD is effective and central to preserve knowledge and limit feature forgetting across model updates.

Loss Components Weights. Finally, in Table 3, we analyze the influence of the stability loss components varying the parameters λ_{KD} and λ_{CSD} of Eq. 6 on CUB-200 with $T = 1$. The table shows the RECALL@1 obtained on the first task, on the final task, and the average between them after training the model. CSD best performs when $\lambda_{KD} = \lambda_{CSD} = 1$, obtaining the highest average RECALL@1.

7   Conclusions

In this paper, we propose Contrastive Supervised Distillation (CSD) to reduce feature forgetting in continual representation learning. Our approach tackles the problem without storing data of previously learned tasks while learning a new incoming task. CSD allows to minimize the discrepancy of new and old features belonging to the same class, while simultaneously pushing apart features from different classes of both current and old data in a contrastive manner. We evaluate our approach and compare it to state-of-the-art works performing empirical experiments on three benchmark datasets, namely CIFAR-100, CUB-200, and Stanford Dogs. Results show the advantages provided by our method in particular on fine-grained datasets where CSD outperforms current state-of-the-art methods. Experiments also provide further evidence that feature forgetting evaluated in visual retrieval tasks is not as catastrophic as in classification tasks.

Acknowledgments. This work was partially supported by the European Commission under European Horizon 2020 Programme, grant number 951911 - AI4Media.

The authors acknowledge the CINECA award under the ISCRA initiative (ISCRA-C - "ILCoRe", ID: HP10CRMI87), for the availability of HPC resources.

References

1. Wan, J.: Deep learning for content-based image retrieval: a comprehensive study. In Proceedings of the 22nd ACM International Conference on Multimedia, pp. 157–166 (2014)
2. Azizpour, H., Sullivan, J., Carlsson, S., et al.: CNN features off-the-shelf: an astounding baseline for recognition. In: CVPRW, pp. 512–519 (2014)
3. Yosinski, J., Clune, J., Bengio, Y., Lipson, H.: How transferable are features in deep neural networks? Adv. Neural Inf. Process. Syst. (2014)
4. Chen, W., et al.: Deep image retrieval: A survey. arXiv preprint arXiv:2101.11282 (2021)
5. Tolias, G., Sicre, R., Gou, H.: Particular object retrieval with integral max-pooling of CNNactivations. In: ICLR 2016-International Conference on Learning Representations, pp. 1–12 (2016)
6. Yue-Hei Ng, J., Yang, F., Davis, L.S.: Exploiting local features from deep networks for image retrieval. In Proceedings of the IEEE Conference on Computer Vision and Pattern Recognition Workshops, pp. 53–61 (2015)
7. Price, W.N., Cohen, I.G.: Privacy in the age of medical big data. Nat. Med. **25**(1), 37–43 (2019)
8. Cossu, A., Ziosi, M., Lomonaco, V.: Sustainable artificial intelligence through continual learning. arXiv preprint arXiv:2111.09437 (2021)
9. McCloskey, M., Cohen, N.J.: Catastrophic interference in connectionist networks: the sequential learning problem. In: Psychology of learning and motivation, vol. 24, pp. 109–165. Elsevier (1989)
10. Ratcliff, R.: Connectionist models of recognition memory: constraints imposed by learning and forgetting functions. Psychol. Rev. **97**(2), 285 (1990)
11. Vijayan, M., Sridhar, S.S.: Continual learning for classification problems: a survey. In International Conference on Computational Intelligence in Data Science, pp. 156–166. Springer (2021)
12. Delange, M., et al.: A continual learning survey: defying forgetting in classification tasks. IEEE Trans. Pattern Anal. Mach. Intell. (2021)
13. Masana, M., Liu, X., Twardowski, B., Menta, M., Bagdanov, A.D., van de Weijer, J.: Class-incremental learning: survey and performance evaluation on image classification. arXiv preprint arXiv:2010.15277 (2020)
14. Parisi, G.I., Kemker, R., Part, J.L., Kanan, C., Wermter, S.: Continual lifelong learning with neural networks: a review. Neural Netw. **113**, 54–71 (2019)
15. Belouadah, E., Popescu, A., Kanellos, I.: A comprehensive study of class incremental learning algorithms for visual tasks. Neural Netw. **135**, 38–54 (2021)
16. Davari, M.R., Belilovsky, E.: Probing representation forgetting in continual learning. In: NeurIPS 2021 Workshop on Distribution Shifts: Connecting Methods and Applications (2021)
17. Chen, W., Liu, Y., Wang, W., Tuytelaars, T., Bakker, E.M., Lew, M.: On the exploration of incremental learning for fine-grained image retrieval. arXiv preprint arXiv:2010.08020 (2020)
18. Pu, N., Chen, W., Liu, Y., Bakker, E.M., Lew, M.S.: Lifelong person re-identification via adaptive knowledge accumulation. In: Proceedings of the IEEE/CVF Conference on Computer Vision and Pattern Recognition, pp. 7901–7910 (2021)
19. Chen, W., Liu, Y., Pu, N., Wang, W., Liu, L., Lew, M.S.: Feature estimations based correlation distillation for incremental image retrieval. IEEE Trans. Multimedia (2021)

20. Li, Z., Hoiem, D.: Learning without forgetting. IEEE Trans. Pattern Anal. Mach. Intell. **40**(12), 2935–2947 (2017)
21. Chen, T., Kornblith, S., Norouzi, M., Hinton, G.: A simple framework for contrastive learning of visual representations. In: International Conference on Machine Learning, pp. 1597–1607. PMLR (2020)
22. Rebuffi, S.-A., Kolesnikov, A., Sperl, G., Lampert, C.H.: icarl: Incremental classifier and representation learning. In: CVPR, pp. 5533–5542. IEEE Computer Society (2017)
23. Hou, S., Pan, X., Loy, C.C., Wang, Z., Lin, D.: Learning a unified classifier incrementally via rebalancing. In: CVPR, pp. 831–839. Computer Vision Foundation/IEEE (2019)
24. Wu, Y., et al.: Large scale incremental learning. In: CVPR, pp. 374–382. Computer Vision Foundation/IEEE (2019)
25. Pernici, F., Bruni, M., Baecchi, C., Turchini, F., Del Bimbo, A.: Class-incremental learning with pre-allocated fixed classifiers. In: 2020 25th International Conference on Pattern Recognition (ICPR), pp. 6259–6266. IEEE (2021)
26. Kirkpatrick, J., et al: Overcoming catastrophic forgetting in neural networks. Proc. Natl. Acad. Sci. **114**(13), 3521–3526 (2017)
27. Van de Ven, G.M., Tolias, A.S.: Three scenarios for continual learning. arXiv preprint arXiv:1904.07734 (2019)
28. Jung, H., Ju, J., Jung, M., Kim, J.: Less-forgetting learning in deep neural networks. arXiv preprint arXiv:1607.00122 (2016)
29. Gretton, A., Smola, A.J., Huang, J., Schmittfull, M., Borgwardt, K.M., Schölkopf, B.: Covariate Shift and Local Learning by Distribution Matching. MIT Press (2009)
30. Shen, Y., Xiong, Y., Xia, W., Soatto, S.: Towards backward-compatible representation learning. In: Proceedings of the IEEE/CVF Conference on Computer Vision and Pattern Recognition, pp. 6368–6377 (2020)
31. Pernici, F., Bruni, M., Baecchi, C., Del Bimbo, A.: Regular polytope networks. IEEE Trans. Neural Netw. Learn. Syst. (2021)
32. Biondi, N., Pernici, F., Bruni, M., Del Bimbo, A.: Compatible representations via stationarity, Cores (2021)
33. Aljundi, R., Kelchtermans, K., Tuytelaars, T.: Task-free continual learning. In: Proceedings of the IEEE/CVF CVPR, pp. 11254–11263 (2019)
34. Pernici, F., Bruni, M., Del Bimbo, A.: Self-supervised on-line cumulative learning from video streams. Comput. Vis. Image Underst. **197**, 102983 (2020)
35. Chopra, S., Hadsell, R., LeCun, Y.: Learning a similarity metric discriminatively, with application to face verification. In: 2005 IEEE Computer Society Conference on Computer Vision and Pattern Recognition (CVPR'05), vol. 1, pp. 539–546. IEEE (2005)
36. He, K., Fan, H., Wu, Y., Xie, S., Girshick, R.: Momentum contrast for unsupervised visual representation learning. In: Proceedings of the IEEE/CVF Conference on Computer Vision and Pattern Recognition, pp. 9729–9738 (2020)
37. Misra, I., van der Maaten, L.: Self-supervised learning of pretext-invariant representations. In: Proceedings of the IEEE/CVF Conference on Computer Vision and Pattern Recognition, pp. 6707–6717 (2020)
38. Khosla, Pet al.: Supervised contrastive learning. In: NeurIPS, Aaron Maschinot (2020)
39. Hinton, G., Vinyals, O., Dean, J.: Distilling the knowledge in a neural network. arXiv preprint arXiv:1503.02531 (2015)
40. Romero, A., Ballas, N., Kahou, S.E., Chassang, A., Gatta, C., Bengio, Y.: Fitnets: Hints for thin deep nets. arXiv preprint arXiv:1412.6550 (2014)

41. Krizhevsky, A., Hinton, G., et al.: Learning multiple layers of features from tiny images (2009)
42. Wah, C., Branson, S., Welinder, P., Perona, P., Belongie, S.: The caltech-ucsd birds-200-2011 dataset. Computation & Neural Systems Technical Report (2011)
43. Khosla, A., Jayadevaprakash, N., Yao, B., Li, F-F.: Novel dataset for fine-grained image categorization. In: First Workshop on Fine-Grained Visual Categorization, IEEE Conference on Computer Vision and Pattern Recognition, Colorado Springs, CO, June 2011
44. Oh Song, H., Xiang, Y., Jegelka, S., Savarese, S.: Deep metric learning via lifted structured feature embedding. In: Proceedings of the IEEE Conference on Computer Vision and Pattern Recognition, pp. 4004–4012 (2016)
45. Wang, X., Han, X., Huang, W., Dong, D., Scott, M.R.: Multi-similarity loss with general pair weighting for deep metric learning. In: Proceedings of the IEEE/CVF Conference on Computer Vision and Pattern Recognition (2019)
46. He, K., Zhang, X., Ren, S., Sun, J.: Deep residual learning for image recognition. In: Proceedings of the IEEE Conference on Computer Vision and Pattern Recognition, pp. 770–778 (2016)
47. Szegedy, C., et al.: Going deeper with convolutions. In: Proceedings of the IEEE Conference CVPR (2015)
48. Jegou, H., Douze, M., Schmid, C.: Product quantization for nearest neighbor search. IEEE Trans. Pattern Anal. Mach. Intell. 33(1), 117–128, 102983 (2010)

A Comparison of Deep Learning Methods for Inebriation Recognition in Humans

Zibusiso Bhango and Dustin van der Haar$^{(\boxtimes)}$ (iD)

University of Johannesburg, Kingsway Avenue and University Road,
Auckland Park, Johannesburg 2092, South Africa
dvanderhaar@uj.ac.za

Abstract. Excessive alcohol consumption leads to inebriation. Driving under the influence of alcohol is a criminal offence in many countries involving operating a motor vehicle while inebriated to a level that renders safely operating a motor vehicle extremely difficult. Studies show that traffic accidents will become the fifth most significant cause of death if inebriated driving is not mitigated. Inversely, 70% of the world population can be protected by mitigating inebriated driving. Short term effects of inebriation include lack of balance, inhibition and fine motor coordination, dilated pupils and slow heart rate. An ideal inebriation recognition method that operates in real-time is less intrusive, more convenient, and efficient. Deep learning has been used to solve object detection, object recognition, object tracking and image segmentation problems. In this paper, we compare deep learning inebriation recognition methods. We implemented Faster R-CNN and YOLO methods for our experiment. We created our dataset of sober and inebriated individuals made available to the public. Six thousand four hundred forty-three (6443) face images were used, and our best performing pipeline was YOLO with a 99.6% accuracy rate.

Keywords: Deep learning · Computer vision · R-CNN · YOLO · Inebriation recognition · Inebriation detection · Drunk driving

1 Introduction

Alcohol abuse is a social issue in need of addressing [3]. Inebriation is a temporary "situational impairment" affecting the subject's interaction with their environment [6]. Driving under the influence of alcohol is a criminal offence in many countries involving operating a motor vehicle after consuming alcohol beyond the legal limit [14]. A WHO study showed that unless measures to reduce inebriated driving are implemented, vehicle accidents will become the fifth leading cause of death globally [15]. Inversely, handling inebriated driving can save 70% of the population.

Consistent excessive alcohol consumption can lead to irregular heartbeats or elevated blood pressure, leading to hypertension. After consuming alcohol,

© The Author(s), under exclusive license to Springer Nature Switzerland AG 2022
S. Sclaroff et al. (Eds.): ICIAP 2022, LNCS 13231, pp. 610–620, 2022.
https://doi.org/10.1007/978-3-031-06427-2_51

physical and physiological changes begin to take places, such as loss of inhibition, caution, finer motor coordination and inability to perform critical hand-coordinated tasks such as operating a motor vehicle [6]. Alcohol consumption also results in slurred speech, poor balance and low heart rate.

Due to these issues, it is essential to recognise inebriation. The short-term effects of inebriation can be used to recognise inebriation in humans. An ideal inebriation recognition method is faster, non-invasive, convenient and practical. Deep learning methods have been used for object detection, semantic segmentation and object recognition problems. We believe they can be used effectively to recognise inebriation in humans.

The rest of the paper aims to outline the problem and compare deep learning methods in literature used to recognise inebriation. The following sections are divided as follows. In the problem background, we outline the inebriation recognition problem in humans using deep learning, followed by a discussion on deep learning methods in the literature. We describe our study's methodology, data sampling, and benchmark in the experimental setup. The implementation is then discussed, where we provide the details of our baseline deep learning pipelines, followed by the results of our methods and a comparison with methods in the literature. In conclusion, we provide our findings from the research and future work.

2   Problem Background

Every 33 min, a person in the world is dying in a road accident instigated by an inebriated driver [17]. In 2014, about 27 people died daily due to inebriated driving in the USA [6]. An estimated 1.24 million people die on the road annually [1], and if this trend does not change, 2.4 million are expected to die on the road by 2030 due to inebriated driving [10]. Road accidents cost USD 500 billion a year, which is between 1% and 3% of the world's GDP [16].

Ten minutes after alcohol consumption, the heart rate increases to filter the toxins from the bloodstream to the kidneys. Initially, alcohol acts as a stimulant, producing intense feelings of warmth, well-being and relaxation, a feeling the alcohol consumer is after. However, inhibition, fine motor coordination, and reaction time begin to suffer, with exponential effects experienced as more alcohol is consumed. The alcohol penetrates the blood-brain barrier, affecting the cognitive neuromotor functions, leading to loss of balance [1]. The effects render operating a motor vehicle extremely difficult. Heavy alcohol consumption meddles with the delicate balance of neurotransmitters, which are in charge of the brain's functionality.

Due to the effects of inebriation, methods exist that can recognise inebriation in humans, such as blood tests, urine tests and breath tests. These methods are very intrusive, and their usage has legal and ethical connotations as blood and urine samples contain sensitive information and deprive privacy. Also, these above mentioned methods can be inefficient, as using a mouthwash with alcohol content can lead to a positive breath test, for example. We believe there are better, faster and more efficient methods to recognise inebriation that are convenient and less invasive. These methods include automated biometrics systems that use deep learning to recognise inebriation. This paper will compare these methods in our experiments and the literature.

3   Related Work

In this section, we look at existing methods in the literature that use deep learning for inebriation recognition.

Mehta et al. [7] developed a dataset called DIF (Dataset of perceived Intoxicated Faces) containing audio-visual data of inebriated and sober individuals curated from social networks such as YouTube. Video titles and captions from the website were used to assign class labels. Due to the nature of the collection procedure of their dataset, they tested their methods in the wild. They had 78 sober subjects and 88 inebriated subjects. They argue that this is a novel approach for non-invasive inebriation detection. They used CNN and Deep Neural Networks to compute the video and audio baselines. 3D CNN was used to exploit the Spatio-temporal change in the video. The audio model used a 2-layer perceptron with ReLU action function using batch normalisation and dropout. Their method performed well, achieving an accuracy rate of 88.39%. Their research involved creating a dataset and making it publicly available, which is a very good contribution to literature. Using both audio and visual data in a video to recognise inebriation makes the system more robust, but multimodal systems can be more computationally expensive.

Lee et al. [5] presented a method to detect abnormal behaviour of inebriated individuals using surveillance videos. They looked for videos of people walking in zigzags, staggering and in a lying posture. They argued that sober people walk straight and upright instead of walking in zigzags or staggering. Using their tracking method, they obtained 890 pedestrian trajectories (846 sober trajectories and 44 inebriated trajectories). They used YOLOv3 for pedestrian detection and tracking, with motion trajectories and pose trajectories evaluated from these detections. Their system achieved 93% recall and 98% precision rates. The authors did not provide their accuracy score to the best of our knowledge. 95% of their dataset was of sober individuals, with 5% of the sample made up of inebriated individuals. We believe this level of data imbalance can affect the experiment's results.

Neagoe and Diaconescu [9] developed a method to recognise inebriation using an ensemble of Deep Convolutional Neural Networks (DCNNs) to process thermal infrared facial images. Two modules (the first with 12 layers and the second with 10 layers) were trained separately using different architectures and sets of parameters, and classification was based on the confidence of both modules. They used 400 thermal infrared facial images belonging to 10 subjects for data sampling. Forty images were used per subject, 20 sober and 20 taken 30 min after drinking 100 ml of whisky. Their experiment had an accuracy rate of 95.75%, which is very high. Although 400 images are a relatively good sample, 10 subjects are a very small sample to use and might affect their results. The usage of ensemble methods is very robust, but it can also be computationally expensive as two deep neural networks are used independently.

Menon et al. [8] developed a system that captures a vehicle driver's face in a thermal image spectrum. The face is recognised using a CNN then classified as inebriated or sober using Gaussian Mixture Model (GMM) with Fischer Linear Discriminant (FDA). Forty-one subjects were used, with samples captured while sober and after consuming alcohol. Their face recognition algorithm had a 97% accuracy, and their classifier had an 87% classification rate, which is a good performance. We believe 41 samples are too small for generalising a classifier, and the sensor used to capture thermal imaging is expensive.

Bhango and van der Haar [2] developed a method to recognise inebriation in humans using computer vision. They used RGB face images made up of 153 inebriated subjects and 101 sober subjects, collated from the internet. Their method used Viola-Jones-based face-detection for the region of interest localisation, and the face image is used as input to a LeNet CNN algorithm to classify inebriation. Their model had an accuracy rate of 84.31%, which was a very good performance, but we believe their dataset was small for the generalisation of their classifier. Also, they had a low recall rate indicating a high number of inebriated cases classified as sober.

4   Experimental Setup

4.1   Methodology

We used the design science research methodology for our study because we believe it is best suited for IT artefacts, is rooted in engineering, and has generally accepted IT practitioners [4]. In particular, we used the design-oriented approach, which aims to develop and provide an artefact as a research contribution or output [11].

The design-oriented method is made up of four phases: analysis, initial design, evaluate and validate & diffuse. During analysis, our research problem is identified, and objectives formulated. During the initial design, the artefact is designed using generally accepted methods. In the evaluation phase, the artefact is produced against the objectives in the analysis phase. During validate & diffuse, validation of the artefact takes place. In our research, we used the quantitative research design because our research is best suited for numerical data that can be statistically analysed to compare deep learning inebriation recognition methods.

4.2   Data Sampling

For data sampling, we could not find a dataset in literature for inebriated and sober individuals; therefore, we are testing in the wild in our research. We created a dataset of sober and inebriated people by collating our data from publicly accessible images on the internet with free usage rights, as shown in Fig. 1. One thousand two hundred fifty-two (1252) images were collated, and data augmentation was performed on each image by rotating the image to the left and right

Fig. 1. Some samples from the database used for data sampling to test our experiments. The top row consists of drunk individuals, while the bottom row consists of sober individuals.

within the 15° range and performing a horizontal flip. The data augmentation resulted in 5008 images (2508 drunk images and 2500 sober images), making up 6443 face images. The augmentation results are shown in Fig. 2 below.

Fig. 2. An augmented sample of a sober individual. Far left is the original image. Middle left is the horizontal flip of the original image. Middle right is the right rotation of the original image and far right is the left rotation of the original image.

Our dataset (which can be found at https://github.com/dvanderhaar/ UJInebriate) is made up of randomly selected inebriated and sober people from different walks of life. Males and females are equally represented, and the races represented include Blacks, Caucasians and Asians. We used 5835 face images for training and 608 face images for testing. For benchmarking, we used the accuracy, precision, recall, f1 score and the area under the curve metrics to measure the performance of our experiments. We also used these metrics to compare our experiments' performance against deep learning methods on inebriation recognition in literature.

5   Implementation

For our experiment, we implemented the Faster R-CNN and YOLO methods. These deep learning methods have been implemented with great results in object detection, object segmentation, object recognition and object tracking problems. The algorithms are discussed in the sections below.

5.1 Faster R-CNN

Faster R-CNN [13], an improvement on R-CNN and Fast R-CNN, uses the region proposal network (RPN) to learn the region proposals. Faster R-CNN consists of two modules: a deep fully connected CNN for region proposals and an SVM classifier for object classification. The RPN is a fully connected layer that predicts object bounds and object probability scores at each position. RPN ranks region boxes called anchors and proposes the ones flagged as containing objects.

We used Roboflow's Faster R-CNN implementation based on the TensorFlow object recognition API. We used the COCO dataset for transfer learning and the Inception v2 CNN algorithm as our architecture because of its high accuracy performance. We used a batch size of 12 for training and L2 regularisation because of its ability to force weights to be small but not zero.

5.2 YOLO

You Only Look Once (YOLO) [12] is an object recognition algorithm that uses a CNN to predict the confidence of the bounding boxes and the class probability for these boxes using an entire image. Object recognition is treated as a regression problem. The method excels at object recognition in real-time.

During training, an image is divided into grids of N × N cells, each cell responsible for predicting possible bounding boxes, which are rectangles surrounding the detected object. These bounding boxes have five values: x, y, w, h and cs. X and y are coordinates of the centre of the bounding box relative to the grid. W and h are width (w) and height (h) of the bounding box relative to image dimension, and cs is the confidence value determining the network's confidence about the object's presence inside the bounding box. The confidence value does not contain object class information.

We used the recently released YOLOv5 method for our experiment, consisting of 3 important parts: model backbone, model neck, and model head. The model backbone is essential for extracting important features from an image. We used Cross Stage Partial Networks as our model backbone, improving our processing time. The model neck is used to generate feature pyramids which help the model generalise well on scaled images. We used a path aggregation network (PANet) as a model neck. The model head is used for final object recognition by applying anchor boxes on features and creating final output vectors containing class probabilities. The Leaky ReLU (rectified linear unit) and sigmoid activation functions are used. ReLU is used in the hidden layers, and the sigmoid function is used in the final recognition layer. We used the stochastic gradient descent (SGD) for the learning optimiser, which computes the network's loss function gradient concerning each weight in the network. We used Pytorch's Binary Cross-Entropy with Logits Loss to classify class probability and object score loss.

6   Results

In this section, we will discuss the results from our experiments. The pipelines' results are discussed below.

6.1   Faster R-CNN

We looked at the pipeline's face detection and inebriation classification for functional requirements. Our Faster R-CNN pipeline performed well, showing promising results. Forty face images used for testing our model were not detected. In Fig. 3 above are a few face images that were both detected and classified by our Faster R-CNN pipeline, including augmented images.

Fig. 3. Face images detected and classified by our algorithm.

After obtaining results from our pipeline, we obtained the confusion matrix shown in Fig. 4:

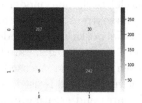

Fig. 4. The confusion matrix for the faster R-CNN pipeline.

We used 608 face images for inferences. Forty face images could not be detected. Two hundred forty-two individuals were correctly predicted as sober, and 287 individuals were correctly predicted as inebriated. However, 30 individuals were wrongly predicted as inebriated, and nine individuals were wrongly predicted as sober. Some of the misclassifications are shown in Fig. 5.

Fig. 5. Misclassified images.

We calculated our accuracy, precision, recall and f1-score from the confusion matrix. These metrics are listed Table 1:

Table 1. Faster R-CNN metric results.

Metrics	Accuracy	Precision	Recall	AUC	F1-score
Accuracy	93.1%	89.0%	96.4%	93.5%	93.0%

Our prototype had an accuracy of 93.1%, the precision of 89.0%, recall of 96.4%, f1-score of 93.0% and area under the curve (AUC) of 93.5%. The precision is lower because the sober cases are misclassified as inebriated. Our recall is high because there were few misclassifications on inebriated cases. We believe this is a good performance on recognising inebriation in humans.

6.2 YOLO

We looked at our pipeline's performance in face detection and inebriation classification for functional requirements. Our YOLO pipeline performed well, showing state-of-the-art results. All face images were successfully detected and classified. Figure 6 shows a sample of face images detected and classified by our YOLO pipeline, including augmented images.

Fig. 6. Face images detected and classified by YOLO.

After obtaining results from our pipeline, we obtained the confusion matrix shown in Fig. 7:

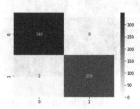

Fig. 7. The confusion matrix for the YOLOv5 pipeline.

We used 608 face images for inferences. Two hundred fifty-eight individuals were correctly predicted as sober, and 348 individuals were correctly predicted as inebriated. However, two individuals were wrongly predicted as inebriated. The two misclassification cases consist of an original image and its augmented version.

We calculated our accuracy, precision, recall and f1-score from the confusion matrix. These metrics are listed Table 2:

Table 2. YOLOv5 metric results.

Metrics	Accuracy	Precision	Recall	AUC	F1-score
Accuracy	99.7%	100%	99.3%	99.6%	99.6%

Our prototype had an accuracy of 99.7%, the precision of 100%, recall of 99.3%, f1-score of 99.6% and area under the curve (AUC) of 99.6%. Our precision and recall are very high because of the low misclassification rate, resulting in a very high F1 score. These are great results because a high F1-score means the system can be used in various environments and use cases that might require high precision, high recall or both. We believe this is a state-of-the-art performance on recognising inebriation in humans.

Table 3 compares the performance of our experiments against similar systems in the literature. Our YOLOv5 pipeline achieved the highest accuracy, F1-score, AUC and recall. We believe our YOLOv5 pipeline had the best results based on the Table 3 comparison.

Table 3. Comparing our model with similar classifiers in literature.

Method	Data sampling	Accuracy	Precision	Recall	F1-score	AUC
YOLOv5	**6443**	**99.7%**	**100%**	**99.3%**	**99.6%**	**99.6%**
Faster R-CNN	**6443**	93.1%	89.0%	96.4%	93.0%	93.5%
Mehta et al. [7]	166	88.39%	85%	99%	–	–
Lee et al. [5]	890	-	98%	93%	–	–
Neagoe and Diaconescu [9]	400	95.75%	–	–	–	–
Menon et al. [8]	41	87%	–	–	–	–
Bhango and van der Haar [2]	153	84.31%	84.38%	71.05%	77.14%	83%

7 Conclusion

Excessive alcohol consumption leads to inebriation. Alcohol abuse is a serious social issue in need of solutions. According to a study by WHO, vehicle accidents will become the fifth highest cause of death if inebriated driving is not mitigated [15]. Inversely, 70% of the world population can be protected by handling drunk driving effectively.

Our research shows enough feature separability between sober and inebriated individuals. Our experiment yielded state-of-the-art results on YOLOv5 and Faster R-CNN implementations on inebriation recognition. We compared results that proved our experiments superior to those in the literature. YOLOv5 and Faster R-CNN algorithms are high performings at recognising inebriation in humans. Although Faster R-CNN struggled with precision, the recognition performance of both Faster R-CNN and YOLOv5 methods was very high.

To the best of our knowledge, no publicly available datasets on inebriated and sober face images to classify inebriation exists. For future work, we believe a scientifically proven dataset consisting of inebriated and sober individuals will be of immense help for benchmarking inebriation recognition methods. Different types of alcohol have varying effects when consumed. Our experiments compared methods that recognise inebriation but not the cause of inebriation. An algorithm that can detect the type of intoxicating substance that caused an individual to be inebriated will be a significant contribution to literature.

References

1. Arnold, Z., Larose, D., Agu, E.: Smartphone inference of alcohol consumption levels from gait. In: 2015 International Conference on Healthcare Informatics, pp. 417–426 (2015)
2. Bhango, Z., van der Haar, D.: A model for inebriation recognition in humans using computer vision. In: Abramowicz, W., Corchuelo, R. (eds.) BIS 2019. LNBIP, vol. 353, pp. 259–270. Springer, Cham (2019). https://doi.org/10.1007/978-3-030-20485-3_20

3. Goffredo, M., Bouchrika, I., Carter, J.N., Nixon, M.S.: Performance analysis for gait in camera networks. In: Proceedings of the 1st ACM Workshop on Analysis and Retrieval of Events/Actions and Workflows in Video Streams. AREA 2008, pp. 73–80. ACM, New York (2008). http://0-doi.acm.org.ujlink.uj.ac.za/10.1145/1463542.1463555

4. Gregor, S., Hevner, A.: Positioning and presenting design science research for maximum impact. MIS Q. **37**, 337–356 (2013)

5. Lee, J., Choi, S., Lim, J.: Detection of high-risk intoxicated passengers in video surveillance. In: 2018 15th IEEE International Conference on Advanced Video and Signal Based Surveillance (AVSS), pp. 1–6 (2018)

6. Mariakakis, A., Parsi, S., Patel, S.N., Wobbrock, J.O.: Drunk user interfaces: determining blood alcohol level through everyday smartphone tasks. In: Proceedings of the 2018 CHI Conference on Human Factors in Computing Systems. CHI 2018, p. 1–13. Association for Computing Machinery, New York (2018). https://doi.org/10.1145/3173574.3173808

7. Mehta, V., Katta, S.S., Yadav, D.P., Dhall, A.: DIF dataset of perceived intoxicated faces for drunk person identification. In: 2019 International Conference on Multimodal Interaction. ICMI 2019, pp. 367–374. Association for Computing Machinery, New York (2019). https://doi.org/10.1145/3340555.3353754

8. Menon, S., Swathi, J., Anit, S.K., Nair, A.P., Sarath, S.: Driver face recognition and sober drunk classification using thermal images. In: 2019 International Conference on Communication and Signal Processing (ICCSP), pp. 0400–0404 (2019)

9. Neagoe, V.E., Diaconescu, P.: An ensemble of deep convolutional neural networks for drunkenness detection using thermal infrared facial imagery. In: 2020 13th International Conference on Communications (COMM), pp. 147–150 (2020)

10. Neagoe, V.E., Carata, S.V.: Drunkenness diagnosis using a neural network-based approach for analysis of facial images in the thermal infrared spectrum, pp. 165–168 (2017)

11. Oesterle, H., et al.: Memorandum on design-oriented information systems research, vol. 20, January 2011. http://www.alexandria.unisg.ch/Publikationen/71089

12. Redmon, J., Divvala, S., Girshick, R., Farhadi, A.: You only look once: unified, real-time object detection (2016)

13. Ren, S., He, K., Girshick, R., Sun, J.: Faster R-CNN: towards real-time object detection with region proposal networks (2016)

14. Wang, W.F., Yang, C.Y., Wu, Y.F.: SVM-based classification method to identify alcohol consumption using ECG and PPG monitoring. Pers. Ubiquitous Comput. **22** (2018)

15. Wu, C.K., Tsang, K.F., Chi, H.R.: A wearable drunk detection scheme for healthcare applications. In: 2016 IEEE 14th International Conference on Industrial Informatics (INDIN), pp. 878–881 (2016)

16. Wu, C., Tsang, K., Chi, H., Hung, F.: A precise drunk driving detection using weighted kernel based on electrocardiogram. Sensors **16**, 659 (2016)

17. Wu, Y., Xia, Y., Xie, P., Ji, X.: The design of an automotive anti-drunk driving system to guarantee the uniqueness of driver. In: Proceedings - 2009 International Conference on Information Engineering and Computer Science, ICIECS 2009, vol. 62, December 2009

Enhanced Data-Recalibration: Utilizing Validation Data to Mitigate Instance-Dependent Noise in Classification

Saeed Bakhshi Germi$^{(\boxtimes)}$ and Esa Rahtu

Tampere University, Tampere, Finland
{saeed.bakhshigermi,esa.rahtu}@tuni.fi

Abstract. This paper proposes a practical approach to deal with instance-dependent noise in classification. Supervised learning with noisy labels is one of the major research topics in the deep learning community. While old works typically assume class conditional and instance-independent noise, recent works provide theoretical and empirical proof to show that the noise in real-world cases is instance-dependent. Current state-of-the-art methods for dealing with instance-dependent noise focus on data-recalibrating strategies to iteratively correct labels while training the network. While some methods provide theoretical analysis to prove that each iteration results in a cleaner dataset and a better-performing network, the limiting assumptions and dependency on knowledge about noise for hyperparameter tuning often contrast their claims. The proposed method in this paper is a two-stage data-recalibration algorithm that utilizes validation data to correct noisy labels and refine the model iteratively. The algorithm works by training the network on the latest cleansed training Set to obtain better performance on a small, clean validation set while using the best performing model to cleanse the training set for the next iteration. The intuition behind the method is that a network with decent performance on the clean validation set can be utilized as an oracle network to generate less noisy labels for the training set. While there is no theoretical guarantee attached, the method's effectiveness is demonstrated with extensive experiments on synthetic and real-world benchmark datasets. The empirical evaluation suggests that the proposed method has a better performance compared to the current state-of-the-art works. The implementation is available at https://github.com/Sbakhshigermi/EDR.

Keywords: Label noise · Classification · Data-recalibration

1 Introduction

Inexperienced workers, insufficient information about samples, confusing patterns, tiresome nature of the work, and other factors make the manual labeling

© The Author(s), under exclusive license to Springer Nature Switzerland AG 2022
S. Sclaroff et al. (Eds.): ICIAP 2022, LNCS 13231, pp. 621–632, 2022.
https://doi.org/10.1007/978-3-031-06427-2_52

Fig. 1. Multiple samples of the same category in different datasets: (A) Number two in MNIST [15], (B) Deer in CIFAR-10 [14], (C) Caesar salad in Food-101N [16], and (D) Underwear in Clothing1M [34]. The images on the top row are more straightforward to label than the images on the bottom row.

of samples in a large dataset prone to errors and noisy labels [10,29]. Unfortunately, deep learning algorithms have the potential to memorize these noisy labels, which leads to poor generalization and lower performance on clean test datasets [37]. Due to the importance of the topic in different sectors, such as safety-critical applications [2] and medical imaging [23], researchers have been developing methods to mitigate such label noise [3,10,27].

Most recent works assume the labels to be affected by a class-conditional noise (CCN) where the noise is instance-independent [20]. This type of noise can be estimated [13] or mitigated by adding extra loss terms in the model [4]. However, Chen utilized visual examples and mathematical analysis to prove that the label noise in a real-world dataset (Clothing1M [34]) is actually instance-dependent. To better understand why this is the case, take a look at Fig. 1. As seen in this figure, two samples of the same category have different complexity of labeling, which suggests that the label noise is instance-dependent.

With the previous assumption of CCN proven wrong, a new mathematical foundation for mitigation methods had to be developed. Therefore, researchers started defining variations of instance-dependent noise (IDN) patterns to represent synthetic noise and propose mitigation approaches based on them. One of the effective strategies used in the state-of-the-art methods is the iterative data-recalibration [18]. These methods use the predictions of a network trained over noisy samples to select and correct samples iteratively.

While recent works on IDN provided theoretical analysis to prove the convergence of their models to an oracle Bayes classifier [5,38], the limiting assumptions in their theories cannot be met in practical implementation, as shown by their empirical findings. Due to these limitations, this paper will focus on empirical experiments on synthetic and real-world datasets to showcase the effectiveness of the proposed method.

This paper proposes an enhanced data-recalibration algorithm that corrects labels affected by instance-dependent noise by utilizing validation set. On each iteration, the proposed method trains a model with the cleansed data from the last iteration to achieve higher performance on a small, clean validation set. Then, the best-performing model is chosen to correct labels in the training set based on the model's confidence for the next iteration. The intuition is that better performance on the clean validation set means a better prediction of training labels than the previous iteration.

The main difference between the proposed method and previous works is utilizing a clean validation set to influence the training stage to help the network approach an oracle model that can predict ground truth labels. Small, clean validation sets can be easily obtained with computer-assisted tools [1]. While previous works often use the validation set as a selector of the final model for accuracy reports, none utilize it any further to the best of the authors' knowledge. The main contributions of this paper are:

- Proposing a practical data-recalibration algorithm that utilizes easy-to-gather clean validation set to enhance the performance over the existing state-of-the-art methods.
- Providing empirical evaluation with extensive experiments on both synthetic and real-world datasets to show the effectiveness of the proposed method.

The rest of the paper is structured as follows. Section 2 covers the related works. Next, Sect. 3 explains the proposed method in detail. After that, Sect. 4 deals with the experiments and the empirical evaluation to show the effectiveness of the proposed method. Finally, Sect. 5 concludes the work.

2   Related Works

Menon provided one of the major theoretical frameworks for IDN in binary problems. This framework provided the basis to construct a loss function with specific criteria to mitigate IDN. While the work was necessary at the time, the method is not extensible to deep neural networks [21]. Chen provided mathematical proof that the label noise in a large real-world dataset called Clothing1M [34] follows the IDN pattern. They proposed a method of generating IDN patterns by averaging the predictions of an oracle classifier over the training session to find complex samples and flip their labels. The mitigation method provided also relies on averaging the predictions of a network, with the intuition that the network can find a soft representation of labels that are closer to ground truth over time. While this work provided essential information about IDN, the mitigation method is cost-heavy with low performance compared to other works [7].

Zhang defined a new family of noise called poly-margin diminishing (PMD). This new noise family follows the same intuition that data points near the decision boundary are more challenging to classify, thus more prone to noise. Based on the previously stated reasons for label noise, this definition seems realistic. To mitigate this family of noise, they proposed an iterative correction method that corrects the labels based on the network confidence over the training set in each iteration. While the work provided theories to prove the effectiveness, their hyperparameter settings and assumption violation in implementing the method contradict their idea [38].

Several state-of-the-art methods managed to reach high performance on real-world benchmarks. Tan combined a supervised and an unsupervised network and co-teach them with the help of an encoder to maximize the agreement between the networks in latent space [28]. Wu utilized the spatial topology of data in the latent space of the network iteratively to collect clean labels and refine the network further [32]. Zhu focused on the second-order approach to estimate covariance terms for IDN with peer loss function [19] and defined a new loss function to change the problem to CCN [40]. Xia eliminated the need for anchor points in estimating the noise transition matrix [33]. Han described a two-stage algorithm where the trained network is used to select multiple class prototypes to represent the characteristics of the data better and correct the noisy labels [11]. Lee focused on reducing human supervision by introducing a method that required a small clean training set to extract the information about label noise [16]. Li divides the training data into labeled clean and unlabeled noisy samples to utilize semi-supervised learning techniques by training two networks and correcting more labels over each iteration [17]. Other methods such as PENCIL [36], ILFC [5], CORES[2] [8], Meta-Weight-Net [24], estimation of transition matrix [35], and JoCoR [31] are also noteworthy.

3   Proposed Method

In this section, we present the details of our proposed method. The proposed method alternates between training the network to find the best performance on the clean validation set and correcting the noisy labels based on confidence scores from the top-performing network. Before the proposed algorithm starts the process, we prepare a deep neural network by training it for a few epochs with a high learning rate, which allows the network to reach a reasonable confidence level without overfitting to noise [37].

3.1   Preliminaries

Let $\mathcal{X}$ be the feature space, $\mathcal{L}$ be the label space, $(x, y), (x, \tilde{y}) \in \mathcal{X} \times \mathcal{L}$ be a clean and a noisy sample respectively, $D = \{(x_i, y_i)\}_{i=1}^n$ be a dataset, $f^t(x) = (C_1, \ldots, C_k)$ be a classifier at the t-th iteration of the algorithm, where C_i is the confidence score of the network for the i-th class (output of softmax layer in this paper), and k is the total number of classes. Finally, let S^t be the performance of the classifier over clean validation set at t-th iteration of the algorithm.

3.2   Iterative Label Correction Method

The overall algorithm is summarized in Algorithm 1. In practice, we use an average of confidence scores from several top-performing networks. Since there is no

---

Algorithm 1: Enhanced Data-Recalibration

Require: Initial training set $\widetilde{D}^0_{train} = \left\{\left(x_i, \tilde{y}_i^0\right)\right\}_{i=1}^n$, Initial classifier f^0,
  threshold value θ, Number of epochs T, Validation set $D_{valid} = \{(x_i, y_i)\}_{i=1}^m$

1: **for** $t \in 1, \ldots, T$ **do**
2:     Train f^{t-1} on $\widetilde{D}^{t-1}_{train}$ to get f^t and get the performance score S^t
3:     Compare S^t to previous scores $\{S^i\}_{i=1}^{t-1}$ to find best performing classifier f^B
4:     **for** $(x, \tilde{y}) \in \widetilde{D}^{t-1}_{train}$ **do**
5:        Get the confidence scores $(C_1, \ldots, C_k)$ of f^B on x
6:        Find the best confidence score C_M and the noisy confidence score C_N
7:        Calculate $Gap = |\log(C_M) - \log(C_N)|$
8:        **if** $Gap \geq \theta$ **then**
9:           Set new label $\tilde{y}^t = M$
10:       **else**
11:          Keep old label $\tilde{y}^t = \tilde{y}^{t-1}$
12:       **end if**
13:     **end for**
14:     **if** $\forall i \in [1, \ldots, n]$, $\tilde{y}_n^t = \tilde{y}_n^{t-1}$ **then**
15:       Decrease θ by a small amount
16:     **end if**
17: **end for**

return Best trained network f^B

---

guarantee of improving the network on every iteration, there might be a random instance where the trained network arbitrarily achieves a high performance score. Averaging multiple confidence scores mitigates the effect of these random encounters as they do not introduce a bias towards any class. Moreover, the top-performing networks are selected from a range of recently trained networks to ensure that the network is not stuck in a loop. In the following subsections, we will describe what happens in the t-th iteration of the algorithm:

3.3 Stage One

In this stage, the algorithm starts training the network for one epoch with the labels acquired from the previous iteration. In other terms, the network from the previous iteration f^{t-1} is trained on the training set with labels generated in the previous iteration $\widetilde{D}^{t-1}_{train} = \left\{\left(x_i, \tilde{y}_i^{t-1}\right)\right\}_{i=1}^n$ to obtain the new network f^t. Then, the performance of the network is evaluated to obtain the top-performing network for the next stage. It is done by evaluating the new network f^t on the clean validation set $D_{valid} = \{(x_i, y_i)\}_{i=1}^m$ to get its performance score S^t. Then, this performance score S^t is compared to all previous scores $\{S^i\}_{i=1}^{t-1}$ to find the best-performing network $\left\{f^B \mid \forall i \leq t : S^B \geq S^i\right\}$.

3.4 Stage Two

In this stage, the algorithm starts collecting the confidence scores of the chosen network on the training set. It is done by predicting the confidence scores

$(C_1, \ldots, C_k)$ of the best-performing network f^B for each sample in training set from the previous iteration $(x, \tilde{y}) \in \widetilde{D}_{train}^{t-1}$. Then, the confidence scores are evaluated to decide the labels for the next iteration. For each sample in the dataset $(x, \tilde{y}) \in \widetilde{D}_{train}^{t-1}$, the highest confidence score $\{C_M \mid \forall i \leq k : C_M \geq C_i\}$ and the confidence score for the noisy label $C_{N=\tilde{y}}$ are considered. If the difference of logarithms between them is greater than a threshold $|\log(C_M) - \log(C_N)| \geq \theta$, then the sample is selected for correction. The intuition behind the process is that a noticeable gap between the prediction of the best-performing network and the current label suggests the label is noisy. After that, the labels for the next iteration are generated. It is done by swapping the label of the selected samples to the prediction of the best-performing network $\tilde{y}_{sel}^t = M$ while keeping the labels of other samples the same as before $\tilde{y}_{rest}^t = \tilde{y}^{t-1}$. Finally, the threshold value is evaluated and reduced if the algorithm cannot select samples anymore. By initializing a high threshold value and lowering it in small steps, the best-performing network gains more trust from the algorithm gradually, which prevents confirmation bias to some degree.

4   Experiments and Evaluation

4.1   Synthetic Datasets

For proof of concept, the public datasets CIFAR-10 and CIFAR-100 [14] are chosen for synthetic experiments. Both datasets contain 50,000 training and 10,000 testing samples over ten categories. In the case of CIFAR-100, each category is further divided into ten subclasses. As argued by the previous works [5,7,38], a realistic noise does not uniformly affect all data space points. The most common solution among previous works to generate reliable IDN is to find challenging samples and then flip their label from the most confident category to the second most confident category. A challenging sample is typically located at the edges of the decision boundary and results in a low network confidence score. Such samples can be found by training an oracle network and selecting the low confidence samples [5] or averaging the network's confidence over the training period and selecting the confusing samples [7]. To generate reliable and comparable IDN, we follow the definition for the PMD noise family [38].

Let $\aleph_{C_1, C_2}(x) = \mathbb{P}[\tilde{y} = C_2 \mid y = C_1, x]$ be the probability of corrupting the label of a sample from the most confident class C_1 to the second-most confident class C_2, and $f^*(x)$ be an oracle classifier trained on clean samples. The three types of IDN used in our experiments are defined as in Eq. 1.

$$
\begin{aligned}
\aleph_{C_1, C_2}^{I}(x) &= \frac{1}{2} - \frac{1}{2}\left[f_{C_1}^*(x) - f_{C_2}^*(x)\right]^2 \\
\aleph_{C_1, C_2}^{II}(x) &= 1 - \left[f_{C_1}^*(x) - f_{C_2}^*(x)\right]^3 \\
\aleph_{C_1, C_2}^{III}(x) &= 1 - \frac{1}{3}\left[f_{C_1}^*(x) - f_{C_2}^*(x)\right]^3 \\
&\quad - \frac{1}{3}\left[f_{C_1}^*(x) - f_{C_2}^*(x)\right]^2 - \frac{1}{3}\left[f_{C_1}^*(x) - f_{C_2}^*(x)\right]
\end{aligned}
\tag{1}
$$

For the sake of completion, we also include the most common CCN noise types in our experiments: uniform and asymmetrical [22]. Let $\beth_{C_1,C_2} = \mathbb{P}\left[\tilde{y} = C_2 \mid y = C_1\right]$ be the probability of corrupting the label of a sample from class C_1 to class C_2, $\mathcal{R}$ be the noise rate and k be the total number of classes. The two types of CCN used in our experiments are defined as in Eq. 2.

$$
\begin{aligned}
\beth_{C_1,C_2}^{\text{Uniform}} &= \begin{cases} \frac{\mathcal{R}}{k-1} & C_1 \neq C_2 \\ 1 - \mathcal{R} & C_1 = C_2 \end{cases} \\
\beth_{C_1,C_2}^{\text{Asymmetrical}} &= \begin{cases} \mathcal{R} & C_1 \neq C_2 \\ 1 - \mathcal{R} & C_1 = C_2 \end{cases}
\end{aligned}
\tag{2}
$$

The ResNet-34 [12] is used for synthetic experiments. All models are trained from scratch for 180 epochs with a batch size of 128 images. Stochastic gradient descent is used as the optimizer with a momentum value equal to 9×10^{-1} and a weight decay rate of 5×10^{-4}. The learning rate is initialized as 1×10^{-2} and gets divided by 2 after 40 and 80 epochs. Standard data augmentations are applied: random horizontal flip, 32×32 random crop after padding 4 pixels, and standard normalizing with mean $= (0.4914, 0.4822, 0.4465)$, std $= (0.2023, 0.1994, 0.2010)$. In each experiment, 10% of the clean training data is reserved as the validation set. Each experiment is repeated 5 times to report the mean and standard deviation for final accuracy. The initial value for θ in Algorithm 1 is set to 7×10^{-1} with a decrement step of 1×10^{-1}. The algorithm averages 5 top-performing networks from the last 30 epochs on each iteration.

Table 1. Final accuracy on the CIFAR datasets for different IDN patterns and rates.

Dataset	Noise info	SL [30]	LRT [39]	PLC [38]	Ours
CIFAR-10	$\aleph_{35\%}^{I}$	79.76×0.7	80.98×0.8	82.80×0.3	$\mathbf{83.60 \times 0.3}$
	$\aleph_{70\%}^{I}$	36.29×0.7	41.52×4.5	42.74×2.1	$\mathbf{46.47 \times 1.1}$
	$\aleph_{35\%}^{II}$	77.92×0.9	80.74×0.3	81.54×0.5	$\mathbf{83.41 \times 0.3}$
	$\aleph_{70\%}^{II}$	41.11×1.9	44.67×3.9	46.04×2.2	$\mathbf{46.24 \times 0.9}$
	$\aleph_{35\%}^{III}$	78.81×0.3	81.08×0.4	81.50×0.5	$\mathbf{83.16 \times 0.3}$
	$\aleph_{70\%}^{III}$	38.49×1.5	44.47×1.2	45.05×1.1	$\mathbf{46.33 \times 1.1}$
CIFAR-100	$\aleph_{35\%}^{I}$	55.20×0.3	56.74×0.3	60.01×0.4	$\mathbf{63.85 \times 0.3}$
	$\aleph_{70\%}^{I}$	40.02×0.9	45.29×0.4	45.92×0.6	$\mathbf{46.38 \times 0.3}$
	$\aleph_{35\%}^{II}$	56.10×0.7	57.25×0.7	63.68×0.3	$\mathbf{63.91 \times 0.3}$
	$\aleph_{70\%}^{II}$	38.45×0.6	43.71×0.5	45.03×0.5	$\mathbf{46.63 \times 0.2}$
	$\aleph_{35\%}^{III}$	56.04×0.7	56.57×0.3	63.68×0.3	$\mathbf{63.92 \times 0.4}$
	$\aleph_{70\%}^{III}$	39.94×0.8	44.41×0.2	44.45×0.6	$\mathbf{46.22 \times 0.2}$

Table 1 holds the results of testing the proposed method on synthetic data affected by three different IDN patterns with 35% and 70% noise rates. The performance of baseline methods is obtained from [38]. As shown in this table, our method outperforms the alternatives in all cases. Judging by the numbers, some alternative approaches have a high standard deviation rate, indicating possible instability of that method.

Table 2. Final accuracy on the CIFAR datasets for different combinations of noise.

Dataset	Noise info	SL [30]	LRT [39]	PLC [38]	Ours
CIFAR-10	$N^I_{35\%}$ + Uniform 30%	77.79 × 0.5	75.97 × 0.3	79.04 × 0.5	**80.94 × 0.2**
	$N^I_{35\%}$ + Asymmetrical 30%	77.14 × 0.7	76.96 × 0.5	78.31 × 0.4	**79.93 × 0.5**
	$N^{II}_{35\%}$ + Uniform 30%	75.08 × 0.5	75.94 × 0.6	80.08 × 0.4	**81.07 × 0.2**
	$N^{II}_{35\%}$ + Asymmetrical 30%	75.43 × 0.4	77.03 × 0.6	77.63 × 0.3	**79.90 × 0.5**
	$N^{III}_{35\%}$ + Uniform 30%	76.22 × 0.1	75.66 × 0.6	80.06 × 0.5	**80.54 × 0.3**
	$N^{III}_{35\%}$ + Asymmetrical 30%	76.09 × 0.1	77.19 × 0.7	77.54 × 0.7	**79.54 × 0.5**
CIFAR-100	$N^I_{35\%}$ + Uniform 30%	51.34 × 0.6	45.66 × 1.6	60.09 × 0.2	**61.46 × 0.4**
	$N^I_{35\%}$ + Asymmetrical 30%	50.18 × 1.0	52.04 × 0.2	56.40 × 0.3	**59.94 × 0.4**
	$N^{II}_{35\%}$ + Uniform 30%	50.58 × 0.3	43.86 × 1.3	60.01 × 0.6	**61.16 × 0.3**
	$N^{II}_{35\%}$ + Asymmetrical 30%	49.46 × 0.2	52.11 × 0.5	**61.43 × 0.3**	59.34 × 0.5
	$N^{III}_{35\%}$ + Uniform 30%	50.18 × 0.5	42.79 × 1.8	60.14 × 1.0	**61.82 × 0.3**
	$N^{III}_{35\%}$ + Asymmetrical 30%	48.15 × 0.9	50.31 × 0.4	54.56 × 1.1	**59.76 × 0.5**

Table 2 holds the results of testing the proposed method on synthetic data simultaneously affected by IDN and CCN patterns. The final noise rate is typically lower than the sum of two individual noise rates due to overlaps in selected samples. As shown in this table, our method still outperforms the alternatives in almost all cases.

4.2   Real-World Datasets

To evaluate the performance of the proposed method on real-world cases, three commonly used datasets were chosen for testing:

ANIMAL-10N [26] – This dataset contains 50,000 training and 5,000 testing samples over ten categories. According to the creators of the dataset, the estimated noise rate is about 8%. Following the authors' work, we chose VGG-19 [25] with a batch normalization for this experiment. The model is trained from scratch for 180 epochs with a batch size of 128 images. Stochastic gradient descent is used as the optimizer with a weight decay rate of 1×10^{-3}. The learning rate is initialized as 1×10^{-1} and gets divided by 5 after 50 and 75 epochs. Standard data augmentations are applied: random horizontal flip and standard normalizing with mean = (0.485, 0.456, 0.406), std = (0.229, 0.224, 0.225). 10% of the training data is manually labeled with the help of [1] and reserved as the validation set. The initial value for θ in Algorithm 1 is set to 7×10^{-1} with a decrement step of 1×10^{-1}. The algorithm averages 10 top-performing networks from the last 30 epochs on each iteration. Table 3 holds the results of testing the proposed method on the ANIMAL-10N dataset. The performance of baseline methods is obtained from their respective papers. As seen in this table, the proposed method outperforms the alternatives.

Table 3. Final accuracy on the Animal-10N and Food-101N datasets.

Dataset	Method	Accuracy	Dataset	Method	Accuracy
Animal-10N	SELFIE [26]	79.40	Food-101N	DeepSelf [11]	79.40
	Co-learning [28]	82.95		PLC [38]	83.40
	PLC [38]	83.40		Ours	86.34
	Ours	**84.47**		**Co-learning** [28]	**87.57**

Food-101N [16] – This dataset contains 310,000 training samples and utilizes the 25,000 testing samples provided by the Food-101 dataset [6] over 101 categories. According to the creators of the dataset, the estimated noise rate is about 10%. Following the authors' work, we chose ResNet-50 with pre-trained weights on ImageNet [9] for this experiment. The model is fine-tuned for 30 epochs with a batch size of 32 images. Stochastic gradient descent is used as the optimizer with a weight decay rate of 1×10^{-3}. The learning rate is initialized as 5×10^{-3} and gets divided by 10 after 10 and 20 epochs. Standard data augmentations are applied: random horizontal flip, 224×224 random crop, and standard normalizing with mean = (0.485, 0.456, 0.406), std = (0.229, 0.224, 0.225). 14% of the labels are verified by the creators of the dataset to be used as the validation set. The initial value for θ in Algorithm 1 is set to 9×10^{-1} with a decrement step of 1×10^{-1}. The algorithm averages 4 top-performing networks from the last 8 epochs on each iteration. Table 3 holds the results of testing the proposed method on the Food-101N dataset. The performance of baseline methods is obtained from their respective papers. This table shows that the proposed method outperforms most of the alternatives but gets beaten by Co-Learning [28].

Clothing1M [16,34] – This dataset contains 1,000,000 samples over 14 categories, out of which 50,000 training, 14,000 validation, and 10,000 testing samples are verified by the creators of the dataset. Following the previous works [16,17,32], the clean training data is discarded. We chose ResNet-50 with pre-trained weights on ImageNet for this experiment. The model is fine-tuned for 20 epochs with a batch size of 32 images. Stochastic gradient descent is used as the optimizer with a momentum value equal to 9×10^{-1} and a weight decay rate of 5×10^{-4}. The learning rate is initialized as 1×10^{-3} and gets divided by 10 after 5 and 10 epochs. Standard data augmentations are applied: random horizontal flip, 224×224 random crop, and standard normalizing with mean = (0.485, 0.456, 0.406), std = (0.229, 0.224, 0.225). The verified validation data is used as the validation set. The initial value for θ in Algorithm 1 is set to 3×10^{-1} with a decrement step of 1×10^{-1}. The algorithm averages 4 top-performing networks from the last 8 epochs on each iteration. Table 4 holds the results of testing the proposed method on the Clothing1M dataset. The performance of baseline methods is obtained from their respective papers. As seen in this table, the proposed method outperforms the alternatives.

Table 4. Final accuracy on the Clothing1M dataset.

Method	Accuracy
CAL [40]	74.17
Reweight [33]	74.18
DeepSelf [11]	74.45
CleanNet [16]	74.69
DivideMix [17]	74.76
Ours	**75.11**

5   Conclusion

This paper proposes a practical iterative label correction method that utilizes clean validation sets to achieve better performance when dealing with instance-dependent noise. The effectiveness of the proposed method is shown with empirical experiments on both synthetic and real-world benchmark datasets. The proposed method outperformed the current state-of-the-art methods in these experiments. The findings suggest that the proposed method's intuition might be correct, and utilizing a clean validation set in iterative label correction methods is helpful.

References

1. Adhikari, B., Huttunen, H.: Iterative bounding box annotation for object detection. In: 25th International Conference on Pattern Recognition (ICPR), pp. 4040–4046 (2021). https://doi.org/10.1109/ICPR48806.2021.9412956
2. Adhikari, B., Peltomäki, J., Germi, S.B., Rahtu, E., Huttunen, H.: Effect of label noise on robustness of deep neural network object detectors. In: Habli, I., Sujan, M., Gerasimou, S., Schoitsch, E., Bitsch, F. (eds.) SAFECOMP 2021. LNCS, vol. 12853, pp. 239–250. Springer, Cham (2021). https://doi.org/10.1007/978-3-030-83906-2_19
3. Algan, G., Ulusoy, I.: Image classification with deep learning in the presence of noisy labels: a survey. Knowl.-Based Syst. **215** (2021). https://doi.org/10.1016/j.knosys.2021.106771
4. Arazo, E., Ortego, D., Albert, P., O'Connor, N., Mcguinness, K.: Unsupervised label noise modeling and loss correction. In: Proceedings of the 36th International Conference on Machine Learning. Proceedings of Machine Learning Research, vol. 97, pp. 312–321 (2019)

5. Berthon, A., Han, B., Niu, G., Liu, T., Sugiyama, M.: Confidence scores make instance-dependent label-noise learning possible. In: Proceedings of the 38th International Conference on Machine Learning. Proceedings of Machine Learning Research, vol. 139, pp. 825–836 (2021)

6. Bossard, L., Guillaumin, M., Van Gool, L.: Food-101 - mining discriminative components with random forests. In: Computer Vision - ECCV, pp. 446–461. Proceedings of Machine Learning Research (2014)

7. Chen, P., Ye, J., Chen, G., Zhao, J., Heng, P.A.: Beyond class-conditional assumption: a primary attempt to combat instance-dependent label noise. In: Proceedings of the AAAI Conference on Artificial Intelligence, vol. 35, no. 13, pp. 11442–11450 (2021)

8. Cheng, H., Zhu, Z., Li, X., Gong, Y., Sun, X., Liu, Y.: Learning with instance-dependent label noise: a sample sieve approach (2021)

9. Deng, J., Dong, W., Socher, R., Li, L.J., Li, K., Fei-Fei, L.: ImageNet: a large-scale hierarchical image database. In: IEEE Conference on Computer Vision and Pattern Recognition, pp. 248–255 (2009). https://doi.org/10.1109/CVPR.2009.5206848

10. Frenay, B., Verleysen, M.: Classification in the presence of label noise: a survey. IEEE Trans. Neural Netw. Learn. Syst. 25(5), 845–869 (2014). https://doi.org/10.1109/TNNLS.2013.2292894

11. Han, J., Luo, P., Wang, X.: Deep self-learning from noisy labels. In: Proceedings of the IEEE/CVF International Conference on Computer Vision (ICCV) (2019)

12. He, K., Zhang, X., Ren, S., Sun, J.: Deep residual learning for image recognition. In: Proceedings of the IEEE Conference on Computer Vision and Pattern Recognition (CVPR) (2016)

13. Hendrycks, D., Lee, K., Mazeika, M.: Using pre-training can improve model robustness and uncertainty. In: Proceedings of the 36th International Conference on Machine Learning. Proceedings of Machine Learning Research, vol. 97, pp. 2712–2721 (2019)

14. Krizhevsky, A.: Learning multiple layers of features from tiny images. Technical report (2009)

15. Lecun, Y., Bottou, L., Bengio, Y., Haffner, P.: Gradient-based learning applied to document recognition. Proc. IEEE 86(11), 2278–2324 (1998). https://doi.org/10.1109/5.726791

16. Lee, K.H., He, X., Zhang, L., Yang, L.: CleanNet: transfer learning for scalable image classifier training with label noise. In: Proceedings of the IEEE Conference on Computer Vision and Pattern Recognition (CVPR) (2018)

17. Li, J., Socher, R., Hoi, S.C.H.: Dividemix: learning with noisy labels as semi-supervised learning (2020)

18. Li, J., Wong, Y., Zhao, Q., Kankanhalli, M.S.: Learning to learn from noisy labeled data. In: Proceedings of the IEEE/CVF Conference on Computer Vision and Pattern Recognition (CVPR) (2019)

19. Liu, Y., Guo, H.: Peer loss functions: learning from noisy labels without knowing noise rates. In: Proceedings of the 37th International Conference on Machine Learning. Proceedings of Machine Learning Research, vol. 119, pp. 6226–6236 (2020)

20. Ma, X., Huang, H., Wang, Y., Romano, S., Erfani, S., Bailey, J.: Normalized loss functions for deep learning with noisy labels. In: Proceedings of the 37th International Conference on Machine Learning. Proceedings of Machine Learning Research, vol. 119, pp. 6543–6553 (2020)

21. Menon, A.K., van Rooyen, B., Natarajan, N.: Learning from binary labels with instance-dependent noise. Mach. Learn. 1561–1595 (2018). https://doi.org/10.1007/s10994-018-5715-3

22. Patrini, G., Rozza, A., Krishna Menon, A., Nock, R., Qu, L.: Making deep neural networks robust to label noise: a loss correction approach. In: Proceedings of the IEEE Conference on Computer Vision and Pattern Recognition (CVPR) (2017)
23. Shi, J., Wu, J.: Distilling effective supervision for robust medical image segmentation with noisy labels (2021)
24. Shu, J., et al.: Meta-weight-net: learning an explicit mapping for sample weighting (2019)
25. Simonyan, K., Zisserman, A.: Very deep convolutional networks for large-scale image recognition (2015)
26. Song, H., Kim, M., Lee, J.G.: SELFIE: refurbishing unclean samples for robust deep learning. In: Proceedings of the 36th International Conference on Machine Learning. Proceedings of Machine Learning Research, vol. 97, pp. 5907–5915 (2019)
27. Song, H., Kim, M., Park, D., Shin, Y., Lee, J.G.: Learning from noisy labels with deep neural networks: a survey (2021)
28. Tan, C., Xia, J., Wu, L., Li, S.Z.: Co-learning: learning from noisy labels with self-supervision. In: Proceedings of the 29th ACM International Conference on Multimedia, pp. 1405–1413 (2019)
29. Veit, A., Alldrin, N., Chechik, G., Krasin, I., Gupta, A., Belongie, S.: Learning from noisy large-scale datasets with minimal supervision. In: IEEE Conference on Computer Vision and Pattern Recognition (CVPR), pp. 6575–6583 (2017). https://doi.org/10.1109/CVPR.2017.696
30. Wang, Y., Ma, X., Chen, Z., Luo, Y., Yi, J., Bailey, J.: Symmetric cross entropy for robust learning with noisy labels. In: Proceedings of the IEEE/CVF International Conference on Computer Vision (ICCV) (2019)
31. Wei, H., Feng, L., Chen, X., An, B.: Combating noisy labels by agreement: a joint training method with co-regularization. In: Proceedings of the IEEE/CVF Conference on Computer Vision and Pattern Recognition (CVPR) (2020)
32. Wu, P., Zheng, S., Goswami, M., Metaxas, D., Chen, C.: A topological filter for learning with label noise (2020)
33. Xia, X., et al.: Are anchor points really indispensable in label-noise learning? In: Advances in Neural Information Processing Systems, vol. 32 (2021)
34. Xiao, T., Xia, T., Yang, Y., Huang, C., Wang, X.: Learning from massive noisy labeled data for image classification. In: Proceedings of the IEEE Conference on Computer Vision and Pattern Recognition (CVPR) (2015)
35. Yang, S., et al.: Estimating instance-dependent label-noise transition matrix using DNNs (2021)
36. Yi, K., Wu, J.: Probabilistic end-to-end noise correction for learning with noisy labels. In: Proceedings of the IEEE/CVF Conference on Computer Vision and Pattern Recognition (CVPR) (2019)
37. Zhang, C., Bengio, S., Hardt, M., Recht, B., Vinyals, O.: Understanding deep learning (still) requires rethinking generalization. Commun. ACM **64**(3), 107–115 (2021). https://doi.org/10.1145/3446776
38. Zhang, Y., Zheng, S., Wu, P., Goswami, M., Chen, C.: Learning with feature-dependent label noise: a progressive approach (2021)
39. Zheng, S., et al.: Error-bounded correction of noisy labels. In: Proceedings of the 37th International Conference on Machine Learning. In: Proceedings of Machine Learning Research, vol. 119, pp. 11447–11457 (2020)
40. Zhu, Z., Liu, T., Liu, Y.: A second-order approach to learning with instance-dependent label noise. In: Proceedings of the IEEE/CVF Conference on Computer Vision and Pattern Recognition (CVPR), pp. 10113–10123 (2021)

DMSANet: Dual Multi Scale Attention Network

Abhinav Sagar[(✉)]

Vellore Institute of Technology, Vellore, Tamil Nadu, India
abhinavsagar4@gmail.com

Abstract. Attention mechanism of late has been quite popular in the computer vision community. A lot of work has been done to improve the performance of the network, although almost always it results in increased computational complexity. In this paper, we propose a new attention module that not only achieves the best performance but also has lesser parameters compared to most existing models. Our attention module can easily be integrated with other convolutional neural networks because of its lightweight nature. The proposed network named Dual Multi Scale Attention Network (DMSANet) is comprised of two parts: the first part is used to extract features at various scales and aggregate them, the second part uses spatial and channel attention modules in parallel to adaptively integrate local features with their global dependencies. We benchmark our network performance for Image Classification on ImageNet dataset, Object Detection and Instance Segmentation both on MS COCO dataset.

Keywords: Attention module · Image classification · Object detection · Instance segmentation

1 Introduction

The local receptive field of the human eye has led to the construction of convolutional neural networks which has powered much of the recent advances in computer vision. Multi scale architecture used in the famous InceptionNet (Szegedy et al. 2016) aggregates multi-scale information from different size convolutional kernels. Attention Networks has attracted a lot of attention recently as it allows the network to focus on only then essential aspects while ignoring the ones which are not useful (Li et al. 2019; Cao et al. 2019).

A lot of problems have been successfully tackled using attention mechanism in computer vision like image classification, image segmentation, object detection and image generation. Most of the attention mechanisms can be broadly classified into two types channel attention and spatial attention, both of which strengthens the original features by aggregating the same feature from all the positions with different aggregation strategies, transformations, and strengthening functions (Zhang et al. 2021).

S. Sclaroff et al. (Eds.): ICIAP 2022, LNCS 13231, pp. 633–645, 2022.
https://doi.org/10.1007/978-3-031-06427-2_53

Some of the work combined both these mechanism together and achieved better results (Cao et al. 2019; Woo et al. 2018). The computational burden was reduced by (Wang et al. 2020) using efficient channel attention and 1×1 convolution. The most popular attention mechanism is the Squeeze-and Excitation module (Hu et al. 2018b), which can significantly improve the performance with a considerably low cost. The "channel shuffle" operator is used (Zhang and Yang 2021) to enable information communication between the two branches. It uses a grouping strategy, which divides the input feature map into groups along the channel dimension.

2   Related Work

There are two main problems which hinders the progress in this field: 1) Both spatial and channel attention as well as network using combination of two uses only local information while ignoring long range channel dependency, 2) The previous architectures fail to capture spatial information at different scales to be more robust and handle more complex problems. These two challenges were tackled by (Duta et al. 2020; Li et al. 2019) respectively. The problem with these architectures is that the number of parameters increased considerably.

Pyramid Split Attention (PSA) (Zhang et al. 2021) has the ability to process the input tensor at multiple scales. A multi-scale pyramid convolution structure is used to integrate information at different scales on each channel-wise feature map. The channel-wise attention weight of the multi-scale feature maps are extracted hence long range channel dependency is done.

Non-Local block (Wang et al. 2018) is proposed to build a dense spatial feature map and capture the long-range dependency using non-local operations. (Li et al. 2019) used a dynamic selection attention mechanism that allows each neuron to adaptively adjust its receptive field size based on multiple scales of input feature map. (Fu et al. 2019) proposed a network to integrate local features with their global dependencies by summing these two attention modules from different branches.

Multi scale architectures have been used successfully for a lot of vision problems like (Hu et al. 2018b; Sagar and Soundrapandiyan 2020). Fu et al. (2019) adaptively integrated local features with their global dependencies by summing the two attention modules from different branches. Hu et al. (2018a) used spatial extension using a depth-wise convolution to aggregate individual features. Our network borrows ideas from Gao et al. (2018) which used a network to capture local cross-channel interactions.

The performance (in terms of accuracy) vs computational complexity (in terms of number of parameters) of the state of art attention modules is shown in Fig. 1:

Our main contributions can be summarized as follows:

- A new attention module is proposed which aggregates feature information at various scales. Our network is scalable and can be easily plugged into various computer vision problems.

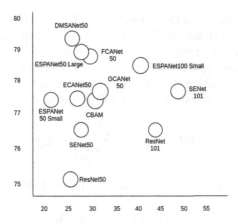

Fig. 1. Comparing the accuracy of different attention methods with ResNet-50 and ResNet-101 as backbone in terms of accuracy and network parameters. The circles reflects the network parameters and FLOPs of different models. Our proposed network achieves higher accuracy while having less model complexity.

- Our network captures more contextual information using both spatial and channel attention at various scales.
- Our experiments demonstrate that our network outperforms previous state of the art with lesser computational cost.

3   Method

3.1   Feature Grouping

Shuffle Attention module divides the input feature map into groups and uses Shuffle Unit to integrate the channel attention and spatial attention into one block for each group. The sub-features are aggregated and a "channel shuffle" operator is used for communicating the information between different sub-features.

For a given feature map $X \in R^{C \times H \times W}$, where C, H, W indicate the channel number, spatial height, and width, respectively, shuffle attention module divides X into G groups along the channel dimension, i.e., $X = [X1, X_G], X_k \in R^{C/G \times H \times W}$. An attention module is used to weight the importance of each feature. The input of X_k is split into two networks along the channel dimension $X_{k1}, X_{k2} \in R^{C/2G \times H \times W}$. The first branch is used to produce a channel attention map by using the relationship of channels, while the second branch is used to generate a spatial attention map by using the spatial relationship of different features.

3.2   Channel Attention Module

The channel attention module is used to selectively weight the importance of each channel and thus produces best output features. This helps in reducing the number of parameters of the network. Let $X \in R^{C \times H \times W}$ denotes the input feature map, where the quantity H, W, C represent its height, width and number of input channels respectively. A SE block consists of two parts: squeeze and excitation, which are respectively designed for encoding the global information and adaptively recalibrating the channel-wise relationship. The Global Average Pooling (GAP) operation can be calculated by the as shown in Eq. 1:

$$GAP_c = \frac{1}{H \times W} \sum_{i=1}^{H} \sum_{j=1}^{W} x_c(i, j) \tag{1}$$

The attention weight of the cth channel in the SE block can be written as denoted in Eq. 2:

$$w_c = \sigma \left(W_1 ReLU \left(W_0 \left(GAP_c \right) \right) \right) \tag{2}$$

where $W_0 \in R^{C \times C}r$ and $W_1 \in R^{Cr \times C}$ represent the fully-connected (FC) layers. The symbol σ represents the excitation function where Sigmoid function is usually used.

We calculate the channel attention map $X \in R^{C \times C}$ from the original features $A \in R^{C \times H \times W}$. We reshape A to $R^{C \times N}$, and then perform a matrix multiplication between A and the transpose of A. We then apply a softmax layer to obtain the channel attention map $X \in R^{C \times C}$ as shown in Eq. 3:

$$x_{ji} = \frac{\exp \left(A_i \cdot A_j \right)}{\sum_{i=1}^{C} \exp \left(A_i \cdot A_j \right)} \tag{3}$$

where x_{ji} measures the ith channel's impact on the jth channel. We perform a matrix multiplication between the transpose of X and A and reshape their result to $R^{C \times H \times W}$. We also multiply the result by a scale parameter β and perform an element-wise sum operation with A to obtain the final output $E \in R^{C \times H \times W}$ as shown in Eq. 4:

$$E_{1j} = \beta \sum_{i=1}^{C} \left(x_{ji} A_i \right) + A_j \tag{4}$$

3.3   Spatial Attention Module

We use Instance Normalization (IN) over X_{k2} to obtain spatial-wise statistics. A $Fc()$ operation is used to enhance the representation of X_{k2}. The final output of spatial attention is obtained by where W_2 and b_2 are parameters with shape $R^{C/2G \times 1 \times 1}$. After that the two branches are concatenated to make the number of channels equal to the number of input.

A local feature denoted by $A \in R^{C \times H \times W}$ is fed into a convolution layer to generate two new feature maps B and C, respectively where $B, C \in R^{C \times H \times W}$. We reshape them to $R^{C \times N}$, where $N = H \times W$ is the number of pixels. Next a matrix multiplication is done between the transpose of C and B, and apply a softmax layer to calculate the spatial attention map $S \in R^{N \times N}$. This operation is shown in Eq. 1:

$$s_{ji} = \frac{\exp(B_i \cdot C_j)}{\sum_{i=1}^{N} \exp(B_i \cdot C_j)} \tag{5}$$

where s_{ji} measures the ith position's impact on j^{th} position. Next we feed feature A into a convolution layer to generate a new feature map $D \in R^{C \times H \times W}$ and reshape it to $R^{C \times N}$. We perform a matrix multiplication between D and the transpose of S and reshape the result to $R^{C \times H \times W}$. We multiply it by a scale parameter α and perform a element-wise sum operation with the feature A to obtain the final output $E \in R^{C \times H \times W}$ as shown in Eq. 2:

$$E_{2j} = \alpha \sum_{i=1}^{N} (s_{ji} D_i) + A_j \tag{6}$$

3.4   Aggregation

In the final part of the network, all the sub-features are aggregated. We use a "channel shuffle" operator to enable cross-group information flow along the channel dimension. The final output of our module is the same size as that of input, making our attention module quite easy to integrate with other networks.

The whole multi-scale pre-processed feature map can be obtained by a concatenation way as defined in Eq. 7:

$$F = \text{Concat}([E_{1j}, E_{2j}]) \tag{7}$$

where $F \in R^{C \times H \times W}$ is the obtained multi-scale feature map. Our attention module is used across channels to adaptively select different spatial scales which is guided by the feature descriptor. This operation is defined in Eq. 8:

$$att_i = \text{Softmax}(Z_i) = \frac{\exp(Z_i)}{\sum_{i=0}^{S-1} \exp(Z_i)} \tag{8}$$

Finally we multiply the re-calibrated weight of multi-scale channel attention a_{tti} with the feature map of the corresponding scale F_i as shown in Eq. 9:

$$Y_i = F_i \odot att_i \quad i = 1, 2, 3, \cdots S - 1 \tag{9}$$

3.5   Network Architecture

We propose DMSA module with the goal to build more efficient and scalable architecture. The first part of our network borrows ideas from (Li et al. 2019) and (Zhang and Yang, 2021). An input feature map X is splitted into N parts along with the channel dimension. For each splitted parts, it has $C_0 = C_S$ number of common channels, and the ith feature map is $X_i \in R^{C_0 \times H \times W}$. The individual features are fused before being passed to two different branches.

These two branches are comprised of position attention module and channel attention module as proposed in (Fu et al. 2019) for semantic segmentation. The second part of our network does the following 1) Builds a spatial attention matrix which models the spatial relationship between any two pixels of the features, 2) A matrix multiplication between the attention matrix and the original features. 3) An element-wise sum operation is done on the resulting matrix and original features.

The operators concat and sum are used to reshape the features. The features from the two parallel branches are aggregated to produce the final output. The complete network architecture is shown in Fig. 2:

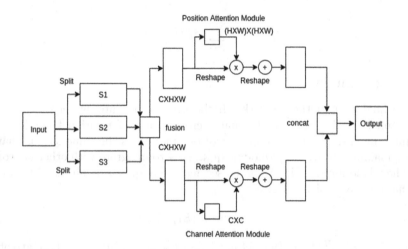

Fig. 2. A detailed illustration of DMSANet

We compare our network architecture with Resnet (Wang et al. 2017), SENet (Hu et al. 2018b) and EPSANet (Zhang et al. 2021) in Fig. 3. We use our DMSA module in between 3×3 convolution and 1×1 convolution. Our network is able to extract features at various scales and aggregate those individual features before passing through the attention module.

The architectural details our proposed attention network is shown in Table 1:

3.6 Implementation Details

We use Residual Networks (He et al. 2016) as the backbone which is widely used in literature for image classification on Imagenet dataset (Deng et al. 2009). Data augmentation is used for increasing the size of the dataset and the input tensor is cropped to size 224×224. Stochastic Gradient Descent is used as the optimizer with learning rate of $1e^{-4}$, momentum as 0.9 and mini batch size of 64. The learning rate is initially set as 0.1 and is decreased by a factor of 10 after every 20 epochs for 50 epochs in total.

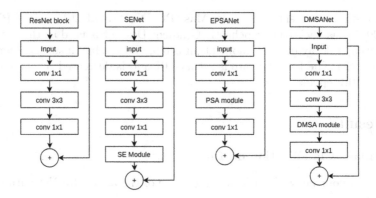

Fig. 3. Illustration and comparison of ResNet, SENet, EPSANet and our proposed DMSANet blocks.

Table 1. Network design of the proposed DMSANet.

Output	ResNet-50	DMSANet
112×112	$7 \times 7, 64$	$7 \times 7, 64$
56×56	3×3 max pool	3×3 max pool
56×56	$\begin{bmatrix} 1\times1, 64 \\ 3\times3, 64 \\ 1\times1, 256 \end{bmatrix} \times 3$	$\begin{bmatrix} 1\times1, \quad 64 \\ DMSA, 64 \\ 1\times1, \quad 256 \end{bmatrix} \times 3$
28×28	$\begin{bmatrix} 1\times1, 128 \\ 3\times3, 128 \\ 1\times1, 512 \end{bmatrix} \times 4$	$\begin{bmatrix} 1\times1, \quad 128 \\ DMSA, 128 \\ 1\times1, \quad 512 \end{bmatrix} \times 4$
14×14	$\begin{bmatrix} 1\times1, 256 \\ 3\times3, 256 \\ 1\times1, 1024 \end{bmatrix} \times 6$	$\begin{bmatrix} 1\times1, \quad 256 \\ DMSA, 256 \\ 1\times1, \quad 1024 \end{bmatrix} \times 6$
7×7	$\begin{bmatrix} 1\times1, 512 \\ 3\times3, 512 \\ 1\times1, 2048 \end{bmatrix} \times 3$	$\begin{bmatrix} 1\times1, \quad 512 \\ DMSA, 512 \\ 1\times1, \quad 2048 \end{bmatrix} \times 3$
1×1	7×7 GAP,1000-d fc	7×7 GAP,1000-d fc

We use Residual Network along with FPN as the backbone network (Lin et al. 2017a) for object detection. The detectors we benchmark against are Faster RCNN (Ren et al. 2015), Mask RCNN (He et al. 2017) and RetinaNet (Lin et al. 2017b) on MS-COCO dataset (Lin et al. 2014). Stochastic Gradient Descent is used as the optimizer with a weight decay of $1e^{-4}$, momentum is 0.9, and the batch size is 16 per GPU for 10 epochs. The learning rate is set as 0.01 and is decreased by the factor of 10 every 10th epoch.

For instance segmentation we use Mask RCNN (He et al. 2017) with FPN (Lin et al. 2017a) as backbone. Stochastic Gradient Descent is used as the optimizer with a weight decay of $1e^{-4}$, momentum is 0.9, and the batch size is 4 per GPU for 10 epochs. The learning rate is set as 0.01 and is decreased by the factor of 10 every 10th epoch.

4   Results

4.1   Image Classification on ImageNet

We compare our network with previous state of the art on ResNet with 50 and 101 layers.

Our network shows the best performance in accuracy, achieving a considerable improvement compared with all the previous attention models along with lower computational cost. The comparison of our network against previous state of the art with ResNet50 as backbone is shown in Table 2:

Table 2. Comparison of various attention methods on ImageNet with ResNet50 as backbone in terms of network parameters (in millions), floating point operations per second (FLOPs), Top-1 and Top-5 Validation Accuracy (%). The best records are marked in bold.

Network	Params	FLOPs	Top-1 Acc (%)	Top-5 Acc (%)
ResNet	25.56	4.12G	75.20	92.52
SENet	28.07	4.13G	76.71	93.38
CBAM	28.07	4.14G	77.34	93.69
ABN	43.59	7.18G	76.90	–
GCNet	28.11	4.13G	77.70	93.66
AANet	25.80	4.15G	77.70	93.80
ECANet	25.56	4.13G	77.48	93.68
FcaNet	28.07	4.13G	78.52	94.14
EPSANet (Small)	**22.56**	3.62G	77.49	93.54
EPSANet (Large)	27.90	4.72G	78.64	94.18
DMSANet	26.25	**3.44G**	**80.02**	**94.27**

The comparision of our network against previous state of the art with ResNet101 as backbone is shown in Table 3:

Table 3. Comparison of various attention methods on ImageNet with ResNet 101 as backbone in terms of network parameters (in millions), floating point operations per second (FLOPs), Top-1 and Top-5 Validation Accuracy (%). The best records are marked in bold.

Network	Params	FLOPs	Top-1 Acc (%)	Top-5 Acc (%)
ResNet	44.55	7.85G	76.83	93.48
SENet	49.29	7.86G	77.62	93.93
BAM	44.91	7.93G	77.56	93.71
CBAM	49.33	7.88G	78.49	94.31
SRM	44.68	7.95G	78.47	94.20
ECANet	44.55	7.86G	78.65	94.34
AANet	45.40	8.05G	78.70	94.40
EPSANet (Small)	**38.90**	**6.82G**	78.43	94.11
EPSANet (Large)	49.59	8.97G	79.38	94.58
DMSANet	42.29	7.11G	**81.54**	**94.93**

4.2 Object Detection on MS COCO

The comparision of our network using Faster RCNN detector against previous state of the art is shown in Table 4:

Table 4. Comparison of object detection results on COCO val2017 using Faster RCNN detector. The best records are marked in bold.

Backbone	Params (M)	GFLOPs	AP	AP_{50}	AP_{75}	AP_S	AP_M	AP_L
ResNet-50	41.53	207.07	36.4	58.2	39.5	21.8	40.0	46.2
SENet-50	44.02	207.18	37.7	60.1	40.9	22.9	41.9	48.2
ECANet-50	41.53	207.18	38.0	60.6	40.9	23.4	42.1	48.0
SANet-50	41.53	207.35	38.7	61.2	41.4	22.3	42.5	49.8
FcaNet-50	44.02	215.63	39.0	61.1	42.3	23.7	42.8	49.6
EPSANet-50 (Small)	**38.56**	**197.07**	39.2	60.3	42.3	22.8	42.4	51.1
EPSANet-50 (Large)	43.85	219.64	40.9	**62.1**	44.6	23.6	44.5	54.0
DMSANet	44.17	222.31	**41.4**	61.9	**46.2**	**25.8**	**44.7**	**55.3**

The comparision of our network using MASK RCNN detector against previous state of the art is shown in Table 5:

Table 5. Comparison of object detection results on COCO val2017 using Mask RCNN detector. The best records are marked in bold.

Backbone	Params (M)	GFLOPs	AP	AP_{50}	AP_{75}	AP_S	AP_M	AP_L
ResNet-50	44.18	275.58	37.2	58.9	40.3	22.2	40.7	48.0
SENet-50	46.67	275.69	38.7	60.9	42.1	23.4	42.7	50.0
Non-local	46.50	288.70	38.0	59.8	41.0	–	–	–
GCNet-50	46.90	279.60	39.4	61.6	42.4	–	–	–
ECANet-50	44.18	275.69	39.0	61.3	42.1	24.2	42.8	49.9
SANet-50	44.18	275.86	39.4	61.5	42.6	23.4	42.8	51.1
FcaNet-50	46.66	261.93	40.3	62.0	44.1	25.2	43.9	52.0
EPSANet-50 (Small)	**41.20**	**248.53**	40.0	60.9	43.3	22.3	43.2	52.8
EPSANet-50 (Large))	46.50	271.10	41.4	**62.3**	45.3	23.6	45.1	54.6
DMSANet	47.23	279.26	**43.1**	61.6	**47.5**	**24.1**	**46.9**	**56.5**

The comparision of our network using RetinaNet detector against previous state of the art is shown in Table 6:

Table 6. Comparison of object detection results on COCO val2017 using RetinaNet detector. The best records are marked in bold.

Backbone	Params (M)	GFLOPs	AP	AP_{50}	AP_{75}	AP_S	AP_M	AP_L
ResNet-50	37.74	239.32	35.6	55.5	38.2	20.0	39.6	46.8
SENet-50	40.25	239.43	37.1	57.2	39.9	21.2	40.7	49.3
SANet-50	37.74	239.60	37.5	58.5	39.7	21.3	41.2	45.9
EPSANet-50 (Small)	**34.78**	**229.32**	38.2	58.1	40.6	21.5	41.5	51.2
EPSANet-50 (Large))	40.07	251.89	39.6	59.4	42.3	21.2	43.4	52.9
DMSANet	41.63	270.17	**40.2**	**59.8**	**44.1**	**23.5**	**44.8**	**54.8**

4.3 Instance Segmentation on MS COCO

We used Mask-RCNN (He et al. 2017) as benchmark on MS-COCO dataset (Lin et al. 2014). The comparision results of our network on instance segmentation using MS COCO dataset against previous state of the art is shown in Table 7:

Table 7. Instance segmentation results of different attention networks by using the Mask R-CNN on COCO. The best records are marked in bold.

Network	AP	AP_{50}	AP_{75}	AP_S	AP_M	AP_L
ResNet-50	34.1	55.5	36.2	16.1	36.7	50.0
SENet-50	35.4	57.4	37.8	17.1	38.6	51.8
GCNet	35.7	58.4	37.6	–	–	–
ECANet	35.6	58.1	37.7	17.6	39.0	51.8
FcaNet	36.2	58.6	38.1	–	–	–
SANet	36.1	58.7	38.2	19.4	39.4	49.0
EPSANet-50 (Small)	35.9	57.7	38.1	18.5	38.8	49.2
EPSANet-50 (Large)	37.1	59.0	39.5	**19.6**	40.4	50.4
DMSANet	**37.4**	**61.1**	**40.7**	19.3	**40.9**	**51.7**

4.4 Ablation Study

The ablation studies of our architecture is shown in Table 8. The results are best obtained using instance normalization. Both removing $F_c()$ and using 1×1 Conv results in reduced performance as compared to the original network. The earlier is because $F_c()$ is used to enhance the performance of individual features while latter is because number of channels in each sub-feature is too few, so it is not important to exchange information among different channels.

Table 8. Performance comparisons of our network using ResNet 50 as backbone with four options (i.e., original, using Batch Normalization, using Group Normalization, using shuffle normalization, eliminating $F_c()$ and using 1×1 Conv to replace $F_c()$ on ImageNet-1k in terms of GFLOPs and Top-1/Top-5 accuracy (in %). The best records are marked in bold.

Methods	GFLOPs	Top-1 Acc (%)	Top-5 Acc (%)
Origin	**3.44**	**80.02**	**94.27**
W BN	3.82	77.37	93.80
W GN	3.56	77.61	92.89
W SN	3.51	78.16	93.48
W/O $F_c()$	4.07	77.64	93.18
1×1 Conv	3.55	78.69	93.71

5 Conclusions

In this paper, we propose a novel Attention module named Dual Multi Scale Attention Network (DMSANet). Our network is comprised of two parts 1) first for aggregating feature information at various scales 2) second made up of position and channel attention modules in parallel for capturing global contextual information. After evaluating our network both qualitatively and quantitatively, we show that our network outperforms previous state of the art across image classification, object detection and instance segmentation problems. The ablation experiments show that our attention module captures long-range contextual information effectively at various scales thus making it generalizable to other tasks. The best part of DMSANet attention module is that it is very lightweight and hence could be easily plugged into various custom networks as and when required.

References

Bello, I., Zoph, B., Vaswani, A., Shlens, J., Le, Q.V.: Attention augmented convolutional networks. In: Proceedings of the IEEE/CVF International Conference on Computer Vision, pp. 3286–3295 (2019)

Cao, Y., Xu, J., Lin, S., Wei, F., Hu, H.: GCNet: non-local networks meet squeeze-excitation networks and beyond. In: Proceedings of the IEEE/CVF International Conference on Computer Vision Workshops (2019)

Deng, J., Dong, W., Socher, R., Li, L.-J., Li, K., Fei-Fei, L.: ImageNet: a large-scale hierarchical image database. In: 2009 IEEE Conference on Computer Vision and Pattern Recognition, pp. 248–255. IEEE (2009)

Duta, I.C., Liu, L., Zhu, F., Shao, L.: Pyramidal convolution: rethinking convolutional neural networks for visual recognition. arXiv preprint arXiv:2006.11538 (2020)

Fu, J., et al.: Dual attention network for scene segmentation. In: Proceedings of the IEEE/CVF Conference on Computer Vision and Pattern Recognition, pp. 3146–3154 (2019)

Gao, H., Wang, Z., Ji, S.: ChannelNets: compact and efficient convolutional neural networks via channel-wise convolutions. arXiv preprint arXiv:1809.01330 (2018)

Gao, S., Cheng, M.-M., Zhao, K., Zhang, X.-Y., Yang, M.-H., Torr, P.H.: Res2Net: a new multi-scale backbone architecture. IEEE Trans. Pattern Anal. Mach. Intell. (2019)

He, K., Zhang, X., Ren, S., Sun, J.: Deep residual learning for image recognition. In: Proceedings of the IEEE Conference on Computer Vision and Pattern Recognition, pp. 770–778 (2016)

He, K., Gkioxari, G., Dollár, P., Girshick, R.: Mask R-CNN. In: Proceedings of the IEEE International Conference on Computer Vision, pp. 2961–2969 (2017)

Hu, J., Shen, L., Albanie, S., Sun, G., Vedaldi, A.: Gather-excite: exploiting feature context in convolutional neural networks. arXiv preprint arXiv:1810.12348 (2018a)

Hu, J., Shen, L., Sun, G.: Squeeze-and-excitation networks. In: Proceedings of the IEEE Conference on Computer Vision and Pattern Recognition, pp. 7132–7141 (2018b)

Li, X., Wang, W., Hu, X., Yang, J.: Selective kernel networks. In: Proceedings of the IEEE/CVF Conference on Computer Vision and Pattern Recognition, pp. 510–519 (2019)

Lin, T.-Y., et al.: Microsoft COCO: common objects in context. In: Fleet, D., Pajdla, T., Schiele, B., Tuytelaars, T. (eds.) ECCV 2014. LNCS, vol. 8693, pp. 740–755. Springer, Cham (2014). https://doi.org/10.1007/978-3-319-10602-1_48

Lin, T.-Y., Dollár, P., Girshick, R., He, K., Hariharan, B., Belongie, S.: Feature pyramid networks for object detection. In: Proceedings of the IEEE Conference on Computer Vision and Pattern Recognition, pp. 2117–2125 (2017a)

Lin, T.-Y., Goyal, P., Girshick, R., He, K., Dollár, P.: Focal loss for dense object detection. In: Proceedings of the IEEE International Conference on Computer Vision, pp. 2980–2988 (2017b)

Ren, S., He, K., Girshick, R., Sun, J.: Faster R-CNN: towards real-time object detection with region proposal networks. arXiv preprint arXiv:1506.01497 (2015)

Sagar, A.: AA3DNET: attention augmented real time 3D object detection. arXiv preprint arXiv:2107.12137 (2021a)

Sagar, A.: AaSeg: attention aware network for real time semantic segmentation. arXiv preprint arXiv:2108.04349 (2021b)

Sagar, A., Soundrapandiyan, R.: Semantic segmentation with multi scale spatial attention for self driving cars. arXiv preprint arXiv:2007.12685 (2020)

Sang, H., Zhou, Q., Zhao, Y.: PCANet: pyramid convolutional attention network for semantic segmentation. Image Vis. Comput. **103**, 103997 (2020)

Szegedy, C., Vanhoucke, V., Ioffe, S., Shlens, J., Wojna, Z.: Rethinking the inception architecture for computer vision. In: Proceedings of the IEEE Conference on Computer Vision and Pattern Recognition, pp. 2818–2826 (2016)

Wang, F., et al.: Residual attention network for image classification. In: Proceedings of the IEEE Conference on Computer Vision and Pattern Recognition, pp. 3156–3164 (2017)

Wang, Q., Wu, B., Zhu, P., Li, P., Zuo, W., Hu, Q.: ECA-Net: efficient channel attention for deep convolutional neural networks (2020)

Wang, X., Girshick, R., Gupta, A., He, K.: Non-local neural networks. In: Proceedings of the IEEE Conference on Computer Vision and Pattern Recognition, pp. 7794–7803 (2018)

Woo, S., Park, J., Lee, J.-Y., Kweon, I.S.: CBAM: convolutional block attention module. In: Proceedings of the European Conference on Computer Vision (ECCV), pp. 3–19 (2018)

Wu, Y., He, K.: Group normalization. In: Proceedings of the European Conference on Computer Vision (ECCV), pp. 3–19 (2018)

Zhang, H., et al. ResNest: split-attention networks. arXiv preprint arXiv:2004.08955 (2020)

Zhang, H., Zu, K., Lu, J., Zou, Y., Meng, D.: EpsaNet: an efficient pyramid split attention block on convolutional neural network. arXiv preprint arXiv:2105.14447 (2021)

Zhang, Q.-L., Yang, Y.-B.: Sa-Net: shuffle attention for deep convolutional neural networks. In: ICASSP 2021–2021 IEEE International Conference on Acoustics, Speech and Signal Processing (ICASSP), pp. 2235–2239. IEEE (2021)

Towards Latent Space Optimization of GANs Using Meta-Learning

Tomaso Fontanini[✉] [iD], Claudio Praticò, and Andrea Prati[iD]

IMP Lab, Department of Engineering and Architecture, University of Parma,
Parma, Italy
{tomaso.fontanini,andrea.prati}@unipr.it,
claudio.pratico1@studenti.unipr.it

Abstract. The necessity to use very large datasets in order to train
Generative Adversarial Networks (GANs) has limited their use in cases
where the data at disposal are scarce or poorly labelled (*e.g.*, in real life
applications). Recently, meta-learning proved that it can help solving
effectively few-shot classification problems, but its use in noise-to-image
generation was only partially explored. In this paper, we took the first
step into applying a meta-learning algorithm (Reptile), to the discriminator of a GAN and to a mapping network in order to optimize the random
noise z to guide the generator network into producing images belonging
to specific classes. By doing so, we prove that the latent space distribution is crucial for the generation of sharp samples when few training
data are at disposal and also managed to generate samples of previously
unseen classes just by optimizing the latent space without changing any
parameter in the generator network. Finally, we show several experiments
with two widely used datasets: MNIST and Omniglot.

Keywords: Generative adversarial network · Meta learning · Few shot

1 Introduction

Training a generative adversarial network (GAN) involves mapping a random
noise z to an image. Nevertheless, in order to obtain results that can not be distinguished from real images, a GAN relies on a long training using huge datasets.
Indeed, at the end of the training, all the information that will be exploited by
the network to produce a new sample is fully contained in z. Nevertheless, the
noise z alone is not enough to control the output of a GAN and for this reason
Conditional GANs [19] (cGANs) were introduced. More in detail, cGANs use
additional information like labels, text or images to guide the outputs of the
generator in the GAN framework.

In addition to that, in order to work properly, GANs require huge datasets
and fail when trying to generate realistic samples using only few images for
training. This fact makes them often unpractical to use in real world applications
where the data at disposal are usually very scarce and/or poorly labelled.

Recently, meta-learning algorithms allowed to achieve outstanding results in few shot image classification using algorithms like MAML [5] or Reptile [21]. Meta-learning divides the training process into a series of task and then is able to learn newly unseen tasks during inference using only few training samples.

In addition to that, previous works combined Reptile and generative models for multi-domain image-to-image translation and image colorization [6,7], reaching interesting and promising results. On the other hand, in the field of noise-to-image generation, being able to generate new sample in a few-shot setting is much more challenging due to the lack of a strong condition (represented by an input image, as in the image-to-image case). In addition to that, the generator of a GAN struggles greatly when it is required to move between different tasks. This is due to the fact that task switching in meta-learning is similar to fine tuning, which can be straightforward in the classification case, but non trivial in the generative one. Here, FIGR (Few-shot Image Generation with Reptile) [4] firstly demonstrated the capability of fusing vanilla GANs and meta-learning for few-shot generation, reaching preliminary results which, nevertheless, suffered from mode collapse and were not sharp enough.

In this paper, we propose a novel way of fusing noise-to-image generative adversarial network and meta-learning (specifically Reptile) and demonstrate its effectiveness for few-shot image generation. In particular, we use Reptile to train only the discriminator of the GAN and guide the generation of specific classes using only the latent code (which is first fed into a fully-connected network in order to be optimized for a specific class). By doing so, we were able to solve the main limitation of combining GANs and meta-learning by removing the generator from the meta-training loop.

The main contributions of this paper are the following:

- A system that combines noise-to-image GAN training with the meta-learning algorithm Reptile and it is able to be train in an end-to-end matter. The model optimizes the input noise z using a linear mapping and then uses it to guide the generator;
- The capability of the proposed network to be trained on a subset of classes from a certain dataset and still manage to generate sharp samples of unseen classes during inference. In addition to that, we were also able to generate samples belonging to a different dataset during inference, showing the good generalization capability of the proposed architecture;
- An extension of the basic system which uses an additional fully-connected network to combine the previously-optimized latent code with information extracted from the discriminator, by permitting to boost the results even more.

2 Related Work

Generative Adversarial Networks. Since a few years, the image generation field is dominated by Generative Adversarial Networks, first introduced by Ian

Goodfellow in [8]. The first version of the GAN model was fully connected, therefore DCGAN [22] proposed a fully convolutional version of the GAN improving its results greatly. Next, conditional GANs (cGANs) [19] were developed in order to allow greater control over the GANs output. Having an additional condition, e.g. the class label, allowed GANs to reach outstanding results in image generation for very large dataset [3]. Recently, StyleGAN models [13,14] were able to produce photorealistyc results in an unconditioned way taking advantage of Adaptive Instance Normalization layers [12]. Nevertheless, GAN training was often prone to failure and, for this reason, WGANs [2,9] were introduced to stabilize the GAN training and avoiding mode collapse.

Meta Learning. Meta-learning is a set of various techniques that are often used to solve few-shot problems by improving the learning process and therefore allowing a neural network to learn better with less data. Some meta-learners work by parameterizing the optimizer of a network [11,23], while in other cases the network itself is used as an optimizer [1,18,25]. In addition to that, often a network is trained on a set of different tasks in order to be able to quickly generalize to previously-unseen ones [17,20,24]. Nevertheless, most meta-learning approaches are dependent on a particular architecture or task. For this reason, MAML [5], which is based on hyper-parameterized gradient descent, was introduced and was designed to be fully model-agnostic. Finally, a simpler and effective version of MAML is represented by Reptile [21] which is used in this paper. The combination of GANs and meta-learning for image generation are a novel topic with the papers [6,7] showing promising results in image-to-image generation and FIGR [4] producing the first basic results for noise-to-image generation.

3   Proposed System

In this section, the proposed architecture, that fuses meta-learning and GANs, will be presented in detail. Also, two different versions of it will be analyzed: the first one uses a single linear mapping for the input noise z, while the second one adds another linear mapping exploiting information extracted from the discriminator network.

3.1   Reptile

As previously mentioned, we choose to employ the Reptile meta-learning algorithm [21]. Reptile works by optimizing the network on a series of different tasks which, in our case, are represented by the different classes in the dataset. Therefore, each training iteration, a task is sampled and the network, which initial weights are defined as θ, is cloned by defining a new set of weights $\tilde{\theta} = \theta$. Then the cloned network performs a series of N gradient steps in order to update its parameters on the current task. Finally, the updated weights $\tilde{\theta}^{(N)}$ are merged with the initial weights θ following the Reptile equation:

$$\theta \leftarrow \theta + \lambda_{ML}(\tilde{\theta}^{(N)} - \theta) \tag{1}$$

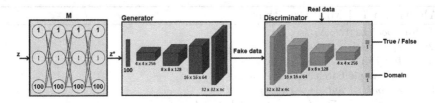

Fig. 1. Network architecture with 1 linear mapping network M. Only the discriminator D and the mapping network M are trained using meta learning.

where λ_{ML} is the meta-learning rate that defines how much of the updated weights needs to be merged with θ. Indeed, if λ_{ML} is set equals to 1, Eq. 1 represents the standard training step with gradient descent. The Reptile equation allows to obtain a network that is able to move very quickly between different tasks using very few samples and gradient steps. In our case, we choose to apply Reptile only to the discriminator and to the mapping network that will be described in the next section.

3.2   Network Architecture

Our GAN architecture follows the DCGAN structure [22] and therefore is composed by one generator G and one discriminator D, both fully convolutional. The generator takes as input the noise z and produces samples almost indistinguishable from real ones trying to fool the discriminator. Then, the discriminator is tasked to discern between real and fake samples. In addition to that, we introduced two additions to this baseline architecture: firstly, we added a linear network M using a Multilayer Perceptron that maps the input noise z to z^* using meta-learning, and, secondly, we added a classification output to the discriminator as in [6] in order to classify if a sample belongs or not to the current task during the meta-learning iterations.

The architecture and the training algorithm for this setting are described in Fig. 1 and in Algorithm 1, respectively. After a random class c_i is chosen, the discriminator and the mapping network are trained for N meta-iterations and their initial weights are updated following the Reptile equation. Therefore, while inside the meta-training loop (rows from 3 to 19 of Algorithm 1), the mapping network M weights are updated and the generator G weights are frozen. Finally, after this step, the generator is updated once. By doing so, the generator learns to produce a specific class using only the optimized noise z^*, without any other condition like class labels.

In addition to this setting, we also experimented with an additional solution that involves the use of a second mapping network M_D that takes as input information from the discriminator. More specifically, features extracted from the second-last layer of D are pooled and then feeded to M_D. The output of M_D is named z_{cls} and is concatenated to z^*. Using this second mapping further enhances the capability of the network to produce results belonging to a specific class without changing the weights of the generator.

Algorithm 1. 1 MLP algorithm

1: **for** *epoch* **in** $0 \ldots N_{\text{epochs}}$ **do**
2:      $c_i \leftarrow$ select_random_class()
3:      **for** j **in** $0 \ldots N_{\text{meta-iter}}$ **do**
4:           sample batch x_{c_i} from class c_i
5:           $out_{adv}, out_{cls} \leftarrow D(x_{c_i})$
6:           $\varepsilon_{D_{\text{real}}} \leftarrow \nabla_{\theta_D} \mathcal{L}_{\text{adv}}(out_{adv}, \text{label\_real})$
7:           $\varepsilon_{D_{\text{cls\_real}}} \leftarrow \nabla_{\theta_D} \mathcal{L}_{\text{cls}}(out_{cls}, \text{label\_real})$
8:           $x_{c_i}^* \leftarrow G(M(z))$
9:           $out_{adv}, \_ \leftarrow D(x_{c_i}^*)$
10:          $\varepsilon_{D_{\text{fake}}} \leftarrow \nabla_{\theta_D} \mathcal{L}_{\text{adv}}(out_{adv}, \text{label\_fake})$
11:          sample batch x from whole dataset without c_i
12:          $\_, out_{cls} \leftarrow D(x)$
13:          $\varepsilon_{D_{\text{cls\_fake}}} \leftarrow \nabla_{\theta_D} \mathcal{L}_{\text{cls}}(out_{cls}, \text{label\_fake})$
14:          $\varepsilon_D \leftarrow \varepsilon_{D_{\text{real}}} + \varepsilon_{D_{\text{fake}}} + \varepsilon_{D_{\text{cls\_real}}} + \varepsilon_{D_{\text{cls\_fake}}}$ ▷ calculates D loss gradients
15:          $\tilde{\theta}_D \leftarrow \tilde{\theta}_D - \lambda_D \varepsilon_D$ ▷ updates D parameters
16:
17:          $\varepsilon_M \leftarrow \varepsilon_{D_{\text{real}}} + \varepsilon_{D_{\text{cls\_real}}}$ ▷ calculates M loss gradients
18:          $\tilde{\theta}_M \leftarrow \tilde{\theta}_M - \lambda_M \varepsilon_M$ ▷ updates M parameters
19:      **end for**
20:      $x_{c_i}^* \leftarrow G(M(z))$
21:      $out_{adv}, out_{cls} \leftarrow D(x_{c_i}^*)$
22:      $\varepsilon_{D_{\text{real}}} \leftarrow \nabla_{\theta_D} \mathcal{L}_{\text{adv}}(out_{adv}, \text{label\_real})$
23:      $\varepsilon_{D_{\text{cls\_real}}} \leftarrow \nabla_{\theta_D} \mathcal{L}_{\text{cls}}(out_{cls}, \text{label\_real})$
24:      $\varepsilon_G \leftarrow \varepsilon_{D_{\text{real}}} + \varepsilon_{D_{\text{cls\_real}}}$ ▷ calculates G loss gradients
25:      $\theta_G \leftarrow \theta_G - \lambda_G \varepsilon_G$ ▷ updates G parameters
26:
27:      $\theta_D \leftarrow \theta_D + \lambda_{ML} \left(\tilde{\theta}_D^{(N_{\text{meta-iter}})} - \theta_D \right)$ ▷ updates initial D parameters
28:      $\theta_M \leftarrow \theta_M + \lambda_{ML} \left(\tilde{\theta}_M^{(N_{\text{meta-iter}})} - \theta_M \right)$ ▷ updates initial M parameters
29: **end for**

The architecture and training algorithm for this setting are described in Fig. 2 and in Algorithm 2, respectively.

After the training on either one of the two settings, an inference step is required to generate a new sample of a specific class. Specifically, few samples of the desired class are extracted from the dataset and a quick meta-learning step is performed in order to allow the mapping network to produce an optimized noise able to push the generator towards the desired class. As it will be shown in the experiments, during inference we were also able to produce samples belonging to class unseen during the training, further proving the effectiveness of the proposed method.

4   Experimental Results

In this section all the experiments performed to validate the proposed algorithm will be presented. Our objective was to move the first step into exploring how meta-learning can be used to optimize the latent space of a noise-to-image GAN in order to generate realistic images using very few samples (20 in our experiments). Consequently, experiments have been carried out by using two well-known datasets like MNIST [16] and Omniglot [15]. More specifically, MNIST is a dataset of 10 handwritten digits, while Omniglot is a dataset of handwritten characters tailored for few-shot learning.

In all the experiments we used $\lambda_{ML} = 0.01$, $N = 20$ meta-iterations and 40000 total iterations.

4.1   Experiments Using MNIST

The first experiment was conducted using the MNIST dataset. Initially, we used all the 10 classes during training and made a comparison between the network with one vs two mapping functions. These results can be observed in Fig. 3, while in Table 1 quantitative results are presented in terms of FID [10]. It is worth emphasizing that the two-mapping version of the proposed system outperforms (by a slight margin) even the baseline (lower bound) of conditional DCGAN in terms of FID (the lower, the better). Moreover, the proposed methods yield the same quality of a standard cDCGAN, but only using the optimised noise to produce results without updating the generator weights.

Finally, we also experimented with generating samples of unseen classes. To do so, we trained the network without using the digit "9" class, then, during inference, leaving the generator weights fixed, we tried to generate this digit only optimizing the mapping network and therefore the input noise. Results of this experiment can be seen in Fig. 4. Pushed by the optimized latent space, the generator was able to utilize the information learned from the other digits to generate the unseen one. In particular, the number "4" was the closest one to "9" in terms of shape and this was exploited by the network in order to produce the new digit.

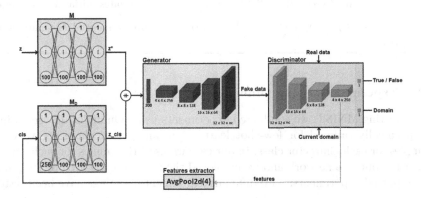

Fig. 2. Network architecture with 2 linear mapping network M and M_D. Only the discriminator D and the mapping networks. M and M_D are trained using meta learning.

Algorithm 2. 2 MLP algorithm

1: **for** *epoch* in $0 \dots N_{\text{epochs}}$ **do**
2:     $c_i \leftarrow$ select_random_class()
3:     **for** j in $0 \dots N_{\text{meta-iter}}$ **do**
4:         sample batch x_{c_i} from class c_i
5:         $out_{adv}, out_{cls} \leftarrow D(x_{c_i})$
6:         $\varepsilon_{D_{\text{real}}} \leftarrow \nabla_{\theta_D} \mathcal{L}_{\mathbf{adv}}(out_{adv}, \text{label\_real})$
7:         $\varepsilon_{D_{\text{cls\_real}}} \leftarrow \nabla_{\theta_D} \mathcal{L}_{\mathbf{cls}}(out_{cls}, \text{label\_real})$
8:         $x_{c_i}^* \leftarrow G(M(z), M_D(AvgPool(D(x_{c_i}))))$
9:         $out_{adv}, \_ \leftarrow D(x_{c_i}^*)$
10:        $\varepsilon_{D_{\text{fake}}} \leftarrow \nabla_{\theta_D} \mathcal{L}_{\mathbf{adv}}(out_{adv}, \text{label\_fake})$
11:        sample batch x from whole dataset without c_i
12:        $\_, out_{cls} \leftarrow D(x)$
13:        $\varepsilon_{D_{\text{cls\_fake}}} \leftarrow \nabla_{\theta_D} \mathcal{L}_{\mathbf{cls}}(out_{cls}, \text{label\_fake})$
14:        $\varepsilon_D \leftarrow \varepsilon_{D_{\text{real}}} + \varepsilon_{D_{\text{fake}}} + \varepsilon_{D_{\text{cls\_real}}} + \varepsilon_{D_{\text{cls\_fake}}}$ ▷ calculates D loss gradients
15:        $\tilde{\theta}_D \leftarrow \tilde{\theta}_D - \lambda_D \varepsilon_D$ ▷ updates D parameters
16:
17:        $\varepsilon_M \leftarrow \varepsilon_{D_{\text{real}}} + \varepsilon_{D_{\text{cls\_real}}}$ ▷ calculates M loss gradients
18:        $\tilde{\theta}_M \leftarrow \tilde{\theta}_M - \lambda_M \varepsilon_M$ ▷ updates M parameters
19:
20:        $\varepsilon_{M_D} \leftarrow \varepsilon_{D_{\text{real}}} + \varepsilon_{D_{\text{cls\_real}}}$ ▷ calculates M_D loss gradients
21:        $\tilde{\theta}_{M_D} \leftarrow \tilde{\theta}_{M_D} - \lambda_{M_D} \varepsilon_{M_D}$ ▷ updates M_D parameters
22:     **end for**
23:     sample batch x_{c_i} from class c_i
24:     $x_{c_i}^* \leftarrow G(M(z), M_D(AvgPool(D(x_{c_i}))))$
25:     $out_{adv}, out_{cls} \leftarrow D(x_{c_i}^*)$
26:     $\varepsilon_{D_{\text{real}}} \leftarrow \nabla_{\theta_D} \mathcal{L}_{\mathbf{adv}}(out_{adv}, \text{label\_real})$
27:     $\varepsilon_{D_{\text{cls\_real}}} \leftarrow \nabla_{\theta_D} \mathcal{L}_{\mathbf{cls}}(out_{cls}, \text{label\_real})$
28:     $\varepsilon_G \leftarrow \varepsilon_{D_{\text{real}}} + \varepsilon_{D_{\text{cls\_real}}}$ ▷ calculates G loss gradients
29:     $\theta_G \leftarrow \theta_G - \lambda_G \varepsilon_G$ ▷ updates G parameters
30:
31:     $\theta_D \leftarrow \theta_D + \lambda_{ML} \left(\tilde{\theta}_D^{(N_{\text{meta-iter}})} - \theta_D \right)$ ▷ updates initial D parameters
32:     $\theta_M \leftarrow \theta_M + \lambda_{ML} \left(\tilde{\theta}_M^{(N_{\text{meta-iter}})} - \theta_M \right)$ ▷ updates initial M parameters
33:     $\theta_{M_D} \leftarrow \theta_{M_D} + \lambda_{ML} \left(\tilde{\theta}_{M_D}^{(N_{\text{meta-iter}})} - \theta_{M_D} \right)$ ▷ updates initial M_D parameters
34: **end for**

4.2 Experiments Using Omniglot

After testing MNIST, we conducted experiments using Omniglot dataset, which is specifically tailored for few-shot learning, meaning that there are very few samples for each character class. In our case we used 16 samples for each class in order to train the network and we employed the two-mapping architecture in all the reported experiments. Two different experimental settings have been tested:

- 30 training classes + 10 unseen classes;
- 100 training classes + 50 unseen classes.

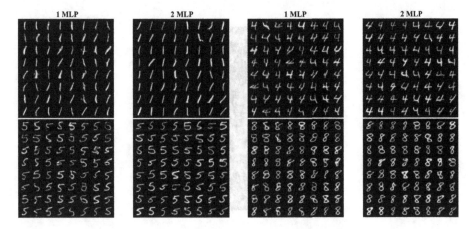

Fig. 3. Comparison between results obtained using a single mapping network and with results obtained using two mapping network.

Table 1. FID score for a conditional DCGAN (considered as a lower bound) and the two proposed settings.

	FID ↓
cDCGAN [19] lower bound	11.92
Ours (1 MLP)	16.53
Ours (2 MLP)	**11.87**

Results for the two experiments can be seen in Figs. 5 and 6. The setting with 100 training classes yields the best results since the network learned to generalize a much higher variety of shapes and, for this reason, was able to learn much quicker the unseen classes. An additional proof of this fact is presented in Table 2. Nevertheless, even with just 30 training classes we were able to obtain reasonably good results for new domains during inference.

Table 2. FID score for the two Omniglot training settings.

	FID ↓
Omniglot 30 + 10	106.61
Omniglot 100 + 50	**77.77**

Finally, as an additional experiment, we pushed the concept of unseen classes to the extreme and, therefore, we tried to generate MNIST digits starting from the network trained on 100 Omniglot characters, with the generator weights kept fixed as usual. Results for this experiment are shown in Fig. 7. In this challenging setting, the network, in order to generate all the ten digits using only the weights

Fig. 4. Results when generating a digit that was unseen during training.

Fig. 5. Results of some of the unseen classes using 30 training classes. Inside the red squares are the real images. (Color figure online)

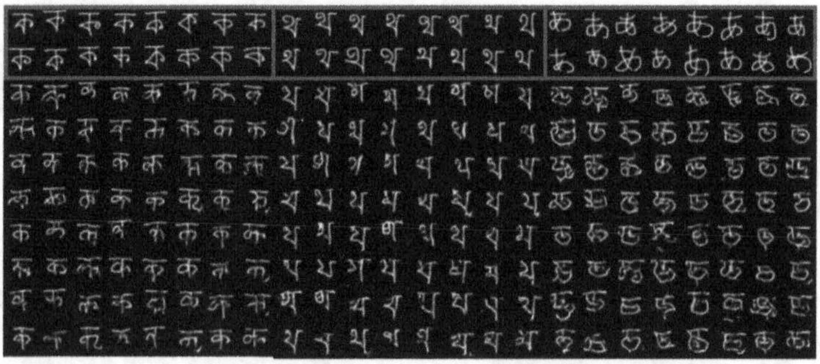

Fig. 6. Results of some of the unseen classes using 100 training classes. Inside the red squares are the real images. (Color figure online)

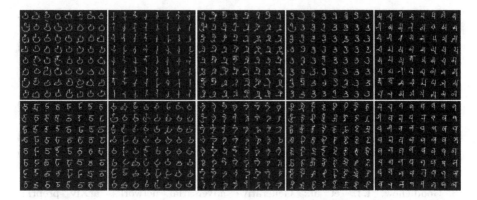

Fig. 7. Results when trying to generate MNIST digits using the network trained with Omniglot.

of a generator trained on alphabets characters, managed to produce characters that resembled closely the desired digits. In other words, the optimization of the latent space alone was able to generate samples very close to the desired output even in the case of an inference performed on a completely different dataset than the one used during training.

5 Conclusions

In this paper, we proved that optimizing the latent space of a GAN with meta-learning can be used to guide the network into generating samples of the desired classes without any additional information (like class labels), allowing to employ fewer training samples. We validate this approach with two widely used datasets like MNIST and Omniglot reaching good results also when dealing with unseen classes. In addition to that, this preliminary work provided a good hint for how to train a few-shot noise-to-image GAN, also using more complex datasets, such as CIFAR10, CIFAR100 or ImageNet.

Acknowledgments. This research has financially been supported by the Programme "FIL-Quota Incentivante" of University of Parma and co-sponsored by Fondazione Cariparma.

References

1. Andrychowicz, M., et al.: Learning to learn by gradient descent by gradient descent. arXiv preprint arXiv:1606.04474 (2016)
2. Arjovsky, M., Chintala, S., Bottou, L.: Wasserstein generative adversarial networks. In: International Conference on Machine Learning, pp. 214–223. PMLR (2017)
3. Brock, A., Donahue, J., Simonyan, K.: Large scale GAN training for high fidelity natural image synthesis. arXiv preprint arXiv:1809.11096 (2018)

4. Clouâtre, L., Demers, M.: FIGR: few-shot image generation with reptile. arXiv preprint arXiv:1901.02199 (2019)
5. Finn, C., Abbeel, P., Levine, S.: Model-agnostic meta-learning for fast adaptation of deep networks. In: International Conference on Machine Learning, pp. 1126–1135. PMLR (2017)
6. Fontanini, T., Iotti, E., Donati, L., Prati, A.: MetalGAN: multi-domain labelless image synthesis using cGANs and meta-learning. Neural Netw. **131**, 185–200 (2020)
7. Fontanini, T., Iotti, E., Prati, A.: MetalGAN: a cluster-based adaptive training for few-shot adversarial colorization. In: Ricci, E., Rota Bulò, S., Snoek, C., Lanz, O., Messelodi, S., Sebe, N. (eds.) ICIAP 2019. LNCS, vol. 11751, pp. 280–291. Springer, Cham (2019). https://doi.org/10.1007/978-3-030-30642-7_25
8. Goodfellow, I.J., et al.: Generative adversarial networks. arXiv preprint arXiv:1406.2661 (2014)
9. Gulrajani, I., Ahmed, F., Arjovsky, M., Dumoulin, V., Courville, A.: Improved training of Wasserstein GANs. arXiv preprint arXiv:1704.00028 (2017)
10. Heusel, M., Ramsauer, H., Unterthiner, T., Nessler, B., Hochreiter, S.: GANs trained by a two time-scale update rule converge to a local NASH equilibrium. Advances in Neural Information Processing Systems, vol. 30 (2017)
11. Hochreiter, S., Younger, A.S., Conwell, P.R.: Learning to learn using gradient descent. In: Dorffner, G., Bischof, H., Hornik, K. (eds.) ICANN 2001. LNCS, vol. 2130, pp. 87–94. Springer, Heidelberg (2001). https://doi.org/10.1007/3-540-44668-0_13
12. Huang, X., Belongie, S.: Arbitrary style transfer in real-time with adaptive instance normalization. In: Proceedings of the IEEE International Conference on Computer Vision, pp. 1501–1510 (2017)
13. Karras, T., Laine, S., Aila, T.: A style-based generator architecture for generative adversarial networks. In: Proceedings of the IEEE/CVF Conference on Computer Vision and Pattern Recognition, pp. 4401–4410 (2019)
14. Karras, T., Laine, S., Aittala, M., Hellsten, J., Lehtinen, J., Aila, T.: Analyzing and improving the image quality of styleGAN. In: Proceedings of the IEEE/CVF Conference on Computer Vision and Pattern Recognition, pp. 8110–8119 (2020)
15. Lake, B.M., Salakhutdinov, R., Tenenbaum, J.B.: Human-level concept learning through probabilistic program induction. Science **350**(6266), 1332–1338 (2015)
16. LeCun, Y., Bottou, L., Bengio, Y., Haffner, P.: Gradient-based learning applied to document recognition. Proc. IEEE **86**(11), 2278–2324 (1998)
17. Lee, K., Maji, S., Ravichandran, A., Soatto, S.: Meta-learning with differentiable convex optimization. In: Proceedings of the IEEE/CVF Conference on Computer Vision and Pattern Recognition, pp. 10657–10665 (2019)
18. Li, K., Malik, J.: Learning to optimize neural nets. arXiv preprint arXiv:1703.00441 (2017)
19. Mirza, M., Osindero, S.: Conditional generative adversarial nets. arXiv preprint arXiv:1411.1784 (2014)
20. Mishra, N., Rohaninejad, M., Chen, X., Abbeel, P.: A simple neural attentive meta-learner. arXiv preprint arXiv:1707.03141 (2017)
21. Nichol, A., Achiam, J., Schulman, J.: On first-order meta-learning algorithms. arXiv preprint arXiv:1803.02999 (2018)
22. Radford, A., Metz, L., Chintala, S.: Unsupervised representation learning with deep convolutional generative adversarial networks. arXiv preprint arXiv:1511.06434 (2015)

23. Ravi, S., Larochelle, H.: Optimization as a model for few-shot learning (2016)
24. Rusu, A.A., et al.: Meta-learning with latent embedding optimization. arXiv preprint arXiv:1807.05960 (2018)
25. Wichrowska, O., et al.: Learned optimizers that scale and generalize. In: International Conference on Machine Learning, pp. 3751–3760. PMLR (2017)

Pruning in the Face of Adversaries

Florian Merkle[✉][ID], Maximilian Samsinger[ID], and Pascal Schöttle[ID]

Digital Business and Software Engineering, Management Center Innsbruck,
Innsbruck, Austria
{florian.merkle,maximilian.samsinger,pascal.schoettle}@mci.edu

Abstract. The vulnerability of deep neural networks against adversarial examples – inputs with small imperceptible perturbations – has gained a lot of attention in the research community recently. Simultaneously, the number of parameters of state-of-the-art deep learning models has been growing massively, with implications on the memory and computational resources required to train and deploy such models. One approach to control the size of neural networks is retrospectively reducing the number of parameters, so-called neural network pruning.

Available research on the impact of neural network pruning on the adversarial robustness is fragmentary and often does not adhere to established principles of robustness evaluation. We close this gap by evaluating the robustness of pruned models against ℓ^0, ℓ^2, and ℓ^∞-attacks for a wide range of attack strengths, several architectures, data sets, pruning methods, and compression rates.

Our results confirm that neural network pruning and adversarial robustness are not mutually exclusive. Instead, sweet spots can be found that are favorable in terms of model size and adversarial robustness. Furthermore, we extend our analysis to situations that incorporate additional assumptions on the adversarial scenario and show that depending on the situation, different strategies are optimal.

Keywords: Security · Neural network pruning · Adversarial machine learning

1 Introduction

Modern deep neural networks (DNNs) are increasingly able to solve sophisticated tasks from computer vision to natural language processing and beyond. This has substantial implications for mankind and thus, we expect DNNs to behave as intended. However, [32] found that adversarial examples, minimally perturbed input samples, can fool DNNs into misclassification. While much of the current research on adversarial robustness is conducted in an artificial, virtual setting, some work has shown that adversarial machine learning is applicable to real-world scenarios, such as road sign classification [10], fooling voice-assistants [6] or face-recognition software with adversarial patterns on eyeglass frames [31].

Although the availability of computational resources drove the recent progress in the field of deep learning, there are many applications where resources

S. Sclaroff et al. (Eds.): ICIAP 2022, LNCS 13231, pp. 658–669, 2022.
https://doi.org/10.1007/978-3-031-06427-2_55

are scarce. Deep learning applications deployed on IoT or mobile devices and real-time applications heavily rely on resource optimization. Previous work [25] has suggested that there is a direct relation between a model's capacity, i.e., the number of parameters of a model, and its respective adversarial robustness. Bigger models are more memory and computationally intensive, and some domains impose restrictions on the resources a model may use. The resources might be bounded by the available hardware or economic aspects. Neural Network (NN) pruning slims down a model's size before deployment, decreasing memory usage and increasing computational efficiency for inference.

Recently, some research on examining the adversarial robustness of pruned NNs has started to emerge. However, much of this work is fragmentary, lacks a clear threat model, suffers from an inadequate choice of attacks, or does not adhere to other principles of a rigorous robustness evaluation as described by [4]. Consequently, existing literature does not provide clear results on the impact of NN pruning on the adversarial robustness.

We conduct an exhaustive study that covers the most relevant attacks, pruning methods, architectures, and data sets. Hereby, we confirm evidence from previous work that NN pruning does not necessarily impact a model's adversarial robustness negatively for various combinations of factors. We show that NN pruning provides a particular space for optimal strategies, balancing clean and robust accuracy.

The remainder of this paper is organized as follows: Sect. 2 introduces the necessary theoretical foundation and covers the current state of the research on the adversarial robustness of pruned NNs. Section 3 presents the design of our experiments. Specifically, we introduce and elaborate on our choice of architectures, attacks, and pruning methods, before we present the experimental results and discuss their relevance on a defender's pruning strategy in Sect. 4. Finally, Sect. 5 concludes this paper and highlights implications of our work for further research and real-world scenarios.

2   Related Work

In this section, we present related work on NN pruning, adversarial machine learning, and the current state-of-the-art in the combination of those fields.

Network pruning refers to the deletion of parameters of a DNN. Modern NNs are typically over-parameterized for the task at hand, leading to extensive redundancy in the model [24]. The goal of pruning is to reduce storage, memory usage, and computational resources. Interestingly, it has been shown that it is possible, by carefully selecting the parameters to be removed, to not only reduce the resource requirements of a model without suffering performance losses but instead to increase the accuracy simultaneously [11,14].

Pruning approaches can be described on five dimensions:

The *structure* describes the granularity of a method. The unstructured approach prunes single weights [2,23] while structured pruning removes entire parts, such as kernels and filters [16], or even whole residual blocks [18]. As the first approach produces sparse matrices of the same size as the unpruned network, dedicated hardware is necessary to accomplish optimizations. The *selection criterion* defines how to select the parameters to be pruned. Many approaches have been proposed: Based on their absolute values [14], the gradients [2], or the ℓ^2 norm of a structure [16]. Network pruning can also be incorporated into the learning procedure [18] or formulated as its own optimization problem [36]. Random pruning can serve as a baseline and sanity check [2,11]. The *scope* determines whether the selection process is performed locally [16], where each layer is pruned separately, or globally where all weights are considered simultaneously for the selection process. *Scheduling* determines when pruning is conducted. Most methods, e.g. [14], apply pruning after training. The network is either pruned in one step to the desired compression rate [24] or as an iterative process of pruning and consequent training [14]. *Finetuning* refers to the training phase after pruning is applied. Traditionally finetuning is conducted with the pre-pruned weight values [14], but recent work explores differences when re-initializing the weights with its initial random values [11] or a set of new random values [24].

Research on **adversarial machine learning** started in 2004 when it was first explored that spam filters utilizing linear classifiers can be fooled by small changes in the initial email that do not negatively affect the readability of the message but lead to misclassification [9]. In 2013 [32] showed that DNNs are just as prone to adversarial examples. Formally, an *Adversarial Example* can be described as follows: A classifier is a function $x \mapsto C(x)$ that takes an input x and yields a class $C(x)$. If a semantic-preserving perturbation δ is added to x, such that the manipulated input $x + \delta = \tilde{x}$ leads to a classification different from the original value $C(\tilde{x}) \neq C(x)$, it is labeled an adversarial example. Usually, distance metrics are used to quantify the difference between x and $\tilde{x}$. In image classification, the most important metrics are [5]:

- The ℓ^0 distance: the number of elements i where $x_i \neq \tilde{x}_i$
- The ℓ^2 norm: $||x - \tilde{x}||_2 = (\sum_{i=1}^{n} |x_i - \tilde{x}_i|^2)^{\frac{1}{2}}$
- The ℓ^∞ norm: $||x - \tilde{x}||_\infty = \max\{|x_1 - \tilde{x}_1|, ..., |x_n - \tilde{x}_n|\}$

Rigorous security evaluations require a precisely stated **threat model**, which encompasses assumptions about an adversary's goal, knowledge and capabilities [4]. In adversarial machine learning, the *adversary's goal* can either be an untargeted attack where $C(x) \neq C(\tilde{x})$ or a targeted attack $C(\tilde{x}) = t$, where t is a defined target class. It is sensible to restrict the *capabilities* of an adversary. Without restrictions, they would be able to manipulate the input pipeline, evade the model at training time, change the semantics of an input image, or even make hard changes on the model's weights. Most works impose constraints so that an adversary can make only small changes to an input. A valid adversarial example $\tilde{x}$ would fulfill $D(x, \tilde{x}) \leq \epsilon$, where ϵ is the upper boundary of the allowed alteration and D is a similarity metric. An adversary has a certain level of *knowledge* of the targeted model regarding the training data, the optimization algorithm, the loss function, hyperparameters, the DNN architecture, or the learned parameters. If an adversary has access to all this information, the setting is labeled a white-box attack. This allows a worst-case scenario evaluation of the examined model. In the black-box setting, the adversary has no knowledge of the model but might have (limited) access to the model to retrieve information [4]. Recently, a wide variety of attack algorithms emerged from the research community. Adversarial attacks can be divided into gradient-based attacks, which require full access to the model (e.g. [3,5,25]) and its weights and black-box attacks that rely either on meaningful model outputs such as logits or probabilities or solely on its final decision. Multiple approaches to defend against adversarial attacks have been proposed, but only few have so far remained unbroken [33]. *Adversarial (re)training*, e.g. [25], enhances a model's robustness by presenting the model adversarial examples during training. Adversarial (re)training reduces the clean accuracy, and due to the necessary additional backward passes, induces high computational costs. Another direction of research aims to achieve *certifiable robustness*. *Randomized smoothing* transforms the problem of classifying under adversarial perturbations into the simpler problem of classifying under random noise. Cohen et al. [7] guarantee a certain level of accuracy under any norm-bounded attack up to a specific attack strength by inducing Gaussian noise at training time, and an additional *smoothed* classifier. Certifiable robustness induces computational complexity and the certified robustness is only a fraction of the empirical robustness gained from adversarial (re)training.

Another string of research aims to unify the act of NN pruning and adversarial training into a single framework. In their work, [35] evaluate the adversarial robustness of an ADMM-based pruning method, implemented for weight, column, and filter-based pruning on the VGG-16 [27] and ResNet-18 [15] architecture. They apply two ℓ^∞-attacks [5,25] and find that for concurrent adversarial training and pruning, the robustness decreases the higher the compression rate is. However, they show that a bigger model that is pruned retrospectively is more robust than an unpruned model with a parameter count similar to the pruned model. Another string of work [30] proposes to align network training and pruning and to make the pruning process aware of the training objective, which can be defined as empirical or verified adversarial robustness. The pruning problem is solved with SGD and assigns an importance score to each weight. They evaluate robustness against PGD [25], and the auto-attack ensemble [8] and claim state-of-the-art clean and robust accuracy. Finally, [13] examine the robustness of a fully connected LeNet 300-100 [21], a LeNet-5 [21], a VGG-like network [27] and a ResNet [15]. They applied two ℓ^∞-attacks [12,34] and two ℓ^2-attacks [5,26] on naturally trained models. The authors find that the sparse DNNs are consistently more robust to FGSM attacks than their respective dense models. As they only evaluated the robustness for one perturbation budget and applied only one unstructured-magnitude pruning approach, no general robustness assertion can be drawn.

3   Experimental Setup

In this section, we elaborate and give a rationale for the design of our experiments. While we are aware that NN pruning is no defense mechanism, we adopt the principles of rigorous robustness evaluation as proposed by Carlini et al. [4], where applicable. In total, we evaluate the adversarial robustness of NNs for a combination of four architectures, three attack methods, four perturbation budgets, nine pruning methods, and seven compression rates, yielding a total of 3 048 data points.

3.1   Threat Model

With regards to the threat model introduced above, we follow the recommendation of [4] and model the strongest adversary possible. As such, the adversary's *goal* is untargeted misclassification, and we assume the white-box scenario in which the adversary has perfect *knowledge*. We grant the adversary different levels and forms of *capabilities*. We choose three attacks, one for each of the ℓ^0, ℓ^2, and ℓ^∞ distance metrics, and perform them in various strengths. Furthermore, we consider an adaptive adversary, so when evaluating the robustness, we use the exact same – possibly pruned – model to craft the adversarial examples.

3.2   Adversarial Attacks

We focus our experiments on gradient-based attacks. We partly follow the recommendations of [4], however as we expect similar results for ℓ^1- and ℓ^2-distortion, we opt to drop ℓ^1 and instead analyze the robustness under an ℓ^0-attack. We considered the ℓ^0-attacks proposed by Carlini and Wagner [5] and [3]. Preliminary experiments showed that both attacks are viable choices to evaluate the ℓ^0-robustness. We choose the Brendel&Bethge attack [3] as it is significantly less computationally expensive. Other than that, we evaluate the attacks proposed by [4]. This gives us the following array of attacks:

- Brendel&Bethge [3] for ℓ^0 perturbations
- Carlini&Wagner [5] for ℓ^2 perturbations
- PGD [25] for ℓ^∞ perturbations

We apply each attack with a set of four ϵ-values, which we have chosen such that the weakest attack does not have any impact on the unpruned model and the strongest attack fools the same model for more than 50% of the test inputs.

3.3   Pruning Methods

The selection of pruning methods is motivated by [2] and [28]. We implement unstructured and structured pruning. For the structured approach, we consider both kernel- and filter-wise pruning. When pruning structures, we only prune the convolutional layers. This leads to a slightly lower total sparsity of the network but no significant reduction of the theoretical speed-up.

Magnitude-based pruning methods are a reliable choice as it is widely adopted in current research, and it has been proven to yield competitive results in comparison with more sophisticated approaches [2]. Additionally, we implement random pruning as a baseline and sanity check. For magnitude-based pruning, we

Table 1. Clean accuracies of the ResNet18 for all pruning methods and compression rates. Bold numbers indicate the highest clean accuracy per row.

Pruning method	Compression rate						
	1	2	4	8	16	32	64
Magnitude global filter	85.37	**86.37**	83.96	79.55	66.62	50.76	37.92
Magnitude global kernel	84.43	86.35	**87.01**	85.75	81.63	75.71	49.44
Magnitude global unstructured	85.57	86.21	86.58	**86.82**	86.01	84.60	82.36
Magnitude local filter	85.79	**86.66**	84.14	80.38	76.91	58.88	42.38
Magnitude local kernel	85.14	86.08	**86.14**	84.93	79.32	53.57	48.03
Magnitude local unstructured	85.44	86.07	**86.68**	86.04	85.24	83.61	81.77
Random local filter	**86.31**	86.26	86.02	83.95	80.35	75.44	61.51
Random local kernel	85.22	86.67	**87.33**	86.11	81.96	39.99	64.12
Random local unstructured	84.96	**87.05**	86.85	86.79	86.20	76.69	48.79

examine local and global pruning. This leaves us with the following nine pruning methods:

- Unstructured {local magnitude|global magnitude|local random} pruning
- Kernel-wise {local magnitude|global magnitude|local random} pruning
- Filter-wise {local magnitude|global magnitude|local random} pruning

As [14] have shown, magnitude-based methods yield better results when conducted iteratively. Thus we refrain from examining one-shot pruning and implement our pruning methods strictly with an iterative pruning schedule. At each pruning step, the network is trained to convergence, and subsequently, half of the remaining weights are removed. For the choice of compression rates, we follow the recommendation of [2] to use the set $2, 4, 8, 16, 32$ and add 64 in order to attain more expressive curves. After the pruning procedure, the remaining weights are retained for retraining. [28] have shown that the results for weight retention [14], re-initialization [24], and rewinding [11] yield comparable results.

3.4 Architectures and Data Sets

We apply all pruning methods and attacks on four different architectures. We have chosen a five-layer convolutional neural network, referred to as CNN5 from here, and a VGG11-like architecture due to its simplicity. Additionally, we expand our experiments to ResNets with pre-activation residual blocks with eight and 18 layers, respectively. We train and evaluate the CNN5 network on the MNIST data set [22], the VGG-11-like and the ResNet-8 on the CIFAR-10 data set [20] and the ResNet-18 on the Imagenette data set [17].

The CNN5 consists of two convolutional layers with a kernel size of five and three dense layers. We adapt the first and last layers of the VGG11-like and the ResNet8 to fit the CIFAR10 data set. On the VGG11-like architecture, we add batchnorm layers for enhanced trainability. On the ResNet18, we adapt the last layer to match the ten classes of the Imagenette data set.

3.5 Evaluation

Our approach aims to observe the adversarial robustness for a selection of network architectures, pruning methods, and adversarial attacks. The compression rate and attack strength are independent variables that are discretely adjusted and the change in adversarial robustness is measured.

We define adversarial robustness as the accuracy of a model f under attack, i.e. the fraction of all images in a data set of size n where the model predicts the correct class for the perturbed image.

$$\text{Acc}_{\text{robust}} = \frac{1}{n} \sum_{i=1}^{n} \chi_{\{y_i\}}(f(\tilde{x}_i)) \tag{1}$$

Note that we treat the attack strength as discrete, which is only partially true for the B&B ℓ^0- and the C&W ℓ^2-attack, which are minimization attacks and return the minimal perturbation that leads to misclassification. We can retrospectively evaluate the robustness for any ϵ. This is necessary to evaluate minimization attacks and fixed-epsilon attacks with the same metric. For each combination of model and attack, we choose a set of ϵ-values such that the weakest attack decreases the accuracy only marginally on the unpruned network and the strongest attack fools the network for more than 50% of the images.

3.6 Implementation

We construct an evaluation pipeline in which for each architecture and pruning method, a separate model with random weights is initialized. We add pruning masks to every layer of the model and set all mask-elements to one. The unpruned network is trained to convergence. Subsequently, we iterate over all compression rates and perform pruning, fine-tuning, and evaluation. We optimize the models with the ADAM algorithm [19] over the categorical cross-entropy loss with an initial learning rate of 0.001. We allow up to 150 epochs for training and implement early stopping observing the validation loss and patience of five epochs for CNN5 and 15 epochs for the other models. For the three bigger models, we apply dynamic learning rate scheduling, multiplying the learning rate by 0.3 after a patience period of twelve epochs. We run all experiments five times with different random seeds and report the average values. All experiments are implemented with Tensorflow 2.2.0 [1] and Foolbox 3.0.0 [29].

4 Results

In this section we present the results of the experiments laid out above. For brevity, we only discuss the experiments we conducted on the ResNet18 architecture. However, we were able to identify the same properties we introduce in this section for the ResNet18 architecture for the other examined architectures.

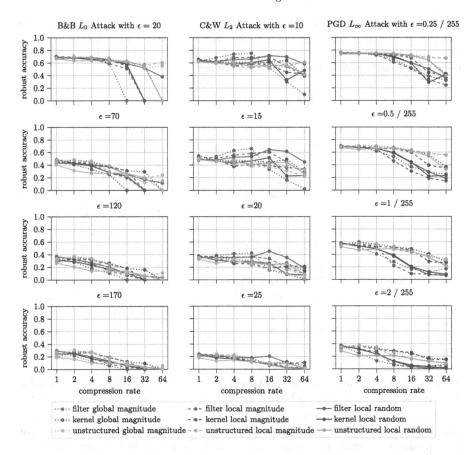

Fig. 1. Adversarial robustness of the ResNet18 models against the B&B ℓ^0- (**left** column), C&W ℓ^2- (**middle**), and PGD ℓ^∞-attacks (**right**) with increasing attack strengths from top to bottom. Each line in each plot depicts a pruning method over the compression rates from one to 64. (Best viewed in color.) (Color figure online)

Clean Accuracies. Table 1 shows the clean accuracies of the ResNet18 for all examined pruning methods and compression rates. Unsurprisingly, our results confirm the findings from previous work [14] that NN pruning can enhance the clean accuracy for mild pruning and yields good results even for higher compression rates, specifically, the unstructured pruning approaches.

Robustness Evaluation. Figure 1 shows the results of our experiments for the ResNet18. Interestingly, the behavior is not consistent for all three attack types. When robustness is evaluated with the PGD attack (rightmost column), as done in most previous work, the expected behavior is observable: For moderate compression rates, the robustness remains stable with small in- and decreases depending on the pruning approach, while more extensive pruning does hurt the

Table 2. Clean and robust accuracies for all attack types and strengths for ResNet18 using unstructured global magnitude pruning with compression rates from 1 (no pruning) to 16. The relative margin (in %) to the unpruned network is displayed in parenthesis. Bold numbers indicate increases in accuracy.

Test data	Compression rate				
	1	2	4	8	16
Benign	85.57	86.21 (**0.64**)	86.58 (**1.01**)	86.82 (**1.25**)	86.01 (**0.44**)
$\ell^0, \epsilon = 20$	68.05	70.62 (**2.58**)	69.30 (**1.25**)	66.56 (-1.48)	62.11 (-5.94)
$\ell^0, \epsilon = 70$	47.66	48.44 (**0.78**)	46.17 (-1.48)	38.67 (-8.98)	25.08 (-22.58)
$\ell^0, \epsilon = 120$	33.91	37.19 (**3.28**)	33.19 (0.00)	24.22 (-9.69)	13.44 (-20.47)
$\ell^0, \epsilon = 170$	27.66	31.02 (**3.36**)	26.44 (-1.33)	18.28 (-9.38)	10.23 (-17.42)
$\ell^2, \epsilon = 10$	61.88	60.94 (-0.94)	62.42 (**0.55**)	61.33 (-0.55)	51.48 (-10.39)
$\ell^2, \epsilon = 15$	47.73	46.41 (-1.33)	48.20 (**0.47**)	47.27 (-0.47)	39.06 (-8.67)
$\ell^2, \epsilon = 20$	34.45	34.30 (-0.16)	32.73 (-1.72)	30.86 (-3.59)	23.59 (-10.86)
$\ell^2, \epsilon = 25$	21.64	24.30 (**2.66**)	18.75 (-2.89)	14.69 (-6.95)	9.22 (-12.42)
$\ell^\infty, \epsilon = .125/255$	74.61	75.00 (**0.39**)	75.23 (**0.62**)	74.61 (0.00)	72.19 (-2.42)
$\ell^\infty, \epsilon = .25/255$	68.52	69.53 (**1.02**)	69.53 (**1.02**)	67.89 (-0.62)	62.66 (-5.86)
$\ell^\infty, \epsilon = .5/255$	55.23	56.88 (**1.64**)	56.25 (**1.02**)	52.42 (-2.81)	42.81 (-12.42)
$\ell^\infty, \epsilon = 1/255$	34.77	37.73 (**2.50**)	34.69 (-0.08)	25.78 (-8.98)	15.78 (-18.98)

robustness considerably for all examined ϵ-values. Evaluation against a B&B ℓ^0-adversary yields comparable results. In contrast, when evaluated with the C&W ℓ^2-attack (middle column), even for higher compression rates, a pruning approach exists that yields better robustness than the unpruned model. The filter-pruning methods (red lines) consistently show superior results where the magnitude-based approaches (dashed and dotted lines) appear to work better for moderate pruning, while pruning random filters (solid line) leads to better robustness for more extensive pruning.

Additionally, we can see that for every attack, there is a pruning approach that keeps the robust accuracy stable or increases it for moderate compression rates. So, if a defender has full information of the adversary with regard to the chosen attack type and strength, it is beneficial to optimize the model to that specific case. I.e., if it can be expected that the adversary will attack with an ℓ^2-attack and a maximum ϵ of 15 (an assumption which might be derived from the nature of the problem) the model should be optimized for this specific case, and filter global magnitude pruning should be applied. Such a level of knowledge about the adversary is unlikely, but motivates to look for sweet spots with favorable trade-offs for the defender regarding clean accuracy, robust accuracy, and the amount of the remaining weights.

Exemplary, we identify such **sweet spots** for the ResNet18 model when applying unstructured global magnitude pruning with a compression rate of two or four: Table 2 shows the absolute accuracies and margins to the unpruned model for all applied attacks and pruning ratios. For a compression rate of two we see a slight decrease of up to 1.33% in robust accuracy compared to the unpruned model, when attacked with the ℓ^2-attack, while for all other combinations of attack type and strength, the robustness is increased by a margin of 3.36%. Even a compression rate of four does not reduce the robustness for any attack by more than 2.89%. For some scenarios, we see an increase of the respective robustness by up to 1.25%. We find these sweet spots for the remaining architectures, too.

5 Conclusion

In this work, we shed some light on the impact of NN pruning in the face of adversaries. We conducted an extensive series of experiments with an ensemble of pruning approaches and attack methods that were carefully selected to provide a broad view. Small increases in robustness for mild pruning were already noticed in small-scaled experiments in prior work and we confirm this for a wide variety of attack-types, attack strengths, pruning approaches, and compression rates. The stronger increase in robustness against an ℓ^2-adversary, observable in the middle column of Fig. 1, is intriguing and calls for further research.

An intuition why robustness might increase with pruning and thus, contradicting the general assumption that capacity helps [25], could be the following: [32] and [12] argue that adversarial examples leverage so-called *blind spots*, which are low-density regions of the training data distribution. Pruning aims to eliminate the least important parts of a DNN, and for lower compression rates, the parts removed contain proportionally more of these *blind spots*.

Our results validate that the method and extent of NN pruning open up additional possible strategies for adversary-aware deep learning practitioners. Furthermore, we show that by making additional assumptions about potential adversaries, we can identify optimal pruning strategies. Factors to consider are possible attack types and strengths. Practitioners should not only think about NN pruning for applications operating under computational and memory constraints, but our findings suggest that security-sensitive use cases might benefit from a carefully selected pruning strategy. NN pruning can simultaneously increase a model's clean accuracy and its robustness against a wide variety of adversarial attack methods and strengths. This is valid for both cases: When resources are not a limiting factor and under resource constraints.

Our choice of perturbation budgets follows a simple heuristic. We can assert that an ℓ^0 and an ℓ^∞-attack are similarly successful but we cannot make any statements about the real strength of an attack. However, the limitations of ℓ^p norms are well known and discussed in the adversarial machine learning community [4]. While we deliberately refrained from incorporating adversarial training methods due to its negative impact on the clean accuracy, future work should examine if a combination of mild pruning, moderate adversarial training, and fine-tuning leads to a significant rise in adversarial robustness while not hurting the clean accuracy compared with a naturally trained, unpruned network. A combination achieving this is considered to be the, so far unreached, "holy grail" of adversarial machine learning research.

Acknowledgement. All authors are supported by the Austrian Science Fund (FWF) under grant no. I 4057-N31 ("Game Over Eva(sion)").

References

1. Abadi, M., et al.: TensorFlow: a system for large-scale machine learning. In: 12th USENIX Symposium on Operating Systems Design and Implementation (OSDI 2016), pp. 265–283 (2016)
2. Blalock, D., Gonzalez Ortiz, J.J., Frankle, J., Guttag, J.: What is the state of neural network pruning? In: Dhillon, I., Papailiopoulos, D., Sze, V. (eds.) Proceedings of Machine Learning and Systems, vol. 2, pp. 129–146 (2020)
3. Brendel, W., Rauber, J., Kümmerer, M., Ustyuzhaninov, I., Bethge, M.: Accurate, reliable and fast robustness evaluation. In: Advances in Neural Information Processing Systems, pp. 12841–12851 (2019)
4. Carlini, N., et al.: On evaluating adversarial robustness. arXiv preprint arXiv:1902.06705 (2019)
5. Carlini, N., Wagner, D.: Towards evaluating the robustness of neural networks. In: 2017 IEEE Symposium on Security and Privacy (SP), pp. 39–57. IEEE (2017)
6. Carlini, N., Wagner, D.: Audio adversarial examples: targeted attacks on speech-to-text. In: 2018 IEEE Security and Privacy Workshops (SPW), pp. 1–7. IEEE (2018)
7. Cohen, J., Rosenfeld, E., Kolter, Z.: Certified adversarial robustness via randomized smoothing. In: International Conference on Machine Learning, pp. 1310–1320 (2019)
8. Croce, F., Hein, M.: Reliable evaluation of adversarial robustness with an ensemble of diverse parameter-free attacks. In: International Conference on Machine Learning, pp. 2206–2216. PMLR (2020)
9. Dalvi, N., Domingos, P., Sanghai, S., Verma, D.: Adversarial classification. In: Proceedings of the Tenth ACM SIGKDD International Conference on Knowledge Discovery and Data Mining, pp. 99–108 (2004)
10. Eykholt, K., et al.: Robust physical-world attacks on deep learning visual classification. In: Proceedings of the IEEE Conference on Computer Vision and Pattern Recognition, pp. 1625–1634 (2018)
11. Frankle, J., Carbin, M.: The lottery ticket hypothesis: finding sparse, trainable neural networks. In: International Conference on Learning Representations (2018)
12. Goodfellow, I., Shlens, J., Szegedy, C.: Explaining and harnessing adversarial examples. In: International Conference on Learning Representations (2015)
13. Guo, Y., Zhang, C., Zhang, C., Chen, Y.: Sparse DNNs with improved adversarial robustness. In: Advances in Neural Information Processing Systems, pp. 242–251 (2018)
14. Han, S., Pool, J., Tran, J., Dally, W.: Learning both weights and connections for efficient neural network. In: Advances in Neural Information Processing Systems, pp. 1135–1143 (2015)
15. He, K., Zhang, X., Ren, S., Sun, J.: Deep residual learning for image recognition. In: Proceedings of the IEEE Conference on Computer Vision and Pattern Recognition, pp. 770–778 (2016)
16. He, Y., Kang, G., Dong, X., Fu, Y., Yang, Y.: Soft filter pruning for accelerating deep convolutional neural networks. In: Proceedings of the 27th International Joint Conference on Artificial Intelligence, pp. 2234–2240 (2018)

17. Howard, J.: The imagenet dataset (2019)
18. Huang, Z., Wang, N.: Data-driven sparse structure selection for deep neural networks. In: Proceedings of the European Conference on Computer Vision (ECCV), pp. 304–320 (2018)
19. Kingma, D., Ba, J.: Adam: a method for stochastic optimization. In: International Conference on Learning Representations (2014)
20. Krizhevsky, A., Hinton, G.: Learning multiple layers of features from tiny images (2009)
21. LeCun, Y., Bottou, L., Bengio, Y., Haffner, P.: Gradient-based learning applied to document recognition. Proc. IEEE **86**, 2278–2324 (1998)
22. LeCun, Y., Cortes, C., Burges, C.: MNIST handwritten digit database. ATT Labs (2010). http://yann.lecun.com/exdb/mnist
23. LeCun, Y., Denker, J.S., Solla, S.A.: Optimal brain damage. In: Advances in Neural Information Processing Systems, pp. 598–605 (1990)
24. Liu, Z., Sun, M., Zhou, T., Huang, G., Darrell, T.: Rethinking the value of network pruning. In: International Conference on Learning Representations (2018)
25. Madry, A., Makelov, A., Schmidt, L., Tsipras, D., Vladu, A.: Towards deep learning models resistant to adversarial attacks. In: International Conference on Learning Representations (2018)
26. Moosavi-Dezfooli, S.M., Fawzi, A., Frossard, P.: DeepFool: a simple and accurate method to fool deep neural networks. In: Proceedings of the IEEE Conference on Computer Vision and Pattern Recognition, pp. 2574–2582 (2016)
27. Neklyudov, K., Molchanov, D., Ashukha, A., Vetrov, D.P.: Structured Bayesian pruning via log-normal multiplicative noise. In: Advances in Neural Information Processing Systems, pp. 6775–6784 (2017)
28. Paganini, M., Forde, J.: On iterative neural network pruning, reinitialization, and the similarity of masks. arXiv preprint arXiv:2001.05050 (2020)
29. Rauber, J., Brendel, W., Bethge, M.: FoolBox: a Python toolbox to benchmark the robustness of machine learning models. arXiv preprint arXiv:1707.04131 (2017)
30. Sehwag, V., Wang, S., Mittal, P., Jana, S.: Hydra: pruning adversarially robust neural networks. Advances in Neural Information Processing Systems (NeurIPS), vol. 7 (2020)
31. Sharif, M., Bhagavatula, S., Bauer, L., Reiter, M.K.: Accessorize to a crime: real and stealthy attacks on state-of-the-art face recognition. In: Proceedings of the 2016 ACM SIGSAC Conference on Computer and Communications Security, pp. 1528–1540 (2016)
32. Szegedy, C., et al.: Intriguing properties of neural networks. In: International Conference on Learning Representations (2014)
33. Tramer, F., Carlini, N., Brendel, W., Madry, A.: On adaptive attacks to adversarial example defenses. Adv. Neural. Inf. Process. Syst. **33**, 1633–1645 (2020)
34. Tramèr, F., Kurakin, A., Papernot, N., Goodfellow, I., Boneh, D., McDaniel, P.D.: Ensemble adversarial training: attacks and defenses. In: 6th International Conference on Learning Representations, ICLR 2018 (2018)
35. Ye, S., et al.: Adversarial robustness vs. model compression, or both. In: The IEEE International Conference on Computer Vision (ICCV), vol. 2 (2019)
36. Zhang, T., et al.: A systematic DNN weight pruning framework using alternating direction method of multipliers. In: Proceedings of the European Conference on Computer Vision (ECCV), pp. 184–199 (2018)

Grad₂VAE: An Explainable Variational Autoencoder Model Based on Online Attentions Preserving Curvatures of Representations

Mohanad Abukmeil[(✉)], Stefano Ferrari, Angelo Genovese, Vincenzo Piuri, and Fabio Scotti

Department of Computer Science, Università degli Studi di Milano, Milan, Italy
{mohanad.abukmeil,stefano.ferrari,angelo.genovese,vincenzo.piuri,
fabio.scotti}@unimi.it

Abstract. Unsupervised learning (UL) is a class of machine learning (ML) that learns data, reduces dimensionality, and visualizes decisions without labels. Among UL models, a variational autoencoder (VAE) is considered a UL model that is regulated by variational inference to approximate the posterior distribution of large datasets. In this paper, we propose a novel explainable artificial intelligence (XAI) method to visually explain the VAE behavior based on the second-order derivative of the latent space concerning the encoding layers, which reflects the amount of acceleration required from encoding to decoding space. Our model is termed as Grad₂VAE and it is able to capture the local curvatures of the representations to build online attention that visually explains the model's behavior. Besides the VAE explanation, we employ our method for anomaly detection, where our model outperforms the recent UL deep models when generalizing it for large-scale anomaly data.

Keywords: Unsupervised learning · VAE · XAI · Anomaly detection

1 Introduction

Explainable artificial intelligence (XAI) is an emerging field in artificial intelligence (AI) and machine learning (ML), and deals with explaining the decisions and behaviors of learned models. XAI models are also associated with unsupervised learning (UL) to learn and visualize the hidden structure of data with limited levels of prior assumptions. Autoencoder models (AEs) are a class of generative UL (UGL) methods, which can reduce dimensionality, visualize and generate data, and perform other ML tasks such as object recognition [1,2]. Many different

This work was supported in part by the Universitá degli Studi di Milano under project 3SUN. We thank the NVIDIA Corporation for the GPU donated.

S. Sclaroff et al. (Eds.): ICIAP 2022, LNCS 13231, pp. 670–681, 2022.
https://doi.org/10.1007/978-3-031-06427-2_56

types of AEs have been introduced recently and they are characterized by a regularization term; such term enforces the AEs to learn with an additional penalty to capture different representations for a better generalization [3,7].

Deep AEs encompass many different encoding and decoding stages, where at each stage diverse layers with associated parameters ($\theta = \{W, B\}$, W and B are weights and biases, respectively) are employed to perform a specific mapping (convolution, deconvolution, dense multiplication, etc.), by utilizing several sets of representations to capture the neurons' responses [6]. Moreover, for each setting among parameters (after each learning iteration or epoch), the gradient is approximated between the input and output by using the first-order partial derivative to optimally fit the model to the data [4]. AEs comprise classic, denoising [27], contractive [23], sparse [20], variational-AE (VAE) [14], and they can also be integrated with other UL models, for example, when combining the generative adversarial networks (GANs) with VAE [17].

Among all AEs, the VAE is regularized by variational inference (VI) [31] to optimize the posterior distribution for large datasets, and it outperforms the others in terms of large-scale generalization (when testing data are larger than the training set). The VAE is utilized in many different fields including image reconstruction and recognition, compression sensing, and other deep learning tasks [14]. However, explaining VAEs did not receive an appropriate interest in the literature, where explaining such a UGL model is considered essential to understanding the behaviors of neurons when new data (normal or anomaly) is generated. [9,24,25]. Thus, different works have been proposed for explaining models through supplementary inputs to carry out specific tasks. However, such works did not explain the behaviors of the models themselves [18,26,32].

The first explainable VAE has been introduced in [16], where it generates offline attention (after learning) by reduplicating the last layer of the encoder, then scaling it up by the global average pooling of the gradient of the latent space concerning that layer. Such attention is seen similar to explaining discriminating models [30]; however, the proposed attention is scaled up by the gradient. The drawback of such an attention lies in unfair scaling, i.e., related and unrelated features in the channels of the filters are scaled with the same factor.

To help to explain VAEs, we propose Grad$_2$VAE, a novel XAI model utilizing online mapping, i.e., after each epoch, visual attention can be produced. Moreover, the Grad$_2$VAE utilizes the second-order (2nd) derivative between the latent and 1st encoder's layers to obtain the 1st derivative of the gradient, which captures the curvatures of neurons responses that are aggregated to show how the VAE learns data without additional scaling. Therefore, our contribution is twofold: (i) introducing a novel method to explain the VAEs employing the gradient derivation, and (ii) expanding our method to accelerate VAE learning (reduced epochs) and one-class anomaly detection. The rest of this paper is organized as follows. Section 2 highlights the VAE and the 2nd derivative interpretation. Section 3 describes the Grad$_2$VAE. The experimental results are given in Sect. 4. The conclusion and future works are reported in Sect. 5.

2   VAE and 2nd Derivative Interpretation

2.1   VAE Model

Similarly to any AE model, the VAE contains two main modules: *(i)* the encoding module (inference side) that is employed to map data $X = \{x_i \mid x_i \in \mathbb{R}^D, i = 1, \ldots, N\}$, D is the original dimensionality, to a latent space $Z = f(X) = \{z_i = f(x_i) \in \mathbb{R}^d, \mid i = 1, \ldots, M\}$; such a module reduces dimensionality where $0 < d < D$, and it is used to infer the model likelihood $P(X|\theta)$; *(ii)* the decoding module (generation side) that is utilized to generate or reconstruct the original data $\tilde{X}$ from the latent space Z [11,12]. For a given data $X \in \mathbb{R}^D$, the encoding module creates a mapping $f : \mathbb{R}^D \to \mathbb{R}^d$, while the decoding module creates a mapping $g : \mathbb{R}^d \to \mathbb{R}^D$, which generates an approximation of the original data: $\tilde{X} = g(Z; \hat{\theta}_d)$ [2]. The AEs are regulated to find the parameters $(\hat{\theta}_e, \hat{\theta}_d)$ that achieve a better generalization [7], and to obtain the minimum loss $\mathbf{L}_{\text{REC}}$:

$$\mathbf{L}_{\text{REC}_{\{\hat{\theta}_e, \hat{\theta}_d\}}} = \min \|X - (f \circ g)X\|_{\text{Er}}^2, \tag{1}$$

where the reconstruction error E_r can be measured by different metrics including mean square error (MSE), Frobenius norm, reconstruction cross-entropy, or β-divergence [1,3].

Among all AE models, VAE is regulated by the VI, and it is optimized based on two different losses that are minimized simultaneously [14]. VI method is one of the Bayesian techniques, which can be utilized to estimate an intractable posterior over a big dataset using a simpler variational distribution to obtain the solution to an optimization problem [31], i.e., the VI approximates probability densities through optimization. By considering the encoder module output, the approximate posterior distribution $Q(Z|X)$ is estimated, which parameterizes the shape of the latent distribution according to the original input data X. Moreover, optimizing $Q(Z|X)$ characterizes the VAE, where it enforces the latent space distribution to follow a unit Gaussian distribution with a certain mean μ (which reflects the center of the Gaussian), and a standard deviation σ (which reflects the Gaussian shape).

Initially, the prior distribution of latent space $P(Z)$ is drawn (simply by copying the unit Gaussian distribution of the data manifold $P(X)$). Thereafter, the approximated distribution $Q(Z|X)$ and the prior $P(Z)$ are compared using the KL divergence [22]. The KL divergence is defined as $\text{KL}(P\|Q) = \Sigma_x P(x) \log \frac{P(x)}{Q(x)}$, which is always positive and tends to zero if and only if P and Q are almost equal. Moreover, appending noise to $Q(Z|X)$ throughout varying σ by a small value ϵ, and then enforcing the AE to reconstruct the data following the true (not varied one) Gaussian $P(Z)$ is called the reparameterization trick; such a trick generates several different distributions (similarly to duplicate the training data with fusion) that are optimized and compared with prior distribution by the KL divergence, thus the model can be better generalized for a large-scale testing stage [14]. Finally, the VAE is optimized to minimize the $\mathbf{L}_{\text{REC}}$ according to Eq. (1), and it is also optimized to minimize the latent loss

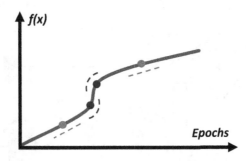

Fig. 1. The neuron activation and gradient over epochs. (Color figure online)

between $Q(Z|X)$ and $P(Z)$ using $\text{KL}(P\|Q)$, which measures to which extent the reparameterized latent distribution can follow a unit Gaussian:

$$\mathbf{L}_{\text{VAE}_{\{\hat{\theta}_e, \hat{\theta}_d, \hat{\mu}_X, \hat{\sigma}_X, \hat{\mu}_Z, \hat{\sigma}_Z\}}} = \min[\mathbf{L}_{\text{REC}} + \text{KL}(P\|Q)]. \tag{2}$$

2.2 The Second-Order (2nd) Derivative Interpretation

The first-order (1st) partial derivative between input and output neurons reflects the 1st gradient, which measures the instantaneous rate of change (velocity or speed) ∂ [21] among model parameters θ that are employed to optimally fit ML model [13]. Moreover, if the gradient sign is negative, then it is decreasing (velocity is reduced), while if the gradient sign is positive then it is increasing (velocity is accelerated).

For a VAE with an encoding layer L_{e1} and a latent layer Z, the 1st gradient of Z with respect to L_{e1} is computed by carrying out the partial derivative of each neuron z_i as $\frac{\partial z_i}{\partial L_{e1}}$; considering that if an additional layer L_{e2} acts between L_{e1} and Z, then the chain rule is introduced as $\frac{\partial z_i}{\partial L_{e1}} = \frac{\partial z_i}{\partial L_{e2}} \frac{\partial L_{e2}}{\partial L_{e1}}$ [5]. The final result of derivations gives all possible rates of changes, which are required to update θ laying between L_{e1} and Z. Because the rate of change (∂) of a neuron's response (activation) is changing during a period of time (over several epochs), thus capturing the variation in the rate of change is essential to obtain the acceleration required for a neuron from the velocity [29], and it is achieved by considering the derivative of the gradient, i.e., 2nd derivative $\frac{\partial^2 z_i}{\partial L_{e1}^2}$ [8].

Graphically, a neuron response that is modeled by a non-linear ReLU function [19] is given according to Fig. 1: the 1st gradient is the slope at a point in the graph (blue curve), whereas the 1st derivative of gradient explains how the slope is changing over time (the red and green points). As it is noticed from Fig. 1, the gradient of a neuron response can be steady at a period of the learning time, i.e., the 2nd derivative around the green points is ≈ 0; however, it can change at a different period, i.e., the 2nd derivative around the red points is $>$ or < 0.

Accordingly, utilizing the 2nd derivative that measures how the 1st gradient of neurons responses is changing (as in deriving the acceleration from speed),

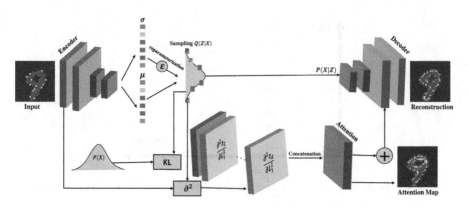

Fig. 2. The Grad$_2$VAE block diagram.

the curvatures of representations and the temporal behavior of neurons when learning data can be captured. Moreover, such a strategy can be represented by a learnable attention map, which aggregates all 2nd partial derivative to explain how the latent neurons of Z are activated to the local curves and edges.

Our Grad$_2$VAE employs the 2nd gradient to visually explain the learned representations of the VAE in an online fashion by reconstructing attention maps, and it exploits such explanations in the application of one-class anomaly detection [24]. Moreover, we will show how such a strategy is able to accelerate the convergence among the learning parameters θ.

3   Grad$_2$VAE

Figure 2 shows our proposed Grad$_2$VAE, where it comprises an encoder, a decoder, and an attention module. Both encoder and decoder contain one stage of down-sampling (convolution with a stride of 2) and up-sampling (de-convolution with a stride of 2), respectively. Moreover, the size of the first two layers of the encoder and the last two layers of the decoder are fixed to uniform the dimensionality. Thus, the obtained attention of the 2nd derivative can be fused with the $L_{d_{n-1}}$ layer (d_n is the total number of the decoder's layers); such a fusion is seen as a form of residual learning [10], which enforces the Grad$_2$VAE to learn the residual of mapping between the encoder and decoder by utilizing the gradient attention. Accordingly, besides the explainability of the Grad$_2$VAE, it also boosts the reconstruction of data by utilizing the curvatures of representations that are combined with the decoder. Therefore, the Grad$_2$VAE optimizes two losses by using Adam [13] as:

$$\mathbf{L}_{\text{Grad}_2\text{VAE}} = \min[\mathbf{L}_{\text{VAE}} + \|X - \theta_{grad}(Z, L_{e1})\|^2_{\text{Er}}], \qquad (3)$$

where the first loss is taken from the vanilla VAE [14] that is depicted at Eq. (2), and the second loss is the reconstruction loss between the data and the aggregated attention that is obtained from the attention module (see Fig. 2). Moreover, θ_{grad} represents the 2nd derivative between each latent neuron z_i with

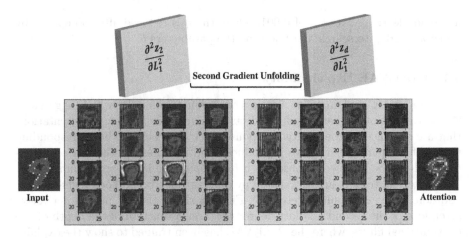

Fig. 3. The unfolding of the tensor that holds all second-order partial derivatives (Grade$_2$) of z_2 (green tensor) and z_{16} (gray tensor) concerning L_{e1}, respectively. (Color figure online)

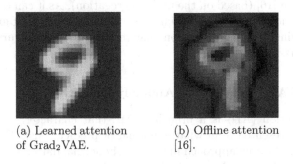

(a) Learned attention   (b) Offline attention
of Grad$_2$VAE.   [16].

Fig. 4. Grad$_2$VAE learnable attention vs. the offline attention proposed in [16] that depends on scaling of the penultimate encoding layer by the gradient of Z respecting that layer when learning the 10th class from MNIST digits.

respect to L_{e1}, i.e., for each z_i there is a corresponding tensor of size of the L_{e1} to allocate all partial derivatives. Additionally, the derivative of gradient of Z can be implemented with respect to all other encoder's layers (as in considering L_{e3}); however, considering more depth layers requires re-scaling the dimensionality which needs more computational time and leads to the loss of global representations.

4   Experimental Results

To show the performance of our Grad$_2$VAE, we employ both MNIST and fashion MNIST datasets [15,28]. Each comprises 60k images for training and 10k for testing, divided in 10 classes with image size of 28×28. Moreover, all quantitative analysis experiments are implemented with a batch size of 128, 100 epochs with a starting learning rate (η) of 0.001, where the η is reduced after 50 epochs by a factor of 10^{-2} to search and fine-tune the parameters θ.

4.1 Grad$_2$VAE Explainability

The Grad$_2$VAE explainability lies in the attention module, which aggregates the derivatives of gradients and it reflects the curvatures among representations that are obtained at the neurons response level. For each z_i, the corresponding tensor of 2nd partial derivative is produced and it represents the neuron attentions, where the number of tensors is a function of the latent space dimensions. Thereafter, all tensors are aggregated by different matrix methods including the addition, mean, convolution, etc. [16]. Figure 3 shows the 2nd partial derivative attentions of the second and last neurons of Z as a function of the depth of the convolutional filters, where the Grad$_2$VAE has been trained to show the explainability for the 9th class of the MNIST, by considering 16 neurons for Z and a depth of 16 filters. Moreover, Fig. 4 shows a comparison between the aggregated attention of the Grad$_2$VAE (based on convolutional aggregation), and the attention proposed in [16] (based on the mean aggregation). As it can be noticed from the figure, considering the 2nd partial derivative (a derivative of gradient) offers a better explainable visual attention that retains all possible curvatures of the representations.

4.2 Grad$_2$VAE in One-Class Anomaly Detection

Anomaly detection (AD) is a branch of ML that characterizes data samples that are misrepresented from what is normal or predicted [25]. One-class AD is referred to as a learning approach in which only normal data is considered at the training stage, where the ML model learns to classify or reconstruct the normal data only [24]. However, at the testing stage, all data samples that are falling out of the normal class distribution of the trained data are considered, and the ML model must be able to distinguish between normal and anomalies samples. The decoder of the Grad$_2$VAE is guided by the curvature of representations from the encoder, thus the reconstruction process is accelerated and optimized. Accordingly, the Grad$_2$VAE is directly applied to the AD due to its reconstruction ability. In the following, we employ the Grad$_2$VAE in the one-class AD, where we benchmark our model on the MNIST and fashion datasets. Moreover, the average area under the receiver operator characteristic curve (AUC-ROC) is considered as a metric to show the performance, where we report the qualitative and quantitative comparisons to the recent works.

Qualitative Analysis Comparison. In this section, we visually compare our work with [16] based on the MNIST dataset. For this analysis, we consider 600 epochs for qualitative comparison. Moreover, we have followed [16] to train our model considering one class from the training set as a normal class, thereafter testing with all classes from the testing set. Hence, our model must be able to

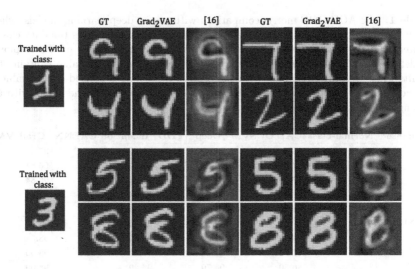

Fig. 5. The qualitative analysis comparison between our Grad$_2$VAE model that produces learnable attention maps (online attention) and the offline attention introduced in [16], where GT represents the ground truth samples. Moreover, the images that appear in the first two rows of the figure have been tested when only the second class of the MNIST digits has been trained, also in the bottom two rows of the figure, the model has been trained considering only the fourth class and tested with all other images from all classes.

produce visual attentions for all testing classes avoiding bias to the previously learned normal class. Figure 5 shows the attention maps comparison between our model and [16] when both are trained with the 2nd and 4th classes from the dataset separately, then considering data from other classes (out of the trained data distribution) as testing samples.

As it can be noticed from Fig. 5, our model produces visual attentions maps that completely retain the curvatures among representations of the input data regardless that they are normal (seen) or anomalies (unseen from other classes). Specifically, our model is able to visually explain the differences between normal and anomalies samples, avoiding further preprocessing such as scaling attentions maps by the gradient as in [16] to partially detect and explain the anomalies.

Quantitative Analysis Comparison. In this section, we compare our model with the recent deep UL models, where we consider the CAE OCSVM, Deep SVDD, and Inception-CAENN [24,25]. Moreover, we followed [25] to train 10 models for each normal class, thus reporting the average AUC-ROC over 10 Grad$_2$VAEs. Additionally, we used 16 and 32 neurons as the bottleneck size for the MNIST and fashion, respectively. Our results are reported in Table 1 considering only 100 epochs, due to the acceleration in learning ability that the Grad$_2$VAE possesses.

Table 1. The AUC-ROC metric comparison with recent deep learning models, where the first column shows the utilized datasets, the second column gives the data classes related to each dataset, and all other columns contain the results of the employed models to evaluate the performance. Moreover, each row in the table contains the results when training the model with the class of data that is labeled by the value of the normal class column and lies in that row, subsequently testing the model with the testing data from all classes.

Dataset	Normal-class	CAE OCSVM	Deep SVDD	Inception CAENN	Grad$_2$VAE
MNIST	0	95.40	99.10	98.70	97.62
	1	97.40	99.70	99.70	96.81
	2	77.60	95.40	96.70	98.84
	3	88.60	95.10	95.20	98.53
	4	83.60	95.90	95.00	97.98
	5	71.30	92.10	95.20	98.70
	6	90.10	98.50	98.30	98.54
	7	87.20	96.20	97.00	98.59
	8	86.50	95.70	96.20	98.09
	9	87.30	97.70	97.00	98.62
	Average	86.50	96.60	96.90	**98.23**
FASHION	T-shirt	88.00	98.80	92.40	96.54
	Trouser	97.30	99.77	98.80	95.88
	Pullover	85.50	93.50	90.00	96.59
	Dress	90.00	94.90	95.00	96.29
	Coat	88.50	95.10	92.00	96.49
	Sandal	87.20	90.40	93.40	96.39
	Shirt	78.80	98.00	85.50	96.65
	Sneaker	97.70	96.00	98.60	96.05
	Bag	85.80	95.40	95.10	96.58
	Boot	98.00	97.60	97.70	96.42
	Average	89.70	95.90	93.90	**96.37**

As it can be noticed from Table 1, the Grad$_2$VAE outperforms the other models under reduced epochs, where all other models have been trained considering 150 epochs [25]. Moreover, the Grad$_2$VAE does not show any bias to a class against all other classes, e.g., the Deep SVDD and Inception-CAENN learn the 2nd class from the fashion dataset perfectly; however, they show minimum accuracies for the 6th and 7th classes, respectively. Finally, the Grad$_2$VAE shows a better mean standard deviation (mstd) among the averaged results (of 10 models), where our maximum mstd for both datasets did not exceed 0.117, whereas it reached 3.8, 3.9 for the Deep SVDD, and Inception-CAENN, respectively [25].

Furthermore, to show the learning complexity and performance convergence (acceleration) of our proposed Grad$_2$VAE model, we compare the accuracy convergence with the baseline convolution AE (Baseline CAE), vanilla VAE, and Inception CAENN models [3,25]. Moreover, under the same experimental setup reported in Sect. 4, the Baseline CAE learns 0.266 M, vanilla VAE learns 3.25 M, and Inception CAENN learns 0.335 M parameters to complete one training

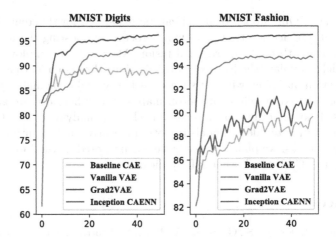

Fig. 6. AUC metric over the first 50 epochs for both MNIST datasets.

epoch. Additionally, our proposed Grad$_2$VAE learns 3.38 M parameters as a consequence of the online attention module and the 2nd order partial derivative parameters.

Figure 6 depicts the learning acceleration ability over the learning epochs of our proposed Grad$_2$VAE model considering the related models in the literature. As it is noticed from the figure, our proposed Grad$_2$VAE shows the highest learning convergence ability at an early stage of the learning period, i.e., our proposed model is able to search and find the suitable learning parameters, ($\theta = \{W, B\}$), at an initial interval of the learning epochs, which accelerate the model learning and prevent the model from overfitting. That is by considering the XAI attention maps, which are fused with the penultimate layer to compensate for the loss caused by encoding and decoding operations. Finally, our model learns a number of parameters that are equal to 12.7×, 1.03×, 10.10× of the Baseline CAE, vanilla VAE, and Inception CAENN models, respectively. However, it converges to an accuracy of greater than 95% for both datasets considering the half number of the learning epochs required to learn the vanilla VAE, and the other remaining models did not reach such accuracy in the first 50 epochs. Accordingly, our proposed model converges to the optimal set of learning parameters under a limited number of learning epochs, and it also outperforms all related models in the literature considering both datasets.

5   Conclusions

We proposed an explainable VAE model termed (Grad$_2$VAE) to be utilized for XAI, image reconstruction, generation, object detection, and anomaly detection applications. We used the 2nd partial derivative of the neuron activation (or responses) between the latent space, Z, and the 1st encoding layer to capture the curvatures of the representations at an early stage by reconstructing visual attention maps. Our proposed model can be expanded for different data types and scales, it also accelerates the learning process by boosting the whole reconstruction process through the residual fusion. Moreover, we employed our proposed model to explain the learned representations through a learnable (online) visual attention mapping, where it shows a better visual explainability than the related works based on offline attention mapping. Furthermore, we generalized our proposed model in the application of one-class anomaly detection. Our model outperforms all related deep models in both qualitative and quantitative analysis. In future works, we plan to investigate our proposed method for other UGL models such as GANs and other ML applications.

References

1. Abukmeil, M., Ferrari, S., Genovese, A., Piuri, V., Scotti, F.: On approximating the non-negative rank: applications to image reduction. In: Proceedings of CIVEMSA (2020)
2. Abukmeil, M., Ferrari, S., Genovese, A., Piuri, V., Scotti, F.: Unsupervised learning from limited available data by β-NMF and dual autoencoder. In: Proceedings of ICIP (2020)
3. Abukmeil, M., Ferrari, S., Genovese, A., Piuri, V., Scotti, F.: A survey of unsupervised generative models for exploratory data analysis and representation learning. ACM Comput. Surv. (CSUR) **54**(5), 1–40 (2021)
4. Abukmeil, M., Genovese, A., Piuri, V., Rundo, F., Scotti, F.: Towards explainable semantic segmentation for autonomous driving systems by multi-scale variational attention. In: Proceedings of ICAS (2021)
5. Ames, W.F.: Numerical Methods for Partial Differential Equations. Academic Press, Cambridge (2014)
6. Baldi, P.: Autoencoders, unsupervised learning, and deep architectures. In: Proceedings of ICML (2012)
7. Bengio, Y., Courville, A., Vincent, P.: Representation learning: a review and new perspectives. IEEE Trans. Pattern Anal. Mach. Intell. **35**(8), 1798–1828 (2013)
8. Fan, K., Wang, Z., Beck, J., Kwok, J., Heller, K.A.: Fast second order stochastic backpropagation for variational inference. In: Proceedings of NIPS (2015)
9. Genovese, A., Piuri, V., Scotti, F.: Towards explainable face aging with generative adversarial networks. In: Proceedings of ICIP (2019)
10. He, K., Zhang, X., Ren, S., Sun, J.: Deep residual learning for image recognition. In: Proceedings of CVPR (2016)
11. Hinton, G.E., Salakhutdinov, R.R.: Reducing the dimensionality of data with neural networks. Science **313**(5786), 504–507 (2006)
12. Hinton, G.E., Zemel, R.S.: Autoencoders, minimum description length and Helmholtz free energy. In: Proceedings of NIPS (1994)
13. Kingma, D.P., Ba, J.: Adam: a method for stochastic optimization. In: Proceedings of ICML (2014)
14. Kingma, D.P., Welling, M.: Auto-encoding variational Bayes. In: Proceedings of ICLR (2014)
15. LeCun, Y., Cortes, C., Burges, C.J.: MNIST handwritten digit database. **7**(23), 6 (2010). https://yann.lecun.com/exdb/mnist
16. Liu, W., et al.: Towards visually explaining variational autoencoders. In: Proceedings of CVPR (2020)

17. Makhzani, A., Shlens, J., Jaitly, N., Goodfellow, I., Frey, B.: Adversarial autoencoders. arXiv preprint arXiv:1511.05644 (2015)
18. Mejjati, Y.A., Richardt, C., Tompkin, J., Cosker, D., Kim, K.I.: Unsupervised attention-guided image-to-image translation. In: Proceedings of NIPS (2018)
19. Nair, V., Hinton, G.E.: Rectified linear units improve restricted Boltzmann machines. In: Proceedings of ICML (2010)
20. Ng, A.: Sparse autoencoder. CS294A Lecture Notes, vol. 72, no. 2011, pp. 1–19 (2011)
21. Pospiech, G., Michelini, M., Eylon, B.-S. (eds.): Mathematics in Physics Education. Springer, Cham (2019). https://doi.org/10.1007/978-3-030-04627-9
22. Rezende, D.J., Mohamed, S., Wierstra, D.: Stochastic backpropagation and approximate inference in deep generative models. In: Proceedings of ICML (2014)
23. Rifai, S., Vincent, P., Muller, X., Glorot, X., Bengio, Y.: Contractive auto-encoders: explicit invariance during feature extraction. In: Proceedings of ICML (2011)
24. Ruff, L., et al.: Deep one-class classification. In: Proceedings of ICML (2018)
25. Sarafijanovic-Djukic, N., Davis, J.: Fast distance-based anomaly detection in images using an inception-like autoencoder. In: Proceedings of DS (2019)
26. Tang, C., Srivastava, N., Salakhutdinov, R.R.: Learning generative models with visual attention. In: Proceedings of NIPS (2014)
27. Vincent, P., Larochelle, H., Bengio, Y., Manzagol, P.A.: Extracting and composing robust features with denoising autoencoders. In: Proceedings of ICML (2008)
28. Xiao, H., Rasul, K., Vollgraf, R.: Fashion-MNIST: a novel image dataset for benchmarking machine learning algorithms. arXiv preprint arXiv:1708.07747 (2017)
29. Zeiler, M.D.: ADADELTA: an adaptive learning rate method. arXiv preprint arXiv:1212.5701 (2012)
30. Zeiler, M.D., Fergus, R.: Visualizing and understanding convolutional networks. In: Fleet, D., Pajdla, T., Schiele, B., Tuytelaars, T. (eds.) ECCV 2014. LNCS, vol. 8689, pp. 818–833. Springer, Cham (2014). https://doi.org/10.1007/978-3-319-10590-1_53
31. Zhang, C., Butepage, J., Kjellstrom, H., Mandt, S.: Advances in variational inference. IEEE Trans. Pattern Anal. Mach. Intell. **41**(08), 2008–2026 (2019)
32. Zhang, J., Bargal, S.A., Lin, Z., Brandt, J., Shen, X., Sclaroff, S.: Top-down neural attention by excitation backprop. Int. J. Comput. Vis. **126**(10), 1084–1102 (2018)

Image Processing for Cultural Heritage

The AIRES-CH Project: Artificial Intelligence for Digital REStoration of Cultural Heritages Using Nuclear Imaging and Multidimensional Adversarial Neural Networks

Alessandro Bombini[1]([✉]), Lucio Anderlini[1], Luca dell'Agnello[2],
Francesco Giaocmini[2], Chiara Ruberto[1], and Francesco Taccetti[1]

[1] INFN, Florence Section, Via Bruno Rossi 1, 50019 Sesto Fiorentino, FI, Italy
bombini@fi.infn.it
[2] INFN CNAF, Viale Carlo Berti Pichat 6, 40127 Bologna, BO, Italy

Abstract. Artificial Intelligence for digital REStoration of Cultural Heritage (AIRES-CH) aims at building a web-based app for the digital restoration of pictorial artworks through Computer Vision technologies applied to physical imaging raw data. Physical imaging techniques, such as XRF, PIXE, PIGE, and FTIR, are capable of exploring a wide range of wavelengths providing spectra that are used to infer the chemical composition of the pigments. A multidimensional neural network, specifically designed to automatically restore damaged or hidden pictorial work, will be deployed on the INFN-CHNet Cloud as a web service, freely available to authenticated researchers. In this contribution, we report the status of the project, its current results, the development plans as well as future prospects.

Keywords: Image processing for cultural heritage · Deep learning · X-ray fluorescence imaging · Web technologies

1 Introduction

In the context of physical technologies applied to pictorial artworks imaging, the digital technologies for image manipulation are becoming increasingly relevant; an example is the application of modern artificial intelligence-based image inpainting techniques to physical imaging [14,16,18,30,35][1], in order to get an automatic digital restoration of damaged or hidden pictorial images. In particular, it is possible to apply a multidimensional, multi-spectral generative deep neural network to digital pre-treated row data of physical images, as a means to

---

[1] For other Machine learning approaches in Cultural Heritage, see [7], and references therein.

S. Sclaroff et al. (Eds.): ICIAP 2022, LNCS 13231, pp. 685–700, 2022.
https://doi.org/10.1007/978-3-031-06427-2_57

digitally reconstruct non-visible pictorial layer, in order to furnish a visual help to restoration professionals through digital technologies.

To optimise the digital restoration process, and for other technical reasons discussed in Sect. 2.2, we will focus on the raw data of non-invasive physical imaging such as the X-Ray Fluorescence.

These imaging techniques provide elemental maps of the pigments forming the pictorial layer(s) by accessing a wide wavelength spectra in the X-ray region; such information may be interpreted by a deep neural network to infer the expected RGB colour from the element composition of the image.

In particular, raw data of physical imaging techniques are counts organised in a rank-3 tensor $\mathcal{I}_{ij;k}$ with dimensions (height, width, energy). The tensor $\mathcal{I}_{i,j}$ contains, for each pixel (i,j), the histogram of the counts in bins of the X photon energies. It is customary to convert the raw data into elemental maps by integrating the per-pixel energy spectra around the characteristic energy peak associated to each element [15,19].

In this context, we refer to a *multidimensional* deep neural network (DNN) because the devised neural network has two branches; a 1D-branch, trained to infer the RGB colour of each pixel independently taking the energy spectrum as an input; and a 2D-branch trained to capture spatial correlations between adjacent pixels to improve the robustness against the Poisson fluctuations of single-pixel counts. The predictions obtained by the two branches are combined by a third neural network as shown in Fig. 1a.

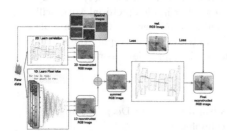
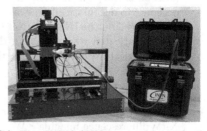

(a) A sketch of the multidimensional DNN.

(b) The INFN-CHNet scanner, with the batteries connected. From [29].

Fig. 1. (a) A sketch of the multidimensional DNN. (b) The INFN-CHNet scanner, with the batteries connected. From [29].

The results obtained with a custom DNN and several well-known architectures for either the 1D and 2D branches have been compared. The results are discussed in Sect. 3.1 and 3.2.

The performance of the algorithm is assessed through several metrics commonly employed in Computer Vision [36], such as the *Peak Signal-to-Noise Ratio*

(PSNR), the *Structural Similarity Index Measure* (SSIM) [32] and its variant, the Multi-Scale SSIM (MS-SSIM) [31][2].

In this letter, we report on the preliminary results obtained in a first, explorative phase of the project, showing that the deepest models developed in the context of Computer Vision do not offer any significant improvement of the algorithm performance over our simpler custom DNN.

An *alpha* version of the recolouring DNN is deployed as a web service in the INFN-CHNet cloud [5], also providing authenticated users with real-time analysis capabilities by processing raw data saved in highly compressed HDF5 format, computing, with few clicks, both the spectra and the integrated image, as well as the 3D scatter plot of the counts.

The ultimate goal of the AIRES-CH project is to develop a DNN capable of inferring the RGB image from an XRF image; this will be obtained by a multi-dimensional DNN, capable of exploiting features of 1D and 2D DNN. Up to now, the two branches were developed and trained on a dataset formed by XRF images of various pictorial artworks (e.g. paintings, drawings, illuminated manuscripts) of different geographical origins and historical periods. This paper reports a study of their performance, demonstrating the feasibility of inferring RGB images from XRF scans. Recolouring networks will play an important role in the future developments of the field. While providing immediate visual support to physicists performing XRF analyses, they represent a necessary building block towards the implementation of algorithms designed to digitally restore hidden pictorial layers, by removing from the XRF scan the components describing the most superficial, visible, layer, in order to furnish a visual aid to CH experts, before the application of any invasive restoration process.

2 X-Ray Fluorescence (XRF) Spectroscopy

X-Ray Fluorescence (XRF) spectrometry is a technique suitable for a wide range of applications in many disciplinary fields (environmental science, industry, geology, etc.), providing a fast, sensitive, multi-elemental analysis. The XRF technique is widely employed in Cultural Heritage (CH) applications, as no sample pre-treatment is required and it allows for non-invasive, non-destructive measurements, thus maintaining the integrity of the sample and the repeatably of the measurements ([15,19], and references therein). In addition, the portable versions can be used in situ also on wide dimension and complex geometry objects [9,29]. Its macro mapping, known as Macro X-Ray Fluorescence mapping (MA-XRF), produces elemental distribution maps of a scanned area of an artwork, proving to be extremely useful for material characterisation, and also for understanding an artist's painting and production techniques [3].

---

[2] For a review on Image Quality measures, their issues and prospects in the fields, see [6].

(a) Visible   (b) Sb

Fig. 2. Visible and Antimony Kα line distribution map of the painting *Sad child*, Unknown author, XIX century circa [8]. It is clearly evident the emergence of a hidden pictorial layer representing two human figures, not visible in the outermost pictorial layer.

2.1   The INFN-CHNet Scanner

The XRF data analysed in this work were generated by three different XRF scanners, all developed by INFN-CHNet for *in situ* analysis; their detailed description is reported in Ref. [29]. The measuring head of the instrument is composed of a X-ray tube (Moxtek - 40 kV maximum voltage, 0.1 mA maximum anode current) and a Silicon Drift Detector (Amptek XR100 SDD). In addition, a telemeter (Keyence IA-100) was also installed on the measuring head to control and adjust the sample-instrument distance during the scan. This measuring apparatus was mounted on three linear motor stages (Physik Instrumente), with $150 \div 300$ mm travel range in the $x - y$ directions (according to the XRF Scanner version), plus a 50 mm stage along the z perpendicular direction.

The whole apparatus was installed on a carbon-fibre box containing, among others, motor controllers and a signal digitiser (CAEN DT5780); for a more detailed description of the apparatus, see [29].

2.2   XRF for Hidden Layers

As explained in the previous section, the X-Ray technique is non-invasive, and can be performed with a portable instrument, allowing for *in situ* analysis. This enables the access to a (relatively) large datasets. In this work, we employed more than 62 XRF scans acquired in measurement campaigns between 2016 and 2021.

Furthermore, XRF is able to detect signal coming from hidden pictorial layers. As an illustrative example, we refer to Fig. 2. XRF analysis conducted upon the artwork *Sad child* (Unknown author, XIX century circa [8]) disclosed the presence of (at least) one currently non-visible layer, highlighting two human figures which are not noticeable in the outermost one (Fig. 2b).

Therefore, being the XRF technique capable of exploring hidden pictorial layers, XRF raw data are suitable to devise a neural network capable of digitally restoring pictorial artworks.

2.3 The Training Dataset

The whole dataset is composed by 62 XRF raw data coming from several XRF analysis on multiple paintings performed both in the LABEC facility in Florence, as well as in situ analysis (the data comes not only from published works, and include some private artworks) [1,2,4,8,20,23,25,26]; for training the 1D models, only a 50% of the pixels where used (being randomly chosen), giving a training dataset of around 2 059 780 [histogram, RGB] pairs, divided into training, test, and validation set. For training the 2D models, 45 XRF scans are used, reserving the remaining as 9 for test, and 8 for validation.

The raw data are obtained by three different devices, all developed, built and assembled by CHNet. The raw data comes from different artwork typologies: multi-layered paintings, drawings without preparatory layers, and illuminated manuscripts, all over different periods and epochs (from middle ages to contemporary art).

It is worth noticing that artworks from different epochs might have been realised using different pigments. Visually similar colour can therefore be associated to completely different XRF spectra depending on the epoch of the painting.

3 Models and Results

In this preliminary stage of the project we focused on the 1D and 2D branches of the network, separately, leaving the combination of their predictions as a following study. We have developed several models, divided in two categories:

- *1D Models*, whose goal is to learn how to associate a 3-vector living in $[0, 1]^3 \subseteq \mathbb{R}^3$, (i.e. the RGB colour) to a vector living in $\mathbb{N}^{500}$ (one for each channels of the detector[3]), the histogram of XRF counts in the range $(0.5, 38.5)$ keV;
- *2D Models*, whose goal is to learn how to associate to 500 grey-scale images a single RGB image.

Following a customary choice in Computer Vision, all the networks infering an RGB colour code are trained using the *Binary Cross Entropy* (BCE) as loss function [10].

---

[3] Actually, the ADC of our XRF detector has $2^{14} = 16384$ channels; we rebinned the histogram down to 500; the number was chosen by trial and error to be the smallest useful for the Deep learning, and the biggest tolerable for RAM memory consumption while training the algorithm.

3.1 1D Models

We have tested seven different 1D models: a simple dense DNN (with dropout and Batch Normalisation layers), a simple few layers Convolutional Neural Network, a 1D Inception-like model [28], a 1D ResNet-like model [11], our custom Multi Input ResNet-based network, a FractalNet-like DNN [17], and a WaveNet-like [21]. We report only the last three, which were found to perform better on our problem.

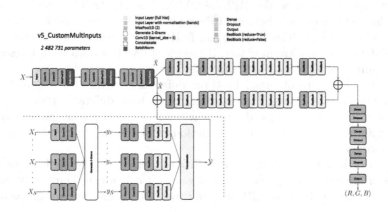

Fig. 3. The architecture of the v5 Custom Multi Input model described in the text. Here, X_i ($i \in [1, N = 4]$) represents the i-th sub-histogram of the relevant element that is fed to the DNN. Similarly, y_α ($\alpha \in [1, \bar{N} = 8]$) represents the α-th elaborated 2-gram described in the text.

Custom Multi Input Model. This model is a small variation of the ResNet model; it employs projection and identity residual blocks for avoiding the vanishing gradient problem [10], as well as Batch Normalisation layers; the final classification is performed by three couples of Dense + Drop Out layers, before the final RGB prediction.

The crucial variant here is that another input is fed to the Neural Network: we pass a set of four element peak histograms (8 bin in size): Tin (Sn) Kα, Potassium (K) Kα, Manganese (Mn) Kα, and Titanium (Ti) Kα. Those were selected because are usually relevant elements in pigments, but usually appears with very small abundance, and thus few counts. So we force the network to look carefully for them, by inserting them into a preliminary network.

This network is divided into 2 stage: the first stage looks each sub-histogram separately (to learn if a peak is actually there or not); after that, processed data are concatenated into 2-grams (all possible independent combination of), and passed to the second stage, which analyse if 2-elements correlations are present, and all the output are concatenated. After that, this information is passed to the original network and combined (by means of an Addition layer) to a pre-processed, full histogram, which underwent an encoding procedure (to extrapolate only relevant information). Also, this pre-processed, full histogram is also used by a different genus-1 topology branch of the network, so that it can be analysed without interference by the network. Finally, the two branches are added together and passed to a final set of three Dense/Dropout couple layers, and finally classified into RGB by the output layer (see Fig. 3), forming our topologically non-trivial genus-2 network.

The model has 2 482 731 trainable parameters, and was trained on a machine having an Intel Core i9-10900K, 10 Core 3.70 GHz 20 MB CPU, 64 (4×16) Gb DDR4 Corsair Vengeance LPX RAMs, and a Aorus nvidia GeForce RTX 2080 super waterforce with 8 Gb internal dedicated memory.

The training, based on the ADAM optimiser [13], proceed quickly towards convergence and is stopped after 14 epochs to prevent overfitting as monitored evaluating the loss function on the validation sample.

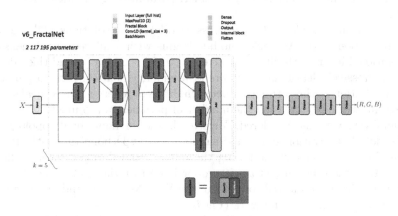

Fig. 4. The architecture of the v6 FractalNet model described in the text.

FractalNet. FractalNet [17] was originally developed to prove that it is possible to eschew residual connections, by replacing them with a *fractal* design. As visible in Fig. 4, it involves a repeated application of a simple expansion rule to generate deep networks whose structural layouts resembles truncated fractals. FractalNet contain sub-paths, even in interaction, but no residual connection. As a side result of this fractal architecture, FractalNet has the property that shallow sub-networks provide a quick answer, while deeper sub-networks provide a more accurate answer. For the construction of the network, we follow both the original paper [17], and Ref. [34]. Our implementation has 2 117 195 trainable parameters.

The training procedure, based on the ADAM optimiser [13], was stopped after 40 epochs, before becoming unstable possibly due to the large learning rate or the emergence of late-stage overfitting.

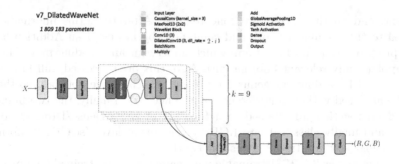

Fig. 5. The architecture of the v7 WaveNet model described in the text.

Dilated WaveNet. WaveNet [21] was originally devised as a DNN for generating raw audio waveforms. Developed by DeepMind, it was able to generate speech which mimics any human voice, reducing the gap with human performance by over 50%. We instead use it as a classifier, paraphrasing the idea suggested in [22,34].

We have also implemented progressive dilated convolutions [33], i.e. each i−th repetition of the WaveNet module begins with a dilated one-dimensional convolutional layer, with dilation rate equal to $2i$; the idea behind this addition is to progressively learn the structure of a wider, global context, in particular the contribution of the correlation between different, energy-spaced peaks in the histogram, without increasing the number of parameters too much.

Our WaveNet has 9 ordered WaveNet blocks of non-trivial topology, due to the double activation functions (sigmoid and hyperbolic tangent) after the Batch Normalisation layer following the dilated convolutional layer, indicated by $i \in [1, 9]$; the dilation factor l is proportional to i via $l = 2i$; each WaveNet block is residually connected to its following WaveNet block, and also connected to the final branch of the network (Fig. 5).

Due to the presence of the dilation factors, this model has 1 809 183 trainable parameters, slightly less with respect to the previous two models. The training, based on the ADAM optimiser [13], converged quickly, before stopping at epoch 150.

Comparison of Results for 1D Models. The comparison of the results obtained for the 1D models is presented in Tables 1a and 1b. We compare both the visual scores (structural similarity index measure, multi-scale structural similarity index measure, peak signal-to-noise ratio), and on the losses (binary cross-entropy, mean squared error). The reported score and loss values were obtained from the validation sample.

Table 1. Performances of the 1D models.

	SSIM	MS-SSIM	PSNR
v5	0.38844	0.68032	20.10390
v6	0.37207	0.67664	20.07558
v7	0.35640	0.67343	19.97972

	BCE	MSE
v5	0.63645	0.01280
v6	0.63306	0.01407
v7	0.62872	0.01456

(a) Comparison table of the visual scores performances of the v5, v6, v7 1D models.

(b) Comparison table of the losses performances of the v5, v6, v7 1D models.

We notice that the Custom Multi Input model described in Sect. 3.1 performs slightly better in all visual scores, and has the lowest (averaged) Mean Squared Error on the validation images. Instead, the model who has the lowest Binary Cross Entropy error - which is also the used loss function - is the one whose training was more long and stable, i.e. the Dilated WaveNet model. This seems to suggest that future training should be performed using different losses (or metrics); we will comment in Sect. 3.2 why MS-SSIM is probably the most relevant metric in the task at hand.

3.2 2D Models

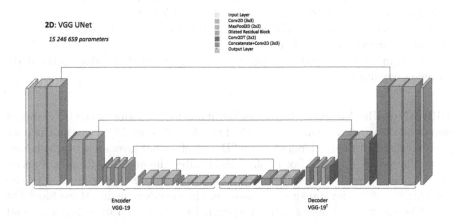

Fig. 6. The architecture of the VGG-like 2D model.

For the 2D branch, we have tested deeply two model architectures of UNet-like shapes [24], whose goal is to learn spatial correlation in the pictures and assign the recolouring to the whole 2D image, having as input $d = 500$ grey-scale images: a VGG-based [27] Encoder-Decoder architecture, and a Dilated Residual UNet Network, based on the *Globally and Locally Consistent Image Completion* (GLCIC) GAN network [12].

However, the exploration of the 2D models is limited by the large memory footprint of our dataset, which requires training in tiny mini-batches making the process extremely slow. So far, we identified two models providing satisfactory results, but further studies on the 2D models using the INFN Cloud resources are planned.

VGG-Like UNet Model. The first 2D model is a UNet with Encoder part a VGG-19 model, and as decoder its transpose (see Fig. 6). The VGG Encoder block is constituted by five blocks: the first two comprises two 2-dimensional convolutional layers with kernel size $(3, 3)$ and a $(2, 2)$-Max Pooling layer to downsample the images; the second three blocks instead have three 2-dimensional convolutional layers with kernel size $(3, 3)$ and a $(2, 2)$-Max Pooling layer.

The Decoder block is symmetric to the Encoder; it has five blocks: the first three blocks instead have three 2-dimensional convolutional layers with kernel size $(3, 3)$ followed by a 2-dimensional transposed convolutional layer to upsample the images, and is finally followed by an Addition layer to glue the upsampled images to the images coming from the encoder via the skip connections; the second two blocks have only two 2D Convolutional layers each.

The skip connections which forms the UNet shape of the network connects the last 2D Convolutional layer of the i-th Encoder's block ($i \in [1, 5]$) to the first 2D Convolutional layer of the $10 - i$-th Decoder's block.

Due to its depth, this model has $15\,246\,659$ trainable parameters. A number extremely larger than the 1D models.

The training converged fast, before stopping at epoch 14, due to the early stopping condition.

DilResNet-like Model. The DilResNet model is a variation of the UNet model described above, and is modelled on the lines described in [12]; it has an Encoder part, consisting of three blocks of two 2D-convolutional layers with $(3, 3)$-kernel size, followed by a $(2, 2)$-Max Pooling layer; after the encoder, there is a series of six Dilated Residual identity blocks. These blocks are thus connected to the decoder, which is the transposed version of the encoder, having three blocks of two 2D-convolutional layers with $(3, 3)$-kernel size, followed by a 2-dimensional transposed convolutional layer for upsampling; the decoder is thus followed by a last convolutional layer producing the 3-channel output (see Fig. 7) (Table 2).

Comparison of Results for 2D Models. The scores of the 2D models appear to be significantly higher w.r.t. the 1D models' scores. But the visual result is less satisfying; as an illustrative example, we refer to Fig. 8, where the two best performing models, the one-dimensional v5 Custom Multi input model and the two-dimensional DilResNet model were applied to on of the validation set raw

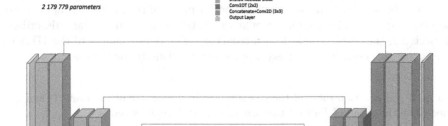

Fig. 7. The architecture of the DilResNet 2D model.

Table 2. Performances of the 2D models.

	SSIM	MS-SSIM	PSNR
VGG	0.73290	0.62626	19.69109
DRN	0.74470	0.66946	21.09683

(a) Comparison table of the visual scores performances of the VGG, DilResNet 2D models.

	BCE	MSE
VGG	0.48037	0.01589
DRN	0.46043	0.01212

(b) Comparison table of the losses performances of the VGG, DilResNet 2D models.

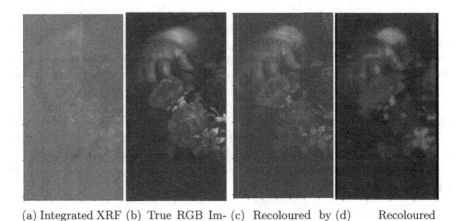

(a) Integrated XRF Image (b) True RGB Image (c) Recoloured by v5 Custom Multi Input (1D branch) (d) Recoloured by DilResNet (2D branch)

Fig. 8. Confront on the results of 1D and 2D best models applied to a detail of the painting *Putto*, Unknown author, 18th-century (circa) [1].

data, which is a detail of the painting *Putto*, Unknown author, 18th-century (circa) [1].

It can be seen that the overall visual performance of the 2D models appears to be poor; even when keeping in account the instrumental issue we face, described in Sect. 3.2, the results are still unsatisfactory. The performances of the 1D networks instead, even if more noisy, appears to be visually more pleasant.

Table 3. Comparison table of the visual scores performances of the best 1D and 2D models on the detail of *Putto*, Unknown author, 18th-century (circa).

	SSIM	MS-SSIM	PSNR
v5 CustomMulInput	0.497	0.840	22.048
DilResNet	0.818	0.810	25.529

For the scores, as reported in Table 3, we see that the Multi-Scale Structural Similarity Index is the only score higher for the 1D model; this seems to suggest that MS-SSIM is the more appropriate score when analysing the performance of 1D models.

4   Deployment in Web Applications

The web-app, offering real-time digital restoration of pictorial artworks employing the models described in Sect. 3, is available on CHNet cloud to all registered users.

It is a back-end app, developed using an open source Python language-based framework, *Dash*; Dash applications are Flask-based web servers and communicating JSON packets over HTTP requests, while Dash's front-end renders components using the JavaScript library React.js. We then used Docker and docker-compose to deploy the application on the cloud.

Users may inspect their raw data by computing XRF images in real time, either manually setting intervals in the XRF spectrum, or selecting an element peak by selecting it in the drop-down menu available on the screen, or resizing the XRF image by selecting a sub-region and computing its XRF spectra.

Finally, users may see immediately the recoloured image directly in app, as a proof-of-concept of the AIRES-CH project.

5   Conclusions and Outlook

In this contribution we outlined the results of the first stage of the AIRES-CH project, whose aim is to build a web-based app for digital restoration of pictorial artworks through Computer Vision technologies applied to physical imaging raw data.

Up to now, a set of 1D and 2D models to perform automatic recolouring were developed, trained and tested upon a relatively small dataset of X-Ray fluorescence imaging data obtained from different pictorial artworks, such as medieval illuminated manuscripts, Flemish and modern multi-layered paintings, and Renaissance drawings. These models are thus employed in the web-app for on-line analysis of XRF raw data hosted in the CHNet cloud.

The models employed are (slight variation of) well know models in the literature, applied to the problem at hand; we also tested a custom model, devised upon insights offered by the physical imaging technique employed. The results of the devised models are extremely promising, offering a starting point for future improvements.

We have also conducted preliminary test of 3D networks, but their results were non-competitive with lower dimensional models; this is most probably due to the fact that, to avoid out-of-memory issues, we employed non-sufficiently deep networks and a rebinned version of the dataset, comprising only 50 Energy channels.

5.1   Next Steps for AIRES-CH

Proven that the 1D and 2D branches are indeed able to provide complementary information to the reconstruction of the acquired image, the next step is to combine them as discussed in the introduction into a fully multidimensional neural network (see Fig. 1a).

It is clear that the small dimension of the dataset constitutes a limit to the learning capabilities of any network. Thus, to improve the scores of both branches by increasing the size of the dataset, a campaign of new XRF imaging measurements on pictorial artworks of the *Biblioteca Marucelliana* is scheduled to begin soon, in the context of the joint project AIRES-CH.

In parallel, we have to develop the technique to take into account the presence of hidden pictorial layers and to recolour them, somehow factoring out the contribution from the outermost layer. To do so, the full AIRES-CH DNN will comprise a *Generative, Adversarial* part [10]; this part associates a guessed XRF image $\mathcal{I}_{ij;k}$ to an RGB image, which is the outermost pictorial layer; this is crucial to "extract" the covered image from the raw data (see Fig. 9).

The generated XRF $\mathcal{I}^{(GAN)}$ will thus be subtracted from the true raw data $\mathcal{I}$, to mimic the outermost layer subtraction, and somehow extract the hidden layer contribution to the XRF image $\mathcal{I}^{(sub)}$.

The generated $\mathcal{I}^{(sub)}$ will be fed to the recolouring architecture, to obtain a recoloured version of the *hidden* layers.

To train this architecture, we can artificially generate training samples of nonvisible pictorial layer by appropriately merge two raw data tensors, regarding the first as the outermost pictorial layer, and the other as the hidden contribution.

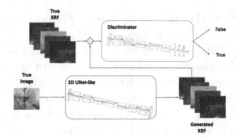

Fig. 9. A sketch of the GAN.

Acknowledgements. This research is part of the project *AIRES-CH - Artificial Intelligence for digital REStoration of Cultural Heritage* (CUP I95F21001120008) jointly funded by Tuscany Region (Progetto *Giovani Sì*) and INFN.

References

1. Ahmetovic, M.: Multi-analytical approach for the study of a XVII century Florentine painting: complementarity and data-crossing of the results of non-invasive diagnostics aimed at attribution and conservation. Master's thesis, University of Florence (2020)
2. Albertin, F., et al.: "Ecce Homo" by Antonello da Messina, from non-invasive investigations to data fusion and dissemination. Sci. Rep. **11**(1), 15868 (2021). https://doi.org/10.1038/s41598-021-95212-2
3. Alfeld, M.: MA-XRF for historical paintings: state of the art and perspective. Microsc. Microanal. **26**(S2), 72–75 (2020)
4. Bochicchio, L., et al.: Chapter 7 "Art is not science": a study of materials and techniques in five of Enrico Baj's nuclear paintings. In: Sgamellotti, A. (ed.) Science and Art: The Contemporary Painted Surface, pp. 139–168. The Royal Society of Chemistry (2020). https://doi.org/10.1039/9781788016384-00139
5. Bombini, A., et al.: CHNet cloud: an EOSC-based cloud for physical technologies applied to cultural heritages. In: GARR (ed.) Conferenza GARR 2021 - Sostenibile/Digitale. Dati e tecnologie per il futuro, Selected Papers. Associazione Consortium GARR (2021). https://doi.org/10.26314/GARR-Conf21-proceedings-09
6. Chandler, D.M.: Seven challenges in image quality assessment: past, present, and future research. ISRN Signal Process. **2013** (2013). https://doi.org/10.1155/2013/905685
7. Fiorucci, M., Khoroshiltseva, M., Pontil, M., Traviglia, A., Del Bue, A., James, S.: Machine learning for cultural heritage: a survey. Pattern Recogn. Lett. **133**, 102–108 (2020). https://doi.org/10.1016/j.patrec.2020.02.017
8. Gagliani, L.: Multi-technique investigations on a XIX century painting for the non-invasive characterization of visible and hidden materials and pictorial layers. Master's thesis, University of Florence (2020)
9. Giuntini, L., et al.: Detectors and cultural heritage: the INFN-CHNet experience. Appl. Sci. **11**(8) (2021). https://doi.org/10.3390/app11083462
10. Goodfellow, I., Bengio, Y., Courville, A.: Deep Learning. MIT Press (2016). http://www.deeplearningbook.org

11. He, K., Zhang, X., Ren, S., Sun, J.: Deep residual learning for image recognition. CoRR abs/1512.03385 (2015)

12. Iizuka, S., Simo-Serra, E., Ishikawa, H.: Globally and locally consistent image completion. ACM Trans. Graph. **36**(4) (2017). https://doi.org/10.1145/3072959. 3073659

13. Kingma, D.P., Ba, J.: Adam: a method for stochastic optimization (2017)

14. Kleynhans, T., Schmidt Patterson, C.M., Dooley, K.A., Messinger, D.W., Delaney, J.K.: An alternative approach to mapping pigments in paintings with hyperspectral reflectance image cubes using artificial intelligence. Heritage Sci. **8**(1), 1–16 (2020). https://doi.org/10.1186/s40494-020-00427-7

15. Knoll, G.F.: Radiation Detection and Measurement, 4th edn. Wiley, Hoboken (2010)

16. Kogou, S., Lee, L., Shahtahmassebi, G., Liang, H.: A new approach to the interpretation of XRF spectral imaging data using neural networks. X-Ray Spectrometry **50**(4) (2020). https://doi.org/10.1002/xrs.3188

17. Larsson, G., Maire, M., Shakhnarovich, G.: FractalNet: ultra-deep neural networks without residuals. CoRR abs/1605.07648 (2016)

18. Licciardi, G.A., Del Frate, F.: Pixel unmixing in hyperspectral data by means of neural networks. IEEE Trans. Geosci. Remote Sens. **49**(11), 4163–4172 (2011). https://doi.org/10.1109/TGRS.2011.2160950

19. Mandò, P.A., Przybyłowicz, W.J.: Particle-Induced X-Ray Emission (PIXE), pp. 1–48. American Cancer Society (2016). https://doi.org/10.1002/9780470027318. a6210.pub3.   https://onlinelibrary.wiley.com/doi/abs/10.1002/9780470027318. a6210.pub3

20. Mazzinghi, A., et al.: MA-XRF for the characterisation of the painting materials and technique of the entombment of Christ by Rogier van der Weyden. Appl. Sci. **11**(13) (2021). https://doi.org/10.3390/app11136151

21. van den Oord, A., et al.: WaveNet: a generative model for raw audio. CoRR abs/1609.03499 (2016)

22. Pandey, S.K., Shekhawat, H., Prasanna, S.: Emotion recognition from raw speech using wavenet (2019). https://doi.org/10.1109/TENCON.2019.8929257

23. Ricciardi, P., Mazzinghi, A., Legnaioli, S., Ruberto, C., Castelli, L.: The Choir Books of San Giorgio Maggiore in Venice: results of in depth non-invasive analyses. Heritage **2**(2), 1684–1701 (2019). https://doi.org/10.3390/heritage2020103

24. Ronneberger, O., Fischer, P., Brox, T.: U-Net: convolutional networks for biomedical image segmentation. CoRR abs/1505.04597 (2015)

25. Ruberto, C., et al.: La rete CHNet a servizio di Ottavio Leoni: la diagnostica per la comprensione dei materiali da disegno. In: Leo, S., Olschki editore, F. (eds.) Accademia toscana di scienze e lettere la colombaria. atti e memorie, vol. LXXXV (2020)

26. Ruberto, C., et al.: Imaging study of Raffaello's La Muta by a portable XRF spectrometer. Microchem. J. **126**, 63–69 (2016). https://doi.org/10.1016/j.microc. 2015.11.037

27. Simonyan, K., Zisserman, A.: Very deep convolutional networks for large-scale image recognition (2015)

28. Szegedy, C., et al.: Going deeper with convolutions. CoRR abs/1409.4842 (2014). http://arxiv.org/abs/1409.4842

29. Taccetti, F., et al.: A multipurpose X-ray fluorescence scanner developed for in situ analysis. Rendiconti Lincei. Scienze Fisiche e Naturali **30**(2), 307–322 (2019). https://doi.org/10.1007/s12210-018-0756-x

30. Wang, M., Zhao, M., Chen, J., Rahardja, S.: Nonlinear unmixing of hyperspectral data via deep autoencoder networks. IEEE Geosci. Remote Sens. Lett. **16**(9), 1467–1471 (2019). https://doi.org/10.1109/LGRS.2019.2900733
31. Wang, Z., Simoncelli, E., Bovik, A.: Multiscale structural similarity for image quality assessment (2003). https://doi.org/10.1109/ACSSC.2003.1292216
32. Wang, Z., Bovik, A., Sheikh, H., Simoncelli, E.: Image quality assessment: from error visibility to structural similarity. IEEE Trans. Image Process. **13**(4), 600–612 (2004). https://doi.org/10.1109/TIP.2003.819861
33. Yu, F., Koltun, V.: Multi-scale context aggregation by dilated convolutions (2016)
34. Zabihi, M., Rad, A.B., Kiranyaz, S., Särkkä, S., Gabbouj, M.: 1D convolutional neural network models for sleep arousal detection (2019)
35. Zhang, X., Sun, Y., Zhang, J., Wu, P., Jiao, L.: Hyperspectral unmixing via deep convolutional neural networks. IEEE Geosci. Remote Sens. Lett. **15**(11), 1755–1759 (2018). https://doi.org/10.1109/LGRS.2018.2857804
36. Zhao, H., Gallo, O., Frosio, I., Kautz, J.: Loss functions for neural networks for image processing. CoRR abs/1511.08861 (2015)

Automatic Classification of Fresco Fragments: A Machine and Deep Learning Study

Lucia Cascone[1] , Piercarlo Dondi[2][(✉)] , Luca Lombardi[2] ,
and Fabio Narducci[1]

[1] Department of Computer Science, University of Salerno, Via Giovanni Paolo II,
132, 84084 Salerno, Italy
{lcascone,fnarducci}@unisa.it
[2] Department of Electrical, Computer and Biomedical Engineering,
University of Pavia, Via Ferrata 5, 27100 Pavia, Italy
{piercarlo.dondi,luca.lombardi}@unipv.it

Abstract. The reconstruction of destroyed frescoes is a complex task: very small fragments, irregular shapes, color alterations and missing pieces are only some of the possible problems that we have to deal with. Surely, an important preliminary step involves the separation of mixed fragments. In fact, in a real scenario, such as a church destroyed by an earthquake, it is likely that pieces of different frescoes, which were close on the same wall, end up mixed together, making their reconstruction more complex. Their separation may be especially difficult if there are many of them and if there are no (or very old) reference images of the original frescoes. A possible way to separate the fragments is to treat this problem as a stylistic classification task, in which we have only parts of an artwork instead of a complete one. In this work, we tested various machine and deep learning solutions on the DAFNE dataset (to date the largest open access collection of artificially fragmented fresco images). The experiments showed promising results, with good performances in both binary and multi-class classification.

Keywords: Machine learning · Deep learning · Classification · Cultural Heritage · Fresco

1 Introduction

The reconstruction of damaged frescoes is an important and challenging task, that requires a great effort from restorers and art experts to be properly handled. Very small fragments, irregular shapes, color alterations and missing pieces are only some of the possible problems that can occur [9]. Moreover, good reference images of the original state of the fresco before the damage are not always guaranteed. It can happen that only very old photos are available [1,13], or that no reference is possible (e.g., for ancient Greek or Roman frescoes [16,20]).

S. Sclaroff et al. (Eds.): ICIAP 2022, LNCS 13231, pp. 701–712, 2022.
https://doi.org/10.1007/978-3-031-06427-2_58

Computer scientists have studied for many years this problem, proposing various approaches to help restorers in their work [3,7,13,16,17]; however, a general solution, able to deal with all the complexities of the problem, has yet to be found.

In this work, we focus on a specific sub-task, namely the proper subdivision of mixed fragments, an important preliminary step in fresco reconstruction. In fact, when a church (or another historical building containing frescoes) is largely damaged, for example by an earthquake or by war bombing, it is likely that pieces of different frescoes end up mixed together, making their reconstruction far more complex. We chose to treat this problem as a particular case of stylistic classification, where the artwork is not intact but split into small pieces, with also some parts missing. Our hypothesis is that established methods able to classify artworks based on artist's style can work well even when we have only small parts of them. This approach has also the additional advantage to be completely independent by the availability of a reference image of the original fresco.

Automatic style and genre classification of artworks was a task difficult to manage until a few years ago, but, in the last years, the deep learning revolution provided new ideas to deal with it. In particular, transfer learning approaches proved to be very effective in this field, achieving good performances, especially for paintings [4,10]. However, since fragments contain a lot less information than the entire artwork, it is possible that deep learning methods have worst performance than expected in the considered scenario. Thus, we tested and compared both traditional machine learning and deep learning classifiers. All the experiments have been conducted on the DAFNE dataset [9], to date the largest collection of artificially fragmented fresco images.

This work has two main goals, one practical and one theoretical. Firstly, we want to provide a useful tool to help restorers in their work. Then, we want to verify whether or not established strategies for stylistic classification can be applied also when we do not have the entire artwork, but only small parts of it, and what is the best approach between machine and deep learning. Considering the relative novelty of the problem of fresco fragments stylistic classification, we think that our study can provide a useful baseline for researchers.

The remaining of the paper in structured as follow: Sect. 2 provides a brief overview of the state-of-the-art in the field; Sect. 3 explains our approach, the tested conditions, and the dataset; Sect. 4 shows and discusses the obtained results; and, finally, Sect. 5 draws some conclusions and presents the possible next steps.

2   Related Works

In the last ten years we assisted to an enormous growth of machine and deep learning applications, that are now widespread in many different fields, such as image processing, healthcare, automotive, robotic, natural language processing, and so on [21,26]. However, the use of these techniques for Cultural Heritage is still relatively limited, mainly due to poor data availability [12]. In fact, even if new datasets are now becoming publicly available (such as Europeana [11], EGO-CH [22], NoisyArt [6] or DAFNE [9]), in most of the cases, data related to artworks are not easily or publicly accessible, for example because they are part of private collections or because there are specific restrictions on their use.

For this reason, it is not a surprise that most of the works dealing with style and genre classification focus on paintings, surely the type of artworks with more publicly available data. Early attempts to apply machine learning in this field include, for example: the work of Li and Wang [18], who in 2004 classified the style of Chinese ink paintings using hidden Markov models; the study conducted in 2005 by Lombardi [19] who tested k-Nearest Neighbor, Hierarchical Clustering and Self-Organizing Maps to discriminate the style of five famous painters; or, more recently, the discriminative and generative Bag-of-Words algorithms employed by Arora and Elgammal in 2012 [2] to classify seven different painting styles (Renaissance, Baroque, Impressionism, Cubism, Abstract, Expressionism, and Pop art). Of course, before the introduction of deep learning, this problem was very difficult to manage. The proposed solutions had rather variable performances or had been applied only to a small set of samples.

To get one of the first applications of deep learning in image style recognition (but not strictly limited to artworks) we had to wait until 2014, when Karayev et al. proved that a Convolutional Neural Network (CNN) pre-trained on Imagenet can be fine-tuned for this task, outperforming traditional machine learning methods [15]. After this preliminary study, in 2016 Tan et al. [25] performed a large-scale paintings classification (using thousands of paintings from the Wikiart dataset), considering style, genre and artist. Their work further confirmed the potentiality of data augmentation and transfer learning in this field. A few years later, in 2018, Cetinic et al. [4] adopted a similar strategy and tested various CNN fine-tuning strategies on five different painting classification tasks (artist, genre, style, time period and nationality). While, in the same year, Elgammal et al. [10] studied the performances of three widely known CNNs (AlexNet, VGGNet and ResNet) for the classification of twenty different painting styles, also obtaining interesting results about connections and evolution of artistic styles. More recently, in 2019, Sandoval et al. [23] proposed an alternative approach using a two-stage deep learning architecture, composed by a CNN, that classifies separately five patches of the image, followed by a shallow network that receives in input the five outputs of the CNN and performs the actual style classification. In 2020, Cetinic et al. [5] focused on a different style classification task, namely they used a CNN to predict the correct Wölfflin's features, five stylistic properties identified by Heinrich Wölfflin in 1915. Finally, in 2021, Dondi et al. [8] focused on the stylistic classification of historical musical instruments rather than of paintings, obtaining interesting results with a modified VGG16 architecture, that showed a behavior comparable to that of human experts.

3   Methods and Data

3.1   Classifiers

As said in the Introduction, we considered both machine and deep learning solutions. Figure 1 shows a simplified scheme of the chosen procedure: after loading the input data and a common Resize step (used to normalize the dimension of the fragments at 150×150 pixels), we run or a neural network classifier or a traditional one.

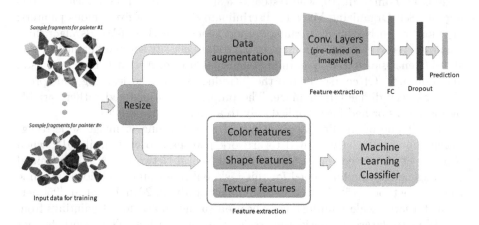

Fig. 1. Simplified scheme of the adopted architectures.

Our neural network architecture is an adaptation of the structure proposed in [8] that is particularly interesting in our scenario, since it was able to obtain good results even with a limited amount of data. In fact, though we have thousands of fragments, they only contain limited information, since they are small parts of large images. As features extractor we tested the convolutional layer of both VGG16 [24] and Resnet34 [14] (pre-trained on Imagenet). The fully connected layer has 1024 nodes and uses a ReLu as activation function. The dropout was set at 0.5. Finally, the prediction layer uses as activation a Sigmoid for the binary classification or a Softmax for the multi-class case. As optimizers we considered Adam, RMSProp and SGD varying the learning rate (0.01, 0.001, 0.0001). As loss function, we used the binary or categorical cross-entropy for the binary and multi-class classification, respectively. The batch size was set at 32. We also inserted a data augmentation step to enlarge the variety of our dataset. More precisely, we applied, to each fragment image, random rotations up to $45°$, horizontal and vertical shifts up to $\pm15\%$ of the original width/height, horizontal and vertical flips, and a $\pm50\%$ random zoom. The training strategy was the following: first of all, we trained only the fully connected layers of the network for 10 epochs, maintaining freezed all the convolutional layers, then, we proceeded with fine-tuning for additional 30 epochs (with early stop) un-freezing only the last convolutional layer (preliminary experiments showed a performance degradation un-freezing more layers).

Instead, as machine learning classifiers, we used Random Forest, Support Vector Machine (SVM) and K-Nearest Neighbor (K-NN). The limited size of the fragments and the consequent lack of a large variation in color and patterns led us to consider a set of three visual features: color, texture and shape. Color histograms are used to obtain color information. Each of the R, G, and B channels has its own color distribution, which is represented by them. Hu moments, on the other hand, are shape descriptors that are particularly useful due to their invariance to translations, rotations, and scaling. The silhouette or outline of an element in an image is used to extract shape element vectors. Basically, the weighted average of the pixel intensities is used to define the image's moments. Finally, Haralick texture vectors collect the data revealed by texture patterns. The Gray Level Co-Occurrence Matrix (GLCM) is used to calculate 13 characteristics in this procedure. It takes advantage of the image's adjacency idea. The core concept is to look for a connection between two nearby pixels at the same time. The selection of hyperparameters is critical for optimal performance. For each method, a set of hyperparameters was fixed, and then the best configuration was discovered by an exhaustive search. Each classifier was trained for each configuration separately, and the top performing hyperparameters were chosen. To avoid overfitting, a k-fold approach was adopted, with "k" set between 2 and 10. As a result, for each exhaustive search, 9 distinct values of "k" were used and the best of them were chosen. For Random Forest we tested the following parameters: n_estimators: (500, 550, 600); min_samples_split: (0.001, 0.1, 1.0); min_samples_leaf: (0.1, 1, 2); max_features: (0.01, 0.1, 1, 5). For SVM we considered: kernels: (rbf, linear, sigmoid); gamma: (scale, auto); degree: (2, 3); c: (0.1, 3.0, 1.0, 65, 100). Finally, for K-NN: metric: (Euclidean, Manhattan, Chebychev, Minkowski); number of neighbors: (5, 10, 16, 17, 18, 25); weights: (uniform, distance); leaf-size: (1, 5, 7, 20, 30, 50).

3.2   Tested Conditions

We focused on two different scenarios: (i) all the fresco fragments are mixed together; (ii) the fragments are only partially mixed together. In both cases we assumed to know the authors of the frescoes (or at least the approximate period in which they have been painted), but no reference images are available.

The first one is, of course, the worst-case scenario. To solve it, we need to completely separate the data, namely the frescoes used from training have to be different from those used for testing (in the following, we referred to this scenario as *Fresco Split*). Thus, in this case, it is possible to properly train the model only having enough samples for each considered author.

The second case is simpler, but also more common. Imagine, for example, a partially collapsed wall having two frescoes near together. It is likely that the only sections mixed together are those originally in the center of the wall, while those belonging to the left and right sides will be more clearly distinguishable. This situation is equivalent to a random split of the dataset, where fragments of the same fresco can be present both in training and in testing set (in the following we referred to this scenario as *Random Split*). It must be stressed that, even if the mixed fragments are few with respect to the first case, they can still be hundreds, thus, it is not a trivial task to manually separate them.

We also considered two possible classification approaches: (i) per-author, in which we try to distinguish fresco fragments based on the style and characteristic of specific artists; (ii) per-period, a more "high-level" classification, where we do not try to associate a fresco to a specific artist, but rather to a century. Even if less precise in the attribution, this second case is still interesting, since it is common to have in the same church frescoes of different epochs, thus, understanding the century can be enough to properly split the fragments, or, at least, to separate the fragments in macro categories, when multiple frescoes of the same period are present. With per-period classification, we are not limited by the availability of frescoes of a given artist, but instead we can train our model on a large number of frescoes made by multiple authors. However, this also means that the per-period classification is useful only for the second scenario (Random Split), since we need at least some samples of each author/fresco considered to properly train the model.

In summary, for the per-author classification we tested both scenarios (Fresco and Random Split), while for the per-period classification we tested only the Random Split case. Both binary and multi-class classification were considered.

3.3   Dataset

DAFNE dataset [9], to date the largest collection of artificially fragmented fresco images, contains images of 62 frescoes, ranging from the 14th to the 18th century. For each fresco, 18 different fragmentations are provided, some containing spurious fragments. For our tests, we chose one fragmentation for each fresco, namely the one with more fragments available and no spurious data (see some samples in Fig. 2). It can be noted that the size of the fragments might be often particularly small, thus greatly lowering the availability of those visual cues that can contribute to successful classification tasks.

Fig. 2. Samples from DAFNE dataset: (a) some fragments of various frescoes mixed together; (b) a fragmented fresco – Giotto, The Ascension of St. John (1318-1322)

Even if the dataset contains frescoes made by 28 different authors, they are not equally distributed, for some of them only one or two artworks are available. Thus, for the per-author classification, we considered only those artists with at least 4 frescoes, and enough fragments per fresco for proper training, namely Fra Angelico, Giotto, and Michelangelo. Regarding instead the per-period classification, we considered only the 14th, 15th, and 16th centuries, since most of the available works of art belong to these periods. Among them, 15th century contains most of the samples/authors of the dataset. Thus, to have a more balanced dataset, we excluded from the 15th century subset the data of two artists (Fra Angelico and Leonardo). Overall, we used a sub-dataset of approximately 13k fragments for per-author classification (4500 for Fra Angelico, 5400 for Giotto and 3100 for Michelangelo) and 26k fragments for per-period (9500 for 14th century, 9800 for 15th and 7600 for 16th).

4   Experimental Results

Here we present the obtained results (for testing) expressed in terms of Accuracy, Precision, Recall and F1-score. All the values are mean over 5 repetitions, with different splits every time. We adopted a standard 60:20:20 split of the dataset between training, validation, and test with the constrain, in the case of Fresco Split, that frescoes present in training and validation cannot be in the test set.

For neural networks, the best outcomes were achieved using the Adam optimizer and a low learning rate (0.0001) for both VGG16 and Resnet34. Instead, for the machine learning classifiers, there was no common best combination of hyper-parameters, that were different in each experiment.

4.1   Per-Author Classification

As said in Sect. 3.2, for the per-author classification we considered both Fresco Split and Random Split. The first one is the most complex scenario. The best classifier was Random Forest, which achieved a F1-score consistently above 80% in all the tests (Table 1). SVM obtained better performances in one case (Fra Angelico vs. Giotto), but Random Forest was still the second best, with very high scores. Neural networks performed poorly, showing particularly low values in Recall. The only exception was the second test (Fra Angelico vs Michelangelo) in which their performance was close to those of Random Forest. This result can be, probably, due to a great difference in the color palette of the available samples of the two authors, however, this is also a clue that the networks can perform well when the styles of the considered artists are very different, even with few samples available for training. This means that we can expect better outcomes in the second scenario, having available for training some fragments of the target fresco.

In fact, with Random Split (Table 2), the neural network classifiers performed far better than with the Fresco Split: we can observe a clear improvement in

Recall, and, in all the tests, F1-score has increased from about 40% of the previous scenario to over 70%. However, despite the improvements, Random Forest remains the best classifier also in this scenario.

Table 1. Results for per-author classification with Fresco Split (from now on, best classifier highlighted in green).

Test	Classifier	Accuracy	Recall	Precision	F1
F. Angel. vs Giotto	VGG16	0.6946	0.2952	0.7876	0.4294
	Resnet34	0.6245	0.2865	0.5773	0.3830
	R. Forest	0.8835	0.8835	0.9094	0.8844
	SVM	0.9364	0.9364	0.9388	0.9367
	K-NN	0.8246	0.8246	0.8774	0.8249
F. Angel. vs Michel.	VGG16	0.8551	0.7196	0.9878	0.8326
	Resnet34	0.8728	0.6988	0.9772	0.8149
	R. Forest	0.8740	0.8740	0.8850	0.8751
	SVM	0.6676	0.6676	0.6963	0.6706
	K-NN	0.6004	0.6004	0.6357	0.6030
Giotto vs Michel.	VGG16	0.6425	0.2878	0.9604	0.4429
	Resnet34	0.6352	0.2656	0.9835	0.4182
	R. Forest	0.9018	0.9018	0.9177	0.9007
	SVM	0.7714	0.7714	0.7802	0.7692
	K-NN	0.6879	0.6879	0.8069	0.6525
F. An. vs G. vs Mic.	VGG16	0.5060	0.5056	0.5059	0.5058
	Resnet34	0.5879	0.5836	0.5896	0.5866
	R. Forest	0.8281	0.8281	0.8507	0.8226
	SVM	0.5895	0.5895	0.6575	0.5242
	K-NN	0.5903	0.5903	0.6255	0.5708

Table 2. Results for per-author classification with Random Split.

Test	Classifier	Accuracy	Recall	Precision	F1
F. Angel. vs Giotto	VGG16	0.7864	0.6731	0.9025	0.7711
	Resnet34	0.7301	0.7133	0.7997	0.7540
	R. Forest	0.8899	0.8899	0.8899	0.8899
	SVM	0.7133	0.7133	0.7264	0.7116
	K-NN	0.6983	0.6983	0.7041	0.6933
F. Angel. vs Michel.	VGG16	0.8908	0.7940	0.9409	0.8612
	Resnet34	0.8717	0.7769	0.9052	0.8362
	R. Forest	0.9377	0.9377	0.9378	0.9376
	SVM	0.6869	0.6869	0.6826	0.6759
	K-NN	0.6390	0.6390	0.6347	0.6020
Giotto vs Michel.	VGG16	0.8829	0.7257	0.9567	0.8253
	Resnet34	0.8418	0.6601	0.8482	0.7424
	R. Forest	0.9439	0.9439	0.9450	0.9436
	SVM	0.6801	0.6801	0.6742	0.6688
	K-NN	0.6994	0.6994	0.7022	0.6798
F. An. vs G. vs Mic.	VGG16	0.7349	0.7328	0.7326	0.7327
	Resnet34	0.6850	0.6814	0.6940	0.6876
	R. Forest	0.8743	0.8743	0.8750	0.8744
	SVM	0.5311	0.5311	0.5964	0.4928
	K-NN	0.5518	0.5518	0.5452	0.5329

Table 3. Per-period classification with Random Split.

Test	Classifier	Accuracy	Recall	Precision	F1
14th vs 15th	VGG16	0.8750	0.8950	0.8580	0.8761
	Resnet34	0.8119	0.7835	0.8267	0.8045
	R. Forest	0.8361	0.8361	0.8401	0.8357
	SVM	0.6530	0.6530	0.6585	0.6508
	K-NN	0.6472	0.6472	0.6478	0.6471
14th vs 16th	VGG16	0.9118	0.9106	0.8930	0.9017
	Resnet34	0.8609	0.8252	0.8567	0.8406
	R. Forest	0.8653	0.8653	0.8674	0.8656
	SVM	0.5799	0.5799	0.6215	0.5691
	K-NN	0.6254	0.6254	0.6383	0.6259
15th vs 16th	VGG16	0.8692	0.9592	0.7934	0.8684
	Resnet34	0.8446	0.8488	0.8140	0.8310
	R. Forest	0.8265	0.8265	0.8277	0.8267
	SVM	0.6074	0.6074	0.6303	0.6042
	K-NN	0.5580	0.5580	0.5662	0.5587
14th vs 15th vs 16th	VGG16	0.8241	0.7949	0.8457	0.8195
	Resnet34	0.7295	0.6746	0.7795	0.7232
	R. Forest	0.7520	0.7520	0.7578	0.7531
	SVM	0.4244	0.4244	0.4638	0.4208
	K-NN	0.4573	0.4573	0.4679	0.4599

4.2   Per-Period Classification

For the per-period classification we can consider only the Random Split scenario (as specified in Sect. 3.2). This time, the neural networks (the modified VGG16, in particular) were able to surpass the traditional machine learning methods (Table 3), showing F1-scores always higher than 80%. However, we must consider that, in this case, the dataset contained double the fragments than in the per-author case and that Random Forest still achieved very good scores, confirming itself as the most stable classifier in all the tested scenarios.

4.3   Discussion

The experimental results revealed that traditional machine learning classifiers were able to achieve a level of performance comparable, or even higher, than that of complex deep learning architectures, by extracting simple features related to color, texture and shape that allow characterizing the content of the pieces. Such a result is justified by the fact that deep learning solutions are in general greedy of features, that they use to separate the samples into different classes. In this case, there are mainly two reasons that make a fragment unfavorable for the classification task: (i) the fragment may be very small, thus covering a negligible portion of the surface of the fresco; (ii) the fragment can be arbitrarily big but represents a pattern-less portion of the fresco (a plain sky, a lake, a grass field, and so on). These conditions lead to a lack of visual features in the input sample, which means no information for the hidden layers of the deep neural model. On the other side, the handcrafted features used as input for machine learning classifiers underscore that, even in case of such challenging conditions, overall classification performance can remain high. More precisely, the outcome for per-author classification, presented in Tables 1 and 2, showed superior performance for Random Forest compared to the other classifiers. Notably, Support Vector Machines suffer from an increasing number of classes to be separated. In this case, even if the total number is not far greater than the binary classification, the limited number of available features does not make it possible to look for a valid separation among the classes. Similarly, the K-NN approach is affected by the same challenging conditions. The results obtained by Random Forest indeed show that ensemble learning techniques may overcome the limitation of the ambiguous representation of the input even when the available set of features is particularly reduced. Neural networks, on the other hand, showed performance inferior to Random Forest until the per-period experiment (Table 3), in which the number of available fragments is far higher than in all the previous tests. Unfortunately, this is also the less precise type of classification, a per-author comparison (either with Fresco Split or Random split) is generally more useful for restorers in a real situation.

From a more theoretical point of view, it is interesting to notice how the modified VGG architecture performed generally better than the modified Resnet architecture. This is the same behavior described in [8], for a different stylistic classification problem but still with a limited amount of data, showing that the VGG convolutional architecture is probably more well suited for extracting stylistic features when few data are available.

5   Conclusions

In this work we presented a study about the problem of mixed fresco fragment classification, an important preliminary step in the reconstruction of broken frescoes, especially useful when hundreds of small fragments must be discriminated. This problem can be considered as a particular, and relatively new, case of stylistic classification, in which we only have small sub-parts of the entire artwork. We proposed a comparison between various machine and deep learning approaches with the objective to find out which are the best performing solutions, and also to provide a valuable baseline for researchers working in the field.

Experimental tests conducted on the DAFNE dataset showed promising results, especially in the case of partially mixed fresco fragments (Random Split), where we obtained a F1-score almost always over 80% and in some cases over 90%, but they also revealed significant limitations, mostly for Fresco Split, further confirming the challenging points of the considered problem.

The outcomes also proved that traditional machine learning classifiers (Random Forest, in particular) can outperform deep learning solutions in this task, especially under the most tough conditions, namely when all fragments are mixed up together and only few samples per-author are available.

Next steps will involve additional tests on a larger number of authors and frescoes, as well as the use of other types of classifiers and neural networks. Fragments with alterations on their surface (like color lacunae or scratches) will be considered, too.

References

1. Abate, D.: FRAGMENTS: a fully automatic photogrammetric fragments recomposition workflow. J. Cult. Herit. **47**, 155–165 (2021). https://doi.org/10.1016/j.culher.2020.09.015

2. Arora, R.S., Elgammal, A.: Towards automated classification of fine-art painting style: a comparative study. In: Proceedings of the 21st International Conference on Pattern Recognition (ICPR 2012), pp. 3541–3544 (2012)

3. Barra, P., Barra, S., Nappi, M., Narducci, F.: SAFFO: a SIFT based approach for digital anastylosis for fresco reconstruction. Pattern Recogn. Lett. **138**, 123–129 (2020). https://doi.org/10.1016/j.patrec.2020.07.008

4. Cetinic, E., Lipic, T., Grgic, S.: Fine-tuning convolutional neural networks for fine art classification. Expert Syst. Appl. **114**, 107–118 (2018). https://doi.org/10.1016/j.eswa.2018.07.026

5. Cetinic, E., Lipic, T., Grgic, S.: Learning the principles of art history with convolutional neural networks. Pattern Recogn. Lett. **129**, 56–62 (2020). https://doi.org/10.1016/j.patrec.2019.11.008

6. Del Chiaro, R., Bagdanov, A.D., Del Bimbo, A.: NoisyArt: a dataset for webly-supervised artwork recognition. In: VISIGRAPP (4: VISAPP), pp. 467–475 (2019)

7. Derech, N., Tal, A., Shimshoni, I.: Solving archaeological puzzles. Pattern Recogn. **119**, 108065 (2021). https://doi.org/10.1016/j.patcog.2021.108065

8. Dondi, P., Lombardi, L., Malagodi, M., Licchelli, M.: Stylistic classification of historical violins: a deep learning approach. In: Del Bimbo, A., et al. (eds.) ICPR 2021. LNCS, vol. 12667, pp. 112–125. Springer, Cham (2021). https://doi.org/10.1007/978-3-030-68787-8_8

9. Dondi, P., Lombardi, L., Setti, A.: DAFNE: a dataset of fresco fragments for digital anastlylosis. Pattern Recogn. Lett. **138**, 631–637 (2020). https://doi.org/10.1016/j.patrec.2020.09.015

10. Elgammal, A., Liu, B., Kim, D., Elhoseiny, M., Mazzone, M.: The shape of art history in the eyes of the machine. In: Proceedings of the AAAI Conference on Artificial Intelligence, vol. 32 (2018)

11. Europeana: Europeana digital library. https://www.europeana.eu/en. Accessed 02 Feb 2022

12. Fiorucci, M., Khoroshiltseva, M., Pontil, M., Traviglia, A., Del Bue, A., James, S.: Machine learning for cultural heritage: a survey. Pattern Recogn. Lett. **133**, 102–108 (2020). https://doi.org/10.1016/j.patrec.2020.02.017

13. Fornasier, M., Toniolo, D.: Fast, robust and efficient 2D pattern recognition for reassembling fragmented images. Pattern Recogn. **38**(11), 2074–2087 (2005). https://doi.org/10.1016/j.patcog.2005.03.014

14. He, K., Zhang, X., Ren, S., Sun, J.: Deep residual learning for image recognition. In: Proceedings of the IEEE Conference on Computer Vision and Pattern Recognition, pp. 770–778 (2016). https://doi.org/10.1109/CVPR.2016.90

15. Karayev, S., et al.: Recognizing image style. In: Proceedings of the British Machine Vision Conference. BMVA Press (2014). https://doi.org/10.5244/C.28.122

16. Koller, D., Trimble, J., Najbjerg, T., Gelfand, N., Levoy, M.: Fragments of the city: Stanford's digital Forma Urbis Romae project. In: Proceedings of the Third Williams Symposium on Classical Architecture, vol. 61, pp. 237–252 (2006)

17. Lermé, N., Hégarat-Mascle, S.L., Zhang, B., Aldea, E.: Fast and efficient reconstruction of digitized frescoes. Pattern Recogn. Lett. **138**, 417–423 (2020). https://doi.org/10.1016/j.patrec.2020.08.006

18. Li, J., Wang, J.Z.: Studying digital imagery of ancient paintings by mixtures of stochastic models. IEEE Trans. Image Process. **13**(3), 340–353 (2004). https://doi.org/10.1109/TIP.2003.821349

19. Lombardi, T.E.: The classification of style in fine-art painting. Pace University (2005)

20. Papaodysseus, C., Panagopoulos, T., Exarhos, M., Triantafillou, C., Fragoulis, D., Doumas, C.: Contour-shape based reconstruction of fragmented, 1600 BC wall paintings. IEEE Trans. Signal Process. **50**(6), 1277–1288 (2002). https://doi.org/10.1109/TSP.2002.1003053

21. Pouyanfar, S., et al.: A survey on deep learning: algorithms, techniques, and applications. ACM Comput. Surv. **51**(5) (2018). https://doi.org/10.1145/3234150

22. Ragusa, F., Furnari, A., Battiato, S., Signorello, G., Farinella, G.M.: EGO-CH: dataset and fundamental tasks for visitors behavioral understanding using egocentric vision. Pattern Recogn. Lett. **131**, 150–157 (2020)

23. Sandoval, C., Pirogova, E., Lech, M.: Two-stage deep learning approach to the classification of fine-art paintings. IEEE Access **7**, 41770–41781 (2019). https://doi.org/10.1109/ACCESS.2019.2907986

24. Simonyan, K., Zisserman, A.: Very deep convolutional networks for large-scale image recognition. arXiv preprint arXiv:1409.1556 (2014)

25. Tan, W.R., Chan, C.S., Aguirre, H.E., Tanaka, K.: Ceci n'est pas une pipe: a deep convolutional network for fine-art paintings classification. In: 2016 IEEE International Conference on Image Processing (ICIP), pp. 3703–3707 (2016). https://doi.org/10.1109/ICIP.2016.7533051

26. Voulodimos, A., Doulamis, N., Doulamis, A., Protopapadakis, E.: Deep learning for computer vision: a brief review. Comput. Intell. Neurosci. **2018**(7068349) (2018). https://doi.org/10.1155/2018/7068349

Unsupervised Multi-camera Domain Adaptation for Object Detection in Cultural Sites

Giovanni Pasqualino$^{(\boxtimes)}$, Antonino Furnari, and Giovanni Maria Farinella

Department of Mathematics and Computer Science, University of Catania, Catania, Italy
giovanni.pasqualino@phd.unict.it, {furnari,gfarinella}@dmi.unict.it

Abstract. Domain adaptation approaches can be used to efficiently train object detectors by leveraging labeled synthetic images, inexpensively generated from 3D models, and unlabeled real images, which are cheaper to obtain than labeled ones. Most of the state-of-the-art techniques consider only one source and one target domain for the adaptation task. However, real world scenarios, such as applications in cultural sites, naturally involve many target domains which arise from the use of different cameras at inference time (e.g. different wearable devices and different smartphones on which the algorithm will be deployed). In this work, we investigate whether the availability of multiple unlabeled target domains can improve domain adaptive object detection algorithms. To study the problem, we propose a new dataset comprising images of 16 different objects rendered from a 3D model as well as images collected in the real environment using two different cameras. We experimentally assess that current domain adaptive object detectors can improve their performance by leveraging the multiple targets. As evidence of the usefulness of explicitly considering multiple target domains, we propose a new unsupervised multi-camera domain adaptation approach for object detection which outperforms current methods. Code and dataset are available at https://iplab.dmi.unict.it/OBJ-MDA/.

Keywords: Cultural sites · Multi-camera · Domain adaptation · Object detection

1 Introduction

Object detection algorithms find their use in different scenarios [1,14,23]. Among these, detecting and recognizing objects from an egocentric point of view can help to model the relationships between the user and the surrounding objects to understand the camera wearer's focus of attention, as well as to reason on current

---

Supplementary Information The online version contains supplementary material available at https://doi.org/10.1007/978-3-031-06427-2_59.

Synthetic   Hololens   GoPro

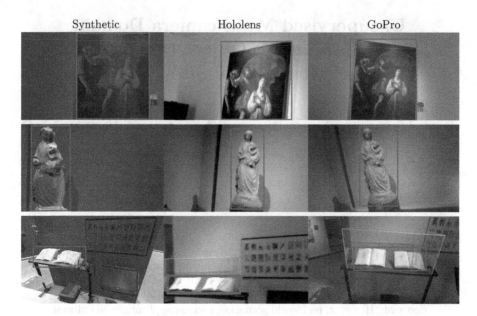

Fig. 1. Example images from the proposed dataset along with the related object annotations. Each row depicts the same object, whereas columns identify the different domains. Note that objects collected with different cameras tend to vary in appearance.

and future interactions. Indeed, past works have proposed to use object detectors in industrial [18], cultural [17] and home [3] scenarios. The use of object detectors on wearable and mobile devices in a cultural site, in particular, can enable useful applications aimed to assist the visitor during the fruition of the cultural goods. Examples of such applications range from the deployment of smart audio guides automatically triggered by object detection [21] to the study of the visitor's behavior [17] from the analysis of a video collected form their point of view. Although several object detectors trained on generic datasets such as COCO [9] are available, these require to be fine-tuned using domain-specific data, which has to be acquired and labeled. Labeling images for object detection requires manual labor, which significantly increases the deployment costs and time. To reduce these costs, previous work [11] has proposed to acquire a 3D model of the cultural site in which object detection algorithms have to be deployed and generate large quantities of synthetic images of artworks by simulating an agent visiting the site. By labeling the 3D model once, specifying the position of each artwork in the 3D space, the process proposed in [11] allows to easily collect large quantities of images automatically labeled for object detection at not cost. While this process can be used to easily generate a large training set for object detection, at inference time, the model is still expected to be used on real images, which may have a significantly different appearance. See for example Fig. 1. The domain difference between training and test images often prevents the model from generalizing. To deal with this gap, domain adaptation methods are generally exploited. These methods assume the existence of a source domain (e.g., the synthetic images) and a target domain (e.g., the real images). Training is performed relying on labeled source images and unlabeled target images with the aim to generalize on the target domain. Although domain adaptation methods allow to significantly reduce the test error on real data, they generally assume the presence of a single target domain. However, in many application scenarios different target domains may naturally arise from the need to deploy a given algorithm in different devices equipped with different cameras, such as a smartphone and a wearable device. While real images collected with different cameras can look similar, they can still present subtle differences in terms of colors, contrast and distortion. For example, Fig. 1 reports synthetic images of artworks, as well as images collected with a Hololens and a GoPro wearable camera. Starting from this observation, we aim to answer two questions: i) are current adaptation methods able to generalize to multiple target domains? ii) can we exploit multiple target domains to improve generalization? In this paper, we study whether the availability of more unlabeled target domains in the training phase can improve the generalization of object detection algorithms trained only on labeled synthetic data. As an application example, we consider the domain of cultural sites. In this context, it is worth noting that unlabeled target data is cheap to obtain when algorithms need to be adapted to a new camera (e.g., a wearable device), which can make the domain adaptation approach scalable to the case of multiple cameras.

To carry out the study, we propose the OBJ-MDA-CH (Object multi-camera Domain Adaptation Cultural Heritage) dataset. The dataset can be used to study the problem of object detection in the presence of one labeled synthetic source domain and multiple unlabeled target domains. The dataset contains images of 16 artworks kept in the "Galleria Regionale di Palazzo Bellomo", generated from a 3D model of the museum (synthetic) and as well as real images collected in the cultural site with 2 different devices: a Microsoft Hololens and a GoPro. We perform experiments to investigate whether current algorithms for unsupervised domain adaptation for object detection can exploit multiple target domains to improve the generalization of the object detector. Motivated by the limited performance of domain adaptation approaches for object detectors, we propose MDA-RetinaNet, a baseline which extends common approaches to unsupervised domain adaptation for object detection to the case of multiple target domains. Despite being conceptually simple, the proposed approach outperforms previous state-of-the-art algorithms, which suggests that explicitly considering multiple target domains can improve generalization in the considered settings.

In sum, the contributions of this paper are as follows: 1) we introduce a dataset to study the problem of unsupervised multi-camera domain adaptation; 2) we study the ability to generalize across multiple target domains of object detectors combined with different forms of domain adaptation mechanisms; 3) we propose an unsupervised domain adaptation baseline which explicitly models multiple unlabeled target domains to improve cross-domain generalization leveraging only synthetic labeled images and real unlabeled images. Code and dataset are available at https://iplab.dmi.unict.it/OBJ-MDA/.

2   Related Work

Our work is related to different lines of research related to egocentric vision in cultural sites, domain adaptation techniques and their use for object detection.

Egocentric Vision in Cultural Sites: Augmented reality is an important application of wearable computer vision. In the context of cultural sites, some authors [15,21] proposed to create a virtual guide application to improve the visitors' experience with the fruition of multimedia materials. These applications are based on the detection and recognition of objects which can trigger the proposal of associated content. The training process of these kinds of algorithms requires many well annotated images that must be acquired and manually annotated affecting development times and costs. Due to the lack of data in this field, the authors of [17] presented a first person dataset acquired using Microsoft Hololens to study fundamental tasks in the context of cultural sites. The authors of [11] proposed a tool to generate synthetic labeled images from a 3D reconstruction of a real cultural site. However, the synthetic images differ in style such as color and shape with respect to the real counterparts. For this reason, an object detection algorithm, which has to work at inference time with the real images, will produces poor results if trained only with synthetic data. In this work, we study the problem of object detection, which is at the core of many applications in cultural sites [15,21]. To avoid the manual labeling of large quantities of real images and make the approach scalable to the introduction of new cameras, we study a multi-camera unsupervised domain adaptation approach.

Domain Adaptation: State-of-the-art domain adaptation methods usually work considering one source labeled domain and one target unlabeled domain. The authors of [5] presented a method to align features extracted in the source and target domains (i.e., make their distributions indistinguishable) using an adversarial learning scheme implemented through a novel gradient reversal layer. The authors of [22] proposed an approach to adapt a backbone model trained on the source domain through adversarial learning. The authors of [28] presented Cycle-GAN, an image translation model that learns a mapping between source and target images in the absence of paired examples. Some works have considered the possibility to exploit many labeled source domains and only one unlabeled target domain. The authors of [27] proposed a method which discriminates between all possible *target-source* domain pairs. The authors of [13] presented a method composed of three components: a feature extractor, a moment matching module and a set of classifiers. Multi-camera domain adaptation has also been studied in the past. The authors of [6] proposed a method based on an autoencoder that finds a latent space capturing domain-invariant and domain-dependent features which can generalize over multiple target domains. The authors of [16]

Table 1. Statistics of the proposed dataset for unsupervised multi-camera domain adaptation. The average occupied area (last column) is the average percentage of the image occupied by the bounding boxes of the considered object class.

Object instances	Synthetic Domain (Source)		Hololens Domain (Target)		GoPro Domain (Target)		Total object instances for each class	Average occupied area
	Training	Test	Training	Test	Training	Test		
Annunciazione	1301	605	191	69	211	74	2451	42.87%
Libro d'Ore miniato	1628	722	105	30	146	42	2673	8.02%
Lastra tombale di Giovanni Cabastida	2313	1181	200	100	247	114	4155	24.58%
Madonna del Cardillo	2345	1264	106	40	166	66	3987	9.74%
Disputa di San Tommaso	2202	965	100	46	155	67	3535	28.17%
Traslazione della Santa Casa	1904	964	161	46	225	71	3371	22.24%
Madonna col Bambino	2135	1044	119	47	161	46	3552	21.93%
L'immacolata Concezione e Dio Padre in Gloria	2557	1139	77	39	100	54	3966	35.70%
Adorazione dei Magi	1517	478	64	36	69	39	2203	30.35%
Sant'Elena e Costantino e Madonna con Bambino in gloria fra angeli	3285	1031	94	44	153	61	4668	33.72%
Taccuini di disegni	1617	513	59	33	75	39	2336	22.34%
Martirio di S. Lucia	3567	2353	106	36	184	45	6291	22.55%
Volto di Cristo	990	519	25	26	50	36	1646	11.74%
Dipinti di Sant'Orsola	2721	1897	83	69	125	86	4981	30.56%
Immacolata e i santi Chiara, Francesco, Antonio, Abate, Barbara e Maria Maddalena	3824	2424	104	69	187	89	6697	32.36%
Storia della Genesi	927	375	55	14	57	15	1443	22.79%
Total object instances for each split	34833	17474	1649	744	2311	944		

extended the idea proposed in [22] by replacing the binary discrimination with a multi-class discrimination between the source and target domains. The authors of [10] proposed a method based on an iterative multi-teacher knowledge distillation from multiple teachers to a common student. While we base our work on previous domain adaptation approaches [5], we consider a different scenario in which there is only one labeled source synthetic domain and two unlabeled real domains produced by two different camera.

Domain Adaptation for Object Detection: Previous works have investigated the application of unsupervised domain adaptation techniques to the problem of object detection. The authors [2] presented DA-Faster RCNN which, exploiting the gradient reversal layer paradigm [5], aligns features at the image and instance level. The authors of [20] proposed to align the feature extracted at the high and low level. The authors of [26] propose the same architecture of [2] adding more discriminators with gradient reversal layers to the Faster-RCNN [19] backbone. The authors of [24] proposed an incremental learning based architecture which gradually stores the knowledge coming from the multiple domains combined with multi-level feature modules that align the features at different levels. We present an architecture based on feature alignment and image to image translation to tackle the problem of multi-camera unsupervised domain adaptation for object detection in cultural sites.

3 Dataset

The proposed OBJ-MDA-CH dataset includes images of 16 artworks kept in the "Galleria Regionale di Palazzo Bellomo"[1].

---

[1] http://www.regione.sicilia.it/beniculturali/palazzobellomo/.

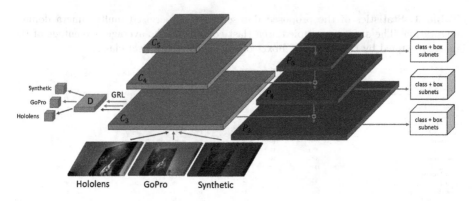

Fig. 2. The proposed MDA-RetinaNet architecture. The diagram follows the original notation of RetinaNet [8].

The dataset covers different types of objects such as books, sculptures and paintings (see Fig. 1) and it is divided into three subsets related to the three different domains, as explained in the following.

Source Synthetic Labeled Images: These images have been generated using the tool proposed in [11]. Given a 3D model of the museum with 3D bounding box annotations around each object, the tool simulates an agent navigating the site, generating images collected from its egocentric point of view. The generated images are automatically annotated with object bounding boxes. This set contains 51284 training and 23960 test images, for a total of 75244 labeled images. Examples of these images are shown in the first column of Fig. 1.

Target Real Images Acquired Using Microsoft Hololens: These images have been collected in the real environment by 10 subjects visiting the site while wearing a Microsoft Hololens device. After the collection, the images have been manually annotated drawing bounding boxes around the 16 selected artworks. This set of images contains 1502 training and 688 test images, for a total of 2190 images. The second column of Fig. 1 reports some examples of these images.

Target Real Images Acquired Using GoPro: These images have been collected by the same 10 subjects visiting the site wearing a GoPro wearable camera. This set contains 1911 training and 796 test images, for a total of 2707 images. The third column of Fig. 1 reports some examples of these images.

Table 1 reports some statistics for each object class included in the proposed dataset. The table also highlights that the proposed dataset is challenging for the task of domain adatation for object detection due to the average size of the objects. In fact, the biggest object present in the dataset occupies only the 42.87% of the images' area while the smallest occupies the 8.02%. The OBJ-MDA-CH dataset for multi target domain adaptation is available at double blind review.

4 Compared Methods and Proposed Approach

Problem Statement: Let $S = \{x_s^i, y_s^i\}_i$ be the source domain, where x_s^i is the *i-th* source image and y_s^i is the corresponding annotation. Let $T = \{T_1, T_2, ..., T_n\}$ be the set of target domains where $T_j = \{x_{t_j}^i\}_i$ corresponds to the *j-th* target. In our study, we set the number of targets to two. The goal is to train the model using synthetic labeled source images and unlabeled real target images and obtain good generalization on the target domains.

Baselines Without Domain Adaptation: To assess the performance of detectors in the absence of domain adaptation, we analyze the behavior of RetinaNet [8] and Faster RCNN [19] when trained on the source domain and tested on the target domains. As an upper-bound, we also report the results of RetinaNet when trained on labeled target images, which we refer to as the "oracle".

Domain Adaptation Based on Feature Alignment: To study whether current domain adaptation approaches can be used for multi-camera domain adaptation, we consider a naive baseline approach in which the two target domains are merged into a single target set.

Specifically, we consider three unsupervised domain adaptation state-of-the-art methods for object detection: DA-Faster RCNN [2], Strong Weak [20] and DA-RetinaNet [12].

Domain Adaptation Through Feature Alignment and Image to Image Translation: We also assess the performance of another popular domain adaptation approach based on image-to-image-translation. To this aim, we train a Cycle-GAN model [28] to translate synthetic images to the domain of real images, obtained by merging images from the two target domains. Hence, we train the previously discussed methods using the translated images and test them directly on the real images. The goal of this approach is to assess the impact of further reducing the domain gap between source and target domains.

Proposed MDA-RetinaNet Architecture: Previously discussed approaches do not explicitly consider the presence of multiple target domain, but can only take advantage of the additional data obtained by merging the two target domains. To assess whether explicitly modeling the existence of multiple target domains is actually beneficial, we propose a simple architecture for unsupervised domain adaptation for object detection based on adversarial learning [5]. Figure 2 reports a diagram of the proposed architecture, which is based on the standard RetinaNet object detector [8]. To achieve domain adaptation, we attach a domain discriminator with a gradient reversal layer to the feature map C_3 obtained from the ResNet backbone [7]. Common domain adaptation techniques use a binary classifier which discriminates between features from the source and target domain. To adapt to the case of multiple target domains, we consider a multi-class classifier D to discriminate between all the targets and source domains, i.e., a 3-class classifier considering the proposed dataset. The discriminator has

Table 2. Results of baseline and feature alignment methods. S refers to Synthetic, H refers to Hololens and G to GoPro. Results are computed using mAP.

Model	Source	Target	Test H	Test G
Faster RCNN [19]	S	–	7.61%	30.39%
RetinaNet [8]	S	–	14.10%	37.13%
DA-Faster RCNN [2]	S	H+G merged	10.53%	48.23%
Strong Weak [20]	S	H+G merged	26.68%	48.55%
DA-RetinaNet [12]	S	H+G merged	31.63%	48.37%
MDA-RetinaNet	S	H, G	**34.97%**	**50.81%**
RetinaNet [8] (Oracle)	H	–	92.44%	77.96%
RetinaNet [8] (Oracle)	G	–	69.70%	89.69%

3 convolutional layers with kernel size equal to 1 and ReLU activation function. Following [5], the model is trained by minimizing the following loss function:

$$L = L_{class} + L_{box} - \lambda(L_D) \tag{1}$$

where L_{class} and L_{box} are the regression losses of RetinaNet, L_D is the loss of the discriminator module and λ is an hyper-parameter that balances the object detection and domain adaptation losses. The gradient reversal layer placed at the input of the discriminator (see Fig. 2) allows to jointly train the discriminator to distinguish between the domains and the model to extract indistinguishable features. We hypothesize that, by providing three classes to the discriminator, the model will learn to extract features which are not only indistinguishable across synthetic and real domains, but also indistinguishable across the different real cameras.

Experimental Settings: All models were trained for 60K iterations starting from ImageNet [4] pre-trained weights. We set the learning rate to 0.0002 for the first 30K iterations. We then multiply it by 0.1 for the remaining 30K iterations. The batch size was set to 4 for RetinaNet [8] and Faster RCNN [19], 6 for DA-RetinaNet [12] (4 source and 2 target images) and 8 for MDA-RetinaNet[2] (4 source, 2 target Hololens and 2 target GoPro images). All these models were implemented using Detectron2 [25]. To reduce the noise arising in the initial training of the Discriminator D, we update the λ hyper-parameter following the update rule proposed by [5]: $\lambda = \frac{2}{1+e^{\gamma*p}} - 1$. DA-Faster RCNN and Strong Weak were trained for the same amount of iterations using the same learning rate as for the previously discussed model. For the other parameters, we adopted the settings proposed by the authors in their respective works [2] and [20]. CycleGAN was trained for 60 epochs using the default parameters. We will make our code public to improve reproducibility of the work.

---

[2] https://github.com/fpv-iplab/STMDA-RetinaNet.

5 Results

Main Results: Table 2 reports the results of the baseline and the methods based on feature alignment. The first two rows highlight that RetinaNet trained with synthetic images and tested on both target domains is less sensitive to the domain gap than Faster RCNN (14.10% vs 7.61% for Hololens and 37.13% vs 30.39% for GoPro). Following this observation, we based our architecture on RetinaNet. The second group of rows reports the results of DA-Faster RCNN [2], Strong Weak [20] and DA-RetinaNet [12] adapted to multi-camera domain adaptation by simply merging the two target domains into one indicated as ("H+G merged" in Table 2). As can be noted, training these methods using this naive approach already allows to improve over the baseline, but it obtains lower results than the proposed MDA-RetinaNet approach (sixth row of the Table 2). In particular, the proposed model achieves better result than DA-RetinaNet with a margin of more than +3% (34.97% vs 31.63%) and +2% (50.81% vs 48.37%) when tested respectively on the Hololens and GoPro target domains. This is probably due to the fact that the multi-class discriminator present in the MDA-RetinaNet architecture allows to extract features that are more domain invariant than the ones obtained with the binary discriminator of DA-RetinaNet [12]. The last two rows of the table report the results of two "oracle" RetinaNet models trained directly on the target domains. These results highlight that there is a domain gap between the two target sets. In particular, the performance drops of about 15–20% when the model is tested on a different target set (92.44% vs 77.96% in the case of training with Hololens images, and 89.69% vs 69.70% in the case of training with GoPro images). A comparison between the best performing domain adaptation approach and oracle performance also highlights that the OBJ-MDA-CH dataset is challenging and there is still space for improvement.

Table 3 reports the results obtained combining the object detectors with CycleGAN. The first two rows show the performance of Faster RCNN and RetinaNet when the training synthetic images are transformed to the merged Hololens and GoPro domain. As can be noted comparing Table 2 and Table 3, CyleGAN allows to increase the performance of Faster RCNN and RetinaNet respectively of more than 7% (7.61% vs 15.34%) and 16% (14.10% vs 31.43%) on Hololens and more than 33% (30.39% vs 63.60%) and 32% (37.13% vs 69.59%) on GoPro. These results show that the performance gap between the Hololens and GoPro target domains is reduced by CycleGAN when the images are translated to the merged domain. The remaining rows of Table 3 confirm that the proposed MDA-RetinaNet achieves better performance than the compared DA-Faster RCNN, Strong Weak and DA-RetinaNet methods (58.11% vs 32.14%, 41.11% and 52.07% on Hololens; 71.39% vs 65.19%, 66.45%, and 71.14% on GoPro).

More experiments of the proposed architecture are available in the supplementary material.

Qualitative Results. Figure 3 compares qualitatively some detection results obtained by the proposed MDA-RetinaNet model with respect to RetinaNet

Table 3. Results of feature alignment methods combined with CycleGAN. H refers to Hololens while G to GoPro. "{G, H}" refers to synthetic images translated to the merged Hololens and GoPro domains.

Model	Source	Target	Test H	Test G
Faster RCNN [19]	{G, H}	–	15.34%	63.60%
RetinaNet [8]	{G, H}	–	31.43%	69.59%
DA-Faster RCNN [2]	{G, H}	H+G merged	32.13%	65.19%
Strong Weak [20]	{G, H}	H+G merged	41.11%	66.45%
DA-RetinaNet [12]	{G, H}	H+G merged	52.07%	71.14%
MDA-RetinaNet	{G, H}	H, G	**58.11%**	**71.39%**

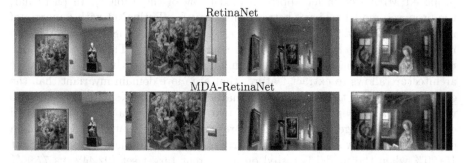

RetinaNet

MDA-RetinaNet

Fig. 3. Qualitative results of RetinaNet and MDA-RetinaNet. The blue box represents ground truth, the red and green boxes denote the predictions. The red box indicates a wrong detection (object localization or classification). The green box represents correct detections. (Color figure online)

baseline (the ground truth is the blue bounding box). We observed that, MDA-RetinaNet perfectly recognize small, large and partially occluded artworks. RetinaNet is not be able to detect some artworks (first column of the first row), it detects some false positive (second and third column of the first row) and it usually predicts incorrect bounding box (last column of the first row).

6   Conclusion

We considered the problem of multi-camera domain adaptation for object detection presenting a new dataset to study the problem which is experimentally shown to be challenging. The study has also proposed the new method MDA-RetinaNet which considers multiple targets domains. Experiments shows that MDA-RetinaNet performs better than state-of-the-art approaches. Future works can focus on the combination of different feature alignment and image to image translation methods to adapt both low- and high-level feature maps.

Acknowledgments. This research has been supported by the project VALUE (N. 08CT6209090207 - CUP G69J18001060007) - PO FESR 2014/2020 - Azione 1.1.5., by Research Program Pia.ce.ri. 2020/2022 Linea 2 - University of Catania, and by MIUR AIM - Linea 1 - AIM1893589 - CUP E64118002540007.

References

1. Chen, X., Ma, H., Wan, J., Li, B., Xia, T.: Multi-view 3d object detection network for autonomous driving. In: Proceedings of the IEEE Conference on Computer Vision and Pattern Recognition, pp. 1907–1915 (2017)
2. Chen, Y., Li, W., Sakaridis, C., Dai, D., Van Gool, L.: Domain adaptive faster R-CNN for object detection in the wild. In: Proceedings of the IEEE Conference on Computer Vision and Pattern Recognition, pp. 3339–3348 (2018)
3. Damen, D., et al.: Scaling egocentric vision: the epic-kitchens dataset. In: Ferrari, V., Hebert, M., Sminchisescu, C., Weiss, Y. (eds.) ECCV 2018. LNCS, vol. 11208, pp. 753–771. Springer, Cham (2018). https://doi.org/10.1007/978-3-030-01225-0_44
4. Deng, J., Dong, W., Socher, R., Li, L.J., Li, K., Fei-Fei, L.: Imagenet: a large-scale hierarchical image database. In: 2009 IEEE Conference on Computer Vision and Pattern Recognition, pp. 248–255. IEEE (2009)
5. Ganin, Y., Lempitsky, V.: Unsupervised domain adaptation by backpropagation. In: International Conference on Machine Learning, pp. 1180–1189 (2015)
6. Gholami, B., Sahu, P., Rudovic, O., Bousmalis, K., Pavlovic, V.: Unsupervised multi-target domain adaptation: an information theoretic approach. IEEE Trans. Image Process. **29**, 3993–4002 (2020)
7. He, K., Zhang, X., Ren, S., Sun, J.: Deep residual learning for image recognition. In: Proceedings of the IEEE Conference on Computer Vision and Pattern Recognition, pp. 770–778 (2016)
8. Lin, T.Y., Goyal, P., Girshick, R., He, K., Dollár, P.: Focal loss for dense object detection. In: Proceedings of the IEEE International Conference on Computer Vision, pp. 2980–2988 (2017)
9. Lin, T.-Y., et al.: Microsoft COCO: common objects in context. In: Fleet, D., Pajdla, T., Schiele, B., Tuytelaars, T. (eds.) ECCV 2014. LNCS, vol. 8693, pp. 740–755. Springer, Cham (2014). https://doi.org/10.1007/978-3-319-10602-1_48
10. Nguyen-Meidine, L.T., Belal, A., Kiran, M., Dolz, J., Blais-Morin, L.A., Granger, E.: Unsupervised multi-target domain adaptation through knowledge distillation. In: Proceedings of the IEEE/CVF Winter Conference on Applications of Computer Vision, pp. 1339–1347 (2021)
11. Orlando, S.A., Furnari, A., Farinella, G.M.: Egocentric visitor localization and artwork detection in cultural sites using synthetic data. Pattern Recogn. Lett. **133**, 17–24 (2020)
12. Pasqualino, G., Furnari, A., Signorello, G., Farinella, G.M.: An unsupervised domain adaptation scheme for single-stage artwork recognition in cultural sites. Image and Vision Computing (2021). https://iplab.dmi.unict.it/EGO-CH-OBJ-UDA/
13. Peng, X., Bai, Q., Xia, X., Huang, Z., Saenko, K., Wang, B.: Moment matching for multi-source domain adaptation. In: Proceedings of the IEEE International Conference on Computer Vision, pp. 1406–1415 (2019)

14. Pirsiavash, H., Ramanan, D.: Detecting activities of daily living in first-person camera views. In: 2012 IEEE Conference on Computer Vision and Pattern Recognition, pp. 2847–2854. IEEE (2012)
15. Portaz, M., Kohl, M., Quénot, G., Chevallet, J.P.: Fully convolutional network and region proposal for instance identification with egocentric vision. In: Proceedings of the IEEE International Conference on Computer Vision Workshops, pp. 2383–2391 (2017)
16. Ragab, M., et al.: Adversarial multiple-target domain adaptation for fault classification. IEEE Trans. Instrum. Meas. **70**, 1–11 (2020)
17. Ragusa, F., Furnari, A., Battiato, S., Signorello, G., Farinella, G.M.: Ego-ch: Dataset and fundamental tasks for visitors behavioral understanding using egocentric vision. Pattern Recogn. Lett. **131**, 150–157 (2020)
18. Ragusa, F., Furnari, A., Livatino, S., Farinella, G.M.: The MECCANO dataset: understanding human-object interactions from egocentric videos in an industrial-like domain. In: IEEE Winter Conference on Application of Computer Vision (WACV) (2021). https://iplab.dmi.unict.it/MECCANO
19. Ren, S., He, K., Girshick, R., Sun, J.: Faster R-CNN: towards real-time object detection with region proposal networks. In: Advances in Neural Information Processing Systems, pp. 91–99 (2015)
20. Saito, K., Ushiku, Y., Harada, T., Saenko, K.: Strong-weak distribution alignment for adaptive object detection. In: Proceedings of the IEEE Conference on Computer Vision and Pattern Recognition, pp. 6956–6965 (2019)
21. Seidenari, L., Baecchi, C., Uricchio, T., Ferracani, A., Bertini, M., Bimbo, A.D.: Deep artwork detection and retrieval for automatic context-aware audio guides. ACM Trans. Multimedia Comput. Commun. Appl. (TOMM) **13**(3s), 1–21 (2017)
22. Tzeng, E., Hoffman, J., Saenko, K., Darrell, T.: Adversarial discriminative domain adaptation. In: Proceedings of the IEEE Conference on Computer Vision and Pattern Recognition, pp. 7167–7176 (2017)
23. Varma, S., Sreeraj, M.: Object detection and classification in surveillance system. In: 2013 IEEE Recent Advances in Intelligent Computational Systems (RAICS), pp. 299–303. IEEE (2013)
24. Wei, X., Liu, S., Xiang, Y., Duan, Z., Zhao, C., Lu, Y.: Incremental learning based multi-domain adaptation for object detection. Knowledge-Based Systems **210**, 106420 (2020). https://doi.org/10.1016/j.knosys.2020.106420, https://www.sciencedirect.com/science/article/pii/S0950705120305499
25. Wu, Y., Kirillov, A., Massa, F., Lo, W.Y., Girshick, R.: Detectron2 (2019). https://github.com/facebookresearch/detectron2
26. Xie, R., Yu, F., Wang, J., Wang, Y., Zhang, L.: Multi-level domain adaptive learning for cross-domain detection. In: Proceedings of the IEEE International Conference on Computer Vision Workshops (2019)
27. Zhao, H., Zhang, S., Wu, G., Moura, J.M., Costeira, J.P., Gordon, G.J.: Adversarial multiple source domain adaptation. Adv. Neural. Inf. Process. Syst. **31**, 8559–8570 (2018)
28. Zhu, J.Y., Park, T., Isola, P., Efros, A.A.: Unpaired image-to-image translation using cycle-consistent adversarial networks. In: Proceedings of the IEEE International Conference on Computer Vision, pp. 2223–2232 (2017)

Robot Vision

Leveraging Road Area Semantic Segmentation with Auxiliary Steering Task

Jyri Maanpää[1,2]([✉]) [ID], Iaroslav Melekhov[2] [ID], Josef Taher[1,2] [ID],
Petri Manninen[1] [ID], and Juha Hyyppä[1] [ID]

[1] Department of Remote Sensing and Photogrammetry, Finnish Geospatial Research Institute FGI, National Land Survey of Finland, 02150 Espoo, Finland
`jyri.maanpaa@nls.fi`
[2] Aalto University School of Science, 02150 Espoo, Finland

Abstract. Robustness of different pattern recognition methods is one of the key challenges in autonomous driving, especially when driving in the high variety of road environments and weather conditions, such as gravel roads and snowfall. Although one can collect data from these adverse conditions using cars equipped with sensors, it is quite tedious to annotate the data for training. In this work, we address this limitation and propose a CNN-based method that can leverage the steering wheel angle information to improve the road area semantic segmentation. As the steering wheel angle data can be easily acquired with the associated images, one could improve the accuracy of road area semantic segmentation by collecting data in new road environments without manual data annotation. We demonstrate the effectiveness of the proposed approach on two challenging data sets for autonomous driving and show that when the steering task is used in our segmentation model training, it leads to a 0.1–2.9% gain in the road area mIoU (mean Intersection over Union) compared to the corresponding reference transfer learning model.

Keywords: Road area semantic segmentation · Multi-task learning · Transfer learning · Domain adaptation · Autonomous driving

1 Introduction

One of the main challenges in fully autonomous driving is that there is a huge amount of different roads and weather conditions in which all perception methods should work within an acceptable accuracy. For example, several areas have gravel roads and regular wintertime, in which the road environment looks remarkably different. Deep learning based perception methods should also be trained with these conditions for safe operation within use in traffic.

Supplementary Information The online version contains supplementary material available at https://doi.org/10.1007/978-3-031-06427-2_60.

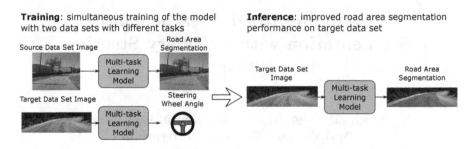

Fig. 1. A simplified summary of our method. We train a multi-task learning model simultaneously with two data sets which have separate road area segmentation and steering wheel angle prediction tasks. This improves the road area segmentation accuracy in a data set which has steering wheel angle information but no ground truth road area segmentation labels during training.

It is relatively easy to collect data from these adverse conditions with cars equipped with sensors. However, it is rather difficult to train the current perception methods without annotated training data. As training data annotation is a laborious task, one could try to tackle this problem with semi-supervised [13] or weakly supervised [2] learning methods as well as with transfer learning approaches, such as unsupervised domain adaptation [26,27]. These methods benefit from the training data with no annotation increasing the accuracy in the original perception task.

One possible solution is to find an auxiliary task which is closely related to the target perception problem and has ground truth labels that can be easily acquired. A multi-task perception model is trained to perform both original and auxiliary tasks on the data set, increasing the accuracy in all tasks if they are related to each other. Autonomous cars have several sensor outputs, such as car control metrics, recorded trajectories, different imaging or lidar sensor modalities, that can be used for auxiliary task learning. If these auxiliary tasks could be used to benefit robust perception model training, this has a potential for large scale implementation as cars equipped with sensors could collect this auxiliary task data constantly from all conditions.

In this work, we improve road area semantic segmentation with steering wheel angle prediction as an auxiliary supportive task. We chose to segment the road area as it is one of the most important semantic classes for staying on the road in adverse environmental conditions. Our hypothesis is that the steering wheel angle relates to the road boundaries in a way that the steering task supports the road area segmentation problem. The results demonstrate that the proposed multi-task model has higher accuracy than the corresponding transfer learning baseline. A simplified summary of our method is illustrated in Fig. 1.

2   Related Work

Semantic segmentation from RGB images is a widely researched topic in pattern recognition and most of the state-of-the-art methods based on convolutional neural networks (CNNs) are also trained and tested on road environment semantic segmentation data sets [1,7,14]. Some of the recently proposed data sets also include gravel road areas and different weather conditions [12,15]. DeepLab V3 Plus [5] is one of the most popular segmentation models as a starting point for different experiments, but more complex state-of-the-art models exist, such as the hierarchical multi-scale attention approach by Tao et al. [17]. We refer an interested reader to the great survey on semantic segmentation models by Lateef et al. [8].

There are several papers on transfer learning in road environment semantic segmentation, mostly related to the simulation-to-real problem. Some of them focus on the adversarial approach, in which a discriminator predicts from the operation of the model if the sample is from the source or from the target data set, thus decreasing the domain gap between the data sets [18,19]. Others may also include self-learning methods in which the target data set is self-labeled by the trained model to support the fine-tuning training to the target data set [27,28]. Some of the works also apply both approaches [26].

Semantic segmentation accuracy can also be improved by using multi-task learning. These methods are used to decrease the computational cost of running several different models for different tasks, but the tasks can also support each other during training and therefore improve the overall accuracy. In road environment semantic segmentation, the depth prediction task has been shown to improve the semantic segmentation accuracy [6,16,24]. There are several approaches to perform multi-task learning as described in the survey of Vandenhende et al. [20], but the most promising deep multi-task models are based on task-specific encoders in the network architecture, such as MTI-Net [21].

In this paper our aim is to improve road area semantic segmentation accuracy with steering wheel angle prediction task as a supportive auxiliary task. Bojarski et al. [3,4] proposed a CNN-based approach for steering wheel angle estimation using front camera images. In our previous work [10] we extended the work by M. Bojarski by utilizing lidar data in addition to camera images and by testing the method in adverse road and weather conditions. Work by Wang et al. [22] and by Xu et al. [25] are relatively close to our method as they improve car control prediction accuracy by utilizing semantic segmentation or object detection as supportive auxiliary tasks. In contrast, our work applies this idea vice versa, as we leverage steering wheel angle estimates to improve road area semantic segmentation performance.

3   Method

Our model is greatly inspired by MTI-Net [21] - a neural network architecture for multi-task learning and the transfer learning scheme proposed by Zheng et al. [26]. MTI-Net is a decoder-based multi-scale task learning model showing

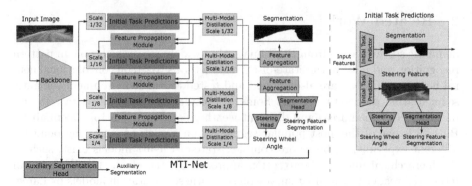

Fig. 2. Our model based on MTI-Net. On the left is the overall model and on the right a graph of a initial task prediction module. The main difference to the original MTI-Net is that one of the auxiliary tasks is predicting a steering feature, which is used to predict steering wheel angle and steering feature based segmentation in different scales. The auxiliary segmentation output is used for transfer learning purposes.

remarkable results on dense prediction problems [20]. We utilize the idea of transfer learning setup [26] since a part of it can be straightforwardly applied in our problem and decreasing the domain gap with transfer learning supports our multi-task approach. It also offers a transfer learning baseline so that the effect of auxiliary steering task could be measured. In this section we describe the details of our implementation and otherwise follow the implementations in [21] and [26].

3.1   Model Architecture

Our model performs two tasks: road area semantic segmentation and steering wheel angle prediction. As presented in Fig. 2, MTI-Net predicts initial estimates for each task output within each scale. This is done using *initial task prediction* modules, each of which uses one feature scale from the backbone network. The sum of these loss predictions for task t is called deep supervision loss $L_{Deep,t}$.

The initial task predictions are converted to task-specific features with *feature propagation modules* and forwarded to the next higher scale in the network structure. This provides information for the next scale initial task predictions so that the predictions could adapt to different task features obtained from the previous scale. Next, the task predictions from each scale are fused with *multi-modal distillation* modules that impose attention for initial task predictions based on all other task outputs. Finally, the output from different scales is combined with *feature aggregation* modules to obtain final output predictions.

However, the steering wheel angle prediction is not an image-spatial task like semantic segmentation as steering wheel angle is a scalar value. Therefore, the initial task prediction module actually outputs a steering feature instead of a steering wheel angle. This feature has the same width and height as the initial prediction of road area segmentation so that they could be fused together. When evaluating deep supervision loss, we apply a corresponding steering head in each of the steering features to predict an actual steering wheel angle. A steering head is also applied in the steering task feature aggregation module output to obtain a final steering wheel angle prediction.

The steering feature could be seen as a prior of the road shape, which is used in the steering. However, we noticed in our initial experiments that the steering feature could not correspond to road area well if it is only directly affected by the steering loss leading to a lower accuracy. Therefore, we also applied separate segmentation heads to each steering feature in the model to produce road area segmentations based on them. Within this way steering features correspond more to road area, even though they mostly contain information for steering prediction. This increases the effectiveness of the transfer from the steering task to segmentation. We call this particular segmentation output as steering feature segmentation in this work.

In our experiments, we decided to use 4 steering feature channels based on our initial tests. Examples of these feature masks are presented in the supplementary material, which also explains the architectures of the steering heads and other specific details of the model.

Note that as the input samples have images and either segmentation or steering ground truth depending on which data set gives the sample, the model is not trained simultaneously to both tasks with a same image. Each input batch during training contain samples from both data sets and the segmentation and steering losses are evaluated depending on the sample type.

3.2   Transfer Learning Scheme

Our transfer learning scheme adopted the adversarial domain adaptation and memory regularisation approaches that were introduced in the work of Zheng et al. [26]. Our method resembles the 'Stage-I' of the original work [26] as we omitted the presented self-learning approach. We consider the Mapillary Vistas data set with road area segmentation masks as the source data set and the FGI autonomous steering data set from [10] as the target data set.

As a part of the transfer learning method, we predict an auxiliary segmentation output with a segmentation head, that operates on the three highest-resolution features from the backbone. This auxiliary segmentation head has otherwise similar architecture to the segmentation feature aggregation module.

We train two discriminators that operate on the primary segmentation output from the MTI-Net and on the auxiliary segmentation output, trying to separate the samples from different data sets from each other. These impose adversarial losses on the segmentation outputs from both segmentation heads. This should improve the model operation on the target set with no ground truth segmentation. The discriminators are trained similarly as in the previous work of Zheng et al. [26], with the exception that the 'real' output used in the real-fake comparison is the primary segmentation on the source data set. The primary segmentation output is also used as the segmentation in the validation and test set results.

A) Sample with segmentation

B) Sample with steering wheel angle

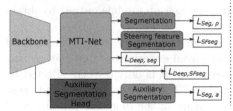
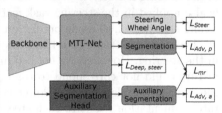

Fig. 3. Our overall training scheme. When the sample is from the Mapillary data set with segmentation (case A) we evaluate the segmentation loss for primary, auxiliary and steering feature segmentation outputs and similarly the deep supervision loss for all segmentations in different MTI-Net scales. When the sample is from the FGI autonomous steering data set with steering wheel angle (case B) we evaluate the primary steering loss, deep supervision loss for steering, adversarial losses for primary and auxiliary segmentation outputs and memory regularisation loss between primary and auxiliary segmentation outputs.

Figure 3 shows our overall training scheme for a single batch that has samples from both of our data sets. Different loss functions are evaluated on the model, depending on the input data set. When the input sample has a road area segmentation mask, we evaluate the binary cross entropy loss for all segmentation outputs in the model: the primary segmentation loss ($L_{Seg,p}$), the auxiliary segmentation loss ($L_{Seg,a}$), and steering feature based segmentation loss (L_{SFseg}). The same loss is also applied for the deep supervision, in which the overall loss is the sum of segmentation losses in each initial prediction scale ($L_{Deep,seg}$ for initial segmentation prediction and $L_{Deep,SFseg}$ for initial steering feature based segmentation). These losses are fully supervised as segmentation masks are available for the Mapillary Vistas data set.

When the input sample is from the FGI autonomous steering data set, we evaluate the mean square error of the steering wheel angle prediction L_{Steer} and a deep supervision loss $L_{Deep,steer}$ that is a sum of the mean squared errors evaluated on initial steering predictions from different scales. Following the current multi-scale discriminator approach in the source code of [26], the predicted segmentation outputs are supervised with the following LSGAN [11] adversarial loss, applied only for the target set:

$$L_{adv,i} = 2 \cdot \mathbb{E}\left[(D_i(F_i(x_t)) - 1)^2\right]. \tag{1}$$

Here x_t is the target input image batch, F_i is the segmentation head output model (primary or auxiliary) and D_i is the corresponding discriminator model. Therefore we get the adversarial losses $L_{adv,p}$ for the primary segmentation from the MTI-Net and $L_{adv,a}$ for the auxiliary segmentation.

In addition, we use the memory regularisation loss as presented in [26], which is the pixel-wise KL-divergence loss:

$$L_{mr} = -\sum_{h=1}^{H}\sum_{w=1}^{W} F_a(x_t)\log(F_p(x_t)) - \sum_{h=1}^{H}\sum_{w=1}^{W} F_p(x_t)\log(F_a(x_t)). \tag{2}$$

Here the sums act pixel-wise and F_p and F_a produce the primary and secondary segmentation masks from the target set input image x_t. This loss enforces the model to be consistent between primary and auxiliary predictions in the target set, also acting in a self-supervising manner.

As a summary, our loss in the source data set with segmentation is

$$L_{source} = L_{Seg,p} + \lambda_{aux}L_{Seg,a} + \lambda_{deep}L_{Deep,seg} \qquad (3)$$
$$+ \lambda_{SFseg}\left(L_{SFseg} + \lambda_{deep}L_{Deep,SFseg}\right)$$

and in the target set with steering wheel angles the loss is

$$L_{target} = \lambda_{steer}\left(L_{Steer} + \lambda_{deep}L_{Deep,steer}\right) \qquad (4)$$
$$+ \lambda_{adv,p}L_{adv,p} + \lambda_{adv,a}L_{adv,a} + \lambda_{mr}L_{mr}.$$

Here λ_{aux}, λ_{deep}, λ_{SFseg}, λ_{steer}, $\lambda_{adv,p}$, $\lambda_{adv,a}$, and λ_{mr} are the weighting coefficients for the auxiliary segmentation, deep supervision, steering feature segmentation, steering, adversarial for the primary and auxiliary segmentation, and the memory regularisation loss respectively. The final loss is the sum of these losses, as each mini-batch contains data from both source and target data sets.

4 Experiments

4.1 Data Sets

We use two data sets: Mapillary Vistas [14] and a data set for autonomous steering in adverse weather conditions from our previous work [10], called the FGI autonomous steering data set in this work. The Mapillary data set has multi-class ground truth segmentations for diverse driving scenarios and the FGI autonomous steering data set has 28 h of camera image sequences with corresponding steering wheel angles in a variety of road and weather conditions. We combined drivable-area classes to one road area class in the Mapillary data set. After preprocessing, the Mapillary data set has 17074 samples and the FGI autonomous steering data set has 990436 samples. More information about the data set preprocessing is provided in the supplementary material. We also annotated the drivable area in 100 images for model validation and 100 images for performance testing in the FGI autonomous steering data set. Most of these samples focus on winter conditions, there are also samples from gravel roads and during night. No training samples are used within 5 s before and after each validation or test set sample in the data set.

4.2 Training Setup

We trained three models for a performance evaluation: *(1)* **single-task (ST) model**: trained only with the segmentation task on Mapillary data set; *(2)* **transfer learning (TL) model**: utilizes the transfer learning setup, MTI-Net and deep supervision on segmentation and steering feature segmentation, but is not trained with the steering task; *(3)* **our multi-task model**: similar to

Table 1. We report road area semantic segmentation performance on the validation and test set in terms of mIoU, precision ($\mathcal{P}$), and recall ($\mathcal{R}$) for different models and backbones. The proposed approach outperforms other methods by a noticeable margin.

Model	Backbone	Validation			Test		
		mIoU	$\mathcal{P}$	$\mathcal{R}$	mIoU	$\mathcal{P}$	$\mathcal{R}$
Zheng et al. [26]	ResNet-18	85.96	92.05	92.67	87.18	94.06	92.47
ST	HRNetV2-W18	84.70	88.21	**94.28**	87.43	91.73	94.64
TL		89.04	95.45	92.70	89.71	**97.36**	92.19
Proposed		**90.44**	**95.68**	94.06	**89.79**	97.21	**92.22**
ST	FPN ResNet-18	80.13	84.62	92.99	83.84	87.56	**94.56**
TL		87.11	93.80	92.25	87.70	94.97	92.01
Proposed		**88.92**	**94.12**	**93.94**	**90.61**	**95.92**	94.38

(2) but is also supervised to predict the steering wheel angle, both with primary output and with deep supervision.

The similar structure of the last two models allow the investigation of the additional impact of the auxiliary steering task. Although the steering segmentation loss is evaluated in the transfer learning model, all actual steering related losses are not evaluated during its training.

We repeated the experiments with two backbones: High-Resolution Network (HRNet) [23] and Feature Pyramid Network [9] applied to ResNet features (FPN ResNet) as in [21]. This was done to confirm the increase in accuracy due to the auxiliary steering task. In addition, we train a stage I model from [26] as a comparison to MTI-Net based architectures, with the exceptions that our implementation has ordinary ResNet instead of dilated ResNet and does not have a dropout layer. For loss function in Eq. 3 and Eq. 4 we used the weights $\lambda_{aux} = 0.5$, $\lambda_{deep} = 1.0$, $\lambda_{SFseg} = 0.3$, $\lambda_{steer} = 0.5$, $\lambda_{adv,p} = 0.001$ and $\lambda_{adv,a} = 0.0002$. The memory regularisation loss was applied after 15000 first training steps with $\lambda_{mr} = 0.1$. Otherwise our training setup mostly follows the work in [26], more details on model training are provided in the supplementary.

4.3   Results

The performance of the models are evaluated on validation and test splits from the FGI autonomous steering data set. The results are reported as road area mean Intersection-over-Union (mIoU) values which means the fraction of the intersection and union between the predicted and actual road area pixels, averaged over validation or test set samples. The results show that although our multi-task model performs best with both backbones, the difference to the transfer learning baseline is not always significant. This is especially seen in the test set mIoU of the HRNet backbone models, in which the steering task causes less than 0.1% increase in mIoU. However, the multi-task model with FPN ResNet backbone has a 0.9% increase in test set mIoU to the HRNet transfer learning model and even 2.9% increase to the corresponding FPN ResNet transfer learning model (Table 1).

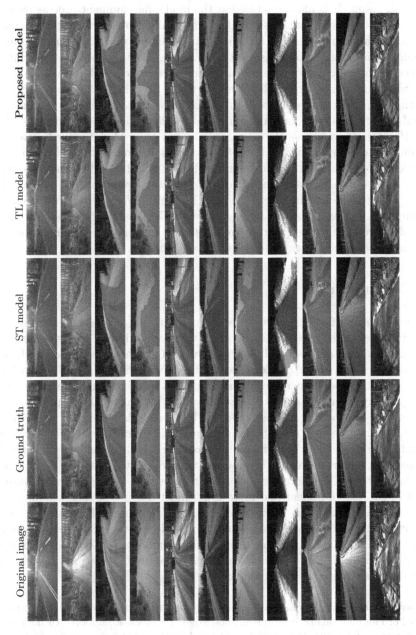

Fig. 4. Example road area segmentation masks evaluated with FPN ResNet based models on test set samples.

The difference between validation and test set results is most probably caused by the small amount of data in validation and test sets. Each set contains 100 sample images and one poorly classified sample leads to almost 1% difference in the mIoU result. It is also possible that the model development process overfitted the HRNet based multi-task model to the validation set with hyperparameter and architecture decisions. In addition, the interpretation of the road area boundaries are often ambiguous even for humans, and therefore it is difficult to observe performance differences when the mIoU accuracy is close to 90%. However, when considering both validation and test set results with both backbones, we can argue that the steering task supports the road area segmentation task, as the mIoU increases systematically and notably when utilizing steering task with the FPN ResNet backbone. We also conclude that using MTI-Net in the transfer learning baseline is justified as the reference transfer model ('Stage-I' from [26]) has smaller or equal accuracy to the FPN ResNet based transfer learning model with a similar ResNet part. Example road area segmentations for each model with FPN ResNet backbone can be seen in Fig. 4. We observe that utilizing steering information makes the segmentation mask follow the road borders more accurately, reducing false positive area outside the road. The segmentation is also more accurate far away and does not contain gaps.

5   Conclusion

We trained a multi-task model to perform both road area segmentation and steering wheel angle prediction in order to transfer road area segmentation performance to a new data set with adverse road and weather conditions. We found out that when the steering task is used in the model training, we gained a 0.1–2.9% percentage point increase in the road area mIoU when compared to the corresponding reference transfer learning model, reaching 90.6% mIoU in the test set with our best multi-task model. This means that steering wheel angle prediction could be used as an useful auxiliary task for improving road area segmentation in a new data set which only has steering wheel angle ground truth. However, there can be plenty of random variation in the results due to small test data set size. Furthermore, one should note that our approach is mostly tested in rural environment and more experiments should be made to optimize the performance of this multi-task setup and to implement it to a multi-class semantic segmentation case. One should also consider if there are other auxiliary tasks that could support road area segmentation more effectively and if there is a limit in which higher road area segmentation accuracy could not be reached due to road area interpretation ambiguity.

Acknowledgement. Academy of Finland projects (decisions 318437 and 319011) and Henry Ford foundation are gratefully acknowledged for financial support. Authors' contribution is the following: Maanpää designed and performed the experiments and wrote the manuscript. Melekhov advised in the model development and provided feedback on the manuscript. Maanpää, Taher, and Manninen took equal shares in instrumenting the autonomous driving platform and developing software. Maanpää and Taher participated in data collection. Hyyppä supervised the project.

References

1. Alhaija, H., Mustikovela, S., Mescheder, L., Geiger, A., Rother, C.: Augmented reality meets computer vision: efficient data generation for urban driving scenes. Int. J. Comput. Vis. (IJCV) **126**, 961–972 (2018)
2. Barnes, D., Maddern, W., Posner, I.: Find your own way: weakly-supervised segmentation of path proposals for urban autonomy. In: 2017 IEEE International Conference on Robotics and Automation (ICRA), pp. 203–210. IEEE (2017)
3. Bojarski, M., et al.: The nvidia pilotnet experiments. arXiv preprint arXiv:2010.08776 (2020)
4. Bojarski, M., et al.: End to end learning for self-driving cars. arXiv preprint arXiv:1604.07316 (2016)
5. Chen, L.-C., Zhu, Y., Papandreou, G., Schroff, F., Adam, H.: Encoder-decoder with atrous separable convolution for semantic image segmentation. In: Ferrari, V., Hebert, M., Sminchisescu, C., Weiss, Y. (eds.) ECCV 2018. LNCS, vol. 11211, pp. 833–851. Springer, Cham (2018). https://doi.org/10.1007/978-3-030-01234-2_49
6. Chennupati, S., Sistu, G., Yogamani, S., Rawashdeh, S.: Auxnet: Auxiliary tasks enhanced semantic segmentation for automated driving. arXiv preprint arXiv:1901.05808 (2019)
7. Cordts, M., et al.: The cityscapes dataset for semantic urban scene understanding. In: Proceedings of the IEEE Conference on Computer Vision and Pattern Recognition (CVPR), pp. 3213–3223. IEEE (2016)
8. Lateef, F., Ruichek, Y.: Survey on semantic segmentation using deep learning techniques. Neurocomputing **338**, 321–348 (2019)
9. Lin, T.Y., Dollár, P., Girshick, R., He, K., Hariharan, B., Belongie, S.: Feature pyramid networks for object detection. In: Proceedings of the IEEE Conference on Computer Vision and Pattern Recognition (CVPR), pp. 2117–2125. IEEE (2017)
10. Maanpää, J., Taher, J., Manninen, P., Pakola, L., Melekhov, I., Hyyppä, J.: Multimodal end-to-end learning for autonomous steering in adverse road and weather conditions. In: 2020 25th International Conference on Pattern Recognition (ICPR), pp. 699–706. IEEE (2021)
11. Mao, X., Li, Q., Xie, H., Lau, R.Y., Wang, Z., Paul Smolley, S.: Least squares generative adversarial networks. In: Proceedings of the IEEE International Conference on Computer Vision (ICCV), pp. 2794–2802. IEEE (2017)
12. Metzger, K.A., Mortimer, P., Wuensche, H.J.: A fine-grained dataset and its efficient semantic segmentation for unstructured driving scenarios. In: 2020 25th International Conference on Pattern Recognition (ICPR), pp. 7892–7899. IEEE (2021)
13. Mittal, S., Tatarchenko, M., Brox, T.: Semi-supervised semantic segmentation with high-and low-level consistency. IEEE Trans. Pattern Anal. Mach. Intell. **43**(4), 1369–1379 (2019)
14. Neuhold, G., Ollmann, T., Rota Bulo, S., Kontschieder, P.: The mapillary vistas dataset for semantic understanding of street scenes. In: Proceedings of the IEEE International Conference on Computer Vision (ICCV), pp. 4990–4999. IEEE (2017)
15. Pitropov, M., et al.: Canadian adverse driving conditions dataset. Int. J. Robot. Res. **40**(4–5), 681–690 (2021)
16. Sener, O., Koltun, V.: Multi-task learning as multi-objective optimization. Adv. Neural. Inf. Process. Syst. **31**, 527–538 (2018)

17. Tao, A., Sapra, K., Catanzaro, B.: Hierarchical multi-scale attention for semantic segmentation. arXiv preprint arXiv:2005.10821 (2020)
18. Tsai, Y.H., Hung, W.C., Schulter, S., Sohn, K., Yang, M.H., Chandraker, M.: Learning to adapt structured output space for semantic segmentation. In: Proceedings of the IEEE Conference on Computer Vision and Pattern Recognition (CVPR), pp. 7472–7481. IEEE (2018)
19. Tsai, Y.H., Sohn, K., Schulter, S., Chandraker, M.: Domain adaptation for structured output via discriminative patch representations. In: Proceedings of the IEEE/CVF International Conference on Computer Vision (ICCV), pp. 1456–1465. IEEE (2019)
20. Vandenhende, S., Georgoulis, S., Gansbeke, W.V., Proesmans, M., Dai, D., Gool, L.V.: Multi-task learning for dense prediction tasks: a survey. IEEE Trans. Pattern Anal. Mach. Intell. 1 (2021)
21. Vandenhende, S., Georgoulis, S., Van Gool, L.: MTI-Net: multi-scale task interaction networks for multi-task learning. In: Vedaldi, A., Bischof, H., Brox, T., Frahm, J.-M. (eds.) ECCV 2020. LNCS, vol. 12349, pp. 527–543. Springer, Cham (2020). https://doi.org/10.1007/978-3-030-58548-8_31
22. Wang, D., Wen, J., Wang, Y., Huang, X., Pei, F.: End-to-end self-driving using deep neural networks with multi-auxiliary tasks. Automot. Innov. **2**(2), 127–136 (2019)
23. Wang, J., et al.: Deep high-resolution representation learning for visual recognition. IEEE Trans. Pattern Anal. Mach. Intell. **43**(10), 3349–3364 (2020)
24. Xu, D., Ouyang, W., Wang, X., Sebe, N.: PAD-Net: multi-tasks guided prediction-and-distillation network for simultaneous depth estimation and scene parsing. In: Proceedings of the IEEE Conference on Computer Vision and Pattern Recognition (CVPR), pp. 675–684. IEEE (2018)
25. Xu, H., Gao, Y., Yu, F., Darrell, T.: End-to-end learning of driving models from large-scale video datasets. In: Proceedings of the IEEE Conference on Computer Vision and Pattern Recognition (CVPR), pp. 2174–2182. IEEE (2017)
26. Zheng, Z., Yang, Y.: Unsupervised scene adaptation with memory regularization in vivo. In: Proceedings of the 29th International Joint Conference on Artificial Intelligence, (IJCAI), pp. 1076–1082. International Joint Conferences on Artificial Intelligence Organization (2020)
27. Zou, Y., Yu, Z., Vijaya Kumar, B.V.K., Wang, J.: Unsupervised domain adaptation for semantic segmentation via class-balanced self-training. In: Ferrari, V., Hebert, M., Sminchisescu, C., Weiss, Y. (eds.) ECCV 2018. LNCS, vol. 11207, pp. 297–313. Springer, Cham (2018). https://doi.org/10.1007/978-3-030-01219-9_18
28. Zou, Y., Yu, Z., Liu, X., Kumar, B., Wang, J.: Confidence regularized self-training. In: Proceedings of the IEEE/CVF International Conference on Computer Vision (ICCV), pp. 5982–5991. IEEE (2019)

Embodied Navigation at the Art Gallery

Roberto Bigazzi[(✉)], Federico Landi, Silvia Cascianelli,
Marcella Cornia, Lorenzo Baraldi, and Rita Cucchiara

University of Modena and Reggio Emilia, Modena, Italy
{Roberto.Bigazzi,Federico.Landi,Silvia.Cascianelli,Marcella.Cornia,
Lorenzo.Baraldi,Rita.Cucchiara}@unimore.it

Abstract. Embodied agents, trained to explore and navigate indoor
photorealistic environments, have achieved impressive results on standard datasets and benchmarks. So far, experiments and evaluations have
involved domestic and working scenes like offices, flats, and houses. In
this paper, we build and release a new 3D space with unique characteristics: the one of a complete art museum. We name this environment ArtGallery3D (AG3D). Compared with existing 3D scenes, the collected space is ampler, richer in visual features, and provides very sparse
occupancy information. This feature is challenging for occupancy-based
agents which are usually trained in crowded domestic environments with
plenty of occupancy information. Additionally, we annotate the coordinates of the main points of interest inside the museum, such as paintings,
statues, and other items. Thanks to this manual process, we deliver a new
benchmark for PointGoal navigation inside this new space. Trajectories
in this dataset are far more complex and lengthy than existing ground-
truth paths for navigation in Gibson and Matterport3D. We carry on
extensive experimental evaluation using our new space for evaluation
and prove that existing methods hardly adapt to this scenario. As such,
we believe that the availability of this 3D model will foster future research
and help improve existing solutions.

Keywords: Embodied AI · Visual navigation · Sim2Real

1 Introduction

In recent years, Embodied AI has benefited from the introduction of rich datasets
of 3D spaces and new tasks, ranging from exploration to PointGoal or Image-
Goal navigation [9,26]. Such availability of 3D data allows to train and deploy
modular embodied agents, thanks to powerful simulation platforms [23]. Despite
the high number of available spaces, though, the topology and nature of the
different scenes have low variance. Indeed, many environments represent apart-
ments, offices, or houses. In this paper, we take a different path and collect and
introduce the 3D space of an art gallery.

Current agents for embodied exploration feature a modular approach [5,10,
20]. While the agents are trained for embodied exploration using deep reinforcement learning, this hierarchical paradigm allows for great adaptability on

© The Author(s), under exclusive license to Springer Nature Switzerland AG 2022
S. Sclaroff et al. (Eds.): ICIAP 2022, LNCS 13231, pp. 739–750, 2022.
https://doi.org/10.1007/978-3-031-06427-2_61

downstream tasks. Hence, models trained to explore the Gibson dataset can solve PointGoal navigation with satisfactory accuracy under the appropriate hypotheses. Furthermore, accurate and realistic simulating platforms such as Habitat [23] facilitate the deployment in the real world of the trained agents [7,15]. While agent architectures and simulating platforms are possible sources of improvement, there is a third important direction of research that regards the availability of 3D scenes to train and test the different agents. Indeed, the nature of the different environments influences the variety of tasks that the agent can learn and perform.

In this work, we contribute to this third direction by collecting and presenting a previously unseen type of 3D space, *i.e.*, a museum. This new environment for embodied exploration and navigation, named ArtGallery3D (AG3D), presents unique features when compared to flats and offices. First, the dimension of the rooms drastically increases, and the same goes for the size of the building itself. In our 3D model, some rooms are as big as 20×15 m, while the floor hosting the art gallery spans a total of 2,000 square meters. However, dimensions are not the only difference with current available 3D spaces. As a second factor, the presented gallery is incredibly rich in visual features, offering multiple paintings, sculptures, and rare objects of historical and artistic interest. Every item represents a unique point of interest, and this is in contrast to traditional scenes where all elements have approximately the same visual relevance. Finally, the museum has sparse occupancy information. Many agents count on depth information to plan short-term displacements. However, when placed in the middle of an open empty hall, depth information is less informative. In our challenging 3D scene, the agent must learn to combine RGB and depth information and not be overconfident on immediately available knowledge on the occupancy map. All these challenges make our newly-proposed 3D space a valuable asset for current and future research.

Together with the 3D model of the museum, we present a dataset for embodied exploration and navigation. For the navigation task, we annotate the position of most of the points of interest in the museum. Examples include numerous paintings, sculptures, and other relevant objects. Finally, we present an experimental analysis including the performance of existing architectures on this novel benchmark and a discussion of potential future research directions made possible by the presence of the collected 3D space.

2   Related Work

Both autonomous robotics [5,14] and embodied AI [6,8,11,12,18,21] have recently witnessed a boost of interest, which has been enabled by the release of photorealistic 3D simulated environments. In such environments, algorithms for intelligent exploration and navigation can be developed safely and more quickly than in the real-world, before being easily deployed on real robotic platforms [2,7,15]. Among the datasets of spaces, the most commonly used are MP3D [9], Gibson [26], HM3D [22], and Replica [25]. These datasets mainly contain house-like and office-like environments, with some environments taken from shops, garages, churches, and restaurants. Rooms in such environments are generally cluttered, and thus, rich of landmarks and texture information that the agent can exploit while navigating. In contrast, the presented AG3D environment has been collected in a museum, with larger, uncluttered spaces.

Algorithms developed in the simulated environments are typically trained with deep reinforcement learning, both for exploration and navigation tasks. The exploration task, which is often tackled to allow other downstream navigation tasks [27,28], consists in letting an agent equipped with visual sensors (*i.e.*, RGB-D cameras) freely navigate the environment to gather as much information as possible, usually in the form of an occupancy map. To this end, intrinsic rewards have been proposed, which can be based on novelty, curiosity, reconstruction enabling, and coverage [11,20,21]. For the navigation tasks, the agent is deployed in an unknown environment (*i.e.*, no map provided) and given some assignments in visual or textual form. These tasks include PointGoal navigation [1], where the robot is expected to reach a coordinates-specified goal, ImageGoal navigation [29], where the robot must reach an observation point in the environment that matches an image-specified goal, and ObjectGoal navigation [4], where the robot is asked to get to any instance of a label-specified object in the environment. Other related tasks involve embodied question answering [13] and vision-and-language navigation [3,16,17], where the robot must follow a natural language instruction to reach the goal. The environment presented in this paper is used for the exploration task and for the PointNav task. In this latter case, we define a variant in which the goal is expressed in terms of both coordinates and orientation.

3   ArtGallery3D (AG3D) Dataset

Existing datasets for indoor navigation comprise 3D acquisitions of different types of buildings, ranging from private houses, that cover the majority of the scenes, to offices and shops. Nevertheless, the focus of these datasets is on private spaces and there is low variance in terms of dimension and contained objects. In fact, to the best of our knowledge, among the publicly available datasets, no acquired indoor environment is composed of large rooms with a low occupied/free space ratio as in a museum. To overcome this deficiency in current literature we release a new indoor dataset for exploration and navigation captured inside a museum environment, called AG3D[1].

Acquisition. To build the 3D model of the art gallery, we employ a Matterport camera[2] and related software. This technology is the same employed to collect Matterport3D and HM3D datasets of spaces [9,22] and is particularly suitable to capture indoor photorealistic environments. We place the camera in the physical

---

[1] The dataset has been collected at the Galleria Estense museum of Modena and can be found at https://github.com/aimagelab/ag3d.

[2] https://matterport.com/it/cameras/pro2-3D-camera.

Fig. 1. On the left: a view of the 3D model of the acquired environment. On the right: images captured during the acquisition of the scene.

environment and capture a 360° RGB-D image of the surrounding. Then, we repeat the same process after moving the camera approximately 1.5 m away. Using consecutive panoramic acquisitions, the software is able to compute the 3D geometry of the space using depth information and the correspondences between the same keypoints in different acquisitions. To capture the entire museum, we make 232 different scans. Thanks to the high number of acquisitions, we are able to reproduce fine geometric and visual details of the original space (see Fig. 1). The resulting 3D model consists of more than 1430 m^2 of navigable space.

Dataset Details. The proposed dataset allows two different tasks: exploration and navigation. Episodes for the exploration task include starting position and orientation of the agent which are sampled uniformly over the entire navigable space. The navigation dataset, instead, extends traditional PointGoal navigation where episodes are defined with a starting pose and a goal coordinate, including an additional final orientation vector. Conceptually, we can consider this setting as the link between PointGoal navigation and ImageGoal navigation since the goal is to rotate the agent towards a precise objective/scene, specifying the goal using coordinates instead of an image. We name this new setting PointGoal++ navigation (PointNav++). To create the navigation dataset we annotate 147 points of interest mostly consisting of paintings and statues. The annotated goal position is around 1 m in front of the artwork and the goal orientation vector is directed to its center. For each point of interest, we define three episodes with different difficulties based on the geodesic distance between start and goal positions: easy (<15 m), medium (>15 m), and difficult (>30 m). In particular, thanks to the dimension of the acquired environment, each difficult episode has a geodesic distance larger than the longest path of MatterPort3D and Gibson datasets. A comparison of the geodesic distance distribution of the episodes of various available PointGoal navigation datasets is presented in Fig. 2. The introduction of AG3D enables the evaluation of agents on long navigation episodes which were previously not possible and highlights the inaccuracy of components of the architecture that accumulate error over time. The exploration task dataset contains 500k, 100, 1000 episodes respectively for training, validation, and test, while the PointNav++ dataset includes 411 annotated navigation episodes.

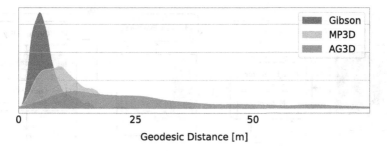

Fig. 2. Comparison of the distribution of the geodesic distances from starting position to goal position of the episode for different datasets.

4   Architecture

We provide an experimental analysis comparing recently proposed approaches on the devised environment, both for exploration and PointNav++ tasks. The evaluated methods are consistent with recent literature on embodied AI [5,10,20] and adopt an architecture shown in Fig. 3, which is composed of a neural mapper, a pose estimator, and a hierarchical navigation policy. The mapper generates a representation of the environment while the agent moves, the pose estimator is in charge of locating the agent in the environment, and the policy is responsible for the movement capabilities of the agent. The core difference between the evaluated approaches resides in the navigation policy, as described in the following. For further details, we refer the reader to the original papers.

4.1   Mapper

The mapper module incrementally builds an occupancy grid map of the environment in parallel with the navigation task. At each timestep, the RGB-D observations (s_t^{rgb}, s_t^d) coming from the visual sensors are processed to extract a $L \times L \times 2$ agent-centric map m_t where the channels indicate, respectively, the occupancy and exploration state of the currently observed region, and each pixel of the map describes the state of an area of 5×5 cm. The RGB observation is encoded using a ResNet-18 followed by a UNet, while the depth observation is encoded using another UNet. The features extracted from the two modalities are combined using CNNs at different levels of the output of the two UNet encoders and a final UNet decoder is used to process the combined features to retrieve the resulting local map m_t. Following the method proposed in [20], our mapper is not limited to predicting the occupancy map of the visible space but tries to infer also occluded and not visible regions of the local map. The global level map of the environment M_t has a dimensionality of $G \times G \times 2$, where $G > L$, and is built using local maps m_t step-by-step. At each timestep the pose of the agent x_t is used to apply a rototranslation to the local map, then, the transformed local map is finally registered to the global map M_t with a moving average.

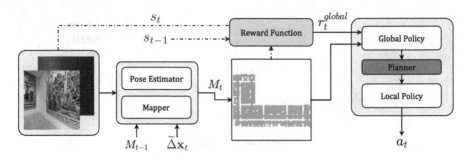

Fig. 3. Overall architecture of the models employed for exploration and navigation on AG3D.

4.2 Pose Estimator

In order to create a coherent representation of the environment during navigation, a precise and robust pose estimation needs to be achieved. To address problems like noise in the sensors and collisions with obstacles, we adopt a pose estimator and avoid the direct use of sensor readings. The pose estimator computes the pose of the agent $\mathbf{x}_t = (x_t, y_t, \theta_t)$ where (x_t, y_t) and θ_t are its position and orientation in the internal representation of the environment. The pose estimate $\mathbf{x}_t$ is computed in an incremental way, adding the displacement $\Delta\mathbf{x}_t$ caused by the action a_t to the current pose estimate $\mathbf{x}_t$. In order to retrieve a first noisy estimate of the displacement $\widetilde{\Delta}\mathbf{x}_t$, we use the difference between consecutive readings of the pose sensor $(\widetilde{\mathbf{x}}_{t-1}, \widetilde{\mathbf{x}}_t)$. To account for errors in the sensor displacement, consecutive local maps (m_{t-1}, m_t) coming from the mapper are used as feedback; m_{t-1} is reprojected using $\widetilde{\Delta}\mathbf{x}_t$ to the same point of view of m_t and the concatenation of the transformed m_{t-1} and m_t is processed using a CNN to retrieve the final robust pose displacement $\Delta\mathbf{x}_t$. At each timestep $\Delta\mathbf{x}_t$ is used to compute the pose of the robot $\mathbf{x}_t$:

$$\mathbf{x}_t = \mathbf{x}_{t-1} + \Delta\mathbf{x}_t, \tag{1}$$

where we assume $\mathbf{x}_0 = (0, 0, 0)$ without loss of generality and $\mathbf{x}_0$ corresponds to the center of the map M_t with the agent facing north.

4.3 Navigation Policy

The navigation policy is the module that determines the movement of the agent in the environment. Its hierarchical design is required in order to allow the agent to uncouple high-level navigation concepts, such as moving across different rooms, and low-level concepts, like obstacle avoidance. The navigation policy is defined by a three-component module consisting of a global farsighted policy, a deterministic planner, and a local policy for atomic action inference.

Global Policy. The global policy is the high-level component of the navigation policy and is responsible for extracting a long-term goal on the global map g_t.

The global policy takes as input an enriched $G \times G \times 4$ current global map M_t^+ retrieved stacking the two-channel global map M_t, the one-hot representation of the current position on the map, and the map of the already visited states. M_t^+ is in parallel cropped with respect to the position of the agent and max-pooled to a lower dimensionality $H \times H \times 4$. These two versions of M_t^+ are stacked together to obtain the final $H \times H \times 8$ input of the global policy. A CNN is used to sample a point of a $H \times H$ grid that is converted to a goal position on the global map g_t. The global policy is trained with reinforcement learning using PPO [24] to maximize different rewards used in literature.

In the experiments, we employ and compare different reward methods, namely Coverage, Anticipation, and Curiosity. The Coverage reward [10, 21] maximizes the information gathered at each time-step, expressed in terms of the area seen. The Anticipation reward [20] is defined by comparing the predicted local occupancy map with the ground-truth considering also occluded areas. The Curiosity reward [19] encourages the agent towards areas that maximize the prediction error of a model trained to predict future states, thus improving the learning of the dynamics of the environment.

Planner. Given the global goal on the map, the planner has the task of computing a short-term goal on the map that the agent should reach. We employ an A* algorithm on the global map M_t to plan a path from the current position of the agent to the global goal and a local goal l_t is computed on the obtained trajectory within a distance D from the agent.

Local Policy. The local policy is the module that allows the movement of the agent in the environment and its objective is to reach the local goal l_t determined by the planner. The input of the local policy, formed by the relative displacement from the position of the agent to the local goal l_t and the current RGB observation s_t^{rgb}, is processed to compute an atomic action a_t. The available actions are: *move ahead 0.25 m, turn left* $10°$, *turn right* $10°$, with the addition of a *stop* action when performing the navigation task. During training with reinforcement learning, the reward of the local policy r_t^{local} encourages the agent to reduce the distance from the local goal:

$$r_t^{local}(\mathbf{x}_t, \mathbf{x}_{t+1}) = d(\mathbf{x}_t) - d(\mathbf{x}_{t+1}), \tag{2}$$

where $d(\mathbf{x}_t)$ is the euclidean distance between the agent and the local goal l_t at timestep t. Following the hierarchical design, the global goal is sampled every N_G timesteps, while the local goal is reset if a new global goal is sampled, if the previous local goal is found to be in an occupied area, or if the previous local goal has been reached.

5 Experiments

We perform experiments on the proposed dataset comparing various models trained with different global rewards on another dataset, with models trained from scratch or finetuned on AG3D on exploration and PointNav++ to evaluate the performance gap between these approaches and highlight the difference between the characteristics of AG3D compared to other datasets. A sample episode of PointGoal++ navigation of ArtGallery3D is shown in Fig. 4.

Fig. 4. An episode of PointNav++ in AG3D where consecutive frames have a distance of 10 timesteps approximately. The red frame indicates the stop action in the traditional PointNav task. The green frame corresponds to the stop action in PointNav++. (Color figure online)

5.1   Experimental Setting

Evaluation Protocol. The baselines are trained with Coverage, Anticipation, and Curiosity rewards on the Gibson dataset for ≈5M frames corresponding to 12 GPU-days on NVIDIA V100. The best performing approach among the baseline is also both trained from scratch and finetuned, but since high-quality textures and memory occupancy of AG3D do not allow training with the same number of environments in parallel as Gibson, we trained the model from scratch on AG3D with the same GPU time for ≈2.8M frames, while the finetuned model is trained for ≈1M additional frames.

For the exploration task we evaluate the following metrics: **IoU** (Intersection-over-Union) between the map built during at end of the episode and the ground-truth map. **Acc** measures the correctly reconstructed map in m^2. **AS** indicates the area seen by the agent during exploration (in m^2). **FIoU**, **OIoU**, **FAS**, and **OAS** measure, respectively, IoU and area seen for free and occupied portions of the environment. **TE** and **AE** are the translation and angular error between estimated and ground-truth pose measured respectively in meter and degrees. PointGoal++ navigation is evaluated considering these metrics: **D2G** (Distance to Goal) and **OE** (Orientation Error) are the mean geodesic distance to the goal and the mean orientation error at the end of the episode. The orientation error is computed considering the vector between the center of the artwork and the position of the agent as ground-truth. **SR** (Success Rate) is the percentage of episodes terminated successfully. In PointNav++ the agent needs to be within 0.2 m to the goal and with an orientation error lower than $10°$ C. **PGSR** (PointNav Success Rate) and **ASR** (Angular Success Rate) consider only one component of SR; respectively D2G and OE. **SPL** and **SoftSPL** are success rates weighted on the length of the trajectory of the agent.

Implementation Details. The experiments are performed extracting 128×128 RGB-D observations from the acquired 3D model using the Habitat simulator. The maximum length of the exploration episodes during training is set to $T = 500$. Regarding the mapping process, we set $L = 101$ and $G = 2881$ for the local and global map dimensionalities. The action space grid $H \times H$ of the global policy is 240×240. The maximum distance of the local goal l_t from the position of the agent is $D = 0.5$ m for exploration and $D = 0.25$ m for PointNav++.

Table 1. Exploration results over the 100 episodes of the AG3D validation split in noise-free and noisy conditions.

Model	Training	IoU ↑	FIoU ↑	OIoU ↑	Acc ↑	AS ↑	FAS ↑	OAS ↑	TE ↓	AE ↓
Noise-Free										
Anticipation [20]	Gibson	0.163	0.170	0.157	294.4	290.6	258.3	32.3	0.0	0.0
Curiosity [19]	Gibson	0.175	0.184	0.166	317.9	317.5	281.7	35.8	0.0	0.0
Coverage [10,21]	Gibson	0.214	0.237	0.191	403.1	384.3	341.1	43.2	0.0	0.0
Coverage [10,21]	AG3D	0.219	0.239	0.200	400.6	354.6	316.6	38.0	0.0	0.0
Coverage [10,21]	Gibson+AG3D	**0.296**	**0.313**	**0.278**	**531.8**	**470.2**	**418.1**	**52.1**	0.0	0.0
Noisy										
Anticipation [20]	Gibson	0.144	0.157	**0.131**	269.6	281.7	249.9	31.9	**0.48**	**2.95**
Curiosity [19]	Gibson	0.119	0.151	0.086	251.1	307.3	272.8	34.5	2.62	15.99
Coverage [10,21]	Gibson	0.148	0.203	0.093	327.0	380.7	337.9	42.8	2.98	12.54
Coverage [10,21]	AG3D	0.144	0.200	0.088	320.0	356.4	317.3	39.2	2.65	13.35
Coverage [10,21]	Gibson+AG3D	**0.191**	**0.266**	0.116	**427.3**	**461.9**	**413.2**	**48.7**	2.68	10.16

5.2 Experimental Results

Exploration Results. As a first experiment, in Table 1 we compare the considered models on the exploration task on the AG3D validation set. Each exploration episode has a length of $T = 1000$ timesteps during which the agent has to disclose the initially unknown environment. Among the baselines trained only on the Gibson dataset, the Coverage-based model achieves the best results in terms of IoU and Area Seen in both noise-free and noisy settings. The model trained with Coverage from scratch obtains competitive results even using fewer training frames (2.8M vs. 5M), showing the importance of adapting the models to AG3D. This conclusion is supported by the fact that the model trained on Gibson and finetuned for 1M frames on AG3D achieves the best results on noise-free and noisy settings, with a significant margin on the second-best model. In both settings, the performance gap in terms of Area Seen (85.9 m^2 and 81.2 m^2) and IoU (0.082 and 0.043) between the best models trained only on Gibson dataset and using AG3D denotes the need of adapting the weight of the models to the different visual characteristics and occupation of AG3D.

Table 2. PointNav++ results on the AG3D navigation episodes under noise-free and noisy settings.

Model	Training	SPL ↑	SoftSPL ↑	SR ↑	PNSR ↑	ASR ↑	Steps ↓	D2G ↓	OE ↓
Noise-Free									
Anticipation [20]	Gibson	0.697	0.780	0.803	0.873	0.808	364.3	4.131	12.2
Curiosity [19]	Gibson	0.625	0.706	0.732	0.803	0.732	416.4	7.934	17.0
Coverage [10, 21]	Gibson	0.760	0.838	0.876	0.954	0.883	314.6	0.700	5.2
Coverage [10, 21]	AG3D	**0.805**	**0.875**	**0.898**	**0.973**	**0.908**	**270.3**	**0.268**	**4.8**
Coverage [10, 21]	Gibson+AG3D	0.793	0.873	0.883	0.964	0.891	273.1	0.323	5.0
Noisy									
Anticipation [20]	Gibson	0.211	0.788	0.224	0.255	0.387	338.6	3.152	32.2
Curiosity [19]	Gibson	0.225	0.655	0.243	0.275	0.341	446.3	9.746	38.7
Coverage [10, 21]	Gibson	0.228	0.783	0.243	0.260	0.392	348.6	3.165	34.1
Coverage [10, 21]	AG3D	0.235	0.832	0.248	0.273	0.445	306.2	2.420	28.8
Coverage [10, 21]	Gibson+AG3D	**0.373**	**0.853**	**0.399**	**0.443**	**0.543**	**283.8**	**1.430**	**19.8**

PointNav++ Results. Moving on to the navigation task, models trained on exploration substitute the global goal with a fixed goal specified by the navigation episode. Experimental results on PointNav++, shown in Table 2, present a similar trend as on the exploration task. In fact, the Coverage model has the best results in terms of SPL and Success Rate related metrics among the models trained only on Gibson. Moving to the Coverage models trained on AG3D, in the noise-free setting, the model trained from scratch achieves the best results even in comparison to the finetuned counterpart which is trained with more than double the total observations (2.8M vs 6M). This behavior can be explained by the performance of its mapper that is trained for more frames using visual observation from AG3D (2.8M vs 1M) and extracts a more detailed map sacrificing robustness and generalization. Accordingly, in the noisy setting, the higher robustness of the Coverage-based model finetuned on AG3D regains the first place with a noteworthy margin on the other models, while the Coverage model trained from scratch goes down to the second position in terms of SPL and SR. As in the case of the exploration task, the performance gap between models trained on Gibson and using AG3D (0.045 and 0.145 for SPL in noise-free and noisy settings) stresses the importance of adapting the parameters to the features extracted from AG3D. Moreover, it is worth noting that the gap of the best model from noise-free to noisy navigation (0.432 for SPL) is a consequence of the length of the navigation episodes of AG3D, and the difficulty of performing precise lengthy trajectories in the presence of noise. This is an interesting aspect that the AG3D dataset offers for exploration in future works.

6   Conclusion

In this work, we introduced the AG3D photorealistic 3D dataset for embodied exploration and PointGoal navigation tasks. The dataset has been collected in an art gallery, which features larger and more uncluttered spaces compared to most of the environments available in commonly used benchmark datasets. For the PointNav task, we propose a variant that is more suitable to the type of environment in the AG3D dataset. The variant entails not only reaching the specified coordinates, as in standard PointNav but also assuming a specified orientation. We also present an experimental comparison of state-of-the-art approaches on the devised dataset, which can serve as baselines for future research on embodied AI tasks performed in museum-like environments.

Acknowledgement. This work has been supported by "Fondazione di Modena" and the "European Training Network on PErsonalized Robotics as SErvice Oriented applications" (PERSEO) MSCA-ITN-2020 project.

References

1. Anderson, P., et al.: On evaluation of embodied navigation agents. arXiv preprint arXiv:1807.06757 (2018)
2. Anderson, P., et al..: Sim-to-real transfer for vision-and-language navigation. In: CoRL (2021)
3. Anderson, P., et al.: Vision-and-language navigation: Interpreting visually-grounded navigation instructions in real environments. In: CVPR (2018)
4. Batra, D., et al.: Objectnav revisited: On evaluation of embodied agents navigating to objects. arXiv preprint arXiv:2006.13171 (2020)
5. Bigazzi, R., Landi, F., Cascianelli, S., Baraldi, L., Cornia, M., Cucchiara, R.: Focus on impact: indoor exploration with intrinsic motivation. RA-L (2022)
6. Bigazzi, R., Landi, F., Cornia, M., Cascianelli, S., Baraldi, L., Cucchiara, R.: Explore and explain: self-supervised navigation and recounting. In: ICPR (2020)
7. Bigazzi, R., Landi, F., Cornia, M., Cascianelli, S., Baraldi, L., Cucchiara, R.: Out of the box: embodied navigation in the real world. In: CAIP (2021)
8. Cascianelli, S., Costante, G., Ciarfuglia, T.A., Valigi, P., Fravolini, M.L.: Full-GRU natural language video description for service robotics applications. RA-L **3**(2), 841–848 (2018)
9. Chang, A., et al.: Matterport3D: learning from RGB-D data in indoor environments. In: 3DV (2017)
10. Chaplot, D.S., Gandhi, D., Gupta, S., Gupta, A., Salakhutdinov, R.: Learning to explore using active neural SLAM. In: ICLR (2019)
11. Chen, T., Gupta, S., Gupta, A.: Learning exploration policies for navigation. In: ICLR (2019)
12. Cornia, M., Baraldi, L., Cucchiara, R.: Smart: training shallow memory-aware transformers for robotic explainability. In: ICRA (2020)
13. Das, A., Datta, S., Gkioxari, G., Lee, S., Parikh, D., Batra, D.: Embodied question answering. In: CVPR (2018)
14. Irshad, M.Z., Ma, C.Y., Kira, Z.: Hierarchical cross-modal agent for robotics vision-and-language navigation. In: ICRA (2021)
15. Kadian, A., et al.: Sim2real predictivity: does evaluation in simulation predict real-world performance? RA-L **5**(4), 6670–6677 (2020)
16. Krantz, J., Wijmans, E., Majumdar, A., Batra, D., Lee, S.: Beyond the Nav-graph: vision-and-language navigation in continuous environments. In: Vedaldi, A., Bischof, H., Brox, T., Frahm, J.-M. (eds.) ECCV 2020. LNCS, vol. 12373, pp. 104–120. Springer, Cham (2020). https://doi.org/10.1007/978-3-030-58604-1_7

17. Landi, F., Baraldi, L., Cornia, M., Corsini, M., Cucchiara, R.: Multimodal attention networks for low-level vision-and-language navigation. CVIU **210**, 103255 (2021)
18. Niroui, F., Zhang, K., Kashino, Z., Nejat, G.: Deep reinforcement learning robot for search and rescue applications: exploration in unknown cluttered environments. RA-L **4**(2), 610–617 (2019)
19. Pathak, D., Agrawal, P., Efros, A.A., Darrell, T.: Curiosity-driven exploration by self-supervised prediction. In: ICML (2017)
20. Ramakrishnan, S.K., Al-Halah, Z., Grauman, K.: Occupancy anticipation for efficient exploration and navigation. In: Vedaldi, A., Bischof, H., Brox, T., Frahm, J.-M. (eds.) ECCV 2020. LNCS, vol. 12350, pp. 400–418. Springer, Cham (2020). https://doi.org/10.1007/978-3-030-58558-7_24
21. Ramakrishnan, S.K., Jayaraman, D., Grauman, K.: An exploration of embodied visual exploration. Int. J. Comput. Vis. **129**(5), 1616–1649 (2021). https://doi.org/10.1007/s11263-021-01437-z
22. Ramakrishnan, S.K., et al.: Habitat-matterport 3d dataset (HM3d): 1000 large-scale 3D environments for embodied AI. In: NeurIPS (2021). https://openreview.net/forum?id=-v4OuqNs5P
23. Savva, M., et al.: Habitat: a platform for embodied AI research. In: ICCV (2019)
24. Schulman, J., Wolski, F., Dhariwal, P., Radford, A., Klimov, O.: Proximal Policy Optimization Algorithms. arXiv preprint arXiv:1707.06347 (2017)
25. Straub, J., et al.: The Replica Dataset: A Digital Replica of Indoor Spaces. arXiv preprint arXiv:1906.05797 (2019)
26. Xia, F., Zamir, A.R., He, Z., Sax, A., Malik, J., Savarese, S.: Gibson env: real-world perception for embodied agents. In: CVPR (2018)
27. Ye, J., Batra, D., Das, A., Wijmans, E.: Auxiliary tasks and exploration enable ObjectNav. In: ICCV (2021)
28. Ye, J., Batra, D., Wijmans, E., Das, A.: Auxiliary tasks speed up learning point goal navigation. In: CoRL (2021)
29. Zhu, Y., et al.: Target-driven visual navigation in indoor scenes using deep reinforcement learning. In: ICRA (2017)

Relaxing the Forget Constraints in Open World Recognition

Dario Fontanel$^{(\boxtimes)}$, Fabio Cermelli, Antonino Geraci, Mauro Musarra,
Matteo Tarantino, and Barbara Caputo

Politecnico di Torino, Turin, Italy
{dario.fontanel,fabio.cermelli,barbara.caputo}@polito.it
{antonino.geraci,mauro.musarra,matteo.tarantino98}@studenti.polito.it

Abstract. In the last few years deep neural networks has significantly
improved the state-of-the-art of robotic vision. However, they are mainly
trained to recognize only the categories provided in the training set
(closed world assumption), being ill equipped to operate in the real world,
where new unknown objects may appear over time. In this work, we investigate the open world recognition (OWR) problem that presents two challenges: (i) learn new concepts over time (incremental learning) and (ii)
discern between known and unknown categories (open set recognition).
Current state-of-the-art OWR methods address incremental learning by
employing a knowledge distillation loss. It forces the model to keep the
same predictions across training steps, in order to maintain the acquired
knowledge. This behaviour may induce the model in mimicking uncertain predictions, preventing it from reaching an optimal representation
on the new classes. To overcome this limitation, we propose the Poly loss
that penalizes less the changes in the predictions for uncertain samples,
while forcing the same output on confident ones. Moreover, we introduce
a forget constraint relaxation strategy that allows the model to obtain a
better representation of new classes by randomly zeroing the contribution of some old classes from the distillation loss. Finally, while current
methods rely on metric learning to detect unknown samples, we propose
a new rejection strategy that sidesteps it and directly uses the model
classifier to estimate if a sample is known or not. Experiments on three
datasets demonstrate that our method outperforms the state of the art.

Keywords: Open world recognition · Robot vision · Deep learning

1 Introduction

Over the last few years, the emergence of deep neural networks has brought significant improvements in the robotic vision, being used in multiple tasks such as

Supplementary Information The online version contains supplementary material
available at https://doi.org/10.1007/978-3-031-06427-2_62.

grasping [9], tool selection [37], depth prediction [29], and autonomous driving [21]. However, modern deep architectures are still trained under the *closed world assumption* (CWA) which assumes that every category the model will need to recognize is fixed and known a priori during the training phase. Clearly, this is a significant limitation since the real-world is continuously changing and the model will likely encounters new classes while operating in new environments. Recognizing the necessity of breaking the CWA, [2] proposed the *open world recognition* (OWR) problem. It consists of two sub-challenges: (i) incremental learning [3,4,36,43], which requires models to extend their knowledge over time without forgetting already learned concepts (*i.e.* incurring into catastrophic forgetting [30]) and (ii) open set recognition [14,38], which requires models to distinguish already seen concepts from unknown ones.

Standard OWR approaches [2,8,13,28] addressed the two challenges separately. To deal with catastrophic forgetting, the state-of-the-art methods [13,28] employ a knowledge distillation loss [18] that prevents changes of the classification outputs for old classes. The model is forced to maintain consistent prediction also when it is not confident, resulting in an overly-regularized training and preventing the model to correctly adapt the feature space when learning novel classes. To overcome this limitation, in this paper we propose a novel distillation loss, *i.e.* the *Poly loss*. It has been designed following two criteria: (i) maintaining the output unchanged when the model is certain, and (ii) letting the model be free to change when the prediction is uncertain. This formulation allows to effectively updates the network to represent novel classes, while also preventing forgetting of the old knowledge. Despite the advantages of the Poly loss, preserving the model unchanged may prevent it from achieving optimal representation on new classes. To this end, we propose the *forget constraint relaxation* strategy. It relaxes the constrain imposed by the distillation loss by randomly removing the contribution from the loss computation of some old classes at each iteration. To address the second challenge of OWR, *i.e.* distinguish between known and unknown samples, the standard approach is to rely on metric learning. In particular, state-of-the-art methods [2,13,28] couple the nearest class mean (NCM) classification strategy [8] with a rejection threshold to categorize a sample into the set of known categories or predict it as unknown. Despite its effectiveness, this approach has two drawbacks: (i) it considers all the features as equally important and (ii) it suffers the curse of dimensionality. In this work, we abandon the metric learning approach in favor of a rejection strategy based on a linear classifier that computes a score for each class as the dot product between the feature representations and a set of class specific learnable weights, implicitly weighting each feature by its importance. As [13], we learn class-specific thresholds on an held-out set.

Following previous works, we benchmark our contributions on Core50 [27], RGB-D Object Dataset [39] and CIFAR-100 [20] datasets, demonstrating the benefits of our new components and outperforming the state of the art.

Contributions. To summarize, in this paper we tackle the challenges of OWR scenario. In particular, we introduce the Poly loss, a novel distillation loss that allows changes in network output when the old model is not confident about the prediction. We propose a forget-constraint relaxation strategy that allows the network to reach an optimal representation of novel classes and a new rejection strategy, abandoning the metric learning approach in favor of a linear classifier. We benchmark our approach on three datasets, showing that it outperforms the previous state of the art.

2 Related Work

Open World Recognition. The necessity of breaking the CWA for robot vision systems [41] has prompted numerous research efforts aiming at equipping models with the capability of both automatically detecting unknown concepts and incorporating them during subsequent learning phases. To that purpose, [2] introduced open world recognition (OWR) as a realistic benchmark for developing agents able to act in the real world. [2] empowers the Nearest Class Mean (NCM) classifier [16,31] with the ability of detecting unknowns, proposing the Nearest Non-Outlier algorithm (NNO). NNO uses a fixed rejection threshold to categorize a test sample as belonging to a known or unknown class. To tackle the OWR scenario, [8] develops the Nearest Ball Classifier which exploits the confidence of the prediction to compute the rejection threshold. [28] extends the NNO method of [2] by incorporating a dynamic updating strategy for the rejection threshold and using a deep neural network as feature extractor. Recently, [13] improves the performances of NCM based classifier introducing two clustering losses and proposing to explicitly learn a specific threshold for each category. Recently, [12] proposed an OWR benchmark considering different visual conditions, showing that current methods struggle in discriminating between unknown and known samples belonging to a different visual domains. In this paper, we go beyond the NCM-based metric learning approach, proposing a simpler but effective rejection strategy that takes advantage of the network outputs confidence.

Knowledge Distillation-based Incremental Learning. Knowledge distillation has been first proposed by [18] as a technique to transfer knowledge from a teacher (cumbersome) model to a student (simple) one. The idea has been then adapted by [24] in incremental learning to alleviate catastrophic forgetting [30]. They considered as teacher the model frozen after the previous learning step and as student the model trained on the new incoming data, and they forced the student to keep its predictions consistent with the teacher. In the following, multiple works proposed different variations this idea in the context of classification [1,5,11,19,25,26,32,36,42,44,45] and only recently in semantic segmentation [6,10,33] and object detection [34,35,40]. Please refer to [7,23] for an extensive survey of incremental learning methods.

While previous works in the OWR setting [13,28] adopted the knowledge distillation strategy presented in [36], we develop a new distillation loss that considers the model uncertainty to prevent forgetting while learning new classes.

3   Method

3.1   Problem Formulation

The open world recognition (OWR) setting is composed of multiple training steps. In the first step, the system is provided with an initial training set $\mathcal{T}_0$ composed by N_0 samples, i.e. $\mathcal{T}_0 = (x_i, y_i)_{i=1^{N_0}}$, where x_i indicate an image and $y_i \in \mathcal{Y}_0$ is the relative class label. In any following step T, the system is provided a new training set $\mathcal{T}_T$, containing samples belonging to a set of novel classes $\mathcal{Y}_T$, where $\mathcal{Y}_T \cap \mathcal{Y}_t = \emptyset \ \forall t \in [0, T-1]$. The goal of OWR is to find a model f that maps an image x to the respective class, if it is known at the step T, or to the unknown class u, i.e. $f : X \to \mathcal{K}_t \cup u$, with $\mathcal{K}^T = \bigcup_{t=0}^{T} \mathcal{Y}_t$ indicates the set of known classes at step T. The model f must be incrementally updated at every training step T to predict new classes but it must still be able to detect unknown concepts. Without loss of generality, we consider a model f made of two components: a feature extractor ω mapping images into a feature space $\mathcal{Z}$ ($\omega : X \to \mathcal{Z}$), and a scoring function ϕ mapping features in $\mathcal{Z}$ to class probabilities ($\phi : \mathcal{Z} \to [0,1]^{|\mathcal{K}^T|}$). We note that, as in [13,28], we consider binary class probabilities obtained from a sigmoid function.

OWR then presents two challenges: (i) learning new classes without forgetting the old ones and (ii) recognizing whether new data falls into previously learned categories or not [2]. In the next section we focus on the former challenge, while in Sect. 3.4 we discuss the latter.

3.2   Learning Without Forgetting

Preliminaries. While learning novel categories without accessing the old data, the model is prone to catastrophic forgetting [15,30], i.e. it gradually forgets the classes it has learned in previous step. To alleviate the catastrophic forgetting issue, previous works [13,28,36] regularize the model using knowledge distillation [18] which forces the current model $\mathcal{M}_T$ to behave like the model of the previous training step $\mathcal{M}_{T-1}$. Practically, this is accomplished by interpreting the outputs of the previous model $\mathcal{M}_{T-1}$ as pseudo-targets within a loss function, so that a sample x can be identified by the the current model $\mathcal{M}_T$ as belonging to previously observed classes with a certain probability.

During the training step T, the model is then trained using sum of two different terms, i.e. the *classification loss* and the *distillation loss*. State-of-the-art methods [13,28,36] employ the binary cross-entropy (BCE) loss for both terms. Formally, the loss is defined as:

$$L = -\frac{1}{|\mathcal{T}_T|} \sum_{(x_i, y_i \in \mathcal{T}_T)} L_C(x_i, y_i) + L_{D_{BCE}}(x_i), \tag{1}$$

with

$$L_C(x_i, y_i) = \sum_{c \in \mathcal{Y}^T} \delta_{c=y_i} \log(\phi(x_i)) + \delta_{c \neq y_i} \log(1 - \phi(x_i)),$$

$$L_{D_{BCE}}(x_i) = \sum_{c \in \mathcal{K}^{T-1}} q_i^c \log(\phi(x_i)) + (1 - q_i^c) \log(1 - \phi(x_i)),$$

where x_i is a sample drawn from the training set $\mathcal{T}_T$, y_i is its ground truth label, $\phi(x_i)$ is the model prediction, q_i^c is the old model probability for class c, with i.e. $q_i = \phi^{T-1}(x_i)$, $\mathcal{K}^{T-1}$ is the set of old classes, and $\mathcal{Y}^T$ the set of new ones.

The distillation loss $L_{D_{BCE}}$ prevents any changes in the model's output, forcing the probability to be equal to the one obtained by the previous model. Indeed, this loss is highly beneficial when the outputs of the previous network are close to a value of maximum certainty (either 0 or 1), preventing any change that may cause the novel network to lose its ability to predict the old classes. However, when the old network is uncertain (outputs values around 0.5), as often occurs when seeing novel classes samples, the network is forced to maintain its uncertainty, preventing to reach an optimal configuration for new classes.

Poly Loss. We therefore aim at finding a distillation loss that prevents the network from modifying the outputs with the maximum certainty, favoring instead the modification of those closer to 0.5, *i.e.* the ones with the highest uncertainty. To formulate a new classification and distillation combination, two major criteria must be satisfied:

1. The total loss L must be globally continuous, differentiable and strictly convex.
2. The distillation loss must have its only minimum in $\phi_T^c = \phi_{T-1}^c$ which means that the first derivative of L must be equal to 0 only when for a certain class c the outputs of both the current and the previous models are equal, i.e. $\phi_T^c = \phi_{T-1}^c$.

To satisfy these criteria, we propose to exploit a *polynomial function* as it is the simplest and most versatile function that can approximate as nearly as needed every continuous function defined on a closed interval.

To prevent the network from changing the outputs with the highest certainty, encouraging instead the modification of those closer to maximum uncertainty status, we formulate Poly loss as follows:

$$L_{D_{POLY}}(x_i) = \sum_{y=1}^{s-1} \frac{1}{4} \Big((2\phi_T^i(x_i) - 1)^4 \tag{2}$$
$$-4(2\phi_{T-1}^i(x_i) - 1)^3 (2\phi_T^i(x_i) - 1) + 3 \Big),$$

where ϕ_{T-1}^i indicates the outputs of the previous model $\mathcal{M}_{T-1}$ interpreted as pseudo-target of class c and ϕ_T^i indicates the outputs of the current model $\mathcal{M}_T$.

Overall, for training the network, we replace $L_{D_{BCE}}$ with the Poly loss $L_{D_{POLY}}$, obtaining the following cost function:

$$L = -\frac{1}{|\mathcal{T}_T|} \sum_{(x_i, y_i \in \mathcal{T}_T)} L_C(x_i, y_i) + L_{D_{POLY}}(x_i). \tag{3}$$

3.3   Forget-Constraint Relaxation

Despite the advantages of the Poly loss, the constraints it imposes may still be too binding. If, on the one hand, forcing the network at step T to not change too much in comparison to the network at step $T-1$ helps to prevent forgetting, on the other hand, this behavior may prevent the network at step T from reaching the *best* possible configuration, which may be very different from what it was at step $T-1$. Indeed, a further improvement would be to relax such constraints, increasing the degrees of freedom of the network. The goal this time is to allow updates in the model configuration not only when the targets are close to 0.5, as obtained using the Poly loss, but also when they are close to 0 or 1. For this reason, we propose the *forget-constraint relaxation (FCR)* strategy that randomly removes some of the old classes from the loss computation, by simply setting their contribution in the distillation loss L_D to 0. More formally,

$$\tilde{L}_D = \sum_{c \in C^{t-1}} \frac{R^c}{p} L_D(x_i),\tag{4}$$

where

$$R^c = \begin{cases} 1 & \text{with } \textit{probability} \quad p, \\ 0 & \text{with } \textit{probability} \quad 1-p; \end{cases}\tag{5}$$

and L_D can be any distillation loss (*e.g.*, $L_{D_{POLY}}$).

When an old class c is removed from the loss computation, *i.e.* $R^c = 0$, the output of the new model ϕ_T^i can take any value for c to minimize the total loss L, allowing it to properly learn the novel class. However, we are not letting the model to forget that class: in the next iterations, it is likely that the class c is again considered in the equation, *i.e.* $R^c = 1$, and the distillation loss will prevent the catastrophic forgetting phenomenon. We remark that this could not be obtained using simple strategies such as multiplying the distillation loss by a

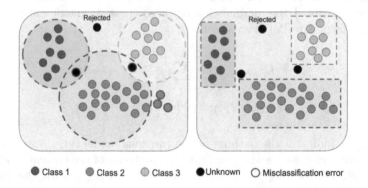

● Class 1   ● Class 2   ○ Class 3   ●Unknown   ○ Misclassification error

Fig. 1. The figure illustrates the metric learning-based (left) and our rejection (right) strategies. Metric learning weights each feature equally, resulting in a sub-optimal rejection strategy. Differently, our strategy considers each feature independently, modeling better the classes distributions.

positive weight below 1 since, while the network would learn new classes more easily, it would also quickly forget the old ones.

3.4 Rejection Strategy

In this section we will analyze the second challenge of OWR, namely the models capability of categorizing as unknown data that does not belong to the set of previously learned classes. The standard approach for detecting unknown samples is to employ a metric learning approach [2,8,13,28]. In particular, previous works assume that the feature extractor ω projects samples in an embedding space where samples of the same class are closer than samples of any other classes. Following this assumption, they compute for each class a centroid, *i.e.* the mean of the feature representation of the samples of that class, and they consider a sample as unknown if its representation is more distant than a threshold η from all the class centroids.

Despite the improvements introduced in B-DOC [13], *i.e.* using class-specific thresholds and learning them rather than computing them, we still identify two important limitations with the metric learning approach:

1. All the features are treated as equally informative to compute per-class thresholds;
2. Computing feature distances on a large scale suffers the curse of dimensionality.

As illustrated in Fig. 1, we argue that considering all the feature as equally important (issue 1) is sub-optimal, since not all the features are meaningful to identify a certain class. Consider, for example, a model having a feature identifying whether or not a *wheel* is present. This feature would be hugely important to classify the *car* class, but it is totally meaningless to classify the *dog* class. Thus, we need to properly consider each feature, weighting its contribution depending on how important it is for a certain class.

To deal with both issues, we propose an approach that completely abandons the metric learning approach. To take into account the different importance of each dimension of the feature vector, we propose to directly use the network classifier weights, that implicitly provide the features importance for each class. The classifier computes the dot product between the sample feature representation and the weights of a certain class, producing a scalar value for each class, *i.e.* the classification score, addressing also the issue 2. Intuitively, the classification score is a value indicating the confidence for a sample x to belong to class c. Thus, as in B-DOC [13], we define a threshold for each class c, η_c, and given an image x_i, we implement the following rejection policy:

$$\begin{cases} \text{accept,} & \text{if } \exists c \in \mathcal{C} : (<\omega(x_i), \mathbf{w}_c>) > \eta_c; \\ \text{reject,} & \text{otherwise.} \end{cases} \tag{6}$$

Chiefly, following B-DOC [13], our training strategy consists of two steps: in the first one, we train the feature extractor on the training set while minimizing

Eq. 3, and in the second one, we learn the thresholds η on a set of samples that we excluded from the training set. Keeping frozen all the network parameters, we learn η_c minimizing the following cost function:

$$\mathcal{L}_{GR}(x, c) = \sum_{c \in \mathcal{C}} \max(0, k \cdot (\eta_c - (<\omega(x), \mathbf{w}_c>))), \tag{7}$$

where k is equal to 1 if $k = y_i$, and -1 otherwise. Intuitively, if the sample belonging to class c has a lower score than η_c, the threshold η_c will decrease. On the other hand, if a sample not belonging to class c obtains a higher score than η_c, it will increase.

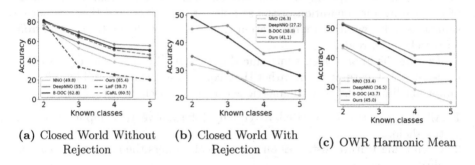

(a) Closed World Without Rejection

(b) Closed World With Rejection

(c) OWR Harmonic Mean

Fig. 2. Comparison of LwF [24], iCaRL [36], NNO [2], DeepNNO [28], B-DOC [13], and our method on Core50 dataset [27]. The parenthesis denote the average accuracy among the different incremental steps.

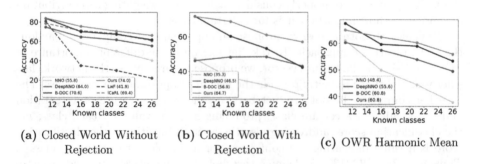

(a) Closed World Without Rejection

(b) Closed World With Rejection

(c) OWR Harmonic Mean

Fig. 3. Comparison of iCaRL [36], NNO [2], DeepNNO [28], B-DOC [13] and our method on RGB-D Object dataset [22]. The parenthesis denote the average accuracy among the different incremental steps.

4   Experiments

Datasets. Following the same evaluation protocol of [13], we evaluate the performance of our model on three datasets: Core50 [27], RGB-D Object [22] and CIFAR-100 [20]. Core50 dataset [27] represents a very challenging benchmark with 50 different objects grouped into 10 semantic categories and captured in 11 distinct sequences under shifting conditions. Following [13], we divide the 10 categories into two splits: 5 are considered as known classes and the remaining 5 as unknown. We use the first 2 known classes as the initial training set and we incrementally add the other classes one by one. The RGB-D Object dataset [22] contains 51 different semantic categories of daily-life objects collected in a controlled scenario. Following previous works [13,28], we divided its categories into two split: the first 26 categories are considered as known classes, while the remaining 25 are considered as unknown ones. Among the 26 categories, the first 11 ones constitute the initial training set and the remaining ones are added incrementally in 4 steps of 5 classes each. CIFAR-100 [20] is a largely adopted benchmark to compare incremental class learning algorithms [36]. It consists of 100 semantic categories with 500 training images and 100 testing images per class. Follow previous works [13,28] we divide the dataset into 50 known and 50 unknown categories. As for Core50 [27] and RGB-D Object dataset [22], we identify an initial training set, which in this case corresponds to 20 classes chosen among the known set. We then incrementally add the remaining ones in steps of 10 classes each.

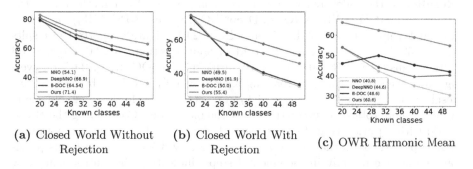

(a) Closed World Without Rejection   (b) Closed World With Rejection   (c) OWR Harmonic Mean

Fig. 4. Comparison of NNO [2], DeepNNO [28], B-DOC [13] and our method on CIFAR-100 dataset [20]. The parenthesis denote the average accuracy.

Networks Architectures and Training Protocols. Following previous works [13] we employ a ResNet-18 architecture [17] as backbone. For each dataset we start by training the network from scratch. For Core50 dataset, we use 5 epochs for the initial training classes and 20 epochs for the incremental steps. For RGB-D Object dataset, instead, we train the network on the initial classes for 4 epochs and then incrementally for 10 epochs. Finally, for CIFAR-100 we set to 70 both the epochs for the initial learning stage and the following incremental steps. We set the learning rate to 0.02 for both the RGB-D Object and Core50 datasets, while we use 0.2 for CIFAR-100. We adapt Stochastic Gradient Descent (SGD) with momentum 0.9 and a weight decay of 10^{-3} for the RGB-D Object and 10^{-4} for both Core50 and CIFAR-100. To learn η_c on the held-out set of samples, we use 20 epochs for the three datasets. We use a learning rate of 0.001 for Core50, 0.07 for the RGB-D Object dataset and 0.01 for CIFAR-100. We also employ the same strategy for memory management of [13], which set the maximum storable samples up to 2000. 40% of the instances from memory are then drawn to construct each training batch. 20% of the stored samples, instead, are not used to directly train the model but only to learn the class-specific thresholds.

Metrics. Following previous works [2,13,28], we use three standard metrics for comparing the performances of OWR methods. In the closed world *without rejection* setting, the models is evaluated only on the known set of classes, with no possibility of considering any sample as unknown. In the closed world *with rejection* scenario, instead, the model may either categorize a sample into one of the known classes or classify it as unknown. This scenario is much more difficult than the preceding one because samples from the known set of classes may be misclassified as unknowns. Overall, for open world evaluations, we use the standard harmonic mean (OWR-H) metric defined in [13].

To compute the method performance, we randomly picked 5 distinct sets of known categories for each dataset and we repeated each experiments 3 times. The final performance is obtained averaging the results of each run and order.

Results. In the following, we report the comparison of our method with the state-of-the-art of OWR (NNO [2], DeepNNO [28], and B-DOC [13]) and incremental learning (IL) (iCaRL [36] and LwF [24]). Figure 2 reports the results on Core50 dataset. On the closed world without rejection (Fig. 2a), our method outperforms the OWR state of the art, surpassing B-DOC [13] by 2.6% on average, and even IL methods, surpassing iCaRL by 4.9% and LwF by 25.7%. This result indicates that the adoption of $L_{D_{POLY}}$ is beneficial for learning, obtaining a model more robust on predictions over the old classes. Considering the closed world with rejection (Fig. 2b), our method rejects less known classes, obtaining higher performances than B-DOC, on average, by 3.1%. It outperforms previous methods especially in incremental steps, indicating that introducing new classes does not reduce the confidence on previous classes. Finally, considering both known and unknown samples, our method is superior to previous works, outperforming B-DOC by 1.6% on the OWR-H (Fig. 2c).

Figure 3 reports the results on RGB-D Object dataset. Similarly to Core50 dataset, our method outperforms IL methods, surpassing iCaRL by 14.1% in the last step and by 4.6% on average (Fig. 3a). It also surpasses OWR methods by a large margin when considering rejection (Fig. 3b), achieving an average accuracy of 64.7%, more than 7% w.r.t. B-DOC and DeepNNO. The effectiveness of our method is also confirmed by the OWR-H metric (Fig. 3c), where it archives performance comparable to B-DOC and outperforms DeepNNO and NNO.

Finally, we report in Fig. 4 the results on CIFAR-100 dataset. As for Core50 and RGB-D datasets, our method outperforms OWR state-of-the-art by a large margin. In particular, in the closed world without rejection, it surpasses DeepNNO by 4.5% and B-DOC by 6.8% on average (Fig. 4a). In closed world with rejection (Fig. 4b) DeepNNO achieves slightly higher performance, reaching up to 52.5%. The reason is that DeepNNO classifies most of the samples into known classes, failing in rejecting them as unknown. This behaviour is confirmed by the OWR-H metric (Fig. 4c) in which our method achieves much higher performance than DeepNNO (56.1% vs 42.8%), benefiting from the rejection strategy based on features importance.

Ablations. Due to lack of space, we report the ablation studies in the supplementary material.

5 Conclusion

In this work, we studied the open world recognition problem in robot vision. We first proposed to relax the forget-constraint imposed by previous methods to prevent catastrophic forgetting. In particular, we proposed a new distillation function, the Poly loss, that enabled changes in the model's output when it was uncertain about the old class prediction. Moreover, we introduced the forget-constraint relaxation strategy to further relax the distillation constraint on certain samples, enabling the network to reach an optimal representation for novel classes without forgetting previous classes. Second, we abandon the metric-learning strategy to detect unknown samples and we propose to directly use the model's classifier. We demonstrate the benefits of our contributions on Core50, RGB-D Object, and CIFAR-100 datasets outperforming the state of the art.

References

1. Belouadah, E., Popescu, A.: Il2m: class incremental learning with dual memory. In: ICCV-19
2. Bendale, A., Boult, T.: Towards open world recognition. In: CVPR-15
3. Camoriano, R., Pasquale, G., Ciliberto, C., Natale, L., Rosasco, L., Metta, G.: Incremental robot learning of new objects with fixed update time. In: ICRA-17
4. Camoriano, R., Traversaro, S., Rosasco, L., Metta, G., Nori, F.: Incremental semi-parametric inverse dynamics learning. In: ICRA-16
5. Castro, F.M., Marín-Jiménez, M.J., Guil, N., Schmid, C., Alahari, K.: End-to-end incremental learning. In: Ferrari, V., Hebert, M., Sminchisescu, C., Weiss, Y. (eds.) ECCV 2018. LNCS, vol. 11216, pp. 241–257. Springer, Cham (2018). https://doi.org/10.1007/978-3-030-01258-8_15
6. Cermelli, F., Mancini, M., Bulò, S.R., Ricci, E., Caputo, B.: Modeling the background for incremental learning in semantic segmentation. In: CVPR-20
7. De Lange, M., et al.: Continual learning: A comparative study on how to defy forgetting in classification tasks. 2(6) (2019). arXiv preprint arXiv:1909.08383
8. De Rosa, R., Mensink, T., Caputo, B.: Online open world recognition. arXiv:1604.02275 (2016)

9. Della Santina, C., et al.: Learning from humans how to grasp: a data-driven architecture for autonomous grasping with anthropomorphic soft hands. RA-L-19

10. Douillard, A., Chen, Y., Dapogny, A., Cord, M.: Plop: learning without forgetting for continual semantic segmentation. In: CVPR-21

11. Douillard, A., Cord, M., Ollion, C., Robert, T., Valle, E.: PODNet: pooled outputs distillation for small-tasks incremental learning. In: Vedaldi, A., Bischof, H., Brox, T., Frahm, J.-M. (eds.) ECCV 2020. LNCS, vol. 12365, pp. 86–102. Springer, Cham (2020). https://doi.org/10.1007/978-3-030-58565-5_6

12. Fontanel, D., Cermelli, F., Mancini, M., Caputo, B.: On the challenges of open world recognition under shifting visual domains. RA-L-20 **6**(2)

13. Fontanel, D., Cermelli, F., Mancini, M., Rota Buló, S., Ricci, E., Caputo, B.: Boosting deep open world recognition by clustering. RA-L **5**(4), 5985–5992 (2020)

14. Fragoso, V., Sen, P., Rodriguez, S., Turk, M.: EVSAC: accelerating hypotheses generation by modeling matching scores with extreme value theory. In: ICCV-13

15. French, R.M.: Catastrophic forgetting in connectionist networks. Trends Cognit. Sci. **3**(4) (1999)

16. Guerriero, S., Caputo, B., Mensink, T.: Deep nearest class mean classifiers. In: ICLR-WS-18

17. He, K., Zhang, X., Ren, S., Sun, J.: Deep residual learning for image recognition. In: CVPR-16

18. Hinton, G., Vinyals, O., Dean, J.: Distilling the knowledge in a neural network. arXiv 1503.02531 (2015)

19. Hu, X., Tang, K., Miao, C., Hua, X.S., Zhang, H.: Distilling causal effect of data in class-incremental learning. In: CVPR-21

20. Krizhevsky, A., Hinton, G.: Learning multiple layers of features from tiny images. Technical report, University of Toronto (2009)

21. Kumar, V.R., et al.: Omnidet: surround view cameras based multi-task visual perception network for autonomous driving. RA-L-21 **6**(2)

22. Lai, K., Bo, L., Ren, X., Fox, D.: A large-scale hierarchical multi-view RGB-d object dataset. In: ICRA-11

23. Lesort, T., Lomonaco, V., Stoian, A., Maltoni, D., Filliat, D., Díaz-Rodríguez, N.: Continual learning for robotics: definition, framework, learning strategies, opportunities and challenges. Inf. Fusion **58**, 52–68 (2020)

24. Li, Z., Hoiem, D.: Learning without forgetting. T-PAMI-17

25. Liu, X., et al.: Generative feature replay for class-incremental learning. In: CVPR-20

26. Liu, Y., Su, Y., Liu, A.A., Schiele, B., Sun, Q.: Mnemonics training: multi-class incremental learning without forgetting. In: CVPR-20

27. Lomonaco, V., Maltoni, D.: Core50: a new dataset and benchmark for continuous object recognition. In: CoRL-17

28. Mancini, M., Karaoguz, H., Ricci, E., Jensfelt, P., Caputo, B.: Knowledge is never enough: towards web aided deep open world recognition. In: ICRA-19

29. Mancini, M., Costante, G., Valigi, P., Ciarfuglia, T.A., Delmerico, J., Scaramuzza, D.: Toward domain independence for learning-based monocular depth estimation. RA-L-17 **2**(3)

30. McCloskey, M., Cohen, N.J.: Catastrophic interference in connectionist networks: the sequential learning problem. In: Psychology of Learning and Motivation, vol. 24, pp. 109–165. Elsevier (1989)

31. Mensink, T., Verbeek, J., Perronnin, F., Csurka, G.: Metric learning for large scale image classification: generalizing to new classes at near-zero cost. In: ECCV-12

32. Michieli, U., Zanuttigh, P.: Continual semantic segmentation via repulsion-attraction of sparse and disentangled latent representations. In: CVPR-21
33. Michieli, U., Zanuttigh, P.: Knowledge distillation for incremental learning in semantic segmentation. CVIU-21 **205**
34. Peng, C., Zhao, K., Lovell, B.C.: Faster ilod: incremental learning for object detectors based on faster RCNN. Pattern Recognit. Lett. **140** (2020)
35. Perez-Rua, J.M., Zhu, X., Hospedales, T.M., Xiang, T.: Incremental few-shot object detection. In: CVPR-20
36. Rebuffi, S.A., Kolesnikov, A., Sperl, G., Lampert, C.H.: iCaRL: incremental classifier and representation learning. In: CVPR-17
37. Saito, N., Ogata, T., Funabashi, S., Mori, H., Sugano, S.: How to select and use tools? Active perception of target objects using multimodal deep learning. RA-L-21 **6**(2)
38. Scheirer, W.J., De Rezende Rocha, A., Sapkota, A., Boult, T.E.: Toward open set recognition. T-PAMI-12 **35**(7)
39. Schwarz, M., Milan, A., Periyasamy, A.S., Behnke, S.: RGB-D object detection and semantic segmentation for autonomous manipulation in clutter. IJRR-18 **37**(4–5)
40. Shmelkov, K., Schmid, C., Alahari, K.: Incremental learning of object detectors without catastrophic forgetting. In: ICCV-17
41. Sünderhauf, N., et al.: The limits and potentials of deep learning for robotics. IJRR-18 **37**(4–5)
42. Tao, X., Hong, X., Chang, X., Dong, S., Wei, X., Gong, Y.: Few-shot class-incremental learning. In: CVPR-20
43. Valipour, S., Perez, C., Jagersand, M.: Incremental learning for robot perception through HRI. In: IROS-17
44. Wu, Y., et al.: Large scale incremental learning. In: CVPR-19
45. Zhao, B., Xiao, X., Gan, G., Zhang, B., Xia, S.T.: Maintaining discrimination and fairness in class incremental learning. In: CVPR-20

Memory Guided Road Segmentation

Praveen Venkatesh, Rwik Rana, and Varun Jain[(✉)]

Indian Institute of Technology, Gandhinagar, India
{praveen.venkatesh,rwik.rana,varun.jain}@iitgn.ac.in

Abstract. In self-driving car applications, there is a requirement to predict the location of the road given an input RGB front-facing image. We propose a framework that utilizes an interleaving strategy of large and small feature extractors assisted via a propagating shared feature space allowing us to realize gains of over 2.5X in speed with a negligible loss in the accuracy of predictions. By utilizing the gist of previously observed frames, we train the network to predict the current road with greater accuracy and lesser deviation from previous frames.

Keywords: Video segmentation · Deep learning · Road detection

1 Introduction

-In recent years, there has been an increasing interest in the development of fully autonomous vehicles. A core component of these systems is the ability of the robot to perceive the world around it. In self-driving car applications, one of the most critical tasks is to accurately predict the location of the road in order to make key driving decisions. While there have been significant strides in developing robust visual perception for autonomous cars, several algorithms utilize techniques that work on a per-frame basis and do not exploit the consistency between frames that is generally the case when an autonomous car moves in a natural environment. For example, inferencing a state-of-the-art segmentation model on each video frame independently can lead to jitter or other unwanted artefacts. By using frames that have already been observed in a given video sequence, a deep learning network can potentially utilize contexts from previous frames leading to an increase in the effectiveness of the predicted segmentation maps.

Typically, video cameras have a frame rate of upwards of 30FPS, meaning that successive frames in a video are spaced at intervals shorter than 1/3rd of a second. In this time, assuming that objects in the video are moving at physically realizable speeds, a given object's displacement in the video frame is negligible,

---

P. Venkatesh, R. Rana and V. Jain—The authors contributed equally.

Supplementary Information The online version contains supplementary material available at https://doi.org/10.1007/978-3-031-06427-2_63.

with an even lower number of objects moving out of the frame between successive images. Recomputing the segmentation at each frame by treating each image independently without prior context becomes computationally inefficient since the similarity between adjacent frames is very high. Hence, we hypothesize that storing an underlying low-level context of each frame throughout inference can drastically improve the speeds at which inference is possible, leading to lesser computational requirements and satisfying stringent time constraints.

In this paper, we use a combination of large and small feature extractors to learn to infer segmentation maps from a shared underlying memory, allowing for fast inference with a low accuracy dip. The proposed method is inspired by how the human eye perceives complex scenes and infers a rich representation based on previously seen scenes. As a result, our framework makes the process of segmenting roads computationally fast, robust, and smooth.

2 Related Work

With the onset of deep learning techniques, semantic segmentation has seen increased interest as a subfield in computer vision. Early works such as those introduced by Long et al. [16] propose FCNs (Fully Convolutional Networks) that aim to replace fully connected layers with fully convolutional layers. This significantly reduces the number of parameters required while also introducing spatial consistency over localized patches of an image. Fisher et al. [18] suggest replacing the convolution layers with dilated convolutions. This allows for an exponential expansion in the receptive field without a loss of resolution or coverage at the same level of computational complexity.

Shao-Yuan Lo et al. [12] introduced EDANet, composed of an aggregation of EDAModules. They exploit the use of pointed convolutions, asymmetric convolutions, and dilated convolutions to increase inference speed at a slightly lower IOU cost. The addition of an extra EDA Module at the start of downsampling acts like a feature selection layer and improves the model's accuracy.

[13] suggests a real-time semantic segmentation network involving a U-net shaped encoder-decoder backbone. The encoder parses the input image and generates a feature vector that is used for processing by the subsequent decoder network. The encoder network allows the network to locally and dynamically adapt to the input frames and predict high-quality outputs. Other networks such as those proposed in [7] extract temporal features by stacking adjacent images for inference.

The closest work to the one proposed in this paper is presented in [10]. The authors suggest a way of generating bounding boxes for object detection in videos. By using a flowing memory that is updated by differently sized models, the context vector is updated to reflect subtle changes between frames.

In our work, we employ an interleaved suite of extractors with a shared underlying memory that allows models with a significant difference in the model size to drastically improve inference speed while sustaining a negligible loss in accuracy. We use models with ResNet-18 and 101 backbones as our feature

extractors and a ConvLSTM and an FCN32 decoder network to predict the
road and segment the input image.

3   Our Approach

In our work, we propose a framework that allows networks to utilize a propagating memory that is updated continually through time by alternating subnetworks. By exploiting the consistency between frames, we can significantly speed
up inference without a considerable sacrifice in segmentation accuracy.

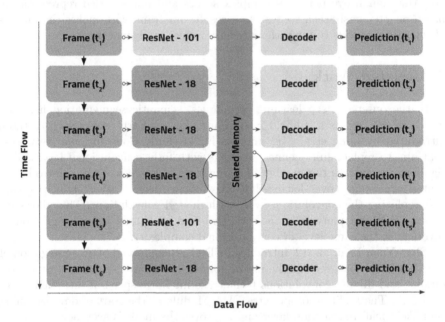

Fig. 1. Pipeline - Successive image frames are passed through alternating networks in
a fixed pattern. The feature extractor stage updates the recurrent shared memory layer
before prediction by the decoder.

We divide our approach into three stages: 1) Feature extraction stage, 2)
Shared Memory stage, 3) Decoder stage. Each of these stages is described in the
following subsections.

3.1   Dataset

There are several large scale datasets available for semantic segmentation. However, they all suffer from the sparsity of fine-grained annotations in video
sequences. Even though fine-grained annotations are available for most available datasets, the annotation rate is generally in the order of 1FPS compared

to the 30FPS of the video sequence. Due to this, there is a significant variation between objects in adjacent annotated frames. If the network is trained on these images, the 30x difference between training and inference timestamps can lead to significant differences between the train and test setup, making the pipeline highly ineffective.

To combat this issue, we create a synthetic dataset that uses video sequences of a car driven in the CARLA simulator [6]. We drive a car in the Town01 map setting the weather to rainy. We collect 8100 front-facing RGB training images as part of 10 video driving sequences and use data augmentation (flipping & rotation) to increase the dataset size to 24000 images.

3.2   Feature Extractors - The Interpreter

A crucial part of any learning-based segmentation technique is the encoder or the feature extraction layer. Here, we utilize two different feature extractors of different sizes to be interleaved. Even though the feature extractors can be chosen from the vast body of work that exists in representation learning, we use two variants of ResNet (18 and 101) in our experiments. The drastic difference in model size (11M vs 44M parameters) significantly changes the speed of inference on both networks, which naturally fits the proposed framework. In our experiments, we remove the last few layers and retain the ResNet backbones until the last convolutional block to obtain a fully convolutional network pipeline. Since the size of the feature vectors in the ResNet-18 and 101 models are different, we also add a trainable convolutional block after the feature extraction layer to equalize the dimensions and ensure uniformity of features. This is important as both the feature extractors pass data directly into the shared memory block, which has a fixed input size.

3.3   Shared Memory - The Time Keeper

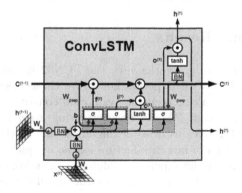

Fig. 2. ConvLSTM cell [17]

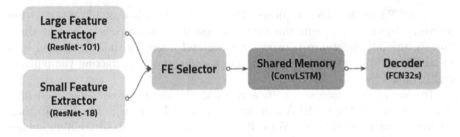

Fig. 3. Framework pipeline

The shared memory block that maintains a floating context of the frame is implemented as a stateful convolutional LSTM (ConvLSTM) block seen in Fig. 2. This cell is essentially a modified LSTM that operates using convolutional gates instead of linear gates. We use a single ConvLSTM layer in our experiments that take in the features predicted by the feature extractor at each time step and produces features that incorporate temporal data. The output of the ConvLSTM block is then passed to the decoder stage for predicting the segmentation maps.

The ConvLSTM executes the role of creating consistency between successive frames for prediction. For example, immediately after the states of the ConvLSTM are cleared, a high-quality context is induced in the memory layer using the large feature extractor. Over subsequent time steps, this context is refined using the smaller feature extractor leading to similar high-quality outputs at lower computational requirements.

3.4   Decoder - Generating Segmentations

In the final stage, the outputs of the ConvLSTM are fed to a fully convolutional upsampling block that predicts the segmentation of the road. In our experiments, we use a decoder block similar to the FCN32 architecture that is composed of 5 transpose convolutional layers and a single convolutional layer at the end.

4   Experimental Results

We tested our various subnetworks with the FCN32 decoder to gain a comparative understanding of their performance. We chose ResNet-18 and 101 as the downsampling network to keep the evaluation consistent while keeping the upsampling network intact. We observe that using a small feature extractor, such as the ResNet18, gives us a faster inference time at the expense of accuracy, whereas using a more intricate feature extractor, such as the ResNet-101, gives us better results at the expense of inference time. We consider the vanilla ResNet models as baselines since we aim to infer results when multiple models are interleaved.

4.1   Training Setup

With the baselines defined, we propose two strategies using which the networks are trained:

1. Sequential-Interleaved: The inputs to the framework are six successive frames from the video sequence passed to either the large or the small feature extractors randomly sampled from a uniform distribution.

2. Batched-Interleaved: The inputs to the framework are six duplicates of the same frame for each frame in the image sequence. They are passed to either the large or the small feature extractor randomly sampled uniformly. Because consecutive frames of a video stream are similar, we hypothesize that passing in the same frame multiple times could be a way to trick the model into thinking it is receiving a consistent video feed.

After obtaining the features from either of the above strategies, they are passed into the ConvLSTM memory stage statefully before the final prediction of the segmentation maps. In both training strategies, the objective function is to minimize the following loss:

$$\text{Loss} = -\sum_{f_i=1}^{6} \sum_{i=1}^{\text{output size}} y_i \cdot \log \hat{y}_i$$

where $\hat{y}$ is the ground truth output, f_i is the ith frame in a sequence and y_i is the prediction.

4.2   Results

During inference, we employ different interleaving strategies between the large and small feature extractors. The state of the ConvLSTM is cleared every time the large feature extractor is called in order to provide a new high-quality context of the frame. Subsequent to the call of the first large, the smaller extractor is called repeatedly until another request for the large extractor is issued. The point at which the large extractor is to be called is determined via one of two strategies: 1) Generate a normally sampled random number ($X \sim N(0,1)$) and set a threshold beyond which the large extractor is called. 2) Call the large extractor once in every n frames. Figure 3 provides a representative example of the order of calling between the large ResNet-101 and small ResNet-18 extractors.

By testing our framework on the above-mentioned training and inference strategies, we tabulate our results in Table 1.

Fig. 4. (a) Vanilla ResNet18 (b) Vanilla ResNet101

Fig. 5. (a) Batched Interleaved (b) Sequential Interleaved

As seen in Figs. 4a, 4b, 5a and 5b, it is evident that the Vanilla ResNet models and interleaved models perform similarly at each frame. From the videos (provided in the supplemental material), it is clear that the interleaved models have a far superior temporal consistency than the vanilla models, which treat each

Table 1. Avg IoU and Avg FPS of various trained models

Name	Strategy	Avg IoU	Avg FPS
Vanilla ResNet-18	–	0.852	**155.4**
Vanilla ResNet-101	–	**0.915**	46.18
Batched Interleaved	randn(0.9)	0.878	132.58
	1 in 6	0.877	135.73
	1 in 10	**0.88**	**146.02**
Sequential Interleaved	randn(0.7)	0.872	117.91
	randn(0.8)	**0.872**	132.12
	randn(0.9)	0.871	**146.53**
Sequential Interleaved	1 in 6	**0.871**	132.42
	1 in 10	0.872	**141.31**
	1 in 12	0.868	139.64

frame as an independent prediction. This validates our hypothesis that introducing temporal consistency can improve the robustness of road segmentation.

The table shows that the interleaved framework performs almost 2.5x as fast as the slowest extractor while having a performance that averages the slow and the fast extractors. There is a high temporal consistency between successive predictions due to the LSTM cell present in the architecture. The strategy employed to use a particular feature extractor affects the performance of the interleaved model. Although the variation between the performance of the different strategies is not very wide, we can optimize the system's abilities by clearing the weights and using the larger extractor depending on changes in the frame.

5   Conclusion and Outlook

In this paper, we present a framework that utilizes an interleaved suite of extractors and exploits the temporal consistency in videos to significantly increase inference speed with a negligible dip in accuracy. By utilizing such a framework, we increase the speed of inference of the network and improve the spatiotemporal consistency of the predictions of the video frames, making them superior in terms of consistency.

The framework presented in this paper can also be extended to include multiple different feature extractors instead of just the two networks shown in this paper. Moreover, we notice that calling the large extractor and clearing the weights when there are large changes in the input video sequence helps improve predictions. This can be further optimized by introducing a feature selector network that decides to choose the model to infer from depending on the context of the frame, potentially improving results drastically.

References

1. Lin, T.-Y., et al.: Microsoft COCO: common objects in context. In: Fleet, D., Pajdla, T., Schiele, B., Tuytelaars, T. (eds.) ECCV 2014. LNCS, vol. 8693, pp. 740–755. Springer, Cham (2014). https://doi.org/10.1007/978-3-319-10602-1_48
2. The Pascal Visual Object Classes (VOC) Challenge. Int. J. Comput. Vis. https://dl.acm.org/doi/10.1007/s11263-009-0275-4
3. Chen, C., Seff, A., Kornhauser, A., Xiao, J.: Deepdriving: learning affordance for direct perception in autonomous driving. In: Proceedings of the 2015 IEEE International Conference on Computer Vision (ICCV). ICCV 2015, pp. 2722–2730. IEEE Computer Society, USA (2015). https://doi.org/10.1109/ICCV.2015.312
4. Deng, J., Dong, W., Socher, R., Li, L.J., Li, K., Fei-Fei, L.: Imagenet: a large-scale hierarchical image database. In: 2009 IEEE Conference on Computer Vision and Pattern Recognition, pp. 248–255. IEEE (2009)
5. Dosovitskiy, A., Ros, G., Codevilla, F., Lopez, A., Koltun, V.: CARLA: an open urban driving simulator. arXiv:1711.03938 [cs] (2017)
6. Dosovitskiy, A., Ros, G., Codevilla, F., López, A.M., Koltun, V.: Carla: an open urban driving simulator. arXiv:abs/1711.03938 (2017)

7. Hu, P., Heilbron, F.C., Wang, O., Lin, Z.L., Sclaroff, S., Perazzi, F.: Temporally distributed networks for fast video semantic segmentation. In: 2020 IEEE/CVF Conference on Computer Vision and Pattern Recognition (CVPR), pp. 8815–8824 (2020)

8. Kim, Y.: Two-step recurnet - younghyun. https://sites.google.com/site/entertain84/research-project/TSRN

9. Kolesnikov, A., et al.: Big Transfer (BiT): general visual representation learning. arXiv:1912.11370 [cs] (2020). version: 3

10. Liu, M., Zhu, M., White, M., Li, Y., Kalenichenko, D.: Looking fast and slow: memory-guided mobile video object detection. arXiv:1903.10172 [cs] (2019)

11. Liu, X., Deng, Z., Yang, Y.: Recent progress in semantic image segmentation. Artif. Intell. Rev. **52**(2), 1089–1106 (2019). https://doi.org/10.1007/s10462-018-9641-3

12. Lo, S.Y., Hang, H.M., Chan, S.W., Lin, J.J.: Efficient dense modules of asymmetric convolution for real-time semantic segmentation. In: Proceedings of the ACM Multimedia Asia (2019)

13. Nirkin, Y., Wolf, L., Hassner, T.: Hyperseg: patch-wise hypernetwork for real-time semantic segmentation. In: CVPR (2021)

14. Rong, G., et al.: LGSVL simulator: a high fidelity simulator for autonomous driving. arXiv:2005.03778 [cs, eess] (2020)

15. Shah, S., Dey, D., Lovett, C., Kapoor, A.: AirSim: high-fidelity visual and physical simulation for autonomous vehicles. arXiv:1705.05065 [cs] (2017)

16. Shelhamer, E., Long, J., Darrell, T.: Fully convolutional networks for semantic segmentation. IEEE Trans. Pattern Anal. Mach. Intell. **39**(4), 640–651 (2017). https://doi.org/10.1109/TPAMI.2016.2572683

17. Xavier, A.: An introduction to convlstm, April 2019. https://medium.com/neuronio/an-introduction-to-convlstm-55c9025563a7

18. Yu, F., Koltun, V.: Multi-Scale Context Aggregation by Dilated Convolutions. arXiv:1511.07122 [cs] (2016)

Learning Visual Landmarks
for Localization with Minimal Supervision

Muhammad Haris[1]([⊠]), Mathias Franzius[2], and Ute Bauer-Wersing[1]

[1] Frankfurt University of Applied Sciences, 60318 Frankfurt, Germany
muhammad.haris@fb2.fra-uas.de
[2] Honda Research Institute Europe GmbH, 63073 Offenbach, Germany

Abstract. Camera localization is one of the fundamental requirements
for vision-based mobile robots, self-driving cars, and augmented reality
applications. In this context, learning spatial representations relative to
unique regions in a scene with Slow Feature Analysis (SFA) has demonstrated large-scale localization. However, it relies on hand-labeled data
to train a CNN for recognizing unique regions. We propose a new approach that uses pre-trained CNN-detectable objects as anchors to label
and learn new landmark objects or regions in a scene using minimal
supervision. The method bootstraps the landmark learning process and
removes the need to manually label large amounts of data. The anchor
objects are only required to learn the new landmarks and become obso-
lete for the unsupervised mapping and localization phases. We present
localization results with the learned landmarks in simulated and real-
world outdoor environments and compare the results to SFA on complete
images and PoseNet. The landmark-based localization shows similar or
better accuracy than the baseline methods in challenging scenarios. Our
results further suggest that the approach scales well and achieves even
higher localization accuracy by increasing the number of learned landmarks without increasing the number of anchors.

Keywords: Localization · Mapping · Landmarks · Service robots

1 Introduction

Visual mapping and localization refer to creating a consistent scene representation and localizing a robot using a camera as the only exteroceptive sensor.
A mobile robot's ability to localize itself in an environment is fundamental to
achieve intelligent behavior. It enables a range of indoor and outdoor applications ranging from household robots (i.e., lawnmowers, vacuum cleaners) to
self-driving cars.

The research in this area encompasses a broad range of methods that address
this challenging task. State-of-the-art simultaneous localization and mapping
(SLAM) algorithms exploit image features [22] or complete image information
[4] to create sparse or semi-dense scene representation. Convolutional neural networks (CNNs) for visual localization have become an appealing alternative to the

S. Sclaroff et al. (Eds.): ICIAP 2022, LNCS 13231, pp. 773–786, 2022.
https://doi.org/10.1007/978-3-031-06427-2_64

traditional methods based on hand-crafted features. In [13,14], the authors have trained PoseNet in an end-to-end way to regress pose from single images. In contrast, there are methods [5,20,21] for mapping and localization that are inspired by neurobiological systems. Earlier learning approaches [5] reproduce the firing characteristics of Place- and Head-Direction Cells [24,28] using a hierarchical model. The model uses the concept of slow feature analysis (SFA) [30], and the intuition behind it is that behaviorally meaningful information changes on a slower timescale than the primary sensory input (e.g., pixel values in a video). Previous work [17] implemented SFA-based localization on a mobile robot. It achieved similar localization accuracy [19] to state-of-the-art visual SLAM methods, i.e., ORB and LSD-SLAM [4,22] in small- to medium-scale environments and demonstrated robustness against changing conditions in outdoor scenarios [9,18].

While the methods mentioned above are sufficient for the localization task, the recent trend has shifted to create semantic maps that will enable the robots to better interact with the world around them. One way to obtain such maps is to incorporate objects into the localization pipeline using deep-learning-based object detection algorithms. Recent work in this direction, i.e., Hybrid-SFA [11], uses a CNN to detect unique objects or regions in a scene and performs localization relative to them. The approach leads to representations similar to those of Spatial View Cells in the hippocampus [5]. The results show a significant improvement in localization accuracy, especially in a large-scale environment. However, it relies on hand-labeled training data to learn unique objects or regions in a scene, which is infeasible for many real-world applications.

This paper's main contribution is a novel approach that uses object instances with pre-trained visual detectors (e.g., MS-COCO objects [16]) as a labeling tool to learn new landmarks for localization. For the sake of simplicity, we will refer to objects or regions with pre-trained detectors as *anchors* and the derived objects or regions to be learned as *landmarks*. The idea is to place an anchor in spatial relation to a landmark and generate labeled training data relative to it. If the scene already contains suitable anchors, they can be used directly. Please note that it is possible to learn spatial representations directly w.r.t anchors. However, these anchors (i.e., pre-learned CNN-object categories) are typically dynamic objects in the scene (e.g., a bicycle, a chair). They thus cannot be used reliably for both indoor and outdoor localization in the long term. Hence, the proposed approach enables selecting long-term stable landmarks for localization with minimal supervision and a faster generation of labeled training data for learning them. Moreover, localization accuracy scales with the number of selected landmarks but without increasing the number of anchors and the amount of supervision.

After landmark learning, the system uses the views of the learned landmarks for mapping and localization phases. This paper presents localization results from simulated and real-world outdoor environments. Most of the available localization methods work online, but only a subset of these are trained in an offline learning phase. To provide fair and straightforward comparisons, we use PoseNet [13] and basic SFA-localization [9].

2   Related Work

The recent performance increase of deep-learning-based object detection algorithms [15] have led the way to incorporate object detection into the traditional SLAM pipeline for creating semantic maps. Earlier work [1] in this field has extended the structure-from-motion (SfM) pipeline for joint estimation of camera parameters, scene points, and object labels. However, its computational complexity limits the method's ability to operate in real-time. Other object-level SLAM methods [6,27] use object detection in a scene to solve the problem of scale uncertainty and drift of monocular SLAM. In [7], the authors used an extensive database of known objects and proposed an algorithm based on bags of binary words [8]. The combined usage of monocular SLAM and object recognition algorithms improves the map and finds its real scale. However, the main limitation of the approach is its dependence on known objects. QuadricSLAM [23] is an object-oriented SLAM that does not rely on prior object models. Rather it represents the objects as quadrics, i.e., sphere and ellipsoids. It jointly estimates a 3D quadric surface for each object and camera position using 2D object detections from images. In [2], the authors were the first to include the inertial, geometric, and semantic information into a single optimization framework. The proposed system performs continuous optimization over the poses while it discretely optimizes the semantic data association. In [12], the authors represented generic objects as landmarks by including an object detector in a monocular SLAM framework. The method exploits the CNN-based objects and plane detectors for constructing a sparse scene representation. The SLAM bundle adjustment includes semantic objects, plane structures, and their completed point clouds. In [29,32], the authors use machine learning-based approaches to perform localization in indoor environments relative to landmarks. CubeSLAM [31] combines 2D and 3D object detection with SLAM pose estimation by generating cuboid proposals from single view detections and optimizing them with points and cameras using a multi-view bundle adjustment. In [25] authors create category-level models with CAD collections for real-time object-oriented monocular SLAM. Their rendering pipeline generates large amounts of datasets with limited hand-labeled data. The proposed system first learns 2D features from category-specific objects (i.e., chairs, doors) and then matches the features to a CAD model to estimate the semantic objects' pose. For obtaining a metrically correct robot pose, the system then combines semantic objects and the estimated robot's pose from VO into an optimizing graph framework. Most of the existing literature on object-SLAM considers indoor scenes or outdoor autonomous driving scenarios. In both cases, it is possible to directly use a pre-trained CNN to identify enough objects in a scene without training a detector on custom objects. However, the problem arises when a scene lacks pre-trained objects. In this scenario, training a detector on custom landmarks would automatically become a necessary pre-condition for most object-based localization approaches. While generating labeled training data for learning new landmarks by hand is cumbersome, the method described in Sect. 3 only requires minimal human supervision for the learning task.

3   Methods

This section introduces our proposed approach for learning new landmarks with minimal human intervention. It further presents Slow Feature Analysis (SFA), the core algorithm we use to extract spatial representation. Finally, it describes the procedure to perform localization using landmarks views.

3.1   Minimal Supervision for Landmark Learning

In this work, we propose to use readily detectable object categories from pre-trained CNNs as anchors to generate labeled data for learning new landmarks. Figure 1 shows the steps of label generation and consequently using the anno-tated image data for landmark learning. A mobile agent explores an environment and records camera images. Here we assume that the CNN-detectable objects (e.g., a bicycle) are already present in the scene, and their location remains fixed during the recording phase. The system then runs the YOLOv3 [26] object detec-tion algorithm on the recorded images to detect the instances of anchors. The next step is to specify the spatial relationship of new landmarks w.r.t anchors. At

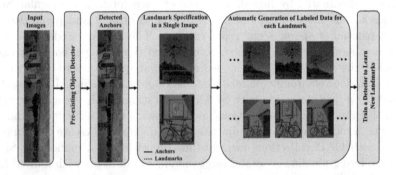

Fig. 1. Label Generation for Learning Landmarks: The input to the pipeline are images collected from a robot recording session. The first step applies a pre-trained visual object detector to these images to identify a pre-learned object's instances, such as a bicycle. The next step is to select new landmarks in a single image (i.e., specification of the new landmarks' spatial relationship w.r.t to the detected anchor). This selection can be made using a human input, setting a fixed offset w.r.t an anchor, or taking a random region around an anchor as a new landmark. Based on the specified relationship, the system then automatically generates labeled data from all available training images. The final step uses the generated data to train a detector for the new landmarks.

this step in the learning phase, minimal human supervision is necessary, i.e., the human has to specify the landmark's spatial relation. There are several possible ways to perform this step. a) A human can cooperatively indicate the location of a new landmark relative to an anchor as a 2D offset in one or a few images. This approach can generate semantic object categories (e.g., a specific tree, a fountain). b) The system autonomously analyzes the regions around the anchor and chooses a visually unique region (e.g., not a section of brick wall from a larger brick wall). c) The system takes a fixed 2D offset w.r.t to an anchor (e.g., above, bottom, besides) to learn new landmarks. We use the fixed 2D offset approach to derive a landmark relative to an anchor in this work. This offset is set only *once (i.e., minimal supervision)* in a single image for each landmark compared to manually annotating thousands of images. The system then uses the instances of detected anchors and a specified offset to automatically annotate the landmarks in the rest of the recorded images. If YOLOv3 fails to recognize the anchors in some images, we run an object tracker to obtain the bounding boxes of anchors in the missing frames. The final step uses the generated labeled data and trains a detector to recognize new landmarks. This step's output is a custom landmark detector, which we use as an independent module in the mapping and localization phases. Please note that the anchor objects in a scene are only temporarily required for landmark learning and can be removed afterwards. The basic implementation of this approach learns one landmark per detected anchor. As an extension, it is possible to scale the system to learn multiple landmarks per anchor, which will improve both robustness against local occlusions and localization accuracy.

3.2   Slow Feature Analysis

To learn the robot's position in 2D space, we use Slow Feature Analysis (SFA) as introduced in [30]. It transforms a multidimensional time series $\mathbf{x}(t)$, in our case images along a trajectory, to slowly varying output signals. The objective is to find instantaneous scalar input-output functions $g_j(\mathbf{x})$ such that the output signals

$$s_j(t) := g_j(\mathbf{x}(t))$$

minimize

$$\Delta(s_j) := \langle \dot{s}_j^2 \rangle_t$$

with $\langle \cdot \rangle_t$ and $\dot{s}$ indicating temporal averaging and the derivative of s, respectively. The Δ-value defines the temporal variation of the output signal, and its minimization is the optimization objective. Thus small Δ-values indicate slowly varying signals over time. There are three optimization constraints: the output signals should have zero mean, unit variance, and are decorrelated. These constraints avoid the trivial constant solution and ensure that different functions g_j code for different aspects of the input.

3.3   Learning of Spatial Representation Using Landmark Views

Acquiring Landmark Views: Figure 2 shows the steps to detect and extract landmark views. The input to the system are images recorded for the mapping and localization phases. The next step applies the trained detector to recognize the instances of the landmarks in images. Afterwards, we resize each landmark's bounding box to have the same size as the biggest bounding box in its category and rescale the extracted image patch to 120×120 pixels. The output of this step generates an image stream for each landmark.

Mapping Phase: We use landmark views to learn camera position regression. We choose SFA to get a compact place representation relative to each landmark and perform light-weight position regression on top. The approach employs a four-layer hierarchical SFA network and has been described recently in [11]. The network learns spatial representations relative to each landmark in an unsupervised learning process. Afterwards, we obtain metric space representation by computing a regression function from the learned spatial representations and odometry data, i.e., the robot's ground truth position (x, y). This step outputs an individual position estimator (x, y) for each landmark.

Localization Phase: The localization phase uses the learned position estimators to obtain the robot's 2D position (x, y) relative to each landmark. Afterwards, it estimates the robot's global 2D position (x, y) by combining each landmarks' position estimation using weighted averaging. We determine the weight of each landmark by taking the inverse of its localization error.

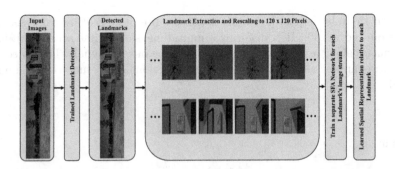

Fig. 2. Landmark-based Learning of Spatial Representation: We use the trained detector from the previous step for recognizing the learned landmarks in the recorded images for the mapping phase. The next step extracts the detected landmarks and rescales the image patches to 120×120 pixels. Afterwards, we train an independent SFA network to learn spatial representation relative to each landmark. The procedure is the same for the test phase (localization), where we pass each test view through its learned network for obtaining the output in the SFA space.

4   Experiments

This section presents the localization results using simulated and real-world data. We have a two-stage system, and the first stage learns new landmarks in a scene using the proposed approach in this work. We derive one landmark per anchor by setting a fixed 2D offset. The system then uses 500 labeled images for each landmark based on the specified relationship and trains a detector to learn these landmarks. The second stage uses the learned landmarks to perform localization relative to them. This stage proves that the learned landmarks in the first stage are well suited for the localization task. To obtain baseline PoseNet [13] results, we use 25% of the data (subsampled from the training set, i.e., every 4th image) for validation and the remaining to train the network. We extract Fourier features to obtain the localization results for SFA localization on complete images, as in [9].

4.1   Simulated Experiments

We perform the experiments in a simulated garden with an area of 18×18 meters. A virtual robot randomly traverses in the environment to record images for the training set and then along a regular grid to collect the test set. We project images from the simulated omnidirectional camera to panoramic views of size 3600×600 pixels. The training and test trajectory consist of $15,000$ and $1,250$ images, respectively. Here, we have used three anchors to learn new landmarks. The anchors include a bicycle (Id_0), a car (Id_1), and an umbrella (Id_2). Table 1 shows the experimental results of localization w.r.t learned landmarks and the baseline methods. All the methods produce localization results in a similar range. However, PoseNet outperforms the SFA-based approaches in this experiment. It achieved good localization accuracy with a large amount of labeled training data, as expected in this case. However, it is infeasible to generate a massive amount of labeled training data in real-world scenarios. On the other hand, we can further improve the accuracy of landmark-based localization by incorporating more landmarks (c.f. Sect. 4.3 on scaling experiments).

Table 1. Localization Results on Simulated Data: Table shows median Euclidean localization accuracy on learned landmarks views, their combination, and the baseline methods. It further reports the detection rates of the learned landmarks. The combined detection rate of 100% indicates that each test image at least contains a single landmark. PoseNet outperforms both Landmark-based and Fourier-SFA localization in this experiment.

Landmark-based localization					Fourier-SFA	PoseNet
Id_1	Id_2	Id_3	Combined	%		
0.53 m [99%]	0.37 m [99%]	0.35 m [97%]	0.33 m	100	0.26 m	**0.21 m**

4.2   Real-World Experiments

We perform the experiments in two garden-like outdoor environments of size $88\,m^2$ and $494\,m^2$, respectively. The autonomous lawn mower robot (Fig. 3a) equipped with a fisheye lens traverses in a scene to record images of size 2880×2880 pixels. Each recording session has two operational phases. In the first phase, the robot traverses the border of an area by using the standard wire guidance technology, while in the second phase, it moves freely within the area defined by the border wire. Figure 3b and 3c show the robot's trajectories in one of the recording sessions from each garden, respectively. During a recording, the robot stores images and the associated odometry information. For the first working phase, we estimate the robot's ground truth metric position (x, y) using a method described by Einecke et al. [3]. The authors used wheel odometry and additional weighted loop closure to get high-quality localization. However, the technique only estimates the metric shape of the boundary. For the second working phase, we estimate the ground truth data (x, y) using commercial photogrammetry software, i.e., Metashape[1]. The obtained ground truth position (x, y) estimates are used to evaluate the metric performance of the localization methods. Instead of using MS-COCO objects [16], we reuse pre-trained manually labeled region detectors (Fig. 3d) from [11] as anchors to learn new landmarks for localization. We show the experimental results using ten recordings collected from both gardens under varying lighting, weather conditions, and dynamic obstacles.

Temporal Generalization. This experiment aims at testing the re-localization ability of the methods in changing conditions over time. Here, we have chosen that the robot traverses on a similar path (border run) and collected three recordings from each garden. We use one dataset to learn the spatial representations and the other two sets to test the localization accuracy. The datasets differ w.r.t dynamic scene variations and changes in lighting conditions. The number of training set images for the small garden is 1138, while the two test sets consist of 1091 and 1109 images. Similarly, the big garden datasets consist of 4336 training images, while the test sets have 4032 and 4050 images. After training PoseNet, its localization accuracy on the validation data from the small and big garden is $0.07\,m$ and $0.41\,m$, respectively. Table 2 reports median localization accuracy of landmark-based localization and the baseline methods. The results obtained with individual landmarks enable coarse localization in an environment. Nevertheless, their combination achieves similar or better localization accuracy than the baseline methods. This effect, however, is more pronounced for the big garden. Fourier-SFA does well on the datasets from the small garden. However, it does not scale to the large environment with the configurations used in [9]. PoseNet produces good localization results when the environmental condition between the training and the test sets is almost identical (e.g., the first dataset

---

[1] https://www.agisoft.com/.

from the small garden). However, it degrades otherwise (e.g., the last dataset from the big garden).

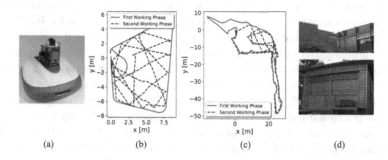

(a)   (b)   (c)   (d)

Fig. 3. Experimental Setup: (a) An autonomous lawn mower robot with a fisheye camera used for the experiments. (b) and (c) show the robot's traversed trajectories in one of the recording sessions from both gardens, respectively. The solid line (red) shows the robot's traversal in the first working phase (border run), while the dashed line (blue) shows its traversal in the second phase (infield run). (d) We used previously trained custom objects (e.g., hut) or regions (e.g., building corner) as anchors to learn new relative landmarks for the real-world experiments. (Color figure online)

Spatial Generalization. This experiment aims at testing the re-localization ability of the methods when the train and test set contain sufficiently different robot trajectories. As described earlier, each robot recording session has two operational phases. Hence, we use the images from the first phase (border run) to learn the spatial representations and the second phase (infield run) to test the localization method. We collected two recordings from both gardens for the experiments. The small garden training sets consist of 1141 and 1131 images, while the test images from infield positions are 158 and 269, respectively. Similarly, the big garden training sets have 4336 and 4011 images, while 234 and

Table 2. Real-World Experiments for Temporal Generalization: Median Euclidean errors of individual landmarks, their combination, and the baseline methods for generalization over time. The drop in detection rates of the landmarks present in the big garden is due to their visibility only in specific parts of the scene. Please note we only use test images for the baseline methods, where at least a single landmark view was available in the corresponding image. Landmarks enable coarse localization in both environments while their combination performs similar or better than the baseline methods, especially in the big garden.

Garden	Test set	Landmark-based localization						Fourier-SFA	PoseNet
		Id_1	Id_2	Id_3	Id_4	Combined	%		
Small	1	0.26 m [99%]	0.31 m [99%]	0.60 m [99%]	–	0.20 m	100	0.19 m	**0.18 m**
	2	0.73 m [99%]	0.93 m [97%]	1.33 m [99%]	–	**0.75 m**	100	1.01 m	0.83 m
Big	1	1.46 m [13%]	3.54 m [25%]	1.74 m [32%]	2.44 m [16%]	**2.22 m**	78	7.50 m	2.99 m
	2	1.83 m [13%]	3.80 m [28%]	2.06 m [32%]	2.97 m [17%]	**2.57 m**	80	8.22 m	6.57 m

180 test set images. The localization accuracy of PoseNet on the validation set for both sets from the small garden is 0.06 m, while the corresponding accuracy on the sets of the big garden is 0.41 m and 0.30 m, respectively. Table 3 reports median localization accuracy of each method. The individual landmarks again show coarse localization accuracy. However, it is slightly worse for landmarks of the big garden. Extreme perspective changes between the train and test images mainly influence this performance drop. Moreover, there is a noticeable change in the lighting conditions between the first and second robot recording phases. Despite that, their combination achieves similar or better accuracy than localization using the baseline methods. There are several ways to improve the current results of landmark-based localization. Firstly, the incorporation of more landmarks leads to a higher localization accuracy (c.f. Sect. 4.3). Secondly, it is possible to filter out the landmarks with the worst performance as a post-processing step and only localize relative to those with mean accuracy better than a specified threshold. Thirdly, the addition of sparse images from infield run during learning can further improve the localization accuracy. Please note that, here, we only need landmark views from the infield that are relatively easy to obtain with a single pass through the pre-trained detector. It contrasts to PoseNet, which requires a computationally expensive structure-from-motion (SfM) step to generate robot poses as labeled data for learning.

Table 3. Real-World Experiments for Spatial Generalization: Table reports localization accuracy when the train and test sets consist of images from different robot trajectories. Similar to temporal generalization experiments, the combination of landmarks achieves similar or better accuracy than the baseline methods.

Garden	Test set	Landmark-based localization						Fourier-SFA	PoseNet
		Id_1	Id_2	Id_3	Id_4	Combined	%		
Small	1	1.41 m [96%]	0.95 m [99%]	1.01 m [96%]	–	0.74 m	100	**0.66 m**	0.85 m
	2	0.82 m [95%]	1.05 m [97%]	1.35 m [95%]	–	0.84 m	100	**0.65 m**	0.82 m
Big	1	4.37 m [12%]	3.95 m [57%]	1.50 m [27%]	5.51 m [44%]	**3.21 m**	95	7.10 m	5.31 m
	2	5.88 m [17%]	4.64 m [61%]	1.82 m [31%]	6.12 m [28%]	3.96 m	96	7.60 m	**3.49 m**

4.3   Scaling Experiments

This experiment aims to analyze the effect of increasing the number of landmarks on localization using simulated and real-world data from the small garden. We use ten different landmarks from each environment by randomly selecting them around a single anchor. The first step learns an independent position estimator for each landmark. The second step processes landmark images from the test set using the estimators and predict the robot's 2D position (x, y). Afterwards, we calculate the test set's median localization error by systematically increasing the number of landmarks. Figure 4 shows the results of 50 random permutations of the ten landmarks for both simulated and real-world data. Both plots initially show improved localization accuracy by increasing the number to three, and adding more landmarks continues to improve the overall localization accuracy until it almost saturates as expected. From an application perspective, a robot could increase the number of landmarks to achieve a certain accuracy level at runtime, depending on the area where high accuracy is required. As we train an independent SFA network for each landmark, the processing time will linearly increase (i.e., $O(n)$) with the number of landmarks. However, SFA-based mapping and localization are drastically faster than one of the fastest state-of-the-art visual localization methods [10].

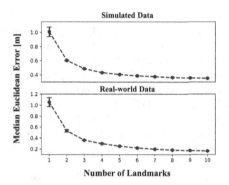

Fig. 4. Effect of Increasing Landmarks on Localization: Simulated data (Top). Real-world data (Bottom). The plot shows the median and variance localization accuracy for 50 random permutations of the ten landmarks. The usage of more landmarks for localization significantly improves the accuracy initially, and eventually, it saturates for a higher number of landmarks.

5   Conclusion

This work proposed an approach to speed up the label generation process for learning new long-term visual landmarks for localization. The method uses instances of readily available CNN objects as anchors to generate labeled data for the unseen imagery based on minimal human supervision. We used a fixed 2D offset to derive new landmarks relative to anchors. When anchor and landmark are within the same plane, perspective changes during recording result in labeling the identical scene part as a landmark in our training paradigm with a fixed 2D offset between anchor and landmark. In the most extreme case, if an anchor is placed such that the robot can go around it, a simple 2D approach may fail and does not capture a semantically meaningful region but a subset of the scene's viewing space. Geometry-based localization methods may fail in such a case. However, unintuitively, these views are classified very well as a landmark with a CNN, and we see no reduction in localization accuracy with pose regression learning (as shown here with SFA). After landmark learning, we used the learned landmarks and performed localization relative to them. The landmark-based localization performed better than the baseline methods, especially in a challenging large-scale outdoor environment. The accuracy can be significantly improved by integrating more landmarks and obtaining a global position estimation relative to them. From an application perspective, our system is suitable for service robots (e.g., lawnmowers and vacuum cleaners), employing a pre-trained visual detector to learn new landmarks in a scene. Thus, the approach enables reliable localization in the long-term even if the anchor objects are no longer present in the scene.

References

1. Bao, S.Y., Bagra, M., Chao, Y., Savarese, S.: Semantic structure from motion with points, regions, and objects. In: 2012 IEEE Conference on Computer Vision and Pattern Recognition, pp. 2703–2710 (2012)
2. Bowman, S.L., Atanasov, N., Daniilidis, K., Pappas, G.J.: Probabilistic data association for semantic slam. In: 2017 IEEE International Conference on Robotics and Automation (ICRA), pp. 1722–1729 (2017)
3. Einecke, N., Deigmöller, J., Muro, K., Franzius, M.: Boundary wire mapping on autonomous lawn mowers. In: Hutter, M., Siegwart, R. (eds.) Field and Service Robotics. SPAR, vol. 5, pp. 351–365. Springer, Cham (2018). https://doi.org/10.1007/978-3-319-67361-5_23
4. Engel, J., Schöps, T., Cremers, D.: LSD-SLAM: large-scale direct monocular SLAM. In: Fleet, D., Pajdla, T., Schiele, B., Tuytelaars, T. (eds.) ECCV 2014. LNCS, vol. 8690, pp. 834–849. Springer, Cham (2014). https://doi.org/10.1007/978-3-319-10605-2_54
5. Franzius, M., Sprekeler, H., Wiskott, L.: Slowness and sparseness lead to place, head-direction, and spatial-view cells. PLoS Comput. Biol. 3(8), 1–18 (2007)
6. Frost, D., Prisacariu, V., Murray, D.: Recovering stable scale in monocular slam using object-supplemented bundle adjustment. IEEE Trans. Rob. 34(3), 736–747 (2018)
7. Gálvez-López, D., Salas, M., Tardós, J.D., Montiel, J.: Real-time monocular object slam. Robot. Auton. Syst. 75(PB), 435–449 (2016). https://doi.org/10.1016/j.robot.2015.08.009
8. Galvez-López, D., Tardos, J.D.: Bags of binary words for fast place recognition in image sequences. Trans. Rob. 28(5), 1188–1197 (2012). https://doi.org/10.1109/TRO.2012.2197158
9. Haris, M., Franzius, M., Bauer-Wersing, U.: Robust outdoor self-localization in changing environments. In: IEEE/RSJ International Conference on Intelligent Robots and Systems (IROS 2019). IEEE (2019)
10. Haris, M., Franzius, M., Bauer-Wersing, U.: Unsupervised fast visual localization and mapping with slow features. In: 2021 IEEE International Conference on Image Processing (ICIP), pp. 519–523 (2021). https://doi.org/10.1109/ICIP42928.2021.9506656
11. Haris, M., Franzius, M., Bauer-Wersing, U., Karanam, S.K.K.: Visual localization and mapping with hybrid SFA. In: Kober, J., Ramos, F., Tomlin, C. (eds.) Proceedings of the 2020 Conference on Robot Learning. Proceedings of Machine Learning Research, vol. 155, pp. 1211–1220. PMLR, 16–18 November 2021. https://proceedings.mlr.press/v155/haris21a.html

12. Hosseinzadeh, M., Li, K., Latif, Y., Reid, I.: Real-time monocular object-model aware sparse slam. In: 2019 International Conference on Robotics and Automation (ICRA), pp. 7123–7129 (2019)
13. Kendall, A., Cipolla, R.: Geometric loss functions for camera pose regression with deep learning. CoRR (2017)
14. Kendall, A., Grimes, M., Cipolla, R.: Convolutional networks for real-time 6-DOF camera relocalization. CoRR (2015)
15. Krizhevsky, A., Sutskever, I., Hinton, G.: ImageNet classification with deep convolutional neural networks. In: Neural Information Processing Systems 25 (2012). https://doi.org/10.1145/3065386
16. Lin, T.-Y., et al.: Microsoft COCO: common objects in context. In: Fleet, D., Pajdla, T., Schiele, B., Tuytelaars, T. (eds.) ECCV 2014. LNCS, vol. 8693, pp. 740–755. Springer, Cham (2014). https://doi.org/10.1007/978-3-319-10602-1_48
17. Metka, B., Franzius, M., Bauer-Wersing, U.: Outdoor self-localization of a mobile robot using slow feature analysis. In: Neural Information Processing (2013)
18. Metka, B., Franzius, M., Bauer-Wersing, U.: Improving robustness of slow feature analysis based localization using loop closure events. In: Villa, A.E.P., Masulli, P., Pons Rivero, A.J. (eds.) ICANN 2016. LNCS, vol. 9887, pp. 489–496. Springer, Cham (2016). https://doi.org/10.1007/978-3-319-44781-0_58
19. Metka, B., Franzius, M., Bauer-Wersing, U.: Bio-inspired visual self-localization in real world scenarios using slow feature analysis. PLOS ONE **13**(9), 1–18 (2018). https://doi.org/10.1371/journal.pone.0203994
20. Milford, M.J., Wyeth, G.F.: Mapping a suburb with a single camera using a biologically inspired slam system. IEEE Trans. Rob. **24**(5), 1038–1053 (2008). https://doi.org/10.1109/TRO.2008.2004520
21. Milford, M.J., Wyeth, G.F., Prasser, D.: RatSLAM: a hippocampal model for simultaneous localization and mapping. In: IEEE International Conference on Robotics and Automation, 2004, Proceedings, ICRA 2004, vol. 1, pp. 403–408, April 2004. https://doi.org/10.1109/ROBOT.2004.1307183
22. Mur-Artal, R., Montiel, J.M.M., Tardós, J.D.: ORB-SLAM: a versatile and accurate monocular SLAM system. IEEE Trans. Robot. **31**(5), 1147–1163 (2015). https://doi.org/10.1109/TRO.2015.2463671
23. Nicholson, L., Milford, M., Sünderhauf, N.: Quadricslam: constrained dual quadrics from object detections as landmarks in semantic SLAM. CoRR abs/1804.04011 (2018)
24. O'Keefe, J., Dostrovsky, J.: The hippocampus as a spatial map. preliminary evidence from unit activity in the freely-moving rat. Brain Res. **34**(1), 171–175 (1971). https://doi.org/10.1016/0006-8993(71)90358-1
25. Parkhiya, P., Khawad, R., Murthy, J.K., Bhowmick, B., Krishna, K.M.: Constructing category-specific models for monocular object-SLAM. In: 2018 IEEE International Conference on Robotics and Automation (ICRA), pp. 4517–4524 (2018)
26. Redmon, J., Farhadi, A.: YOLOv3: an incremental improvement. CoRR abs/1804.02767 (2018)
27. Sucar, E., Hayet, J.: Bayesian scale estimation for monocular slam based on generic object detection for correcting scale drift. In: 2018 IEEE International Conference on Robotics and Automation (ICRA), pp. 5152–5158 (2018)
28. Taube, J., Muller, R., Ranck, Jr, J.: Head-direction cells recorded from the postsubiculum in freely moving rats. I. Description and quantitative analysis. J. Neurosci. **10**, 420–35 (1990). https://doi.org/10.1523/JNEUROSCI.10-02-00420.1990
29. Thrun, S.: Bayesian landmark learning for mobile robot localization. Mach. Learn. **33**(1), 41–76 (1998). https://doi.org/10.1023/A:1007554531242

30. Wiskott, L., Sejnowski, T.: Slow feature analysis: unsupervised learning of invariances. Neural Comput. **14**(4), 715–770 (2002)
31. Yang, S., Scherer, S.: CubeSLAM: monocular 3-D object slam. IEEE Trans. Robot. 1–14 (2019). https://doi.org/10.1109/TRO.2019.2909168
32. Zhao, Z., Carrera, J., Niklaus, J., Braun, T.: Machine learning-based real-time indoor landmark localization. In: Chowdhury, K.R., Di Felice, M., Matta, I., Sheng, B. (eds.) WWIC 2018. LNCS, vol. 10866, pp. 95–106. Springer, Cham (2018). https://doi.org/10.1007/978-3-030-02931-9_8

Correction to: Improving Colon Carcinoma Grading by Advanced CNN Models

Marco Leo⊙, Pierluigi Carcagnì, Luca Signore, Giulio Benincasa⊙,
Mikko O. Laukkanen⊙, and Cosimo Distante⊙

Correction to:
Chapter "Improving Colon Carcinoma Grading by Advanced CNN Models" in: S. Sclaroff et al. (Eds.):
Image Analysis and Processing – ICIAP 2022, **LNCS 13231,
https://doi.org/10.1007/978-3-031-06427-2_20**

In the originally published version of chapter 20, the name of the author Pierluigi Carcagnì contained a spelling mistake. This has been corrected.

The updated version of this chapter can be found at
https://doi.org/10.1007/978-3-031-06427-2_20

Correction to: Improving Autoencoder Training Performance for Hyperspectral Unmixing with Network Reinitialisation

Kamil Książek[ID], Przemysław Głomb[ID], Michał Romaszewski[ID],
Michał Cholewa[ID], Bartosz Grabowski[ID], and Krisztián Búza[ID]

Correction to:
Chapter "Improving Autoencoder Training Performance
for Hyperspectral Unmixing with Network Reinitialisation"
in: S. Sclaroff et al. (Eds.): *Image Analysis*
and Processing – ICIAP 2022, **LNCS 13231,**
https://doi.org/10.1007/978-3-031-06427-2_33

In the originally published version of chapter 33, table 2 included an error. This has been corrected.

The updated version of this chapter can be found at
https://doi.org/10.1007/978-3-031-06427-2_33

Author Index